# THE AMERICAN SEA

**HARTE RESEARCH INSTITUTE**

**FOR GULF OF MEXICO STUDIES SERIES**

*Sponsored by the Harte Research Institute for Gulf of Mexico Studies, Texas A&M University–Corpus Christi*

JOHN W. TUNNELL JR., General Editor

*A list of titles in this series appears at the end of the book.*

The

# American Sea

## A NATURAL HISTORY OF THE GULF OF MEXICO

### Rezneat Darnell

TEXAS A&M UNIVERSITY PRESS  College Station

Copyright © 2015 by Rezneat Milton Darnell
Manufactured in China by Everbest Printing
through FCI Print Group
First edition

This paper meets the requirements of
ANSI/NISO Z39.48–1992 (Permanence of Paper).
Binding materials have been chosen for durability.

Library of Congress Cataloging-in-Publication Data
Darnell, Rezneat M., author.
The American sea: a natural history of the Gulf of Mexico /
Rezneat Darnell. — First edition.
pages cm — (Harte Research Institute for Gulf of Mexico
Studies series)
Includes bibliographical references and index.
ISBN 978-1-62349-282-3 (printed case: alk. paper) —
ISBN 978-1-62349-301-1 (ebook) 1. Natural history—
Mexico, Gulf of. 2. Marine biology—Mexico, Gulf of.
3. Marine ecology—Mexico, Gulf of. 4. Oceanography—
Mexico, Gulf of. I. Title. II. Series: Harte Research Institute
for Gulf of Mexico Studies series.
QH92.3.D37 2015
578.7709163'64—dc23
2015001353

Line art by Heather Prestridge

Cover photo: Elkhorn coral (*Acropora palmata*) on the upper
windward forereef of Topatillo Reef, June 1984. This iconic
Gulf of Mexico and Caribbean Sea coral reef species was
given "threatened" status under the Endangered Species
Act in 2006. Photo by John W. Tunnell.

Frontispiece: A view to the northeast, over the leeward side
of Isla de Enmedio Reef platform and island. The Isla de
Enmedio Reef is the most studied reef in the Veracruz Reef
System. Photo by © Kip Evans, National Geographic Society.

*This book is dedicated to the West Indian monk seal* (Monachus tropicalis), *which thrived in the Gulf of Mexico before the arrival of European explorers, declined during the colonial period, and finally became extinct about the time I finished graduate school. It was formerly found from the Bahamas and Cuba, through the islands of the Caribbean, and along the east coast of Middle America down at least as far as Honduras. It also occurred in the Florida Keys and on the islands off Yucatán (Alacrán, Triángulos, etc.), and it occasionally wandered along the shorelines of the Gulf as far as the coast of Texas. As an island species, it was naive of terrestrial predators and showed no fear of humans; thus it became easy prey for the colonists, who slaughtered the seals in vast numbers for the valuable oil that could be rendered from their fat. Already scarce by the mid-1800s, this species was last seen alive in 1952, apparently killed throughout its range by fishermen who considered this animal to be a competitor for the local fish supply. The demise of this gentle species stands as a monument to the ignorance, neglect, and greed of civilized humanity. Our society can and must do better, and we ignore Nature at our own peril. The case of the West Indian monk seal cannot be ignored or forgotten.*

# CONTENTS

Supplemental tables can be found online at http://harteresearchinstitute.org/hri-media/books/hri-studies-series

# SERIES EDITOR'S FOREWORD

Rez Darnell and I reconnected late in life. Although he was 20 years my senior, we both had a connection to and passion for the Gulf of Mexico, and we both liked writing and summarizing information about the Gulf. My focus was coral reef and coastal ecology, seashells, and oil spill impacts, whereas his expertise was in estuarine and continental shelf ecology, fish and crustacean life history and ecology, as well as food relations and trophic structure of marine ecosystems. I summarized information on some of my favorite regional places, *Laguna Madre of Texas and Tamaulipas* (Tunnell and Judd 2001) and *Coral Reefs of the Southern Gulf of Mexico* (Tunnell et al. 2007), and regional organisms, *Encyclopedia of Texas Seashells* (Tunnell et al. 2010) and *Texas Seashells: A Field Guide* (Tunnell et al. 2014), but Rez decided to do it all, the entire Gulf of Mexico.

There are few generalists today, if any, who would take on the challenge of summarizing and synthesizing all of the information about an ecosystem as large as the Gulf of Mexico. Rez started this massive project in 1994, the year before his retirement as professor of oceanography at Texas A&M University. He continued it as professor emeritus until his death in December 2009. All 18 chapters had been completed and most of the 189 figures and 183 tables were done, and it had been recommended for publication to Texas A&M University Press after two external reviews. Rez was in the midst of adding suggestions or adjusting the manuscript per the reviewer comments when he passed away. You can read a brief obituary of his productive life in the pages that follow this foreword. Below is a brief history of this project and the story of how friends and colleagues pulled together to honor the work and memory of Dr. Rezneat M. Darnell Jr., teacher, researcher, mentor, and friend by bringing the project to culmination.

Dr. Darnell was a professor of oceanography at Texas A&M University (TAMU) in the early 1970s when I was working on my PhD in the department of biology. Although I did not take any of his classes, I knew him through seminars on campus and my college buddies who were in oceanography. He was graduate advisor to some of my good friends back then, and I kept track of his career after I left, particularly where he worked extensively with NOAA (National Oceanic and Atmospheric Administration) and MMS (Minerals Management Service) on large scale ecosystem projects in the Gulf of Mexico.

Fast forward to 1998, when he contacted me for the first time regarding his Gulf book. It was the first of many long, hand-written letters about the book and its progress or needs that I would receive over the coming years. Here is how the long yellow legal pad letter opened "Dear Wes, Please excuse the informality of this means of communicating. I don't mess with computers, and in the age of computers, the department no longer stocks stationary." At his request, I sent him a big stack of publications, technical reports, and theses/dissertations covering our research in coastal South Texas and Mexico. He would devour them like a child eating candy, and then summarize them for his book in the various chapters where appropriate. Everything was always returned when he was finished with them.

After our university, Texas A&M University-Corpus Christi (TAMU-CC), was blessed with the $46 million endowment in 2000 from Ed Harte to establish the Harte Research Institute for Gulf of Mexico Studies (HRI), Rez asked in 2002 if we might fund some of the typing and illustration work needed for his book. Our early HRI Advisory Council and TAMU-CC Management Team thought it was a good idea for us to get involved, since there had not been a summary work on the entire Gulf of Mexico since Galtsoff's *Gulf of Mexico: It's Origins, Waters, and Marine Life* in 1954. We requested and received a short proposal for his work and funded him with $10,000 to keep the project moving. He was working on chapter 15 of the proposed 16 chapters at this time. Rez had previously received a $5,000 grant for typing and other book related work from the Texas Sea Grant Program.

A letter from late February 2002 noted that he had finished chapter 15, and it contained 215 (hand written) pages of text and 218 references. He noted that the heavy work for the chapter was done, but that he still needed to polish it a bit and write a summary. Two months later he indicated he had completely finished that chapter and the final numbers were 243 pages and 224 references. He was happy with his progress and noted, "They say that an author never finishes a book, he simply abandons it after a while."

In early March 2002, he checked in with a short letter to update his progress. He started the letter with a paragraph that was typical of his thoughtfulness and humor: "Wes, I sometimes worry that in the isolation of South Texas you don't have enough material to read. In partial remedy of this situation I am enclosing a number of items which may be of some interest and which should help you endure the long hours in which you have little to do. It's nice to have such thoughtful friends. I also have a couple of heavy bio-atlases and my *New Orleans Guide* to hand you if and when we do get together."

Monthly updates continued as Rez worked to complete the manuscript. He frequently mentioned his colleagues in College Station who lent a hand, such as Norman Guinasso, Bob Stewart, Gil Rowe, and others. You can see a much longer and complete list of people he thanked in the acknowledgments section of this book.

As time progressed, his eyes continued to get weaker, and he had to frequently use a magnifying glass to do his work. He mainly worked in his office in the department of oceanography in the evenings when it was quiet. Barbara Childers, an executive assistant in the oceanography department, typed the entire manuscript and proofed it along the way. Then, unfortunately, his wife Carol died in late 2003, from injuries sustained in an auto accident near campus. This was a substantial setback. Due to Rez's eye problems, he was legally blind and could not drive. Carol was Rez's key to independent living in College Station. But Rez continued writing from home. Norman Guinasso would drive him places occasionally when he needed help with something.

In late 2004 Rez had to have gall bladder surgery and that again slowed his progress, but in 2005 he received a grant from MMS, and this gave him a boost to continue. During these middle years of the 2000s, Rez was having each chapter peer-reviewed by his colleagues in the department of oceanography primarily, but sometimes external to TAMU.

During 2006 into 2008, I heard little from Rez, but late in 2008 he called and said he had finished the narrative of the book and had added two more chapters, including human issues and management issues of the Gulf. He also said that he had moved to Minnesota to be near his daughter Molly who would help take care of him.

In early 2009, two TAMU Press selected reviewers rec-

ommended publication of the book with minor to moderate adjustments. There was concern that some of the material had become out of date, due to the long book writing process, but there was also support for publishing pretty much everything as it was, since it was so inclusive of many topics and it was very well written. It was during the final updating of material that Rez passed away in Minneapolis after a brief illness of pneumonia and suspected lung cancer.

Even though there was still much work to be done from an author's standpoint before the book could be published, we knew that the manuscript had come too far and with expectations from too many people to let it die at this phase. Shannon Davies, then Louise Lindsey Merrick Editor for the Natural Sciences at TAMU Press and now editor-in-chief at the Press, and I discussed options for moving forward with the project. Since the size of the book was equivalent to four normal books moving through their system, Shannon said that it would be too large of a book project and too time consuming for them to just complete it internally without outside help.

After wracking my brain for a while on who we might get to help finish up the book, it came to me that two of his former students who worked at the MMS Gulf of Mexico office in New Orleans had just retired. Rick Defenbaugh and Bob Rogers would be perfect for the task, and they fortunately accepted. Linda Pequegnat also stepped back into the picture and helped greatly with final details.

Finally, all of Rez's colleagues who reviewed the manuscript after his death were reluctant to significantly alter or try to rewrite or add to the unique brilliance and genius of his writing—in spite of the fact that events occurred after his death in December 2009 that had effects on the Gulf of Mexico. Among these were:

- the BP oil spill in the northern Gulf in 2010.
- the fact that MMS was reorganized in 2010 to BOEMRE (Bureau of Ocean Energy, Management, Regulation, and Enforcement) and that BOEMRE was subsequently reorganized in 2011 into two agencies: BSEE (Bureau of Safety and Environmental Enforcement) and BOEM (Bureau of Ocean Energy Management). Rez's colleagues chose to keep the original MMS designation in this book.
- Rez understood that some information in the book would be dated by the time it reached publication, as evidence by statements in his preface: "In a

few cases the author has decided to stick with old, widely-used scientific names rather than use their recent updates (e.g., in the penaeid shrimps, *Penaeus* rather than *Farfantepenaeus, Litopenaeus,* etc.). Some simplification has also occurred in the higher classification of the sharks and some of the bony fishes." We agree that this is a sensible way to handle the constantly changing taxonomy and nomenclature of marine organisms.

· Near the end of his preface, he states: "Finally, it is recognized that new knowledge about the Gulf is accumulating rapidly and that some of the information in this book is being replaced as we write. Obsolescence in some details is the price which must be paid for painting the grand picture." Dr. Darnell's colleagues who reviewed the manuscript agreed with his wise statement.

In closing, I am pleased to offer this unique book in our Harte Research Institute for Gulf of Mexico book series at TAMU-CC, as it is truly a "grand picture" of the entire Gulf of Mexico told by one of the grand masters of this grand ecosystem.

*Wes Tunnell*
July 2014

# OBITUARY
## DEATH OF DR. REZNEAT DARNELL, TEXAS A&M UNIVERSITY OCEANOGRAPHER AND ECOLOGIST

Rezneat Milton Darnell Jr., well-known and respected ecologist and marine biologist, died on December 22, 2009, in Minneapolis, Minnesota, after a brief illness with pneumonia and suspected lung cancer.

Rez was born in Memphis, Tennessee, on October 14, 1924. His parents were Rezneat Milton Darnell Sr. and Matilda Millen Darnell.

Dr. Darnell was professor of oceanography at Texas A&M University from 1968 until his retirement in 1995 as professor emeritus.

Darnell graduated from Southwestern College in Memphis, Tennessee (now known as Rhodes College), with a bachelor's degree in zoology in 1946. He received his master's degree in biology and genetics from Rice University in Houston in 1948 and his PhD in ecology from the University of Minnesota in 1953.

Before coming to Texas A&M Darnell served three years as instructor at Tulane University in New Orleans, Louisiana, and thirteen years as assistant professor at Marquette University in Milwaukee, Wisconsin.

Dr. Darnell published numerous scientific papers on the ecology of the Gulf of Mexico. His last work, a comprehensive book on the ecology of the Gulf of Mexico, has been eagerly awaited by colleagues across the nation. His many graduate students have praised his lectures and his ability to clearly and concisely explain the complex environmental problems of our planet.

Dr. Linda Pequegnat, former research scientist in the Oceanography Department at Texas A&M and a friend and colleague of Dr. Darnell's, said: "Rez has been called 'The Great Synthesizer' because of his ability to take detached scientific information and organize it into meaningful overviews that explain the 'big picture' of ecological relationships in the natural world. He was also a 'Renaissance Man' with extensive knowledge and experiences in such diverse areas as music, languages, and history—in addition to his vast scientific knowledge. His book on the history, biology, ecology, and management of the Gulf of Mexico pulls together more information about the Gulf of Mexico than has ever before been amassed in one volume."

Dr. Darnell was preceded in death by his parents and by his older brother, Rowland Jones Darnell. He is survived by his loving daughter and caregiver, Molly Marie Darnell of Minneapolis, Minnesota; his brother J. Millen Darnell of Memphis, Tennessee; his first wife, Jeanne Hellberg Darnell, of Minneapolis, Minnesota; and many nieces and nephews.

Private memorial services and entombment were held January 22, 2010, at 2 p.m. at Lakewood Chapel Cemetery in Minneapolis.

# FOREWORD
## REZNEAT DARNELL: BRIDGING THE GULF

It might seem overly ambitious for one individual to undertake the synthesis of information gathered over centuries of exploration of the Gulf of Mexico and compress it within a single tome. After all, "America's Sea" has inspired hundreds of scholarly and technical works from weighty atlases, regional analyses, and hundreds of specialized papers to the recent multi-year, multi-volume, multi-author series — *The Gulf of Mexico, its Origin, Waters and Biota*. But Rezneat M. Darnell Jr. was happily blessed with the intellectual skills, personal depth and breadth of experience, professional stature, and sense of purpose lightened with an irrepressible sense of humor that combined to make possible this astonishingly comprehensive — and wonderfully comprehensible — tribute to Earth's tenth largest body of water, the fifth largest sea. Darnell conveys here an underlying, urgent call to action, a message that it is not too late to use our powers to reverse significant damage to one of the planet's most economically, socially, and biologically valuable liquid assets.

This assessment of the Gulf has come at the best possible time. A hundred years before, the ability to access the depths of the Gulf did not exist, nor did the ways and means of gathering data and seeing connections among seemingly unrelated topics. A hundred years later, changes that are sweeping the world will have obliterated or significantly altered the intact ecosystems Darnell witnessed, recorded, and celebrated in this monumental volume.

When Darnell was born in 1924, the world population was fewer than two billion, automobiles were just beginning to displace equestrian-powered transportation, whale oil and gas-lamps were giving way to electric light bulbs, few homes were equipped with amazing new party-line telephones, manuscripts were written by hand on sheets of paper, kids learned long division, sailors navigated using sextants and dead reckoning, and touching the moon or the deepest parts of the sea were dreams yet to be realized.

The decades of Darnell's life were marked by an unprecedented ferment of discovery for the nature of the world and the growing prosperity of humankind, which coincided with the depletion of terrestrial and marine systems that underpin our existence. He witnessed breakthroughs in technology that have made possible an awareness that in all of the universe, only Earth has the characteristics suitable for life as we know it. He could see in the Gulf of Mexico changes that reflect global trends of loss caused by human actions, inexorably altering the nature of nature. And, he was inspired to find ways to embrace the complex processes that shape the Gulf and through deep understanding of the geological, biological, physical, and human realms find ways to heal the harm.

In his introduction to this volume, Darnell expresses his concern, suggesting that Earth's life support system, governed by the ocean, ". . . cannot survive intact without human assistance." Large ecosystem management, he believed, was the key to maintaining conditions that until recently could be taken for granted. During most of the 20th century, it seemed the ocean was limitless in its ability to yield whatever we took out of it and accept whatever noxious things we wanted to put into it. But Darnell witnessed growing evidence of serious impacts in the Gulf; as he put it, a system ". . . under siege by a host of human-induced maladies including among others overdevelopment of coastlines, destruction of estuarine habitats, loss of coastal marshlands, chemical pollution, nutrient over-enrichment, physical damage to shorelines and ocean bottoms, accumulation of solid wastes, overexploitation of fishery stocks, intrusion by alien invasive species, and the effects of global warming."

His motive for dedicating a major part of his life to bridging disciplines, assembling and synthesizing data, making correlations, and communicating the results to his fellow scientists, students, decision makers, and the public at large was driven by a sense of urgency that it is vital to understand what the problems are in order to craft policies that would lead to healthy ecosystems — and therefore a planet that works in favor of humankind. For the Gulf of Mexico, he said ". . . we cannot manage and protect it without knowing and understanding it."

In the Gulf and globally, since the 1950s about half of the coral reefs have disappeared or suffered serious damage and 90 percent of many kinds of commercially exploited marine species have been extracted. Oxygen-producing and carbon dioxide-absorbing marshes, sea grasses, mangroves, and ocean phytoplankton have undergone serious decline while coastal "dead zones" have developed, fueled by fertilizers and toxic chemicals flowing from land-based sources. Increasing discharge of carbon dioxide into the atmosphere has driven global warming, sea level rise and alarming changes in ocean chemistry, notably increasing acidification.

Darnell lived to see Earth host more than three times the number of people than when he was born—though far fewer horses—and the advent of billions of automobiles, millions of aircraft, and thousands of high-flying spacecraft and satellites. He witnessed newly-harnessed fossil fuels revolutionizing agriculture, global communications, astonishingly accurate navigation, weather forecasting, and an unprecedented ability to acquire and analyze data. Patterns, connections, and relationships beloved by big-thinking scientist-ecologists such as Darnell came into focus, and in this volume, Darnell enhances those perceptions with his special insights in language, music, art, history, and human culture.

What Darnell missed seeing was the largest accidental marine oil spill in history, the *Deepwater Horizon* disaster of 2010 when roughly five million barrels of oil, 500,000 metric tons of natural gas, and about two million gallons of toxic dispersants were released into the Gulf. He had observed and taken into account the effects of drilling for oil and gas at more than 30,000 places in the Gulf of Mexico, the thousands of miles miles of pipelines lacing the sea floor and hundreds of miles of canals carved into coastal marshes and waterways. Many pages in this volume are devoted to the impacts of extracting offshore minerals, petroleum, and natural gas, including the noise from seismic surveys, the release of drilling muds, cuttings, and deep "fossil" water and the introduction of an archipelago of new metal "islands" in places where few hard surfaces had existed before. He noted the consequences of the massive 1979 *Ixtoc I* wellhead blowout and discharge of three million barrels of oil as well as numerous smaller spills that have continued to alter the nature of marine systems decades after they occurred. The *Deepwater Horizon* blowout added enormous stress to an already afflicted ocean.

Many who once thought that the Gulf and the ocean beyond were "too big to fail" are recognizing the wisdom of Darnell's mantra that active management based on a sound base of knowledge is necessary in light of ongoing human impacts. An "ecosystem" of institutions, government agencies, industries large and small, and individuals is looking at the Gulf with new respect, and many are working together to grapple with issues such as pollution, shoreline restoration, increasing population, growing ship traffic, offshore aquaculture, and expanding oil and gas operations. Proposals for a network of protected areas are being considered, and there is growing support for restoration of damaged reefs, beaches, seagrasses, marshes, and mangroves as well as enhanced protection for sea birds, turtles, sharks, tunas, and other marine life.

Future changes, good and bad, will demonstrate the importance of this book as a trusted source of knowledge about what the Gulf of Mexico has been in the past, and with care, can become again. In so doing, it will fulfill Darnell's wish that ". . . this volume will not only inform but will also aid in the preservation of the environment and living resources of this, the American Sea."

*Sylvia Earle*

# PREFACE

During the past few decades it has become apparent that the life support systems of this planet cannot survive intact without active human assistance. Broad appreciation of this fact has stimulated interest in a new endeavor, the science of large ecosystem management. But such ecosystems cannot be managed intelligently without a sophisticated input of technical information about the structure, composition, and functional processes of the systems being managed. The Gulf of Mexico itself is now under siege by a host of human-induced maladies including, among others, overdevelopment of coastlines, destruction of estuarine habitats, loss of coastal marshlands, chemical pollution, nutrient overenrichment, physical damage to shorelines and ocean bottoms, accumulation of solid wastes, overexploitation of fishery stocks, intrusion by alien invasive species, and the effects of global warming. In the face of these challenges we must attempt to manage the Gulf, but we cannot manage and protect it without knowing and understanding it. With these thoughts in mind, the author here endeavors to provide a broad perspective of the Gulf of Mexico ecosystem and its various subsystems and to present it in relatively non-technical language so that the information will be available to a wide audience of students, scientists, managers, and educated citizens of the public at large.

This work is an outgrowth of two graduate courses, one called Ecology of the Continental Shelf and the other Organic Cycles of the Sea, which the author taught for many years. It builds on his own research and that of his graduate students, and it brings together knowledge accumulated in various literature summary and data synthesis reports and research reports submitted to government agencies by the author and his colleagues and co-workers.

This book is intended to be used as a text in advanced undergraduate and beginning graduate-level courses, and the writing has been pitched toward this level. However, it should also serve as a reference work for scientists, resource managers, and other professionals who may need greater depth and detail. Chapter 2, "An Introduction to Marine Science," has been included to provide background for those readers previously untrained in marine science. New terms and concepts are defined where first presented in the text, and from the index these definitions may be easily located.

It is important that all readers be able to grasp the complexity of the web of life and the diversity of species in the many biological groups inhabiting the Gulf of Mexico, and the author has gone to great lengths to assemble accurate lists and provide illustrations of representative plants and animals that dwell in the Gulf. However, for the most part the taxonomic names and other specialized information have been relegated to tables. Where present in the text, they are generally given as parenthetical notes so as not to compromise the narrative. The metric system is used throughout, but equivalent data in English units are sometimes provided.

The information on which this book is based is widely scattered in technical books and journal articles, but much of it is hidden in the so-called gray literature. This includes unpublished material located in various project progress and final reports submitted to government agencies, numerous government documents, and graduate student theses and dissertations. There is also considerable information in Spanish, which is not generally available to audiences in the United States because of difficulty in locating the sources or because of the language barrier. Throughout the chapters of this book, extensive literature references have been given to document the information presented and to aid readers seeking to pursue certain topics in greater detail. Much of the literature about the Gulf treats specialized subjects or local areas, so that it is often difficult to grasp the larger picture. However, in recent years several major works have appeared summarizing broad sweeps of this knowledge (see table S5).

Although new information is accumulating at a rapid pace, there are still large gaps in our knowledge of the Gulf of Mexico. Estuaries and other marginal waters and wetlands have been the most thoroughly studied. Continental shelves, especially those of the northern and eastern Gulf, are fairly well known, but our knowledge of the bottoms and water column of the deep Gulf is still in its infancy. Primarily because of the interest in the location and exploitation of petroleum resources, we know a great deal about the topography, surface sediments, and underlying geology of the Gulf. Shipboard observations and satellite data have contributed to our knowledge of circulation patterns in surface waters, but we still do not fully understand the patterns and processes of circulation in deep, subsurface waters of the Gulf. Detailed chemistry of the waters and sediments is not well known. We are fairly knowledgeable about the biology of organisms associated with the sediments, both the infauna and the epifauna, as well as the larger organisms that swim or hover near the bottom, but much less is known about inhabitants of the open water column (the phytoplankton, zooplankton, and larger, free-swimming animals, the nekton). We have accumulated long lists of species that occur in the various habitats, but we are woefully ignorant about how most of the species live and interact with one another. Against this background of spotty information the author has attempted to provide a coherent and balanced picture of various aspects of the Gulf ecosystem. This effort to describe and explain such a vast and complex subject has necessarily resulted in more of a summary of knowledge than a true synthesis. Yet the work does provide the most complete and comprehensive view currently available of the environment, biology, and ecology of this remarkable system, based on our present knowledge.

The book is profusely illustrated with tables and figures. The tables provide detailed information that should be useful to the specialist but that should also prove informative to the nonspecialist. General readers need not dwell on the individual scientific names, but they cannot fail to be impressed with the variety of categories and diversity of species inhabiting the Gulf. They will be interested to learn that this body of water is home to nearly 100 species of squids and octopuses, more than 1,500 species of fishes (about one-tenth of all the world's known marine fish species), 5 species of sea turtles, and more than 25 different types of marine mammals (whales and dolphins, a manatee, and formerly a seal). The figures include various maps and charts as well as shaded line drawings of representative plants and animals to give visual meaning to many of the scientific names and to create additional interest in the details. Many of the species inhabiting the Gulf are of tropical origin. Some are northern species, typical of cooler climates, and are residual in the Gulf, relicts of the recent Ice Ages when Gulf waters were cooler than

at present. Many are cosmopolitan species, being widely distributed through both warm and cool marine environments. However, in most groups a few of the species are endemic to the Gulf, having never been found elsewhere in the world oceans.

A few caveats are in order. About 3,300 different species of microorganisms, plants, and animals are mentioned in this volume. To keep track of such diversity, scientists have developed classification systems that include a hierarchy of categories from the most inclusive (kingdom and phylum) on down to the most specific (genus and species). These systems are arranged so that the most closely related organisms are in the same or nearby categories. However, even the specialists do not always agree on some of the categories, and this is especially true for the microorganisms, which are often difficult to study. For purposes of the present volume, the author has adopted a somewhat simplified classification scheme (see table 2.4), which is more or less consistent with that given at www.itis.gov and with Ruppert et al. (2004) for the invertebrates and with McEachran and Fechhelm (1998, 2005) for the fishes.

In a few cases the author has decided to stick with old, widely used scientific names rather than use their recent updates (e.g., in the penaeid shrimps, *Penaeus* rather than *Farfantepenaeus, Litopenaeus,* etc.). Some simplification has also occurred in the higher classification of the sharks and some of the bony fishes.

Modern meteorology and physical oceanography depend heavily on mathematical equations and computer models of the processes involved. These are considered to be advanced topics and are not included in the present discussions.

Finally, it is recognized that new knowledge about the Gulf is accumulating rapidly and that some of the information in this book is being replaced even as it is written. Obsolescence in some details is the price that must be paid for painting the grand picture. Most of the information herein is up to date.

Conceptually, the book is divided into five sections: Introduction (two chapters), Physical Environment (five chapters), Biology (seven chapters), Ecological Processes (two chapters), and Human Relations (two chapters). When originally conceived, the book was to cover only the science (i.e., the first sixteen chapters). However, during the writing it became clear that enough information was available to support the two additional chapters and that the volume would really be incomplete without them. Having worked closely with various state and federal agencies during the past four decades, the author was in a unique position and therefore had an obligation to discuss and explore human impacts and management perspectives. In our present world, science and management must play complementary roles in protecting the environment and our natural resources. Hopefully, this volume will not only inform but will also aid in the preservation of the environment and living resources of this, the American Sea.

For an interesting manifesto of the author's strong belief in the interconnectedness of all living things, see his "Declaration of Dependence," published in *BioScience* (vol. 20, no. 17) September 1, 1970, available at http://bioscience .oxfordjournals.org/content/20/17/945.extract.

# ACKNOWLEDGMENTS

The present volume, over a decade in preparation, addresses many dimensions of the Gulf of Mexico, and through these years the author has received assistance from numerous colleagues who have aided in the location of literature; permitted examination and use of unpublished material; read and commented upon portions of the manuscript; provided photographs, maps, or drawings; and assisted in numerous other ways. All of this help is deeply appreciated and gratefully acknowledged here.

Assistance with the vast and scattered literature has been provided by Douglas C. Biggs, Luis A. Cifuentes, Steven F. DiMarco, Sayed Z. El-Sayed, Wilford D. Gardner, Stefan Gartner, Ann E. Jochens, John W. Morse, Worth D. Nowlin, Jr., Vita Pariente, Bobby Joe Presley, Maureen E. Reap, Rehilla Shatto, Niall C. Slowey, and John W. Wormuth of the department of oceanography; Dusan Djuric, Robert Duce, and Gerald North of the department of atmospheric sciences; John D. McEachran of the department of wildlife and fisheries sciences; Norman L. Guinasso Jr. and Terry L. Wade of the Geochemical and Environmental Research Group (GERG); and John Firth of the Ocean Drilling Program (ODP); all of Texas A&M University at College Station.

Literature help was also provided by G. Fain Hubbard and Gilbert T. Rowe of Texas A&M University—Galveston; David W. Hicks and John W. Tunnell Jr. of Texas A&M University—Corpus Christi; Edward J. Buskey, Paul A. Montagna, Dean Stockwell, and Tracy A. Villareal of the University of Texas Marine Laboratory in Port Aransas; F. Quay Dortch, Michael J. Dagg, and Nancy N. Rabalais of the Louisiana Universities Marine Consortium (LUMCON) in Chauvin, Louisiana; and Elva G. Escobar-Briones of the Instituto de Ciencias del Mar y Limnología, Universidad Nacional Autónomo de México (UNAM), México D.F., México.

Additional literature assistance was furnished by James D. Simons, Texas Parks and Wildlife Department in Corpus Christi; Thomas Minello of the NOAA Fisheries Laboratory in Galveston, Texas; Patricia A. Tester of the NOAA Fisheries Laboratory in Beaufort, North Carolina; Richard E. Defenbaugh, Robert M. Rogers, and James Sinclair of the Minerals Management Service in New Orleans, Louisiana; and Chris Dorsett of the Ocean Conservancy Gulf of Mexico Field Office in Austin, Texas.

Permission to use unpublished manuscript material was generously granted by Bernd Würsig of Texas A&M University—Galveston; David J. Schmidly of the office of graduate studies at Texas Tech University in Lubbock; Thomas A. Jefferson of Clymene Enterprises; William J. Wiseman Jr., Department of Ocean and Coastal Studies, Louisiana State University in Baton Rouge; Steven E. Lorenz, Alan M. Shiller, and Dennis Wiesenburg of the Center for Marine Sciences, University of Southern Mississippi at the Stennis Space Center in Slidell; A. Arnone, Naval Research Laboratory, Stennis Space Center, Slidell, Mississippi; and Wilton Sturges, department of oceanography, Florida State University, Tallahassee.

The author is also indebted to the following individuals who read and commented upon whole chapters or portions of chapters: Stefan Gartner (chapter 3), William R. Bryant and Niall C. Slowey (chapter 4), Worth D. Nowlin Jr. and Edwin Shaar Jr. (chapter 6), John W. Morse (chapter 7), and John H. Wormuth (chapters 9 and 13), all of the department of oceanography, Texas A&M University. Dusan Djuric of the department of atmospheric sciences read chapter 5. Gilbert T. Rowe of Texas A&M University—Galveston read chapters 14, 15, and 16. Robert B. Abel, Stevens Institute of Technology, Hoboken, New Jersey, and Ralph Rayburn of the Texas Sea Grant College Program in College Station read and commented on chapter 18. In addition, Linda H. Pequegnat of La Jolla, California, read most of the manuscript and commented in detail on chapters 13, 14, 15, and 16.

Thanks are due to many colleagues for detailed discussions which helped to clarify complex matters concerning the Gulf, and the following deserve special mention. Stefan Gartner and John Firth aided in explaining the effects of and evidence for the extraterrestrial bolide that crashed into the Yucatan Platform 65 million years ago. Niall C. Slowey aided in understanding the most recent information regarding the decay of $^{14}$Carbon and other isotopes and their use in dating the post-glacial rise in sea level. Worth D. Nowlin Jr. and Steven F. DiMarco discussed recent information concerning surface circulation in the

Gulf. Gilbert T. Rowe and Linda H. Pequegnat helped elucidate the environmental conditions and the abundance and diversity of life in the deep Gulf. Gerald North and Ping Chang discussed potential effects of global warming on the Gulf area. Thomas Minello, Ralph Rayburn, Robert B. Abel, and Chris Dorsett provided important perspectives on problems associated with marine fisheries management in the Gulf of Mexico. To all these persons the author expresses his deep appreciation.

Typists, artists, and computer experts were important in the production of this volume. The entire text, including the original and revised versions, was typed by Barbara C. Childers. The tables and revisions were typed by Silvia Mendez-Pitts, Barbara C. Childers, Sandra K. Drews, and Diana Vance. Linda Lantz typed the figure captions, and Diana Vance typed most of the front material. Most of the hand-drawn art work was completed by G. Fain Hubbard and Heather Prestridge, but some of the drawings were made by Linda Leatherwood and Marilyn S. Yeager. Computer generated or computer-enhanced figures were created by Heather Prestridge, Leigh Holcombe, and Don Johnson. Ruth Mullins, Heather Prestridge, and Diana Vance aided in obtaining permissions to use published materials.

During the preparation of this book financial support has generously been provided in the form of grants from the Texas Sea Grant College Program, the Harte Research Institute for Gulf of Mexico Studies of Texas A&M University—Corpus Christi, and the Minerals Management Service of the US Department of the Interior. The college of geosciences of Texas A&M University supplied funds for student assistance and art work, and the department of oceanography of Texas A&M University facilitated the project in numerous ways, including unrestricted use of the photoduplicating equipment. For all of this support the author expresses his sincere thanks.

Finally, it is important to acknowledge my deep appreciation to individuals who have been especially important in conceiving this work and in bringing it to fruition. Willis E. Pequegnat played a major role by introducing me to the field of oceanography and by broadening my perspective from the estuaries and continental shelves into the deep Gulf. Linda H. Pequegnat helped in numerous ways, as acknowledged above. She also provided identifications of the deep-water shrimp species, made available bottom photographs of the deep Gulf, and prepared a taxonomic guide and index for the bacteria, fungi, algae, higher plants, and invertebrates. Her continued inspiration, support, and encouragement have been instrumental in seeing this project to completion. Heather Prestridge of the Texas Cooperative Wildlife Collections did much of the art work and computer drawings, as noted above. She also completed a taxonomic guide and index for the vertebrate animals, and she reformatted and retyped the combined guide and index. For her expertise, dedication, and persistence the author is especially appreciative. Thanks are also due to the author's daughter, Molly M. Darnell, for her continued support and help with numerous chores associated with production of this volume, especially transportation and the vast amount of photoduplication.

The sources for the color plates are acknowledged individually in the plate captions. However, the author here gives thanks to all the individuals and organizations who supplied the photos and gave permission for their use.

# THE AMERICAN SEA

# Part I

**Figure 1.1.** Reproduction of a pre-Hispanic wall painting from the Temple of Warriors, Chichén Itzá, Yucatán, depicting a Mayan coastal village. Shown in the water are three dugout canoes and several types of snails, clams, crabs, and fishes, as well as a sea turtle. Dried fishes are also hanging in the shed at the center of the scene. Although some of the animals are represented by stylized drawings, this and other paintings and sculptures emphasize the familiarity of early coastal inhabitants with the marine environment and aquatic life. (From Morris, Charlot, and Morris 1931. Reproduced by courtesy of the Carnegie Institution of Washington.)

# 1 : HISTORICAL BACKGROUND

Before we address the more technical aspects of the Gulf of Mexico, it will be of interest to briefly examine the fascinating history of the area. The story begins with the indigenous peoples and the discovery, exploration, and conquest by Europeans five centuries ago. It continues through a period of national rivalries and wars of independence. It concludes with political stability and the advancement of scientific knowledge. The discussion of the early history of the Gulf area relies heavily on the summary by Galtsoff (1954a).

## PRE-COLUMBIAN PERIOD

Archaeological evidence, mainly from shell middens, reveals that during the pre-Columbian period Native Americans living along shores of the Gulf exploited fishes, shellfishes, and other resources of the bays and estuaries and very shallow waters of the Gulf itself, but most apparently did not venture far from shore. Exceptions include the more highly organized societies of the southern Gulf. Evidence from archaeological remains, wall paintings (fig. 1.1), and sculptures of the Mayas and Aztecs, as well as written accounts left by early Spanish explorers such as de Landa, indicate that these peoples were very active in exploiting marine fishery resources and that this activity, although concentrated largely in shallow coastal waters, extended well offshore. Shells and other marine products were used for ornamentation and were featured in festivals and certain religious ceremonies. Large quantities of marine animals were used as food, and these ranged from mollusks, crustaceans, and echinoderms (starfishes and their relatives) to fishes, sea turtles, and marine mammals. Fishes were captured by a variety of means including wicker traps, set nets, seines and tow nets, hooks and lines, and probably poisoning by rotenone derived from the sap of local plants. Most of the fishes were consumed locally, but some (preserved by drying, salting, smoking, and roasting) were traded inland. Manatees, taken by spears, were consumed at festivals and were particularly prized for their fat. Many types of marine fishes were utilized, and Baughman (1952) attempted to identify some of the species listed in de Landa's account (from a translation by Tozzer 1941). Included were various shallow-water sharks and rays, tenpounders (ladyfish), anchovies, snooks, mullets, and flounders as well as some species of more open waters, such as mackerels and barracudas.

There is also evidence of considerable marine transportation and commerce (Andrews 1998; Healy, McKillop, and Walsh 1984; Romero 1998). Over-water trade routes extended from Guatemala, Belize, and points south to Yucatán and the mainland coast at least as far north as Veracruz. Navigation was enhanced by a series of high masonry towers built along the coast and on offshore islands, and illuminated with fires; these served as lighthouses to guide the marine travelers at night. One of the Spaniards recorded that the Aztec king Montezuma presented Cortés with a map of the Mexican coast from the Río Pánuco to the southern Gulf showing the rivers and bays with great accuracy. Also, Columbus reported that in 1502 he sighted a large Indian ship the size of a Spanish galleon about 130 km east of the Yucatán coast. Clearly, these more advanced native peoples possessed some sophisticated knowledge of mapmaking and marine navigation as well as wind patterns and water currents.

## EARLY EXPLORATION AND MAPPING

Following discovery of the New World in 1492, explorations proceeded rapidly, but details of the early events cannot be fully reconstructed. Among the competing nations of Europe there was much secrecy, and documents detailing some of the earliest activities have not been located. Mapmakers of most European nations eagerly sought information about new discoveries, and maps of the time suggest some knowledge of the existence of the Gulf by 1500. Although the chief explorers of the Gulf and New World, in general, were of Spanish origin, Portugal, France, Holland, and Great Britain were also heavily involved. The islands of the West Indies were quickly overrun, and Cuba became the staging area for most of the Spanish explorations of the Gulf, although some expeditions were also sent from Haiti.

In 1517 Francisco Hernández de Córdoba left Havana, Cuba, with three ships bound for Honduras. Blown off course by a violent storm, they finally landed at Cabo Catoche on the northeast tip of Yucatán. From there they sailed westward along the northern coast and turned south along the northwestern coast of Yucatán to Champotón, just inside the Bay of Campeche. At every landing the natives were fiercely hostile, and most members of the land parties were killed or wounded. Córdoba himself was mortally wounded at Champotón, and the small fleet was forced to return to Cuba. This expedition was the first to enter the Gulf of Mexico, and it returned with samples of gold ornaments worn by the natives as well as information concerning the fine dress and high civilization of the indigenous people.

In Cuba news of the gold spread like wildfire. The following year Juan de Grijalva, nephew of Córdoba, left Havana with four ships and retraced Córdoba's route along the northern coast of Yucatán, but they continued along the coast of the southern and southwestern Gulf to near the present site of Tampico. Some of the Indians proved hostile, but in other instances they were friendly, and the Spaniards were able to barter for gold objects. This expedition expanded geographic knowledge of the coastline and native inhabitants, and it provided further evidence of the wealth of the native civilization. By then it was fairly clear that Yucatán was not an island, as first supposed, but part of a much larger land mass lying to the west.

Hernán Cortés was then commissioned by the governor of Cuba to continue the explorations and to barter with the Indians for more gold objects. Leaving port in February 1519 with a flotilla of 11 ships and more than 800 men (including 553 heavily armed soldiers), as well as an array of cannons and smaller arms and ample ammunition, Cortés was prepared for conquest, not just exploration. Retracing the route of Grijalva, this expedition explored the coastline more thoroughly all the way to the mouth of the Río Pánuco at Tampico. Establishing a base among the friendly Indians of the Veracruz area, Cortés marched inland to conquer the Aztec empire, which he accomplished in 1521. Details of his dealings with the Indians of the southwestern Gulf and of the conquest itself are provided in Prescott (1964).

In 1519 Álvarez de Piñeda, sailing from Jamaica, explored the northern Gulf and discovered the mouth of the Mississippi River, which he named Río de Espíritu Santo. Continuing westward along the Louisiana and Texas coasts and southward along the Mexican shoreline, he arrived at Veracruz, where he encountered the large fleet of Cortés. Retracing his route along the northern Gulf, he eventually returned to his home port in Jamaica. This expedition mapped the shorelines of the northern and western Gulf.

As seen in table S1, within 30 years of the discovery of the New World, Spanish explorers had completely navigated the shorelines of the Gulf. By then they knew that Florida and Yucatán were not islands, as previously thought, but peninsulas, and that a large, previously unknown continent surrounded the Gulf. By 1519 they had viewed the major rivers of Mexico, and Piñeda's discovery of the Mississippi River in the same year indicated a large land mass to the north. By the time he had conquered the Aztecs, Cortés was well aware that this area was part of a continent. Much credit for the early mapping of the Gulf of Mexico goes to Antonio de Alaminos, chief navigator for the expeditions of Córdoba in 1517, Grijalva in 1518, and Cortés in 1519–1521. On the return voyage of the Córdoba expedition, Alaminos sailed northeast from the western tip of Yucatán to Florida and then south to Cuba. On this and succeeding expeditions he mapped over 800 km of the Mexican coastline, and, based on astronomical observations, some of his positions were accurate to within one degree of latitude. His maps give place names for many of the rivers, bays, islands, and other coastal features observed. He provided hydrographic information from depth soundings and observations of wind and water currents, all essential for navigation. Of particular importance for the Spanish, he determined that the shortest route for ships sailing from Veracruz to Spain was north of Cuba through the Straits of Florida.

In 1527 an expedition led by Pánfilo de Narváez to explore the northeastern Gulf resulted in disaster. In a violent storm, apparently off Mississippi Sound, all of the ships and most of the men were lost. One of the crew, Cabeza de Vaca, and two companions survived with the local Indians. During a period of seven years they traveled westward along the northern Gulf coast and overland before eventually encountering another group of Spaniards on the Mexican west coast near the Gulf of California. Cabeza de Vaca and his companions were apparently the first Europeans to view many of the lands bordering the northern Gulf and in northern Mexico. In his journal, published in Spain in 1542, Cabeza de Vaca described in detail the lands traversed and the nature of the inhabitants. Most of the natives encountered eked out a meager existence by hunting and gathering, and there was no evidence of the great riches and high civilization characteristic of the Indians farther south.

By present standards the early maps were crude, and

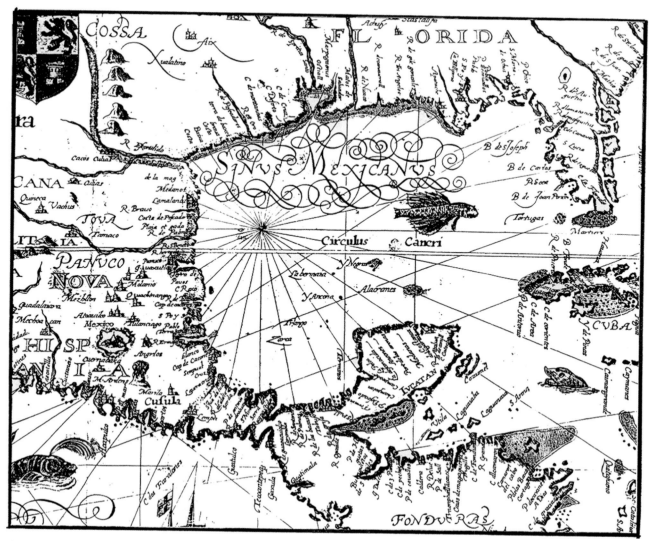

**Figure 1.2.** Map of the Gulf of Mexico and surrounding areas drawn by Gabriel Tatton and engraved by Benjamin Wright in 1600. (Reproduced by *American Heritage* magazine, from original in the American collection of Thomas W. Streeter of Morristown, New Jersey.)

place names did not become fixed for some years. On these early maps the Gulf of Mexico was variously called Mar del Norte, Golfo de Florída, Golfo de Cortés, Sinus Magnus Antilliarum, Mare Cathaynum (Chinese Sea!), and Golfo de Nueva España. The name "Gulf of Mexico" first appeared on a world map in 1550. Following the conquest of the Aztec empire in 1521, treasure fleets flowed from Veracruz to Spain. Dependent on the winds and lacking storm warnings, most of the fleets followed the coastline of the western and northern Gulf before putting in for provisions at Havana. Spanish maps of the time show the Mississippi River mouth labeled Cabo de Lodo (mud cape). During the following century Spain made only modest progress in improving maps of the Gulf,

although details were added concerning local harbors, depths, and water currents. A 1600 map of the Gulf and Caribbean shows the state of geographical knowledge at that time (fig. 1.2).

## REDEFINITION OF POLITICAL BOUNDARIES

From the time of first discovery, all lands surrounding the Gulf of Mexico were claimed by Spain. Until the late 1600s these claims were not seriously challenged, but the situation would soon change. Francis Drake and other British and Dutch privateers attacked Spanish shipping and forced the Spanish ships to travel in large well-guarded fleets. In 1588, with Dutch help, Britain defeated the great Spanish Armada. Thereafter, Britain challenged Spain's domination of the high seas and placed Spanish colonies in jeopardy. In the 1600s Britain established its own colonies on the eastern seaboard from Georgia through New England and raided Spanish outposts in Florida. France established settlements in Canada, where it pursued the lucrative fur trade. In 1682 La Salle descended the Mississippi River from Canada to its mouth and claimed all surrounding lands for France. Two years later, when he returned to found a French colony, he missed the mouth of the Mississippi River and landed at Matagorda Bay on the central Texas coast. After two years of futile searching for the river's mouth, he and some of his band headed northeast overland for Canada, and he was murdered en route. New Orleans was eventually founded in 1718, giving the French a foothold in the Gulf and control of the Mississippi River. Subsequent rivalry among the European powers led to increased interest in mapmaking for military purposes, and much information was accumulated concerning natural harbors around the Gulf.

Events in the establishment of political boundaries in lands surrounding the Gulf of Mexico are shown in table S2. Louisiana passed from French to Spanish and back to French hands before it was sold to the United States in 1803. Florida, which then extended westward to the Mississippi River, passed from Spanish to British and back to Spanish control before the United States purchased it in 1819. Texas, which won its independence from Mexico in 1836, joined the Union in 1845, but its southern boundary was not fixed at the Rio Grande until the end of the US-Mexican War in 1849. Meanwhile, Mexico had succeeded in winning its independence from Spain in 1821, and Cuba finally became independent in 1898. Thus, Spain, which had claimed all lands surrounding the Gulf until 1718, was largely out of the picture by 1821 and completely out by 1898. The United States, which gained its first foothold in 1803, controlled the entire northern Gulf by 1849.

## EARLY SCIENTIFIC STUDIES

With the advent of political stability, and especially with the desire to advance the commercial and scientific interests of the United States, knowledge of the Gulf of Mexico grew steadily. Prominent scientific studies during the early period (1839–1939) are shown in table S3. In 1807 Congress established the US Coast Survey (now the National Ocean Survey). Under the aegis of this office, hydrographic mapping of the Gulf began in 1839 and has continued ever since. By the mid-1800s, as an aid to navigation, attention had focused on the Gulf Stream and re-

lated currents within the Gulf of Mexico. Early studies were concerned with mapping the surface currents, but by the 1930s investigations were underway to understand the three-dimensional structure of the basin water and its relationships with the surface currents. By the late 1930s efforts were being made to map the submarine topography and geological characteristics of the deep Gulf basin.

Major biological studies in the Gulf began with the investigations of Florida coral reefs by Louis Agassiz between 1850 and 1880. During the period 1867–1885 several cruises were sent to the Gulf to ascertain the nature of the deepwater fauna. These studies, carried out by Louis François de Pourtalés, Henry Mitchell, Alexander Agassiz, and others, laid the foundation for future investigations of the continental slope and abyss and demonstrated the existence of a thriving deepwater animal community to a depth of over 3,600 m. It had been widely believed that the deep sea was devoid of animal life, and the Gulf findings stirred considerable interest in the scientific community. The British expedition of the HMS *Challenger* (1872–76) was shortly to confirm the existence of deepwater animal life throughout the Atlantic and Pacific Oceans and at much greater depths (over 8,000 m).

In 1871 Congress established the US Commission of Fish and Fisheries (now NOAA Fisheries), and in 1884 the agency commenced studies of the commercial fishery potential of bays, estuaries, and the nearshore continental shelf of the northern Gulf. Beginning with investigations of oyster and sponge grounds, this work expanded to include shrimp and bottom fish harvesting areas. Several marine laboratories were set up to examine problems such as diseases and parasites of oysters and other commercially important species. An exception was the Carnegie Laboratory in the Dry Tortugas, which operated from 1904 through 1939. Here long-term studies were carried out on many invertebrate and fish species of little or no commercial importance. By the onset of the Second World War considerable knowledge had been gained concerning the flora and fauna of the bays and estuaries. The importance of the continental shelf in the life histories of many estuarine species was understood, primarily as a result of the work of John Pearson. Some information had also accumulated about coral reefs, sponge grounds, and other nearshore areas of the continental shelf, but for much of the Gulf, especially the outer continental shelf and deep basin, information was quite sketchy.

## WARTIME AND POST-WAR EXPANSION

During the early 1940s considerable wartime activities centered around ports of the northern Gulf. Notable were those of several ship-building yards near New Orleans, which turned out patrol torpedo (PT) boats, landing craft, and other vessels at a rapid pace, and the petrochemical industries of the lower Mississippi River and along the Houston Ship Channel above Galveston, Texas, which provided fuel for the armed forces. Shore batteries with gun emplacements were set up to protect the major ports, and naval and air bases extended from Pensacola, Florida, to Corpus Christi, Texas. Nevertheless, German U-boats lurking off the mouth of the Mississippi River took a heavy toll on coastal shipping before they themselves were sunk in the "battle of the Gulf."

In the years since the end of the Second World War, knowledge of the Gulf of Mexico has grown at an accelerating pace as a result of a number of factors, chief of which are expanded economic activities (especially the petroleum and commercial fishing industries) and interest by federal and state agencies, which have provided increased funding for marine studies. During the same period much new marine technology has been developed, and there has been a steady increase in scientific personnel and facilities for studying both the shallow and deeper portions of the Gulf. These factors are taken up below.

### Economic expansion

The postwar increase in economic activities is illustrated by the petroleum and commercial fishing industries. The petroleum industry first ventured into the Gulf in 1938 with the opening of the Creole Field 2 km offshore from the Louisiana coast. In 1947 the Ship Shoal Field came in 19 km (12 mi) offshore from Terrebonne Parish, Louisiana. This was followed in 1948 by the Main Pass Field near the mouth of the Mississippi River, and the following year by three major discoveries: the Eugene Island, Bay Marchand, and Vermilion Fields, all off Louisiana. Subsequent discoveries came quickly, and by the early 1980s more than 70 million acres had been leased, more than 20,000 km of pipeline had been laid, and more than 6 billion barrels of oil and 60 trillion cubic feet of natural gas had been produced by wells on the outer continental shelf. These were primarily off Louisiana, but they extended from Texas through northern Florida.

The commercial fishing industry expanded after the war largely because of better ships, more powerful engines and deck winches, refrigeration, better fishing gear, and improved processing facilities on shore. The exploratory fishing program of the National Marine Fisheries Service, operating out of Pascagoula, Mississippi, pioneered many of these developments and also located new offshore fishing grounds. Although many species were harvested in the bays and estuaries and on the continental shelf, the primary target species soon became the Gulf menhaden (a type of herring) and three species of commercial shrimp. From very small prewar catches, by the 1980s over a million metric tons of fishes and shellfishes were being landed Gulfwide.

The petroleum and commercial fishing industries have stimulated growth in other economic activities such as manufacturing, processing, shipping, and service industries. Additional activities are associated with the recovery and processing of sulfur, salt, shell, and construction materials. At the present time, over one-sixth of the US population resides in Gulf coastal states. Ninety percent of the nation's offshore oil and gas comes from the Gulf, and 45 percent of the country's shipping passes through Gulf ports. Between 1956 and 1984, income from outer shelf oil and gas leasing brought $76 billion to the US Treasury, making it the nation's largest single revenue source outside of federal income taxes. Forty percent of the country's commercial fish landings and one-third of the recreational marine fishing activities in the continental United States now occur in Gulf waters.

Increased human intrusion has been accompanied by a series of problems. Chemical pollutants from municipal, industrial, and agricultural sources have found their way into estuaries and coastal lagoons and eventually into the open Gulf. Shipping accidents and oil spills have contaminated beaches and coastal waters. Discarded objects have accumulated as debris in the inside waters and on the open beaches, and to protect public health from a series of contaminants and infections, fishing grounds have been closed. Economic progress has come at a price.

### Increased interest by state and federal agencies

Lands bordering the Gulf and the enclosed coastal waters have long been considered to fall primarily within state jurisdiction, although the federal government has retained an interest in such matters as maintenance of navigable waterways and public health issues. Jurisdiction in the open Gulf is another matter. Laws relating to land and inland waters are inadequate to deal with complex legal issues beyond the shoreline. It needed to be determined just how far seaward federal jurisdiction extended and how much within this area could be considered to fall within the jurisdiction of coastal states.

In the postwar years, as it became clear that significant recoverable petroleum and other resources lay beneath continental shelf waters, federal and state action was swift. In 1945 President Truman issued a proclamation declaring that the United States considered natural resources of the seabed and subsoil of the continental shelf to be owned by the nation and that the exploration and development of the natural resources of this area should come under the authority of the US Department of the Interior. As a follow-up, in 1953 Congress passed the Submerged Lands Act and the Outer Continental Shelf Lands Act. These laws and subsequent litigation established the seaward limits of state jurisdiction and confirmed the role of the US Department of the Interior in managing all marine submerged lands beyond state jurisdiction, that is, what has come to be known legally as the Outer Continental Shelf (OCS). In most cases state jurisdiction was considered to extend three nautical miles from the coastline. However, in the case of Texas and Florida, whose historical titles went back to Spanish rather than English law, within the Gulf the states were given jurisdiction three marine leagues (about nine nautical miles) seaward from the coastline.

In 1969, in response to growing problems of pollution and other forms of environmental degradation, Congress passed the National Environmental Policy Act, which, among other things, directed all federal agencies to plan their policies and actions in such a way as to ensure that environmental factors receive adequate consideration and that environmental effects are understood in advance as demonstrated through preparation of environmental impact statements. This Act generated a compelling need to know about the environment of the US Gulf coast and continental shelf, which could be degraded through federal actions. At a time when the National Science Foundation was providing funds for basic research, various federal agencies began funding mission-oriented studies of the Gulf. Involved in a major way were the US Navy (through the Office of Naval Research), US Army (through the Corps of Engineers), US Department of the Interior (through the Bureau of Land Management and Geological Survey, portions of which were later joined to form the Minerals Management Service), and US Department of Commerce (through the National Marine Fisheries Service, Sea Grant Program, and National Ocean Survey), as well as the US Department of Energy, National Aeronautic

and Space Administration (NASA), and others. All states bordering the Gulf had established coastal laboratories with very active marine programs that contributed to the growing knowledge of bays and estuaries and the nearshore continental shelf.

## Development of marine research technology

During the postwar years, scientists and engineers worldwide turned their attention to the development of technology for exploration of the sea. Knowing the precise location of a ship's position at sea had always been a problem. In the early postwar years, radar triangulation from known targets on shore gave the approximate position of a ship, but precise location (within a few meters) became possible with the advent of satellite navigation. In the 1920s the echo-sounder was developed to determine the depth of water, and it was used aboard the German research vessel *Meteor* to map the ocean bottom between South America and Africa. An improved version known as the precision depth recorder (PDR) is now standard equipment on all research vessels, and it provides a continuous record of water depths traversed during each cruise. In order to understand the physical structure of the water column, scientists must measure very accurately the vertical distribution of such properties as temperature, salinity, and oxygen content. This was formerly accomplished by laboriously taking water samples from various depths at each station and analyzing each sample in laboratories aboard ship or on shore. Now this is all accomplished electronically by sensors on lowered wires and with shipboard displays of real-time data, electronic recording, and computer analysis. To measure the dynamic three-dimensional structure of the water column, current-meter arrays may be deployed with built-in recording devices. The International Ocean Discovery Program (IODP) operates a deep-sea drilling ship (*JOIDES Resolution*), which is capable of taking bottom core samples over 2,000 m long and from water depths in excess of 6,000 m. When analyzed, such sediment columns can reveal the past history of the earth, oceans, and atmosphere. Recoverable packages have been developed for recording deep-sea panoramas, events, and processes. These may include waterproof and pressure-resistant batteries, lights, triggers, still and motion-initiated picture cameras, laboratories for measuring respiration and other processes, data recorders, signal transmission devices, and other apparatus.

For taking quantitative biological samples, many forms

of collecting devices are now available including some that can be opened and closed at a specified depth, thus avoiding contamination by specimens from other depths. Mapping of wide swaths of sea bottom is now carried out with side-scan multiple-channel sonar, and the sea floor can also be investigated with color television cameras mounted on remotely operated vehicles (ROVs). Satellites now carry instrumentation for mapping clouds and other meteorological phenomena, sea surface water temperatures, concentrations of the plant pigment chlorophyll, and other features, and they have the advantage of viewing wide expanses of the ocean surface. In a few hours the surface of the entire Gulf of Mexico can be mapped. A variety of small submersibles is available for taking human observers into the sea, and these have operated at depths of over 6,000 m. Finally, development of the Aqua-Lung and scuba gear has permitted humans to swim freely among kelp beds, through coral reefs, and under the Antarctic ice and to make observations and carry out shallow-water studies that previously would have been impossible. These and many other technological developments have greatly increased our ability to investigate and understand the Gulf and the world oceans in general.

## Recent scientific studies

During the postwar years, with expanded economic interest and the availability of new technology, scientists have focused increased attention on the Gulf of Mexico. Early effort was confined largely to the bays, estuaries, and coastal lagoons, that is, areas that could be investigated with small boats and limited shore facilities. Research on the continental shelf and open Gulf became possible in the 1960s with the advent of larger and better-equipped ships and higher levels of funding, primarily from the National Science Foundation and Office of Naval Research. However, the era of large research projects on the Gulf shelf and slope came in the 1970s with the great expansion of offshore petroleum activities. Mandated by the National Environmental Policy Act of 1969 to prepare comprehensive preleasing environmental impact statements, the US Department of the Interior began funding major multidisciplinary studies of the continental shelf to provide the necessary background information. Other agencies including the National Marine Fisheries Service, Department of Energy, US Army Corps of Engineers, NOAA, NASA, and EPA have been involved to a lesser extent. Table S4 provides information on some of the major broad-scale and multidisciplinary projects carried out in

the Gulf since 1960. Table S5 lists some of the more important books, technical reports, and atlases that have appeared since 1950. Trends and specific studies are discussed below.

In 1951 the Woods Hole Oceanographic Institution published a very detailed map of the submarine topography of the northwest quarter of the Gulf based on its own investigations and data supplied by the US Coast and Geodetic Survey. That year the US Fish and Wildlife Service initiated a long-term exploratory fishing program out of Pascagoula, Mississippi, to study gear development and the oceanography and fishery potential of the continental shelf and deeper Gulf. The same year, prompted by the oil industry's interest in coastal and marine environments, the American Petroleum Institute began a seven-year program (Project 51) to study shallow-water sedimentary environments of the northwestern Gulf. A component of this project investigated factors associated with oyster mortality in coastal environments to determine to what extent this might be the result of petroleum activities. A major work appeared in 1954 entitled *Gulf of Mexico: Its Origin, Waters, and Marine Life.* Edited by Paul Galtsoff, this volume, the work of many authors, was an effort to summarize existing knowledge of the Gulf. It remains the single best source of information on many topics and the only scientific overview of early literature relating to the entire Gulf.

By 1960 Project 51 had been completed, and some of the results appeared in book form (Shepard, Phleger, and van Andel 1960). To obtain a clearer understanding of the environment and seasonal distribution patterns of commercial shrimp and associated fish species, in 1961–65 the National Marine Fisheries Service completed a five-year survey of the hydrography, plankton, and demersal shrimp and fish fauna of the northern Gulf shelf from the Rio Grande to Mobile Bay. This remains the most thorough general survey of the northern Gulf yet undertaken. Results of the shrimp and fish surveys are given in Darnell, Defenbaugh, and Moore (1983) and in Darnell and Kleypas (1987). In 1965–67 the state of Florida carried out a two-year field program to investigate the hydrography and biology of the central Florida continental shelf. These were the Hourglass Cruises, so called because the shape of each monthly cruise tract on a map resembled an hourglass (west from Tampa Bay, southeast to the mouth of Charlotte Harbor, west again, and then northeast to Tampa Bay). Results of these studies have been appearing as *Memoirs of the Hourglass Cruises.*

Through the late 1950s and 1960s scientists of the department of oceanography at Texas A&M University, sponsored largely by grants from the Office of Naval Research and the National Science Foundation, conducted basic research on the continental shelf and open waters of the Gulf of Mexico. During the early 1970s this group published three volumes detailing their accumulated knowledge of the biology, physical oceanography, and geological and geophysical oceanography of the Gulf (W. Pequegnat and Chace 1970; Capurro and Reid 1972; and Rezak and Henry 1972). With collaborators they also put out two atlases (W. Pequegnat, L. Pequegnat, et al. 1971; and El-Sayed et al. 1972) summarizing our knowledge of the deep-sea crustacean and other fauna as well as the chemistry, productivity, and benthic algae of the Gulf. A book edited by Thomas Bright and Linda Pequegnat appeared in 1974 and reported on studies carried out on the West Flower Garden Bank, an elevated hard bank on the outer continental shelf southeast of Galveston, Texas, which is capped by a living coral reef community (Bright and L. Pequegnat 1974). Of the many theses and dissertations that appeared during this period, that of Defenbaugh (1976) is of particular interest. He reported on the invertebrates identified from 146 trawl collections (representing more than 50,000 specimens) from soft bottoms of continental shelves around most of the Gulf of Mexico (excluding Florida and Yucatán). In the late 1970s a major work edited by Arnold Bouma, George Moore, and James Coleman (1978) provided much new information on the physiography, structural characteristics, and sediments of the continental shelves and slopes of the Gulf, summarizing knowledge gained from recent high-resolution seismic profiling studies of the Gulf bottoms.

During the 1970s federal agencies actively supported major studies of the continental shelves to provide background information useful for preparation of environmental impact statements and to provide a basis for informed management of petroleum production and other human activities in the Gulf. In this period, as shown in table S4, the Bureau of Land Management sponsored several multidisciplinary investigations on continental shelves of the northern Gulf including those off Mississippi, Alabama, and Florida; south Texas; and Louisiana. Other more localized studies off Louisiana and Texas were supported by several agencies. In the 1970s the US Department of Commerce initiated the "Mussel Watch" program, which is still active. It had long been known that large mollusks inhabiting bays and estuaries accumulate pollutants, especially heavy metals such as mercury, cadmium, and lead. Therefore, harvesting and chemically analyzing a few mussels or oysters from each bay on a regular basis provides a long-term record of chemical pollution. This has proven to be a very successful method of keeping track of the quality of the chemical environment in coastal waters of the Gulf and of the world oceans, in general.

In the 1980s progress in the development of new knowledge about the Gulf of Mexico proceeded much as it had in the previous decade. Governmental funding of large research programs resulted in the production of detailed reports, data summaries, atlases, and books. The period was also marked by the appearance of several major works by Mexican scientists studying the ecology and fishery resources of the southern Gulf. Outstanding among the large research projects were those sponsored by the Minerals Management Service on the southwest Florida and Mississippi-Alabama continental shelves and on the northeastern Gulf continental slope. Another project, sponsored by the US Department of Energy, dealt with the Strategic Petroleum Reserves (SPR) to be stored off Texas and Louisiana. The plan was to dissolve enormous cavities in salt domes for petroleum storage and to discharge the resulting brines onto the continental shelf. The SPR project was designed to determine what effects, if any, this discharge might have on the environment and marine life near the discharge outlets. During the 1980s two data syntheses appeared, the Tuscaloosa Trend Study, dealing with the continental shelf off Mississippi and Alabama, and the Offshore Texas and Louisiana Study, relating to the shelf area from the Mississippi Delta to the level of Corpus Christi, Texas. Both projects entailed literature surveys and synthesis of previous studies and other published information. Summarizing results of a detailed investigation of topographic features of the northern Gulf, a book was published providing information on the distribution, hydrography, geology, biology, and zonation of these hard banks (Rezak, Bright, and Mc-Grail 1985). Three atlases appeared, two sponsored by the Minerals Management Service, and one by NOAA of the US Department of Commerce. The former two addressed the distribution of demersal fishes and penaeid shrimps of the US Gulf shelf from the Rio Grande to the Florida Keys (Darnell, Defenbaugh, and Moore 1983; Darnell and Kleypas 1987). The latter provided maps of the entire Gulf showing general information on the environment, living marine resources, economic activity, environmental quality, and jurisdictions in the Gulf (NOAA 1986). Scientists at the Institute of Marine Sciences and Limnology of the University of Mexico produced three books relating to the southern Gulf (Yáñez-Arancibia 1985a, 1985b; Yáñez-Arancibia and Sánchez-Gil 1986). These described the environments, fishes, and fisheries of the southern Gulf shelf from Yucatán to above Veracruz including a large estuarine area, the Laguna de Términos, where four and a half centuries earlier Grijalva and Cortés had learned that the local Indians called their land Mexico.

In the 1990s the Mississippi-Alabama Marine Ecosystems Study was completed, and several new investigations were undertaken. One of these concerned petroleum and natural gas seeps on the continental slope off Louisiana and Texas. Another, the Louisiana-Texas Shelf Physical Oceanography Program (LATEX) dealt with water circulation and other physical processes on the Louisiana-Texas continental shelf. A third, the Gulf Offshore Operators Monitoring Experiment (GOOMEX), was designed to determine long-term, subtle effects of offshore petroleum operations on the environment and biota of the northwest Gulf shelf. A symposium sponsored by the EPA and other federal and state agencies was published dealing largely with socioeconomic issues of the northern Gulf (Flock 1990). Another symposium sponsored by the Minerals Management Service also appeared in print. This included a series of technical articles summarizing results of major multidisciplinary studies of the Gulf during the two previous decades (Darnell and Defenbaugh 1990). In the past thirty years or so several books have appeared summarizing our knowledge of various aspects of the Gulf. These treat the geological history of the Gulf of Mexico basin (Salvador 1991a), estuarine ecology (R. Stickney 1984), the Laguna Madre of south Texas and northeastern Mexico (Tunnell and Judd 2002), shoreline ecology (mainly of the western and southern Gulf) (Britton and Morton 1989), fishes (McEachran and Fechhelm 1998, 2005), and marine mammals (Würsig, Jefferson, and Schmidly 2000).

Thus, during the past three decades there has been an explosion of new knowledge about the Gulf of Mexico in general, but particularly about the continental shelf and slope of the northern Gulf. However, much of this information lies buried in large technical reports to government agencies, theses and dissertations at various universities, and in the Spanish language, where it is not readily available to scientists, environmental managers, and the general public. It is the aim of the present volume to summarize this knowledge and to place it in perspective of what is known about the Gulf in general.

The Gulf of Mexico is much smaller than the great oceans of the world. In effect, it is an ocean in miniature. To understand it one must have some grasp of marine systems in general, as well as their compositions and dynamic processes. Hence, before proceeding to describe the Gulf of Mexico, it is appropriate to provide a foundation of marine science. This discussion will take up, in order, the structure and surface sediments of ocean basins, the physical and chemical properties of the water within the basins, and the composition and characteristics of the biological systems that make up the living communities of the oceans.

**THE CRUST OF THE EARTH AND MARGIN OF THE SEA**
The earth is a roughly spherical body with a radius of 6,378 km (3,963 mi) and a circumference of 40,074 km (24,900 mi) at the equator. As shown in table 2.1, the earth is composed of four primary layers with densities increasing toward the center of the sphere. The thin, solid outer crust consists of seven major plates (Eurasian, North American, South American, African, Pacific, Indo-Australian, and Antarctic) and at least that number of minor ones (Caribbean, Cocos, Nazca, Juan de Fuca, Scotia, Philippine, Arabian, etc.). These plates are separated by great cracks in the crustal material and are free to slide slowly (a few centimeters per year) over the surface of the denser liquid mantle that lies beneath. Areas where the plates are pulling apart from one another are called spreading centers or divergent boundaries, and areas where the plates are pushing together are referred to as convergent boundaries. The latter are sites of great compression where the crust may be cracked, buckled, and elevated into mountain ranges, and the leading edge of one plate may be forced beneath that of the other. Compressional boundaries are often the scene of extensive earthquakes and volcanic activity. The early history of the Gulf of Mexico basin was marked by both convergence and divergence of continental plate boundaries.

The continent itself consists of one or more large blocks of granite often overlain with a blanket of sedimentary rocks. Beneath both the continents and the sea lies a layer of basalt derived from the upper mantle, and the

**Table 2.1.** *Layers of the earth, their depth ranges from the surface, thicknesses, and average densities.*

| Layer | Depth range (km) | Thickness (km) | Average density g/cm³ | Average density lb/ft³ |
|---|---|---|---|---|
| Crust | 0–33 | 33 | 2.8 | 175 |
| Mantle | 33–2,900 | 2,867 | 4.5 | 231 |
| Outer core | 2,900–5,100 | 2,200 | 11.8 | 735 |
| Inner core | 5,100–6,370 | 1,270 | 17.0 | 1,060 |
| Total earth | 0–6,370 | 6,370 | 5.5 | 343 |

basalt is much thicker (ca. 40 km) beneath the continent than it is beneath the sea (< 10 km) (fig. 2.1). The continental margin, where it meets and underlies the edge of the sea, is characterized by several zones. These include, in order from land seaward, the continental shelf, continental slope, and continental rise. Beyond this stretches the abyssal plain of the deep ocean. The approximate depth ranges and slopes of each of these zones are given in fig. 2.2. The marine environment includes the neritic zone, which lies over the continental shelf, and the oceanic zone, which includes all the ocean beyond the shelf. The latter is divided on the basis of depth (epipelagic, mesopelagic, bathypelagic, and abyssopelagic).

**SURFACE SEDIMENTS**
The rocky basement of the ocean floor is carpeted with layers of particulate sediments. Based on their modes of origin, these sediments fall into five primary classes: relict, terrigenous, biological, volcanic, and authigenic. Each will be described briefly.

**Relict sediments**
During the great Ice Ages of the past million and a half years, so much water was locked up in continental glaciers that several times the level of the world oceans dropped by 100 m or more, and the coastline stood at or near the outer edge of the continental shelves. At such times ma-

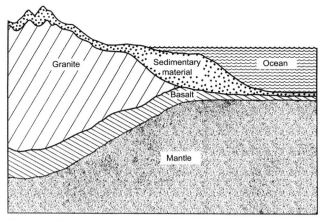

**Figure 2.1.** Cross section of the continental margin showing major layers and their proportional thickness.

terial composing the shelves was exposed to subaerial erosion and weathering. Sand grains became pitted and stained red with iron oxide, and shells and other calcareous materials sometimes became cemented into crusts and nodules. As the sea rose again to its present level, many of these relict deposits were covered by more recent sediments, particularly around river mouths, but some are still exposed and recognizable, primarily on outer portions of continental shelves.

### Terrigenous sediments

Terrigenous sediments include those clastic materials (gravel, sand, silt, and clay) derived from land sources and transported to the sea primarily by streams and to a lesser extent by winds. Chemical compositions vary depending on the source rocks. Throughout the world river-borne terrigenous sediments dominate the continental shelves.

### Biological sediments

Biological sediments are made up of the remains of plants and animals that once lived in association with the bottom or as free-floating organisms of the water column, called plankton. Chemically, the biological remains are of two types, calcareous (formed of calcium carbonate) and glassy or siliceous (formed of silicon dioxide). Calcareous remains of bottom-dwelling organisms are derived from a variety of algae, protozoans, sponges, corals, bryozoans, mollusks, echinoderms, and other groups. Such deposits tend to accumulate on continental shelves of tropical and subtropical seas. Siliceous deposits of bottom organisms are rare and derive largely from skeletal remains of certain sponges. Planktonic deposits may also be calcareous or siliceous. The former predominate on the bottoms of warm shallow seas and include the remains of planktonic algae (coccolithophores), protozoans (foraminiferans), and tiny planktonic mollusks (pteropods). Planktonic siliceous deposits, which are dominant on the bottoms of deep cold seas, are derived from planktonic algae (diatoms and silicoflagellates) and some protozoans (radiolarians).

### Volcanic sediments

Volcanic deposits are derived from volcanic ash or from the weathering of lava. This material is often rich in iron and when oxidized produces reddish deposits. Some volcanic material is also quite rich in silicon.

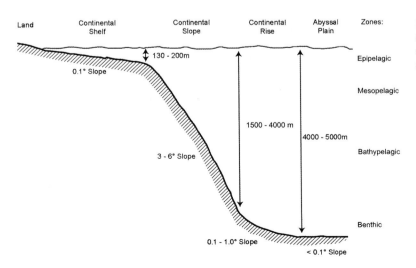

**Figure 2.2.** Idealized diagram of zonation in the sea showing depth zones in the open water column as well as primary regions of the sea bottom with average slopes and approximate depths.

## Authigenic sediments

Authigenic sediments are produced in place by chemical precipitation from seawater. They typically occur on outer edges of continental shelves in areas of upwelling, where they form deposits of the minerals glauconite and phosphorite. Authigenic sediments may also form in warm, shallow, marginal lagoons where partial evaporation of seawater can lead to precipitation of calcium carbonate and other materials.

After deposition, marine sediments are often transported from one place to another, and sometimes they are altered chemically. In the case of mixed sediments brought in by streams, the larger and denser particles (primarily sand) are dropped out first, and the finer material remains in suspension longer to be deposited farther from the river mouth. Materials laid down near the shore are transported parallel to the coastline by alongshore currents. In the high-energy surf zone, waves extract the silts and clays for further transport, leaving behind the sand that forms beaches, dunes, and nearshore sea bottoms. Silts and clays come to rest on the middle to outer shelf and are often swept out to be deposited in the deeper ocean. In warm, clear, tropical and subtropical seas the abundant calcareous remains are moved around by bottom currents and fragmented into smaller particles. Skeletal remains of planktonic organisms are present in all marine sediments, but on nearshore and middle continental shelves and around the mouths of rivers they are often overwhelmed by larger volumes of silt and clay. However, on outer shelves and in deeper water planktonic sediments often dominate.

## PHYSICAL CHARACTERISTICS OF THE WATER COLUMN

### Density

Density is one of the most important properties of seawater, and it is determined primarily by the temperature and salt content, or salinity. The colder the water, down to about 4°C (39.2°F), the greater its density. Freshwater is much less dense than seawater, and the greater the salt concentration, the higher the density. In the open sea, where salinity varies only slightly from place to place, temperature is the chief factor governing density. Most oceanic waters contain about 3.6 percent salt by weight, or 36 parts per thousand (ppt). In tropical seas, where evaporation is prominent, surface water salt concentration may exceed 37 ppt. Near shore, especially near mouths of rivers, the salinity may be much lower because of the diluting effect of freshwater entering the sea.

Examination of a cross section of an oceanic water column, from the surface to the bottom, reveals that the column consists of several distinct layers, or water masses, with the densest layer on the bottom and the least dense at the surface. All the water masses move horizontally, generally in different directions, and each has its own prehistory and future path. As the layers flow past one another they tend to remain distinct, with little mixing between them. Bottom water of the world oceans is formed in the polar or near-polar regions (especially in the North Atlantic). Here, because of the intense chilling by frigid winds and also because of the rise in salinity from the freezing out of water crystals, surface water becomes extremely dense. This cold saline seawater sinks to the bottom and begins circulating around the globe. Obtaining heat from the sea bottom, it gradually warms, becomes less dense, and rises to become an intermediate layer. After a few thousand years it eventually becomes the surface layer.

### The surface mixed layer

The topmost layer of the sea, often referred to as the surface mixed layer, extends to a depth of around 100 m, although the depth varies from one place to another and with the season of the year. It is important because it represents the interface between the ocean and the atmosphere and because much of the life in the sea is limited to this layer. Friction caused by winds blowing across the sea surface creates waves, foam, and sea spray, which facilitate the transfer of both heat and gases (oxygen, nitrogen, and carbon dioxide) between the air and the sea. This layer is important in controlling regional climates of the world as well as concentrations of atmospheric gases. Wind stirring of surface water sets up turbulent motions that stir and mix the entire layer. High levels of oxygen throughout the layer are important in supporting living systems. Sunlight that enters from above penetrates to a depth of 150–200 m in the open sea, providing ample light to support photosynthesis by marine phytoplankton. Below the surface layer lies the "twilight" zone, which receives less than 1 percent of the surface illumination, and below this lies the zone of total darkness. Representing over 95 percent of the ocean's volume, this zone receives no light except that produced by living organisms (bioluminescence).

**Table 2.2.** *Approximate rotational speed of the earth at different latitudes.*

| Position | Circumference of earth (km) | Rotational speed of earth (km/hr) |
|---|---|---|
| Equator | 42,000 | 1,750 |
| 30°N or S | 37,300 | 1,560 |
| 60°N or S | 20,000 | 833 |
| 90°N or S | 0 | 0 |

## Water circulation

The various water masses of the sea are in a constant state of horizontal movement or circulation. This movement is powered primarily by friction at the sea surface caused by winds blowing across the ocean in tropical latitudes. As the earth spins on its axis from west to east, the air lags behind. Surface winds that flow from east to west in the tropics force surface waters westward. In the Atlantic Ocean surface waters flow from Africa to the Americas, and in the Pacific they flow from the Americas toward the Philippines and southern Asia. An important steering mechanism for oceanic currents is the Coriolis effect. This is most easily explained by an analogy. Quito, Ecuador (on the equator), lies almost directly south of Miami, Florida (at about 25.5°N latitude). Suppose an airplane takes off from Miami and points its nose directly south. When it reaches the equator it will not be over Quito, but hundreds of miles to the west over the Pacific. Due to the earth's rotation (which is faster at the equator than at 25.5°N), the target has moved eastward, and to the pilot it will appear that the plane has veered to the right. Conversely, a plane heading directly northward out of Quito will reach the latitude of Miami east of the target, well out in the Atlantic. Before the plane took off, it was traveling eastward at the speed of the earth's rotation at the equator (1,750 km/hr) (table 2.2). In the air it retains this eastward rotational speed, which is greater than that at the latitude of Miami. Hence, without a course correction, the plane heading due north appears to veer toward the east. These factors, the westward flow in the tropics and the Coriolis effect thereafter, account for the major oceanic gyres, clockwise in the Northern and counterclockwise in the Southern Hemisphere (fig. 2.3). Closer to shore, major oceanic currents may induce smaller circulation cells on the continental shelf. Thus, along the east coast of the United States, where the Gulf Stream heads generally northward, counterrotating cells bring water southward along the shoreline. Nearshore water currents also respond to wind direction, velocity, and persistence, as well as to basin configuration and other factors.

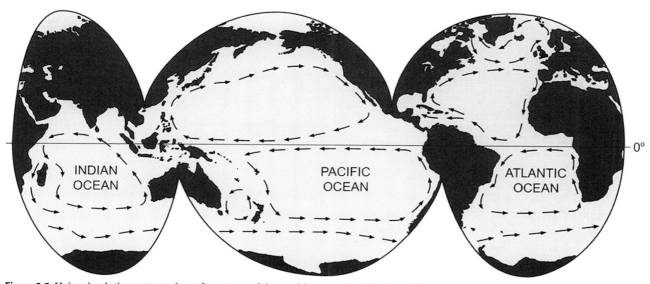

**Figure 2.3.** Major circulation patterns in surface water of the world oceans. Note the clockwise gyres in the Northern Hemisphere and counterclockwise gyres in the Southern Hemisphere.

## Upwelling

In addition to mixing within water masses and horizontal circulation around the earth, under certain conditions water masses may undergo vertical movement. An example of this is upwelling. Outside the tropics, when the wind blows across the surface of the sea it causes surface waters to move at a right angle to the wind direction. Since this is due to the Coriolis effect, surface water movement is to the right in the northern and to the left in the southern hemisphere. As the surface water moves away it must be replaced, and this is generally accomplished by water rising from a deeper layer below. For example, suppose that off the coast of Spain and Portugal the wind is blowing from the north. This causes the surface mixed layer to move westward away from shore, and this, in turn, is replaced by deeper water rising to become the new near-shore surface mixed layer. In effect, the wind is pumping deep water to the surface. This type of pumping may also occur when one water mass moves across another.

## The pressure factor

The atmospheric pressure at the surface of the earth represents the weight of the air column above. At sea level this is about 1 kg/cm² (14.7 lb/in²), which is referred to as 1 atmosphere. Water is denser than air, and in the ocean the pressure increases regularly about 1 atmosphere for every 10 m of depth increase. For example, the pressure at 4,000 m depth is around 400 atmospheres.

## CHEMICAL CHARACTERISTICS OF SEAWATER
### Composition

The most abundant chemical in seawater is, of course, water itself. This makes up about 96.5 percent of seawater, the remainder being primarily a variety of salts. Properties of the water molecule are such that seawater readily dissolves acids, bases, salts, most organic molecules, and a great variety of other chemical types. Practically every chemical element known to occur naturally on earth has been found in seawater, although most occur in very low concentrations. Seawater is salty, and salts are composed of charged particles called ions. The most abundant of these are shown in table 2.3. Cations are positively charged and anions are negatively charged, and the association of a cation and an anion produces a salt. Sodium and chloride are the most abundant ions, and together these make up about 85 percent of the salts in the sea. Magnesium and sulfate are next, and together these four ions constitute 97 percent of the ions in sea-

**Table 2.3.** *Approximate concentrations of the major ions in seawater expressed as a percentage of total dissolved materials.*

| Ion | Symbol | Percentage by weight |
|---|---|---|
| Cations | | |
| Sodium | $Na^+$ | 30.6 |
| Magnesium | $Mg^{++}$ | 3.7 |
| Calcium | $Ca^{++}$ | 1.2 |
| Potassium | $K^+$ | 1.1 |
| Strontium | $Sr^{++}$ | 0.02 |
| Anions | | |
| Chloride | $Cl^-$ | 55.1 |
| Sulfate | $SO_4^{--}$ | 7.7 |
| Bicarbonate | $HCO_3^-$ | 0.4 |
| Bromide | $Br^-$ | 0.2 |
| Borate | $H_2BO_3^-$ | 0.01 |
| Fluoride | $F^-$ | 0.01 |

water. The list of ions known from the sea is very long, but most are present in only trace amounts, that is, in parts per million (ppm) or less.

In addition to the chemical elements and charged particles, seawater includes many organic, or carbon-containing, compounds including long lists of sugars, amino acids, vitamins, proteins, lipids, enzymes, and so forth. Of particular importance are the nutrient elements nitrogen, phosphorous, and silicon, which are required for growth by the phytoplankton. When one or more of these is in short supply, the growth and production of marine algae is limited. The element carbon is also required for growth, but it is always present in excess supply as carbon dioxide, bicarbonate, or carbonate. The chief dissolved gases include nitrogen, oxygen, and carbon dioxide, which are readily exchanged with the atmosphere at the sea surface.

The chemical composition of seawater is regulated by a number of physical, chemical, and biological factors. The water content is increased by river runoff, precipitation, and the melting of ice. It is decreased by surface evaporation and the formation of sea ice. Exchange with the atmosphere influences the concentration of the major gases. Biological activity plays a major role in regulating the concentrations of biologically active chemicals, especially in the surface layer of the sea. These include some gases (oxygen and carbon dioxide) and nutrients (phosphorus, silicon, and inorganic nitrogen sources—nitrate, nitrite, and ammonium) as well as certain other chemicals (sulfates, vitamins, some trace metals, etc.). Most other chemicals are controlled by their ability to adsorb to the surfaces of clay and other particles suspended in seawater. The actual concentration of any chemical in seawater rests on a balance between the rate of entrance and the rate of removal. Biologically active chemicals are generally removed very quickly, whereas others may persist for very long periods of time.

## Buffer systems

The pH, or acidity, of seawater is regulated by several chemical buffer systems that prevent much change in the acidity of the water. An example of this is the inorganic carbon system illustrated by the following set of equations.

$$CO_2 + H_2O \leftrightarrow H_2CO_3 \leftrightarrow H^+ + (HCO_3)^- \leftrightarrow H^+ + (CO_3)^{--}$$

This may be read as follows: carbon dioxide gas may combine with water to produce carbonic acid, which may dissociate to produce hydrogen and bicarbonate ions, which may further dissociate to produce more hydrogen ions plus carbonate ions, which (when combined with calcium, magnesium, etc.) tend to precipitate out of solution. The entire set of transformations is reversible and may go either way depending on conditions. Under acid conditions, the shift is toward the left, producing an excess of carbon dioxide, some of which may be lost to the atmosphere. Under alkaline conditions and in the presence of an excess of cations (sodium, potassium, calcium, magnesium, etc.), the equilibrium shifts toward the right, and some of the carbonate may precipitate out of solution. This ability to shift either way tends to retain the acidity of seawater within rather narrow limits. It also has major atmospheric implications because it suggests that the oceans may be able to take up a great deal of carbon dioxide, thus partially buffering the atmosphere against a build-up of a major "greenhouse" gas.

## BIOLOGICAL ORGANIZATION OF THE SEA
## Classification of marine organisms

The sea contains a bewildering array of life forms, and classification systems have been developed to help understand and keep track of this diversity. One such system is based on natural relationships. It constitutes a family tree as best as can be determined by careful study of anatomy,

biochemistry, and so forth. This is the taxonomic classification. The second system is an ecological one. It is based on how and where the organisms live, or their modes of life. Both systems are addressed below.

TAXONOMIC CLASSIFICATION

The taxonomic classification is made up of a hierarchy of categories. From highest to lowest these include kingdom, phylum, class, order, family, genus, and species. Sometimes prefixes are added, as in "superfamily" or "subfamily," to denote larger or smaller groupings. Most commonly encountered by the nonspecialist are genus and species names, such as *Callinectes sapidus* for the blue crab or *Micropogonias undulatus* for the Atlantic croaker. Since technical names receive international usage, these are normally of Latin or Greek derivation, and they generally describe characteristics of the particular organisms. The name of the blue crab translated means "beautiful swimmer that tastes good," whereas that of the croaker means "small-bearded one with wavy lines."

As shown in table 2.4, all of the phyla of living organisms are grouped within just five kingdoms: Monera, Protista, Fungi, Metaphyta, and Metazoa. The Monera, containing bacteria and blue-green algae, are the most primitive and lack clearly defined nuclei within their cells. Nuclear (genetic) material is distributed throughout the cytoplasm. In the more highly developed Protista all nuclear material is confined within a nucleus, which is surrounded by a nuclear membrane. Included here are most algae and all protozoans. These are single-celled or multicellular organisms with only limited differentiation of cell types. Fungi were probably derived from algae but have lost all photopigments, and most survive by decomposing living or dead organisms. Metaphytes include all higher plants. These have well-developed organ systems and contain photopigments that permit them to carry out photosynthesis. The Metazoa include all animals above the level of the Protozoa. These contain specialized organ systems, but because they lack photopigments, they are unable to carry out photosynthesis. A brief survey of some of the major plant and animal groups follows, and a phylogenetic tree showing the relationships of the kingdoms and phyla is given in fig. 2.4. In order to simplify the discussions, throughout the present volume we will revert to an older classification scheme that includes all algae, fungi, and higher plants in the plant kingdom and the protozoa and all higher animals in the animal kingdom.

Bacteria are found throughout the sea. Some are

**Table 2.4.** *Higher classification categories containing common marine organisms. This classification scheme is somewhat simplified and omits relatively rare groups.*

| | | |
|---|---|---|
| Kingdom | MONERA | |
| Phylum | Schizophyta – bacteria | |
| Phylum | Cyanophyta – blue-green algae | |
| | | |
| Kingdom | PROTISTA | |
| Phylum | Pelagophyceae – microalgae | |
| Phylum | Chrysophyta – golden-brown algae | |
| Class | Chrysophyceae – golden algae | |
| Class | Coccolithophorida – coccolithophores | |
| Class | Diatomophyceae – diatoms | |
| Class | Silicoflagellida – silicoflagellates | |
| Phylum | Pyrrophyta – dinoflagellates | |
| Phylum | Protozoa – protozoans | |
| Class | Ciliophora – ciliates | |
| Class | Mastigophora – flagellated protozoans | |
| Class | Sarcodina – amoeboid protozoans | |
| Class | Sporozoa – sporozoans | |
| | | |
| Kingdom | FUNGI | |
| Phylum | Mycophyta – fungi | |
| | | |
| Kingdom | PLANTAE | |
| Phylum | Chlorophyta – green algae | |
| Phylum | Phaeophyta – brown algae | |
| Phylum | Rhodophyta – red algae | |
| Phylum | Tracheophyta – vascular plants | |
| Class | Angiospermae – flowering plants | |
| | | |
| Kingdom | METAZOA | |
| Phylum | Porifera – sponges | |
| Phylum | Cnidaria (Coelenterata) – hydroids, medusae, corals | |
| Phylum | Ctenophora – comb jellies | |
| Phylum | Platyhelminthes – flatworms | |

| | | |
|---|---|---|
| Phylum | Nematoda – roundworms | |
| Phylum | Annelida – segmented worms | |
| Phylum | Mollusca – mollusks | |
| Phylum | Arthropoda – arthropods | |
| Subphylum | Chelicerata – horseshoe crabs, sea spiders | |
| Subphylum | Mandibulata – mandibulates | |
| Class | Crustacea – copepods, barnacles, shrimps, crabs, lobsters, and relatives | |
| Subphylum | Uniramia – insects | |
| Phylum | Bryozoa – moss animals | |
| Phylum | Brachiopoda – lamp shells | |
| Phylum | Chaetognatha – arrow worms | |
| Phylum | Echinodermata – starfish, brittle stars, sea cucumbers, sea urchins, crinoids | |
| Phylum | Hemichordata – acorn worms | |
| Phylum | Chordata – chordates | |
| Subphylum | Urochordata – salps, sea squirts | |
| Subphylum | Cephalochordata – lancets | |
| Subphylum | Vertebrata – backboned animals | |
| Class | Agnatha – jawless fishes | |
| Class | Chondrichthyes – cartilaginous fishes | |
| Class | Osteichthyes – bony fishes | |
| Class | Amphibia – frogs and their relatives | |
| Class | Reptilia – sea turtles, crocodiles, snakes, lizards | |
| Class | Aves – birds | |
| Class | Mammalia – mammals: seals, sea lions, walruses, sea otters, whales, dolphins, manatees | |

capable of photosynthesis, and a few obtain energy by special chemical transformations, but the great majority act as decomposers, breaking down formerly living tissues to the primary chemical elements. Blue-green algae are capable of photosynthesis, and they are especially important in estuaries and nearshore portions of the sea.

Among the protists are several groups of algae that differ from one another primarily in which photopigments they contain. By far the most important groups making up the phytoplankton (microscopic free-floaters) of the open sea are the chrysophytes and pyrrophytes. The former includes the extremely diverse group known as the diatoms. These are encased in siliceous or glassy structures known as "frustules." Diatoms form the chief food for many small animals of the sea. A few are known to produce toxins harmful to both marine life and to humans who con-

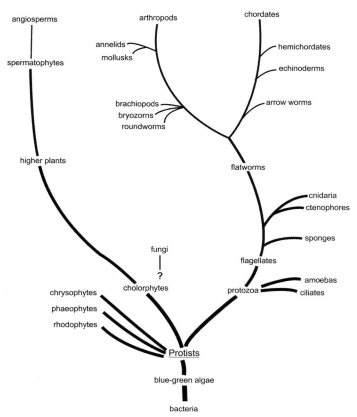

**Figure 2.4.** A simplified phylogenetic tree showing relationships of the various plant and animal groups. See text for details.

sume marine products (causing amnesic shellfish poisoning, or ASP). Pyrrophytes include the dinoflagellates. Abundant in the plankton of all oceans, they are particularly important in subtropical and tropical seas. Several species are known to produce extremely toxic chemicals and to cause massive die-offs of marine life (so-called red tides, etc.). Humans consuming seafood from such areas may suffer serious illness or death (from paralytic shellfish poisoning or PSP, ciguatera, etc.).

Brown and red algae are generally nonplanktonic and most grow attached to hard substrates of the continental shelf. Brown algae include the giant kelps and a host of smaller attached species primarily of colder waters. Two species, *Sargassum fluitans* and *S. natans,* contain buoyant structures and form unattached free-floating patches in the open ocean. When stranded on beaches they are recognized as "seaweeds." Red algae are anchored to hard substrates in warmer seas. Both brown and red algae are harvested commercially to obtain agar and carrageenan, which are used as thickening agents in cosmetics and certain foods.

Among the protozoa the Sarcodina, or amoeba-like forms, are of particular interest. Some such as the Foraminifera produce calcareous shells, while others including the Radiolaria have glassy shells. These protozoans may live either in the water column or on the bottom, and over the millennia their shells accumulate and become major components of bottom sediments.

Metaphytes appear to be derived from green algae. Mosses and ferns do not occur in the sea, and only a few flowering seed plants are found there. Two groups that do occur are the mangroves and seagrasses. Mangroves are bushy or treelike plants occurring on muddy, semiprotected areas in tropical and subtropical coasts around the world. To survive in such environments they have developed specialized structures that allow them to tolerate salt water, anoxic sediments, and the force of wave action. Some have also developed specialized reproductive structures. Seagrasses occur in bays and other relatively protected areas of continental shelves. Unlike mangroves, they are found in cold as well as warm waters of the world. These are rooted plants with long, bladelike leaves, and they grow in large patches or submarine meadows. Among their many specialized structures are subsurface runners, or rhizomes, which aid in anchoring the plants against the force of water movement. Both mangroves and seagrasses will be discussed in some detail in later chapters.

Metazoans apparently developed from flagellated protozoans. Included here are many phyla, of which only the most important will be considered. All have marine representatives, and several of the phyla are strictly marine. The lowest group, the Porifera, includes the sponges. These may be shaped like vases, baskets, cones, and so forth, but they all possess a central cavity open at the top. The walls contain pores connected with sometimes very elaborate systems of canals. Seawater is pumped through these canals to the central chamber and out the top, permitting the animal to filter out small food particles. To aid in providing support, sponges produce skeletal frames of calcareous, siliceous, or proteinaceous material. These animals are found attached to the bottom in waters of all depths. Their canals and central cavities afford living quarters for a great variety of fishes, shrimps, crabs, and other types of marine animals.

The bodies of Cnidaria, or coelenterates, are made up of two cell layers surrounding a central cavity connected to the outside by a single opening. Surrounding this "mouth" is a set of movable tentacles armed with stinging cells used to stun and capture prey. Included here are the hydroids, jellyfishes (medusae), corals, sea anemones, sea fans, sea pens, sea whips, and their relatives. Medusae and some of their relatives (siphonophores) are free floating in the sea and occur at all depths. The remainder are attached to hard substrates. Corals produce calcareous structures and over time construct massive reefs in warmer seas. Ctenophores, or comb jellies, are close relatives of the Cnidaria but lack stinging cells, and like medusae, they are free floating in the ocean.

Flatworms, or platyhelminths, are more advanced in that they have three body layers and definite front and back ends. There is one body cavity with a single opening. Flatworms are a relatively unimportant group in the sea, where they live in association with the bottom sediments.

Roundworms, or nematodes, have two openings to the digestive tract, a mouth and an anus, as well as a second body cavity developed in part from the middle tissue layer. Roundworms are very abundant in marine sediments at all depths.

Annelids and mollusks do not look alike, but similarities in their larval forms and basic body structure show them to be closely related. Annelids are segmented worms, that is, the body consists of a linear series of compartments, each an almost exact duplicate of the ones before and behind. Several of the front compartments have sense organs and modified appendages to sense and manipulate the environment ahead. In the major marine group, the polychaetes, each segment has a pair of lateral oar-like structures with bristles that aid in locomotion and respiration. Evidence of segmentation is seen in the most primitive mollusks, but it is lost in the more advanced groups. Mollusks include the snails, clams, oysters, tusk shells, cuttlefishes, squids, octopuses, and some lesser-known groups. Some have a hard, calcareous shell on the outside, and in others this is reduced to an internal supporting structure or is lost completely. Both annelids and mollusks are important inhabitants of the sea at all depths.

Bryozoans (moss animals) and brachiopods (lamp shells) appear to be distant relatives of annelids, and both are survivors of ancient groups, formerly much more numerous. Some bryozoans produce calcareous structures and crusts and contribute to the formation of reefs. Lamp shells look superficially like bivalve mollusks, but the two shells are quite different in detail. Neither group (moss animals or lamp shells) is of great importance in present oceans.

The phylum Arthropoda includes the joint-legged animals. These are clearly derived from segmented worms. Appendages of the various segments have been highly modified to form antennae, mouth parts, walking legs, and so forth. Included here are the crustaceans, spiders and their relatives, and insects. Of these, the crustaceans form by far the most diverse and widespread group in the sea. Although relatively unknown to the nonspecialist, such crustacean groups as copepods, isopods, amphipods, ostracods, and mysids are quite diverse, abundant, and important in marine food chains. Also included are such well-known forms as shrimps, lobsters, and crabs. Marine relatives of spiders include horseshoe crabs and pycnogonids, or "sea spiders." There are no insects living in the sea, but this group, so successful on land and in the air, does have several representatives that live in brackish water. One group, the water striders, has a few members that dance over the surface of the sea. Covered with a waxy coating that protects them from the salt water, these are called "grease bugs."

Chaetognaths (arrow worms) and echinoderms (starfishes and their relatives) do not look alike as adults, but similar larval forms show them to be related. Arrow worms are slender semitransparent animals a centimeter or so in length with clawlike bristles around the mouth. This is a very successful group in the plankton, where the individuals prey on tiny crustaceans such as copepods. Echinoderms include starfishes, brittle stars, sea urchins, and sea cucumbers, all of which move around from place to place on the bottom. Also included are stalked forms, the crinoids, most of which are attached to the substrate. Both chaetognaths and echinoderms are very important groups in the ocean.

Hemichordates, or acorn worms, embrace a small group of wormlike animals that burrow in the sediments. Their larvae are similar to those of echinoderms, and the adults display anatomical characters linking them with the more advanced phylum, the chordates. Although relatively unimportant in the economy of the sea, this group is significant in showing relationships with the other phyla.

Members of the phylum Chordata typically possess a dorsal tubular nerve cord, a longitudinal strengthening rod, or notochord, and gill slits in the throat or pharyngeal region. In higher forms these appear as embryonic structures. Three subphyla are recognized, the Urochordata, Cephalochordata, and Vertebrata. In the urochordates the larvae look like tadpoles and possess the characteristics of the phylum, but as adults they become highly

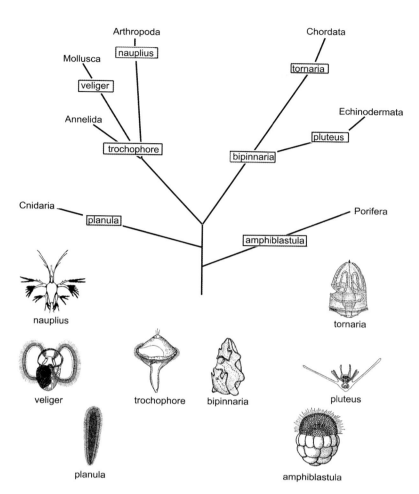

**Figure 2.5.** Some examples of marine larval types and their positions in the phylogenetic tree.

modified for a free-floating or sedentary life. Included here are salps, sea squirts, and similar forms. Cephalo-chordates are slender animals a few centimeters in length that live in sandy bottoms of shallow seas.

Vertebrates are animals with backbones made up of cartilage, bone, or both. Seven classes are recognized, and each will be addressed briefly. The most primitive are the Agnatha, or jawless fishes. Two small groups occur in the sea, the hagfishes and the sea lampreys. The Chondrich-thyes, or cartilaginous fishes, include the sharks, skates, rays, and their relatives. The Osteichthyes, or bony fishes, embrace a diverse and abundant array of forms and in-clude most of the fishes of the sea. The fourth group, the Amphibia, includes frogs, toads, and salamanders. These are not represented in salt water, although a few do live in brackish water areas. The next group is the Reptilia. In-cluded here are about 50 species of sea snakes (all highly venomous and distributed primarily through the Indian and South Pacific Oceans), 7 species of sea turtles (found worldwide in warm seas), the saltwater crocodile (of the East Indies and southwest Pacific), and the marine iguana (of the Galapagos Islands). The Aves, or birds, have many marine representatives that swim in the water, float on top, or fly above and dart or dive for marine life. The final class, Mammalia, embraces several groups of marine mammals including the whales and dolphins, sea cows and manatees, seals, sea lions, elephant seals, walruses, and sea otters.

This brief review has provided some idea of the great variety of marine life. Adding to this diversity is the fact that many marine animals produce larval forms, some-times a sequence of several larval forms, which may differ from one another and from the adults in anatomy, habi-tat, and lifestyle. A few of the major types of marine larvae are depicted in fig. 2.5.

CLASSIFICATION BASED ON MODES OF LIFE

Another method of classifying marine organisms is based on where and how they live. Here they all fall into just four categories: *plankton, neuston, nekton,* and *benthos.*

The *plankton* includes all microscopic plants and animals of the water column that possess only limited powers of locomotion. They are at the mercy of the water currents and are swept around with the water masses. The *phytoplankton* includes the small suspended algae, primarily diatoms and dinoflagellates, but also includes coccolithophores, silicoflagellates, and a few others. These carry out photosynthesis, and their production supports the remaining marine life of the open sea. The *zooplankton* includes all the animal plankton forms. Within this group are protozoans, small worms, mollusks, crustaceans, and arrow worms together with a wide variety of larval forms. Some groups that spend their entire lives as zooplankters are known as *holoplankton,* whereas others, including most larval forms, are planktonic during only portions of their lives. These make up the *meroplankton.*

A second group of marine organisms is the *neuston,* which lives in intimate association with the surface layer of the sea. Some dwell upon the surface, some within the surface layer, and some just below the surface. Most are microscopic, and many planktonic forms are temporary or permanent members of the neuston. This group is distinguished because the sea surface represents a special and generally hostile environment for most marine organisms. It is subject to extremely high light intensities and much mechanical buffeting by wave action, and since many chemical pollutants float, it is a damaging chemical environment. Eggs and larvae of many species of invertebrates and fishes contain oil droplets that cause them to float. Hence, these are also important components of the neuston.

The *nekton* consists of the larger free-swimming animals of the sea. Here one finds the medusae, larger worms, squids, cuttlefishes, some swimming shrimps and crabs, salps, and many cartilaginous and bony fishes, as well as sea turtles, whales, dolphins, and other higher vertebrates.

The *benthos* includes all organisms intimately associated with the sea bottom. These may be mobile or attached. They may live within the sediments, on the surface, or in the waters just above the bottom. They may be classified on the basis of size, type of substrate inhabited, or relationship with the substrate. Most major groups of plants and animals found in the sea are represented. Many produce floating eggs and planktonic larvae.

Classification of marine organisms on the basis of modes of life has real meaning in an ecological sense because it groups those organisms that live together and interact with one another. It also has a very practical value because gear for collecting and studying marine life is designed to sample one habitat or another.

## MARINE ECOSYSTEMS

All marine ecosystems are based on the production of organic matter from inorganic chemicals and the subsequent transfer of organic matter through food chains. The present section will lay out basic concepts and provide examples of these systems and how they function.

### Concepts of mass and energy

Within the context of living systems all things exist in one of two states: mass or energy. Mass includes all chemicals and things made up of chemicals. Energy, defined as the ability to do work, is what makes things happen. In dealing with energy three principles are recognized.

1. Energy exists in many forms: light, electrical, kinetic (the energy of chemical reactions), and heat (the random movement of molecules).
2. All forms of energy may be transformed into one another. Electrical energy may be used to produce light; chemical energy may be used to produce motion, and so on.
3. At each transformation some of the energy is lost as heat. No transformation is 100 percent efficient except that which results in all being transformed to heat. Since all energy can be converted to heat, the heat equivalent, or *calorie,* is often used as a measure of energy content or energy transfer. In biological systems energy is generally bound as chemical energy.

### Photosynthesis and respiration

Plants that possess photopigments such as chlorophyll (among others) have the remarkable ability to transform sunlight into chemical energy. In this process photons from the sun are captured by electrons, which then become high-energy electrons capable of powering chemical reactions. In photosynthesis carbon dioxide is combined with water to produce sugar and the by-product oxygen, as shown by the following equation:

$$CO_2 + H_2O + energy \rightarrow sugar + O_2$$

The sugar, in turn, provides energy for a series of subsequent chemical reactions by which the plant cell produces all the complex carbohydrates, proteins, lipids, and other compounds required by the plant. Along the way the plant must pick up nitrogen, phosphorus, and other elements required for its own metabolism. In addition to photosynthesis, all plants must engage in the chemical process of respiration to provide energy to carry out their metabolic activities. Chemically, this involves oxidizing or burning a portion of the sugar produced. The equation for respiration is as follows:

$$sugar + O_2 \rightarrow energy + CO_2 + H_2O$$

Thus, respiration is the exact reverse of photosynthesis, and the by-products of respiration are the raw materials of photosynthesis. Lacking photopigments, animals can carry out only respiration. They obtain their energy from preexisting organic matter, that is, they consume plants or other animals. The need to replace energy supplies that are constantly dwindling through respiratory loss is what drives marine ecosystems.

## Production and productivity

Production of organic material by photosynthetic plants is referred to as *primary production* because it involves making organic from inorganic chemicals. The total amount of organic matter produced, measured in terms of calories or of carbon fixed, represents *gross primary production.* However, as mentioned above, plants must expend some of the captured energy to support their own metabolism, and thus, they lose some of the energy and carbon through respiration. The excess of gross production over respiration is what they have left over, and this remaining or stored energy and organic carbon is referred to as *net production.* Both gross production and respiration can be measured, and net production can be determined by subtraction. *Productivity* refers to the rate of production, for example, g/m² per hour or tons/acre per year.

A comparison of gross primary production in major ecosystems of the world reveals that although the oceans cover well over twice the surface area that the land does, the total production of the land and sea is roughly the same. This is largely because most of the open ocean is only about as productive as deserts (ca. seven tons of organic carbon fixed per square kilometer per year). Nutrient chemicals are locked up in deeper water layers but are very scarce in the surface mixed layer. However, on continental shelves, in areas of upwelling, and in areas of water mass convergence where nutrients do rise into the sunlit surface mixed layer, primary production may equal or exceed that of forests (250 tons/km² per year). Overall, taking into account areas of both high and low production, the world oceans average around 46 tons/km² per year, a figure comparable to the 43 tons/km² per year fixed by grasslands (see table 8.7).

## Food chains, webs, and pyramids

Since animals do not carry out photosynthesis, they must obtain nourishment by consuming preexisting organic material of plants and other animals. Therefore, the production of animal matter is called *secondary production.* In the sea, as on land, animals are linked to one another in nutrient transfer systems, involving eat-and-be-eaten sequential steps called food chains. These may be pelagic or benthic or a combination of the two. A pelagic food chain may appear as follows:

diatoms→copepods→anchovies→flying fishes→ barracudas

A benthic food chain might look like this:

bacteria→amphipods→crabs→groupers→sharks

Each level in a food chain is known as a *trophic level.* The concept of food chains is helpful, but it is a great oversimplification of what really takes place. The copepod may consume a variety of diatoms as well as dinoflagellates and other types of algae. The anchovy may take in copepods, isopods, amphipods, fish larvae, and a variety of other small creatures. Putting all these details on paper creates a complex network of lines, many crossing each other, and such a realistic picture is referred to as a *food web.* Most food chains or webs contain only four or five trophic levels, and the reason is their inefficiency. Clearly the size of the food supply must exceed the size of the groups feeding on it, and it has been estimated that only about 10 to 20 percent of the carbon and calories of one trophic level passes on to the next. At the molecular level some energy is lost as heat. Digestion is not perfect, and some food and energy are lost in fecal material. The consumer species must support their own metabolism and expend a great deal of their chemical and energy resources in locating and consuming their food. In addition, much of the potential food supply is never consumed. Thus, each succeeding trophic level has only about 10 percent of the biological mass (*biomass*) of the one before, and this progressive loss limits the length of the food chain. It does not pay to have another level above barracudas and sharks.

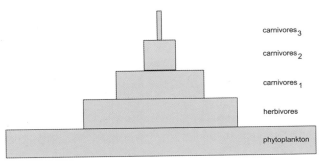

carnivores$_3$

carnivores$_2$

carnivores$_1$

herbivores

phytoplankton

**Figure 2.6.** Example of a marine food pyramid showing relative sizes of the succeeding trophic levels.

A *food pyramid* is a diagram showing the trophic levels as blocks stacked one upon the other, with each higher block reduced to a fraction of the width of the one below it (fig. 2.6). This provides a visual image of the five levels (producer, herbivore, and three carnivore levels) in proportion to their actual abundance in the sea. If sunlight had been included as the lowermost block, the diagram would have been much wider, because photosynthetic plants capture only about 1 percent of the sun's energy. Some interesting facts are correlated with the food pyramid concept. Proceeding from lower to higher levels there are fewer individuals, fewer species, less biomass, and less energy. Organisms at higher levels tend to be larger than those at lower levels.

Pelagic and benthic food chains, as discussed above, are the most obvious, but there are other food chains in the sea. An important one is the *decomposer chain,* whereby dead organic matter (often called *organic detritus*) is broken down to its component chemical elements. Involved here is a series of scavengers and decomposers such as bacteria and yeasts. They return carbon dioxide, nitrogen, phosphorus, and other chemicals to the system, to be reused by the phytoplankton. Another type of food chain in the deep sea derives its energy through *chemosynthesis* rather than photosynthesis. In some chemosynthetic communities the bacteria derive energy by oxidizing inorganic substances such as iron or sulfur. Others oxidize petroleum or natural gas seeping from the bottom sediments. In such communities the bacteria themselves become food for invertebrates, and these support a series of predators. Thus, whether the nutrients and energy start through photosynthesis or chemosynthesis, their transfer through food chains, webs, or pyramids provides the organizational basis for all marine ecosystems.

# Part II

## THE PHYSICAL ENVIRONMENT

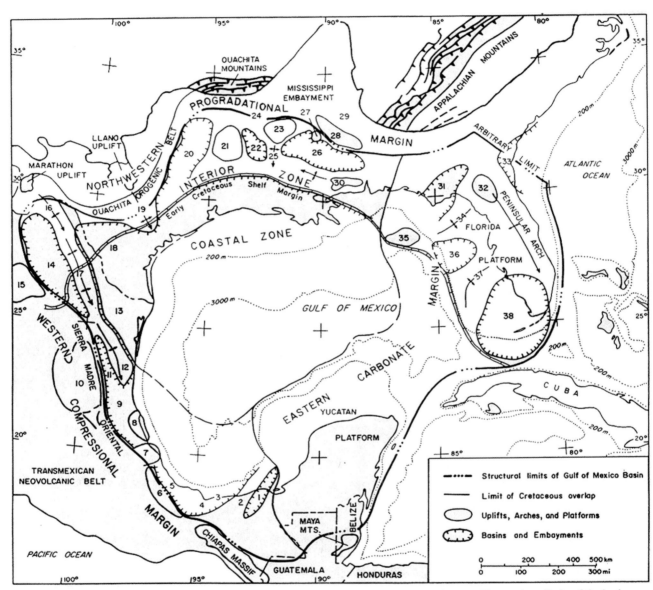

**Figure 3.1.** Map showing approximate limits of the Gulf of Mexico structural basin and related features. The northern limits of the basin are formed by the southern margins of the Appalachian and Ouachita Mountains. The western margin is the Sierra Madre Oriental and (on the south) the Chiapas Massif. The eastern margin, essentially the edge of the continent, represents old fault zones that have widened to form channels, the Caribbean Sea, and the Atlantic Ocean. Subsurface geology reveals ancient elevations (uplifts, arches, and platforms) and depressions (basins and embayments). These numbered features are named and discussed in the original source. (From Salvador 1991b.)

# 3 : ORIGIN AND HISTORY OF THE GULF OF MEXICO BASIN

The Gulf of Mexico occupies one of the world's deepest basins. How and when this basin was formed and its subsequent development are the subjects of this chapter. Attention will focus on the underlying rock structures and the sediments brought in by rivers draining the North American continent.

## DEVELOPMENT OF KNOWLEDGE ABOUT THE GULF BASIN

Speculation about the origin of the Gulf of Mexico basin began over a century ago, and a discussion of early theories is provided by Lynch (1954). Of particular interest are the ideas of Schuchert (1935), which were widely held for several decades. He concluded that the Gulf floor is part of a large continental plate extending from Illinois to Tabasco, Mexico, and from eastern Mexico to the Bahamas. Formerly a shallow sea, the Gulf floor began to subside by the Cretaceous period and has continued to do so to the present time. The presumed mechanisms include isostatic adjustment (i.e., sinking in response to sediment loading) and subcrustal flow of molten rock away from the central basin. Subsequent authors provided further evidence for this scenario, citing the widespread salt beds, thought to have been formed in a very shallow basin but now present in sediments under the deep Gulf, and near-vertical scarps off the continental shelves west of the Florida Peninsula and off Yucatán, considered to be evidence of faulting that could have occurred as the Gulf floor was being depressed.

At the time these ideas were popular, it was assumed that the positions of all continents had been fixed early in the earth's history. Earthquakes, volcanic activity, mountain building, erosion, and similar phenomena were known to cause vertical changes, but lateral drift of the continents with respect to one another was not considered possible. However, by the late 1960s sophisticated geophysical studies of the land and ocean basins had revealed that the earth's crust is composed of large plates that are slowly moving with respect to one another, sometimes colliding, sometimes drifting apart. Alternative interpretations of the origin of the Gulf basin were now possible. During the past four decades, spurred largely by the search for onshore and offshore petroleum resources, geologists and geophysicists have greatly added to our knowledge of the subsurface structure of the Gulf margins and the deep Gulf basin itself. Seismic reflection and refraction surveys have provided information about the earth's crust and overlying rocks and sediments, and drill cores from land, the shallow Gulf, and the deep Gulf basin have aided in the interpretation of the seismic data. Based on plate tectonic considerations and detailed local geophysical information, a new scenario has emerged. Although knowledge gaps remain, the general picture of the origin and development of the Gulf of Mexico basin appears to be firmly established, and the details are provided by several authors in Salvador (1991a).

## DEFINITION OF THE STRUCTURAL BASIN

As seen in fig. 3.1, the structural basin of the Gulf of Mexico is a roughly circular depression approximately 1,500 km (900 mi) in diameter. It includes the present area of the Gulf and surrounding lands that fall within the structural framework of the basin. In Mexico it is bounded by the Chiapas Massif, the Sierra Madre Oriental, and the eastern edge of the Coahuila platform. Within the United States its borders include the Marathon Uplift and Ouachita orogenic belt of Texas, the Ouachita Mountains, central Mississippi deformed belt, and southern extensions of the Appalachian Mountains. Contained within the basin are the Florida and Yucatán Peninsulas, as well as the coastal plains of the southern United States and eastern Mexico, all of which have filled in or built up during development of the Gulf basin.

## FORMATION OF PANGAEA

Major events in the development of the Gulf of Mexico basin and the geological periods and approximate time frames in which they occurred are presented in table 3.1. Little is known about the history of the area during the early Paleozoic because of the paucity of appropriate outcrops and deep-drill well cores. However, there is ample evidence for great disturbance of the margins of the basin during the late Paleozoic, that is, from the late Mississippian and Pennsylvanian of the Carboniferous period

**Table 3.1.** *Major events in the development of the Gulf of Mexico basin and their times of occurrence. Mybp = millions of years before present.*

| Geological periods and stratigraphic units | | | | Mybp | Major events |
|---|---|---|---|---|---|
| Cenozoic | Quaternary | | Holocene | 0.02 | Continental glaciers. |
| | | | Pleistocene | 1.6 | Variable sea levels. |
| | Tertiary | | Pliocene | 5.3 | Southern Rocky Mountain and Colorado Plateau uplift. Mississippi River progradation. |
| | | | Miocene | 23.7 | Volcanism in western Sierra Madre. |
| | | | Oligocene | 36.6 | Rio Grande progradation. |
| | | | Eocene | 57.8 | Laramide orogeny. Houston embayment progradation. |
| | | | Paleocene | 66.4 | Chicxulub event. |
| Mesozoic | Cretaceous | U | Maastrichtian | 74.5 | |
| | | | Campanian | 84 | |
| | | | Santonian | 87.5 | High sea level stand. Basin margins flooded. Gulf of Mexico connected with western interior seaway. |
| | | | Coniacian | 88.5 | |
| | | | Turonian | 91 | |
| | | | Cenomanian | 97.5 | Yucatán platform submerged. |
| | | L | Albian | 113 | |
| | | | Aptian | 119 | Reefs and shoals develop around rim of Gulf basin. |
| | | | Barremian | 124 | |
| | | | Hauterivian | 131 | |
| | | | Valanginian | 138 | Florida peninsula partially submerged. |
| | | | Berriagian | 144 | |
| | Jurassic | U | Tithonian | 152 | Gulf of Mexico first in contact with Atlantic Ocean. |
| | | | Kimmeridgian | 156 | Formation of oceanic crust under central Gulf of Mexico. |
| | | | Oxfordian | 163 | Mississippi River first in evidence. |
| | | M | Callovian | 169 | |
| | | | Bathonian | 176 | Pacific Ocean flows across Mexico into Gulf basin. |
| | | | Bajocian | 183 | Louann and Campeche salt layers deposited. |
| | | | Aalenian | 187 | |
| | | L | Toarcian | 193 | |
| | | | Pliensbachian | 198 | Crustal expansion. |
| | | | Sinemurian | 204 | Breakup of Pangaea. |
| | | | Hettangian | 208 | Africa–South American plate moves away. |
| | Triassic | | | 245 | |
| Paleozoic | Permian | | | 286 | Crustal compaction. |
| | Carboniferous | | | 360 | Assembly of Pangaea. |
| | Devonian | | | | |

through Permian time (ca. 330–280 mybp). Major continental folding, thrusting, subduction, uplift, and formation of deep foreland basins all point to compressional events associated with the collision of continental plates around the northern and western perimeters of the Gulf of Mexico basin. This evidence appears in geological formations from the southern edge of the Appalachians in Alabama westward through the orogenic and uplift areas of Texas and possibly into northeastern Mexico. A string of granitic intrusive bodies from northeastern Mexico down to Central America of about the same age also suggests strong compressional activity. These formations have been interpreted as evidence of the collision of the African–South American plate with the North American plate, which resulted in the formation of the supercontinent Pangaea. Although the northern and western margins of the basin show evidence of the collision, not much is known about contemporaneous land masses within the basin or those to the east and south.

## BASIN STABILITY FOLLOWED BY RIFTING

Following the compressional events, for a period of over 60 million years during the late Permian and most of the Triassic (ca. 280–215 mybp), the Gulf of Mexico basin and immediately surrounding areas appear to have been part of a large stable land mass subject to erosion but with little evidence of marine deposition. However, by the late Triassic the African–South American plate began to pull away from the North American plate, initiating a period of crustal extension that continued for over 50 million years to the end of the middle Jurassic or the earliest part of the late Jurassic (ca. 215–160 mybp). It was during this interval that the Gulf of Mexico basin was formed, achieving the general configuration recognizable today.

## CHANGES IN THE UNDERLYING CRUST

The earth's crust, which underlies all land and marine areas, is composed of igneous rock called basalt. Although this lies beneath subsequently deposited sedimentary rocks, the top of the crust and its thickness can often be determined by deep seismic refraction studies. Within the Gulf of Mexico basin three crustal types are recognized: continental (ca. 40 km thick), oceanic (6–8 km thick), and transitional crust, the latter divisible into thick transitional (ca. 20–30 km thick) and thin transitional (ca. 10–20 km thick) types. As the continents pulled apart, the intervening crust was stretched and thinned,

with the greatest thinning taking place beneath the deep basin of what is now the Gulf of Mexico.

As presently understood, the evolution of the crust beneath the Gulf basin involved four stages: (1) graben formation, (2) crustal stretching and thinning, (3) formation of oceanic crust and subsidence, and (4) sediment loading and continued subsidence. In the earliest phase (late Triassic through early Jurassic), as the continental plates separated, the underlying crust began to pull apart along ancient fault lines or cracks in the brittle crust. Large blocks of land bounded by faults subsided, creating valleys or grabens that, over time, were filled with terrestrial and volcanic debris of iron-rich deposits known as "red beds." These grabens are evident today as deep subsurface features along the eastern coast of North America and along the northern edge of the Gulf of Mexico basin into Mexico. During the second phase (middle Jurassic), stretching and thinning created transitional crust throughout the area and widened some faults into basins. Specifically, the faults between the east-west-oriented Yucatán block and the northern and western margins of the Gulf widened and began to deepen, forming the early Gulf of Mexico itself. In the process the Yucatán block started to rotate slowly in a counterclockwise direction. In the third phase (late Jurassic), these processes continued. Widening of the Gulf was accompanied by stretching, thinning, and slow subsidence of what is now the floor of the deep central and western Gulf, creating true oceanic crust in this area. The Yucatán block continued to rotate toward its present north-south orientation. The final phase (all post-Jurassic time) has been marked by crustal subsidence throughout the entire Gulf of Mexico, but especially along the northern margins, under the burden of increasingly thick layers of sediment brought in by streams draining the United States and northern Mexico. The present distribution of major crustal types beneath the Gulf of Mexico basin is shown in fig. 3.2.

## ASSOCIATED EVENTS DURING THE JURASSIC

Throughout the early and middle Jurassic, central Mexico was covered by a shallow sea, an embayment of the Pacific Ocean. Toward the end of the middle Jurassic (beginning in the Callovian or somewhat earlier), the sea flowed eastward into the newly formed Gulf of Mexico depression, and this connection persisted intermittently through the late Jurassic. Repeated evaporation of the hypersaline Gulf waters produced widespread layers of salt (halite).

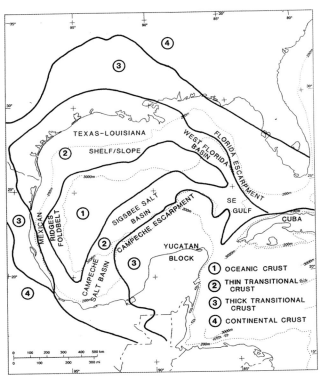

**Figure 3.2.** Map showing the distribution of major crustal types lying beneath the Gulf of Mexico structural basin. (From Buffler 1991.)

Over millions of years the total amount of salt deposited was enormous. Today the Louann salt bed beneath the Texas-Louisiana continental slope is over 3,000 m thick, and in the southern Gulf the Campeche salt layer is over 2,000 m in thickness. Coastal plains surrounding the Gulf were intermittently flooded, and evaporative deposits of anhydrite of approximately the same age are widespread.

During most of the late Jurassic the sea level was low, and shallow lagoonal deposits continued to form around margins of the basin. Seismic studies of the deep Gulf suggest quiet deposition of fine-grained pelagic shales and limestones over the older salt beds. Toward the end of the upper Jurassic the Gulf of Mexico first became connected with the Atlantic Ocean, and at about this time the level of the Gulf began to rise, flooding most of the lands bordering the Gulf depression. Some deposits in the Oxfordian suggest the existence of the ancestral Mississippi River, and by the end of the upper Jurassic, evidence of the river is clear. During most of the Jurassic both the Florida and Yucatán platforms were above water, but by the end of the period the southern portion of Florida had become submerged by the rising sea.

## CRETACEOUS EVENTS

Throughout the lower Cretaceous the continued rise in sea level ultimately flooded most of the Gulf of Mexico basin. Florida became totally submerged, but the Yucatán Peninsula remained above water. Carbonate reefs and banks developed in shallow water around the rim of the deep Gulf, with landward lagoons and near-vertical seaward scarps that dropped off to deep water. Partly because of the increased weight of accumulated sediments, the floor of the deep Gulf continued to subside. The distribution of sedimentary environments during the middle part of the lower Cretaceous is shown in fig. 3.3.

In the upper Cretaceous, after a brief drop, the sea level rose to new heights and inundated the entire basin, including the Yucatán Peninsula. Flooding of lands to the northwest created a shallow inland seaway connecting the Gulf of Mexico, through Texas and northward through the Great Plains, with the Arctic Ocean, a condition that persisted until near the end of the Cretaceous. Along the northern Gulf, deposits of terrigenous clastic materials (sand, silt, and clay) accumulated, whereas on the Florida and Yucatán platforms and along eastern Mexico carbonate deposition continued.

Toward the end of the Cretaceous, as a result of compressional forces from the west, midcontinental mountain building began, and this process was to last through the Eocene. This was the great Laramide orogeny, which lifted the Sierra Madre Oriental of Mexico and the Rocky Mountains of the United States. The early stages of this vast elevation broke the marine connection with the Arctic Ocean, and rivers draining these uplifted lands brought large volumes of terrestrial sediments into the northern and western sections of the Gulf. The slowly subsiding Florida and Yucatán platforms continued to build up layers of carbonate materials (limestone), and in the deepening Gulf basin calcareous muds were deposited over most of the floor except along the Mexican coast, where river-borne clastics dominated. Thus, toward the end of Cretaceous time, three zones of depositional activity were evident. In the east and south, carbonate layers were being laid down. In the west, the rise of the Sierra Madre Oriental was accompanied by erosion and local marine deposition of terrestrial clastics. In the northwest and north, large volumes of terrestrial sediments were being laid down, filling the basin margins and prograding the Gulf edge seaward.

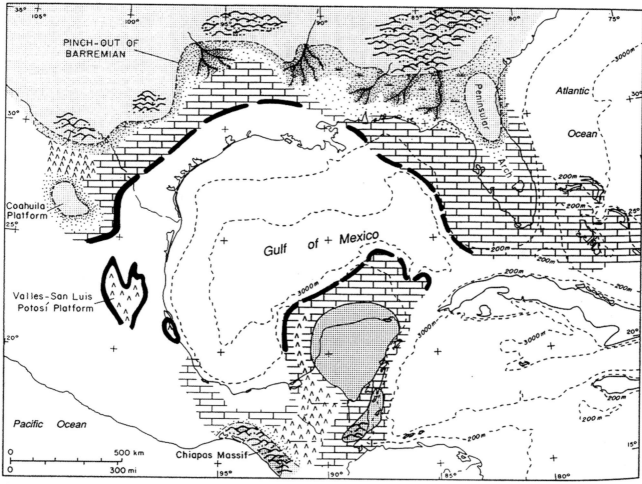

**Figure 3.3.** Map of the Gulf of Mexico basin during the lower Cretaceous (Barremian) showing the edge of the sea with river distributaries, clastic sediments, reef and shoal complexes producing shallow back lagoons, and open Gulf waters. In northern Mexico and Texas through Florida, the edge of the lower Cretaceous sea is marked by a thin dashed line. Stippled areas represent clastic deposits, some with submerged river deltas, on the shallow shelf. The brick-like pattern depicts shallow shelf carbonate deposits. The bold dashed lines mark the positions of carbonate shoals, banks, and reefs. Inverted v's represent very shallow evaporative basins. (From Salvador 1991c.)

### TERTIARY PROCESSES

Depositional processes begun in the Cretaceous continued through the Cenozoic period. Carbonates accumulated in the east and south and terrigenous clastic sedimentation was active elsewhere, particularly in the northwest and north. As shown in table 3.2, this deposition occurred in three phases, each correlated with major orogenic events on land. Alternately three basins were filled in: the Houston embayment of central and east Texas, the Rio Grande embayment of south Texas and northeastern Mexico, and the Mississippi embayment of Louisiana and Mississippi. Thick sedimentary sections that accrued in the Cenozoic have prograded the Gulf margins of northern Mexico, Texas, and Louisiana about 300 km basinward of the old Cretaceous reef-dominated carbonate shelf edge. Sediment loading has accentuated subsidence in the underlying transitional and oceanic crust to the extent that the sedimentary wedge just offshore of Texas and Louisiana is now over 15 km (9 mi) thick. Elsewhere marginal sediment accumulation rates have been only a fraction of those of the north.

### QUATERNARY PROCESSES

During the Quaternary (1.6–0.0 mybp), sedimentation in the Gulf has been dominated by the Mississippi River. In this period alone deposits 3,600 m thick have accumulated offshore of Louisiana and Texas, and the Mississippi Fan has been extended across the deep Gulf to the base of the Yucatán platform. Most of the recent deposition has been associated with events of the Pleistocene Ice Ages.

**Table 3.2.** *Areas of major sediment deposition in the northwestern Gulf of Mexico basin during the Tertiary, and associated orogenic events.*

| Area of sediment buildup | Geologic period | Correlated orogenic events |
| --- | --- | --- |
| Houston embayment | Paleocene-Eocene | Buildup of southern Rocky Mountains |
| Rio Grande embayment | Oligocene | Volcanism in west Texas and Sierra Madre Occidental |
| Mississippi River embayment | Miocene-Pliocene | Reactivation of southern Rockies, uplift of Colorado Plateau, eastward tilting of Great Plains, and renewed uplift of southern Appalachians |

During the past million and a half years or more, the northern part of the North American continent has been subject to repeated episodes of ice advance and retreat, and these ice sheets stood several miles thick in some areas. The four most recent episodes have occurred during the past 400,000 years. From the oldest to the youngest these include the Nebraskan, Kansan, Illinoian, and Wisconsin. At the time of each advance so much water was locked up in the continental ice that sea levels around the world dropped 120 m or more. During interglacial periods the sea level rose to its present stand or somewhat higher. As ice accumulated on land and sea levels dropped, major rivers became more deeply entrenched in their alluvial valleys, scouring these deeper and transporting the resulting sediments to the Gulf, where they rapidly prograded the river deltas seaward and built out the continental margins. On the other hand, during periods of sea level rise, the rivers deposited most of their sediments in the floodplain valleys and on the inner portions of the continental shelf, building these up and transporting to the outer shelf and upper slope only small volumes of sediments. The alternate scouring and filling of the lower Mississippi River valley is illustrated in fig. 3.4.

Of particular interest are the levels of the Gulf from near the end of the last Ice Age until the present. Figure 3.5 shows the levels of the world oceans during this period according to the best information currently available. Eighteen thousand years ago sea level stood at about 120 m (394 ft) below its present stand. Since then it has risen at an average rate of about 0.67 m (2.2 ft) per century. This general increase has been marked by two periods of very rapid rise when the sea level rose 20 m or more in about a thousand years, or 2 m (6.6 ft) per century. These rapid rises were centered around 14,000 and 11,500 years before the present. During these two periods a great deal

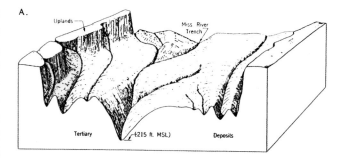

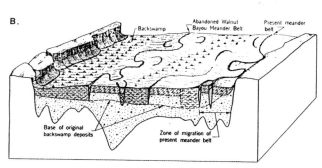

**Figure 3.4.** Block diagrams showing (A) the evacuated Mississippi River valley during a period of lowered sea level and (B) the partially filled valley during an early stage of sea level rise. (From Coleman, Prine, and Garrison 1980; Fisk 1944.)

of cold freshwater entered the world oceans, and a large percentage of it was delivered to the Gulf of Mexico by the Mississippi River. The actual volumes and the environmental and biological effects on the Gulf have not been fully investigated. During the past 8,000 or 9,000 years the rate of sea level rise has slowed considerably.

Parenthetically, it should be noted that prior to the late 1980s most of the recent sea level curves for the Gulf of Mexico were based on $^{14}$C dating of coral skeletons and other carbonate materials. However, it is now known that in some cases this technique produces results that are not entirely accurate, particularly for ages beyond about

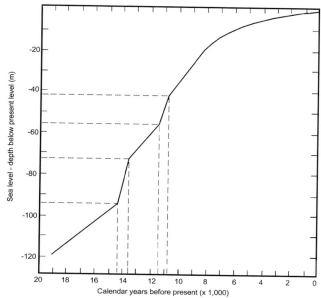

**Figure 3.5.** Rise in world sea levels during the past 19,000 years based on [230]Thorium / [234]Uranium dating of corals from Barbados. Note two periods of rapid rise centered around 14,000 and 11,500 years before the present and a slowing of the rise during the past 8,000 or 9,000 years. The present sea level is shown here at two or three meters below its actual level. This is not an artifact but the result of data based on samples from elkhorn coral, *Acropora palmata*, which generally grows at a depth of less than five meters. (Modified from Bard, Hamelin, and Fairbanks 1990; Fairbanks 1989.)

9,000 years ago. Carbon-14 is produced when high-energy cosmic rays from outer space collide with nitrogen atoms in the earth's atmosphere. This cosmogenic [14]C then begins to disintegrate at a regular rate, losing 50 percent of the remaining [14]C every 5,730 years. Radiocarbon dating is based on the assumption that the rate of formation of [14]C in the earth's atmosphere (from cosmic rays colliding with nitrogen atoms) has remained constant during the past, which is now known not to have been the case. The earth's magnetic field acts as a shield that deflects some of the incoming radiation, and the stronger the magnetic field, the more cosmic rays are deflected. It has recently become clear that the earth's magnetic field has varied over the past several thousand years, thus changing the rate of formation of cosmogenic [14]C in the atmosphere and thereby introducing errors in the [14]C-derived dates. These errors increase in magnitude the further back in time one goes. Before 9,000 years ago the [14]C dates are systematically younger than the actual calendar ages as determined by the more accurate [230]Thorium / [234]Uranium technique, resulting in a difference of 3,500 years for materials actually 20,000 years old. (Bard, Hamelin,

and Fairbanks 1990; Fairbanks 1989). Fortunately, the old [14]C measurements can easily be calibrated.

Past sea levels have left records of their occurrence. At lower stands of the sea much of the continental shelf was exposed to subaerial erosion, terrestrial animals roamed the shelf, and at maximum sea level lowering, the edge of the Gulf lay at or near the outer margin of the continental shelf, where it was subject to more intense erosion by the surf. Submarine terraces carved by wave action mark paleoshorelines, evidence of earlier stillstands of the Gulf. Cemented beach ridges and dunes and drowned coral reefs also provide evidence of earlier stillstands. During periods of exposure the limestone platforms of Florida and Yucatán became riddled with caverns leached out by rainfall and subterranean rivers, and occasionally the roofs caved in. Now sinkholes, some filled with sand and other debris, are found on the carbonate shelves. Although sometimes far from the present shoreline, some of these off west Florida still flow with freshwater, and sea bottoms here may be populated by estuarine rather than marine flora and fauna.

On the Yucatán shelf Logan et al. (1969) found terraces and other shallow-water features at depths of 128, 91.5, 51.6–64.0, and 30.5–36.6 m (428, 300, 170–210, and 100–120 ft) below the present level of the Gulf. Off western Louisiana, at a depth of 38 m (124.6 ft), the writer has collected river-washed gravel and shells of the American oyster (*Crassostrea virginica*), an inhabitant of estuaries. Thayer, La Roque, and Tunnell (1974) determined that Seven and One-Half Fathom Reef, located in 14 m (46 ft) of water on the south Texas shelf, is most likely an old lake bottom deposit, and from this formation these authors and Tunnell and Causey (1969) reported terrestrial and freshwater snails as well as bones and teeth of the Columbian mammoth (*Mammuthus columbi*), American mastodon (*Mammut americanum*), and bison (*Bison* sp.). Active formation of barrier islands and spits along the coasts of the northern and northwestern Gulf began about 5,000 years ago as the rate of sea level rise diminished. Future changes in sea level will depend on local and distant geophysical events as well as on alterations in global temperature regimes that control the formation and melting of glaciers and the polar ice caps.

## SALT TECTONICS

Conditions of high compression and elevated temperature cause salt to become viscous and to flow slowly, somewhat like toothpaste in a tube. Under the great

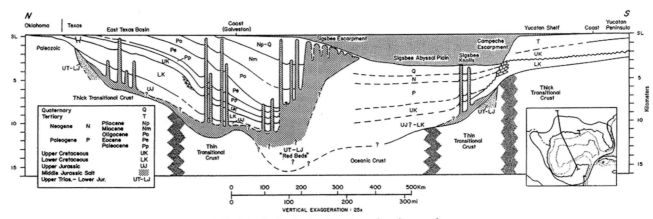

**Figure 3.6.** Generalized cross section of the Gulf of Mexico basin showing the various layers of sedimentary deposits and the present distribution of Jurassic salt. (From Salvador 1991b.)

weight of Cretaceous and Cenozoic sediments the old Jurassic salt beds have flowed laterally as well as vertically along faults and other weaknesses in the overlying sedimentary blanket. Vertical columns or salt *diapirs* occur beneath the coastal plain, continental shelf, and continental slope of the northern Gulf and the southern Gulf west of the Yucatán platform. They also are found beneath the deep Gulf west and north of Yucatán. Vertical expressions of salt diapirs, often called "salt domes," may take the form of columns, walls, or massifs. The latter are most common beneath the middle and outer continental slope, where the overburden is thinnest.

As they near the surface, the rocks that are pushed up form low mounds or hills, and these are evident on the coastal plain, the continental shelf and slope, and in the bottom of the deep Gulf, where they form a line of low mounds referred to as the Sigsbee Knolls. Where a salt column actually penetrates the sea bottom of the continental shelf or slope, it is gradually dissolved by the seawater, creating a very concentrated brine-filled basin. A cross section of the Gulf of Mexico basin showing the configuration of the salt beds and other features in relation to the sedimentary deposits is given in figure 3.6. Somewhat like salt, shale under compression may also flow. The continental slope of the western Gulf of Mexico is thrown into a system of ridges and valleys running more or less parallel to the coastline. These apparently represent the surface expressions of underlying shale deposits deformed by compression and gravity slumping into a system of long ridges and troughs.

## PETROLEUM DEPOSITS

Petroleum and natural gas generally develop from marine deposits where large amounts of organic material are buried by layers of sediments. This may occur in hypoxic or anoxic basins with poor circulation and little or no oxygen in the lower part of the water column. It may also take place on the open continental shelf or slope where layer upon layer of sediments build up through the millennia, each containing only a small percentage of organic matter. Both types of environment are often encountered around the deltas of major rivers.

Large streams entering the sea bring dissolved and finely particulate organic material derived from algae, leaf litter, and other upstream sources, and much of the particulate matter settles on the nearby continental shelf and slope. Rivers also bring to the sea quantities of dissolved nutrients (nitrates, phosphates, and silicates) that stimulate the growth of marine phytoplankton, and much of this also falls to the nearby sea bottom. As sediments near the river mouth containing 1 or 2 percent organic matter become buried, they are shut into an environment totally devoid of free oxygen. Here the organic materials gradually undergo chemical transformations that convert them to petroleum compounds. Over millions of years the continual build-up, burial, and transformation create large underground deposits that may migrate laterally through sand layers and pore spaces or upward toward the surface. Some of the petroleum and gas escape directly into the sea, where they are swept away. In other cases the materials become trapped beneath impermeable layers of rock or clay and form large subsurface reservoirs. Salt columns that penetrate such layers facilitate the upward migration, and reservoirs often develop around the sides of salt diapirs. In the Gulf of Mexico area major oil and gas reservoirs are located around the deltas of present or ancient rivers including those along the southern margin of the Bay of Campeche, Veracruz, Tampico, and Tuxpan; on shore and offshore along the northwestern and northern Gulf into the Florida panhandle; and possibly in an ancient basin off the Everglades of south Florida. Natural oil and gas seeps have been identified on lands surrounding the Gulf, on the continental shelf, and on the continental slope. There is also evidence of petroleum deposits in association with the Sigsbee Knolls and elsewhere in the deep Gulf. Given the long and extensive depositional history of the Gulf of Mexico, it is not surprising that the areas of great sediment accumulation harbor some of the world's major deposits of petrochemicals.

## THE CHICXULUB EVENT

A footnote to the history of the Gulf of Mexico basin involves an event that took place 65 million years ago. This event, which wiped out most terrestrial animal groups, changed the nature of terrestrial vegetation and killed off nearly three-quarters of the earth's shallow-water marine life, marking the end of the Cretaceous period. The scientific community now generally agrees that at that time the earth was struck by a bolide (comet or meteorite) 8 to 10 kilometers in diameter and with a mass of around a trillion tons. Traveling at an estimated velocity of 20 km/sec, it came in from a southerly direction and struck the earth on the northern coast of Yucatán, its epicenter being north of Merida where the coastal village of Chicxulub now stands. The energy released from this impact ($10^{24}$ joules) was enormous, equivalent to that of 100 million megatons of TNT. The force excavated a crater 30 km (18 mi) deep, and after collapse of the side walls the rim now measures about 180 km (108 mi) across (fig. 3.7). Although now buried under a mile-thick blanket of Tertiary limestone, the crater has been successfully mapped by its magnetic anomaly signature (fig. 3.8). Much effort has gone into determination of the effects of this impact on the earth, the atmosphere, and the sea, and good discussions may be found in the works of Frankel (1999) and Ryder, Fastovsky, and Gartner (1996), among others. A general scenario of events associated with the impact follows, but the details are subject to revision as new evidence accumulates.

The initial impact generated powerful atmospheric shock waves beginning at a velocity of 20 km/sec but diminishing rapidly as they spread out in all directions.

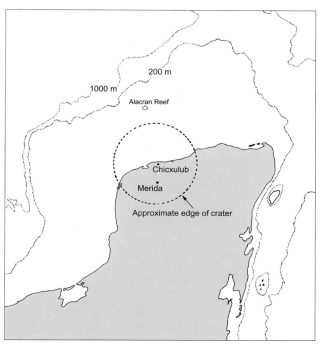

**Figure 3.7.** Location and approximate extent of the Chicxulub crater on the northern portion of the Yucatán platform.

Across the Gulf of Mexico and surrounding lands this was accompanied by winds of several hundred km/hr, far greater in force than any associated with present-day hurricanes. Eventually, as the shock waves dissipated, the winds reversed direction and swept back toward the center from which they had come. Great tidal waves were generated that probably exceeded 50–100 m in height as they inundated coastal lowlands around the Gulf. Shock waves through the solid earth must have reached at least 10 on the Richter scale, triggering massive earthquakes, landslides, and fault movements in lands surrounding the Gulf and in the Gulf floor itself. An estimated 200,000 km³ of target rock was blasted out from the crater area, and this was added to blast debris from the mass of the bolide itself. The ejecta included vast amounts of solid, molten, and vaporized rock sent out at velocities of up to several km/sec. Some must have gone halfway to the moon before falling back to the earth. The fiery reentry of such rocks and smaller particles heated the atmosphere perhaps several hundred degrees, igniting global wildfires that ultimately burned up much of the preexisting terrestrial biomass over large portions of the earth. Rocky material from the blast, some of it molten or very hot, bombarded much of the earth's surface, but the fallout was asymmetrical, most falling to the west of the impact

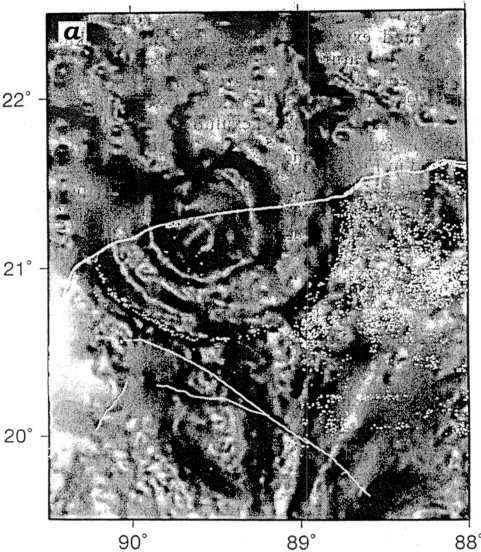

**Figure 3.8.** Gravity anomaly map of the northern portion of the Yucatán platform showing the circular structure of the Chicxulub crater now covered by over a thousand meters of limestone deposits. At least six concentric rings have been identified. Ejecta is splayed out in a northwesterly direction, showing that the bolide struck from the southeast. The white line traversing the crater's center marks the present shoreline of Yucatán. White lines in the lower portion of the image are earthquake faults. White dots represent sinkholes, or "cenotes." Their clustering around the periphery of the crater suggests that their formation is closely related to the presence of slump faults near the rim of the crater. (From Hildebrand et al. 1995.)

as the earth rotated beneath the cloud of flying debris. Finer particles remained suspended in the stratosphere and atmosphere for months to years following the initial blast. Together with smoke from the global conflagrations, they blocked out the sunlight so that the earth was in total darkness for perhaps six months and under twilight conditions for additional months or years. Thus, following the firestorms the earth's temperature must have dropped dramatically, with chill conditions prevailing until much of the dust settled out. Since the target rocks of Yucatán contained large deposits of sulfur-rich anhydrite, a great deal of sulfur (probably about 100 billion tons) must have entered the atmosphere as sulfur dioxide,

which, when combined with atmospheric water, was converted to sulfuric acid. In addition, the blast itself and ensuing events must have converted much atmospheric and terrestrial nitrogen to nitrous oxide, which, with atmospheric water, became nitric acid. The ensuing acid rains greatly reduced the pH of soils, fresh waters, and surface waters of the sea and leached large amounts of heavy metals (aluminum, antimony, arsenic, mercury, nickel, selenium, etc.) from soils and ashes and deposited them in lakes, estuaries, and the sea. The target rocks of Yucatán were composed largely of limestone, which was rich in carbon, and perhaps as much as a trillion tons of carbon were converted to carbon dioxide gas. Together with the carbon dioxide released from combustion, the atmosphere must have been enriched in carbon dioxide by four or five times its original level. This, in turn, through the greenhouse effect, greatly elevated global temperatures for a thousand years or more.

The sequence of events, particularly for the Gulf of Mexico region, appears to have been as follows:

1. Following the initial impact, atmospheric shock waves produced very high winds, and terrestrial shock waves generated earthquakes, landslides, faults, and fault slips within and around the basin.

2. Great tidal waves spread out, inundating marginal lowlands.

3. Heating of the earth's atmosphere by the initial blast, subsequent falling of hot debris, and friction due to reentry of large particles led to raging wildfires that burned up much of the earth's terrestrial organic material.

4. During and following the conflagration, the earth was blanketed by layers of fine particulate material suspended in the atmosphere and stratosphere. Thus, insulated from solar radiation, the earth descended into total darkness for several months, followed by perhaps a year or more of twilight. All terrestrial, freshwater, and marine photosynthesis ceased during this period, and the temperature of the atmosphere, soil, and surface water of the sea plunged sharply.

5. During and possibly following the period of darkness, large volumes of acid rain fell to earth, acidifying terrestrial and marine systems and concentrating toxic metals in lakes, estuaries, and marine sediments.

6. These events were followed by a thousand years or more in which the temperature of the earth was greatly elevated due to high atmospheric concentrations of greenhouse gases, particularly carbon dioxide.

7. Gradually thereafter, photosynthetic plants returned to the land and surface waters of the sea. These eventually reduced the levels of atmospheric carbon dioxide, and world temperatures returned to normal. Surviving plant and animal species underwent evolutionary radiation, giving rise to grasses and other flowering plants and the great diversity of land and marine mammals.

The question naturally arises as to where and how any plants and animals survived the devastation, and only partial and speculative answers are available. It is likely that most if not all of the flora and fauna of estuaries, continental shelves, and surface waters of the Gulf of Mexico was annihilated by the blast and its aftermath. The fact that many types of plants and animals did survive, however, attests that there were refuges elsewhere. Dense clouds or heavy local rainfall may have protected some areas of the earth from wildfires. Some terrestrial plants may have survived as resistant seeds, spores, or deeply buried root systems, perhaps located in the lee of protective barriers such as mountains or canyon walls; and small amphibians, reptiles, birds, and mammals may have survived in such areas or in subterranean burrows, although apparently no terrestrial animal larger than 25 kg (55 lb) made it through. Freshwater species may have persisted in deeper portions of alkaline lakes located in limestone areas where the carbonates could neutralize acid runoff. Animal species that did survive were probably omnivores that could subsist on a diet of dead plant and animal material. Some species of plankton and other marine life may have survived in deeper water far from shore, and photosynthetic algae may have made it through as cysts or resting spores. What we do know from the fossil record is that ferns (which have a resistant spore stage) were the first major plant group to repopulate the burned terrestrial landscape, and these were gradually replaced by other species of higher vegetation. In the oceans the diatoms and radiolarians (both with acid-resistant silicon-based tests) came back relatively quickly, but foraminiferans (which have calcareous tests) took much longer, presumably due to the acidity of surface waters of the sea. Most Cretaceous species of foraminiferans became extinct, and those that made it through were primarily tiny species with small tests. Research currently underway should eventually paint a clearer picture of which groups persisted, which did not make it, and the life history and environmental factors associated with the differential survival patterns. Effects on bottom sediments of the deep Gulf are discussed in the next chapter.

# 4 : PHYSIOGRAPHY AND SEDIMENTARY ENVIRONMENTS

The Gulf of Mexico today is, in many respects, unique among the world's ocean systems. It lies in a semienclosed Mediterranean-type basin bounded by the North American continent and the island of Cuba, but it is connected with the Caribbean Sea by the Yucatán Channel and with the Atlantic Ocean through the Straits of Florida. From its large drainage basin it annually receives a great deal of river-borne freshwater and sedimentary material. Compared to the Atlantic and Pacific, it is much smaller and shallower, and a large portion of its margin is bounded by estuaries and coastal lagoons. It is characterized by broad continental shelves, and relatively speaking, only a small fraction of its surface area is underlain by very deep water. The bottom in some areas is hard and rocky; in others it is soft and smooth. It may be flat and featureless or punctuated by outcrops, hills, and valleys. In the following pages these and related topics will be examined in greater detail. Special attention will be given to the historical and ongoing processes that have shaped the present structure and composition of the subbottoms and sediment surfaces.

## PHYSIOGRAPHY

Basic physical data concerning the Gulf are provided in table 4.1 and are given in greater detail in Moody (1967) and Martin and Bouma (1978). Situated between 18° and 31°N latitude, the Gulf is bisected by the Tropic of Cancer (ca. 23.5°N). Thus, the environment of the southern portion of the Gulf is subtropical, whereas the northern half is subtropical to temperate. The Gulf occupies a surface area of over 1.5 million km², of which 35.2 percent is underlain by continental shelf (0–200 m depth), and only 24.3 percent is underlain by very deep water (> 3,000 m). The maximum depth of 3,750 m, known as the Sigsbee Deep, is located in the west-central sector. Compared to the major world oceans, the Gulf is characterized by a very large percentage of continental shelf and the absence of really deep water (> 4,000 m). The deepest point in the Gulf of Mexico is only about one-third as deep as the greatest known ocean depth of 11,022 m in the Mariana Trench of the western Pacific. The Gulf occupies 1.8 percent of the surface area of the Atlantic Ocean, but because

**Table 4.1.** *Physical data concerning the Gulf of Mexico. (Modified from Moody 1967.)*

| | |
|---|---|
| Geographic location | |
| longitude | 81° – 98°W |
| latitude | 18° – 31°N |
| Dimensions | |
| surface area | $1.58 \times 10^6$ km² |
| maximum east-west length | 1,575 km |
| maximum north-south width | 900 km |
| Depth | |
| maximum | 3,750 m |
| mean | 1,485 m |
| Yucatán Channel | 1,500 – 1,900 m |
| Straits of Florida | ca. 800 m |
| Freshwater inflow | |
| total Gulf | ca. $1.60 \times 10^{11}$ m³/yr |
| US streams only | $1.26 \times 10^{11}$ m³/yr |
| Mississippi River input | |
| annual freshwater discharge | $1.03 \times 10^{11}$ m³/yr |
| annual sediment load | $4.54 \times 10^{11}$ kg/yr |
| Saltwater outflow (Florida Current) | $> 8.0 \times 10^{14}$ m³/yr |

of the large extent of shallow water, it contains only 0.7 percent of the volume of the Atlantic Ocean.

Major physiographic and topographic features of the present Gulf of Mexico basin are shown in figure 4.1. The Gulf is remarkable in that over half its perimeter is bounded by shallow bays, estuaries, and coastal lagoons. These are especially prominent along the northern and northwestern coasts, where they provide protected harbors and nursery areas for the production of fish and shellfish. Broad continental shelves occur off Florida and Yucatán as well as in the northern Gulf from Alabama to south Texas, the latter shelf being interrupted by the Mississippi River Delta. Prominent also are the steep scarps off the Florida and Yucatán platforms and the hummocky continental slope areas off Texas and Louisiana and west of the Yucatán platform. The Mississippi Fan, which stretches southward from the northern Gulf slope, extends to the

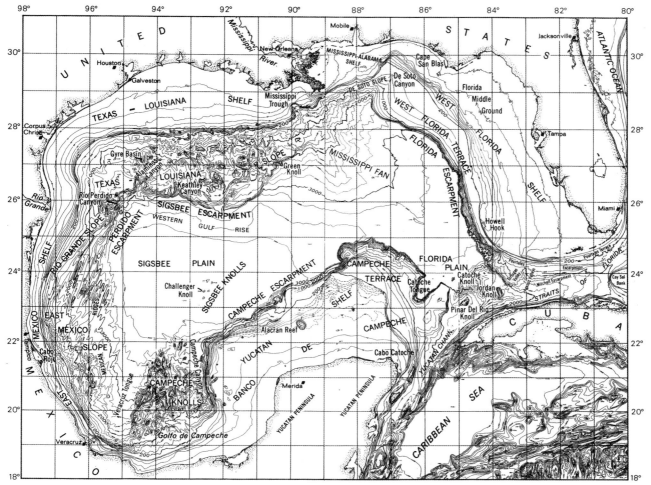

**Figure 4.1.** Bathymetric map of the Gulf of Mexico showing major provinces and topographic features. (From Martin and Bouma 1978.)

base of the Yucatán rise. Although not obvious from the map, the continental slope of the western Gulf consists of a series of ridges and troughs running more or less parallel to the Mexican coastline. The continental slope and, to some extent, the shelf are dissected by DeSoto Canyon off the Florida panhandle, the Mississippi Trough west of the Mississippi River Delta, and Campeche Canyon west of the Yucatán platform. The deep Gulf is a flat plain broken only by a series of low hills in the west known as the Sigsbee Knolls and a few hills in the small eastern section. The sill beneath the Yucatán Channel lies at a depth of 1,500–1,900 m, but the controlling depth of the Straits of Florida is only about 800 m.

The Gulf of Mexico drainage basin includes about two-thirds of the coterminous United States, half of Mexico, and small portions of Canada, Guatemala, and Cuba (fig. 4.2). From this vast area rivers annually bring to the Gulf around 160 billion m³ of freshwater. Seventy-nine percent

is from streams entering the northern Gulf (Rio Grande through Florida), and 82 percent of this, or 65 percent of the total, comes from the Mississippi River drainage. The average monthly discharge rate of the Mississippi-Atchafalaya system is shown in figure 4.3, but this is subject to wide annual and seasonal variation. Most of the freshwater is brought to the Gulf during the spring months (March–May), whereas very low flow rates characterize late summer and fall (August–November). Discharge velocities may exceed 90 cm/sec. Each year, the Mississippi River transports to the Gulf 454 billion kg (500 million tons) of sediments representing about a half ton of topsoil per acre of drainage. The bedload, containing fine sands, represents 10–20 percent of the total, and the remainder is primarily suspended silts and clays. Deposition at the river mouth has prograded the delta toward the outer shelf so that Southwest Pass is now less than 20 km from the 200 m isobath and 45 km from the

**Figure 4.2.** The surface freshwater drainage basin of the Gulf of Mexico involving portions of the United States, Canada, Mexico, Guatemala, and Cuba. Subsurface drainage also enters from the Florida and Yucatán Peninsulas and probably elsewhere. The Mississippi River drainage basin is shaded. (From Moody 1967.)

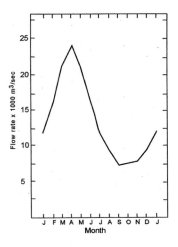

**Figure 4.3.** Average monthly discharge rates for the Mississippi River system.

1,000 m isobath. More than 800 trillion m³ of salt water flow through the Straits of Florida annually, about 8,000 times the yearly discharge of the Mississippi River.

Moody (1967) estimated that at the present rate of sedimentary input the Gulf of Mexico basin could be completely filled in less than seven million years, assuming that the added weight does not cause further subsidence of the Gulf floor and that no sediments are lost to the connecting oceans. As Moody recognized, neither of these assumptions is valid, but this does raise the question of the future of the Gulf. One scenario is that sediments brought in by the Mississippi-Atchafalaya system will gradually build up an underwater dam or sill connecting the continental slopes of Louisiana and Yucatán, thus dividing the deep Gulf into two basins. With restricted internal circulation, bottom waters of the western basin would become anoxic and devoid of higher life. With continued build-up of the sill, the level of the oxygen-free water would rise, and before the western Gulf itself was finally filled with sediment it would become an anoxic basin somewhat like the Black Sea. Meanwhile, sediments would also accumulate in the eastern Gulf, but these would build up only to the controlling depth of the Straits of Florida. Because of the magnitude and strength of the Yucatán Current and associated mixing of water masses, the eastern Gulf basin would never become anoxic. Frictional resistance in the shallower basin could reduce the volume and velocity of the current, but without major changes in world ocean circulation patterns the modified eastern basin would persist.

## REGIONAL ENVIRONMENTS AND TOPOGRAPHIC FEATURES

Consideration of environments around the Gulf of Mexico will be facilitated by dividing the Gulf into different sectors. The first will include the marginal environments and continental shelves beginning with Florida and the northeastern coast and moving counterclockwise around the Gulf to Yucatán and Cuba. This will be followed by examination of environments of the deeper water. Emphasis is placed on the surface sediments, but prominent coastal structures, outcrops, canyons, and other features will receive some attention. For orientation a general map of the surface sediments is presented in figure 4.4, and rock, mineral, and sediment types important in the Gulf are described in table 4.2.

### Continental shelves

EASTERN AND NORTHEASTERN GULF

This section extends from the Florida Keys to the Mississippi River Delta (fig. 4.5). Here the coastline is quite complex. The southern end is bounded by the Florida Reef Tract and a chain of low-lying limestone islands,

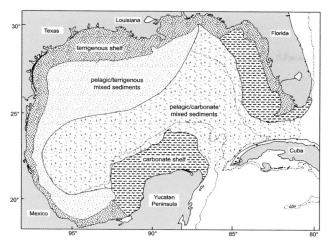

**Figure 4.4.** General surface sediment map of the Gulf of Mexico. (Modified from Uchupi and Emory 1968.)

the Florida Keys, which lie offshore from Florida Bay and the Everglades. The northward-trending coastline of the Florida Peninsula is indented by several estuaries, notably Charlotte Harbor and Tampa Bay. Apalachee and Apalachicola Bays lie at the northern end of the peninsula, and the coast of the panhandle is indented by St. Andrew, Choctawhatchee, and Pensacola Bays. A string of barrier islands lies offshore of the coasts of Alabama, Mississippi, and eastern Louisiana. From east to west, these include Dauphin, Petit Bois, Horn, Ship, Cat, and the Chandeleur Islands. These bound Mississippi, Chandeleur, and Breton Sounds. Coastal low-salinity habitats inland of these sounds include Mobile Bay and Lakes Borgne and Pontchartrain. Since most streams draining the Florida Peninsula are fairly short, they bring only

**Table 4.2.** *Rock, mineral, and sediment types important in the Gulf of Mexico basin.*

aragonite – soft, chalky white crystalline form of calcium carbonate that forms under marine conditions and is the most common form produced by marine plants and animals.

basalt – dense, igneous rock or condensed lava that constitutes the basement rock of ocean bottoms, generally rich in iron, silicon, and aluminum combined with base elements such as sodium, potassium, and/or calcium.

calcareous material – solid rocky material such as limestone, mollusk shells, etc., that is rich in calcium.

calcite – hard, glassy crystalline form of calcium carbonate that forms primarily under atmospheric or freshwater conditions.

carbonate rocks – rocks such as limestone that are rich in calcium carbonate and often other related materials such as magnesium carbonate.

clastic material – material formed primarily from the weathering or breakup of terrestrial rocks, e.g., gravel, sand, silt, and clay.

    gravel – small rocky materials with particle size greater than 2.00 mm in diameter that are formed from the breakup of larger rocks.

    sand – granular material from the breakup of rocks, often composed of carbonate or quartz, with grains in the size range of 0.062–2.00 mm in diameter.

    silt – fine particulate material from the breakup of rocks, finer than sand and with grains in the size range of 0.004–0.062 mm in diameter.

    clay – very fine particulate material from the breakup of rocks, made up largely of hydrated aluminum silicates, finer than silt and with particle sizes of less than 0.004 mm in diameter.

evaporite – rocky mineral material formed by the evaporation of seawater or brine: includes anhydrite, gypsum, and related materials.

    anhydrite – mineral formed by the evaporation of seawater and composed of calcium sulfate.

    gypsum – mineral formed by the evaporation of seawater and composed of hydrated calcium sulfate.

halite – rock salt.

limestone – rock composed wholly or chiefly of calcium carbonate.

quartz – silicon dioxide, often occurring in crystalline form, and a chief component of most beach sands.

sandstone – rock formed by the consolidation of sand and held together by a cement of silica, lime, gypsum, or clay.

shale – weak rock formed by the consolidation of clay.

siltstone – rock formed by the consolidation of silt.

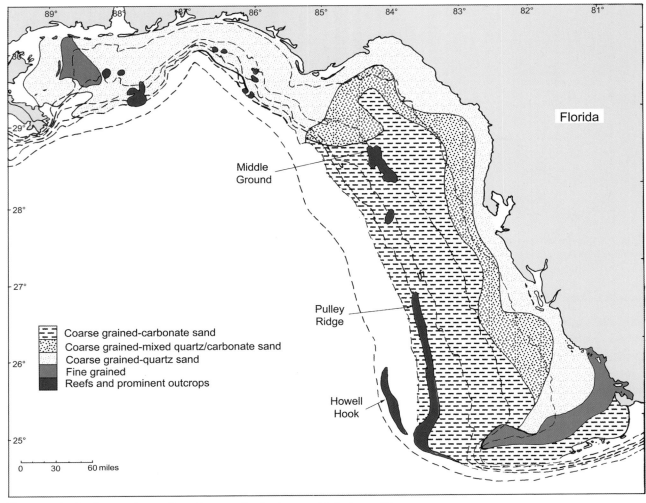

**Figure 4.5.** Predominant surface sediment types and major topographic features of the eastern and northeastern Gulf shelf. (Modified from Darnell and Kleypas 1987.)

limited quantities of freshwater and sediment to the Gulf, and here the estuaries are separated by long open stretches of saline water. However, from Mobile Bay to the Mississippi River Delta there is greater freshwater input and protection by the barrier islands, and these factors allow for lower salinity conditions in Mississippi Sound and related coastal waters.

Off peninsular Florida the broad continental shelf is underlain by a limestone platform dipping gently to the west. This is covered by a thin sedimentary blanket made up of bands running parallel to the shoreline. The nearshore bottom out to a depth of 10–20 m consists largely of quartz sand derived ultimately from weathering of the southern Appalachian Mountains and distributed along the coasts by alongshore currents. The outer portion of the shelf, from 20–40 m to the shelf break, is floored pri-

marily by biologically derived coarse carbonate sand and rubble. This material includes fragmental remains of calcareous algae, foraminifera, sponges, corals, mollusks, and echinoderms in various proportions. Between the two bands lies a strip made up of a mixture of quartz sand and carbonate materials. In the south a tongue of fine-grained carbonate sediments runs seaward from the Everglades and extends the full length of the Florida Keys. From about the level of Apalachicola Bay westward the shelf is blanketed by a massive quartz sand sheet, which, west of Mobile Bay, becomes mixed with finer silts and clays derived from the Mississippi River and nearby streams. Sediments tend to become finer toward the outer shelf, and along the eastern rim of DeSoto Canyon lies a thin band of fine-grained carbonate deposits.

A number of significant topographic features appear

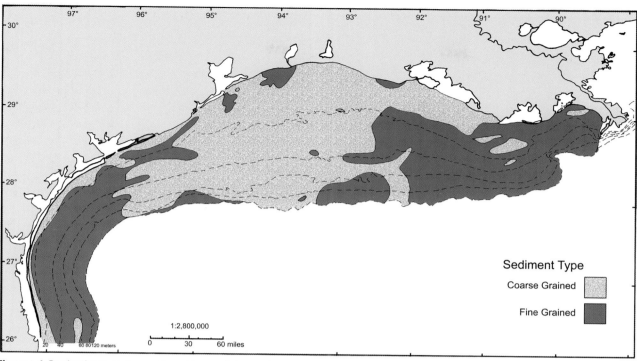

**Figure 4.6.** Predominant surface sediment types of the northern and northwestern Gulf shelf. (From Darnell, Defenbaugh, and Moore 1983.)

on the continental shelf of the eastern and northeastern Gulf. Extending south and west from the Florida Peninsula and stretching to the Dry Tortugas lies the Florida Reef Tract. This limestone arc includes the Florida Keys and associated submerged outcrops and coral reefs. Seaward of the Everglades and extending northward to the level of Charlotte Harbor at a depth of around 80 m lies another limestone trend known as Pulley Ridge, and this is paralleled by an even deeper ridge at a depth of 130–150 m known as Howell Hook, and yet another at a depth of 210–235 m. These ridges appear to be drowned reefs or consolidated beach dunes built up during lower sea level stands of the Pleistocene. The Florida shelf displays occasional sinkholes, generally filled with sedimentary debris. Also present are numerous low limestone outcrops and terraces as well as massive living and dead coral heads. Two carbonate features of particular prominence are the Elbow, at around 60 m off Tampa Bay, and the Middle Ground, at 40–50 m south of Apalachicola Bay. The latter is a complex of limestone outcrops, terraces, ridges, and pinnacles with vertical relief of up to 15 m and covering an area of about 1,500 km². Flat limestone outcrops and broken slabs have been reported around the head of De-Soto Canyon, and patches of low-relief rocky outcrops are known from about 20 m off Mobile Bay. South and south-

west of Mobile Bay in the depth range of 38–130 m lies a series of topographic features including parallel linear ridges, patch reefs, and pinnacles ranging to over 20 m in height. These may occur singly or in clusters, and they often trend along isobaths. These features appear to represent eroded remnants of ancient shorelines and ridges of sand, gravel, and shell cemented together. The three major groups include ridges and scarps at about 60 m, patch reefs around 65–70 m, and tall pinnacles at about 105 m.

The eastern and northeastern Gulf area is marked by its high diversity of habitat types. These include mangrove swamps, seagrass meadows, coral reefs, "live-bottom" development on rocky outcrops, and various combinations of soft bottoms made up of carbonate rubble, quartz sand, and/or silt.

## NORTHERN AND NORTHWESTERN GULF

This section stretches from the Mississippi River Delta westward and southward to the Mexican border (fig. 4.6). Well over two-thirds of this coastline is bounded by bays and coastal lagoons. Prominent are those of Louisiana (Barataria, Terrebonne, Timbalier, Atchafalaya, and Vermilion Bays) and Texas (Galveston, Matagorda, San Antonio, Copano, and Corpus Christi Bays and the long

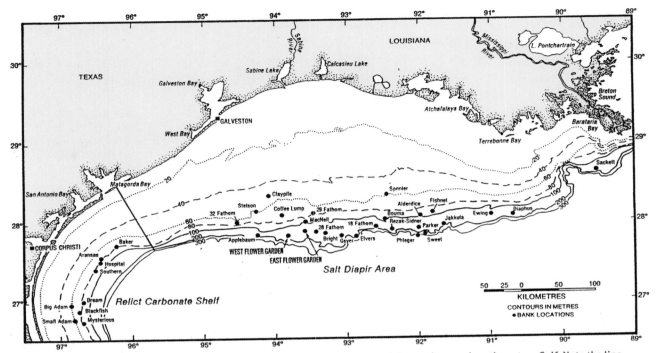

**Figure 4.7.** Location of major hard banks on the continental shelf and upper slope of the northern and northwestern Gulf. Note the line drawn across the shelf below Matagorda Bay that separates the salt diapir province to the east from the relict carbonate shelf to the west. (From Rezak, Bright, and McGrail 1983.)

Laguna Madre). Streams entering the upper portions of this coast bring in large volumes of freshwater so that the estuaries from the Mississippi River Delta through Matagorda Bay are of fairly low salinity. Here freshwater input exceeds the loss due to evaporation. Below the level of Corpus Christi the reverse is the case, and the Laguna Madre and related waters of south Texas tend to be hypersaline, with salinities well above that of seawater. Because of the large volumes of freshwater discharged by the Mississippi-Atchafalaya and other Louisiana streams, the nearshore continental shelf off Louisiana and upper Texas is often bathed by water of less than full marine salinity. This is especially true during the spring months when the rivers are in full spate.

The entire continental shelf from the Mississippi River Delta to the Rio Grande is blanketed by a layer of clastic sediments that are distributed in complex patterns with much local variation. Sand lies adjacent to the coastline from central Louisiana to the Rio Grande. Coarser sediments also make up the dominant type from shore to the outer shelf throughout the entire central third of this sector. Finer sediments, silt and clay, cover the eastern third (dominated by the Mississippi and Atchafalaya Rivers) and the western third of the shelf (influenced by

the present and ancestral Rio Grande). Fine sediments also occur along the outer edge of the continental shelf.

Many topographic features occur on the northern and northwestern shelf. About three dozen of the more prominent hard banks are depicted in figure 4.7. As pointed out by Rezak, Bright, and McGrail (1985), those banks off south Texas below the level of Matagorda Bay represent drowned coral/algal reefs that developed on a carbonate shelf during periods of lower sea level stand. East of Matagorda Bay the middle and outer shelf banks, with vertical relief of up to 50 m, are all outcrops of Tertiary limestones, sandstones, claystones, and siltstones associated with salt domes. Some are capped with active coral and/or algal reefs. The most prominent and highly developed bank/reef complex includes the East and West Flower Garden Banks, located on the outer continental shelf 190 km southeast of Galveston, Texas, and rising to within 15 m of the surface of the water. In addition to these major banks, many minor rocky outcrops are known throughout this section of the shelf. Another feature of interest is the Mississippi Trough, which indents the outer shelf just west of the Mississippi River Delta and extends southeastward across the continental slope. This represents an old distributary canyon of the Mississippi River from a time of lower sea level stand. As in the eastern and northeastern Gulf, the continental shelf west of the Mississippi River Delta is a broad plain that slopes gently seaward but is punctuated with numerous topographic features that reflect the past history of the Gulf basin.

WESTERN AND SOUTHERN GULF

This portion of the Gulf extends from the Rio Grande, on the north, around the eastern shoreline of Mexico to the base of the Yucatán Peninsula. Coastal lagoons here include the long hypersaline Laguna Madre de Tamaulipas (just below the Texas border), Laguna de Tamiahua (south of Tampico), and in the southern Gulf below Veracruz, the Lagunas Alvarado, Carmen, Machona, and Términos. Except in the very southern Gulf, the coastal lagoons are widely separated. Mention should also be made of the major rivers entering this stretch of coastline. Draining the high Sierra Madre Oriental during the summer monsoonal season, these streams bring to the Gulf large volumes of freshwater and fine, mostly calcareous, sedimentary material derived from the limestones and shales of the mountains and coastal plains. In order, from north to south, these include the Soto la Marina, Tamesí and Pánuco (which enter the Gulf together), Tuxpan, Papaloapan, Coatzacoalcos, and Usumacinta and Grijalva (which also enter together). This coastline is bounded by one of the narrowest continental shelves of the Gulf; it is only about 40 km wide off Tampico but widens at the base of Yucatán to about 130 km.

Surface sediments and topographic features of most of this shelf appear not to have been mapped in detail, but enough is known to provide a general picture. The majority of the sediments are clastic materials of terrestrial origin. Along most of the coastline quartz sand dominates in a nearshore band, and away from major river mouths sand may extend across the middle shelf, as well. Off the mouths of the major rivers finer silts and clays, often rich in carbonates, extend across the middle shelf. As elsewhere, the outer shelf is covered with a blanket of fine sediments containing both terrestrial materials and pelagic ooze. Two volcanic salients are noted. One lies a few kilometers north of the city of Veracruz (related to the volcanoes Orizaba and Cofre de Perote in the Jalapa transvolcanic mountain belt), and the other is situated well south of Veracruz (related to the volcanoes Santa Marta, San Martín Tuxtla, and San Martín Pajapan). Offshore from each of these areas some minerals of volcanic origin are mixed with the normal terrigenous sediments. About 120

km south of Tampico lies a small island, Isla de Lobos, and several small islands occur just offshore from Veracruz. In addition, a series of shallow-water coral reefs occupies submerged prominences near Isla de Lobos, off Veracruz, and about 30 km farther south off the fishing village of Antón Lizardo. Altogether about 27 of these reefs are known, and all lie within 25 km of the coastline. Along the continental shelf of the western Gulf, particularly in the stretch from just above Tampico to below Veracruz, rocky outcrops are numerous, some affording attachment sites for deepwater corals, but their location and composition are poorly known. Such outcrops are less abundant in the southern Gulf between Coatzacoalcos and the base of the Yucatán Peninsula. A number of relict features have been described on the middle and outer shelf of the western Gulf.

YUCATÁN

The Yucatán Peninsula is a flat, low-lying platform of Tertiary limestone with a maximum elevation of just over 200 m. Despite high annual rainfall there is little surface drainage into the Gulf of Mexico. The water sinks through cracks and the porous limestone into a cavernous underground network and flows seaward subsurface. Along the north coast a long spit fronts a canal-like feature that connects in the east with the Laguna de Lagartos. The continental shelf itself is the seaward extension of the limestone platform that slopes gently to the shelf break at a depth of 170–270 m. The shelf varies in width from 130 km in the southwest to 300 km in the north, and in the east it narrows to less than 10 km (fig. 4.8). The inner and middle shelf are covered with a blanket of fine and coarse carbonate materials derived largely from plant and animal skeletal remains but also including fragments of limestone. Much of this material has been cemented and refragmented repeatedly. Toward the outer shelf these materials give way to fine pelagic carbonate ooze.

Rocky outcrops are numerous, particularly in the 60–90 m range. Along the western and northern portions of the shelf just inside the 55 m contour lies a series of islands, reefs, and rocky outcrops. Proceeding clockwise from the southwest, these include the Arcos Reef group; Obispo Shoals; Banco Nuevo and Banco Pera; West, South, and East Triangulos; Nuevo Reef; Banco Igleses; Cayo Arenas; a series of unnamed hard banks; and Alacrán Reef. The latter is a large island/reef of about 600 km², arcuate in shape and encompassing a lagoon with a maximum depth of about 23 m. Active coral reefs are well

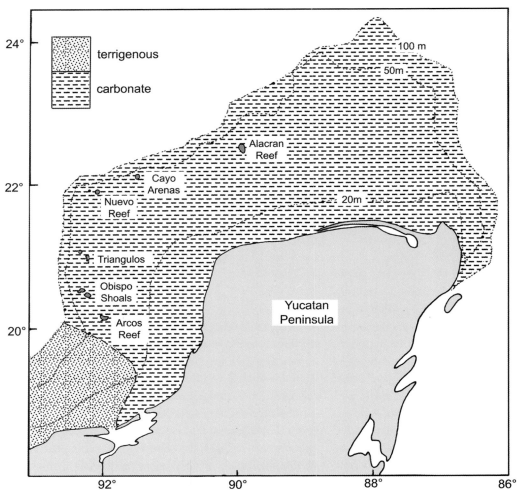

**Figure 4.8.**
Predominant surface sediment types and major topographic features of the Yucatán shelf.

developed around most of the islands, banks, and outcrops that approach within 20 m of the water's surface. A number of relict features are known on the Yucatán shelf. Broad terraces in the limestone occur at depths of around 30, 55, and 100 m, and these appear to represent wave-cut erosional features associated with stillstands of the sea during the Pleistocene. The islands, reefs, and hard banks inland of the 50 m contour have been interpreted as relict shoreline features, and the 30 m terrace is bounded landward by a long, low ridge possibly representing a series of old consolidated beach dunes, now submerged. As on the west Florida shelf, filled or partially filled sinkholes represent the surface expressions of the collapsed roofs of caverns.

CUBA

For a distance of over 500 km the northern coast of Cuba bounds the southeastern portion of the Gulf of Mexico

and Straits of Florida. Along this stretch most of the continental shelf is only a few kilometers wide, but two sizable bays occur here. The Golfo de Guanahacabibes occupies much of the western tip of the island, and toward the east lies the Bahía de Santa Clara, the latter being fronted by a long spit and numerous small islands. This is largely a carbonate shelf, and on submerged prominences it supports the growth of numerous coral reefs. Calcareous oozes blanket most of the outer shelf. Although relict features must be present, they apparently have not been documented in detail.

**Deepwater environments**

Deepwater environments of the Gulf of Mexico have been discussed by many investigators, and good summaries are found in Martin and Bouma (1978) and Bryant, Lugo, et al. (1991).

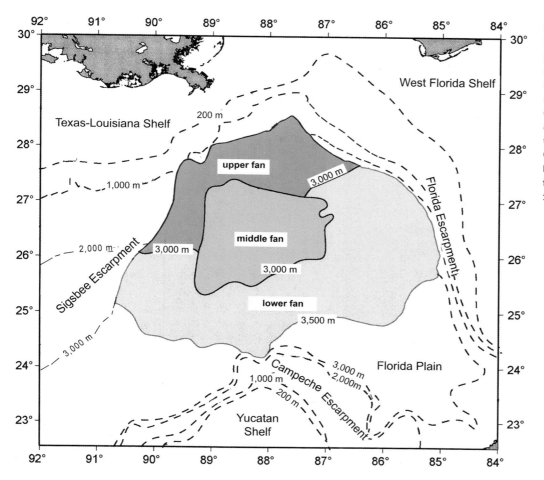

**Figure 4.9.**
Bathymetry and physiography of the eastern Gulf of Mexico showing the upper, middle, and lower fans of the Mississippi River and the relationship of the fans to the Florida, Campeche, and Sigsbee Escarpments. (Modified from G. Moore et al. 1978.)

## CONTINENTAL SLOPE

The continental slope is often conceived of as a gently sloping section of the seafloor extending from the edge of the continental shelf to the upper limit of the rise, that is, in the general depth range of 200–2,800 m. However, in the Gulf the slope in some areas is quite steep (escarpments), and it may be interrupted by many topographic features. As seen in figure 4.4, its surface is blanketed by pelagic oozes, often rich in carbonates (from tests of foraminiferans and other planktonic forms) and variously mixed with terrigenous silts and clays.

Off the Florida shelf the slope angles gently downward to a depth of around 800 m before plunging steeply toward the abyssal plain. To the north this slope is incised by DeSoto Canyon, an ancient erosional feature extending sinuously from the shelf to deep water. Both the upper and lower sections of the canyon are buried by sedimentary filling.

Toward the west lies the Mississippi Fan, which is a broad sedimentary apron deposited by the ancestral Mississippi River mainly during the Pleistocene. From its apex near the Mississippi River Delta this fan widens as it spreads across the slope, rise, and abyssal plain. To the east it abuts the Florida Escarpment, to the south the Campeche Escarpment, and on the west it extends to the Sigsbee Escarpment of the lower Texas slope (fig. 4.9). The surface of the Mississippi Fan is marked by erosional gullies, fault scarps, and low hills overlying salt domes. The Louisiana–upper Texas slope is gentle and featureless in some areas, but in other sections, where it is underlain by complex diapiric structures, it displays rugged hill and valley terrain. The outermost portion of this slope features the Sigsbee Escarpment, a minor steepening of the slope just above the continental rise. Off south Texas and northeastern Mexico lies the Rio Grande slope, with an upper portion that grades gently seaward with only low relief features but a lower slope that is quite complex. Here steep hills and troughs appear to be surface expressions of buried salt and/or shale structures.

The east Mexican slope extending southward to near

Veracruz (20°–26°N latitude) features the remarkable Mexican Ridge System. This consists of a series of regular ridges and valleys that run more or less parallel to the coastline for about 500 km and are the most well developed along the middle continental slope. The ridge crests lie 5–12 km apart and exhibit vertical relief of up to 500 m. From north to south the system is composed of three zones that differ in the number and directional orientation of the ridges, width of the field, and other characteristics (Zones 1, 2, and 4 in fig. 4.10). In the northernmost group (Zone 1) there are fewer ridges, and these trend from southwest to northeast. In the middle group (Zone 2) the ridges trend from south-southwest to north-northeast, and the number of ridges has more than doubled. In the third group (Zone 4) the ridges trend more or less from south to north, but they are arcuate in shape, and here the number of ridges also exceeds a dozen. The gap between Zones 2 and 4 (Zone 3), which lies off Tampico, has not been well explored. Here the ridge systems appear to be interrupted, and it has been suggested that this area may be underlain by a major fault zone. In each of the zones the uppermost ridges lie buried beneath continental shelf sediments. The several ridge systems appear to be surface expressions of buried shale deposits that have undergone gravity slumping and faulting during the mid-Tertiary, and this has probably continued to the present time. Most river-borne clastic sediments have been trapped shoreward of the first and second ridges so that seaward the surface sediments are largely pelagic in origin.

South of the east Mexican province lies the Veracruz Tongue (between Zones 4 and 5 in fig. 4.10). Here the bottom lacks salient topographic features as it slopes gently seaward. Surface sediments are largely terrigenous clastics that have been funneled southward by the ridges and eventually seaward beyond the south end of the ridge systems.

The Campeche Knoll section of the continental slope extends eastward from the Veracruz Tongue to Campeche Canyon (see fig. 4.1). Here the topography is very irregular, and the complex hills and valleys are in many respects similar to the slope topography off Louisiana and Texas. As in the northern Gulf, the Campeche slope is underlain by a massive body of salt deformed into massifs, pillars, and walls with up to 1,500 m of vertical relief. This is part of a salt formation that stretches from well inland out beneath the continental margin and under the abyssal plain of the southern Gulf. To the east lies Campeche Canyon, a narrow gentle plain sloping seaward along the western face of the Campeche Escarpment.

As in the case of western Florida, the Yucatán slope consists of an upper terrace and escarpment. The terrace off northeastern Yucatán is quite broad, having a maximum width of about 200 km, and is a gradually sloping plain extending downward from the shelf. The Campeche Escarpment, which stretches for 1,100 km, borders the western, northern, and part of the eastern edge of the Yucatán platform and has a maximum vertical relief of 3,600 m. The western and eastern walls are relatively smooth, but canyon cutting and massive slumping characterize the northwestern section.

The southeastern Gulf slope includes the Cuban Escarpment and the bottoms of the Yucatán Channel and Straits of Florida as well as the seafloor between the Campeche and Florida Escarpments (fig. 4.1). This region has a very complex geological history that involves faulting, crustal thinning, and volcanism. Present seafloor topography is quite rugged and features knolls up to 700 m in relief, plateaus, and sedimentary aprons. The knolls appear to be remnants of ancient dikes or volcanoes, and one knoll off south Florida is capped with an early Cretaceous reef complex (fig. 4.11).

CONTINENTAL RISE

The continental rise is that portion of the seafloor lying between the continental slope and the abyssal plain, which in the Gulf of Mexico occupies the approximate depth range of 2,800–3,500 m. For the most part the rise is made up of a wedge of materials that have slumped or flowed downslope and piled up at the base of the continental slope. Hence, its width varies depending on the environments and processes occurring at shallower depths. In the Gulf the northern section of the rise, from the Florida Escarpment westward to just beyond the Mississippi River Delta, is quite broad and consists of the lower section of the Mississippi Fan. West of this area and extending south into the Bay of Campeche the rise is generally narrower, reflecting the reduced amount of sediment reaching this depth. Around the base of the Yucatán, Cuban, and south Florida Escarpments very little continental rise is evident, although local debris piles do occur. For the most part the continental rise is a gently sloping plain of low relief, but low hills, banks, and channels do occur in areas underlain by deformed salt deposits, that is, off Louisiana and Texas in the north and off the Campeche slope in the south.

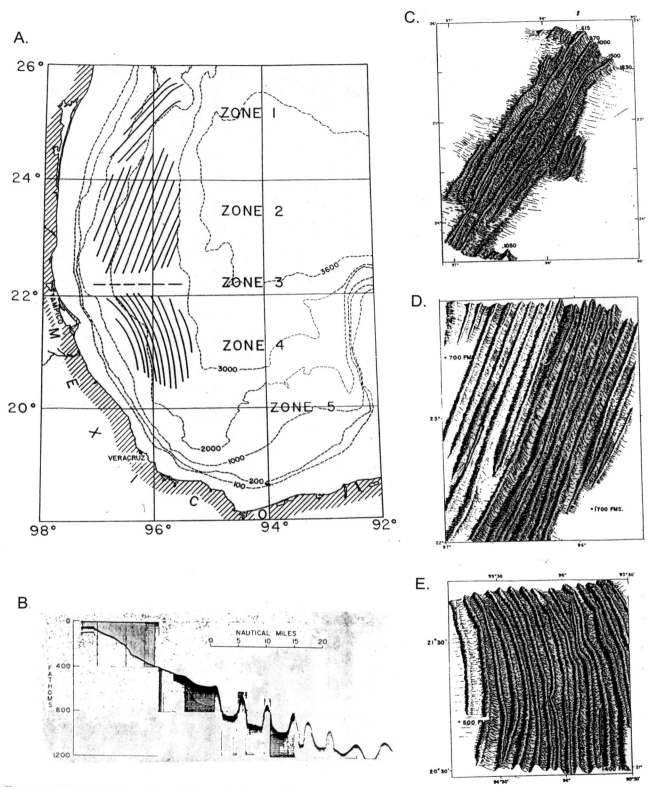

**Figure 4.10.** Structural ridge systems of the east Mexican continental shelf and slope. (A) Trends of ridge systems and locations of physiographic zones. (B) Seismic profile across shelf break and slope along 21°40′ N latitude (in Zone 4). (C) Perspective drawing of Zone 1. (D) Perspective drawing of Zone 2. (E) Perspective drawing of Zone 4. In (A), depths are given in meters. In (B)–(E), depths are given in fathoms. (From Bryant, Antoine, et al. 1968.)

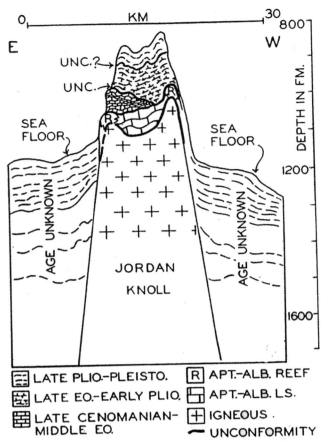

0 |————— KM —————| 30

E

UNC.? →

UNC. →

SEA FLOOR

SEA FLOOR

AGE UNKNOWN

AGE UNKNOWN

JORDAN KNOLL

800

1200

1600

DEPTH IN FM.

W

▤ LATE PLIO.-PLEISTO.        Ⓡ APT.-ALB. REEF

▧ LATE EO.-EARLY PLIO.      ⊞ APT.-ALB. LS.

▦ LATE CENOMANIAN-          ⊞ IGNEOUS .
  MIDDLE EO.                ⌇ UNCONFORMITY

**Figure 4.11.** Cross section of Jordan Knoll, located between the Yucatán and Florida continental shelves, as interpreted from seismic reflection data. Note the basic igneous structure capped by Cretaceous limestone and unclassified Tertiary deposits. The presence of the Aptian- and Albian-age reefs, which must have developed in shallow water, attests to considerable subsidence since the lower Cretaceous time. Geological periods are as given in table 3.1. Depths are in fathoms. (From Bryant, Meyerhoff, et al. 1969.)

ABYSSAL PLAIN

The abyssal plain is the flattest portion of the ocean floor. In most of the world oceans it occurs at 4,000–5,000 m, but in the Gulf of Mexico it lies in the depth range of 3,400–3,750 m. In the Gulf this plain is bisected by the lower portion of the Mississippi Fan. The larger western section, known as the Sigsbee Plain, is flat and featureless except for a belt of low hills parallel to the west and north-west faces of the Campeche Escarpment. These hills, known as the Sigsbee Knolls, are the surface expressions of salt domes arising from a tongue of salt that extends north and east from the Campeche Knolls area of the southern Gulf (fig. 4.1). About 13 prominences, with elevations of up to 200 m, make up the Sigsbee Knolls area,

and some have been shown to contain hydrocarbon deposits. East of the lower Mississippi Fan lies a small section of the abyssal plain known as the Florida Plain. This small, flat area is featureless except for several knolls, the largest of which has a vertical relief of about 700 m. These are not salt diapirs but appear to be ancient igneous, possibly volcanic, structures.

SEDIMENTARY PROCESSES

The processes by which sedimentary materials are formed, transformed, transported, and deposited are quite complex. They deal with the nature of the particles themselves; physical agents of transport, settlement, and resuspension; aspects of chemical and biological transformation; and the temporary and ultimate depositional sinks. These processes may be active in the water column, at the bottom surface, or in the subsurface environment. Among the latter are compaction and lithification of the buried sediments, intrusions from below by salt or petroleum hydrocarbons, faulting, and gravity slumping. Of primary importance is the nature of the material itself, its chemical and physical makeup, density, and particle size distribution.

Some sedimentary materials are formed within the sea. This is particularly true in continental shelf environments of tropical and subtropical areas where the calcareous tests and skeletal remains of various marine plants and animals contribute greatly to bottom deposits, and it is also true on the outer continental shelves and in open oceanic areas where the rain of planktonic remains is so important. However, the ultimate source of much of the sedimentary material, especially in the Gulf of Mexico, is the uplands of the continental interior. Here mountains, rocks, and soils undergo processes of erosion, and the resulting products are carried by streams and rivers to the shoreline and shelf, where they become subject to the various marine processes.

Once in the marine environment the particles may undergo repeated cycles of sorting, transport, settlement, and resuspension until they eventually come to rest in an area little disturbed by water movement or become deeply buried. The marine physical forces responsible for transport and resuspension are the water currents generated by winds, tides, spin-offs from major oceanic currents, and thermohaline density currents, although biological agents may also be involved. Extreme wind events such as those associated with storms and hurricanes generate strong currents that are particularly important in sediment transport, and the high-energy environments of continental shelves are where much of this activity takes place. In fact, most of the terrigenous material brought in from the uplands is eventually moved to the outer shelf, slope, or continental rise, although a small fraction also winds up on the abyssal plain.

Loose bottom deposits of the outer shelf and upper slope become compacted as they are weighed down from above by subsequent deposits. Under such stress they may undergo faulting and gravity slumping and flow downslope to be redeposited in deeper water. Under more stable conditions, as the sediments become deeply buried, they may solidify into layers of rocky material such as claystone (shale), siltstone, sandstone, or limestone, depending on the initial particle composition. Subsequent intrusion by salt columns rising from below may deform the bedding structure of the layers, and liquid petroleum or natural gas (methane) may rise along faults or around the margins of salt intrusions.

The topographic and other features we observe upon and within the various bottoms of the Gulf of Mexico today have resulted, in part, from processes that were more active in the past, especially during the periods of continental glaciation and deglaciation. When sea levels were much lower, large portions of the continental shelves were exposed to the atmosphere, and streams and rivers were more deeply entrenched, with their mouths located at or near the outer margins of what are now the continental shelves. So, interpretation of present-day features must take account of both the ongoing processes and the prehistory of each area. Among the world oceans the Gulf is remarkable for the diversity of its present sedimentary environments, active processes, and prior sedimentary events. These topics will be addressed in some detail in the following sections. Largely because of interest in the location and development of the Gulf's vast petroleum reserves, a great deal of attention has been devoted to the study of the sedimentary processes and the surface and subsurface geology of the Gulf of Mexico.

## Carbonate shelves and slopes

As mentioned earlier, the broad carbonate platform off northern and western Yucatán consists of flat-bedded limestone dipping gently seaward and covered with a veneer of unconsolidated carbonate rubble. This blanket tends to be thinnest on the inner shelf, and it gradually thickens with depth, from a few inches near shore to a few feet on the outer shelf and upper slope. It is composed of

the carbonate tests and skeletal remains of many types of planktonic and benthonic algae and invertebrates together with nonbiological carbonate particles (Harding 1964). The latter may be roughly classified into four categories: *ooids, calcareous pellets, lithic fragments,* and *aggregates.* The ooids are tiny ovoid or egg-shaped particles a few millimeters in diameter, which in cross section are seen to be composed of a nucleus surrounded by concentric layers of carbonate deposits. They are formed in high-energy environments (where water is moved by waves or currents) in less than about 9 m (30 ft) of water. Calcareous pellets represent fecal material deposited by gastropods, crustaceans, and other marine animals. They also are very small, but they occur in various shapes and show no internal structural organization. Lithic fragments are simply small pieces of limestone that have broken from the bedrock, either in shallow water by wave cutting of terraces or, in somewhat deeper water, by natural processes of erosion or biological activity. These particles are irregularly shaped and often sharp and angular, although in some cases they have been worn and rounded by abrasion. Aggregates result from cementation of one or more of the above types into larger particles.

The processes involved in cementation of small carbonate particles into larger particles, limestone, beachrock, and so forth are complex and only partly understood. Along the shoreline, beach dunes composed mostly of sand-sized grains of calcium carbonate are pelted by tropical rains. Passing through the atmosphere, the raindrops pick up some carbon dioxide, which combines with the water to produce a weak solution of carbonic acid. Falling on the beach dunes, this acid dissolves some of the carbonate of the dune sand, and when the water evaporates the carbonate resolidifies, cementing the particles together. These processes may be augmented by certain algae, particularly blue-green algae, which reside in spaces between the particles. As a result of this dissolution and recementation, solid beachrock forms fairly rapidly, but the presence of uncemented spaces facilitates dissolution of the rock if it subsequently becomes submerged. Some of the same processes occur in unconsolidated sediments of the sea bottom once they become buried beyond the influence of water currents and burrowing animals, and here too the presence of algae and perhaps bacteria may enhance the cementation of particles. The presence of organic material may also hasten the processes. Lithification (the formation of limestone or other hard, stonelike material) of loose carbonate deposits may be further enhanced by compaction or the physical pressing of the particles closer together, thereby reducing the amount of interparticle space. However, in the formation of limestones, cementation is far more important than compaction because of the high solubilities of the carbonate components and the ease with which water can penetrate the interparticle spaces.

Calcium carbonate exists in a number of crystalline forms, of which *calcite* and *aragonite* concern us here. These two crystalline states are easily recognized. Calcite is a hard, clear, glass-like substance that is the stable form in subaerial and freshwater environments. Beachrock, for example, is cemented by calcite. By contrast, aragonite is a softer, opaque, white, chalky material, and this is the stable form of calcium carbonate in seawater. It forms the tests and shells of foraminiferans, mollusks, and other marine animals as well as the cementing matrix of marine aggregates. The history of marine carbonate particles is often written in the crystalline nature of the particles. For example, the conversion of aragonite to calcite indicates marine-formed particles that have been exposed to freshwater or the atmosphere. Recrystallization of calcite to aragonite indicates material formed subaerially that is now subject to marine chemical processes.

The distribution of unconsolidated sediments of the Yucatán shelf is, at first glance, somewhat enigmatic. Ooids and recemented ooid aggregates, which should be found only near shore, occur at most depths of the shelf, and their abundance on the outer shelf is evidence of a time when the outer shelf was a shallow marine environment. This area is also marked by the presence of calcite particles, wave-cut terraces, and around the 55 m (180 ft) contour a series of islands and submerged shoals (many capped with living coral reefs) that apparently represent old beach dune formations. Confirmation of the relict nature of these middle and outer shelf formations and deposits comes from radiocarbon dates of about 13,780 ± 200 years before the present (Logan et al. 1969). The inner shelf off Yucatán is blanketed largely by shallow-water deposits characteristic of the environment in which they now reside and are taken to represent recent postglacial deposits. Relict sediments and related topographic features characterize middle to outer shelf environments around the Gulf and around the world, and they are not limited to carbonate shelves. Seaward, on the upper Yucatán slope, the relict sediments give way to deep deposits of fine, bluish, calcareous muds, pelagic oozes derived from the tests of planktonic foraminiferans but also containing the remains of diatoms, coccolithophores, and other planktonic inhabitants of the open Gulf.

The continental shelf west of the Florida Peninsula also consists of a flat limestone platform dipping gently seaward, but its thin sedimentary blanket is a bit more complex in that it contains both carbonate and noncarbonate deposits. Unlike the northern Yucatán Peninsula, which has virtually no surface drainage, Florida has several rivers that drain portions of the southeastern states. Over the years these have brought to the Gulf quantities of terrigenous materials, mostly quartz sand composed largely of silicon dioxide. Parenthetically, it should be noted that terrigenous sediments, derived originally from the rocky uplands, tend to be rich in silicon, often bound with aluminum, iron, lithium, sodium, potassium, calcium, and/or magnesium, as well as other chemical elements. Thus, the quartz sand itself may contain grains of other minerals such as feldspar, glauconite, and mica by which the source rocks may be identified. The sand on the Florida shelf is derived ultimately from the southern Appalachians. After arriving on the shelf this sand was distributed north and south by alongshore currents, and it now lies in a band extending from the shoreline to a depth of about 10 m. Seaward it becomes mixed with typical coarse carbonate materials, and this mixed band covers the bottom out to the 20 or 30 m isobath (see fig. 4.5). Beyond this zone the shelf sediments consist mostly of unconsolidated coarse carbonate rubble that is largely of biological origin but contains some ooids, pellets, aggregates, and limestone fragments, as off Yucatán. Toward the very outer shelf these grade into fine-grained pelagic ooze. At the southern tip of the Florida Peninsula a tongue of fine-grained carbonate mud extends seaward from the Everglades, and below this, along the Florida Reef Tract, coarse carbonate rubble carpets the bottom in a zone that extends westward to the middle or outer shelf, where it gives way to deeper-water carbonate algal nodules and pavement. As elsewhere around the Gulf, the middle to outer shelf deposits and topographic structures are relict features developed during periods of lower sea level stands.

The west Florida shelf is also characterized by numerous flat bedrock exposures and low-relief limestone outcrops. Wave-cut terraces, 1 to 3 m in height, are found on the middle and outer shelf. Large prominences also occur. North of Tampa Bay in the depth range of 38–44 m (125–144 ft) lies the Florida Middle Ground, which consists of a series of north-south-trending ridges with vertical relief of up to 5 m (16.4 ft) (fig. 4.5). The two primary ridges are about 12–15 km long and are separated by a flat plain 5–9 km wide, which at the southern end is partially blocked by a short third ridge. Seaward of the westernmost major ridge lie at least two lower ridge systems. Relief of the top of the ridge systems consists mostly of steep isolated pinnacles. The Middle Ground structures probably began as long consolidated beach dune ridges, which during periods of submergence became reef systems capped by the growth of coralline algae and possibly hermatypic corals. During subsequent periods of exposure they underwent considerable dissolution and erosion. Now that the structures are resubmerged, reef building is again active. Along the outer shelf of southwest Florida lie three more or less parallel ridges that appear to represent Pleistocene or even pre-Pleistocene reef formations. Pulley Ridge lies at a depth of 70 m (230 ft). The Howell Hook formation is 65 km long and resides at about 150 m (492 ft), and this is paralleled by yet another, which is probably older, in somewhat deeper water. These ancient reef tracts are now partially buried by a veneer of more recent sediments.

The continental slopes off Yucatán and Florida are steep scarps that drop off to the continental rise; that off Florida is the steeper of the two. At the base of the Campeche Escarpment, particularly along the northwest corner, lies a talus formation consisting of limestone blocks, stones, and rocks that have fallen from above as the scarp face has eroded back.

## Terrigenous shelves and slopes

The terrigenous bottom province extends from north Florida around the northern, western, and southern Gulf to the base of the Yucatán Peninsula. For the most part, continental shelves of this vast area are smooth, flat plains sloping gently seaward. Surface sediments consist primarily of quartz sand, silt, and clay, clastic materials brought to the Gulf by rivers and subsequently sorted and redistributed by water currents. Locally, particularly on the middle and outer shelf, the soft bottoms are punctuated by hills, rocky outcrops, and erosional gullies, many of which represent relict features of a time of lower sea level stand. Calcareous deposits of biological origin are found around some of the rocks and banks, and some are capped by living coral/algal reefs. These occur on the outer shelf off the Texas-Louisiana border and on the inner shelf south of Tampico and around Veracruz, Mexico.

In the DeSoto Canyon area the flat limestone platform supporting the west Florida shelf terminates and is re-

placed by deposits of riverine sediments in a band that extends westward through the Texas shelf. Here the basement rocks generally lie deeply buried and are capped by hundreds to thousands of feet of alluvial deposits brought to the Gulf by streams draining lands to the north, the most important of which is the Mississippi River. When the sea level was much lower these streams were deeply entrenched, and they transported enormous volumes of sedimentary material to the Gulf. Rising sea levels were accompanied by filling of the river valleys, and during the past 5,000 years, as sea level approached its present stand, the Mississippi River has formed and abandoned a series of alternate deltaic fans as it built out onto the continental shelf (fig. 4.12).

A number of major processes occurring within the deep alluvial sediments of the northern Gulf should also be noted. Burdened by the weight of thousands of feet of sediments, the underlying crust of the earth has sunk to greater depths. The sediments themselves, composed largely of sand, silt, and clay particles, were originally laid down with a fairly high water content, but over time and with great pressure from above much of the water has been squeezed out, resulting in greater compaction of the sediments. This has led to lithification (in this case, the formation of sandstone, siltstone, and claystone layers) within the deeper deposits. The force of gravity and unequal pressures on different sections of the sediments has resulted in frequent faulting and slumping (fig. 4.13), particularly of those deposits on the outer shelf and continental slope.

The basement layer of salt, subject to conditions of elevated temperature and high pressure, has become plastic and has flowed laterally as well as vertically, presumably along weaknesses in the sediments such as fault lines. Approaching the surface, the salt has solidified into isolated columns, walls, and, toward the outer slope, massive ramparts (fig. 4.14). Once solidified in place, these salt structures provide a measure of stability to the sediments of the shelf and slope, where they act as dams behind and above which the loose sediments accumulate. As the salt moves vertically, it distorts the rocky layers, and these often wind up tilted at an angle and leaning against the salt columns. Petroleum and natural gas deposits deep within the sediments are under great pressure from the load above, and these substances seek the surface through weaknesses such as cracks and faults. Oil and gas from deposits penetrated by salt columns also move vertically and frequently become trapped in pockets beneath

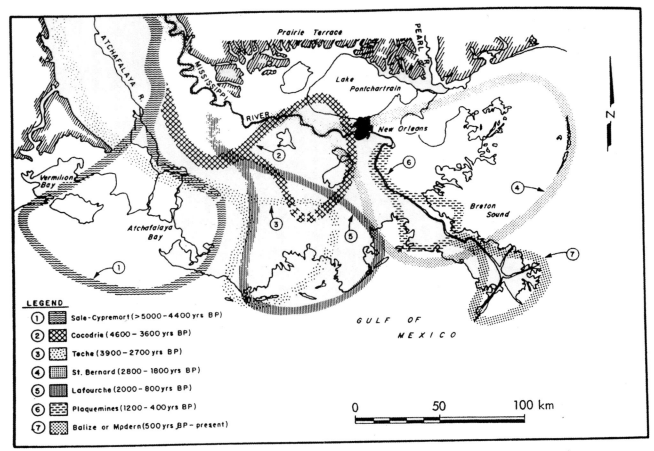

**Figure 4.12.** Locations of major deltas of the Mississippi River during the past 5,000 years. (From Kolb and Van Lopek 1958.)

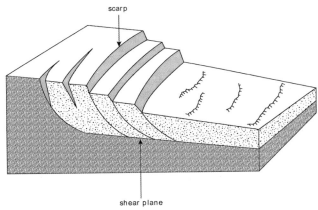

**Figure 4.13.** Profile of a minor sediment slump on the continental slope off the Mississippi River Delta.

the tilted rocks around the periphery of salt intrusions, although some escapes into the water column above. As it nears the sediment surface, the salt itself is subject to dissolution. Seawater penetrating through cracks and faults dissolves some of the salt, and very saline brine may exit

through seeps along the sides of the formation (fig. 4.15). Although initially covered by "caprock" pushed up from below, over long periods of time upper portions of the salt columns may dissolve away, causing the top of the formation to collapse and creating crater or caldera-like structures or, in some cases, brine lakes (fig. 4.16) on the continental slope.

Despite continued postglacial sedimentation, in some areas the outer margin of the northern Gulf continental shelf is irregular and incised by a series of erosional cuts or gullies. Most are remnants of lower stands of the sea, but some are still active as conduits for the passage of sedimentary materials to the continental slope. On the slope itself there are many minor and several major troughs and canyons by which sediments are funneled to the deep Gulf. Chief among these are DeSoto Canyon in the east and the Mississippi Trough, located a few kilometers southwest of the Mississippi River Delta (fig. 4.1). The latter is a relict distributary of the glacial or postglacial Mississippi River that still serves as a sluiceway

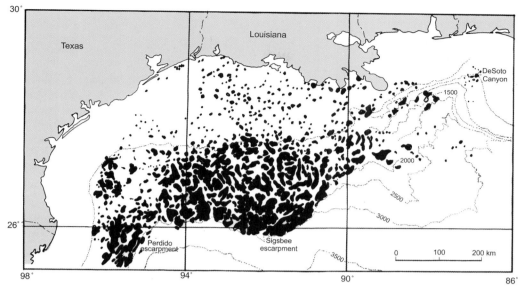

**Figure 4.14.**
Map showing the locations of major salt diapirs and massifs in offshore sediments of the northern Gulf and indicating their approximate sizes and shapes. (Modified from Martin and Bouma 1978.)

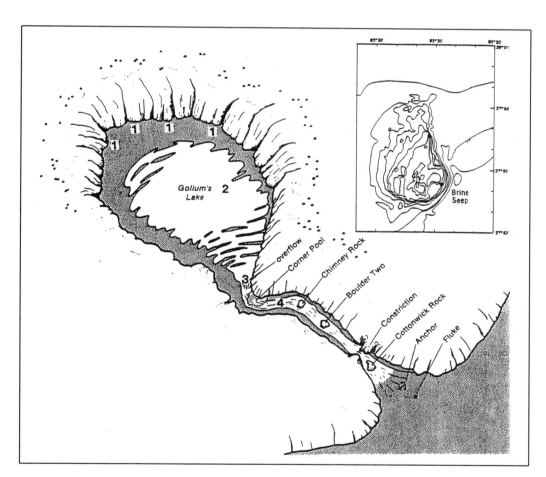

**Figure 4.15.**
Brine seep complex on the East Flower Garden Bank south of the Texas-Louisiana border. 1 = brine seeps, 2 = brine lake, 3 = overflow from lake, and 4 = canyon with mixing stream. (From Rezak, Bright, and McGrail 1983.)

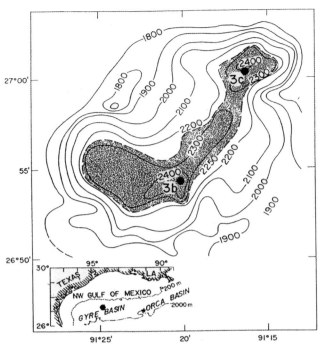

**Figure 4.16.** Bathymetry of Orca Basin. The contour interval is 100 m. The anoxic brine pool is shaded. The inset map shows the location of both Orca and Gyre Basins; 3b and 3c mark sites where bottom sediment samples were taken for analysis. (From Trabant and Presley 1978.)

for passage of sediments to the deep Gulf. A smaller but still important channel is Bryant Canyon, somewhat to the west of the Mississippi Trough.

The physical structure and sedimentary processes described for the northern Gulf apply fairly well to the continental shelf and slope of the southern Gulf, which is also underlain by massive salt deposits. Here the slope is hummocky and incised by erosional gullies, and Campeche Canyon to the east channels sediments to the deep Gulf. The continental shelf and slope of the western Gulf are not underlain by salt, but the ridge systems paralleling the coastline act as barriers to the seaward passage of river-borne sediments. The uppermost troughs are partially filled with loose alluvial deposits, and they also serve to channel the sediments southward to the Veracruz area. Here, at the south end of the ridge systems, these sediments extend gently seaward into deeper water.

## READING THE DEEP GULF

The history of the Gulf of Mexico is recorded in the sediments and topography of the deep Gulf, and much of what has been said in the present and previous chapter about this history has come from examination of this record. To obtain and interpret this information, scientists have relied on some of the older standard methods now supplemented with newer technology. In the present section we will describe a few of the methods and provide illustrations of some of the more interesting results. The techniques include bathymetric analysis, seismic profiling, piston coring, and deep sea drilling, as well as computer analysis and enhancement of the results.

In order to map the bottom of the ocean, one must be able to obtain two pieces of information: the position of a ship on the surface of the sea and the depth of the water column below. Before the advent of satellites, ship positions could be approximated only by triangulation with shore stations or through a log of the ship's speed and direction of travel. Now satellite global positioning systems permit very accurate determination of a ship's position (within a few feet). In the early days of oceanography, depth was determined by measuring the length of a weighted rope or cable lowered to the bottom, with the assumption that it went straight down. However, during the past half century, with development of sonar and other echolocation methods, it is now a simple matter to bounce sound signals off the bottom and deduce depth from the time interval between sending the signal and receiving the echo. Now both the ship's position and water depth can be recorded automatically throughout the cruise, thus providing the information needed for production of very accurate maps of the sea bottom and submerged topographic features, such as that shown in figure 4.1. With the aid of computers, bathymetric data can be converted to images; the topography can be viewed from any angle, and the various elevations or depths can be assigned false colors. In fact, subsea prominences can be colored to appear as though they were lighted from an angle by the sun. The results of such data manipulation are shown in figure 4.17, which exposes the continental slope of the entire northern Gulf, much of the eastern Gulf, and the tip of the Campeche Escarpment. Immediately obvious are DeSoto Canyon and the Mississippi Trough as well as the hummocky bottom of the salt dome province off Louisiana and Texas. The steep scarp off the Florida shelf and portions of the Yucatán shelf are also clear, as are the upper and middle portions of the Mississippi Fan.

Out beyond the main slope off central Louisiana stands a large hill known as Green Knoll (for location see fig. 4.1). This feature, enlarged in figure 4.18, is the surface expression of a major salt dome of which the summit has

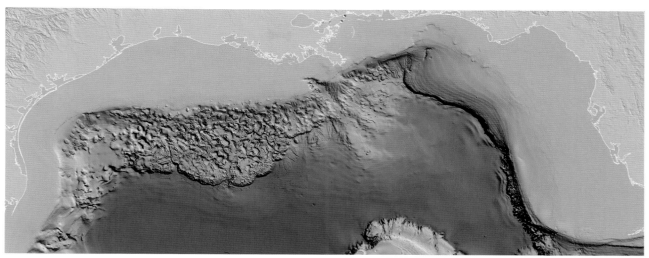

**Figure 4.17.** False-color images of subsea prominences in the northern Gulf of Mexico.

**Figure 4.18.** Green Knoll, a prominent large hill off the main slope of central Louisiana.

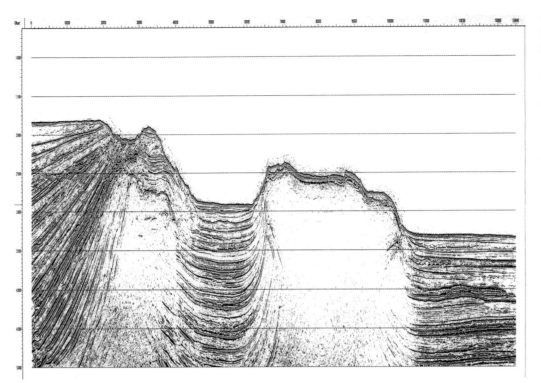

**Figure 4.19.**
North-south seismic profile through (left to right) the edge of the Sigsbee Escarpment, intervening valley, Green Knoll, and abyssal plain. Note the flat-bedded sedimentary layers of the abyssal plain and their distortion and up-tilting in association with the diapirs of the Sigsbee Escarpment and Green Knoll. Also note the irregular surface of the top of Green Knoll, apparently a result of the dissolution of salt. (Print provided by W. Bryant.)

eroded away, giving it a caldera-like appearance. Portions of the Sigsbee Escarpment are also shown in the foreground and on the upper left, themselves also salt dome features. Of considerable interest to physical oceanographers are the long fields of parallel furrows in the bottom sediments behind, around, and in front of Green Knoll, evidence of strong and sustained bottom currents previously unknown and still only partially understood.

If stronger signals are sent out from the ship, the sound waves may penetrate the bottom and bounce off reflective layers within the sediments, and by this means it is possible to map subsurface features, providing cross sections of the various sediment layers within the sea bottom. An example of such a seismic profile is shown in figure 4.19, which is a north-south section through the Sigsbee Escarpment (left), Green Knoll (center), and abyssal plain (right). Immediately obvious are the salt columns, rising from below, which form the cores of Green Knoll and the outer edges of the Sigsbee Escarpment. In both cases some caprock lies above the salt. The flat-bedded sediment layers characteristic of the abyssal plain have been greatly distorted by the salt intrusions, particularly by the salt column of the Sigsbee Escarpment. Note that the summits of both salt columns are irregular and partially cratered, evidence of the dissolution of exposed portions

of the salt. This section also shows how loose sediments moving down the slope tend to accumulate on the landward side of the salt intrusions, and the clear view of sediment tilting shows why petroleum and natural gas tend to accumulate around margins of the salt columns.

One of the standard ways of obtaining actual samples of the bottom sediments is through the use of a piston corer. This device consists of a weighted metal pipe that is driven into the ocean floor. When retrieved it brings up a cylindrical core of sediments 7–13 cm in diameter and 40 m (131 ft) or more in length. This material is extruded onto a length of plastic (PVC) piping that has been cut in half lengthwise. Back in the laboratory the sediment core itself is sliced in half lengthwise, and the exposed flat section can be visually examined, photographed, and x-rayed, and subsamples can be removed for examination of physical and chemical properties and biological remains. The structure, composition, and bedding patterns of the sediment core can reveal a great deal about the source of the materials and circumstances surrounding their deposition. Subsamples can be isotopically dated. Oxygen isotope ratios provide information about past surface-water temperatures. Populations of microfossils (skeletal remains of planktonic organisms such as diatoms, coccolithophores, foraminiferans, radiolarians, etc.) reveal much about the ages and environments in which the organisms lived. These and other types of analyses permit scientists to put together a chronological picture of the recent and more distant sedimentary history of the Gulf.

Although piston coring is a rather simple procedure that can be carried out from the decks of most oceanographic vessels, the short length of the core samples limits the time span covered by the sediments. In order to study deeper and older sediments, one needs much longer cores, some taken from very deep water. To solve this problem, drilling rigs have been mounted on special ships that can be very accurately positioned on the sea surface. By adapting oil field drilling technology, scientists of the internationally sponsored Ocean Drilling Program have been able to drill in water depths of over 8,000 m and to extract cores over 2,000 m long. The samples cover time spans of hundreds of millions of years, and some have penetrated through the sediments to the underlying crust of the earth. Since its inception in 1968 this program has retrieved cores from more than 1,200 sites located in all the world's major oceans, providing a remarkable record of the earth's history.

### EVIDENCE OF THE K/T BOUNDARY EVENT

Of particular interest here are several long cores taken in deep water (> 2,500 m) in the southeastern Gulf of Mexico between the Yucatán and Florida platforms, which cover the entire Quaternary and Tertiary periods and penetrate the Cretaceous (see table 3.1). Thus, they cover the K/T boundary, the period when the extraterrestrial bolide struck Yucatán. For comparison, a core taken from the deep Atlantic Ocean off northern Florida, well away from the impact site, shows the following simple sedimentary sequence (fig. 4.20): in a section of core less than 45 cm (18 in) long there is a series of definite bands representing (from bottom to top) (A) sediments deposited during the late Cretaceous prior to the impact, (B) material ejected by the impact, (C) dust that finally settled following the impact (rich in the exotic chemical element iridium), (D) a dead ocean from which no fossil remains were produced, and (E) sea bottom sediments containing fossilized remains of planktonic organisms that were beginning to repopulate the world oceans in the early Tertiary period.

By contrast, the K/T boundary sediments from the deep Gulf off Yucatán show a chaotic pattern that has been referred to as the K/T boundary "cocktail" (Alvarez et al. 1992; Bralower, Paull, and Leckie 1998). In some places

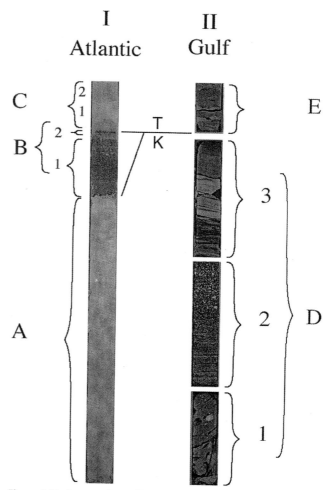

**Figure 4.20.** A comparison of bottom sediment core samples from the Atlantic Ocean and Gulf of Mexico straddling the K/T (Cretaceous/Tertiary) boundary and showing samples of the K/T boundary "cocktail" from the deep Gulf sediments. (I) Core taken from the deep Atlantic Ocean off the coast of northern Florida (ODP Leg 171B, Hole 1049A, North Atlantic). (II) Core taken from the deep Gulf of Mexico at the base of the south Florida Escarpment (ODP Leg 77, Hole 540, Gulf of Mexico). (A) Material laid down in a quiet sea during the late Cretaceous (clayey nannofossil ooze with some soft sediment deformation). (B) Impact-related deposits. (B-1) Impact ejecta (green layer containing glass spherules). (B-2) Dark iridium-rich layer (postimpact dust fall). (C) Early Tertiary deposits (clayey carbonate oozes). (C-1) Deposits from a dead sea (no nannofossils). (C-2) Deposits from a recovering sea (planktonic nannofossils present). (D) Representative sections from the highly disturbed upper and lower Cretaceous deposits. (D-1) Reworked Cenomanian pebbly limestone with Albian claystone and limestone fragments. (D-2) Reworked Cenomanian and Maastrichtian pyritic sandstone with Albian limestone rock fragments. (D-3) Reworked Cenomanian and Maastrichtian laminated chalk. (E) Early Tertiary (Paleocene) pebbly chalk with claystone rock fragments. (Photos of core samples and geological interpretations provided by J. Firth, Ocean Drilling Program.)

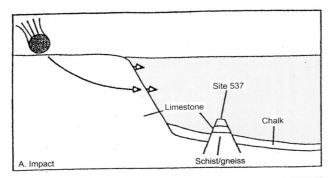

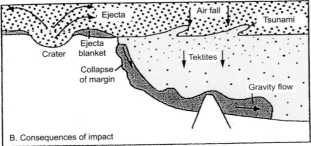

**Figure 4.21.** A reconstruction of processes associated with the Chicxulub event leading to formation of the Cretaceous/Tertiary (K/T) boundary "cocktail." A = Situation of the Yucatán platform and southeastern Gulf just prior to the bolide impact. B = Processes taking place during and following the impact. Note especially the massive downslope gravity flow of continental shelf deposits. (After Bralower, Paull, and Leckie 1998.)

this sedimentary mixture is over 50 m thick. It consists of materials originating in shallow as well as deep water and contains mud mixed with chalk and rock fragments, cross-banded sandstones, and fine ooze with Cretaceous-age fossils. Included among these mixed sediments are tiny spherules (*tektites,* formed when ejected particles reentered the earth's atmosphere), glass, shocked quartz crystals, and fragments of chalk and limestone—all this

from an environment normally characterized by undisturbed layers of finely particulate sedimentary material. Examples of the K/T boundary sediments from the southeastern Gulf are shown in figure 4.20-B, C, and D). Factors associated with deposition of these chaotic sediments are discussed below and illustrated in figure 4.21.

When the bolide struck Yucatán, it hit with such force that continental margins around the Gulf collapsed, resulting in great downslope gravity flows of mud and rocky materials from previous shelf and slope deposits. The collision itself and gravity flows, in turn, generated large tidal waves, or *tsunamis,* which swept back and forth through the Gulf basin creating currents that caused further gravity flows and reworking of the bottom sediments, even in deep water. To these deposits were added ejecta that landed directly in the Gulf (shocked quartz, rock fragments, etc.) together with material sent out of the earth's atmosphere and reentering from space (*tektites*). The gravity flows and bottom roiling were so extensive that some of the suspended material was even deposited on the crests of the knolls of the southeastern Gulf (composed of igneous material called *schist* and *gneiss*). After the shaking, slumping, and roiling settled down, fine dust particles, rich in iridium, accumulated on the surface of the disturbed sediments. This iridium-rich layer, clearly identifiable in the sediment cores, marks the upper end of the K/T boundary sequence in the Gulf, as elsewhere around the earth. Thus, long sediment cores obtained by the International Ocean Discovery Program have in effect provided the "smoking gun," direct evidence of what happened in the deep Gulf basin during and following the Chicxulub event, which killed off the dinosaurs, marked the end of the Cretaceous period, ushered in the Age of Mammals, and changed life on earth forever.

# 5 : METEOROLOGY

Air masses that impinge upon or traverse the Gulf of Mexico in large measure control the climate and local weather. Through wind stress, precipitation, evaporation, and heat exchange they also influence the temperature, density, circulation patterns, and other characteristics of the surface mixed layer of the Gulf itself. Our knowledge of meteorological conditions over the open water is based on inference from shore stations, direct observations taken from buoys or recorded in ship logs, and radar and satellite data. Although some limitations are associated with each of these data sources, together they provide a reasonable picture of climate and weather conditions over the expanse of the Gulf. Important historical references include Franceschini (1961), Jordan (1973), Leipper (1954a), and Orton (1964), as well as publications by the US Department of Commerce (1959, 1968), US Naval Oceanographic Office (1963, 1967), and US Naval Weather Service Command (1970).

## AIR MASSES

The climate and local weather of the Gulf are controlled largely by two air masses, maritime tropical and continental polar, although others including maritime polar and continental Arctic are sometimes important (fig. 5.1).

## Maritime tropical

Over the North Atlantic Ocean stands a permanent high pressure area known as the Bermuda High. This area expands and contracts seasonally, being small in the winter and much enlarged during the summer. Wind circulation around this area is always clockwise, or anticyclonic. During the winter months, when the area is small, winds circulating around the southern portion of the high blow across the Atlantic from the east and enter the Gulf from an easterly or southeasterly direction. With the onset of warmer weather in spring and summer, the Bermuda High expands, and east winds cross the Atlantic closer to the equator. During this period tropical air enters the Gulf primarily from the south. In the fall the situation reverses. Weakening of the Bermuda High results in reestablishment of easterly and southeasterly wind patterns. Thus, maritime tropical air masses dominate the Gulf through-

out the year, although this dominance is frequently interrupted during late fall, winter, and early spring by the arrival of colder air masses from the north.

## Continental polar

Air masses originating in the Great Plains of North America push southward into the Gulf at all seasons but are most frequent from October through April. In the summer months they are weak and affect primarily the northeastern Gulf. During the winter, stronger cold fronts may affect the entire northern Gulf. Overriding warmer waters, they often stall out over the continental shelf of the northern Gulf, but some make it all the way to the southern Gulf, and occasionally all the way to South America.

## Maritime polar

Air masses entering from the Pacific Ocean during the fall and early spring traverse the Great Plains and enter

**Figure 5.1.** General paths of major air masses affecting the Gulf of Mexico.

the northern Gulf, where their effects are much the same as those of continental polar fronts. During the summer months they are often too weak to cross the Rockies.

## Continental Arctic

In the winter months, but particularly during January and February, very cold air descending from Canada may push well into the Gulf as "blue northers." Extending into the southern Gulf they are locally characterized as "nortes."

Thus, from March through September winds circulating around the Bermuda High control weather patterns of the Gulf. Easterly and southeasterly winds of spring give way to southerly winds in summer, and they quickly return to southeasterly in the fall. By October and November, with strengthening of the polar air masses, winds enter the Gulf more frequently from a northerly direction accompanying cold fronts. These are particularly intense during January and February in association with Arctic air masses. About 15 to 20 cold fronts enter the Gulf each year, and they often stall out over open water. The leading edge is often accompanied by low clouds, fog, and rainy weather. When especially strong, these cold fronts may be accompanied by large temperature changes, torrential rains, gale-force winds, high waves, and long sea swells. A few days after passage of a cold front, the wind returns from the east or southeast, and the weather clears up.

## WIND

Prevailing wind directions and speeds during a period of time are conveniently displayed in a figure called a wind rose, and monthly wind roses for the northwest Gulf of Mexico are shown in figure 5.2. These roses are constructed using the oceanographic convention for wind direction, that is, the wind is blowing from the center out toward the tip of the petal. Further explanation is provided in the figure legend. Examination of the roses reveals that during the spring months (March through May) the most persistent winds are from the southeast. During the summer (June through August) they come primarily from the south. In September they switch abruptly and blow from an easterly direction, a condition that persists through the winter. However, in the late fall and winter (November through February), northerly winds are also in evidence. During this period wind frequencies are about the same from the north, northeast, east, southeast, and south.

Monthly mean wind speeds are lower during the sum-

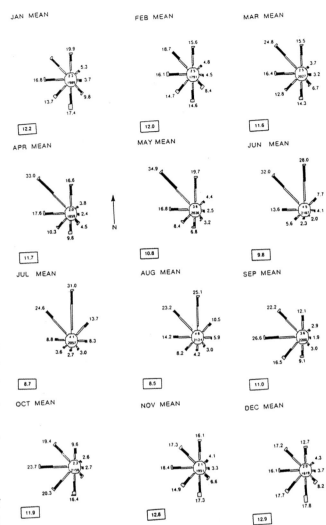

**Figure 5.2.** Monthly wind roses for the northwest Gulf of Mexico. Numbers inside the central part of each rose represent the percentage of calm (upper) and the number of observations (lower). The point of each petal designates the direction the wind is blowing, and the number at the end of each petal gives the percentage of time in that direction. The segments of each petal represent the percentage of the time the wind is blowing at a given speed. From the center toward the point these represent 0–3.99 kts, 4.0–10.99 kts, 11.0–21.99 kts, and 22.0–33.99 kts. The box at the lower left of each rose gives the monthly mean wind speed. (From Kelly 1988.)

mer (June through August), when they average less than 10 kts, and highest in the late fall and winter (November through February), when they average over 12 kts. High winds often accompany passage of winter polar fronts. However, the highest temporary wind speeds, sometimes exceeding 100 kts (about 115 land miles/hr), occur during the period of June through October in association with major cyclonic disturbances such as tropical storms and hurricanes. Wind plays a major role in setting up surface wave fields and in establishing water circulation patterns, especially over continental shelves, and the magnitude of these effects is proportional to the wind speed.

### WIND STRESS

Wind blowing across the sea surface produces a frictional force, or wind stress, the magnitude of which is determined by the density of the air, a value called the "drag coefficient," and the square of the wind speed. The direction of the force is, of course, the direction in which the wind is blowing. Deformation of the sea surface produces surface waves, familiar to all mariners. Pressure gradients are created between areas where water is blown out and where water is blown in. These gradients, in turn, drive flow in the surface water that is only indirectly coupled with the wind.

Seasonal wind stress fields across the entire Gulf have been calculated by several investigators employing different data sets and different levels of resolution. Figure 5.3 shows results of calculations based on $1°$ squares for summer and winter. In the summer, when air circulation over the Gulf is dominated by the western arm of the Bermuda High, over most of the Gulf the directional orientation of the vectors in the stress field is largely toward the west or northwest. The highest values (indicated by the longest arrows) occur in the west and northwest sectors, and the lowest values are seen in the east and northeast sectors. During the winter, when the Bermuda High weakens and strong polar fronts invade the Gulf from the north, the direction of the stress field shifts, and the vectors are now oriented toward the southwest. The highest values occur along the far western Gulf as the polar air is funneled southward by the wall of the Sierra Madre of eastern Mexico. The data that form the basis for such vector maps provide input for complex computer models of surface water circulation in the Gulf of Mexico.

### AIR PRESSURE

Sea level air pressure over the Gulf is highest during the winter months because of the intrusion of continental high-pressure anticyclones from the north. A maximum air pressure of about 1,019 mb occurs in December. Air pressure tends to be lower during the summer because of the northward migration of the equatorial trough. A minimum of about 1,013 mb occurs in May, and a second mini-

mum of 1,014 mb appears in August and September. A slight rise in the pressure is observed during extension of the Bermuda High in June.

### AIR TEMPERATURE

Mean monthly air temperature for selected stations around the Gulf is given in table 5.1. Here it is seen that mean winter temperatures (in degrees Fahrenheit) around the northern Gulf lie in the upper 50s and lower 60s, whereas around the southern Gulf they range in the upper 60s and lower 70s. However, such data do not provide reliable indications of conditions over the Gulf itself because air temperature over land varies with respect to season, latitude, and distance from the Gulf and because heating and cooling are different over land and water. Over the water, air temperatures show narrower limits of both daily and seasonal variations, especially during the winter. Air temperature over the center of the Gulf in the winter averages around $17.0°–23.0°C$ ($62.6°–73.4°F$), and in the summer around $29.0°C$ ($84.2°F$). In the northern Gulf, which is most affected by winter cold fronts, during the period of December through February air over the inner shelf averages about $15.5°C$ ($59.9°F$) and over the outer shelf around $18.0°C$ ($64.4°F$). In this season variability is fairly high, and 90 percent of the observations fall within $6.0°C$ ($10.8°F$) of the mean. By May, air temperature over both the inner and outer shelves has climbed to $24.5°C$ ($76.1°F$), and the highest value of $29.0°C$ ($84.2°F$) is observed in July and August. Low variability characterizes temperature regimes during summer, when 90 percent of the observations fall within $3.0°C$ ($5.4°F$) of the mean. Air temperatures begin to drop in September, and by October differences appear between air over the inner and outer shelves. During this month air over the inner shelf averages $24.0°C$ ($75.2°F$), while that over the outer shelf averages $26.0°C$ ($78.8°F$).

### RELATIVE HUMIDITY

Relative humidity remains high over the Gulf of Mexico throughout the year. It is lowest in late fall and winter in response to the passage of cold, dry continental air masses, and it reaches its maximum in spring and summer when the Bermuda High dominates the circulation. In the northern and eastern Gulf, coastal fog is produced in late fall and winter when warm, moist air overrides cooler water and land masses. Radiation fog lasts for only three or four hours, but dense advection fog, or sea fog,

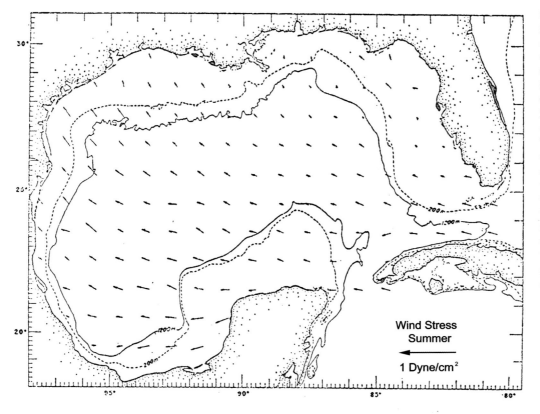

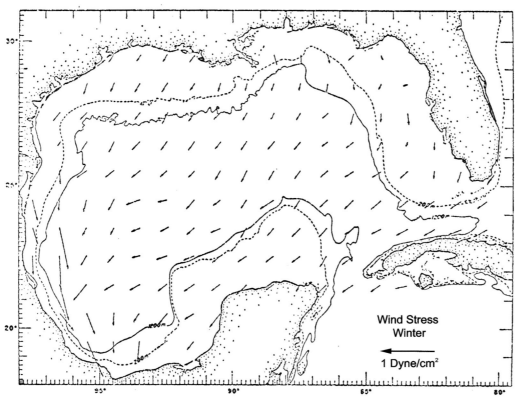

**Figure 5.3.**
Mean seasonal wind stress fields for the Gulf of Mexico during the summer (June, July, and August) and winter (December, January, and February) calculated by 1° squares. The direction of each arrow represents the direction of the stress force in that square, and the length of the arrow is proportional to the magnitude of the force (in dynes/cm$^2$). (Note: 1 dyne = 1.45 × 10$^{-5}$ pounds per square inch.) (From Elliott 1979.)

**Table 5.1.** *Monthly mean temperature and precipitation data for selected cities around the periphery of the Gulf of Mexico. The cities include Tampa, Florida; New Orleans, Louisiana; Corpus Christi, Texas; Tampico, Tamaulipas, Mexico; and Merida, Yucatán, Mexico.*

| Month | Temperature (°F) | | | | | Precipitation (inches) | | | | |
|---|---|---|---|---|---|---|---|---|---|---|
| | Tampa | New Orleans | Corpus Christi | Tampico | Merida | Tampa | New Orleans | Corpus Christi | Tampico | Merida |
| January | 59.8 | 52.4 | 56.3 | 67.1 | 73.2 | 2.17 | 4.97 | 1.63 | 0.78 | 1.01 |
| February | 60.8 | 54.7 | 59.3 | 68.7 | 75.0 | 3.04 | 5.23 | 1.55 | 0.72 | 1.05 |
| March | 66.2 | 61.4 | 65.9 | 71.6 | 78.4 | 3.46 | 4.73 | 0.84 | 0.52 | 0.56 |
| April | 71.6 | 68.7 | 73.0 | 76.6 | 81.1 | 1.82 | 4.50 | 1.99 | 1.03 | 0.46 |
| May | 77.1 | 74.9 | 78.1 | 80.1 | 82.4 | 3.38 | 5.07 | 3.05 | 1.20 | 3.35 |
| June | 80.9 | 80.3 | 82.7 | 82.0 | 81.9 | 5.29 | 4.63 | 3.36 | 5.40 | 5.78 |
| July | 82.2 | 82.1 | 84.9 | 81.3 | 81.0 | 7.35 | 6.73 | 1.96 | 7.52 | 4.89 |
| August | 82.2 | 81.7 | 85.0 | 82.8 | 81.7 | 7.64 | 6.02 | 3.51 | 3.50 | 6.56 |
| September | 80.9 | 78.5 | 81.5 | 80.4 | 80.8 | 6.23 | 5.87 | 6.15 | 13.11 | 8.66 |
| October | 74.5 | 69.2 | 74.0 | 77.7 | 78.6 | 2.34 | 2.66 | 3.19 | 5.92 | 4.38 |
| November | 66.7 | 60.0 | 65.0 | 75.0 | 76.1 | 1.87 | 4.06 | 1.55 | 2.63 | 0.90 |
| December | 61.3 | 54.6 | 59.1 | 67.1 | 73.8 | 2.14 | 5.27 | 1.40 | 0.70 | 1.01 |
| Average or total | 72.0 | 68.2 | 72.1 | 75.6 | 78.6 | 46.73 | 59.74 | 30.18 | 43.04 | 38.62 |

may be quite widespread and may persist for several days. The former occurs from November through March, but sea fogs develop primarily from November through February. In the northern and eastern Gulf, river discharges in the winter and spring may produce cold surface water, resulting in local fog conditions near river mouths.

### PRECIPITATION

In the southern Gulf, precipitation is highly seasonal and is limited largely to the summer months, that is, the period of tropical cyclones. Here the rainy and dry seasons are quite pronounced. Data in table 5.1 show that during the monsoonal months of May through October, Tampico and Merida both receive over 80 percent of their annual rainfall. In the northern Gulf seasonal differences are less pronounced, and precipitation may occur during any month due to instabilities associated with frontal passages in the winter and tropical cyclones in the summer. Here the rainiest months are July through September, and the month of least precipitation is May. Most of the water falls as rain, and the little frozen precipitation that does fall melts upon contact with the ground or water surface. In contrast with the southern Gulf, New Orleans

receives only 52 percent of its annual precipitation during the period of May through October, and Tampa and Corpus Christi both receive around 70 percent.

As seen in table 5.1, the north-central, eastern, and southern Gulf all receive considerably more precipitation than does south Texas. Most winter rains and tropical cyclones bypass the extreme northwestern sector of the Gulf. It is important to note that average annual precipitation values do not tell the whole story because variability from one year to another can be quite great. For example, Pensacola, Florida, with a mean annual precipitation of 1,468 mm (58 in) has recorded years with as little as 737 mm (29 in) and as much as 2,286 mm (90 in).

Since there are no rain gauges in the open Gulf, estimates of precipitation over open water must be derived by combining information from shore stations, ship logs, and satellite observations. Using such data sources, Dorman and Bourke (1981) calculated the average annual precipitation for the entire Gulf to be about 530 mm (20.9 in), with the maximum amount of 190 mm (7.5 in) being received in the fall and the minimum of 60 mm (2.4 in) coming in the spring. However, this is not uniformly distributed over the Gulf, and separate estimates suggest

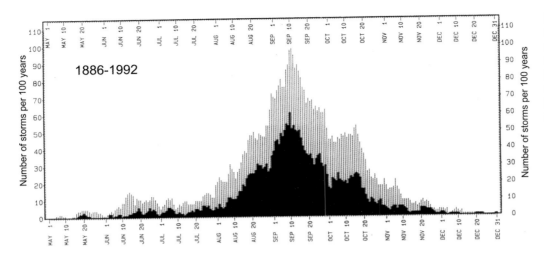

**Figure 5.4.**
Seasonal distribution of tropical cyclones in the North Atlantic region based on a record from 1886 through 1992. Lower bars are for hurricanes only. Upper bars are for hurricanes plus tropical storms. (From Neumann et al. 1993.)

that the northwestern Gulf annually receives about 1,016 mm (40 in) of precipitation.

## TROPICAL CYCLONES

Tropical cyclones are nonfrontal, low-pressure, large-scale systems that develop over tropical or subtropical waters and show definite organized circulation. They are classified on the basis of sustained surface wind speeds as follows: tropical depressions, 0–64.9 km/hr (0–38.9 mi/hr); tropical storms, 65–122.9 km/hr (39–73.9 mi/hr); and hurricanes, 123 km/hr (74 mi/hr) or more. Tropical cyclones draw their energy from the latent heat of condensation of water vapor over warm water. Hurricanes typically have a central "eye" with a radius of 10–30 km (6–18 mi) or more where winds are light and air pressure is at a minimum. The eye is surrounded by a wall of rapidly rotating winds, and the entire system may be hundreds of miles in diameter. High winds associated with hurricanes generate large waves and strong water currents. Sea level rises dramatically, especially over shallow continental shelves. There is intense vertical mixing within the water column, and bottom sediments over much of the continental shelf may be stirred and transported elsewhere. Making landfall, hurricanes often reshape barrier islands and other coastal features and cause extreme property damage.

The seasonal distribution of tropical storms and hurricanes in the North Atlantic based on a 100-year record is shown in figure 5.4. Although such cyclonic events have been known to occur as early as May and as late as December, the official "hurricane season" stretches from the beginning of June through the end of October; and within the Gulf, hurricanes occur with greatest frequency

and intensity during the period of July through October. Based on the 100-year record, monthly compilations of hurricane tracks through the Gulf are available (fig. 5.5). Although many complex factors are involved in steering individual hurricanes, some sense can be made from the tracks of large numbers occurring throughout the season. In the month of June some of the tracks are erratic, but most trend in a northerly direction, and few enter the western Gulf. In July the tracks trend northerly or northeasterly, and some activity takes place in the western Gulf. In August the predominant directions are toward the west and northwest, and many hurricanes make landfall along the coasts of Mexico, Texas, and Louisiana. In September, although the frequency of hurricanes increases, the same westerly and northwesterly patterns persist. Many of the hurricanes make landfall near the Mississippi River Delta. In October, the general direction changes. Although most hurricanes still enter the Gulf from the south or southeast, most now exit in a northeasterly direction. Furthermore, most of the tracks are concentrated in the eastern Gulf, with few passing inland through Mexico, Texas, or Louisiana. By November, the number of hurricanes has dropped dramatically. Their tracks are erratic, and regardless of initial direction, they wind up heading south. By this season, surface water temperatures are too low to sustain strong cyclonic activity, and high pressure builds over the northern Gulf.

The vast amount of energy associated with hurricanes modifies both the open ocean and coastal areas where they make landfall. Over open water, low air pressure in the center of the storm causes the level of the sea to rise, forming a broad dome of water, and wind stress from the rapidly rotating cyclonic winds sets up a rotating system

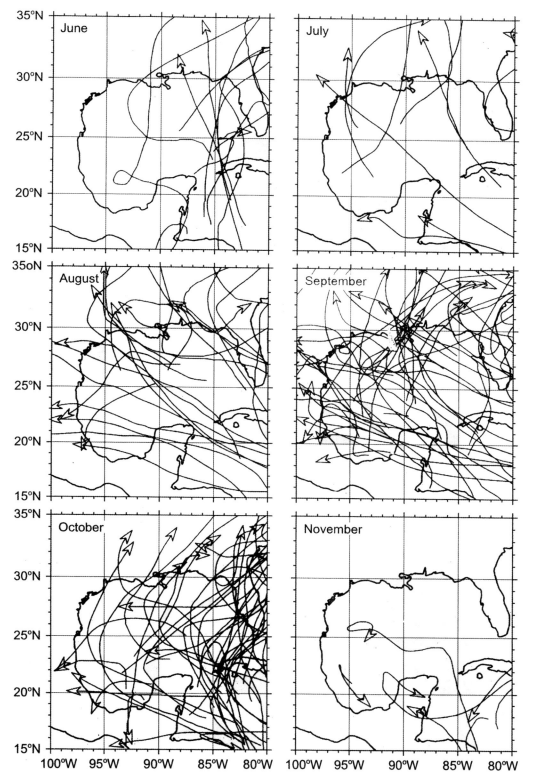

**Figure 5.5.**
Paths taken by hurricanes that entered the Gulf of Mexico and originated during the first 10 days of each month during the hurricane season, based on the record from 1886 through 1992. (From Neumann et al. 1993.)

in the water column. Together these factors act as a pump, moving surface water to the periphery and bringing deep cooler water to the surface. In the Gulf such upwelling areas may be as much as 5.0°C (9.0°F) cooler than the surrounding surface water. Approaching landfall, the hurricane is preceded by a storm surge, a local elevation in sea level resulting from high winds and lowered air pressure. Storm surges of 4.5 m (15 ft) are not uncommon, and the great hurricane of 1969 (Camille), with wind speeds of 320 km/hr (200 mi/hr), was accompanied by a storm surge estimated to be at least 7.2 m (24 ft). Damage to coastal property and loss of human life come primarily from the high winds, floods, and storm surges. In the great Galveston hurricane of 1900, more than 6,000 people lost their lives, and most of the buildings of the city were leveled by the 4.5 m (15 ft) storm surge. Now, with greatly improved weather forecasting and emergency management procedures, such loss of human life is not likely to be repeated, but property damage can still be quite extensive. However, during the past two decades, as global temperatures have risen, the northern Gulf of Mexico has been lashed by four major hurricanes, as follows (based on data provided by NOAA):

1992 – Hurricane Andrew first struck the southern tip of Florida and then moved on to the central coast of Louisiana, with a loss of 26 human lives and $26 billion in property damage.

2005 – Hurricane Katrina struck central Florida and then Mississippi and eastern Louisiana (including New Orleans), with a loss of 1,833 lives and $81 billion in property damage.

2005 – Hurricane Rita struck western Louisiana and eastern Texas, with a loss of 180 lives and $11.3 billion in property damage.

2008 – Hurricane Ike struck Galveston, Texas, with a loss of 103 lives and $19.3 billion in property damage.

## COLD FRONTAL PASSAGES

The maritime tropical atmosphere over the Gulf of Mexico may respond to cold fronts as much as 48 hours before their actual passage. During this period, surface air pressure falls, temperature increases, and the dew point rises. Frontal passages into the Gulf are accompanied by dramatic events such as gale-force winds, heavy rains, and large temperature fluctuations. After passage of a cold front, about 48 hours are required for the polar

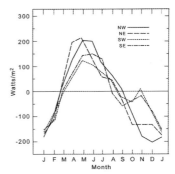

**Figure 5.6.** Annual variation in the rates of heat storage (positive values) and heat loss (negative values) for the four quadrants of the Gulf of Mexico. These include the northwest (NW), northeast (NE), southwest (SW), and southeast (SE) sectors. Units used are watts/m². (Note: 1 watt = 0.01433 kilogram-calorie/min.) (From Etter 1975.)

air to reach new equilibrium with the warmer, moister atmosphere over the Gulf. Such passages occur with high frequency during the winter months (ca. nine passages/month) and low frequency during the summer (ca. two passages/month). The transition is gradual during the spring but quite sharp during the fall (between September and October).

## HEAT AND MASS EXCHANGE AT THE WATER'S SURFACE

Gulf waters gain heat by absorption of solar radiation, which, of course, is greater in the summer than in the winter. Some of the heat is stored by the water, some is lost to the atmosphere (by radiation and evaporation), and some is transferred elsewhere through mixing and horizontal circulation. The annual progression in the rate of heat storage and loss for different quadrants of the Gulf of Mexico is shown in figure 5.6. Here heat gain (in watts/m²) is represented by positive values and heat loss by negative values. For all four sectors the rate of gain is greatest in the summer and the rate of loss is greatest in the winter. However, there is a basic difference between the two northern and the two southern sectors. In the northern regions the single maximum occurs around May and the minimum in December–January, but the two southern regions exhibit a double maximum (May–June and November) and a single minimum (January). Also, during the spring and early summer the northern Gulf gains more heat than the southern Gulf because its waters are colder to begin with, and it loses more heat in the fall and winter as a result of the passage of cold fronts.

Each cold air outbreak cools and deepens the surface

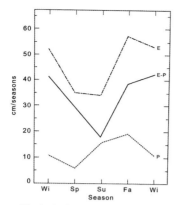

**Figure 5.7.** Seasonal hydrologic balances for the Gulf of Mexico. E = evaporation rates, P = precipitation rates, E-P = hydrologic balances (evaporation minus precipitation). (From Etter 1983.)

mixed layer of the Gulf. Calculations suggest that during January the mean monthly heat loss of the northern Gulf is about 310 cal/m². The effects of a strong early-season (November) polar front crossing the Gulf of Mexico have been analyzed by Henry and Thompson (1976). During its 24-hour passage from the northern Gulf coast to Yucatán, the cold air extracted about 1.44 cal/m² per minute from the surface waters. Over the 24-hour period the surface layer of the Gulf would have lost almost 2,100 cal/m², and most of this loss must have occurred from waters of the northern Gulf. This passage also resulted in the formation of a cloud that stretched most of the way to Yucatán.

A freshwater budget for the Gulf of Mexico has been calculated by Etter (1983). He first determined that the annual loss of freshwater through evaporation is about 180 cm (70.9 in), with a maximum in the fall of 56 cm (22.0 in) and a minimum in the spring of 35 cm (13.8 in) (fig. 5.7). Subtracting the amount gained through precipitation (as discussed earlier) from that lost through evaporation, he arrived at a net annual loss of 127 cm (50.0 in) of freshwater. Rivers bring in freshwater at an average rate of $31.6 \times 10^6$ kg/sec, which computes to 62 cm (24.4 in) of water depth per year. This partially makes up the deficit, but it still leaves a freshwater loss of 65 cm (25.6 in) for the year. Since the Gulf of Mexico is not a closed system, the lost volume is made up by marine water entering through the Yucatán Channel, but it also means that water exiting through the Straits of Florida is more saline than that entering the Gulf, as has been confirmed by direct measurements.

Since air and water are both fluids, physical oceanography has a great deal in common with atmospheric science. However, water is much denser than air, its responses to physical forces are often slower, and the effects of these forces may persist for much longer periods. Furthermore, being a liquid, water dissolves many substances, most notably salts, which in turn affect its density. In dealing with waters of the Gulf, we will be concerned primarily with surface phenomena such as air-sea interactions, the origins and properties of the various water masses, and patterns of horizontal and vertical circulation. Considerable research has been carried out on the physical oceanography of the Gulf of Mexico, and there is a large amount of literature on the subject. Much of the early work has been summarized by Galtsoff (1954a) and Leipper (1954b). Through the early years, ships' officers accumulated observational information concerning weather, water temperature, and surface currents. On the basis of these data, Soley (1914) was able to map the general seasonal circulation patterns of surface waters of the Gulf with fair accuracy (fig. 6.1). During the past three or four decades, under sponsorship of the US Navy, NOAA, NASA, and other agencies, the pace of physical research has sharply accelerated. Equipped with sophisticated environmental sensing and recording devices, satellites, and advanced computers, scientists have greatly increased the body of observational data, and they have been working to establish a mathematical basis for data interpretation. Important references include Capurro and Reid (1972), Cochrane and Kelly (1986), Ichiye, Kuo, and Carnes (1973), and Kelly (1988). Important summaries of the more recent literature have been provided by Nowlin et al. (2000), Schmitz (2005), and Sturges and Lugo-Fernandez (2005). The following discussion provides a descriptive overview of the physical oceanography of the Gulf of Mexico. Mathematical and theoretical treatments are reserved for the advanced reader.

### TIDES AND SEA LEVEL

Around the surface of the earth, marine waters are attracted by the gravitational pull of the moon, sun, and various planetary bodies, and the resulting rise and fall of sea level is referred to as the astronomical tide. The difference between consecutive high and low tides is the tidal range, and the midway point is the mean sea level. Half the tidal range is the tidal amplitude. Mean sea level may be calculated from tidal observations of a few hours to days, months, or years.

Technically, at least 30 astronomical components are known to have some effect on the tides, but most of these are inconsequential. Only 6 components have significant effects on tidal amplitude, and only 4 are important in the Gulf of Mexico. These are due to the moon and sun, and both have semidiurnal (ca. 12-hour) and diurnal (ca. 24-hour) components. These 4 components are given as follows:

$K_1$ = diurnal luni-solar tide
$O_1$ = diurnal lunar tide
$M_2$ = principal semidiurnal lunar tide
$S_2$ = principal semidiurnal solar tide

There are fortnightly variations in the magnitude of these tidal components as the moon goes through its phases relative to the sun. Maximum amplitudes occur when the moon and sun are aligned or in phase (i.e., on the same or opposite sides of the earth), and minimum amplitudes take place when they are 90° out of phase. These changes produce, respectively, spring (maximum) and neap (minimum) tides.

It can be seen that the quantity $K_1 + O_1$ represents the sum of the components affecting diurnal tides, and $M_2 + S_2$ is the sum of components affecting semidiurnal tides. Thus, as a rule, when $M_2 + S_2$ equals or exceeds $K_1 + O_1$, there is a semidiurnal or mixed semidiurnal tide. When $K_1 + O_1$ exceeds $M_2 + S_2$, there is a diurnal or mixed diurnal tide. Note that there are four basic types of tide: semidiurnal, mixed semidiurnal, diurnal, and mixed diurnal. The diurnal tide has one high and one low tide per day, whereas the semidiurnal has two highs and two lows per day. In both cases succeeding highs and lows have about the same amplitude, but in the case of mixed tides there are large inequalities between succeeding highs and/or lows. In any given basin the tide may be driven directly by gravity, as noted above, or it may be driven, in part,

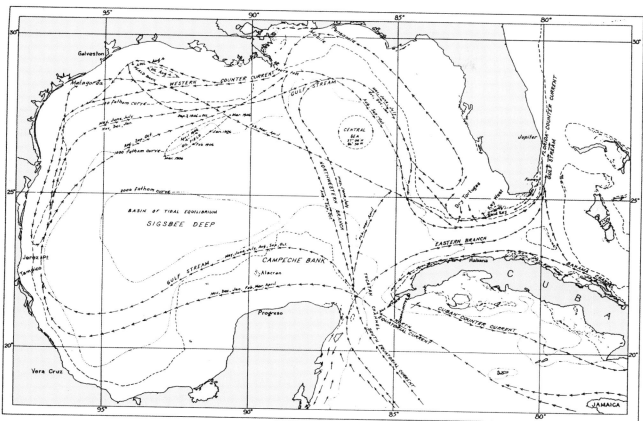

**Figure 6.1.** General seasonal circulation patterns of surface waters of the Gulf of Mexico based on early ship observations. (From Soley 1914.)

by tidal waves entering from another basin. The latter is known as a co-oscillating tide.

With respect to the world oceans, the Gulf of Mexico occupies a relatively small basin with two ports of entry: the Straits of Florida and the Yucatán Channel. The tides of the Gulf vary from diurnal to mixed, and the latter include both mixed semidiurnal and mixed diurnal. Plain semidiurnal tides do not occur here. Representative tide records for stations around the Gulf are shown in figure 6.2. Note the contrast between the diurnal and mixed tides of the various Gulf stations and those of Miami, Florida. Miami lies on the Atlantic coast, where plain semidiurnal tides are the rule. Note also the small tidal range of Gulf stations. The records show that the Gulf of Mexico is a microtidal environment with a maximum tidal range of less than one meter. Calculations have shown that in the Gulf of Mexico the diurnal portion of the tide is primarily a co-oscillating tide driven by that of the Atlantic Ocean and Caribbean Sea. For the diurnal component, direct lunar and solar forcing accounts for only 15 percent of the tide, but for the semidiurnal component, such direct forcing accounts for 65 percent of the tide. Also, the semidiurnal tide appears to rotate counterclockwise around a node or amphidromic point located just north of the Yucatán Peninsula.

In addition to being affected by the astronomical tides, whose amplitudes and timing are periodic and predictable, the actual level of the sea at any given time and place may be influenced by a number of other factors. Over longer periods, uplift or subsidence of the land may change the level of the water at a given site. Formation and melting of ice in polar regions can lower or raise sea levels worldwide. On a more temporary and local level, flooding from rivers, changes in hydrology (temperature-salinity-density characteristics) or barometric pressure of the atmosphere, wind-driven currents, and storm surges may be important factors. Therefore, the actual height of the sea at any given time and place is determined by the astronomical tide as modified by a host of geographical and weather-related phenomena.

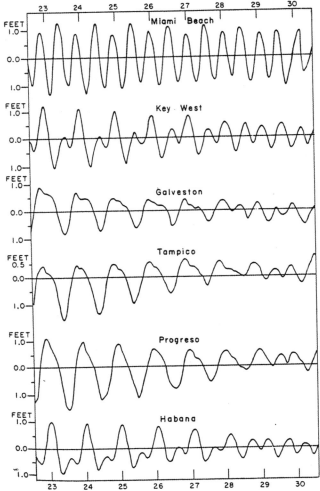

**Figure 6.2.** Tidal curves at various locations around the margin of the Gulf of Mexico, 23–30 June 1948. (From Marmer 1954.)

**Table 6.1.** *Definition of terms relating to gravity waves at the sea surface.*

fetch – distance traveled over open water by a wave after its generation.

seas – waves under the direct influence of the wind in a wave-generating field.

swell – waves outside a generating field and no longer subject to significant wind action.

wave amplitude – one-half the height of a wave.

wave height – vertical distance from the trough to the crest of a wave.

wave length – horizontal distance between two successive wave crests.

wave period – time between the passage of two successive wave crests.

wave velocity – distance traveled by a wave in one second.

## SURFACE WAVES AND SWELL

As noted earlier, winds blowing across the sea deform the surface and produce surface waves. Terminology relating to waves is provided in table 6.1. Wind waves are often called gravity waves because gravity is the principal restoring force. Gravity waves fall into two categories, seas and swell. The former are produced under the direct influence of the wind in a wave-generating area such as a storm. Swell refers to gravity waves after they have left the area of wave generation and are no longer subject to significant wind action. Seas generally have steeper waves with short periods and wave lengths, whereas waves associated with swell have relatively long periods and wave lengths, and until they reach shallow waters they have low amplitudes. In general, such wave characteristics as height, period, and length are determined by fetch, wind speed, wind duration, and decay distance. In shallow water the depth can also affect wave properties.

In the Gulf of Mexico mean monthly wave heights are generally lowest in the spring and summer. Winds entering the Gulf from the southeast and south may be quite persistent, but the speed is generally low to moderate. Such winds usually generate waves in the range of 0.5 to 1.5 m in height. Waves are often higher during the winter months as a result of the frequent passage of fronts. Along the northern Gulf coast, winter wave heights may be small, but the average height increases seaward as the fetch becomes greater. Northers typically generate 2–3 m waves near shore and 4–6 m waves toward the outer shelf. Since many of these fronts stall out over open water, they send long swells to the southern Gulf. The greatest average wave heights occur during September, the month with the most frequent storms and hurricanes. Here wave heights may exceed 10 m. The most common periods associated with nonhurricane waves in the Gulf are between 4 and 6 sec, with very few greater than 10 sec. However, the extreme wind speeds of hurricanes generate waves with periods of 9 to 13 sec.

Wave characteristics of an area are of great importance to commercial ship traffic and to engineers who design fixed structures such as oil platforms to be placed on the continental shelf. Hence, a great deal of attention has been given to the wave environment of the northern Gulf. Table

**Table 6.2.** *Annual percentage frequency of wave heights in various categories based on ship operations in statistical areas 27, 28, and 29 representing the Texas-Louisiana continental shelf. Asterisks indicate values less than 0.05 but greater than 0.0. (From Quayle and Fulbright 1977.)*

| Area | Wave height (m) | | | | | | | | | | | | |
|---|---|---|---|---|---|---|---|---|---|---|---|---|---|
| number | 0 | ½ | 1 | 1½ | 2 | 2½ | 3 | 3½ | 4–4½ | 5–5½ | 6–6½ | 7–7½ | 8–9½ |
| 27 | 14.2 | 25.6 | 29.7 | 17.3 | 7.1 | 3.4 | 1.3 | 0.7 | 0.6 | 0.1 | 0.1 | * | * |
| 28 | 13.9 | 26.1 | 30.4 | 17.1 | 6.7 | 3.3 | 1.2 | 0.6 | 0.5 | 0.1 | 0.1 | * | * |
| 29 | 8.3 | 24.2 | 33.3 | 18.3 | 8.2 | 4.2 | 1.7 | 0.9 | 0.7 | 0.1 | 0.1 | * | 0 |

6.2 provides information on the annual percentage frequencies of wave heights for three statistical areas covering the Texas-Louisiana continental shelf. Here it can be seen that waves of 1 m in height or less occur between 65 and 71 percent of the time, and waves of 3 m or greater are encountered less than 4 percent of the time. The very highest waves in the 7–9.5 m range occur less than 0.1 percent of the time. Data from a moored buoy located 200 km south of the shelf break provide a 30-month record of wave statistics near the central Gulf. Here monthly mean wave heights typically fall in the 1–1.5 m range, and monthly maximum heights are usually around 3–4 m, but seasonally these may rise to 6–8 m. Average wave periods at the buoy range from 4.2 to 5.4 sec.

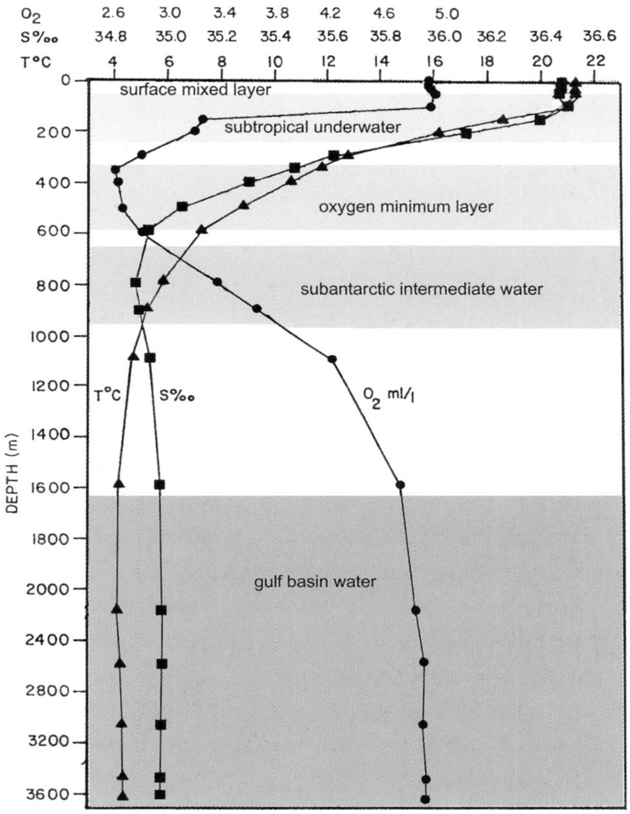


**Figure 6.3.** Physical characteristics and water mass designations from a west-central Gulf hydrographic station. The deepest water sample was taken four meters above the bottom.

## WATER MASSES

The principal inflow of water into the Gulf of Mexico is from the Caribbean Sea through the Yucatán Channel, where the sill depth is estimated to be between 1,650 and 1,900 m. Under normal circumstances this sill determines the greatest depth from which Caribbean water can enter the Gulf. Most of the outflowing water passes through the Straits of Florida into the North Atlantic. The latter passage has a sill depth of around 800 m. Waters entering the Gulf through the Yucatán Channel are a mixture of South Atlantic water (transported northwestward by the Guinea and equatorial current systems) with North Atlantic water (from the west Sargasso Sea). The ratio of South Atlantic to North Atlantic water has been estimated to be between 1:2 and 1:4.

Within the Gulf five vertically layered water masses are recognized. From top to bottom these include: (1) the surface mixed layer, (2) the subtropical underwater layer, (3) the oxygen minimum layer, (4) Antarctic Intermediate Water, and (5) Gulf basin water. Each can be distinguished in the Gulf by distinct temperature, salinity, and oxygen characteristics, that is, by the different values, gradients, and relative maxima and minima of these aspects. The actual water depth distribution and physical and chemical characteristics of each layer tend to vary somewhat with time and place. In figure 6.3 the approximate depth range of each of the five water masses is shown. Also depicted are the temperature, salinity, and dissolved oxygen profiles in relation to depth for a March hydrographic station in the west-central Gulf. Distinguishing characteristics of each of the water masses are given below.

### Surface mixed layer

This is the upper isothermal layer, where the temperatures depend on the heat budget, and the salinity distribution depends on evaporation (minus precipitation), runoff, and horizontal advection (lateral transport) by water currents. The thickness of this layer may vary from a fraction of a meter to over 125 m depending on location, time of year, and local influences. In the figure the depth of the surface mixed layer is shown as approximately 75 m.

### Subtropical underwater

This layer is characterized by an intermediate salinity maximum in the depth range of 50–200 m. This water mass is present throughout the Caribbean, but in the Gulf its salinity maximum becomes eroded outside the Loop Current (to be discussed in a later section). The horizontal distribution of the core salinity maximum and the depth distribution of the core in the Gulf are shown in figure 6.4. The source of the subtropical underwater in the Caribbean and the Gulf is probably the North Atlantic in the area of 20°–25°N and 30°–50°W.

### Oxygen minimum layer

This layer is characterized by minimum oxygen values in the depth range of approximately 300–600 m. It is not associated with temperature or salinity extremes. This layer within the Gulf is clearly continuous with that of the Caribbean. In the eastern Gulf a secondary oxygen minimum layer is present in water bounded by the Loop Current, but this is almost completely absent in the western Gulf. Representative dissolved oxygen curves for various sections of the Gulf are shown in figure 6.5. In each curve the oxygen minimum layer(s) is apparent.

### Antarctic Intermediate Water

In the Gulf this water mass is characterized by a salinity minimum of 34.86–34.89 ppt at depths between 550 and 900 m. It has its origin at the Antarctic Convergence in the South Atlantic, where cold, low-salinity water sinks and spreads to the north. This core enters the Caribbean with salinities slightly less than 34.7 ppt, but by the time it reaches the Gulf, mixing with other water masses has raised the salinity to 34.88–34.89 ppt. The depth distribution of the core layer is shown in figure 6.6.

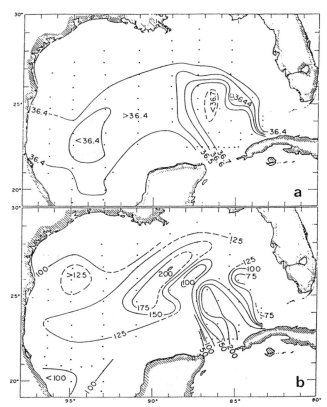

**Figure 6.4.** Core of salinity maximum in the subtropical underwater. (a) Core salinity at intervals of 0.1 ppt. (b) Core depth in 25 m intervals. (From Nowlin 1972.)

### Gulf basin water

This mass is defined as those waters below a depth of 1,650–1,900 m (Yucatán sill depth). At around 2,000 m the mean temperature and salinity are 4.23°C and 34.97 ppt, respectively. Temperature-salinity relationships of deep water on both sides of the Yucatán sill are quite similar, suggesting the present-day exchange of Gulf basin and deep Caribbean waters.

### HYDROGRAPHY

This section describes the temperature, salinity, and density structure of the Gulf of Mexico as well as the dissolved oxygen characteristics. In horizontal perspective the greatest variability in these properties is seen on the continental shelf of the northern Gulf and in the eastern Gulf in the area of the Loop Current. In vertical perspective the greatest variability occurs in the surface mixed layer (which is most affected by solar heating, the atmosphere, river runoff, and other local factors) and in waters associated with the Loop Current.

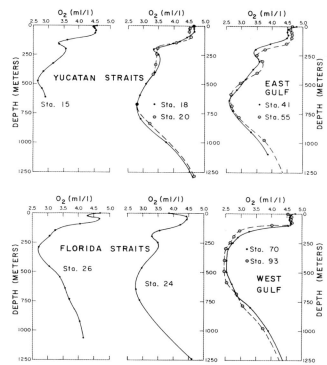

**Figure 6.5.** Dissolved oxygen concentration versus depth for hydrographic stations in various sections of the Gulf of Mexico. Station numbers refer to Hidalgo Cruise 62-H-3. (From Nowlin 1972.)

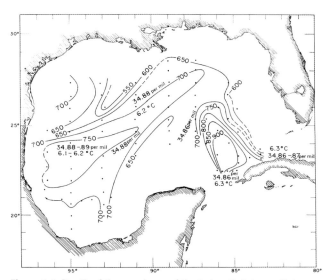

**Figure 6.6.** Depth of the core layer of the Antarctic Intermediate Water in 50 m intervals. Selected values of temperature and salinity at core depth are shown. (From Nowlin 1972.)

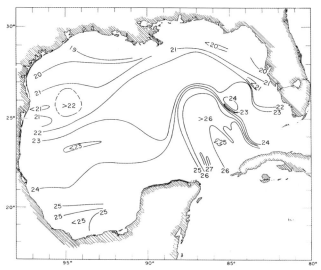

**Figure 6.7.** Distribution of surface water temperatures (in degrees Centigrade) of the Gulf of Mexico in midwinter. (From Nowlin 1972.)

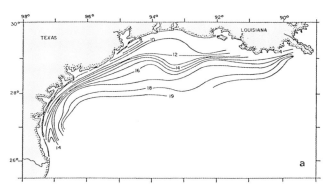

**Figure 6.8.** Distribution of surface water temperatures (in degrees Centigrade) on the continental shelf of the northwestern Gulf in January. (Based on data from Temple, Harrington, and Martin 1977.)

## Temperature

During the summer months, surface water temperatures over the entire Gulf of Mexico fall in the range of 28°–30°C. This is true for nearshore and offshore waters and for the northern as well as the southern Gulf. However, much greater variation is seen during the winter when a latitudinal gradient is established and when waters of the northern Gulf are most affected by cold fronts. The distribution of surface water temperature in open waters of the Gulf in midwinter is shown in figure 6.7. Here it is seen that over the entire shelf of the northern and northeastern Gulf from south Texas to the level of Tampa Bay, Florida, surface water temperatures are 20°C or lower. Actually, as seen in figure 6.8, nearshore temperatures may drop below 10°C. Temperatures throughout the open Gulf remain above 20°C, and most of the southern third stays at 24°C or above. In the winter the highest temperatures occur around the Loop Current, where values remain at 26°C or above. This water, having recently arrived from the Caribbean, has been less affected by chilling winds from the north.

The depth of the surface mixed layer varies with the season. During the spring and summer solar radiation heats the uppermost layer, which, being less dense, remains on top. Light winds cause little disturbance or vertical mixing. In the summer the mixed layer extends to a depth of only about 10 m on the continental shelf and 25–50 m in the open Gulf. During the winter, as cold winds chill the surface waters, they become slightly more dense and sink. This process, together with greater wind velocities, induces deeper mixing within the surface layer. Therefore, during the winter the surface mixed layer on the northern Gulf shelf may extend to a depth of 25 m, and in the open Gulf it can exceed a depth of 100 m.

Figure 6.3 provides a vertical temperature profile taken in the west-central Gulf during March. From a surface mixed layer temperature of 21.5°C the temperature drops rapidly with depth. At 400 m it has reached 10.0°C, and by 1,000 m it is down to 5.0°C. Thereafter, to the bottom of the Gulf it remains just above 4.0°C. With slight variation, this temperature profile is characteristic of most of the open Gulf, and of the world oceans in general.

The situation is somewhat different in the eastern Gulf in areas affected by warm waters of the Loop Current heading northward through the Yucatán Channel. Figure 6.9 shows a cross section of the southeastern Gulf in a line running diagonally southwest from Tampa Bay to a point on the Yucatán shelf. In the heart of the Loop Current 25.0°C water extends to a depth of about 180 m, 10.0°C water goes down to 600 m, and 5.0°C water extends below a depth of 1,100 m. Bottom water ranges between 4.32° and 4.38°C. The depression of isotherms in the core of the Loop Current is accompanied by a rise in the isotherms on the outer continental shelf of the Florida Peninsula, where 14.0°C water bathes the outer shelf. Note that on the west side of the Loop Current, 20.0°C water bathes the edge of the Yucatán shelf. The depression of isotherms by the Loop Current in the southeastern Gulf can also be inferred from information given in figure 6.6, which shows the depth of the core layer of Antarctic Intermediate Water.

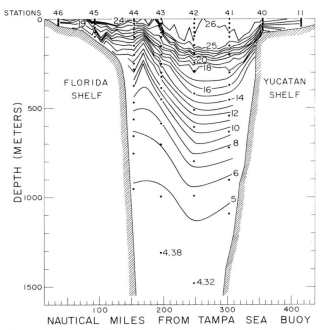

**Figure 6.9.** Hydrographic transect showing isotherms (in degrees Centigrade) from Tampa Bay southwest to a point on the Yucatán shelf in January. Numbers along the top represent cruise station numbers. (From Nowlin 1972.)

## Salinity

As seen in figure 6.10, surface salinities of offshore waters over the entire Gulf fall in the range of 36.0–36.4 ppt. Slightly higher salinity sometimes occurs locally off Yucatán (where there is no surface drainage), and lower salinity values generally characterize continental shelf waters off the mouths of large rivers, particularly in the northern Gulf. Salinity of the offshore waters basically represents oceanic salinity modified by rainfall and evaporation as well as some mixing from shelf waters. Salinity values on the continental shelf reflect rainfall, evaporation, and mixing as well as runoff, river discharge, and, in some localities, upwelling.

As noted in an earlier chapter, the Mississippi-Atchafalaya system brings vast amounts of freshwater to the northern Gulf, and this discharge is highly seasonal, with a maximum of about 23,000 m³/sec in March and April and a minimum of around 5,000 m³/sec in the fall months of September–November. In the spring, when prevailing winds are primarily from the southeast, much of the high-volume discharge of the Mississippi-Atchafalaya system is swept westward over the Louisiana and Texas continental shelves, and river-influenced, low-salinity water has been recorded as far south as the Rio Grande (fig. 6.11). During the remainder of the year the in-

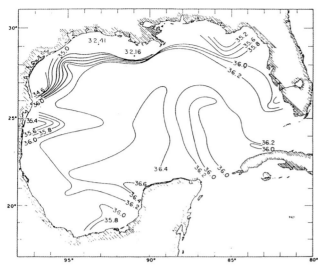

**Figure 6.10.** Surface salinities throughout the Gulf of Mexico for the month of March. (From Nowlin and McLellan 1967.)

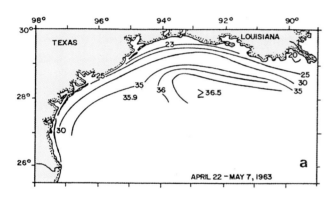

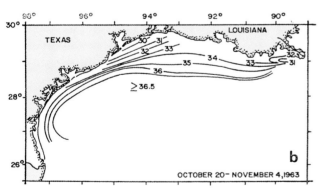

**Figure 6.11.** Surface salinities (in parts per thousand) for the northwest Gulf of Mexico continental shelf during (a) spring and (b) fall. (Based on data from Temple, Harrington, and Martin 1977).

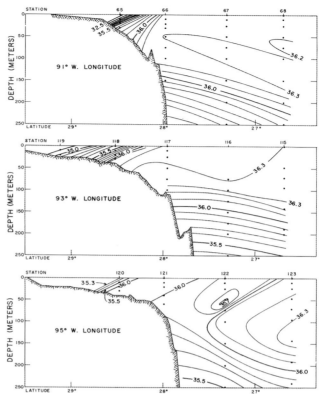

**Figure 6.12.** Vertical sections of salinity (in parts per thousand) over the continental shelf and upper slope of the northwest Gulf of Mexico in March. (From Nowlin and McLellan 1967.)

fluence of the Mississippi River on surface water salinity over the Texas-Louisiana shelf is not as great. As seen in figure 6.12, the low salinity influence in this area is limited largely to the upper 50 m of water. On the continental shelf east of the Mississippi River the effects of freshwater discharge vary with wind direction, induced circulation effects brought about by the upper arm of the Loop Current, and other factors, and they are not as strictly linked with season of the year. Here lower inshore salinity is not limited to the spring months (fig 6.13).

As in the case of temperature, salinity shows its greatest variation in the surface mixed layer, and it is less variable with depth. A typical temperature-salinity profile for the west-central Gulf is provided in figure 6.3. Here it is seen that the salinity in the surface mixed layer is about 35.50 ppt. Below this layer the salinity decreases through the subtropical underwater and oxygen minimum layer, reaching a minimum salinity value of about 34.90 ppt at 800 m in the core of the Antarctic Intermediate Water. Thereafter the salinity rises, and in the Gulf basin water it is around 34.97 ppt.

A.

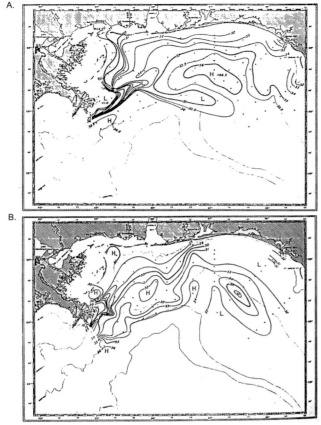

B.

**Figure 6.13.** Surface salinities (in parts per thousand) for the northeast Gulf of Mexico continental shelf during (a) spring and (b) fall. (From Drennan 1968.)

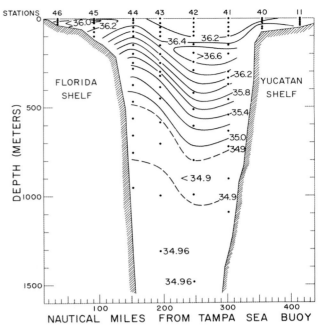

**Figure 6.14.** Hydrographic transect showing lines of equal salinity (in parts per thousand) from Tampa Bay southwest to a point on the Yucatán shelf in February. (From Nowlin 1972.)

In the eastern Gulf the Loop Current affects the depth distribution of salinity values. Figure 6.14 provides a cross section view of depth-related salinity values in a southwest line from Tampa Bay to a point on the Yucatán shelf. Toward the western side the surface salinity is 36.20 ppt, and this increases to a maximum of 36.60 ppt at around 200 m. Below this depth salinity decreases, reaching a minimum value of below 34.90 ppt in the depth range of 800–1000 m. Thereafter, the salinity rises, and in the deepest water values range around 34.96 ppt. Toward the Florida shelf the salinity values parallel those to the west but are found 200–300 m shallower than those described above. The depression of the salinity minimum in the core of the Antarctic Intermediate Water in the southeastern Gulf shows clearly in figure 6.6.

## Density

The density of seawater is determined by three factors: temperature, salinity, and pressure. Since water is nearly incompressible, large changes in pressure cause only slight changes in density, and even the great pressures of the deep sea (several thousand tons/m²) cause only a 2–2.5 percent increase in its density. Since it is important to determine all seawater densities at the same pressure, it is customary to do so aboard ship or in the laboratory (that is, at atmospheric pressure). However, when a water sample is brought up from the deep sea, the slight expansion is accompanied by a small drop in temperature (0.1°–0.2°C). The temperature of such a reading is referred to as the *potential temperature,* as opposed to a reading at depth (*in situ temperature*).

Since all the readings are carried out at the same (atmospheric) pressure, the density of the seawater depends only on the temperature and salinity of the sample. Under standard conditions freshwater has a density of 1.000 g/cm³, and the densities of most ocean waters fall in the range of 1.024–1.028 g/cm³. The first two digits are always the same, and it is customary to drop the 1, move the decimal point three places to the right, and drop the units (g/cm³). The resulting value for density is called sigma-t (or $\sigma_t$). For example, if the density of a seawater sample is 1.0275 g/cm³, its sigma-t value is 27.5.

In practice, if one plots the temperature versus the salinity of a sample on a special graph, one can read off the sigma-t value, and a series of such plots from surface-

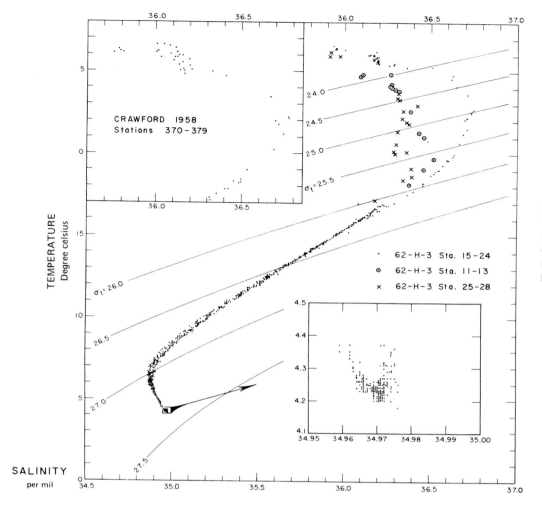

**Figure 6.15.**
Temperature versus salinity profile for waters in the southwestern Gulf of Mexico during the winter. Sigma-t values are indicated by the scale of curved lines trending from the upper right to the lower left. This profile, discussed in the text, is based entirely on stations 15–24 of cruise 62-H-3. Other points in the diagram represent nearby areas. (From Nowlin 1972.)

to-bottom samples provides a density profile of the entire water column. Such a profile for the Gulf of Mexico north of the Yucatán Channel is given in figure 6.15. In the present case the curve is S-shaped due to changing salinity values. As the temperature drops the salinity increases to a maximum value of around 36.7 ppt, then diminishes to a minimum value of about 34.9 ppt, and finally increases slightly at the lowest temperatures. Note that despite the curvature, the sigma-t value increases rather regularly as the temperature decreases (that is, from a value of about 23.5 at a temperature of 26.5°C to maximum of 27.3 at a temperature of 4.0°C). Examination of the temperature and salinity values in figures 6.9 and 6.14 shows that the salinity maximum coupled with a temperature of around 25.0°C occurs at a very shallow depth (ca. 100 m or less), whereas the minimum salinity value coupled with a temperature of slightly above 5.0°C occurs at a depth of about 800–900 m. So the density profile is strongly related to depth, and this is because the sea temperature is the pri-

mary factor controlling seawater density. As the temperature decreases with depth, the density increases.

### Dissolved oxygen

As pointed out in an earlier chapter, dissolved oxygen can be added only to the upper layer of the sea through the processes of atmospheric exchange and photosynthesis of marine plants such as diatoms and dinoflagellates. Oxygen can be lost from the system through atmospheric exchange, respiratory consumption by bacteria and higher organisms, and nonbiological chemical oxidation of certain substances. In the surface mixed layer the oxygen content is generally high and uniformly distributed. The depth of this layer often corresponds more or less with the *euphotic zone,* that is, the topmost layer of the sea that receives sufficient sunlight to support photosynthesis. Below lies the *aphotic zone,* where photosynthesis is not possible. The upper portion of the aphotic zone receives a rain of dead organic material from above, and respiration by bacteria and small zooplanktonic animals partially depletes the oxygen supply of this layer, contributing to the formation of the oxygen minimum layer of the sea. The degree to which this core of low dissolved oxygen is depleted depends mainly on three factors: the original dissolved oxygen content when the water mass was formed (i.e., its prehistory), the amount of oxidizable matter reaching the layer from above, and the residence time of the water mass.

A typical oxygen profile for the west-central Gulf of Mexico is given in figure 6.3. Here dissolved oxygen values in the upper 100 m are around 5.0 mL/L. Below 100 m, the oxygen content drops sharply, reaching 3.2 mL/L by a depth of 200 m and minimum values of around 2.6 mL/L in the depth range of 350–600 m. Below this depth the oxygen content slowly rises, and in the Gulf basin water the dissolved oxygen level stands at 4.5 mL/L or greater. As noted earlier, the oxygen minimum layer is already present in Caribbean water entering the Gulf, and the core minimum value of this water is 2.6 mL/L. However, Antarctic Intermediate Water entering the Gulf has higher oxygen values, and any organic material raining from above is oxidized very slowly in this cold, deep layer. Hence, this organic material creates little oxygen demand in deeper waters of the Gulf.

An interesting sidelight is that the oxygen minimum layer bathes bottom sediments of the outer edge of the continental shelf and the upper portion of the continental slope. With less oxygen available to oxidize the organic matter, it might be expected that the organic content of the sediments in this depth range would be higher than at other depths. As seen in figure 6.16, this is indeed the case, at least in the northwest Gulf of Mexico where it has been investigated. When dissolved oxygen values drop to 2.0 mL/L and below, the water is said to be *hypoxic* (i.e., deficient in oxygen), and at this level some marine animals find it difficult to obtain enough oxygen to support respiration. Hypoxic conditions occur regularly on the continental shelf of Louisiana and sometimes on the Texas shelf. This phenomenon will be discussed in the next chapter.

## CIRCULATION

The circulation of oceanic waters is influenced by a number of factors. Chief among these are the rotation of the earth, winds, tidal forces, and density fields, the latter being determined primarily by temperature and salinity gradients. In the real world these factors work in combi-

A.

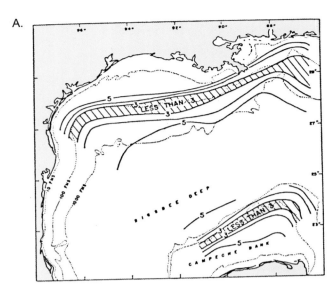

B.

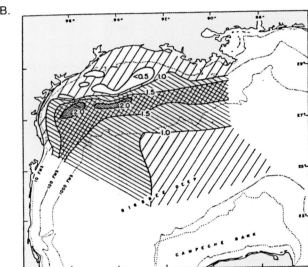

**Figure 6.16.** Bottom waters and surface sediments of the continental shelf and slope of the western Gulf. (a) Dissolved oxygen content (mL/L) of water contiguous with the bottom. (b) Organic content (in percent) of surface sediments. (From Richards 1957.)

nation and their interaction is often very complex, even in the open ocean. In a semienclosed basin such as the Gulf of Mexico, one-third of which is underlain by continental shelves, the nature and depth of the bottom as well as coastal topography become major factors in confining and directing the currents. Here, to achieve a true picture of the actual circulation patterns, theoretical calculations must be modified by observational data (from current meters, satellite observations, ship drift, and surface drifters).

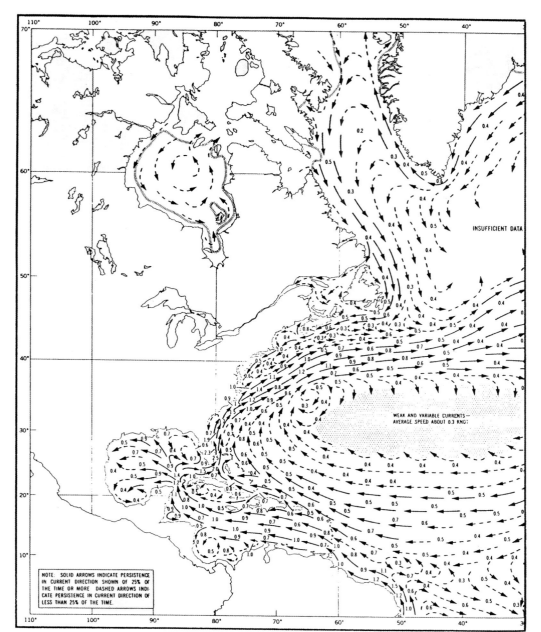

**Figure 6.17.**
Average surface currents in the western North Atlantic and adjacent waters in November. (US Navy Hydrographic Office 1959.)

## Surface circulation

### CIRCULATION IN THE EASTERN GULF

The dominant feature of surface circulation in the Gulf of Mexico is the Loop Current, which reaches a depth of over 500 m. As seen in figure 6.17, surface water entering the Gulf through the Yucatán Channel is a small branch of the much larger, clockwise-rotating North Atlantic gyre. Once in the Gulf this branch flows in a clockwise loop and exits the Gulf through the Straits of Florida. The extent of penetration and the location of the loop are quite vari-able, and large current rings or eddies are known to separate from the Loop Current, often migrating to other sectors of the Gulf. Some of the variations observed in this current are shown in figure 6.18. In an extreme case (fig. 6.18a), the main current shows very little northward extension, and shortly after entering the Gulf it passes right out again, Note, however, the residual clockwise ring to the north, evidence of further penetration at an earlier time. Figure 6.18b gives an example of deeper penetration almost to the northern Gulf continental shelf. The Loop

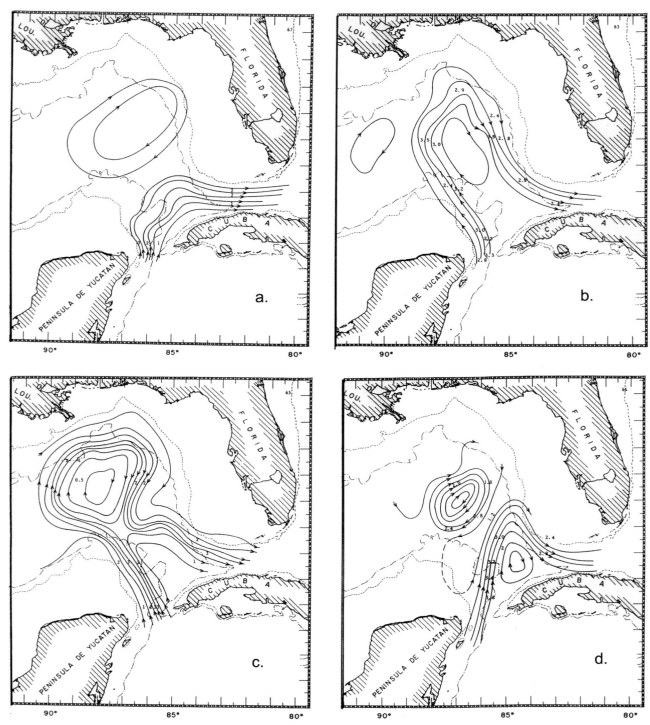

**Figure 6.18.** The eastern Gulf of Mexico showing several variations in the Loop Current and related rings. The Loop Current is shown as a series of streamlines with the greatest current speed in the core of the flow and the least toward the edges. (a) Loop Current with little northward extension. (b) Loop Current with great northward extension. (c) Loop Current with northern portion beginning to pinch off. (d) Loop Current after ring has pinched off. (From Ichiye, Kuo, and Carnes 1973.)

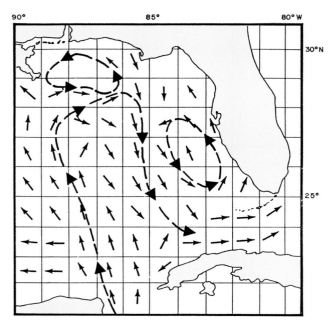

**Figure 6.19.** Generalized picture of circulation in the eastern Gulf showing the Loop Current and counterclockwise eddies on the continental shelves of west-central Florida and the northern Gulf. (From Ichiye, Kuo, and Carnes 1973.)

Current is known to transgress the northern Gulf shelf and has been documented to come within 8 km of the shore between Pensacola and Panama City, Florida. Note the central clockwise ring inside the northern loop of the current and the clockwise ring to the west. Further development (fig. 6.18c) shows the upper portion of the Loop Current with its own circulation that is about to pinch off. The final stage (fig. 6.18d) shows the upper portion rotating clockwise and totally separated from the main loop, which now extends only a short distance into the Gulf. These four diagrams illustrate only a few of the possible configurations of this large and powerful current. Speeds in the center of the Loop Current may range up to 3.5 kts, with lower speeds toward the inner and outer edges.

As shown in figure 6.19, counterclockwise eddies are sometimes present on the continental shelves of west-central Florida and the northern Gulf east of the Mississippi River Delta. Counterrotating shallow-water currents of the eastern Gulf may be induced, in part, by the large off-shelf Loop Current. However, shallow shelf waters around the Gulf out to a depth of 50–60 m are strongly influenced by the wind. Also of importance is that such waters may be highly stratified due to freshwater outflow (on the Mississippi-Alabama shelf) or to summertime heating of the surface layer (on most of the west Florida

shelf). In such cases circulation systems may be set up whereby surface waters are wind-driven off the shelf, with replacement water from deeper layers moving shoreward near the bottom.

CIRCULATION IN THE WESTERN GULF

A permanent feature of circulation in the western Gulf is the strong, persistent westerly flow of water between 21° and 24°N. Approaching the Mexican coast, the water tends to spread and flow northward above 24°N and southward below 21°N, although the alongshelf flow pattern varies seasonally, being strongly influenced by the prevailing winds. In the southern Gulf, below 21°N, there is a permanent counterclockwise eddy in the Bay of Campeche. Along the northeast Mexican coast the northerly flow of the western boundary current is weak in the wintertime and strongest during the warmer months.

Circulation on the continental shelf of the northwestern Gulf has been studied in some detail (fig. 6.20). During the fall, winter, and spring, under the influence of east and southeast winds, shelf currents flow down the coast, that is, west. As shown in the figure, they may set up a counterclockwise eddy, although evidence for the east-flowing southern arm of the eddy is somewhat tenuous. During the cooler months the downcoast current may extend well south of the Rio Grande. In late spring and summer, when the prevailing wind is from the southeast and south, the nearshore current along the Mexican coast pushes northward past the Rio Grande. By July it has reached the upper Texas coast in the vicinity of Galveston, where it flows offshore toward the south. Thereafter, the Mexican current quickly weakens, and by September the winter condition has begun to develop.

Two semipermanent rings are often situated over the continental slope off northeastern Mexico and far south Texas (fig. 6.21). The southernmost, off Mexico, rotates clockwise and is derived from eddies that have spun off the Loop Current and migrated westward to the Mexican coast. The northernmost, off south Texas, rotates in the opposite direction, that is, counterclockwise. Its origin has not been completely established, but it may be induced by the southern ring.

EDDIES AND SMALLER-SCALE FEATURES

As noted earlier, when the Loop Current penetrates far northward into the Gulf, its path becomes unstable and large eddies (diameter ± 400 km) are shed. These generally spin westward or southwestward and ultimately

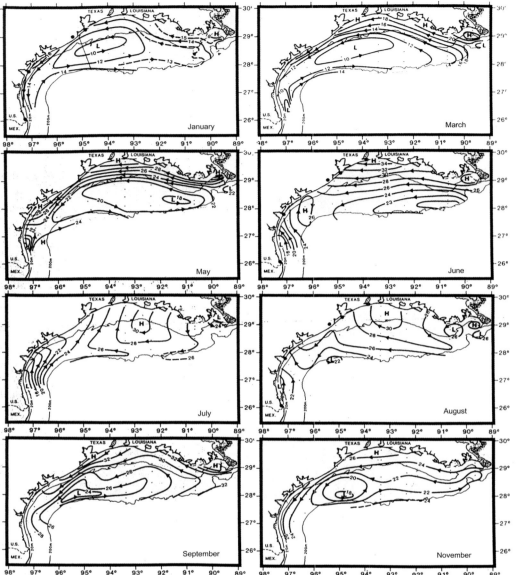

**Figure 6.20.**
General patterns of circulation on the Louisiana-Texas continental shelf during several months of the year. Note the counterclockwise circulation from September through May and the intrusion of a northward-flowing nearshore current from Mexico in June and July. (From Cochrane and Kelly 1986.)

reach the Mexican slope. Then they tend to migrate northward to a point just south of the Rio Grande where they gradually dissipate, largely because of frictional resistance with the bottom. Some of the eddies follow a more northerly course along the continental slope of the northern Gulf, and some head more to the south, eventually reaching the Bay of Campeche, where they gradually spin down. Large eddy detachment occurs as frequently as once every 6 months, although the interval may be as long as 14 months or more. The average is about once every 11 months.

Filaments often develop along the edges of the Loop Current and largest eddies. Filaments from the northern limb of the Loop Current may entrain parcels of Mississippi River water and transport this water out through the Straits of Florida. Smaller eddies (diameter 100–150 km) are often spun off from the Loop Current and from large eddies, and these smaller rings are found throughout the Gulf. They are responsible for the transport of large volumes of surface water from the eastern to the western sectors of the Gulf. Details of their shedding and movements around the Gulf may be followed by satellite-based measurements of the dynamic height of the surface waters.

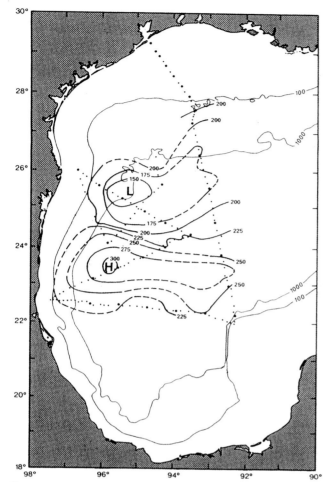

**Figure 6.21.** The western Gulf of Mexico showing depth contours (in meters) of the 15°C isotherm in April. Centers of the clockwise and counterclockwise eddies are denoted by H (high) and L (low), respectively. Note that the 15°C isotherm is 150 m shallower in the center of the counterclockwise eddy than in the clockwise eddy. This is due to upwelling of deep cool water in the former. (From Merrell and Morrison 1981.)

AVERAGE ANNUAL NEAR-SURFACE CIRCULATION

DiMarco, Nowlin, and Reid (2005) analyzed the records of a large series of surface drifters released at various places in the Gulf of Mexico and tracked between 1989 and 1999. The drifters were generally drogued at 50 m below the sea surface, and the data represent the mean current direction and velocity at that depth calculated for 1.5º squares (bins). Clearly depicted are the gross seasonal and overall annual field patterns. Figure 6.22 shows the near-surface direction and velocity field of the Gulf averaged for the entire year. From this figure the following features are noted:

- dominant flows through the Yucatán Channel and the Straits of Florida;
- persistent westward zonal flow across the Gulf between 21° and 24°N;
- definite cyclonic circulation in the Bay of Campeche, south of 21°N;
- a northward-flowing western boundary current off northeastern Mexico between 95° and 97°W and 24° and 26°N;
- mean westward flow on the Louisiana-Texas shelf;
- highly variable flow on the shelves and slopes of the northeastern Gulf;
- mean southward flow on the lower west Florida shelf; and
- a large area of high variability over the deep regions of the central Gulf.

The main driving force for much of the circulation, particularly in the deep water, appears to be the Gulf Loop Current and its associated eddies. Other current features can be directly correlated with seasonal wind driving. These include the westerly current on the Louisiana-Texas shelf, the western boundary current off northeastern Mexico, and the cyclonic circulation in the Bay of Campeche. Further details are given in Nowlin et al. (2005) and Vázquez et al. (2005).

## Upwelling

Upwelling refers to the vertical movement of subsurface water into the surface layer. It may be caused by wind displacing surface water away from the coast or by water currents impinging on one another or on the edge of the continental shelf. Where wind and/or water currents vary, upwelling may be a seasonal phenomenon. Deeper water brought to the surface is characterized by low temperatures and high levels of dissolved nutrients (nitrates, phosphates, and silicates). Warm, moist air passing over cool upwelled water can generate widespread sea fog. Also, the nutrients brought into the upper sunlit zone of the sea stimulate the growth of phytoplankton in areas of upwelling.

Within the Gulf of Mexico there are three main areas of upwelling, all associated with the outer edge of the Loop Current. These include the eastern and northern edges of the Yucatán continental shelf, DeSoto Canyon (off Pensacola, Florida), and the western edge of the continental shelf of peninsular Florida. In the open Gulf upwelling oc-

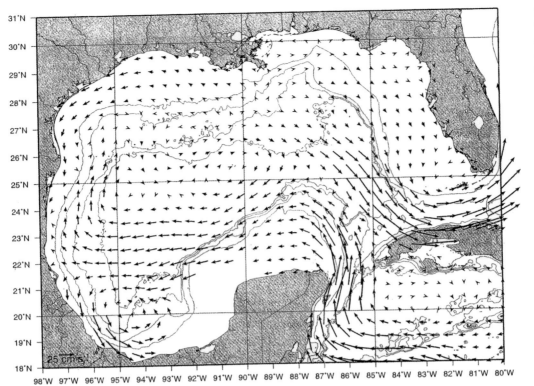

**Figure 6.22.**
The current direction and velocity field of near-surface waters of the Gulf of Mexico averaged over the entire year. (From DiMarco, Nowlin, and Reid 2005.)

curs along the frontal boundaries of large eddies, and in all counterclockwise eddies some upwelling takes place in the center of the ring. Localized upwelling has been reported at the outer edge of the south Texas shelf (Cedar Breaks) and off Veracruz, Mexico. It is likely that other small areas of upwelling exist, at least on a seasonal basis.

## Suspension and transport of continental shelf sediments

In clear waters of the open Gulf there is very little suspended particulate material, either organic or inorganic, and sunlight penetrates 100 m or more. However, near the coast the load of suspended matter is much larger, and light penetration may be greatly reduced. This is particularly true for continental shelves with terrigenous sediments, that is, bottoms composed of sand, silt, and/or clay particles. Except for the carbonate platforms off peninsular Florida and Yucatán, all continental shelves of the Gulf of Mexico are terrigenous.

The relationship between water circulation and sediment suspension has been studied in greatest detail on the continental shelf of the northwestern Gulf. Mississippi River water entering this area contains particles of many sizes, but near its mouth most of the sand and silt drop to the bottom, leaving primarily clay particles in sus-

pension. During the spring and summer a layer of low-density fresher water, cloudy with clay particles, sweeps down along the coast of Louisiana and Texas, and this turbid surface layer may extend from the shoreline a third of the way to the shelf break. During this season the surface mixed layer is very shallow and winds are gentle, so there is little stirring of the bottom sediments. Even tropical storms have only a temporary effect.

However, in the winter the surface mixed layer is deeper, and cold fronts are often accompanied by high wind speeds. During this season the soft bottoms may be stirred to water depths of 60 m or more, and resuspended sediments rise into the water column, creating an inky layer that may rise 10 m or more from the bottom. As cold fronts chill the surface waters they become denser, sink to the bottom, and flow out across the shelf into the deeper Gulf, transporting a load of resuspended sediments to the upper slope.

These processes are reflected in cross-sectional diagrams of continental shelf and upper slope waters running south from Galveston Island, Texas, in March and May (fig. 6.23). Included are lines of equal density (sigma-t), and the most turbid waters are shaded. In the March transect, reflective of winter conditions, the turbid layer extends to the surface only a short distance out

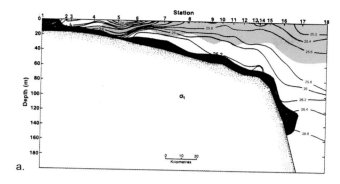

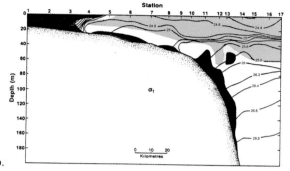

**Figure 6.23.** Cross-sectional diagrams of water over the continental shelf and upper slope running south from Galveston Island, Texas, in (a) March and (b) May. Lines of equal density (sigma-t) are shown. The most turbid waters are heavily shaded, and the least turbid waters are lightly shaded. (From Rezak, Bright, and McGrail 1983.)

from the shoreline, but a continuous layer of turbid bottom water, 5–10 m thick, extends across the shelf onto the upper slope, largely in water with a density of 26.2 or less. This, in effect, represents sediment transport from shallow to deeper water. In May, representing the onset of summer conditions, as Mississippi River water flows westward along the inner shelf it carries a load of fine clay particles, and at this time the surface turbid layer extends from shore at least a third of the way across the shelf. Meanwhile, the bottom turbid layer is discontinuous. Toward the outer shelf and upper slope, where it is probably stirred by oscillations and deepwater currents, the layer thickens. The surface and bottom turbid layers are often referred to as "nepheloid" layers (from Latin *nephelus,* meaning "cloud").

Continental shelf waters east of the Mississippi River Delta are also generally turbid, and this is probably a result of wind stirring as well as strong currents associated with the upper limb of the Loop Current. By contrast, the carbonate shelves off Florida and Yucatán, although well stirred, lack fine particulate sediments and are characterized by very clear water. Storms may temporarily raise

coarser particles, but these quickly drop out of suspension after the storm passes. Off south Florida attached green algae are known to thrive at depths of at least 100 m.

**Subsurface circulation**

Circulation in subsurface waters of the Gulf of Mexico is surprisingly complex and only partially understood. However, a number of points may be made with some confidence. In the Straits of Florida surface water moves from the Gulf into the Atlantic Ocean, but there is a bottom countercurrent that brings Atlantic water into the Gulf. Current speeds have been estimated at 2–4 kts. In the Yucatán Channel the Loop Current flows northward along the western side of the channel and includes water to a depth of about 800 m. Here also there is a countercurrent. Water along the western side below 1,000 m and at all depths along the eastern side of the channel flow southward from the Gulf into the Caribbean at an average speed of about 0.07 kts. Along the continental slope off Louisiana and Texas both clockwise and counterclockwise eddies stir the water column and sediments to depths of 800 m, and these long-lived eddies may move back and forth along the slope. Basically the same phenomenon occurs throughout open waters of the western Gulf and along the Mexican slope.

Circulation in deeper waters of the Gulf (between 1,500 and 3,000 m) appears to be dominated by a large clockwise gyre that fills the entire deep basin. Superimposed on this system are large, long-period horizontal waves that appear to be generated by pulsations of the Loop Current near the west Florida shelf. With wave lengths of 150–200 km, these waves progress westward at an average velocity of around 9 km/day, and as they move into the western Gulf they become progressively more independent of the surface eddies, which also propagate toward the west. The energy of these deepwater waves tends to dissipate along the continental slopes of the western Gulf basin. Finally, an interesting bottom current has been described from the eastern Gulf beneath the Loop Current at depths of over 3,000 m. First identified in photographs of ripple marks on the bottom, this current has been measured to have speeds of 0.11–0.37 kts. Most current directions were toward the east, but some were toward the south and southwest. Clearly, waters of the Gulf are restless at all depths, with the greatest current velocities in the east and with energy propagating from there into the western basin.

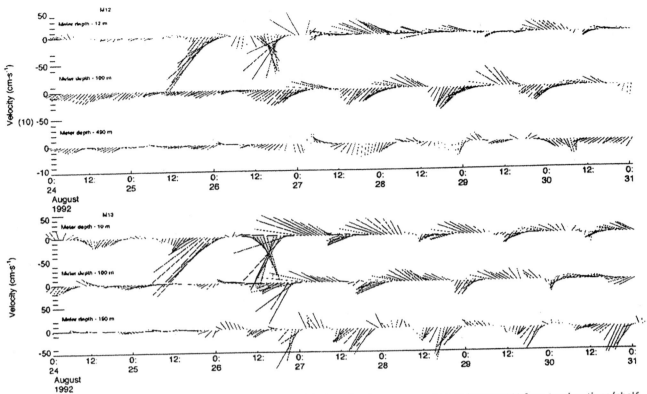

**Figure 6.24.** Hourly values for horizontal current vectors recorded during a hurricane passage in late August 1992 from two locations (shelf edge and upper slope) off central Louisiana. Across-shelf components are oriented up and down (up being toward the shore), and alongshelf components are oriented horizontally (eastward being toward the right). Moorings 13 (bottom) and 12 (top) were located in water depths of 200 and 504 m, respectively, and current meter depths are indicated. The eye of Hurricane Andrew passed on a northeastward track approximately 85 km north of mooring 13 at about 0000 hours (local time) on 26 August and induced water current velocities (at 10 m) at mooring 13 of 163 cm/sec (3.7 mi/hr). (From Nowlin et al. 2000.)

## THE RESTLESS GULF

So far in this chapter we have reviewed what might be thought of as the "classical" physics of the Gulf of Mexico, based largely on studies from shipboard and a few moored instruments. However, during the past dozen years or so data from more extensive studies, especially those involving satellites and mathematical modeling, have painted an exciting new picture. Waters of the Gulf are in constant motion and are far more active than previously realized. This applies not only to currents at the surface but also to the movement of water at greater depths. The energy for this motion derives from two primary sources, wind-induced water movements (especially energetic episodic events such as winter cold fronts and summer storms and hurricanes) and the Gulf Loop Current (together with its spin-off clockwise eddies and induced counterclockwise rings). Much of this new information is summarized in Nowlin et al. (2000).

Easterly winds blowing across the Gulf ultimately encounter the Sierra Madre of eastern Mexico. Those passing across the southern Gulf are deflected toward the south, and recent studies suggest that this wind curl is responsible for setting up and maintaining the semipermanent counterclockwise eddy in the Bay of Campeche (fig. 6.22). In the northern Gulf, data from an array of moored instruments have provided information on the effects of the passage of a major hurricane (Andrew) on current directions and velocities at several water depths. Figure 6.24 shows the hourly directions and velocities at several depths for two moorings off central Louisiana. Just after passage of the eye, near-surface instruments recorded a strong surge of water directed to the left of the storm's track with water velocities of 163 cm/sec (3.7 mi/hr) at the 10 m depth. Following this initial surge an oscillation was set up in the water column that penetrated, with diminished force, to the deepest instrument on the moor-

**Table 6.3.** *General characteristics of rings or eddies spun off from the Gulf Loop Current. (Modified from Nowlin et al. 2000.)*

Shedding frequency – Intervals range from 4 to 16 mo; mean is about 10 mo.

Horizontal migration rate – Range = 1–14 km/day; average = 5 km/day.

Diameter when first closed – Initially ≥ 250 km; average = ca. 350 km. Decrease by 45 percent within 150 days and 70 percent within 300 days.

Swirl speeds – For newer rings: at the surface the speed is 150–250 cm/sec (2.7–5.7 mi/hr); at a depth of 500 m it is 10 cm/sec.

Depths – For new rings rotary motions extend below 1,000 m. Filament motions confined to upper 50–300 m.

Lifetimes – 300–400 days.

Decay modes – Interactions with boundaries (surrounding water or bottom areas), spin-off of filaments, ring shedding, ring-ring interactions.

---

ing (190 m), arriving approximately 24 hours later. Here the maximum speed still exceeded 100 cm/sec. A coherent but weak response was also recorded at mooring 12 (several miles to the south), which occurred at all recorded depths, even down to 490 m. Inertial oscillations continued for about a week with diminishing amplitudes.

Forcing by the Gulf Loop Current is also an important factor, and much energy is transferred through the eddies that spin off toward the west and southwest. A summary of our present knowledge of ring characteristics has been provided by Nowlin et al. (2000) (table 6.3). Insight into ring dynamics in surface waters of the Gulf comes from a report by Biggs, Fargion, et al. (1996). As seen in figure 6.25, during the April–May period of 1992 (panel a) the clockwise rotating ring, T, is being cleaved by a counterclockwise eddy. A month later the separation is complete, with the larger fragment remaining in place and the smaller fragment, t, having migrated southward to about latitude 22°N. By the June–July period the counterclockwise eddy has weakened, and the smaller fragment, t, is moving northward. By July–September it has moved north to 24° and is beginning to merge again with T, and by the following period the smaller fragment has dis-

appeared as a recognizable entity. Meanwhile, another clockwise eddy has been shed by the Loop Current and is moving in a westerly direction. It first appears in the May–June picture (panel b), and by July–September (panel d) it is fully developed as eddy U. This eddy elongates, and by October–November (panel f) it too has been cleaved into two distinct fragments, eddies U and V. Using satellite data, for the first time we can obtain an instantaneous view of the entire Gulf, and repeated imagery presents, in effect, a motion picture of the surface properties and dynamics of this restless body of water.

Attention is also being focused on movements of subsurface waters, and several classes of deepwater currents have been identified. As mentioned earlier, the Loop Current and its eddies have effects on water motion that extend to depths of several hundred meters. Below this level (from about 800 m to the bottom of the Gulf), the waters appear to be set in motion by oscillations of the Loop Current, which generate waves (Rossby waves) that spread westward at about 9 km/day and expend their energy on the continental slopes of the western Gulf. In some cases, at least, the associated currents tend to intensify near the bottom. Subsurface high-speed currents (jets) have been observed at several locations on the continental slope of the northern Gulf in water depths of 1,500 m or less, and they generally occur within the depth range of 100–300 m. Maximum speeds of 150 cm/sec have been reported, although causal mechanisms have not yet been identified.

Along the base of escarpments in the northern Gulf south and west of the Mississippi River Delta in the depth range of 2,000–3,000 m, there have recently been discovered rows of parallel furrows extending for tens of kilometers or more. These are oriented along depth contours. The furrows have depths of 5–10 m, widths of several tens of meters, and they are spaced about 100 m apart. Bottom currents responsible for these furrows have not been identified for certain, but they appear to be ultimately generated by the Loop Current and steered along the face of the Sigsbee Escarpment.

## NUMERICAL MODELING

Since about 1980 efforts have been underway to simulate patterns of water circulation in the Gulf of Mexico by means of mathematical models. These consist of sets of differential equations that take account of the various forces acting on the water masses, as discussed throughout the present chapter. Of particular importance are the

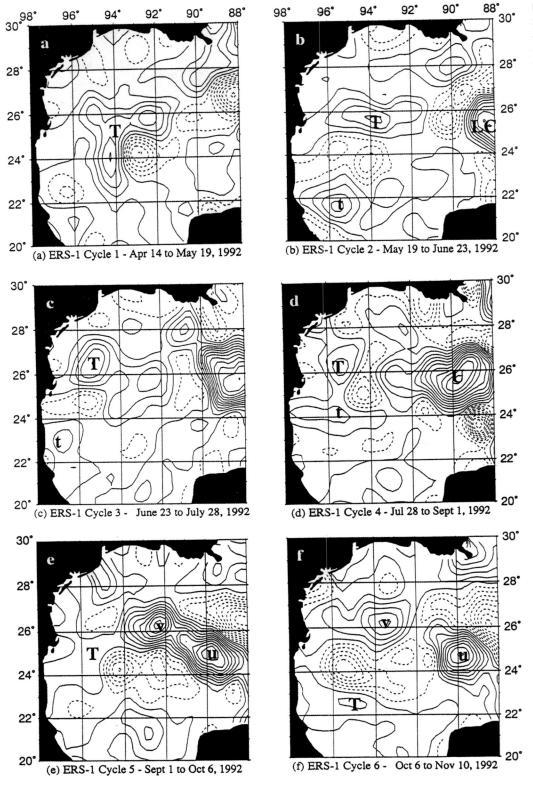

(a) ERS-1 Cycle 1 - Apr 14 to May 19, 1992

(b) ERS-1 Cycle 2 - May 19 to June 23, 1992

(c) ERS-1 Cycle 3 - June 23 to July 28, 1992

(d) ERS-1 Cycle 4 - Jul 28 to Sept 1, 1992

(e) ERS-1 Cycle 5 - Sept 1 to Oct 6, 1992

(f) ERS-1 Cycle 6 - Oct 6 to Nov 10, 1992

**Figure 6.25.**
Successive patterns of surface water circulation in the western Gulf of Mexico at approximately monthly intervals (with a gap in August) from mid-April to mid-November 1992, based on (ERS) satellite data. Clockwise rings are shown in solid lines, counterclockwise circulation by dashed lines. The major fragment of ring "Eddy Triton" is denoted by "T" and the minor fragment by "t." Eddies U and V are denoted by "u" and "v." (From Biggs, Fargion, et al. 1996.)

flow of water into the Gulf through the Yucatán Cannel, wind stress, and water density (as determined by local temperature and salinity conditions). Since the models are three-dimensional, they must also take into consideration the size and shape of the Gulf basin as well as bottom topographic features that constrain the water flow. During the past decade, with data input from satellite imagery of the sea surface and from major physical oceanographic studies of the water column itself, and with the enhancement of such data by powerful computers, these models have achieved considerable sophistication. They are now capable of making fairly accurate forecasts of surface and subsurface circulation patterns of water within the basin. Such information is useful to mariners, and it is critical to engineers who must design oil drilling and recovery structures able to withstand predicted water forces even in the very deep environments of the Gulf. For reviews of some of the recent modeling efforts, the reader is referred to Nowlin et al. (2000) and Sturges and Lugo-Fernandez (2005).

# 7 : CHEMICAL OCEANOGRAPHY

In its broadest sense, chemical oceanography deals with the types of chemicals present in seawater and bottom sediments as well as in the tissues of living marine organisms. It is also concerned with where these chemicals come from and the transformations they undergo while in the marine system. Considering the great variety of natural and human-made chemicals now in the sea, by its very nature the subject is quite complex and involved. Literature concerning the chemistry of the Gulf is uneven and heavily weighted toward the chemistry of hydrocarbons, particularly petroleum hydrocarbons. Trace metals and other potential chemical pollutants have received fair treatment, as have nutrient elements. Other areas, although not completely neglected, have received far less attention. Moreover, chemical studies have addressed primarily the continental shelves of the northern Gulf, areas most affected by Mississippi River effluents and by petroleum exploration and development. Other continental shelves and the open Gulf have not received similar treatment. Despite the complex nature of the subject and the literature bias, the present chapter attempts to provide a balanced overview of the chemistry of the Gulf. For the most part, the discussion will treat chemicals as they exist in the water column, but in the case of the hydrocarbons, trace metals, and a few other forms, information is also provided on concentrations in bottom sediments and in tissues of living organisms. Important literature summaries include the earlier reports by El-Sayed (1972), Fredericks (1972), Jeffrey (1972), and Sackett (1972) and the more recent reports of Vittor and Associates (1985), Kennicutt (1991, 2000), Kennicutt, Brooks, et al. (1988), Presley (1991), and Presley and Brooks (1988). Chemistry of the west Florida shelf is summarized in a report by Environmental Science and Engineering, Inc., LGL Ecological Research Associates, Inc., and Continental Shelf Associates, Inc. (1987).

## GENERAL BACKGROUND

The basic chemical composition of water in the Gulf of Mexico is established by the characteristics of water flowing in from the Caribbean Sea. The Gulf also receives chemical inputs from several other sources including streams, the atmosphere, submarine seeps, and human marine activities. From its vast drainage basin, streams bring large volumes of freshwater to the Gulf. This contains dissolved and particulate material derived from natural weathering of rocks and soils as well as a host of other chemicals resulting from human activities (i.e., from agricultural, municipal, and industrial sources). From the atmosphere come particles of dust, pesticides, aerosols, and other substances. Into bottom waters of the Gulf salt and hydrocarbons seep from subsurface deposits. Human marine activities such as shipping and oil and gas exploration and development add additional chemicals, particularly petroleum hydrocarbons and some metals. Once in the Gulf the chemicals may react with one another in various ways, and they are subject to a number of physical processes such as adsorption to clay particles, mixing, transport, sedimentation, and upwelling. In addition, biological activity involves the uptake of nutrients, the uptake and release of gases such as oxygen and carbon dioxide, the release of various metabolites (through secretion and excretion), the production and decomposition of particulate organic materials, and the precipitation and sedimentation of silicates and carbonates.

In dealing with a specific chemical or group of chemicals, one must always consider three aspects of its Gulf adventures: where it comes from, what changes it undergoes, and where it winds up (i.e., its sources, transfer processes, and ultimate sinks). One must also take into consideration the states in which it might exist (organic or inorganic, inanimate or biological, dissolved or suspended, or sedimented particulate). Many chemicals may exist in more than one of these states, and some are transferred back and forth from one to another. In practice, particulate materials are defined as those substances retained by filters with pore diameters of 0.5 micron, whereas dissolved materials pass through such filters. Most anthropogenic chemicals that we think of as pollutants are brought into the Gulf by the Mississippi River and are derived from marine ship traffic or oil and gas development activities, but some enter from the atmosphere and other sources.

## SUSPENDED MATERIAL

Suspended material includes both the inorganic and organic particulate material found in seawater. Its presence interferes with the transmission of light and sound, provides a substrate for bacterial attachment, and by adsorption and settling provides a mechanism for the removal of various inorganic and organic materials from the water column. Most of the suspended inorganic particles are fine clays introduced to the Gulf by river outflow, the Loop Current, and to some extent by wind transport. Biologically derived materials include pollen grains, skeletal remains of diatoms and other algae, fragments of dinoflagellates and foraminiferans, spicules of unknown origin, woody fibers of land plants, and tiny spherical bacteria.

As seen in table S6, the levels of total suspended material and of the inorganic and organic components are several times greater over the continental shelf than in the open Gulf. This is because of the proximity of river outflows, resuspension of bottom sediments, and higher levels of plankton growth in shelf waters. Offshore the level of particulate organic matter in surface waters exceeds that of deeper levels because of higher plankton production in the euphotic zone. The level of particulate inorganic material is about the same at all depths. Particulate inorganic material makes up a little less than half of the suspended matter over the shelf and in surface waters of the open Gulf, but in deeper waters it accounts for about two-thirds of the total suspended matter. The clay fraction includes a variety of mineral types such as chlorite, feldspar, illite, kaolinite, mica, and montmorillonite, which contain aluminum silicates, often hydrated, and various amounts of calcium, iron, magnesium, potassium, and other elements. Also included are fine particles of quartz and other minerals.

The distributions and concentrations of clay minerals provide insight into the sedimentary history of the Gulf. The Mississippi, Mobile, and Apalachicola Rivers supply most of the clay minerals entering the northern Gulf, and montmorillonite makes up about 60 percent of this load. Montmorillonite is also a major constituent of sediments of the continental shelf of the northwest Gulf as well as sediments off the Mississippi River Delta, on the Mississippi Cone, and throughout the deep basin of the western Gulf. However, this mineral is in very low concentration in open waters of the western Gulf. Apparently, montmorillonite clays brought in from the north are flocculated and deposited near the river mouth, and they enter the deep Gulf through sediment slumping and are widely transported by bottom turbidity currents. Waters entering from the Caribbean have low concentrations of montmorillonite and high concentrations of mica and illite, which also characterizes waters in the Straits of Florida exiting the Gulf. Waters of the western Gulf also show high concentrations of mica and illite, but the minor constituents are different from those of the Loop Current, indicating the relatively distinct nature of western Gulf waters.

### GENERAL INORGANIC CHEMISTRY

This section treats the inorganic chemicals dissolved in Gulf waters. Included are the major ions, minor elements, nutrients, and isotopes.

### Major ionic constituents

Concentrations of the most abundant cations and anions of the Gulf and in Mississippi River water are given in table S7. The four cations and three anions together are present in a little over 36,000 parts per million (ppm), or 36 parts per thousand (ppt), and account for most of the salinity of seawater. In this respect waters of the open Gulf differ little from oceanic waters around the world. By contrast, Mississippi River water contains only about 210 ppm or 0.210 ppt of dissolved ions. Freshwater in general contains less than 1.0 ppt of dissolved salts. Whereas the ionic concentration of the open Gulf runs close to 36 ppt, it may run a little higher in the western Gulf and on the Yucatán shelf where evaporation is high, and it is often lower on continental shelves where seawater is diluted by river runoff. The seven ions listed account for over 99 percent of all the salts in Gulf water. All the remaining ions together constitute less than 1 percent of the dissolved inorganic load of seawater.

### Minor elements

Minor elements (or trace elements, as they are sometimes called) are present in seawater in parts per million, that is, in micrograms rather than milligrams per liter. Values of some of the common minor elements reported from the Gulf of Mexico are given in table S8. Because these elements occur in such low concentrations, their levels are very difficult to determine, and the early measurements shown in the table are somewhat suspect and should be reevaluated. Nevertheless, they do serve to illustrate general trends. None of the values exceeds 4.0 µg/L. In most cases the dissolved form is more abundant than the particulate form. Concentrations in continental shelf waters exceed those of the open Gulf, and in the open Gulf surface values are higher than those of deeper water. Note that there is some individuality in the behavior of each chemical species. For example, manganese does not behave exactly like zinc or copper. The primary source of these minor elements is the rivers that flow into the Gulf, but additional quantities result from human activities such as shipping and petroleum development. Hence, there are regional differences in local concentrations of some elements.

## Nutrient elements

In order to meet their metabolic needs, all phytoplankton organisms require the elements carbon, hydrogen, nitrogen, and phosphorus in some quantity and a variety of trace elements in minute amounts. In addition, diatoms, silicoflagellates, and some other forms require silicon for construction of their skeletal material. Carbon, hydrogen, and oxygen are present in most marine waters in quantities well in excess of phytoplankton needs, but nitrogen, phosphorus, and silicon are in smaller supply, and the scarcity of one or another of these elements may become a factor limiting phytoplankton growth. Hence, these three chemicals are referred to as the nutrient elements.

Concentrations of nutrient elements in seawater are sometimes expressed in terms of weight, that is, as micrograms per liter (µg/L) of water. However, it is of interest to understand the actual ratio of the atoms of each nutrient to the others. Chemical analysis of plankton reveals a rather constant ratio of 40 g of carbon to 7 g of nitrogen to 1 g of phosphorus, and this ratio often holds for surface waters of the sea, as well. Expressed in terms of atoms, this ratio becomes 106 atoms of carbon to 16 of nitrogen to 1 of phosphorus. The conversion is easy, since one simply multiplies the atomic ratio number of each element by its atomic weight to obtain the concentration in terms of actual weight. Thus, the ratio number of carbon is multiplied by 12, that of nitrogen by 14, and that of phosphorus by 32. Hence, (106 × 12): (16 × 14): (1 × 32) gives 1272: 224: 32, which simplified becomes 40: 7: 1. In much of the technical literature the nutrient elements are given as microgram-atoms per liter (µg-at/L), which expresses concentrations in terms of atoms rather than weight.

Nutrients enter the Gulf through river runoff, the Yucatán Current, and, to a lesser extent, the atmosphere. In the euphotic zone, nutrients are depleted largely by phytoplankton uptake. Hence, they are locally and sea-sonally variable. They are regenerated through decomposition in the surface waters, oxygen minimum layer, and bottom sediments. Various physical factors transport nutrients from the continental shelf to the open Gulf and from deeper layers into the surface waters. The distribution and dynamics of each of the nutrient types will be examined below.

### NITROGEN

A simplified diagram of the nitrogen cycle in the sea is presented in figure 7.1. Plankton and other organisms release both organic and inorganic nitrogen compounds that become dissolved in seawater, and the inorganic ammonium becomes progressively oxidized to nitrite and finally nitrate. All these forms of dissolved nitrogen compounds may be taken up by phytoplankton, so within the euphotic zone dissolved nitrogen compounds do not remain long in the water column except for some relatively unavailable (refractory) types. It has been estimated that over two-thirds of the new nitrogen added to surface waters of the Gulf comes from exchange between the Loop Current and open Gulf waters. Nearly a quarter derives from upwelling and upward diffusion from deeper waters, and somewhat less than 10 percent is contributed by outflow from the Mississippi-Atchafalaya system. The total nitrogen output of the Mississippi has greatly increased since the turn of the century and especially since the early 1930s, when application of nitrogenous fertilizers became widespread in the Great Plains and Ohio River Valley. The river's contribution now exceeds 32 million tons per year, with the highest concentrations during peak discharge in the spring months and the least during the fall period of low flow.

In surface waters of the open Gulf, nitrate nitrogen values are quite low, ranging between 0.05 and 2.20 µg-at/L and averaging about 0.29 µg-at/L. Waters of the continental shelves average around 0.22 µg-at/L of nitrate nitrogen. In cyclonic or counterclockwise rings, nutrients

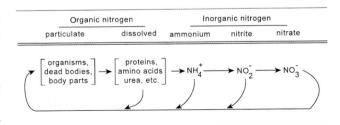

**Figure 7.1.** Generalized diagram of the nitrogen cycle in the sea.

are pumped upward from deeper water, and at a depth of 100 m nitrate nitrogen may run around 10 μg-at/L, whereas in the center of anticyclonic or clockwise rings, the corresponding value may be only 0.2 μg-at/L. It has been estimated that the depth-integrated nitrate nitrogen value in the open Gulf averages around 832 mg-at/m². This represents the total amount of nitrate nitrogen in a column of water 200 m deep underlying one square meter of ocean surface. A typical profile of the vertical distribution of nitrate nitrogen in the upper 2,000 m of the water column in the open Gulf is shown in figure 7.2. Here it is seen that in the surface mixed layer (100–150 m), nitrate is quite low. Between 200 and 800 m there is a considerable increase in the nitrate level, and this depth range corresponds roughly with the oxygen minimum layer, where remineralization of the sinking plankton is taking place. Below 800 m the concentration decreases somewhat and finally levels off below about 1,100 to 1,200 m. Note the correspondence of nitrogen concentrations with vertical water mass distribution described in the previous chapter (fig. 6.3). Detailed studies around the mouth of the Mississippi River point to the importance of dissolved organic nitrogen. Concentrations are quite high near the outflow and westward down the Louisiana-Texas continental shelf. Here values range from 12 to 50 μg-at/L. These high levels of organic nitrogen reflect intense phytoplankton activity in shelf waters downstream from the Mississippi River Delta.

PHOSPHORUS

As in the case of nitrogen, both inorganic and organic forms of phosphorus are present in seawater. Furthermore, the distribution of phosphorus in the Gulf generally parallels that of nitrogen and for the same reasons. Phytoplankton reduces the level of phosphorus in surface waters, but it is reintroduced through recycling and various mixing and upwelling processes. In open waters of the Gulf phosphate values range from near 0 to 2.66 μg-at/L, with few values above 1.0 μg-at/L. They average around 0.26 μg-at/L and range somewhat higher in the central and western sectors than they do in the eastern Gulf. Waters of the continental shelf generally average about 0.27 μg-at/L. As seen in figure 7.2, depth-integrated phosphate values (0–200 m) for the open Gulf average around 65.36 mg-at/m². Seasonally they range from 50 to 78 mg-at/m², with higher values in the fall. As indicated in figure 7.2, in the open Gulf the vertical profile for phosphate is generally similar to that for nitrate. Annually the Missis-

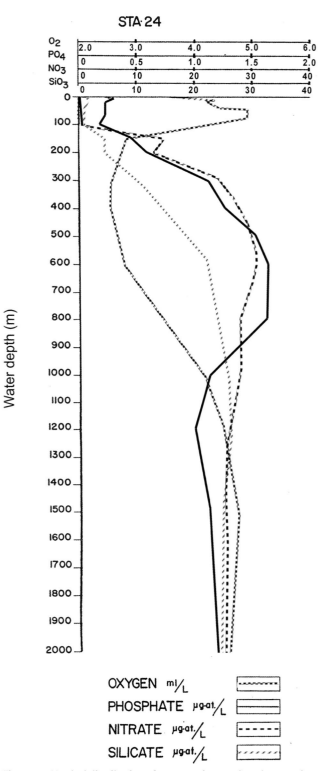

Figure 7.2. Vertical distribution of oxygen, nitrate, phosphate, and silicate at a summer station in the west-central Gulf of Mexico. (From El-Sayed et al. 1972.)

sippi River brings about 7.6 million tons of phosphorus to the Gulf, and much of this reflects fertilizer use in the upstream drainage area. Near the river's mouth phosphorus concentrations are about eight times higher than in the open Gulf. High phosphorus values persist in nearshore waters of Louisiana and upper Texas as the effluent flows down the coast from the Delta.

SILICON

The element silicon is present in seawater only in inorganic forms, either as silicon dioxide or as silicates. The distribution of this element generally parallels that of the other nutrients, but local concentrations are more variable since silicon is required by only a portion of the phytoplankton. Surface values in the open Gulf range from 0.50 to 20.0 µg-at/L, with an overall average of 3.87 µg-at/L. Seasonally, surface values are low during the spring and summer months and quite high in the winter (fig. 7.3). Concentrations on the continental shelf average about 5.47 µg-at/L, somewhat higher than in the open Gulf. Integrated values (0–200 m) average around 427 mg-at/m² and are lowest in the spring and summer and highest in the fall. The pattern of the vertical distribution of silicon in the open Gulf is similar to that of the other nutrient elements. Silicon values in deep waters of the Gulf run somewhat higher than at corresponding depths of the northwest Caribbean, apparently because of more rapid regeneration in the deep Gulf.

The silicon concentration in the lower Mississippi River averages 108.2 µg-at/L, which is much higher than values characteristic of the open Gulf. Annually, the Mississippi River contributes about 24.2 million tons of silicon to the Gulf, somewhat less than the contribution of nitrogen. Once in the Gulf, silicon is rapidly picked up by the local diatom population. Upon death, the diatoms fall to the bottom, where they deposit their siliceous frustules. Since silicon is regenerated very slowly, downcoast diatom populations may be silicon-limited.

## Isotopes

A given chemical element is made up of atoms, all having the same number of protons and electrons and approximately the same number of neutrons. Isotopes of the element have different numbers of neutrons and, thus, slightly different atomic weights. Two classes of isotopes are recognized, stable (nonradioactive) and unstable (radioactive). Stable isotopes remain unchanged through time, but unstable isotopes break down to stable forms, in

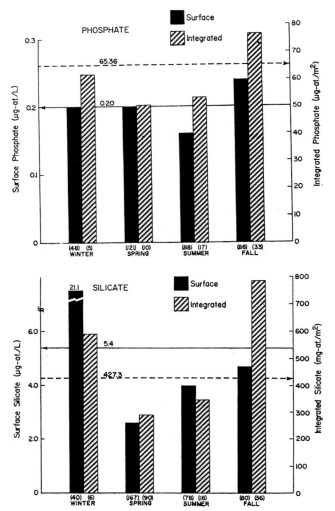

**Figure 7.3.** Seasonal variation in the concentration of phosphate and silicate in the euphotic zone of the Gulf of Mexico. Solid bars represent surface values, and shaded bars represent depth-integrated values (0–200 m). Solid horizontal lines are annual averages for surface values, and dashed horizontal lines are annual averages for integrated values. Numbers in parentheses along the bottom of the graphs indicate the numbers of observations on which the averages are based. (From El-Sayed et al. 1972.)

the process releasing small particles or radiation that can be measured. Since the rate of disintegration of a given isotope is absolutely constant over time, it may sometimes be used to measure the rates of current processes or the elapsed time since past events. Isotopes are formed naturally by thermonuclear processes in the earth's interior and by cosmic ray bombardment of atoms in the earth's atmosphere. Those formed within the earth reach the surface by volcanic output as gases, ashfalls, and lava flows. From terrestrial deposits they may reach the sea through erosion and stream runoff or wind transport of dust particles. Those formed in the atmosphere enter the sea largely by gas equilibration at the air/sea interface. During the past half century, both stable and radioactive isotopes have been added to the environment in quantity through human activities including atomic weapons production and testing, nuclear power plant operations, and reprocessing of nuclear fuel, as well as through nuclear accidents such as the Chernobyl meltdown in Russia in the mid-1980s. By studying concentrations of certain isotopes and comparing isotope ratios, oceanographers have been able to date water masses, examine physical and chemical processes and their rates of occurrence, and assess impacts of human activities on marine ecosystems. Examples of relevant studies in the Gulf of Mexico are provided below.

### STABLE CARBON ($^{13}$C)

Most elemental carbon has an atomic weight of 12, but other isotopes exist, and two are of particular interest: $^{13}$carbon, which is stable, and $^{14}$carbon, which is radioactive. Each is present in the atmosphere in relatively constant percentages, and each enters the Gulf through river runoff and through carbon dioxide equilibration with the atmosphere. In the process of photosynthesis, phytoplankton organisms in the euphotic zone take up carbon atoms from carbon dioxide and bicarbonate. Although both $^{12}$C and $^{13}$C are included, the organisms preferentially remove $^{12}$C, thereby enriching the remaining carbon pool in $^{13}$C. As the phytoplankton and their predators die and sink out of the euphotic zone, decomposition releases much of the carbon in the oxygen minimum layer, enriching these waters with $^{12}$C. Thus, a vertical profile of open Gulf waters shows the following pattern for the ratio of $^{12}$C to $^{13}$C. At the very surface the ratio reflects that of the atmosphere. By the bottom of the euphotic zone the percentage of $^{13}$C has increased and is about 1 percent greater than in the very deep Gulf. Below the euphotic zone the percentage of $^{12}$C increases in the depth range of 100–1,000 m. Below this depth the percentage of $^{12}$C decreases somewhat, since different water masses occupy the deeper Gulf. Many terrestrial and freshwater vegetation species preferentially take up $^{12}$C, and this material and its decompositional derivatives exhibit characteristic $^{12}$C : $^{13}$C signatures distinct from that of marine phytoplankton. Such signatures can often be detected in bays and estuaries as well as on the nearby continental shelf such as off Mobile Bay. This ability to distinguish carbon sources permits determination of the importance of land- and freshwater-derived organic matter in supporting nearshore marine food chains.

### RADIOACTIVE CARBON ($^{14}$C)

The isotope $^{14}$C is produced in the earth's atmosphere by cosmic ray bombardment of nitrogen atoms. Since this process has been occurring at a rather steady rate throughout the recent past, the concentration of $^{14}$C in carbon dioxide of the lower atmosphere has remained relatively uniform around the world prior to the nuclear age. This isotope enters the Gulf through atmospheric equilibration and through river runoff. $^{14}$C disintegrates to the stable form $^{12}$C at a steady and known rate. Thus, the ratio of $^{14}$C to $^{12}$C provides a yardstick for measuring the elapsed time since its formation. Using this measure, it has been estimated that deep waters of the western Gulf have a residence time of around 270 years. The isotope $^{14}$C is also produced in thermonuclear detonations, and both the atmosphere and surface waters of the Gulf have become greatly enriched in $^{14}$C beginning in the early 1960s, following a period of atmospheric nuclear weapons testing. Early in the 1960s, when activity in surface waters was very high, low $^{14}$C values were reported for surface water along the edge of the continental shelf northeast of Yucatán, demonstrating the upwelling of deep water not yet contaminated by weapons-derived $^{14}$C. Thus, studies of the ratio of $^{14}$C to $^{12}$C in various water masses can provide information concerning residence times and water mass mixing rates.

### THORIUM ISOTOPES

Thorium isotopes enter the Gulf primarily through river runoff and, to some extent, by wind transport, and they have been used to study particle dynamics in the northern Gulf. Surface water concentrations of both dissolved and particulate thorium decrease from near shore to offshore. Particulate material scavenges thorium from the water column. Colloidal organic material, especially diatom-derived photopigments, may also play a significant role in scavenging. Benthic resuspension and lateral transport are major processes in removing this material from the continental shelf to deeper water. In the open Gulf both dissolved and particulate thorium concentrations are elevated in surface waters. They decrease to a minimum in the oxygen minimum layer and thereafter increase with depth. An increase in particulate thorium in near-bottom waters suggests scavenging in the bottom nepheloid layer. The residence time of $^{230}$Th in the

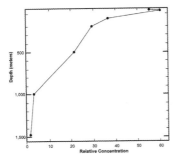

**Figure 7.4.** Concentration of $^{129}$Iodine in surface sediments at various depths relative to its concentration at a depth of 1,506 m. The data are from two stations over the continental slope south of Galveston, Texas. (Based on data from Schink et al. 1995.)

water column over the northern Gulf continental slope has been estimated to be about five years.

### $^{129}$IODINE

Radioactive iodine is produced by cosmic ray bombardment of the atmosphere and by thermonuclear processes within the earth. In recent years, it has been released in greater quantities through nuclear weaponry and power generation, and this recent increase has led to its use as a tracer for ocean mixing processes. Data on the vertical distribution of $^{129}$I in the water column for a transect across the continental slope of the Gulf of Mexico south of Galveston, Texas, are presented in figure 7.4. The value at a depth of 1,506 m (near the bottom) is considered to represent the normal oceanic value due to natural causes, and much of the enrichment in shallower waters is regarded as contamination by anthropogenic $^{129}$I. Near-surface concentrations are around 60 times the prenuclear value, and they suggest that airborne contamination may be significant. However, new $^{129}$I is clearly found down to at least 500 m and likely to 1,000 m. Much of the surplus $^{129}$I is probably released into the eastern North Atlantic. From there it is transported southward and westward in the North Atlantic Gyre, and subsequently delivered into the Caribbean Sea and then into the Gulf of Mexico. Lower concentrations at depths of 500 and 1,000m can be attributed to slower water movement at these depths and to admixture of Antarctic Intermediate Water, which has a lower concentration of $^{129}$I.

### GENERAL ORGANIC CHEMISTRY

Organic chemicals contain the elements carbon, hydrogen, and oxygen, and often other elements such as nitrogen and phosphorus. These chemicals are important components of living organisms, and they enter seawater through secretion and excretion by living marine plants and animals and through decomposition of dead bodies. Such chemicals may be produced in the sea or they may be brought in by rivers and, to a limited extent, by atmospheric fallout. In recent years sewage and other organic wastes have become major components of river effluents. Except for petroleum hydrocarbons, which will be treated in a later section, the organic chemistry of the Gulf of Mexico has not been well studied.

### Particulate and dissolved organic matter

Particulate organic material includes the bodies of living and dead organisms as well as body parts such as the cellulose cell walls of plants and shed exoskeletons of crustaceans and other animals. Its concentration in seawater is often expressed in terms of particulate organic carbon (POC) or particulate organic nitrogen (PON). Table S9 shows the general distribution of POC in the Gulf of Mexico. The highest values are observed in waters of the continental shelf, which have an average of 0.214 mg C/L. Exceptional values of nearly 2.0 mg C/L are encountered off the mouth of the Mississippi River. In the open Gulf, surface waters average 0.050 mg C/L, about one-fourth that of the shelf, and deeper waters average 0.028 mg C/L, or about half the surface value. Vertical profiles exhibit a sharp drop in POC below the euphotic zone, some fluctuation in the upper section of the oxygen minimum layer, and only slight variation below a depth of 500 m. Concentrations of particulate organic nitrogen are much lower than for POC, but otherwise the vertical profile looks much the same, that is, high and somewhat variable in the upper portion of the water column and fairly constant (around 1.5 µg/L) below 800 m.

As mentioned earlier, dissolved organic matter is defined as material that passes through 0.5 µm filters. Included here are the truly dissolved molecules as well as subparticulate colloidal materials. As shown in Table S9, the general distribution pattern for dissolved organic carbon (DOC) is similar to that of POC, but the values run about an order of magnitude higher. Concentrations on the continental shelf average 1.08 mg C/L, in open Gulf surface water 0.79 mg C/L, and in deeper water 0.52 mg C/L. The vertical profile is similar to that of POC. High levels of POC and DOC in waters of the continental shelf are the result of river enrichment, resuspension, and high plankton production. In the surface layer of the open Gulf, plankton production is the primary source. Lower levels of organic material in the deep Gulf reflect residual organic matter derived from source waters of the Atlantic as well as small amounts sinking from local surface waters of the Gulf itself.

## Organic compounds

Seawater contains both refractory and nonrefractory organic compounds. Refractory compounds, due to their chemical makeup, are generally unavailable for use by most marine organisms. These appear to be related to or derivatives of humic acids such as are found in soils, marshes, swamps, and other organic-rich stagnant waters on land. Many contain phenolic or ring-shaped structures that are difficult to break down and metabolize. Nonrefractory or labile compounds include proteins, amino acids, sugars and other carbohydrates, lipids, vitamins, hormones, enzymes, and related compounds, mostly derived from marine organisms and most available for further use by other organisms. Many of the nonrefractory compounds are quickly recycled in the euphotic zone, so their concentrations in the water column are kept low. The dissolved organic matter of seawater includes thousands of compounds, most in very low concentrations. Of these, probably less than a quarter have been specifically identified. Only a few studies have been carried out to determine the presence and concentrations of dissolved organic compounds in the Gulf of Mexico and associated bays and estuaries, and some of the more reliable information is presented in table 7.1. For total carbohydrates the following concentrations have been reported: bays and estuaries 0.0–19.4 mg/L, continental shelf 0.0–20.0 mg/L, and surface waters of the open Gulf 0.05–0.13 mg/L. Around a score of amino acids have been reported from offshore waters, the most abundant of which are given in table 7.1. A vertical profile of the concentrations of free and combined amino acids (proteins) from the west-central Gulf produced values in the range 8.5–13.65 μg/L and showed only slight variation from the surface to a depth of 3,500 m. The ratio of combined to free amino acids was about three to one.

A number of studies have been carried out on lipids in Gulf waters without distinguishing between recently formed and petroleum-derived compounds. Chloroform extracts showed that lipids made up 12–20 percent of the total dissolved organic carbon. In nearshore waters (< 50 m deep) lipid concentrations were relatively high, ranging from 0.12 to 0.95 mg C/L, and were significantly higher near oil rigs. In surface waters of the open Gulf, concentrations were in the range of 0.11–0.16 mg C/L, and

**Table 7.1.** *A sampling of organic chemical compounds identified from the Gulf of Mexico and adjacent bays and estuaries. (From Jeffrey 1972.)*

Amino acids
  Aspartic acid, glutamic acid, threonine, serine, alanine, valine, leucine, isoleucine
Carbohydrates
  Total carbohydrates only
Lipids
  Free and esterified (mono- and di-) unsaturated fatty acids, substituted phenols, fatty alcohols, sterols (cholesterol, beta sitosterol, stigmasterol), paraffins
Vitamins and growth factors
  Biotin, adenine, uracil
Organometallic complexes
  Complexes with boron, copper, manganese, and zinc

in deep waters of the western Gulf the range was 0.09–0.16 mg C/L. A vertical profile down to 900 m showed an apparent decrease in the length of the carbon chains with depth. Petroleum hydrocarbons in the Gulf are discussed in a later section. Examination of surface slicks showed alcohols and fatty acids that appeared to be less water soluble and of higher molecular weight than lipids of underlying waters. Such slicks modify wave heights and various physical processes normally occurring at the surface. Additional studies have demonstrated the presence of vitamins and growth factors as well as organometallic complexes. However, a great deal of work remains to be done before the organic chemistry of the Gulf is really understood.

## CHEMISTRY OF THE SEDIMENTS

As noted earlier, seawater generally contains a large supply of oxygen, even down to the bottom. Hence, the surface layer of bottom sediments throughout the Gulf also contains a certain amount of free oxygen. However, a short distance below the surface free oxygen is totally absent, and this anoxic zone continues downward through the entire length of the sediment column. Chemically speaking, the surface layer is an oxidizing zone, and the anoxic layer is a reducing zone. Between these two zones lies a thin intermediate layer where the oxygen content is very low and somewhat variable over time. In shallow water very close to the shoreline, where wave action is intense and the bottom is composed of coarse sediments (mostly sand), the oxidizing layer may be as much as a meter thick, but in less energetic environments, where the bottom is made up of silts and clays, oxygen penetrates only slightly, and the oxidizing layer may be only a few millimeters thick. This latter condition characterizes the sediments underlying all of the world oceans. The difference between oxidizing and reducing environments is of extreme importance in terms of chemistry as well as biology. The present section will examine the effects of these environments on sulfur and several trace metals that have been studied in the Gulf of Mexico.

## Sulfur

The chemical element sulfur is remarkable in having a wide range of oxidation states, from −2 as in sulfides ($S^{2-}$) to +6 as in sulfates ($SO_4^{2-}$), including various intermediate states. Some are solid, some liquid, and some gaseous. Sulfate, the most oxidized state, is the third most abundant ion in seawater, making up nearly 8 percent of the dissolved solids. Sulfur is an important component of all living systems, and animals require sulfur-containing compounds such as cysteine, cystine, methionine, thiamine (vitamin $B_1$), and biotin (vitamin H). Within the sediments many of the chemical transformations of sulfur are mediated by specialized bacteria, and such bacteria have been recovered from over a mile deep in the sediments, where they derive energy from petroleum hydrocarbons and transform sulfur compounds to elemental sulfur and the gas hydrogen sulfide ($H_2S$). Sulfur-transforming bacteria are present in surface sediments as well as the overlying water column. Most anoxic marine sediments are rich in hydrogen sulfide and other reduced sulfur compounds.

In the Gulf of Mexico the distribution and transformation of sulfur have been investigated in a series of sediment sample transects across the continental shelf into deep water. These transects were located off the Mississippi River Delta, Louisiana and Texas coasts, and off the Mexican coast below the Rio Grande, below Tampico, and in the southernmost Gulf near the mouth of the Rio Usumacinta. It was found that at all stations the oxidized form, sulfate, was present in pore waters of the sediments for considerable depths. It was most abundant at the surface and decreased with depth in the sediments due to chemical reduction of the sulfate to other compounds, most notably to the relatively insoluble iron disulfide ($FeS_2$), or pyrite.

Results of a transect off the Mississippi River Delta are given in table S10. Here it is seen that the rates of sulfate reduction and pyrite formation decreased exponentially with the depth of the overlying water column, and the percentage retention of reduced sulfur decreased from 26.4 percent at 59 m to 0.3 percent at a depth of 574 m. For all the transects around the Gulf, the pyrite concentration was highest on the continental shelf and decreased offshore with increasing water depths. In this study the organic carbon content of the sediments varied between 0.5 and 1.0 percent at the surface, and it tended to decrease with increasing depth of the overlying water. However, there was always a maximum of organic carbon in surface sediments at depths between 200 and 500 m, that is, where the oxygen minimum layer impinges on the bottom surface. Within the sediments, the organic carbon concentration decreased with depth in the sediment column.

It was concluded that the rate of sulfate reduction and pyrite formation is directly influenced by the organic carbon concentration of the sediments, which in turn depends strongly on the sedimentation rate. Nonrefractory organic carbon is probably the controlling factor. Thus, the major areas of sedimentation, the Mississippi River Delta and the Texas-Louisiana shelf, are the primary areas of pyrite formation and account for around 96 percent of the pyrite buried in the study area (northwestern, western, and southern Gulf) and about 1.5 percent of worldwide pyrite formation. Sedimentation and organic carbon burial rates are quite low in the western and southern Gulf, and iron is lacking on the carbonate platforms of Florida and Yucatán. Hence, the rate of pyrite formation is quite low in these areas. Although the actual mechanisms of pyrite formation were not addressed, it appears likely that the reduction of sulfates to sulfides and the subsequent formation of pyrite depend on the availability of labile organic carbon to serve as an energy source for sulfate-reducing bacteria, which carry out the actual chemical transformations.

## Heavy metals

A number of studies have been carried out on the transport of heavy metals to the Gulf of Mexico by the Mississippi River and the subsequent distribution of these metals in suspended matter and plankton as well as in sediments of the northern Gulf continental shelf. Additional studies have reported on the distribution of such metals in bays and estuaries of the northern Gulf. Most of this work has been undertaken to determine the extent to which our industrialized society is contaminating the Gulf and the rich supply of seafood derived from the Gulf and adjacent waters. Annually, the Mississippi River brings to the Gulf of Mexico an average of $1.3 \times 10^{11}$ m³ of water containing $4.5 \times 10^{11}$ kg of suspended particulate matter, but the flow and sediment load vary greatly from one year to another. The suspended matter is composed of about 57.6 percent clay (< 2 μm in diameter) and 42.1 percent silt particles (2–62.5 μm in diameter), with a very low percentage of sand. Along with the water and suspended particles, the river brings to the Gulf a suite of heavy metals, of which more than 90 percent are associated with the particulate load, and less than 10 percent are dissolved in the water itself (table S11). Iron and aluminum are by far the most abundant, but a variety of other metals are involved (manganese, zinc, lead, copper, chromium, nickel, cadmium, arsenic, cobalt, etc.). The ratio of the various metals to one another in the suspended river matter is roughly similar to their natural concentrations in rocks of the continent.

Table S12 gives the concentrations of a series of metals in suspended matter of the lower Mississippi River, suspended matter of the Gulf near the river mouth, and sediments around the Delta. This table shows the actual concentrations as well as the abundance of each metal in relation to that of iron. Iron is a relatively conservative element, and normalizing the concentrations of other metals to that of iron reduces the effects of overall variability in samples from different areas. Note that in the table the concentrations of the two abundant metals, iron and aluminum, are given as percentages (parts per hundred), whereas the other metals are expressed as micrograms per gram of sediment (parts per million). Upon mixing with seawater, the suspended matter concentrations of most metals (iron, aluminum, cobalt, nickel, and chromium) remained similar to those in the river, but manganese decreased seaward strongly, suggesting gradual loss by desorption from the silt and clay particles. On the other hand, zinc, lead, copper, and cadmium increased in concentration, suggesting uptake from the dissolved forms. In sediments off the Delta the proportions of the metals zinc, lead, cobalt, chromium, and cadmium remained similar to their concentrations in river-suspended matter, but manganese, copper, and nickel were significantly less abundant, suggesting desorption or migration of these metals. On a broader scale, for sediments throughout the northwest Gulf continental shelf, many of the metals have remained in the same metal/iron

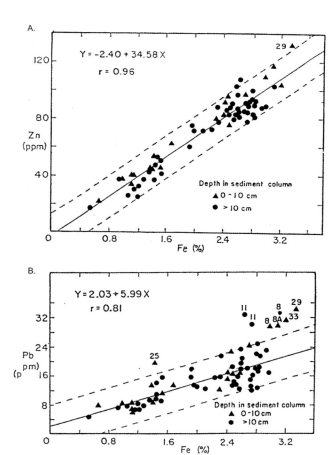

**Figure 7.5.** Scatter diagram plotting the distribution of (a) zinc versus iron and (b) lead versus iron in sediments off the Mississippi River Delta and on the continental shelf of the northwestern Gulf. The 95 percent prediction intervals are indicated by dashed lines. Note the values for lead lying outside the 95 percent interval. (From Trefry and Presley 1976.)

ratio as in the Mississippi River, as, for example, in the plot of the zinc/iron ratio shown in figure 7.5a. However, lead (fig. 7.5b) and cadmium display some values above the normal ratio. The depths of maximum concentrations of these elements in the sediments demonstrate that the increase in lead and cadmium has taken place during the past half century, that is, their increase has been associated with industrialization and widespread use of leaded gasoline following the Second World War.

In some areas of the continental shelf, phytoplankton appears to be enriched with lead and cadmium, and zooplankton is enriched with lead, cadmium, and copper. Within bays and estuaries of the northern Gulf, localized high concentrations of a series of metals in bottom sediments have been documented. These include zinc, cadmium, chromium, mercury, molybdenum, and arsenic.

Within estuaries some of the oysters and blue crabs were found to contain excess quantities of zinc, cadmium, silver, mercury, and/or selenium. Despite the widespread use of heavy metals by society and the occasional finding of locally high concentrations of some metals, sediments and waters of the northern Gulf of Mexico are generally not contaminated to levels that pose a threat to human health, although there are exceptions.

In the past half century, more than 30,000 oil wells have been drilled on the continental shelf and upper continental slope off east Texas and Louisiana, and the effect of these activities on the trace metal concentrations is of interest. During the drilling process large volumes of drilling muds are used, and these contain vast amounts of the element barium, a fair amount of chromium, and small concentrations of other metals. The particulate residues are dumped onto the seafloor around the bases of the drilling rigs. Furthermore, the wells themselves bring to the surface large volumes of brines that are charged with heavy metals, some of which are radioactive. Studies have been carried out to determine the fate of these metals. Only a small fraction of the barium-enriched sediments remains in place. Over time they become widely distributed across the entire Texas-Louisiana shelf before finally being swept out to deeper water. Some of the barium may also be lost by passing into solution in the water column. Metals contained in the brines are already in solution, and as the brines are discharged into the sea, apparently they are rapidly diluted and dispersed by the water currents. Thus, surprisingly, with some exceptions, there appears to be little long-term excess accumulation of heavy metals in this area of the continental shelf. However, metal contamination of fishes close to the platforms has been documented, and of course, this is just where many sportsmen like to fish.

## HYDROCARBONS

Hydrocarbons are chemical compounds consisting primarily of carbon and hydrogen but often containing oxygen, nitrogen, sulfur, trace metals, or other elements. Many thousands of individual hydrocarbon compounds are known, and there is a vast literature on the topic. The present discussion focuses on studies carried out on hydrocarbons in the Gulf of Mexico, and of necessity, it represents a simplification of a very complex and technical subject. Chlorinated and other industrial hydrocarbons, which are major pollutants, will be covered in a later chapter.

In general, hydrocarbons consist of two basic types, chains and rings. The chain compounds (alkanes or aliphatic hydrocarbons) may include a single carbon atom or two, three, or many carbon atoms linked together in either simple or branched chains with one or more hydrogens attached to each carbon atom. Shorter-chained aliphatic compounds such as methane, ethane, propane, and butane are quite soluble in seawater, and they are volatile, that is, they are gases under normal temperature and pressure conditions. As the carbon chains become longer this group progresses through gasolines, kerosenes, light and heavy oils, then greases and waxes (including paraffins). These longer-chained molecules become virtually insoluble and nonvolatile. From a biological standpoint some of the shorter-chained aliphatics may be toxic in high concentrations, but in seawater they are rapidly diluted and eventually escape to the atmosphere. The longer-chained aliphatics can gum up gills, fur, and feathers, but chemically they are not toxic.

Ring compounds (cyclic or aromatic hydrocarbons) generally consist of a closed hexagonal ring of six carbon atoms. Such compounds may be monocyclic (with a single ring) or polycyclic (with two or more rings attached). The latter are referred to as polynuclear or polycyclic aromatic hydrocarbons. Many of these compounds are oily, but the lighter ones, at least, are fairly soluble in seawater. Aromatic hydrocarbons are often quite toxic, and some are known to induce tumorous growths in marine life and humans. From a biological standpoint this is a dangerous group of compounds. From a general chemical mixture such as crude petroleum, analytical techniques permit the separation and identification of hydrocarbons containing fewer than 30 or 40 carbon atoms. However, longer-chained aliphatics and more complex polycyclic aromatics are difficult to remove with solvents, and they are often lumped together in a general category called "unresolved complex mixture." Much of this material would be popularly recognized as tar and asphalt.

Hydrocarbons from several sources occur in the Gulf of Mexico. As indicated in table 7.2, these include petroleum, anthropogenic, and biogenic hydrocarbons, the latter including compounds derived from both terrestrial vegetation and marine plankton. Bottom sediments of the northern and southern Gulf contain rich deposits of natural gas (methane) and petroleum, and on the continental slope of the northern Gulf numerous active gas and oil seeps have been documented. Bottom sampling gear has also revealed bottom surface deposits of tar and

**Table 7.2.** *General hydrocarbon groups present in the water column and sediments of the Gulf of Mexico and their sources.*

## Hydrocarbon groups

Gaseous – Hydrocarbons volatile at normal temperatures. Alkanes with 1–4 carbon atoms (methane, ethane, propane, and butane).

Very light – Hydrocarbons with 5–14 carbon atoms. Alkanes and light aromatics (pentanes, hexanes, heptanes, benzene, toluene, xylene, etc.).

Polyaromatic – Aromatic hydrocarbons with two or more carbon rings. Not produced by living organisms. Formed at elevated temperatures. Characteristic of petroleum and pyrogenic hydrocarbons.

High molecular weight – Hydrocarbons with 15–32 or more carbon atoms. Includes both alkanes and aromatics.

Unresolved complex mixture – "Gunky" material remaining after lighter hydrocarbons have been removed by solvents. Includes very high molecular weight hydrocarbons difficult to identify chemically (tars, asphaltenes, etc.)

## Hydrocarbon sources and characteristics

Petroleum – Crude oil from natural sources (seeps and oil wells). Contains a mixture of many low- to high-molecular-weight hydrocarbons. Includes a complete homologous series of alkanes ($C_{16}$ through $C_{46}$), polyaromatic hydrocarbons, and unresolved complex mixture.

Anthropogenic – Hydrocarbons produced by human activities. Includes fuel oils, pyrogenic hydrocarbons, pesticides, industrial hydrocarbons, etc.

Pyrogenic – Hydrocarbons resulting from combustion of wood and fossil fuels (from forest fires, internal combustion engines, coal gasification, etc.). Includes aromatics, especially polyaromatics.

Biogenic – Hydrocarbons produced by living organisms. Rich in odd-numbered carbon chains. Includes few light aromatics and no polyaromatics.

Terrestrial – Hydrocarbons produced by land plants. Rich in bio-waxes with a preponderance of long, odd-numbered carbon chains of 23, 25, 27, 29, and 31 carbon atoms.

Marine planktonic – Hydrocarbons produced by marine planktonic algae. Rich in shorter odd-numbered chains of 15, 17, and 19 carbon atoms. Pristane often present.

---

asphalt that are probably remnants of earlier seeps. When crude petroleum moves upward through the sediments, generally along geological faults, and reaches the water column, lighter-weight hydrocarbons may bubble upward and/or become dissolved in the water, leaving behind the higher-molecular-weight compounds. These also may float into the water column, but as they accumulate sand grains and other particles, they become denser and sink to the bottom. Here the weathered tars and asphalts create a special habitat for marine organisms seeking attachment sites or semisoft material for boring and tunneling. Some species apparently consume weathered tar, but it is not known whether they receive nourishment from it. In addition to the petroleum that enters the Gulf from oil seeps and from spills, leaks, and other incidents associated with petroleum recovery operations, hydrocarbons may enter the Gulf by other avenues. Various industrial and agricultural hydrocarbons are brought to the Gulf by the atmosphere and by the Mississippi and other rivers. They may be released in tanker and barge accidents and through bilge washing of ships at sea. Biogenic hydrocarbons derived from terrestrial plants and marine plankton are also present in Gulf waters and sediments. Analytical chemical techniques are now available that often make it possible to identify the source of petroleum in a given sample.

It is conservatively estimated that the Gulf of Mexico receives from half a million to more than a million metric tons of hydrocarbons annually from various sources. Natural seepage from sediment deposits probably makes up only 1 percent or less, the remainder being a result of human activities. Although there is considerable annual variation, rough estimates suggest that on average the following partitioning of hydrocarbon inputs generally applies: oil and gas development (20–40 percent), transportation activities (20–40 percent), and Mississippi River effluent (17–50 percent). Additional quantities from bays, other rivers, the Yucatán Current, and the atmosphere, although difficult to quantify, are probably not of major importance. Input from oil and gas development includes spillage, dumping of formation waters (fossil brines found with oil and gas deposits), and underwater venting of natural gas containing various hydrocarbons. Furthermore, drilling muds contain a great deal of oil additives to aid in lubricating the drilling pipes and to facilitate removal of rock chips from the drill hole. Transportation hydrocarbon loss is largely a result of ship and barge accidents, spillage, tanker cleaning, and bottom pipeline leakage due to anchor damage. One of the largest oil spills in the Gulf and one of the largest in the world was the 1979 Ixtoc 1 blowout on the Campeche shelf of the southern Gulf. This single event, which lasted for months, introduced an estimated 0.44 to 1.4 million tons of crude oil into the southern Gulf, and western boundary currents carried some of this along the Mexican coast and into the northern Gulf. Thus, through the years the Gulf of Mexico has been bathed in hydrocarbons from both natural and anthropogenic sources. These hydrocarbons include hundreds of specific chemical compounds representing both aliphatic and aromatic groups. They involve gases, liquids, oils, greases, waxes, and tars. Present concentrations of certain hydrocarbon groups in the water column and sediments are discussed below.

### Hydrocarbons in the water column

As a result of the various inputs, waters of the open Gulf now contain about 0.1–75 mg/L of hydrocarbons, around twice the level for Caribbean waters, indicating chronic low-level hydrocarbon pollution of the open Gulf. Gaseous hydrocarbons, derived from seeps and other sources, are present throughout the Gulf, but they are particularly concentrated on the Louisiana shelf, which receives outflow from the Mississippi River and which supports a major offshore oil field. For example, in the open Gulf the background levels of methane and propane run around 50 ng/L and 1.0 ng/L, respectively, whereas in waters of the Louisiana shelf they average 3,100 ng/L and 22 ng/L. (Note: a nanogram is one millionth of a milligram.)

Very light hydrocarbons, alkanes and aromatics with 5–14 carbon atoms, average 60 ng/L in the open Gulf, about twice the concentration in the Caribbean Sea, but in Louisiana shelf waters they reach 500 ng/L. About 60–85 percent of these are the very toxic light aromatic hydrocarbons. Light alkanes run around 15 ng/L in the open Gulf and 40 ng/L on the Louisiana shelf. Near the Ixtoc 1 blowout they were measured as high as 400 ng/L. Major sources of very light hydrocarbons in the northern Gulf are brine discharges and underwater venting of gases from production platforms, as well as Mississippi River effluent.

Higher-molecular-weight hydrocarbons in Gulf water include tiny tar balls that float at or near the surface. Throughout the Gulf they average around 1.35 mg/m$^2$ of surface area, with the highest concentrations in the western and southwestern Gulf. The total standing crop throughout the Gulf has been estimated at about 2,000 metric tons. Chemical analysis suggests that around 30 percent of this tar is derived from tanker residues (sludge from bilge washing), 65 percent from crude petroleum (from many sources), and 2 percent from fuel oil residues. Possibly 60 percent does not originate in the Gulf, and the high sulfur content suggests that these tars come from foreign crude oils. Most of the tar balls are concentrated at the surface, where they make up 20 percent of the organic matter in the top 5 cm, and their concentration decreases rapidly with depth. Below 10 m the values are generally less than 1.0 mg/L.

### Hydrocarbons in the sediments

Concentrations of hydrocarbons in surface sediments of the Gulf are in the parts per million range, that is, several orders of magnitude higher than in the water column. The distribution of extractable sediment hydrocarbons for selected areas of the northern and eastern Gulf continental shelves is shown in table S13. Relative to other areas of the northern and eastern Gulf, contamination in Louisiana sediments is quite high. However, the summits of offshore banks, which rise many meters above the surrounding bottom surface, are still relatively pristine. Although many sources are involved, most of the sediment hydrocarbon pollution of the Louisiana shelf derives from the vast quantity of hydrocarbon additives to the drilling muds, which are simply discarded around the drilling rigs or shunted to nearby dump sites. Special studies carried out close to active drilling sites have shown hydrocarbon concentrations of up to 940 mg/g of sediments, but the concentration decreases rapidly with distance from the platforms. Furthermore, great day-to-day variation in concentration relates in part to changes in drilling operations and in part to resuspension and transport of sediments by bottom currents. A strong correlation has been found between sediment type and hydrocarbon content. Sandy sediments have the lowest levels (25–50 mg/g), silty sediments next (100–200 mg/g), and clayey sediments the highest levels (200–500 mg/g).

A.

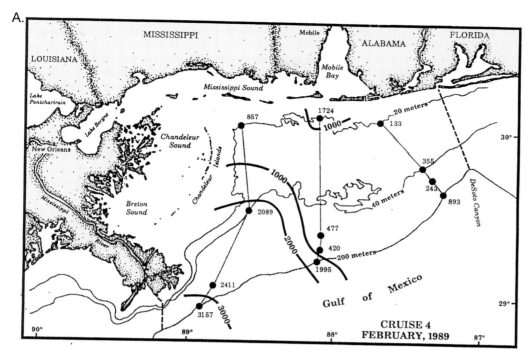

B.

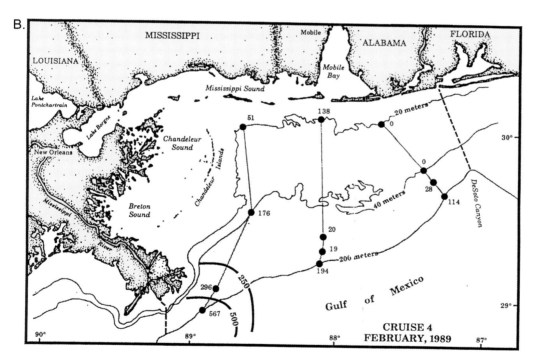

**Figure 7.6.**
Distribution of hydrocarbons in surface sediments of the Mississippi-Alabama continental shelf during February 1989. A = distribution of n-alkanes. B = distribution of polycyclic aromatic hydrocarbons. In both cases concentrations are highest near the Mississippi River Delta and decrease with distance from the Delta. Concentrations are given in parts per billion. (From Kennicutt 1991.)

Special studies of sediments off Mississippi and Alabama have shown the effects of Mississippi River outflow and possibly eastward transport of hydrocarbons from the Louisiana shelf. Figure 7.6a shows the distribution of n-alkanes with concentrations in excess of 3,000 ppb near the Mississippi River Delta diminishing to less than 1,000 ppb off Alabama, although a slight increase is noted just off Mobile Bay (possibly a result of the dumping of dredged materials). A somewhat similar picture is shown by the distribution of polycyclic aromatic hydrocarbons (fig. 7.6b). From a concentration of over 500 ppb near the river mouth, the values quickly drop to less than 100 ppb across most of the shelf. However, values in excess of 100 ppb persist at the deeper stations (ca. 200 m) on all transects up to the head of DeSoto Canyon. On the Louisiana shelf hydrocarbons are a grand mixture of petroleum and anthropogenic types with a lower percentage of biogenic hydrocarbons. Away from Louisiana and the influence of the Mississippi River the reverse is true. Off south Texas and Florida biogenic hydrocarbons dominate, with anthropogenic inputs limited largely to areas near shore and off bays and estuaries. Studies have been carried out to measure the hydrocarbon content of sediments of the continental slopes off Louisiana and the upper Florida Peninsula (table S14). Even at depths greater than 500 m, hydrocarbon concentrations off Louisiana were found to be three or four times higher than off Florida. These included a mixture of petroleum and biogenic (both terrestrial and marine planktonic) hydrocarbons.

## Hydrocarbons in marine organisms

Despite the widespread distribution of hydrocarbons in the water column and in sediments of the northern Gulf, the biota has been found to be relatively free of hydrocarbon contamination. Exceptions do occur near oil and gas platforms, in brine disposal sites, and in bays and estuaries in localized areas of known pollution such as marinas, urban runoff outfalls, and industrial sites. In continental shelf waters of the northwestern Gulf the zooplankton has been reported to be contaminated with high-molecular-weight hydrocarbons, presumably from ingesting tiny tar balls. These may be from tanker traffic, petroleum seeps, or a combination of the two.

Around active drilling rigs surface zooplankton organisms contain variable amounts of hydrocarbons, but some display fairly high levels of contamination. On legs of the platforms barnacles at a depth of three meters have high levels of weathered petroleum. Like the zooplankton mentioned above, barnacles are filter feeders, and they probably take in small tar balls. Associated with the fouling mat on legs of the platform are small fishes called blennies. These reside in shells and crevices and feed on small crustaceans and other organisms living on the mat. The blennies have been found to contain fresh crude oil in their tissues. Larger fishes called sheepshead live in the water column and feed on barnacles and other attached species, and their bodies contain weathered oil apparently picked up from the barnacles consumed. Spadefish, which also live in the water column, pick at the fouling mat and take smaller organisms, and their bodies have lower concentrations of both fresh and weathered crude oil. Red snappers live around the base of the oil platforms.

Many remain in the vicinity of the rigs, but some move around from one area to another. Hydrocarbon levels in their tissues are variable but generally high. Shrimp live directly on the bottom but move around a great deal, and specimens captured below the platforms show low levels of hydrocarbon contamination. Near shore around brine disposal sites shrimp show higher levels of contamination, averaging around 10–30 mg of hydrocarbons per gram of shrimp tissue. At some platforms aromatic hydrocarbons are present in tissues of both sheepshead and spadefish.

Within bays and estuaries throughout the US Gulf coast, oysters have been found to be relatively uncontaminated except at certain localized sites, particularly in Galveston Bay, Texas, and Barataria Bay, Louisiana. Where waters and sediments are highly polluted, the oysters show high levels of contamination, with tissue levels of hydrocarbon two or three orders of magnitude above the levels in those from uncontaminated sites. In these studies tissue contamination levels ranged from less than 20 to 18,600 ppb. Particularly disturbing was the presence of significant levels of low-molecular-weight polyaromatic hydrocarbons. Fortunately, oysters from such areas are regularly monitored to reduce the possibility of contamination of human food resources. These and related studies show that marine organisms can pick up hydrocarbons from the water column and through their food supplies. However, the actual levels of contamination depend a great deal on the behavior of the organisms, what they eat, and where they are located in relation to potential contamination sources. Fortunately, levels of tissue contamination are generally well below the levels of acute toxicity to the organisms, but the effects of long-term exposure to low levels of hydrocarbon pollution are not well understood. Of particular concern are the possibilities of interference with chemoreception, modification of behavior, or inhibition of reproduction. For the most part, human food supplies appear not to be seriously affected by hydrocarbon pollution in the northern Gulf and associated bays and estuaries.

### HYPOXIA

In surface waters of the Gulf of Mexico the level of dissolved oxygen at any locality reflects a balance between the processes of oxygen input and removal. Input takes place through atmospheric exchange and photosynthetic release of oxygen by the phytoplankton. Removal occurs through biological oxygen demand for respiration as well as the nonbiological chemical oxygen demand. These, of course, are modified by mixing and water mass transport processes as well as the tendency for organic particles to sink. For most areas oxygen input exceeds removal so that the oxygen level in the water column remains high. For example, surface waters of the open Gulf average around 5 mg/L, and surface waters of the continental shelf range between about 6 and 8 mg/L. Bottom waters of the northern Gulf continental shelf generally run slightly lower due to higher oxygen demand in the turbid layer. However, under special circumstances oxygen is removed faster than it is generated. When the oxygen level falls below 2 mg/L the condition is referred to as "hypoxia" (low oxygen), and when it reaches 0 mg/L the water is said to be anoxic (without oxygen). Hypoxia and anoxia can have a profound effect on the ability of marine organisms to survive in an area.

## Hypoxia on the continental shelf

Hypoxic events are regularly recorded on the northern Gulf continental shelf west of the Mississippi River, where they have been studied in some detail. They develop annually and are associated with warm weather. Although they generally develop in late spring or early summer, they have been recorded as early as April. These events reach their greatest areal extent in July and August and generally disappear in September, but they have been known to persist into October. The area most affected is the Louisiana shelf west of the Mississippi River Delta, but in exceptional years the hypoxic area may extend along the upper Texas coast as far west as Freeport (just west of Galveston Bay). The mass of hypoxic bottom water generally lies between the 10 and 30 m isobaths, but it may extend both shallower and deeper (6–40 m). It generally lies within 50 km of the coastline and may stand 10–15 m thick. At maximum development during mid or late summer the hypoxic water mass may cover 6,000–10,000 km² of bottom surface area (fig. 7.7).

The basic mechanisms of hypoxia development are well understood, although the details are still under investigation. During the spring and early summer, the Mississippi and Atchafalaya Rivers discharge a large volume of freshwater, which flows westward along the nearshore Louisiana and upper Texas continental shelf. Contained in this water is a great deal of suspended inorganic and organic particulate matter as well as high levels of dissolved nutrients. This surface layer of freshwater flowing over more dense salt water sets up a strong density

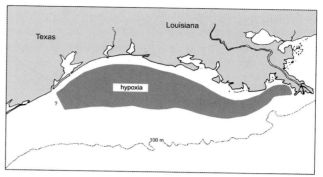

**Figure 7.7.** General distribution of hypoxic bottom water during summer months off the coasts of Louisiana and upper Texas. In any given year only a portion of the shaded area is hypoxic.

stratification in the water column, which, as the season progresses, becomes reinforced by the rising temperature of the surface water. After particulate material drops out near the river mouths, sufficient light can penetrate to support a heavy growth of phytoplankton downstream.

This explosive growth of phytoplankton is so great that zooplankton grazing cannot keep up with the bloom, and much of the phytoplankton sinks into the bottom layer, where, lacking sufficient light, the organisms die and decompose. Particulate organic material brought in by the rivers also settles to the bottom and undergoes decomposition. As a result, demand for oxygen in the lower layer increases rapidly and soon outstrips the supply, which, because of stratification, cannot be renewed. Then the bottom layer becomes hypoxic and eventually anoxic. Furthermore, summer winds from the south often induce some upwelling in this area, sending deeper shelf waters closer to the coast and blocking off-shelf circulation of the nearshore bottom waters. Thus, trapped by density stratification and blocked by a wall of deep-shelf water, the coastal bottom waters stagnate as the summer sun raises the temperature and intensifies decomposition processes that remove the last vestiges of free oxygen. This situation prevails until the winds associated with September storms are strong enough to induce vertical mixing and horizontal transport, causing oxygen values in nearshore bottom waters to rise. At this time the hypoxic event terminates.

Hypoxia is one of several factors that cause mass mortality among the biological inhabitants of the northern Gulf coast. Although some mobile species are able to detect and avoid the hypoxic areas, many do not and are overcome by the lack of oxygen or the accompanying increase in hydrogen sulfide of the anoxic waters. Both the number of species and the abundance of individuals are dramatically reduced, although a few tolerant species do survive. Following the fall period of reaeration of bottom waters, population levels in the area quickly return to normal, probably as a result of reinvasion from other areas.

Along the northern Gulf coast, hypoxia may have occurred as a natural phenomenon in preindustrial times, but there is strong evidence that such events have greatly intensified during the past century and particularly in the past few decades. The primary cause appears to be increased use of agricultural fertilizers in the nation's heartland. Great quantities of nitrate and phosphate are now regularly washed from farm fields and transported by the Mississippi-Atchafalaya system to the Gulf, where they stimulate the massive phytoplankton growths in shallow coastal waters downstream from the river mouths. Hypoxia occurs off the mouths of other major streams such as the Po River of northern Italy, which drains into the upper Adriatic Sea, but it is not known to occur elsewhere on continental shelves of the Gulf of Mexico. Parenthetically, it has been argued that hypoxic basins are necessary for the accumulation of the vast quantities of reduced organic matter that form the basis of major petroleum deposits, and it is certainly true that many deposits were formed under such conditions. However, the annual burial of particulate material containing 1 or 2 percent organic carbon in hypoxic sediments, such as normally occurs off the mouth of the Mississippi River, should accomplish the same thing regardless of the oxygen content of marine waters overlying the sediments.

## Hypoxia on the continental slope

On the continental slope southwest of the Mississippi River Delta at a depth of 1,700–1,900 m lies an L-shaped depression known as the Orca Basin (fig. 4.16). Located in hummocky topography, this basin is approximately 30 km long and covers an area of about 400 km². The surrounding ridges and hills appear to be surface expressions of subsurface salt diapirs. Within this depression lies a pool of concentrated brine 200 m deep whose surface stands at a depth of 2,300 m. Salinity of the brine is around 260 ppt (near the saturation level), or about seven times that of seawater. Chemical analysis indicates that the brine was derived from leaching of the adjacent salt diapirs. Although oxygen in the overlying water stands at 5 mg/L, the brine itself is anoxic and is charged with hydrogen sulfide. There appears to be little mixing between the brine pool and the overlying water column. Because of

the high salinity, the brine is totally devoid of higher life, but some bacteria live at the surface of the brine. Considering the rough topography and widespread distribution of diapirs among continental slopes of the Gulf of Mexico, the discovery of other hypoxic basins would not be surprising. Detailed study of the chemistry of such basins and their subsurface sediments could shed light on conditions under which the massive salt deposits were laid down in the distant past.

## Hypoxia in bays, estuaries, and related waters

On a hot evening in late summer on the western shore of Mobile Bay, shrimps, crabs, and fishes begin crawling or jumping out of the water onto the shore. Word quickly spreads among the local inhabitants, and people arrive with shovels and washtubs to harvest the seafood that is suddenly and inexplicably available. Such "Jubilees," as they are called, have been known to occur since early colonial times. This strange occurrence, so bewildering to the early settlers, is now fairly well understood. The forcing factor is hypoxia.

Currents in the northern Gulf of Mexico move sediments along the shoreline, and these often settle and form shallow shoals or sandbars across the entrance of Mobile Bay. During summer months when river flow is low, the water column of the bay becomes highly stratified, and bottom waters, blocked by the bay-mouth bar, remain trapped in the basin. Decomposition of organic matter depletes oxygen and results in hypoxic conditions in the bottom layer. After a very hot day in August a breeze comes in from the west and blows the surface waters eastward, piling them up along the eastern side of the bay. Here the depth of the surface oxygenated layer deepens. However, as more of the surface water is swept eastward, along the western side of the bay the level of the hypoxic bottom layer rises closer to the surface, reducing the width of oxygenated bottom habitat. Finally, when the hypoxic layer approaches the surface along the western side, respiration becomes impossible, and marine life is forced to abandon the water in an effort to obtain oxygen. This natural phenomenon may have been intensified by human activities. Organic and inorganic pollution from agricultural practices, increased urban runoff, and industrial development have greatly increased the load of oxidizable material in bay-bottom waters and sediments. Although deep navigation channels now cut the bay-mouth bar, bottom circulation is still limited and insufficient to permit full oxygenation of the bottom layer.

Hypoxic conditions have also been reported for portions of Galveston and Tampa Bays and other Gulf coastal areas where organic loading is high and bottom circulation is restricted. A deadly situation can arise in shallow coastal marshlands where deep navigation channels have been cut. These permit the penetration of saline Gulf waters into areas rich in organic peats. As hypoxic conditions develop during the warmer months, sulfate in the seawater is converted to the highly toxic hydrogen sulfide. This not only kills marine life in the immediate area, but it blocks the normal migratory route of species seeking to return to the Gulf to spawn. The situation is particularly acute in coastal Louisiana where extensive marshlands have been crisscrossed with canals and channels.

## OXYGEN LEVELS IN THE DEEP GULF

Although hypoxia regularly occurs on the continental shelf of the northern Gulf and has been recorded from special areas of the continental slope and in marginal bays and estuaries, most of the Gulf is well oxygenated. Except in the oxygen minimum layer, where the level may fall below 3.0 ml/L, dissolved oxygen levels in most open Gulf waters remain in the range of 4.5–5.0 ml/L (or even higher in the euphotic zone). Jochens et al. (2005) undertook a recent study to determine the processes that affect dissolved oxygen levels in deep waters of the Gulf, their rates of activity, and the balance that maintains the existing levels.

Well-oxygenated western Atlantic water is transported into the Gulf from the Caribbean Sea through the Yucatán Channel, which has a sill depth of around 2,000 m, but only the surface water can exit through the Straits of Florida, where the sill depth is only about 800 m. Oxygen can be added to the euphotic zone through atmospheric exchange at the sea surface as well as through photosynthesis by the phytoplankton, but the only sources of oxygen for the deep Gulf are the input of Caribbean water and vertical mixing of surface and deep waters within the Gulf. Oxygen is removed by respiration of living organisms, decomposition of sinking organic matter (dead bodies, secretions, fecal material, etc.), and oxidation of petroleum hydrocarbons derived from natural seeps, mineral extraction activities, shipping, and so forth. Additional demands on the oxygen supply could come from detritus and pollutants brought in by streams and from airborne pollutants. All of these processes are known to decrease exponentially with depth.

Analysis has shown that the two primary controls of oxygen levels in the deep Gulf are the quantity and quality of water entering from the Caribbean (which already has a high level of dissolved oxygen as well as an oxygen minimum layer) and vertical mixing processes within the Gulf itself. All other factors are clearly secondary and of relatively small magnitude. Most respiration and oxidation of organic material takes place in the upper 800 m of the water column. Continental shelf waters transported to the open Gulf could affect oxygen concentrations of surface waters, but they are not dense enough to sink into the deep Gulf. Most of the organic material brought in by the Mississippi River appears to drop out before reaching the deep Gulf. Oxygen consumption by most bottom sediments of the deep Gulf is a slow process, and the effect shows up only in the near-bottom waters. It has been estimated that between 80,000 and 200,000 metric tons of hydrocarbons enter the Gulf each year from natural oil and gas seeps in the bottom sediments. These reduce oxygen levels locally, but the overall effect on the Gulf is quite small. Around 10,000 metric tons of hydrocarbons are added to the Gulf annually by human activities. Again, there are significant local effects, but there is little impact on the total amount of dissolved oxygen. Clearly, the dissolved oxygen levels in the deep Gulf depend largely on the nature and volume of water transported into the Gulf from the Caribbean and on vertical mixing processes within the Gulf basin.

# Part III

---

**BIOLOGY**

Marine phytoplankton consists of a diverse group of primitive plants, each made up of a single cell or a colony of more or less identical cells and lacking multicellular tissues or organs. They are unattached and live a free-floating existence drifting around in the surface layer of the sea. Most are capable of photosynthesis, and since this process requires sunlight they normally flourish only in the lighted layer, or euphotic zone. In clear waters of the open Gulf this zone may extend from the surface to as deep as 200 m, but in areas of high turbidity, such as on the inner and middle continental shelf, the zone may be restricted to only a few meters. To facilitate the process of photosynthesis, phytoplankton organisms have developed a series of photopigments (chlorophylls, carotenoids, etc.) that mediate the conversion of the radiant energy of sunlight into chemical energy. These pigments impart color to the cells. Hence, different groups of marine algae are often recognized and referred to by their color: blue-green, green, or golden algae, and so forth. Classically, marine phytoplankton has been considered to include primarily diatoms, dinoflagellates, coccolithophores, and silicoflagellates, that is, algae 20 $\mu$m or more in diameter. However, in recent years it has been determined that many of the smaller species, nanoplankton (2–20 $\mu$m) and picoplankton (< 2 $\mu$m), are abundant and extremely important in the economy of the sea. These include some blue-green algae, microflagellates, and other forms. Most of these tiny algae contain photopigments and carry out photosynthesis, but some lack photopigments and apparently subsist by taking up dissolved organic chemicals or even particulate material. The smaller phytoplankton (< 20 $\mu$m in diameter) appears to be responsible for three-fourths of the cell counts and at least two-thirds of the photosynthetic production in the Gulf, and in this respect the blue-green alga *Trichodesmium* is very important.

To carry out photosynthesis, marine phytoplankton organisms must take in water (as a source of hydrogen) and either carbon dioxide or bicarbonate (as a source of carbon). Since water and carbon sources are present in ample quantities, neither is a limiting factor. Also required for growth and metabolism are the major nutrient elements (nitrogen, phosphorus, and sometimes silicon), and one or more of these is frequently in short supply and limits phytoplankton growth. In addition, phytoplankton organisms may require certain trace elements (iron, copper, nickel, cobalt, manganese, etc.) in minute amounts, and some species require specific organic compounds such as certain vitamins ($B_{12}$, thiamine, biotin, niacin, etc.) and growth factors. Many phytoplankton species are known to release organic substances, and they may discard into the water column up to 50 percent of the organic matter produced through photosynthesis. Such materials include carbohydrates, oils and waxes, amino and other acids, vitamins, various nitrogenous compounds, growth stimulators, antibiotics, toxins, and so forth.

In temperate and boreal areas of the world oceans, there are marked seasonal peaks when phytoplankton populations increase very rapidly and achieve high levels, or "blooms." At other seasons the populations are quite reduced. These highs and lows reflect periods when sunlight and nutrients are readily available and when they are not. In tropical marine areas where the water column remains stratified throughout the year, vertical mixing is limited, and despite the year-round availability of sunlight, standing crops are low, and only slight seasonal changes in phytoplankton abundance take place because there is no mechanism for pumping nutrients into the euphotic zone. In subtropical waters of the Gulf of Mexico, seasonal changes in phytoplankton abundance have been noted, especially in the northern Gulf, where surface waters are most affected by seasonal wind-induced mixing and where coastal waters seasonally receive peak runoff of nutrient-rich waters from the Mississippi and other major rivers.

In both oceanic and coastal waters, regular seasonal shifts in species composition have been documented, and these reflect the differential responses of individual species to changes in the physical environment, levels of nutrient availability, and the presence or absence of biochemical growth stimulators and inhibitors. Each species has certain requirements and tolerances, and these vary somewhat from one species to another. An inhibitor for one species may be a stimulator for another. So, as the season progresses, the chemical environment keeps changing, and species groups flourish for a time only to be replaced by others, and then others. For example, high levels of dissolved silicon favor diatoms, whereas low silicon limits diatoms and permits dinoflagellates and coccolithophores to bloom. In the higher-nutrient areas of coastal waters, blue-green algae, diatoms, and/or dinoflagellates dominate. In the relatively nutrient-poor offshore waters, silicoflagellates and coccolithophores are more in evidence. Microalgae are ubiquitous.

Many recent studies of marine phytoplankton populations have concentrated on two aspects, standing crop and rate of photosynthesis. Standing crop is a measure of the biomass present. It can be expressed in terms of cell density (number of cells per volume of water) or as the amount of photopigment present. Although photopigment concentration is only loosely correlated with cell density, it is much quicker and easier to determine, and surface pigment concentrations can be measured by satellite imagery. So, photopigment concentration is often employed as an imperfect but convenient index of standing crop. Photosynthesis involves the creation of organic matter from inorganic, and thus, it is often referred to as "primary production." Since it involves the uptake of inorganic carbon to produce organic material and the release of free oxygen, it can be measured by ascertaining the amount of radiocarbon ($^{14}$C) taken up or the amount of oxygen produced per volume of water during a given period. It may also be expressed in terms of the depth-integrated value, that is, the amount of photosynthesis taking place in a column of water lying beneath one square meter of sea surface and extending to the bottom of the euphotic zone (generally expressed as the depth to which only 1 percent of the sunlight penetrates).

In the Gulf of Mexico, studies have been carried out to determine species composition and abundance, horizontal and vertical distribution of photopigments, and rates of photosynthesis in relation to various physical factors such as Mississippi River outflow, upwelling, position and dynamics of the Loop Current, and the cyclonic and anticyclonic rings. Much work has also been done on the distribution and ecological relations of certain species that release highly toxic chemical compounds into the water column and cause outbreaks of the "red tide" and other phenomena. Such chemicals are not only lethal to other marine life, but they contaminate seafood products and pose distinct health hazards for humans. Major literature references to phytoplankton studies in the Gulf of Mexico include the earlier works of Balech (1967a, 1967b),

Björnberg (1971), Bushnell (1972), Conger, Fryxell, and El-Sayed (1972), El-Sayed (1972), Gaardner and Hasle (1971), Iverson and Hopkins (1981), Saunders and Fryxell (1972), Saunders and Glenn (1969), Steidinger (1972a, 1972b; 1973), and Steidinger and Williams (1970). More recent works include a number of unpublished theses and dissertations and summary reports by Biggs and Ressler (2000, 2001), Darnell and Schmidly (1988), Flint and Rabalais (1981), and Vargo and Hopkins (1990), as well as a series of recent studies around the mouth of the Mississippi River such as those of NOAA (1992) and Turner and Rabalais (1991).

## PHYTOPLANKTON COMPOSITION

The phytoplankton of the Gulf of Mexico is conveniently divided into six major groups, each of which is described briefly below, and many of the common genera are illustrated in figures 8.1–8.3.

### Blue-green algae (Cyanophyta)

Lacking a defined cell nucleus, the blue-greens are the most primitive of the algal groups. As seen in table 8.1, these algae lack chlorophyll-c, but they do possess several auxiliary photopigments. Although widespread in fresh waters and in bays and estuaries, few species are prominent in oceanic waters. Two species are important in the Gulf, a coccoid form, *Synechococcus* sp., and the filamentous species *Trichodesmium thiebautii* (fig. 8.3). Both are often widespread off the Louisiana coast, where they sometimes make up over 90 percent of the phytoplankton.

### Diatoms

Diatoms are encased in paired frustules composed of glass or silicon dioxide. They exist as single cells, long chains, or colonial aggregates. Resting spores are found in some species, particularly those that inhabit shallow nearshore waters. Residing primarily in bottom sediments, these spores are dormant life history stages that permit the species to persist through unfavorable environmental conditions. Diatoms are widespread in fresh and brackish environments as well as in marine waters throughout the world. In the Gulf of Mexico they constitute a major component of the phytoplankton, and nearly 1,000 species have been reported from Gulf waters. Diatoms often dominate the phytoplankton over the continental shelf. In general, they are more abundant inshore than offshore, but on an annual basis species diversity is greater in the open Gulf. As seen in table 8.1, diatoms

**Table 8.1.** *Major phytoplankton groups of the Gulf of Mexico and their characteristic photopigments.*

| Group | Chlorophyll types | Other photopigments |
|---|---|---|
| Blue-green algae | a | beta carotene, echinenone, zeaxanthin, phycocyanin, phycoerythrin, phycourobilin |
| Diatoms | a, c | beta carotene, fucoxanthin, diatoxanthin, diadinoxanthin, other carotenoids |
| Dinoflagellates | a, c | beta carotene, fucoxanthin |
| Coccolithophores | a, c | beta carotene, fucoxanthin, diatoxanthin, other carotenoids |
| Silicoflagellates | a, c | carotenes, xanthophylls |
| Microflagellates | variety (a, b, c, etc.) | variety, including alpha and beta carotene, lutein, alloxanthin, neoxanthin, zeaxanthin, etc. |

contain both chlorophyll-a and -c as well as several other photopigments. Diatom species are most abundant in areas with high nutrient concentrations, and populations tend to be reduced in nutrient-poor waters, particularly if the silicon concentration is low. Important diatom genera in the Gulf are listed in table 8.2, and some of the most important genera are shown in figure 8.1. Species of one genus, *Pseudo-nitzschia,* are known to secrete toxic chemicals that can be a health hazard for humans who consume contaminated seafood.

## Dinoflagellates

Dinoflagellates exist as single cells, chains, or filaments, and many species have hard external coverings, or "tests," composed of cellulose-like carbon compounds. Many possess a pair of whiplike or ribbonlike flagella and are capable of locomotion. Some species undergo diurnal vertical migrations, rising toward the surface in the morning and moving back down in the evening. The average rate of vertical movement is about 1.5 m/hr. Such migrations place the cells higher in the water column during the daytime (facilitating photosynthesis) and lower in the water (closer to the nutrient supply) at night. All dinoflagellate species form resting spores or cysts. However, not all species are photosynthetic. Some take up dissolved organic material from the water, while others are capable of ingesting particles such as bacteria. Hence, dinoflagellates are sometimes classified as zooplankton. The group reaches its greatest diversity in tropical marine waters. Within the Gulf of Mexico this is a very important component of the phytoplankton, and around 400 species and varieties have been recorded from the Gulf. Like diatoms, dinoflagellates are often more abundant in nearshore waters but more diverse offshore. However, unlike diatoms dinoflagellates do not require silicon and often do well in nutrient-poor conditions. A number of species are bioluminescent. Photochemicals produced by dinoflagellates are given in table 8.1, and important genera of the Gulf are listed in table 8.2. Some of the common types are shown in figure 8.2. Some species of dinoflagellates secrete chemicals into the water that are highly toxic to marine species and to humans who consume them. When certain dinoflagellate blooms are extremely dense, they may color the waters of the continental shelf or the bays and estuaries, producing the condition known as "red tide."

## Coccolithophores

Coccolithophores are single-celled algae possessing a pair of flagella and containing strange calcium carbonate structures (rings, plates, spines, bulbs, "trumpets," etc.) called coccoliths. Although some species are large, many are quite small (< 10 µm in diameter) and difficult to identify. Coccolithophores occur primarily in offshore waters of tropical and subtropical oceans. Around 200 species are known worldwide and nearly 100 have been reported from the Gulf of Mexico. They are rare in coastal waters and are most abundant and diverse in the open Gulf, where they are often the numerically dominant group. Their calcareous skeletal remains accumulate on the ocean bottom, and in some areas coccolithophore sediments may be tens to hundreds of meters thick. Photopigments are given in table 8.1, and important genera are listed in table 8.2. Common genera are shown in figure 8.3. In the open Gulf four species are of particular importance: *Coccolithus* (= *Emiliania*) *huxleyi, Umbellosphaera irregularis, U. tenuis,* and *Florisphaera profunda.*

**Table 8.2.** *Important phytoplankton genera reported from the Gulf of Mexico. The list includes some brackish-water diatoms abundant off the mouths of rivers. Asterisks denote genera with the largest number of species.*

| Coccolithophores | Diatoms | Dinoflagellates | Misc. algae |
|---|---|---|---|
| *Acanthoica** | *Achnanthes* | *Amphidium* | Blue-green algae |
| *Alisphaera* | *Amphora** | *Ceratium** | *Synechococcus* |
| *Calyptrolithina* | *Bacteriastrum* | *Cochlodinium* | *Trichodesmium** |
| *Calyptrosphaera* | *Biddulphia** | *Corythodinium* | |
| *Chrysochromulina* | *Campylodiscus** | *Dinophysis** | Silicoflagellates |
| *Coccolithus* | *Cerataulina* | *Exuviaella* | *Dictyota* |
| *Discosphaera* | *Chaetoceros** | *Glenodinium* | |
| *Emiliaria* | *Cocconeis* | *Gonyaulax** | Microalgae |
| *Florisphaera* | *Coscinodiseus** | *Gymnodinium* | *Bodo* |
| *Gephyrocapsa* | *Detonula* | *Heterodinium* | *Carteria* |
| *Helicosphaera** | *Diploneis* | *Noctiluca* | *Chlamydomonas* |
| *Michaelsarsia* | *Fragilaria* | *Ornithocercus* | *Chrysomonas* |
| *Phaeocystis* | *Hemiaulis* | *Oxytoxum* | *Dinobryon* |
| *Platychrysis* | *Leptocylindrus* | *Peridinium** | *Pavlova* |
| *Pontosphaera* | *Mastogloia** | *Podolampus* | |
| *Poricalyptra** | *Navicula** | *Pronoctiluca* | |
| *Prymnesium* | *Nitzschia* | *Prorocentrum* | |
| *Rhabdosphaera* | *Plagiogramma* | *Protoperidinium* | |
| *Scyphosphaera** | *Pleurosigma* | | |
| *Syracolithus* | *Rhizosolenia** | | |
| *Syracosphaera* | *Surirella* | | |
| *Umbellosphaera* | *Synedra* | | |
| *Zygosphaera* | *Thalassionema* | | |
| | *Thalassiosira* | | |
| | *Thalassiothrix* | | |
| | *Triceratium** | | |

## Microalgae

These organisms make up a very diverse catchall category representing several different groups of planktonic marine algae. They are all extremely small (< 20 μm) single-celled organisms, some of which possess flagella. Although 4 is the most common number, there may be as few as 1 or as many as 16 flagella. Some species are photosynthetic, some take up dissolved organic material, and others apparently ingest particles. Many combine the above methods of nutrition. Being so small, these algae pass through the mesh openings of most collecting nets, and being very delicate, they require special methods of capture and study. Hence, they have been poorly inves-

tigated in the Gulf of Mexico. However, they are apparently quite important components of the phytoplankton. Recent studies show that they are abundant in both the inshore and offshore waters of the Gulf, and one species is responsible for the "brown tide" of south Texas. Photopigments are given in table 8.1, and important genera are listed in table 8.2. A few examples are shown in figure 8.3.

## Silicoflagellates

Silicoflagellates are very small flagellated marine algae that are sometimes included with the microflagellates. However, they are unique in possessing a reticulate external or internal skeleton of siliceous material. Virtu-

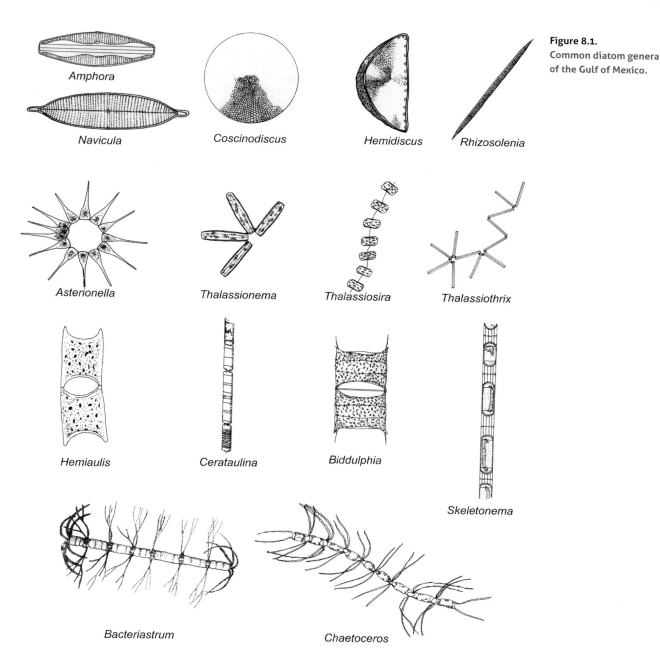

**Figure 8.1.**
Common diatom genera
of the Gulf of Mexico.

*Amphora*

*Navicula*

*Coscinodiscus*

*Hemidiscus*

*Rhizosolenia*

*Asterionella*

*Thalassionema*

*Thalassiosira*

*Thalassiothrix*

*Hemiaulis*

*Cerataulina*

*Biddulphia*

*Skeletonema*

*Bacteriastrum*

*Chaetoceros*

ally no work has been done on this group in the Gulf of Mexico. Photopigments of this group are given in table 8.1. A single genus reported from the Gulf is given in table 8.2 and shown in figure 8.3.

## DISTRIBUTION PATTERNS

Examination of phytoplankton data from the Gulf of Mexico reveals patterns of horizontal and vertical distribution that reflect both the ambient environmental conditions and the requirements of individual species.

### Horizontal distribution

The most obvious pattern of regional differentiation is seen in the distinction between the neritic and oceanic phytoplankton flora. Many species and certain genera are more heavily represented in shelf waters, on the one hand, or open ocean waters, on the other. Table 8.3 provides representative examples of diatoms, dinoflagellates, and coccolithophores characteristic of each realm. However, some species are more cosmopolitan and are found with similar frequency in both habitats, and ex-

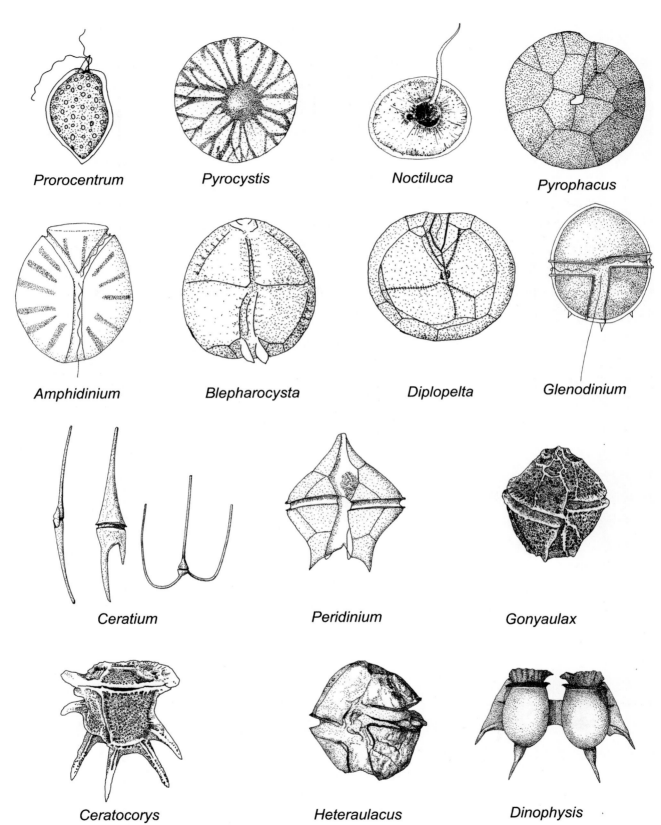

**Figure 8.2.** Common dinoflagellate genera of the Gulf of Mexico.

*Prorocentrum* *Pyrocystis* *Noctiluca* *Pyrophacus*

*Amphidinium* *Blepharocysta* *Diplopelta* *Glenodinium*

*Ceratium* *Peridinium* *Gonyaulax*

*Ceratocorys* *Heteraulacus* *Dinophysis*

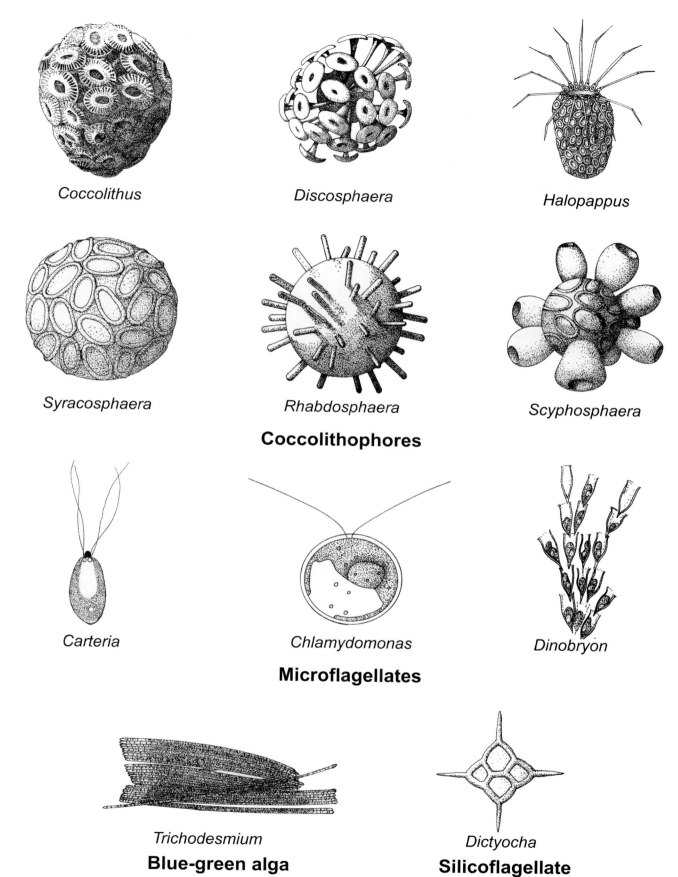

*Coccolithus*     *Discosphaera*     *Halopappus*

*Syracosphaera*     *Rhabdosphaera*     *Scyphosphaera*

**Coccolithophores**

*Carteria*     *Chlamydomonas*     *Dinobryon*

**Microflagellates**

*Trichodesmium*
**Blue-green alga**

*Dictyocha*
**Silicoflagellate**

**Figure 8.3.** Common blue-green algae, coccolithophore, silicoflagellate, and microflagellate genera of the Gulf of Mexico.

**Table 8.3.** *Onshore-offshore patterns of phytoplankton distribution. (Data from several sources.)*

**Groups that are primarily neritic**

Diatoms – *Asterionella japonica, Leptocylindrus danicus, Skeletonema costatum, Thalassionema nitzschioides, Lithodesmium* spp., *Streptotheca* spp.

Dinoflagellates – *Diplopelta asymmetricus, Gymnodinium splendens, Peridinium brochii, Prorocentrum micans, Pyrophacus horologium*

Coccolithophores – *Umbellosphaera irregularis*

**Groups that are primarily oceanic**

Diatoms – *Ethmodiscus* spp., *Gossleriella* spp., *Planktoniella* spp.

Dinoflagellates – *Ceratium teres, Ceratocorys horrida, Pyrocystis pseudonoctiluca, Amphisolenia* spp., *Cladopyxis* spp., *Heterodinium* spp., *Kofoidinium* spp., *Murrayella* spp., *Ptychodiscus* spp., *Tripsolenia* spp.

Coccolithophores – *Florisphaera profunda*

**Groups that occur in both neritic and oceanic areas**

Diatoms – *Chaetoceros coarctatum, Guinardia flaccida, Hemiaulis hauckii, Plagiogramma vanheuckii, Rhizosolenia imbricata, R. robusta, Thalassiothrix frauenfeldii*

Dinoflagellates – *Blepharocysta splendomaris, Ceratium carriense, C. furca, C. fusus, C. massiliense, C. trichoceros, C. tripos, Heteraulacus polyedricus, Peridinium* spp., *Podolampus* spp.

Coccolithophores – *Coccolithus huxleyi*

---

amples of such species are also provided in the table. Furthermore, on the Louisiana continental shelf, which has been more thoroughly studied, it is tentatively suggested that there is also some differentiation between diatom species characteristic of the inner, middle, and outer shelf. Inner shelf species include *Rhizosolenia denticulata, Skeletonema costatum,* and *Thalassionema nitzschioides.* Middle shelf forms include *Chaetoceros affinis* and *C. breve.* On the outer shelf *Chaetoceros decipiens* is more prominent. Trans-shelf species include the diatom *Leptocylindrus danicus* and the blue-green alga *Trichodesmium thiebautii.*

Another pattern of horizontal differentiation shows up in the coastwise distribution of phytoplankton groups, and this is best seen along the continental shelf of the northwestern Gulf from the Mississippi River Delta to the Rio Grande (table 8.4). Near the mouth of the Mississippi River diatoms are numerically dominant, and these include a number of freshwater or brackish-water species. Proceeding westward diatoms still dominate, but the freshwater species drop out. In the nutrient-poor waters off south Texas the number of important diatom groups is diminished, and dinoflagellates are more prominent. In both the onshore-offshore and downcoast patterns nutrient availability is clearly one of the key factors. Shelf

waters tend to be nutrient rich, whereas open Gulf waters are generally nutrient poor. Nutrient levels are quite high near the Mississippi River mouth and relatively depleted off south Texas. This general paradigm, although somewhat simplistic, appears to hold up on even finer geographical scales.

**Vertical distribution**

As discussed earlier, sunlight enters the sea from above, and its intensity decreases with depth in the water column. Likewise, the quality of the light changes with depth. Near the surface most wavelengths of the visible spectrum are present, and the light is rich in the longer wavelengths, including the yellow-green portion of the spectrum. With increasing depth the longer wavelengths become depleted, and in the lower portion of the euphotic zone light is relatively richer in the shorter wavelengths, or blue end of the spectrum. Both the intensity and quality of the light are important in determining phytoplankton species distribution with respect to depth. The types of photopigment present in phytoplankton cells play a role in determining preferred light levels. Although chlorophyll-a appears to be the primary light-trapping pigment in most, if not all, phytoplankton species, the auxiliary pigments (table 8.1) also play a role.

**Table 8.4.** *Numerically dominant phytoplankton genera reported from several stations along the Louisiana-Texas continental shelf. The "fw" in parentheses refers to genera in which freshwater species are dominant. (Data from several sources.)*

| Genera | Louisiana East | Central | West | South Texas |
|---|---|---|---|---|
| **Diatoms** | | | | |
| *Thalassiothrix* | X | | | |
| *Navicula* (fw) | X | | | |
| *Melosira* (fw) | X | X | | |
| *Cyclotella* (fw) | X | X | X | |
| *Asterionella* | X | X | X | |
| *Chaetoceros* | X | X | X | |
| *Nitzschia* | X | X | X | X |
| *Skeletonema* | X | X | X | X |
| *Thalassionema* | X | X | X | X |
| *Cerataulina* | | X | X | |
| *Leptocylindrus* | | X | X | |
| *Rhizosolenia* | | X | X | X |
| *Thalassiosira* | | X | X | X |
| *Achnanthes* | | | X | |
| *Coscinodiscus* | | | X | |
| *Guinardia* | | | X | |
| **Dinoflagellates** | | | | |
| *Gonyaulax* | | X | | X |
| *Gymnodinium* | | | X | |
| *Katodinium* | | | X | |
| *Ceratium* | | | | X |
| *Prorocentrum* | | | | X |

Some phytoplankton species tolerate high light levels and often flourish near the ocean surface, whereas others are subject to photoinhibition and survive well only some distance below the surface. Examples of light-tolerant as well as shade-adapted species from waters of the open Gulf are given in table 8.5. The blue-green algae are particularly tolerant of high light levels, as are some diatom, dinoflagellate, and coccolithophore species, but most of the oceanic phytoplankton species are shade adapted and are likely to be absent or poorly represented in near-surface water samples.

Maintenance of proper depth in the water column is a problem that has been resolved in a number of different ways. Some phytoplankton species, particularly many diatoms, have developed spines and other projections as well as chains and colonies that increase their surface area and retard the rate of sinking. Many species maintain a low specific gravity by the inclusion of oil droplets or certain low-density salts that enhance their ability to float. For example, in the phosphorescent dinoflagellate *Noctiluca,* large intracellular vacuoles are rich in ammonium chloride, which has a lower specific gravity than that of the sodium chloride of seawater. Some coccolithophores are able to increase or reduce their number of coccoliths, thus changing their density and enhancing their ability to sink or rise. Flagellated forms are motile and can achieve appropriate depths through directed movement. Finally, it should be mentioned that some phytoplankton species are found below the euphotic zone, where photosynthesis should be impossible. Not all species are photosynthetic, but it is possible that some photosynthetic species periodically move down into this zone to replenish their supplies of nutrient elements.

Since each water mass has characteristic species, phytoplankton can sometimes provide evidence of water mass movements. For example, indicator species mark the intrusion of Loop Current waters across the Florida continental shelf. From time to time shade-adapted species appear in phytoplankton collections made near the surface, and their presence is taken as an indication of upwelling of water from deeper layers. The most striking example is the finding of typical Antarctic diatom species (*Chaetoceros atlanticum, C. concavicorne, C. convolutum,* and *Dactyliosolen antarcticus*) in surface waters off Yucatán. These apparently represent dead cells carried northward in the Antarctic Intermediate Water and brought to the surface from a depth of several hundred meters by very strong upwelling currents.

Many species of diatoms, dinoflagellates, coccolithophores, and so forth normally dwell in the benthos for part or all of their life histories. Yet, particularly in shallow neritic areas, these often appear in plankton samples, having been suspended by waves or water currents. Such transients are not considered to be true components of the plankton since their presence is accidental. In freshwater samples these are referred to as *pseudoplankton* ("false plankton"), but the marine term is *tychoplankton* ("accidental plankton").

**Table 8.5.** *Examples of phytoplankton species and genera from the Gulf of Mexico that are known or presumed to be light tolerant or shade adapted. (Data from several sources.)*

**Light-tolerant species**

Blue-green algae – *Trichodesmium thiebautii*
Diatoms – *Chaetoceros compressum, C.decipiens, Guinardia flaccida, Hemiaulis membranaceus, Thalassiosira subtilis*
Dinoflagellates – *Ceratium trichoceros, C. massiliense*
Coccolithophores – *Umbellosphaera irregularis*

**Shade-adapted groups**

Diatoms – *Biddulphia tuomeyi, Coscinodiscus eccentrica, Rhizosolenia imbricata*
Dinoflagellates – *Ceratium arietinum, C. coarctatum, C. hexacanthum, C. lunula, C. tenue, Dinophysis hastata, D. odiosa, D. schuettii, Gonyaulax birostris, G. pacifica, G. tamarensis,* genera *Amphisolenia, Centrodinium, Heteroschisma, Murrayella, Tripsolenia*
Coccolithophores – *Florisphaera profunda*

## Seasonal distribution

Phytoplankton investigators have been impressed by the short-term temporal variation in phytoplankton composition. In nearshore areas influenced by river or bay water, particularly where there are strong currents, the phytoplankton composition at a given location can change radically on a daily, even an hourly, basis. This reflects the fact that different water masses sweeping through the area are each characterized by distinct phytoplankton communities. This is particularly true around the mouth of the Mississippi River. However, the phenomenon is more or less widespread and reflects the microscale transport of water parcels along the continental shelf as well as in the open Gulf. Swirls, filaments, rings, eddies, and loops all result in much short-term variability in phytoplankton composition. Parcels of water containing Mississippi River phytoplankton have even been recorded off southwest Florida, apparently having been transported there along the edge of the Loop Current.

The problem is compounded when attempts are made to ascertain seasonal trends in phytoplankton composition. Analysis of repeated samples taken on a regular monthly basis throughout the year, however, does provide considerable insight into seasonal changes. Such a series is available for dinoflagellates taken in samples from the west-central coast of Florida (off Tampa Bay and Charlotte Harbor) (fig. 8.4). Of the 230 species and varieties identified, 89 (38.7 percent) were present throughout the year. The remainder were seasonal in occurrence. Among the seasonal species 54.6 percent were present during the summer, but only 29.8 percent occurred during the winter

months. Among the seasonal species 51.8 percent were present during only a single season, and 22.7 percent each occurred during two or three seasons. Thus, even in this subtropical environment the marine dinoflagellates clearly show strong seasonal changes in species composition, with the greatest diversity in summer and the least during the winter. Seasonal shifts in phytoplankton species composition have also been demonstrated on the continental shelf of the northern Gulf, but here the summer/winter pattern is modified by Mississippi River outflow (which is generally high in the spring and low in late summer and fall) as well as water column stratification and bottom hypoxia. Elsewhere the seasonal composition

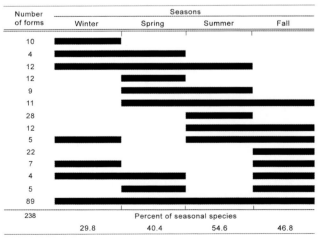

**Figure 8.4.** Seasonal occurrence of individual dinoflagellate species (and varieties) at offshore and inshore stations off the west-central Florida coast. (Data from Steidinger and Williams 1970.)

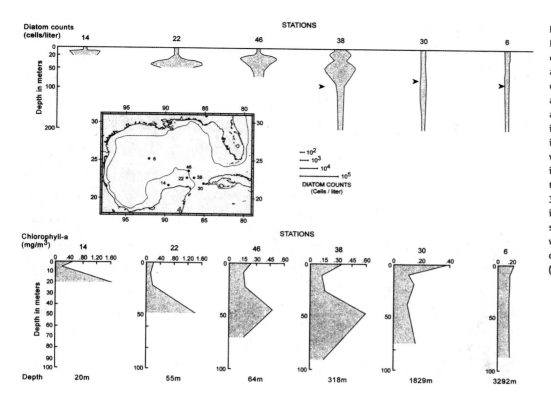

**Figure 8.5.**
Depth distribution of diatom densities (cells/L) and chlorophyll-a concentrations (mg/m³) at stations made on or around the Yucatán shelf in late summer. Arrows indicate the depths at which light intensity is 1 percent of that at the surface. Stations 6, 30, and 38 were made in deep water, whereas stations 14, 22, and 46 were made over the continental shelf. (From E1-Sayed 1972.)

may be influenced by upwelling or the aperiodic intrusion of the Loop Current and its filaments or rings and eddies.

## PHYTOPLANKTON STANDING CROP

The standing crop is the amount of living phytoplankton present at any one time. Thus, it is the quantity of living plant material available as food for the grazers or herbivores of the sea. Two primary methods have been employed to estimate standing crop: cell counts and chlorophyll density. These methods have both theoretical and practical problems, and neither method is entirely satisfactory in depicting the actual living plant matter present, but both have been widely used.

### Abundance of phytoplankton cells

In most of the older and some of the newer literature, standing crop has been expressed in terms of the number of cells per liter of water. Here it is important to distinguish between living and dead cells. Since phytoplankton cells vary greatly in size (from the tiny microflagellate and blue-green algal cells to the giant cells of some diatoms and coccolithophores), some investigators have expressed their results in terms of cell volumes or estimated carbon content. Also, since many phytoplankton species occur as chains or colonies, some workers have reported their results as "unit counts" (where, for example, 10 cells in a chain equal 1 solitary cell). Furthermore, since most microalgae require special methods of collection and study, they have often been overlooked or ignored in routine phytoplankton counts. Despite these problems, cell density does provide a reasonable estimate of phytoplankton abundance, especially since the difference between low and high density is often several orders of magnitude.

From the available information it is clear that throughout the open Gulf phytoplankton cell densities in the upper few meters of surface waters are quite low. Because of the high light intensity and low nutrient content, there are only a few hundred cells per liter, but cell densities generally increase severalfold lower in the euphotic zone (fig. 8.5). In areas of upwelling the surface concentration may rise locally to more than 10,000 cells per liter. On continental shelves, where nutrients are more readily available, cell densities often average in the tens of thousands, and in areas of very high nutrients, densities may rise into the tens of millions per liter. Stated more succinctly, cell densities in the open Gulf run around $10^2$ cells/L, in upwelling areas and on fairly nutrient-rich shelves they range around $10^4$–$10^5$ cells/L, and in areas of highest nutrient concentrations densities of $10^7$–$10^8$ cells/L are at-

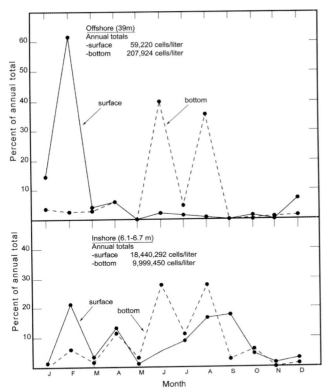

**Figure 8.6.** Monthly diatom population levels expressed as a percentage of the annual total, for surface and bottom collections in offshore and inshore waters off the west-central Florida coast, based on data from August 1965 through July 1966. (Data from Saunders and Glenn 1969.)

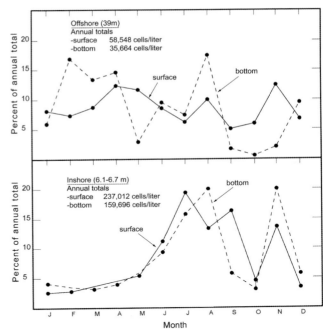

**Figure 8.7.** Monthly dinoflagellate population levels expressed as a percentage of the annual total, for surface and bottom collections in offshore and inshore waters off the west-central Florida coast, based on data from August 1966 through July 1967. (Data from Steidinger and Williams 1970.)

tained. Such high densities are often the result of extraordinary blooms of one or a few species of blue-green algae, diatoms, or dinoflagellates.

Studies carried out on the west coast of Florida based on cell densities provide information on seasonal changes in diatom and dinoflagellate populations (figs. 8.6 and 8.7). Densities of surface diatom populations averaged over 300 times greater at nearshore than at midshelf stations, but densities at bottom stations were only about 50 times greater at the nearshore locations. The bulk of the surface diatoms at the offshore sites appeared in February, and the numbers were quite low throughout the summer months. However, in bottom waters of the offshore stations about three-quarters of the annual cell density occurred in July and August. In the nearshore surface waters diatom densities were moderate during both summer and winter, but in bottom waters they achieved the highest levels in the summer.

For dinoflagellates cell densities in both surface and bottom waters were four or five times greater at the in-

shore stations than they were offshore. At the inshore sites there were strong peaks in summer and fall. At the offshore stations dinoflagellates of the bottom water showed peaks in winter and early spring as well as late summer, but there was only minor seasonal variation at the surface. These data indicate differences between phytoplankton populations at nearshore and midshelf locations, especially for diatoms. They show the generally low-density surface populations of diatoms offshore and the dramatic effect of a major bloom. Although dinoflagellate populations behaved differently, they too gave clear evidence of seasonality, especially at the inshore locations.

Recent studies on the inner continental shelf of the northwestern Gulf include examination of cell densities. Data from a given cruise (fig. 8.8a) show that diatoms > 8 μm in diameter made up from 60 to over 80 percent of the surface phytoplankton population for most of the Louisiana and upper Texas coast. However, below the level of Galveston Bay diatom abundance dropped below 40 percent of the total population. Figure 8.8b shows the surface distribution of a single species, *Skeletonema costatum,* from the same cruise. All along the coast this diatom was fairly abundant (> 1 × 10^6 cells/L), but its highest

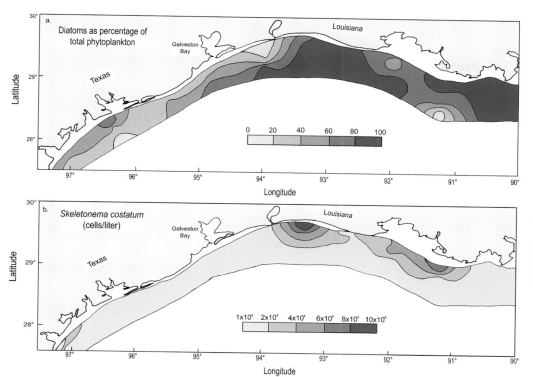

**Figure 8.8.** Phytoplankton distribution patterns in surface waters of the inner continental shelf off Louisiana and upper Texas as shown by data from a single cruise. (a) Diatoms as a percentage of the total phytoplankton population of cells > 8 μm in diameter. (b) Distribution of the diatom species *Skeletonema costatum,* given as the number of cells/L. (From Dortch 1996.)

densities were evident close to shore, often near sources of fresher water. Such patterns change with respect to season and in response to freshwater outflow and nutrient availability.

Additional studies reveal the effects of Mississippi River floodwaters on phytoplankton composition and cell densities at a single station off the central Louisiana coast (table 8.6). The flood that struck the shelf during the summer of 1993 brought high levels of nutrients, including silicates. In response, phytoplankton cell density increased by a factor of nearly 15. This was primarily because of a bloom of blue-green algae that increased almost 23-fold, and to a lesser extent because of the elevation of the diatom population, which rose only 3.6-fold. In this area blue-greens are often more numerous than diatoms, but because they are quite small their cells make up a smaller fraction of the biomass. However, during the flooding event blue-green algae were so abundant that their biomass (expressed as picograms of carbon/liter) exceeded that of the diatoms.

## Concentration of chlorophyll

A second method of estimating the standing crop of phytoplankton involves determining the chlorophyll concentration. It is recognized that this is not a perfect in-

**Table 8.6.** *Comparison of phytoplankton density (cell number and estimated carbon biomass) averaged over July–September for the period 1990–1992 vs. 1993 (flood year) at a single station off the central Louisiana coast. The carbon values are estimated by assuming cell volumes (cyanobacteria and diatoms assumed to be spherical cells of diameter 2 μm and 20 μm, respectively) and converting cell volumes to carbon content (assuming a carbon content of 0.185 pg C/μm³ of cell volume). (From Dortch 1994.)*

| Measure | 1990–1992 | 1993 | Ratio 1993/ 1990–1992 |
|---|---|---|---|
| Cell numbers (cells/L) | | | |
|   Total phytoplankton | $1.26 \times 10^8$ | $1.85 \times 10^9$ | 14.7 |
|   Cyanobacteria | $8.07 \times 10^7$ | $1.84 \times 10^9$ | 22.8 |
|   Diatoms | $4.49 \times 10^5$ | $1.64 \times 10^6$ | 3.6 |
|   Dinoflagellates | $5.63 \times 10^4$ | $7.90 \times 10^4$ | 1.4 |
| Organic carbon | | | |
|   Cyanobacteria | $4.68 \times 10^7$ | $1.07 \times 10^9$ | 22.8 |
|   Diatoms | $2.63 \times 10^8$ | $9.53 \times 10^8$ | 3.6 |

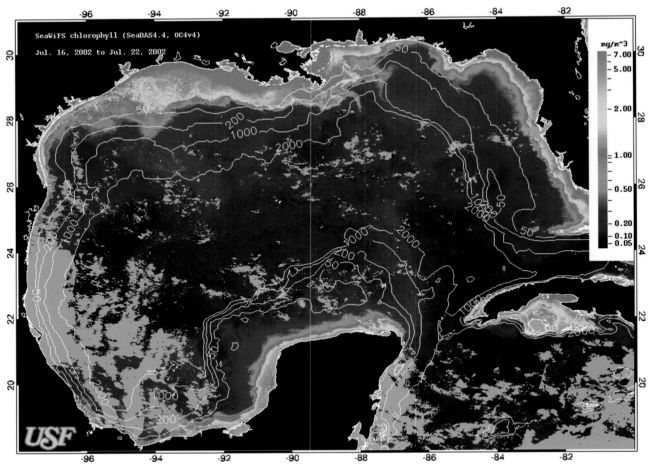

**Figure 8.9.** Distribution of chlorophyll-a in surface waters of the entire Gulf of Mexico in January 2002, as determined from satellite data. The picture is printed in false colors selected for contrast. (Courtesy University of South Florida Institute of Marine Remote Sensing.)

dicator, since the concentration of this photopigment is only loosely correlated with living phytoplankton biomass. However, the method is quick and easy, and many samples can be collected and analyzed in a short period. Furthermore, the concentration of chlorophyll in surface waters can be estimated from certain satellite images. By this means the surface concentration of chlorophyll over large areas can be measured in a relatively short period, providing a near-instantaneous picture of the standing crop in surface waters of the entire Gulf. The method is particularly useful for comparative purposes. (See figs. 8.9 and 8.10.)

In offshore waters of the Gulf of Mexico surface chlorophyll values range from 0.09 to 0.28 mg/m³ and average around 0.15 mg/m³ (fig. 8.11). In upwelling areas they are much higher than those of the normal open Gulf and may exceed 2.0 mg/m³. Over continental shelves surface values average about three times those of the open

Gulf, but they vary locally depending on nutrient availability. For example, exceptionally high levels of around 10.0 mg/m³ sometimes occur in the vicinity of the Mississippi River Delta. Downcoast the concentrations decrease to around 1.2 mg/m³ in nearshore waters of south Texas. Toward the outer shelf these values decrease to about 0.3 mg/m³ off Louisiana and 0.2 mg/m³ off south Texas.

Vertical profiles show that in waters of the open Gulf and continental shelves the highest chlorophyll concentrations generally are not found at the surface but reach maximum levels somewhat deeper in the water column, often near the bottom of the euphotic zone or the bottom of the surface mixed layer. Here concentrations may be several times higher than those at the surface. In fact, fairly high concentrations often extend well below the bottom of the euphotic zone, and this may be a result of the sinking of dead phytoplankton cells from above, the temporary presence of living cells, or both.

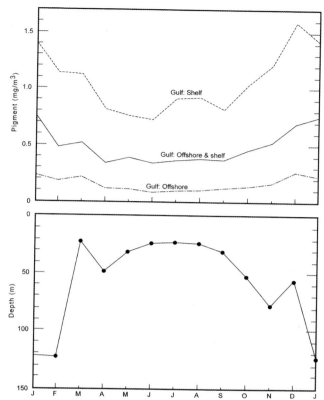

**Figure 8.10.** Average monthly pigment concentration in surface waters of the Gulf of Mexico—total Gulf (offshore + shelf), continental shelves (shelf), and open Gulf waters (offshore)—based on satellite data, compared with the average monthly depth of the surface mixed layer for the entire Gulf. (From Müller-Karger et al. 1991.)

In order to express the entire standing crop of phytoplankton within the euphotic zone, the photopigment is often given in terms of depth-integrated values, that is, the total amount of chlorophyll in a column of water lying beneath one square meter of surface and extending to the bottom of the euphotic zone. Depth-integrated values in the open Gulf range from about 5 to 21 mg/m² and average around 10 mg/m², but in upwelling areas the level may be more than double that. Depth-integrated chlorophyll levels on continental shelves average two or three times those of the open Gulf. Around the mouth of the Mississippi River depth-integrated values are quite high, sometimes exceeding 90 mg/m². Downcoast the values decrease to less than 20 mg/m² off south Texas. Depth-integrated values for selected stations in the open Gulf during June are shown in figure 8.12.

A seasonal comparison of average surface and depth-integrated chlorophyll values is provided in figure 8.13.

The surface concentrations are highest (> 0.2 mg/m³) during the winter and lowest (ca. 0.160–0.165 mg/m³) in the spring and summer. Depth-integrated values also reach maximum levels (ca. 12.6 mg/m²) in the winter, but the summer values are nearly as high. The lowest levels are seen in the fall months. How closely these surface chlorophyll values track the actual seasonal phytoplankton standing crops is not certain.

## PRIMARY PRODUCTION

As noted earlier, primary production involves the uptake of water and carbon dioxide and, in the presence of sunlight and photopigments, the production of organic matter and the release of oxygen. This process can be measured by determining the rate of carbon fixation or the rate of oxygen production. Most studies in the Gulf of Mexico have employed the former method, which involves the transformation of inorganic to organic carbon using $^{14}C$ as a tracer. Results of such studies may be expressed as production per cubic meter of surface water or as depth-integrated values, and they may be given as hourly, daily, or annual rates of production. For simplicity, only annual production will be discussed below. (Note: the hourly rate × 12 gives the daily rate, and that × 360 gives a conservative estimate of the annual rate at this subtropical latitude.)

Primary production in the Gulf of Mexico is characteristic of subtropical marine waters worldwide, that is, fairly low annual production with only modest seasonal variability. In surface waters of the open Gulf production rates run around 1.0 g C/m³ per year, and coastal waters average two and a half to three times this rate. Patterns of vertical distribution of $^{14}C$ uptake at selected offshore stations in June are shown in figure 8.12. Here it is seen that maximum uptake rates do not occur at the surface, apparently because of photoinhibition at intense light levels. Rather, maximum uptake rates are observed at depths where there is 25–50 percent of surface light intensity. Thereafter, uptake diminishes to the bottom of the euphotic zone, and it is negligible below that level. As seen in the figure, there is no clear correlation between the quantity of chlorophyll and the rate of carbon fixation at a given depth. Furthermore, high chlorophyll levels often continue into waters below the euphotic zone, where carbon fixation is essentially absent.

Depth-integrated primary production values for low-nutrient open ocean systems of the world generally fall in the range of 50–160 g C/m² per year. El-Sayed (1972)

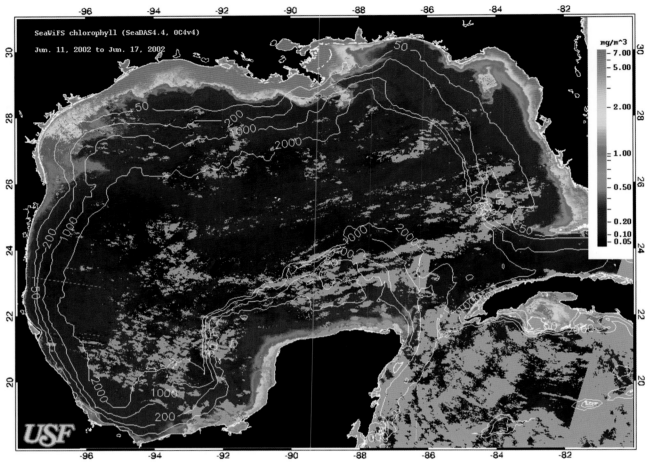

**Figure 8.11.** Distribution of chlorophyll-a in surface waters of the entire Gulf of Mexico in July 2002, as determined from satellite data. The picture is printed in false colors selected for contrast. (Courtesy University of South Florida Institute of Marine Remote Sensing.)

reported the average value for the open Gulf of Mexico to be 25 g C/m$^2$ per year, equivalent to about 5.79 mg C/m$^2$ per hour, which is quite low even for open oceans. In upwelling areas values of up to 90 g C/m$^2$ per year have been reported. Production on the continental shelves averages about 34 g C/m$^2$ per year, or 1.4 times that of the open Gulf, but it varies in relation to nutrient availability. For example, very high values of 250–300 g C/m$^2$ per year characterize the Louisiana coast, and values of more than 1,000 g C/m$^2$ per year have been recorded off the mouth of the Mississippi River. Much lower values are found downcoast, and annual phytoplankton production off south Texas is around only 20 g C/m$^2$ per year. In the eastern Gulf primary production rates have been reported from the outer edge of the Florida shelf. Along the northern third, values fall around 30 g C/m$^2$ per year, but along the southern third, which is more affected by upwelling, values are around 90 g C/m$^2$ per year. On the continental

shelf of the Bay of Campeche in the southern Gulf, summer studies revealed very low values averaging only 14.7 g C/m$^2$ per year. Taking into account both the continental shelves and open Gulf, El-Sayed (1972) concluded that the average rate of primary production for the Gulf of Mexico was about 27 g C/m$^2$ per year (ca. 6.25 mg C/m$^2$ per hour).

## Seasonal changes

The seasonal pattern of $^{14}$C uptake in the open Gulf, as presented by El-Sayed (1972), is shown in figure 8.13. Surface values were found to average about 0.328 mg C/m$^3$ per hour (= 1.4 g C/m$^3$ per year), with the highest values in the winter and the lowest in the fall. Depth-integrated values averaged 6.25 mg C/m$^2$ per hour (= 27 g C/m$^2$ per year), with the highest levels in the winter and the lowest in the spring, winter values being about twice those of the spring.

Satellite images showing surface chlorophyll concen-

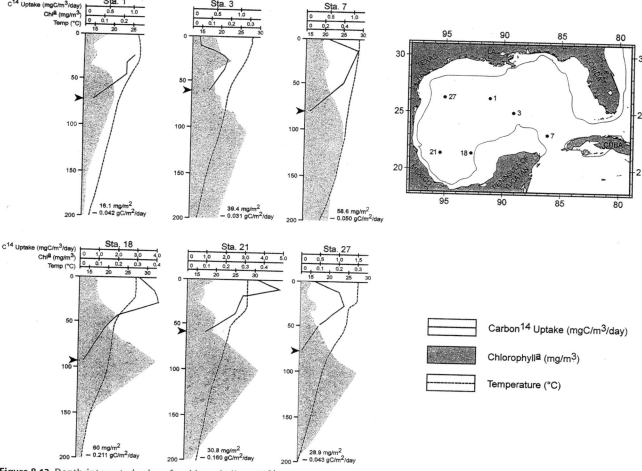

**Figure 8.12.** Depth-integrated values for chlorophyll-a and ¹⁴carbon uptake for selected stations in the open Gulf of Mexico in June. (From El-Sayed 1972.)

trations throughout the Gulf reveal a slightly different picture. Here the seasonal cycle appears to be well defined and generally synchronous throughout the Gulf. The highest standing crop values (ca. 0.2 mg/m$^3$) occur in the winter (December–February) and the lowest values (0.06 mg/m$^3$) appear in the late spring and early summer (May–July), a threefold difference. The surface pigment concentrations appear to be controlled largely by the depth of the surface mixed layer. Whether the winter value in the Gulf is two or three times that of the late spring, this difference is small compared to the seasonal variation in temperate or boreal marine waters.

## Deepwater "hot spots"

It is generally true that the open Gulf of Mexico is essentially a biological desert where production is restricted by the paucity of nutrients, but within this desert there are oases or "hot spots," localities where production rates are

much higher than those in surrounding areas, and these owe their existence to physical factors that operate locally to increase nutrient levels within the euphotic zone. Four such mechanisms will be discussed. These include (1) classical wind-driven upwelling, (2) entrainment and off-shore transport of freshwater, (3) vertical pumping along frictional boundaries of strong currents, and (4) vertical pumping within the core of cyclonic rings and eddies.

### WIND-DRIVEN UPWELLING

Wind blowing across the ocean pushes the surface layer of water ahead of it, but because of the Coriolis effect, in the Northern Hemisphere this layer tends to veer to the right. When water is transported from the continental shelf, it must be replenished by deeper water from below. Being richer in nutrients, such upwelled water becomes the location for increased phytoplankton production and standing crop. Thus, a strong wind from the east would

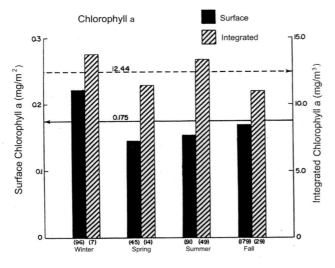

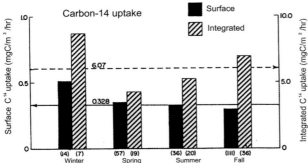

**Figure 8.13.** Seasonal variation in chlorophyll-a and ¹⁴carbon uptake (surface and depth-integrated values) for the Gulf of Mexico. Solid horizontal lines indicate average surface values, and dashed horizontal lines indicate average integrated values. Numbers in parentheses below the vertical bars represent the numbers of observations from which the averages were derived. (From El-Sayed 1972.)

push Yucatán shelf waters toward the north, resulting in upwelling along the outer shelf and upper slope of this bank. In similar fashion, wind-induced upwelling occurs along the outer shelf and upper slope of other areas of the Gulf, the actual locations depending on the orientation of the shelf edge and wind speed and direction. As pointed out earlier, primary production in upwelling areas may be triple the rate in nearby undisturbed waters.

### ENTRAINMENT AND OFFSHORE TRANSPORT OF FRESHWATER

Freshwater brought to the Gulf by major rivers tends to be rich in nutrients, but for the most part such water is swept downcoast and remains on the continental shelf. However, under special circumstances the river water may be carried offshelf into the deeper water of the open Gulf, where it becomes the focus for elevated phytoplankton production. Patches of low-salinity water are frequently drawn offshore from the mouth of the Mississippi River. This happens when there are two counterrotating rings near the river mouth and when the cyclonic ring lies north or east of its partner. The parcel of freshwater is drawn out by the strong currents between the two rings. Such freshwater parcels have been reported from open water in the northern Gulf and even from south Florida, having been entrained and transported along the margin of the Loop Current.

### VERTICAL PUMPING ALONG FRICTIONAL BOUNDARIES OF STRONG CURRENTS

Along the outer margins of high-velocity surface currents, some vertical mixing of water masses takes place. In the Gulf this occurs along the outer edge of the Loop Current and in the margins of large rotating eddies, and this appears to be a very important mechanism for introducing nutrients into the euphotic zone. In such areas phytoplankton standing crop and productivity may be two or three times the background level. Here surface chlorophyll values of 0.4–0.5 mg/m³ are observed, and integrated primary production values of 79 g C/m² per year have been reported (with maximum values of over 900 g C/m² per year). In satellite images of the Gulf, elevated chlorophyll levels often mark edges of the Loop Current and the various eddies.

### VERTICAL PUMPING WITHIN THE CORE OF CYCLONIC RINGS

In the Northern Hemisphere, when water is rotating in a cyclonic (counterclockwise) ring, surface water is spun from the center to the periphery and then to the outside of the rotating system. To replenish this loss, cold, nutrient-rich water from deeper layers rises or domes up in the core of the ring, and as this water enters the euphotic zone locally it stimulates the growth of phytoplankton. Thus, in the core of cyclonic rings in the Gulf, measurements reveal average integrated chlorophyll values of 38 mg/m² and integrated production values of 90 g C/m² per year, these values being two or three times those encountered outside the rings. Parenthetically, it should also be noted that tropical storms and hurricanes, which are powerful, low-pressure, cyclonically rotating systems, also cause doming of core water, resulting in phytoplankton blooms along the storm tracks.

**Table 8.7.** *Primary production in different areas of the Gulf of Mexico compared with primary production in world ecosystem types based on estimates from various sources.*

| Ecosystem type | Annual production | |
|---|---|---|
| | metric tons C/km²/yr | lb C/A/yr |
| **Gulf of Mexico** | | |
| Open Gulf (avg.) | 25 | 225 |
| Upwelling areas (avg.) | 90 | 810 |
| Continental shelves | 34 | 306 |
| Coastal Louisiana | 250–300 | 2,250–2,700 |
| South Texas | 20 | 180 |
| Entire Gulf (avg.) | 27 | 243 |
| **Other Ecosystems** | | |
| Forests | 250 | 2,250 |
| Cultivated land | 147 | 1,323 |
| Grasslands | 43 | 387 |
| Deserts | 7 | 63 |
| World oceans (avg.) | 46 | 414 |

## Phytoplankton production in perspective

To place the above values in perspective, table 8.7 provides summary information on primary production rates for several areas around the Gulf and for other types of ecosystems. The data are presented in terms of tons of organic carbon produced per square kilometer per year as well as pounds per acre per year. (Note: $1 \text{ g/m}^2 = 1,000$ kg or $1.2 \text{ tons/km}^2 = \text{ca. } 9 \text{ lb/A}$.) Except for high-nutrient coastal areas and open-water "hot spots," most Gulf values fall between those of grasslands and deserts. Although typical of subtropical marine areas in general, the open Gulf production rates fall somewhat below the average of the world oceans. However, if the "hot spot" production were factored in, total Gulf production would likely approach the average world ocean values. In high-nutrient coastal waters around the mouth of the Mississippi River, annual production is similar to that of terrestrial forests. All values are given in terms of the amount of organic carbon produced, but for some purposes it would be interesting to know the amount of dry weight or total living organic matter produced. As a rule of thumb, carbon represents about 50 percent of the dry weight of plant and animal biomass, and dry weight is about 20 percent of the wet weight. So, to convert from carbon to wet weight, multiply by a factor of 10.

It is now possible to provide a rough estimate of the amount of organic material produced annually by the Gulf of Mexico. Multiplying the average Gulf phytoplankton value of $27 \text{ g C/m}^2$ per year by the surface area of the Gulf (1.6 million km²), El-Sayed (1972) estimated that the phytoplankton produces about 43 million tons of organic carbon per year. However, as we saw earlier, the figure of $27 \text{ g C/m}^2$ per year is undoubtedly too low. Taking into account the continental shelves and various "hot spots," the total phytoplankton production is probably closer to 50 million tons of organic carbon, or 500 million tons of living phytoplankton biomass per year. Phytoplankton, of course, accounts for only a portion of the primary production in the Gulf area. Bays and estuaries as well as coastal salt marshes, mangrove swamps, benthic algae, and seagrass meadows are all much more productive than open-water phytoplankton populations. If all these were added in, the total production of the Gulf of Mexico ecosystem would likely approach 60 million tons of organic carbon, or 600 million tons of living plant biomass per year.

### NOXIOUS PHYTOPLANKTON SPECIES

During recent years, worldwide and in the Gulf of Mexico, many reports of toxic and other noxious marine algal blooms have appeared in the technical literature. This is in part a result of increased awareness of the marine environment, but there seems to be a definite increase in the frequency and severity of such occurrences. Human activities are suspected of causing or intensifying at least some of these cases. The causative species themselves are normal constituents of coastal or oceanic phytoplankton communities, which, under exceptional environmental conditions, explode to concentrations of a million or more cells per liter. At such densities, photopigments within the cells often impart color to the water, leading to the terms "red tide," "brown tide," and so forth. For half a century it has been known that certain dinoflagellate species are associated with such events, but recently diatoms and microalgae have also been implicated. Examples of some of these forms are illustrated in figure 8.14.

### Diatoms

Present evidence indicates that toxic diatoms are limited to the genus *Pseudo-nitzschia,* and several species of this group produce a potent neurotoxin known as *domoic acid.* In affected areas shellfishes, such as oysters, accumulate

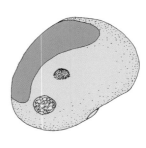

*Pseudo-nitzschia sp.*
**Diatoms**

Figure 8.14.
Examples of noxious diatoms and dinoflagellates reported from the Gulf of Mexico.

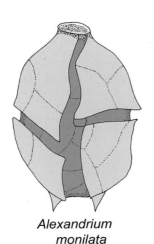

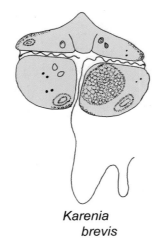

*Alexandrium monilata*

*Karenia brevis*

*Aureoumbra lagunensis*

**Dinoflagellates**

**Pelagophycean**

the substance in their tissues. Humans who consume the contaminated shellfishes can develop severe neurological symptoms, and on the East Coast several deaths have been reported. One of the major effects is permanent loss of short-term memory; hence the syndrome is called *amnesiac shellfish poisoning* (ASP). Although no deaths or major symptoms have so far been reported from the Gulf, at least two of the suspected species (*P. multiseries* and *P. pseudodelicatissima*) are known from bay and shelf waters of Louisiana and upper Texas, and they are likely more widespread.

### Dinoflagellates

Toxic dinoflagellate species are found in tropical and subtropical waters around the world. Within the Gulf of Mexico the primary toxic planktonic species is *Karenia brevis* (= *Gymnodinium breve*), and this species is largely responsible for major outbreaks of the "red tide." Others, such as *Alexandrium monilatum* (= *Gonyaulax monilata*) and several other forms, are also known or suspected to have toxic properties. Red tide outbreaks have been reported throughout the northern Gulf from the Florida Keys to south Texas and northeastern Mexico and around

portions of the Yucatán shelf. However, the most frequent, severe, and long-lasting outbreaks occur on the continental shelf off southwest Florida, where they appear, on average, every five or six years. Most of the events last only 2 to 4 months, but they have been known to persist for up to 11 months.

*Karenia brevis* and some of the other dinoflagellates produce the toxic chemical *brevetoxin,* which causes mass mortality in fish populations and a series of human neurological symptoms (as well as death in areas outside the Gulf). Fish mortality results from poisoning, gill clogging, and oxygen depletion in the water column. Surprisingly, few invertebrate species are affected. Humans may suffer from respiratory irritation due to aerosols from sea spray or dermatitis from swimming in water containing dinoflagellate blooms. However, the neurological symptoms result only from consumption of contaminated shellfish. The syndrome is called *neurological shellfish poisoning* (NSP).

### Microalgae

Along the coastline of south Texas and stretching from Corpus Christi Bay to the Rio Grande is the long, shallow lagoon known as the Laguna Madre of Texas. With little freshwater input, a high evaporation rate, and poor circulation, salinities here generally stand well above those of the Gulf of Mexico, often exceeding 60 ppt. During the severe winter of 1990–91, following two intense cold snaps that resulted in massive local fish and invertebrate mortality, there developed in the upper reaches of the lagoon a massive bloom of a tiny (0.5–0.6 µm) microalga. Previously unknown, this species is now called *Aureoumbra lagunensis*. After a few months of buildup, populations reached concentrations of more than two million and later six million cells per milliliter. At high concentrations, since it imparts a brownish color to the water, it is referred to as a "brown tide." Unlike the dinoflagellates, *A. lagunensis* apparently does not produce toxins, but since it is very tiny, this alga appears to outcompete the other algal species, and under bloom conditions it exists in almost pure cultures. By screening out the light it shades the seagrass beds, causing local die-offs of the rooted vegetation. It is a poor food supply, and when it is abundant zooplankton populations fall, larval fishes die, and benthic invertebrate populations are drastically reduced. Thus, although nontoxic, it does severely affect the ecosystem of the lagoon. After persisting in high concentrations for several years, the bloom was finally reduced in 1998 by heavy rains, lowered salinity, and flushing of the lagoon. In subsequent years there have been local short-term episodic outbreaks.

In summary, certain toxic planktonic dinoflagellates have the potential to cause moderate and severe ailments, and in some cases, death of humans exposed to aerosols, waters, or shellfishes from contaminated areas. Some planktonic diatoms of the Gulf are related to known toxic species, but so far no cases of diatom poisoning have been reported from the Gulf. Neither the dinoflagellates nor the diatoms appear to cause any long-term damage to the marine ecosystems. The reverse is true for microalgae of the Laguna Madre of south Texas. The causative species is nontoxic, and human populations are not directly affected, but long-term persistence of the bloom and widespread destruction of seagrass beds can have extended ecological consequences. The economic impact of noxious algal blooms may be locally severe. Millions of dollars are lost because of the closure of coastal fisheries and the loss of tourist trade when seafood becomes contaminated and when beaches are littered with rotting carcasses of fishes and invertebrates. These topics will be examined in greater detail in chapter 15.

# 9 : ZOOPLANKTON

The zooplankton includes an extremely diverse group of animals that live in the water column and have only limited powers of locomotion. They are not strong swimmers and are basically at the mercy of the water currents. Most are quite small. Many, such as the protozoa, are microscopic, but some, such as large medusae or jellyfish, may reach a diameter of several feet. Included in the zooplankton are representatives of all the animal phyla, although some groups are much more abundant than others, and a number of phyla are represented only by larval forms.

On the basis of life history characteristics, two categories of zooplankton are recognized. The *holoplankton* includes those species that pass their entire lives within the plankton. This group is present in both neritic (coastal) and oceanic waters, and it is dominant in the open ocean. *Meroplankton,* on the other hand, includes those species in which only a portion of the life history (generally the larval stages) is planktonic, the adults being either benthic (associated with the bottom) or nektonic (free swimming in the water column). Meroplanktonic forms are present in neritic and open ocean waters, but in tropical and subtropical areas they are often dominant in waters over the continental shelf.

Zooplankters are often classified on the basis of size, as follows:

| | |
|---|---|
| ultrazooplankton | «{ 5 μm |
| nanozooplankton | 5 μm–76 μm |
| microzooplankton | 76 μm–1 mm |
| macrozooplankton | }» 1 mm |

(Note: 1 μ or micron = 1 μm or one micrometer = one thousandth of a millimeter.)

Such a classification is convenient because the primary means of capturing zooplankton is by using towed nets, and the mesh size of the net determines which sizes are retained and which pass through the mesh openings. Ultrazooplankton includes many protozoa and some tiny larval forms. Macrozooplankton embraces large medusae, comb jellies, worms, crustaceans, larval fishes, and the like. Between these two extremes lies a complex mixture of larval and adult animal groups of great taxonomic diversity. Because of this variety and because each group of organisms requires special methods of study, most zooplankton workers have become specialists in one or a few groups. The term *nekton* refers to larger, free-swimming forms such as fishes, squids, and so forth, and between the zooplankton and the nekton lies a gray area of intermediate-sized animals called *micronekton.* When captured in zooplankton nets, these are often included with the zooplankton.

The larval forms constitute a special class of problems because for a given species there may be as many as 10 or more larval stages, most of which bear little or no resemblance to one another or to the adults. Hence, it is necessary to rear each larval form in the laboratory to determine just what it turns into. Historically, many larval types have been described as new species, whereas the adults have been described as different species. Needless to say, the study of marine larvae has lagged far behind the study of adults.

Zooplankton distribution patterns have received considerable attention. At the species level there is often a fairly clear differentiation between the neritic and oceanic groups, and among the oceanic species there is often a vertical stratification within the water column, each species occupying a preferred depth range. For many species the preferred depth changes with the time of day. During daylight hours most vertically migrating species are found at deeper, darker levels, whereas at night they move up into surface waters to feed on phytoplankton and smaller zooplankton organisms. Seasonal patterns of abundance and reproductive activity have also been reported.

Considering the food chains of the sea, the zooplankton stands intermediate between the phytoplankton, or producers, and the larger consumers such as fishes and mammals. Feeding strategies are variable, and details of the food webs are extremely complex. Of particular interest are the effects of zooplankton on deeper layers of the water column and on the sea bottom. Fecal pellets, still containing organic matter, are produced in streamlined packets that sink fairly rapidly, providing nourishment for organisms in the deeper layers. Skeletal structures of some protozoans, composed of inorganic material (especially carbonate and silicate), sink and accumulate in bottom sediments.

Individual zooplankton species or groups of species may be associated with particular water masses, and their presence is considered to be an indication of the water masses themselves. Individual behavior patterns and special species associations should be mentioned. A number of fish species form associations with gelatinous zooplankton such as jellyfishes and comb jellies. Floating seaweed, or *Sargassum,* forms the basis of a complex community of invertebrates and small fishes, some of which are seldom found anywhere else. Organisms associated with surface layers of the sea are referred to collectively as *neuston,* and this group will be discussed subsequently. The present chapter is concerned primarily with zooplankton in waters overlying the continental shelves and in surface waters of the open Gulf. Zooplankton and micronekton of deeper layers of the open Gulf are examined in chapter 14, which deals with the deep Gulf of Mexico.

Literature dealing with the zooplankton of the Gulf of Mexico has treated all the topics mentioned above, some in greater detail than others. Much of the work has been primarily taxonomic, with information on species composition, distribution, and abundance. It is heavily weighted toward estuaries and coastal marine waters, but a fair amount of work has also been done on zooplankton of the open Gulf. References to general zooplankton studies and literature reviews include Berkowitz (1976), Biggs and Ressler (2000, 2001), Björnberg (1971), Darnell and Schmidly (1988), Flint and Rabalais (1981), T. L. Hopkins (1973, 1982), Iverson and Hopkins (1981), Perry and Christmas (1973), and Vargo and Hopkins (1990), among others. References pertaining to more specific groups are given in the sections dealing with each group.

## ZOOPLANKTON COMPOSITION

As noted above, all phyla of the animal kingdom are represented in the marine zooplankton, but in terms of numbers of species and/or numbers of individuals some groups are far better represented than others. Many occur only as larval forms (fig. 9.6) (table 9.2). Table 9.1 provides a much simplified classification of the animal kingdom that includes only those groups important in the zooplankton of the Gulf, and it indicates whether a particular group is represented by larvae, adults, or both. Reference to this table will aid in understanding the discussion of individual groups within the zooplankton.

**Table 9.1.** *Simplified classification of the animal kingdom including only those groups important in marine zooplankton during the larval (l) or adult (a) stage or both. Note: In some of the Cnidaria, where there is alternation of sexual and asexual generations, the concepts of larva and adult are meaningless.*

| Phylum, subgroups, and common names | Representation in zooplankton |
|---|---|
| Protozoa | |
|   Ciliophora – Tintinnoida (tintinnids) | -, a |
|             – Others (aloricate ciliates) | -, a |
|   Sarcodina – Foraminifera (foraminiferans) | -, a |
|             – Radiolaria (radiolarians) | -, a |
| Porifera (sponges) | l, - |
| Cnidaria | |
|   Hydrozoa – Siphonophora (siphonophores) | l, a |
|          – Hydromedusae (hydromedusans) | l, a |
|   Scyphozoa (jellyfishes) | l, a |
| Ctenophora (comb jellies) | l, a |
| Mollusca | |
|   Gastropoda – Heteropoda (heteropods) | -, a |
|           – Pteropoda (pteropods) | -, a |
|   Bivalvia (bivalve mollusks) | l, - |
| Annelida – Polychaeta (polychaetes) | l, a |
| Arthropoda | |
|   Crustacea – Cladocera (cladocerans) | l, a |
|          – Ostracoda (ostracods) | l, a |
|          – Copepoda (copepods) | l, a |
|          – Cirripedia (barnacles) | l, - |
|          – Malacostraca | |
|             – Mysidacea (mysids) | l, a |
|             – Amphipoda (amphipods) | l, a |
|             – Euphausiacea (euphausiids) | l, a |
|             – Decapoda | |
|               – Natantia (swimming shrimp) | , a |
|               – Macrura (lobsters, etc.) | l, - |
|               – Anomura (hermit crabs) | l, - |
|               – Brachyura (true crabs) | l, - |
| Chaetognatha (arrow worms) | l, a |
| Echinodermata (starfish, brittle stars, etc.) | l, - |
| Hemichordata (acorn worms) | l, - |
| Chordata | |
|   Urochordata (tunicates, salps, pyrosomes, etc.) | l, a |
|   Vertebrata (fishes, etc.) | l, - |

## Protozoa

Protozoa are all microscopic single-celled or colonial animals that are abundant in both freshwater and marine environments. Among the marine forms many live in the water column, some reside in the bottom sediments, and others are parasitic on other animals and plants. All major groups are represented in the Gulf. Flagellated photosynthetic forms (most dinoflagellates, euglenoids, etc.) have been included with the phytoplankton. Two groups of nonphotosynthetic protozoans are listed here, the ciliates (Ciliophora) and the amoeboid forms (Sarcodina).

### CILIOPHORA

As implied by the name, all members of this category possess cilia, hairlike structures used for locomotion and feeding. Two groups of ciliates are important in the Gulf, those that form and live within a capsule or lorica (tintinnids) and those that do not form such capsules (aloricates).

### TINTINNOIDA

The lorica of tintinnids may be tubular, conical, or keyhole shaped, but the form is absolutely constant for each species and can be used for identification. The lorica may be gelatinous or chitinous, and in many species tiny inorganic particles (sand grains, diatom frustules, coccoliths, etc.) may be glued to the outside. Tintinnids range in size from about 20 to 640 μm. Common genera found in the Gulf are listed in table S15, and examples are given in figure 9.1. Important references include Balech (1967a, 1976b) and Casey (1979a).

### ALORICATE CILIATES

This complex group of protozoans has received scant attention in the Gulf of Mexico, yet the group is probably widespread, abundant, and important in marine food chains. A few common genera are listed in table S15, and some are shown in figure 9.1. The presence and abundance of some species off south Texas are given by Casey (1979b).

### SARCODINA

This group of protozoans lacks flagella or cilia (except in some sexual stages), and movement is by means of protoplasmic extensions, or *pseudopodia*. Two groups are important in the Gulf, the Foraminifera and Radiolaria.

### *Foraminifera*

Forams are almost exclusively marine, where they live in bottom sediments as well as the water column. Most species produce shells of calcium carbonate. These may have a single chamber or several chambers, each succeeding one slightly larger than the previous one. Some are spiral shaped and look like tiny snails. Although most are less than 5 mm in diameter, a few are much larger. During reproduction they often discard their old shells and construct new ones. Hence, within the sea there is a constant rain of discarded foraminiferan shells that build up in bottom sediments. In temperate and tropical latitudes thick deposits may be formed in depths shallower than 4,000–5,000 m. Below this depth range, calcium carbonate dissolves as fast as it accumulates, and deeper ocean bottoms are floored by siliceous oozes and/or abyssal clays. Among the Foraminifera some species are coastal, whereas others are oceanic. Most planktonic species live in the euphotic zone, where they feed largely on phytoplankton. Most of the work in the Gulf has dealt with benthic species, but some planktonic forms have been reported by Casey (1979b), Jones (1968), Phleger (1960), and Phleger and Parker (1954a, 1954b), among others. Important genera are listed in table S15, and examples are shown in figure 9.1.

### *Radiolaria*

Radiolarians are amoeboid protozoans that produce needlelike spicules radiating out from the center of the cell or sometimes a highly ornamented rigid framework. These hard structures are generally composed of silicon. This group is confined to marine waters, where the species are all planktonic, but they do occupy all depths of the sea. They feed on nanoplankton such as diatoms, protozoans, and small crustaceans. Species that reside in near-surface waters are generally capable of limited vertical migration, and they often sink to deeper layers to avoid stormy seas or heated surface water. Important species in the Gulf are listed in table S15, and examples are given in figure 9.1. References to Gulf studies include Casey (1979b), Leavesley et al. (1978), and McMillen and Casey (1978).

## Porifera

Sponges occur in fresh and brackish waters, but they are most abundant and diverse in the sea. Adults are entirely sessile and may live on the bottom or attached to solid substrates projecting into the water column. Only the

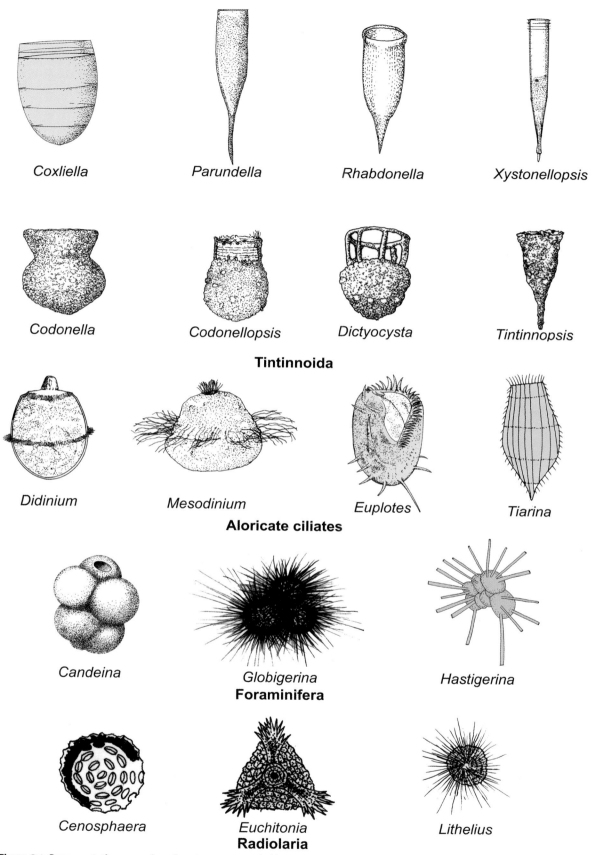

**Figure 9.1.** Representative examples of protozoans recorded in zooplankton of the Gulf of Mexico. Included are tintinnids and aloricate ciliates, foraminiferans, and radiolarians.

larval forms (*amphiblastulae,* etc.) (fig. 9.6) appear in the zooplankton.

## Cnidaria

The cnidarians, or coelenterates, make up a diverse group of primitive, often gelatinous animals possessing a sac-like digestive tract with a single opening that serves as both mouth and anus. The mouth is surrounded by a number of tentacles, each armed with powerful stinging cells for stunning prey. Two basic body forms are present, the medusa, which is generally free living, and the hydroid (anatomically like an upside-down medusa), which in most groups is sedentary and attached to a firm substrate, although some are free floating. Many of the hydroids are colonial. The typical life history involves alternation of generations, with a sexually reproducing medusa stage followed by an asexually reproducing (by budding) hydroid stage. However, among the different species there is great variation on this general life history pattern. Most medusae and some hydroids are planktonic. Two larval forms are planktonic, the *planula,* which develops from a sexually fertilized egg, and the *ephyra,* which develops asexually by budding (fig. 9.6). Many of the planktonic cnidaria are bioluminescent.

Two major taxonomic groups are important in the zooplankton: the Hydrozoa, in which the hydroid stage is dominant, and the Scyphozoa, in which the medusa stage is dominant. The Hydrozoa are represented in the plankton by two groups, the *hydromedusae* and the *siphonophores.* In the hydromedusae the hydroid stage is normally attached to a solid substrate, but many species produce small planktonic medusae. Among the siphonophores the hydroid stage itself is planktonic. Some are solitary individuals, but many consist of complex colonies dangling in the water column beneath a float. Examples include spectacular forms such as the Portuguese man-of-war (*Physalia*) and the by-the-wind sailor (*Velella*). Scyphozoans include the commonly recognized jellyfishes and sea nettles, some of which cause rashes among swimmers. The world's largest zooplankters are medusae, which may reach diameters of over 1.8 m (6 ft). Examples of cnidarians important in the Gulf of Mexico are listed in table S16, and some are shown in figure 9.2. Important references for the Gulf include Biggs, Bidigare, and Smith (1981), Biggs, Smith, et al. (1984), Hedgpeth (1954g), P. Phillips (1972), P. Phillips, Burke, and Keener (1969), Sears (1954b, 1954c), Segura-Puertas (1992), and D. Smith (1982).

## Ctenophora

Ctenophores, or comb jellies, are gelatinous zooplankton related to the Cnidaria. However, they lack stinging cells and do not undergo alternation of generations. All are marine, although some species invade estuaries. Most are planktonic, and locomotion is accomplished by means of ciliated bands. All species are carnivorous, feeding on zooplanktonic forms such as small crustaceans and larval fishes. Swarms of ctenophores, such as occur from time to time in coastal waters, may temporarily wipe out most of the zooplankton from an area. Reproduction involves the *cydippid* larva. As in the case of the cnidaria, many are bioluminescent. Important genera in the Gulf of Mexico are listed in table S16, and examples are shown in figure 9.2. Important references include Biggs, Bidigare, and Smith (1981), Biggs, Smith, et al. (1984), P. Phillips, Burke, and Keener (1969), and Sears (1954a).

## Mollusca

Mollusks are unsegmented soft-bodied animals with complete digestive tracts and definite front and hind ends. Typically the body is more or less surrounded by a thin tissue called a mantle. In some groups the body is sheltered in a carbonate shell, but in others the shell may be internal or absent. The most commonly recognized forms are in the classes Gastropoda (snails, etc.), Bivalvia (clams, oysters, etc.), and Cephalopoda (squids, octopuses, etc.). Mollusks occur on land, in fresh waters, and in the sea. In the ocean most of the adults are limited to the benthic environment, but many of these produce planktonic larvae. In addition, some groups are entirely planktonic. Several larval types occur, of which the *trochophore* and *veliger* (fig. 9.6) are most prominent in the zooplankton.

### GASTROPODA

Typically the gastropods have a broad, flat foot, well-developed head, and single shell that may be twisted into a spiral, as in snails. Most of the planktonic species are greatly modified from this general pattern. Three groups are important in marine zooplankton: the Janthinidae, Heteropoda, and Pteropoda.

#### *Janthinidae*

This group consists of a single genus, *Janthina,* called the "violet snail" because of its color. This animal, which is recognizable as a snail with a coiled shell, is often found floating at the surface of the sea and is particularly abun-

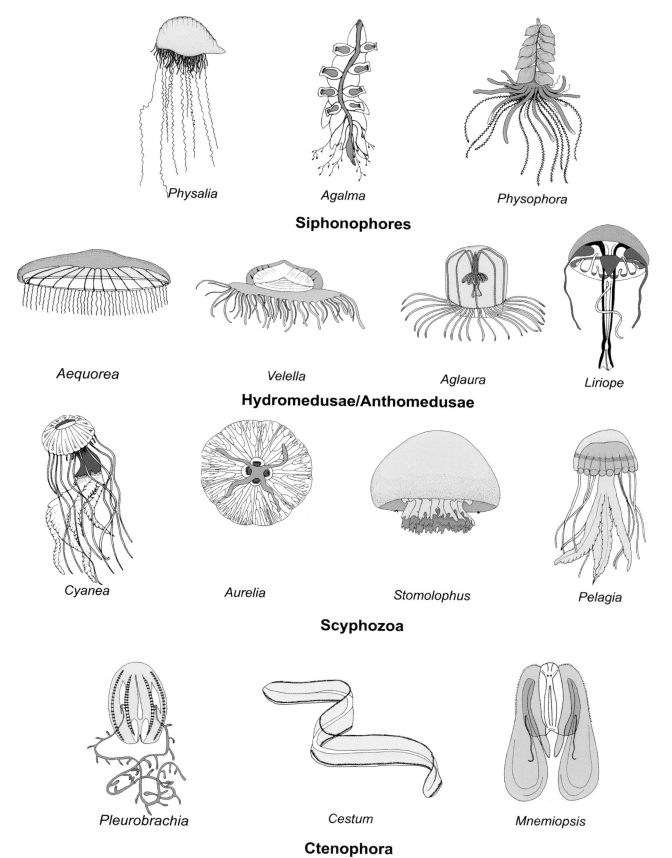

*Physalia*  *Agalma*  *Physophora*

**Siphonophores**

*Aequorea*  *Velella*  *Aglaura*  *Liriope*

**Hydromedusae/Anthomedusae**

*Cyanea*  *Aurelia*  *Stomolophus*  *Pelagia*

**Scyphozoa**

*Pleurobrachia*  *Cestum*  *Mnemiopsis*

**Ctenophora**

**Figure 9.2.** Representative examples of gelatinous zooplankton groups recorded from the Gulf of Mexico. Included are siphonophores, hydromedusae, scyphozoans, and ctenophores.

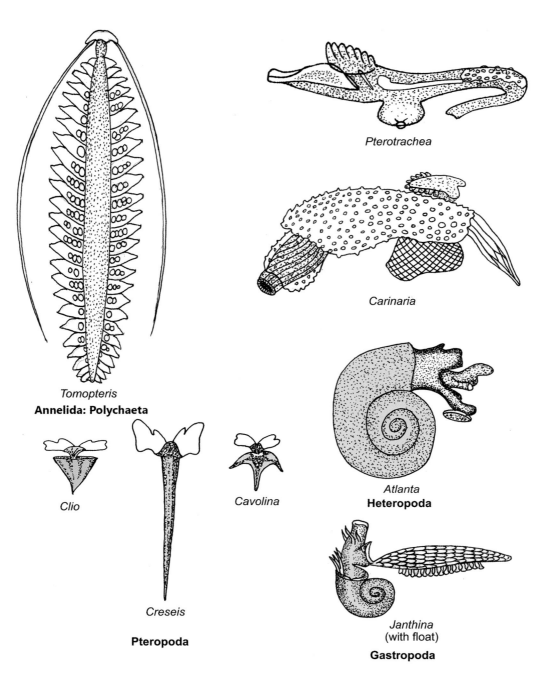

**Figure 9.3.** Representative examples of annelids and mollusks recorded in zooplankton of the Gulf of Mexico. Included are a polychaete as well as heteropods, pteropods, and the floating snail *Janthina*.

*Pterotrachea*

*Carinaria*

*Tomopteris*
**Annelida: Polychaeta**

*Clio*

*Creseis*

*Cavolina*

**Pteropoda**

*Atlanta*
**Heteropoda**

*Janthina* (with float)
**Gastropoda**

dant in the eastern Gulf, where it drifts in on the Yucatán Current. This snail produces a long gelatinous raft of gas-filled bubbles that enables it to float at the sea surface and live an entirely pelagic existence. Eggs are attached to the bottom of the raft. *Janthina* is illustrated in figure 9.3.

### Heteropoda

The heteropods constitute a small group of marine gastropods highly modified for a free-swimming pelagic existence. There is a definite head and tail, and a flattened ventral fin used for locomotion. The shell, when present, is much reduced, and a visceral mass is positioned on the dorsal surface. Sixteen species belonging to seven genera have been reported from the Gulf of Mexico. The genera are listed in table S17 and representatives are shown in figure 9.3. The major references to Gulf heteropods are D. Taylor (1969) and D. Taylor and Berner (1970).

## Pteropoda

The pteropods, or "sea butterflies," are marine pelagic gastropods also highly modified for a planktonic existence. The foot has developed into two large flat fins for locomotion and flotation. The thin calcareous shell is often uncoiled. Pteropods may be quite abundant in surface waters, where they feed primarily on phytoplankton. The bottoms of shallow tropical and subtropical seas may contain thick oozes of pteropod shells. About 30 species belonging to 12 genera have been reported from Gulf waters. Representative genera are listed in table S17, and examples are presented in figure 9.3. Important literature references include Berkowitz (1976), Hughes (1968), and Snider (1975).

### BIVALVIA

All adult bivalve mollusks are benthic. The group is represented in the zooplankton only by the *trochophore* and *veliger* larvae (fig. 9.6).

### CEPHALOPODA

Most squids and octopuses are nektonic, although the young of many species and the adults of some (such as the paper nautilus) could be included in the zooplankton. For simplicity, the entire group will be considered in chapter 13, on nekton.

## Annelida

The body of annelid worms is elongated and always divided into segments. The first segment, which contains the brain and is known as the head, may bear sense organs and tentacles. In most groups there is little differentiation of the segments behind the head. Annelid worms occur in terrestrial soil as well as freshwater and marine environments. Within the sea the adults of most species are benthic, but the adults of a few and the larvae of many occur in the zooplankton. A number of different larval types are produced, of which the *trochophore* (quite similar to that of the mollusks) and the *polytroch* (fig. 9.6) are the most common. All annelids in which the adults are planktonic belong to the Polychaeta.

### POLYCHAETA

Pelagic marine polychaetes generally have long bodies composed of many segments. Each segment behind the head bears a pair of large fleshy appendages called *parapodia* that are used for locomotion and respiration. Only four families have planktonic adults, and these are seldom abundant. In nearshore waters certain benthic species may appear briefly in the plankton in spawning swarms. Representative genera of pelagic polychaetes of the Gulf are listed in table S17, and one is illustrated in figure 9.3. Little work has been done on planktonic polychaetes of the Gulf, but Berkowitz (1976) provides a list of pelagic forms.

## Arthropoda

In body development, the arthropods are basically advanced annelids. Some of the anterior segments are fused and more specialized in function. The fleshy parapodia have become jointed appendages with a variety of functions. Furthermore, the entire body is encased in a plastic-like material called *chitin,* which is tough but flexible. In some forms this integument is impregnated with insoluble salts such as calcium carbonate, in which case it may be more brittle. Although a number of arthropod groups occur in the sea, only the crustaceans are important in the zooplankton as adults.

### CRUSTACEA

Crustaceans are arthropods that breathe by means of gills. They have two pairs of antennae, and the segmented appendages are generally *biramous,* that is, they are divided into an inner and outer branch. Beyond this, there is great diversity in body form and specialization of appendages for different functions. Several groups of crustaceans are important in the marine zooplankton.

## Cladocera

In cladocera, or water fleas, the body is compressed laterally, and the body and most appendages are covered by a rigid bivalved shell open at the bottom. The head remains uncovered. The first antennae are reduced in size, but the second antennae are large and branched and used for swimming. In marine forms there is a single median eye. Developing eggs are carried inside the hind portion of the body valves. All species are basically planktonic. Cladocera are abundant in fresh waters, but only a few species occur in the sea. Three genera have been recorded from the Gulf (table S18), and these are illustrated in figure 9.4. Darnell and Schmidly (1988) recorded these genera from coastal waters of the northern Gulf.

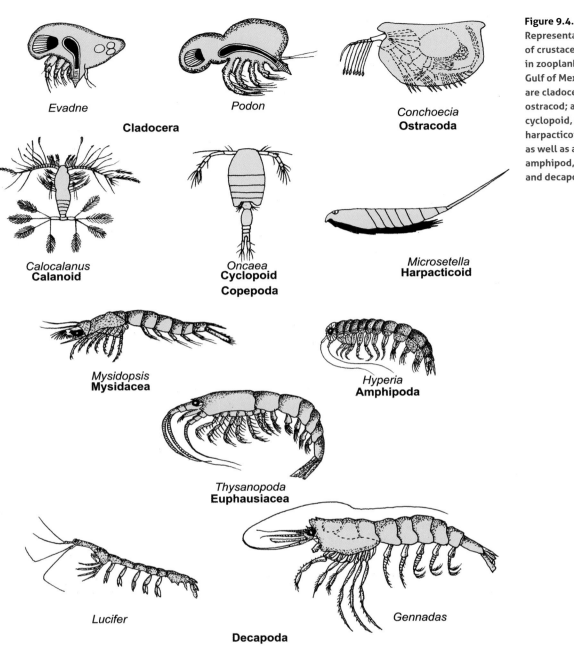

**Figure 9.4.**
Representative examples of crustaceans recorded in zooplankton of the Gulf of Mexico. Included are cladocerans; an ostracod; and calanoid, cyclopoid, and harpacticoid copepods; as well as a mysid, amphipod, euphausiid, and decapods.

*Evadne*          *Podon*

**Cladocera**

*Conchoecia*
**Ostracoda**

*Calocalanus*
**Calanoid**

*Oncaea*
**Cyclopoid**
**Copepoda**

*Microsetella*
**Harpacticoid**

*Mysidopsis*
**Mysidacea**

*Hyperia*
**Amphipoda**

*Thysanopoda*
**Euphausiacea**

*Lucifer*

*Gennadas*

**Decapoda**

*Ostracoda*

In the ostracods, or clam shrimps, the entire body is enclosed in a pair of valves hinged along the top. Enlarged antennae used for locomotion can, in some species, be withdrawn into the valved shell. Ostracods occur in both fresh and marine waters. Most are benthic, but in the ocean several species are planktonic. Two genera are known from the Gulf zooplankton. These are listed in table S18, and one is shown in figure 9.4. Primary references include Darnell and Schmidly (1988) and Tressler (1954).

*Copepoda*

The copepods constitute a large and diverse group of small crustaceans that are abundant in fresh waters, estuaries, and the world oceans. Some scientists consider them to be the most abundant animals on earth. Most are microscopic, but some species reach a length of about one centimeter (half an inch). The body is divided into three sections, the head, thorax, and abdomen, but the first two may be fused into a single part known as the cephalothorax. Typically, each segment bears a pair of appendages that may be specialized for feeding, locomotion, or reproduction. The body is not encased in valves. The first pair of antennae may be enlarged and used for locomotion. Some forms live in the bottom sediments, but most are planktonic. All are extremely important in aquatic food chains. Among the nonparasitic forms, three groups occur in the sea: the calanoids, cyclopoids, and harpacticoids. In the calanoids the first antennae are very long, and there is a movable joint between the fifth and sixth thoracic segments. In the cyclopoids the antennae tend to be somewhat shorter, and the movable joint is between the fourth and fifth thoracic segments. Furthermore, the thorax is much wider than the abdomen. In the harpacticoids the first antennae are quite short. As in the cyclopoids, the movable joint is positioned between the fourth and fifth thoracic segments. The body is nearly cylindrical, and the abdomen is almost as wide as the thorax. More than 100 species and 60 or 70 genera of copepods have been described from the Gulf. Representative genera of the three copepod groups are listed in table S18, and several are shown in figure 9.4. Copepods of coastal and pelagic waters of the Gulf have been studied by many investigators, and important references include the following: Alvarez-Cadena and Segura-Puertas (1997), Berkowitz (1976), Cummings (1982), Ferrari (1973), Fleminger (1956), González (1957), Grice (1960, 1969), Park (1970), and Schmitt (1954), among others.

### Mysidacea

The mysids are fairly primitive crustaceans with slender bodies and a shrimplike appearance. The thoracic appendages all have two branches, and eggs are carried in a special brood pouch beneath the abdomen of the female. Although they occur in fresh waters, most species are estuarine or marine. Within the ocean most live on or just above the bottom, but a number appear in the plankton. Important genera from the Gulf are listed in table S18, and an example is shown in figure 9.4. Major references include Banner (1954), T. L. Hopkins and Lancraft (1984), W. W. Price (1976), W. W. Price et al. (1986), Stuck, Perry, and Heard (1979a, 1979b), and Tattersall (1951).

### Amphipoda

These small crustaceans do not have a carapace or covering over the thorax, and all thoracic segments are free jointed. The body is compressed from side to side. The eyes are flat on the head and not on stalks. There are five pairs of large unbranched thoracic walking legs and three pairs of abdominal swimmerets. Amphipods occur in fresh waters, but they are much more diverse in the sea, where they are mostly associated with the bottoms. A number, however, are planktonic. Several genera reported from the Gulf are given in table S18, and a single example is shown in figure 9.4. Important references include Berkowitz (1976), McKinney (1977), and Morée (1979).

### Euphausiacea

Euphausiids are shrimplike crustaceans in which the head and all thoracic segments are covered by a well-developed carapace. Eyes are mounted on eyestalks. The thoracic legs have two branches, the inner being fan shaped and functioning as gills. The last two thoracic legs are much reduced in size. The abdomen bears five pairs of well-developed swimmerets. Luminescent organs may be present on several parts of the body. Euphausiids are exclusively marine and planktonic or nektonic throughout their life history. In their development the larvae pass through a complicated series of stages. Several genera found in the Gulf are listed in table S18, and one species is shown in figure 9.4. Important references include Banner (1954), James (1966, 1970), and Schroeder (1971).

### Decapoda

As the name implies, the decapods have five pairs of thoracic legs. Some are modified as pincers for grasping, and others are walking legs. The head and thorax are covered by a carapace. The body may be more or less cylindrical, as in shrimps and lobsters, or flattened, as in crabs. Included among the decapods are the shrimps and prawns, lobsters, crayfish, hermit crabs, and true crabs. Although represented on land and in fresh waters, the decapods are primarily marine, where they are generally associated with the bottom environment. Most, if not all, marine decapods produce planktonic larvae. Among the species that inhabit the water column as adults, most are strong swimmers, but a few are weak swimmers and are considered to be planktonic. These include representatives of the families Sergestidae and Penaeidae. Planktonic genera of these two families found in the Gulf are listed in table S18, and representatives are shown in figure 9.4. Important references include Berkowitz (1976), Cruise (1971), Huff and Cobb (1979), and T. Roberts (1970). Members of the caridean shrimp family Oplophoridae are also planktonic, but these are largely inhabitants of very deep water and are taken up in chapter 14.

## Chaetognatha

The chaetognaths, or arrow worms, make up a small phylum of exclusively marine animals ranging in length from one to about eight centimeters (one-half to four inches). The body is slender and transparent and equipped with lateral and tail fins that aid in stabilization in the water column. Locomotion is effected by longitudinal muscles within the body. The most striking feature is the sets of sickle-shaped spines or hooks on either side of the mouth, attached to powerful muscles, which are used for grasping prey. Chaetognaths are voracious predators on small zooplankters including protozoa, copepods, larval crustaceans, small fishes, and other chaetognaths. They often appear in swarms, and they are considered to be extremely important intermediate carnivores in food chains of the sea. Worldwide there are only seven genera, all but one of which are strictly planktonic. About 30 species are known, of which more than half have been recorded from the Gulf. Genera reported from the Gulf are listed in table S19, and one is illustrated in figure 9.5. Important references for the Gulf of Mexico include Adelman (1967), Every (1968), Mille-Pagaza, Reyes-Martínez, and Sánchez-Salazar (1997), and Pierce (1954).

## Echinodermata

Echinoderms include the starfishes, brittle stars, sand dollars, sea cucumbers, and their relatives. All are marine, and most are associated with the benthic environment. However, they produce a variety of larval forms, most of which are planktonic. Examples are given in figure 9.6. Their embryonic development shows that they are related to the chaetognaths, on the one hand, and the hemichordates, on the other.

## Hemichordata

The hemichordates, or acorn worms, are a small group of burrowing marine worms that produce planktonic larvae, one of which is illustrated in figure 9.6. Embryonic development as well as adult body structure shows them to be related to echinoderms and chordates.

## Chordata

The phylum Chordata is characterized by three primary body features: a dorsal tubular nerve cord, a longitudinal strengthening rod, or *notochord,* and paired pharyngeal gill slits. Among the lower chordates these features may be present in the adults, but in the higher forms some appear only during embryonic development. Three subphyla are recognized: the Urochordata, Cephalochordata, and Vertebrata. Only the urochordates and vertebrates are well represented in the zooplankton. The cephalochordates include the lancets, which are small marine burrowing forms. These do produce ciliated larvae, which are generally quite rare in the zooplankton of the sea.

### UROCHORDATA

Members of this subphylum are all marine, and they all feed by straining water through a basketlike pharyngeal cavity lined with what amounts to a plankton net. Adults may be free living or sedentary, and they may be solitary or colonial. In a few forms (salps) there is alternation of sexual and asexual generations. Thus, the adults of some groups and the larvae of all are planktonic. In general body form, the larva (fig. 9.6) much resembles a tadpole and clearly displays the characteristics of the phylum. In the adult some of these features become obscured or lost entirely. Three groups have adults represented in the zooplankton: the Larvacea, Thaliacea, and Ascidiacea. In the Larvacea (represented by *Oikopleura*), the adult more or less resembles the larval form. In the Thaliacea, including the salps and their relatives, the adults are transparent barrel-shaped organisms open at both ends, with several rings of muscle running around the periphery of the barrel. Successive constriction of these rings of muscle draws water into one end and squirts it out the other. Locomotion is by jet propulsion. Examples include *Doliolum* and *Salpa*. Most members of the Ascidiacea are solitary and are known as sea squirts. However, members of one group, the pyrosomes, are free living and planktonic. These colonial organisms form hollow transparent tubes, open at one or both ends, which vary in length from a few inches to several feet. Pyrosomes are brilliantly luminescent. An example is the genus *Pyrosoma*. Representative genera from the Gulf of Mexico are listed in table S19, and several are illustrated in figure 9.5. The main reference for the Gulf is Van Name (1954).

### VERTEBRATA

The vertebrates are characterized by the development of an internal skeleton of cartilage or true bone. The brain is encased in a cranium. In adults the notochord is replaced by a column of vertebrae made of cartilage or bone. In most groups there are two sets of paired appendages. Within the zooplankton the vertebrates are represented by fish eggs and larvae, which are both abundant and quite diverse. Of particular note are the leaflike *lepto-*

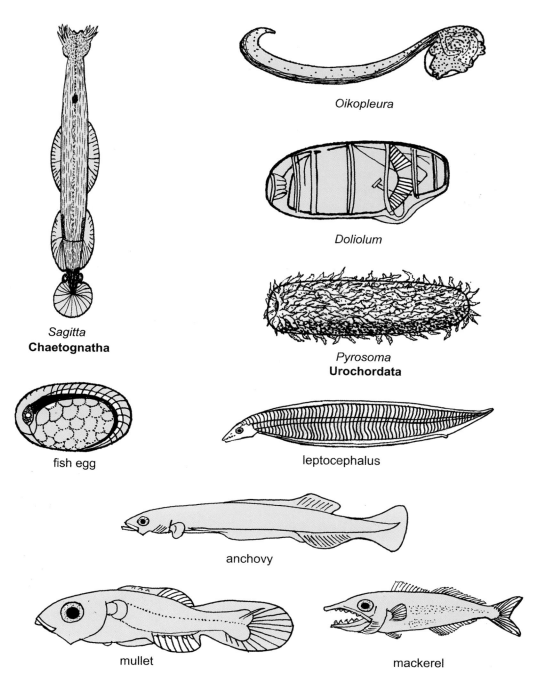

*Oikopleura*

*Doliolum*

*Pyrosoma*
**Urochordata**

*Sagitta*
**Chaetognatha**

fish egg

leptocephalus

anchovy

mullet

mackerel

Fish egg, larvae, and young
**Vertebrata**

**Figure 9.5.** Representative examples of chaetognaths and chordates recorded in zooplankton of the Gulf of Mexico. Included are a chaetognath, urochordates, and vertebrates.

*cephalus* larvae, which bear no resemblance to the adult eels into which they eventually develop. Representative fish eggs and larvae are illustrated in figure 9.5, and a list of representative larval forms found in the zooplankton is given in table 9.2.

**ZOOPLANKTON DISTRIBUTION AND ABUNDANCE**
Standing crop represents the abundance of all zooplankton species present in an area at a given time. It may be expressed in terms of volume, wet weight, dry weight, or total organic matter (ash-free dry weight). It may also be expressed numerically as direct counts of each species or

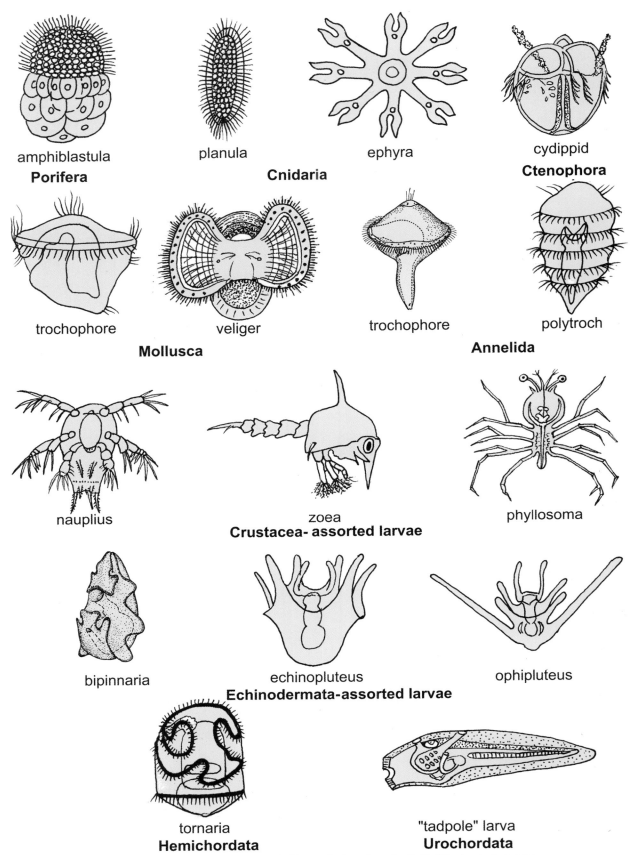

amphiblastula
**Porifera**

planula  ephyra
**Cnidaria**

cydippid
**Ctenophora**

trochophore  veliger
**Mollusca**

trochophore  polytroch
**Annelida**

nauplius  zoea  phyllosoma
**Crustacea- assorted larvae**

bipinnaria  echinopluteus  ophipluteus
**Echinodermata-assorted larvae**

tornaria
**Hemichordata**

"tadpole" larva
**Urochordata**

**Figure 9.6.** Representative examples of larval forms recorded in zooplankton from the Gulf of Mexico. Included are larvae of poriferans, cnidarians, ctenophores, mollusks, annelids, crustaceans, echinoderms, and chordates.

**Table 9.2.** *Representative larval types occurring in marine zooplankton.*

| Phylum | Representative larvae |
| --- | --- |
| Porifera | amphiblastula, parenchymula |
| Cnidaria | ephyra, planula |
| Ctenophora | cydippid |
| Mollusca | trochophore, veliger |
| Annelida | trochophore, polytroch |
| Arthropoda | cypris, calyptopis, furcilia, megalops, mysis, nauplius, phyllosoma, protozoea, zoea |
| Echinodermata | auricularia, bipinnaria, brachiolaria, doliolaria, echinopluteus, ophiopluteus, pluteus |
| Hemichordata | tornaria |
| Chordata | tadpole larva, leptocephalous |

each general taxonomic group present in a given volume of water. Although the concept is straightforward, determination of standing crop is not a simple matter, and all determinations are considered to be estimates, biased one way or another. The first source of bias relates to distribution patterns of the organisms themselves. Considerable evidence indicates that for many, if not most, coastal and oceanic species the individuals tend to be distributed in a patchy fashion, that is, in areas of high and low species density, and these may reflect inequalities in the physical environment or behavioral factors (schooling or feeding and breeding aggregates) or both. Thus, nearby samples or repeated samples from a given area may show a high degree of variability depending on whether the samples include species swarms or aggregates. Average values based on two or three samples provide the best estimate of the local standing crop.

Another set of problems relates to the sampling gear and its method of use. Most zooplankton samples are taken by means of towed nets. Experience has shown that the size and composition of the catch is affected by the diameter of the mouth opening of the net and by the aperture size of the mesh. However, depth in the water column, towing speed, and duration of the tow are also important. For those species that exhibit vertical migration, the time of day can be a major factor determining the catch. For general survey work, many marine zooplankton investigations have employed nets with 1 m openings

and a mesh size of 333 μm, towed below the surface for 15 minutes at a speed of about 2.5 kts. For smaller organisms such as protozoans and many larval forms, smaller mesh openings are required, and for gelatinous organisms (medusae, ctenophores, salps, etc.), which can be damaged or destroyed in collecting nets, special techniques must be used. During the past few decades the development of nets that can be opened and closed at a given depth has provided valuable information on the depth distribution of many marine zooplankton species.

## Coastal studies

A number of important zooplankton investigations have been carried out on continental shelves of the northern Gulf including those off Mississippi (Perry and Christmas 1973), Louisiana (Marum 1979; Wolff and Wormuth 1984), and south Texas (Casey 1979a, 1979b; Park 1975c), among others. Extensive species lists are available for several taxonomic groups including tintinnids, foraminiferans, heteropods, pteropods, copepods, and a few other groups. All told, around 2,000 species have been recorded from the northern Gulf continental shelf, and many of the larval types could not be identified beyond major group. The zooplankton of the northern Gulf shelf is rich in both holoplanktonic and meroplanktonic forms. Unfortunately, a variety of gear types and at least eight different mesh sizes have been employed in these studies, making quantitative comparisons all but impossible. The range of reported standing crop values is given as follows:

- wet weight — 0.66 to 2.5 g/m³
- dry weight — 3.4 to 122.8 mg/m³
- volume — 0.1 to 3,500 μL/m³
- density — 166 to 1,539,373 individuals/m³

LOUISIANA-TEXAS SHELF

Along the Louisiana-Texas shelf there is a definite downcoast gradient from high zooplankton densities along Louisiana to low densities off south Texas. For example, in monthly collections off the mouth of the Calcasieu River of western Louisiana, the numerical abundance of total zooplankton ranged from 365 to 1,539,373 individuals/m³, with an average value of 85,408 individuals/m³. By comparison, total zooplankton values at inshore stations off south Texas averaged only 2,671 individuals/m³, even though gear types and mesh sizes in the two studies were roughly comparable. The Louisiana average was elevated due to one month of very high density (see below), but the general nature of the downcoast gradient is clear.

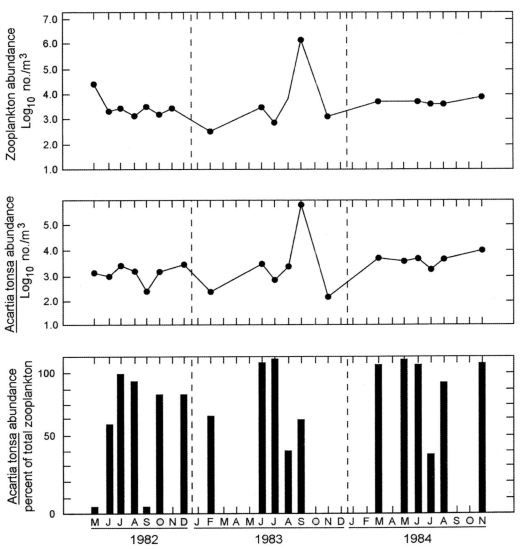

**Figure 9.7.**
Monthly zooplankton abundance at a shallow-water station off western Louisiana. (From Wolff and Wormuth 1984.)

Seasonal trends in some zooplankton parameters for the western Louisiana study are presented in figure 9.7, which shows monthly values for zooplankton abundance, numerical abundance of the dominant copepod species, *Acartia tonsa,* and the abundance of *A. tonsa* as a percentage of total zooplankton. Here it is seen that for total zooplankton most monthly values ranged between 1 and 10,000 individuals/m³, but during September 1983, the average density exceeded 1.5 million organisms/m³, apparently because of the passage of a major storm and outwelling from the estuary. During the three-year study period, the density values for *A. tonsa* ranged from 144 to 861,926 individuals/m³, with a mean value of 47,610 individuals/m³. This single species constituted from 4.8 to 98.3 percent of the total zooplankton population, with a mean value of 55.7 percent. In this study, zooplankton volumes ranged from 0.1 to 3,500 μL/m³ (= 0.01 to 350 mL/100 m³).

SOUTH TEXAS ZOOPLANKTON STUDIES

The south Texas zooplankton investigations are the most thorough yet carried out anywhere in the Gulf of Mexico, and since they illustrate the major points about zooplankton distribution, these results will be discussed in some detail. As seen in figure 9.8, there were four transects extending across the continental shelf and three stations along each transect: inshore (Station 1, 18–27 m), midshelf (Station 2, 42–65 m), and outer shelf (Station 3, 91–134 m). During each of two years, collections were made seasonally: winter (December–January), spring (April–May), and late summer (August–September). At

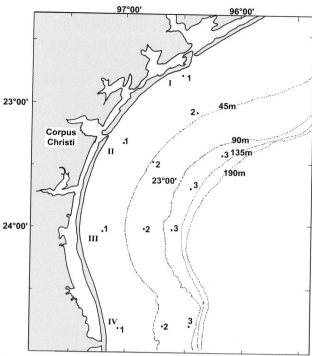

**Figure 9.8.** South Texas continental shelf showing the location of transects (roman numerals) and collecting stations (arabic numerals) for the south Texas zooplankton study. (From Park 1975c.)

each station day and night collections were made, and each collection was taken in duplicate. All plankton collections were made with a 1 m net (mesh openings of 233 µm) mounted with a flowmeter to measure the amount of water filtered. All were oblique tows lasting 15 minutes, with a towing speed of 2.5 kts. Results of the first year's collections are presented below.

On a seasonal basis, the highest average catch (1,937 individuals/m³) occurred during the spring, the next highest (1,642/m³) in late summer, and the lowest (1,439/m³) in winter. Day/night comparisons reveal that during the spring, night catches exceeded day catches in 83.3 percent of the cases, and during the summer, night catches were greater in 75.0 percent of the cases. However, for the winter cruises night catches were greater in only 25.0 percent of the cases. Biomass values are presented in table S20. On average, zooplankton volumes were greatest (304.8 µL/m³) on the inner shelf, intermediate (186.0 µL/m³) at midshelf, and least (83.9 µL/m³) on the outer shelf. The same pattern appeared on all transects. Both dry weight and ash-free dry weight followed this general pattern.

Data for zooplankton numerical abundance are presented in table S21. On average, zooplankton densities were highest inshore (2,671/m³), intermediate at mid-

shelf (1,590/m³), and lowest on the outer shelf (760/m³). This pattern held up on all except the southernmost transect (Transect IV), where the midshelf density exceeded that of the inner shelf (largely because of a huge swarm of ostracods in late summer). Mean densities of the two northernmost transects (Transects I and II) were somewhat higher than those of the southernmost transects, apparently reflective of the fact that the shelf off south Texas receives very little outflow of freshwater and nutrients now that most of the water of the Rio Grande is diverted for agricultural irrigation. Copepods, which are the major component of the zooplankton, showed high densities on the inner shelf and much lower densities on the outer shelf, the average ratio being about four to one. Although this ratio varied somewhat between transects, the general pattern held up all along the coast. Copepods were found to make up around two-thirds of the total zooplankton individuals taken. The highest percentage (81.3) occurred on the inner shelf, and the lowest percentage (53.1) was seen at midshelf. Calanoids constituted about three-fourths and cyclopoids about one-fourth of the copepods, with harpacticoids making up only a fraction of a percent. Interestingly, calanoids decreased from a high of 81.1 percent near shore to 68.1 percent offshore, whereas cyclopoids increased from 18.8 percent on the inner shelf to 31.3 percent on the outer shelf.

GULF-WIDE COMPARISON OF SHELF ZOOPLANKTON
Further insight into the south Texas neritic zooplankton populations derives from analysis of taxonomic groups in relation to distance from shore and comparison of these data with comparable information from nearshore collections taken off Mississippi and Yucatán (table S22). In south Texas coastal waters, copepods were clearly the numerically dominant group, followed by ostracods (primarily *Euconchoecia* and some *Conchoecia*), and then mollusks (mainly gastropod larvae and fair numbers of heteropods and pteropods, and an occasional cephalopod). Polychaetes, mysids, and euphausiids were generally rare and never averaged more than 6 individuals/m³. Sergestids (*Lucifer faxoni*), although more abundant near shore than offshore, were also rare. Medusae and urochordates (larvaceans, *Doliolum,* and salps) showed only slight changes in mean density across the continental shelf. All the remaining groups except ostracods showed highest densities inshore and lowest densities offshore. Surprisingly, the ostracods were least abundant inshore and most abundant at midshelf.

The Mississippi data represent the average of monthly collections taken throughout a year in the pass between Ship and Horn Islands, barrier islands off Mississippi Sound. This is a very fertile area influenced by nutrient-rich outflow from streams draining eastern Louisiana and southern Mississippi. At the collecting station the water is shallower, more fertile, and closer to shore than the inshore stations off south Texas. Here the total zooplankton density was much higher, and this was reflected in most taxonomic groups. Copepods constituted slightly less than half the total zooplankton. Mollusks, crustacean larvae, chaetognaths, and urochordates were also relatively abundant, but ostracods, mysids, amphipods, and euphausiids were barely represented or absent. Echinoderm larvae, which accounted for most of the "other" category, were seasonally abundant. Thus, in addition to displaying a much higher total density, the Mississippi populations showed distinct differences in relative composition of the various taxonomic groups.

The Yucatán data also represent the annual average of monthly collections made at a single nearshore station. This was located at a depth of four meters and situated on the continental shelf three kilometers off the northeast tip of Yucatán. Since the area is virtually unaffected by nutrient-rich surface runoff and too close to the shore to be affected by upwelling, the zooplankton populations here are greatly impoverished, having less than a tenth the average density of inshore stations off south Texas. Even copepods were quite rare, and the only taxonomic group of consequence was echinoderm larvae making up most of the "other" category.

## Oceanic studies

Open waters of the Gulf of Mexico are inhabited by an extremely diverse array of zooplankton species, including both larval and adult forms. Consequently, most investigations of oceanic zooplankton in the Gulf have concentrated on particular taxonomic groups. Most thoroughly studied have been the copepods, but much information has also been obtained on other groups including the gelatinous forms (medusae, salps, and ctenophores), as well as pteropods, euphausiids, and chaetognaths. Little work has been done on the larval forms. To date, a fair amount of information has been amassed on the taxonomic composition, horizontal and vertical distribution patterns, zooplankton abundance, and species that undergo vertical migrations. These topics are addressed below.

SPECIES DISTRIBUTION PATTERNS

Some zooplankton species are largely confined to coastal waters. These include species found here year-round as well as certain estuarine forms (such as the copepods *Acartia tonsa, Oithona nana, Paracalanus crassirostris,* and *Pseudodiaptomus coronatus*) that occur only under low-salinity conditions off the mouths of rivers and estuaries often associated with the spring outflow. Other species are essentially limited to open Gulf waters, but some are widely distributed in both neritic and oceanic environments. Representative examples of neritic, oceanic, and ubiquitous species are given in table 9.3. From available sources, this list could be greatly expanded. For most groups the taxonomic diversity is far greater offshore than in coastal waters. Within the offshore waters there is some differentiation between the eastern and western Gulf, some species being more characteristic of one area or the other. Seasonal differences in species composition and abundance have also been noted.

In addition to the horizontal patterns discussed above, some information is also available on the vertical distribution of species in a number of oceanic zooplankton groups. Table S23 provides examples of epipelagic, mesopelagic, and bathypelagic species of medusae, heteropods, pteropods, copepods, euphausiids, and chaetognaths recorded from the Gulf. For most groups species abundance and diversity are greatest in the epipelagic zone, and both abundance and diversity decrease with depth. However, some zooplankton species are present even in the deepest waters of the Gulf.

ZOOPLANKTON ABUNDANCE

Lacking chlorophyll, zooplankton cannot be imaged by satellites, and our information on general zooplankton abundance in the open Gulf must rely on samples taken from surface ships. Fortunately, two Gulf-wide surveys have been carried out, one by the US Fish and Wildlife Service in the 1950s and the other by Soviet scientists as part of the Soviet-Cuban fishery investigations of the 1960s. In addition, a major zooplankton and fish larva survey has been underway since 1983, the Southeast Area Monitoring and Assessment Program (SEAMAP-Gulf), under the auspices of the National Marine Fisheries Service.

The 1950s survey was designed to determine the distribution and abundance of total zooplankton as well as fish eggs and larvae throughout the entire Gulf of Mexico, and the results are summarized in table S24. Here it is seen that zooplankton volumes throughout the open Gulf

**Table 9.3.** *Onshore and offshore patterns of zooplankton distribution. (Data from Every 1968; Park 1970; P. Phillips 1972; Snider 1975; and other sources.)*

**Groups that are primarily neritic**

Medusae – *Chiropsalmus quadrumanus, Chrysaora quinquecirrha, Cyanea capillata, Rhopalonema verrilli, Tamoya haplonema*

Pteropods – *Limacina trochiformis*

Copepods – *Centropages furcatus, Eucalanus pileatus, Labidocera aestiva, Paracalanus parvus, Temora turbinata*

Chaetognaths – *Krohnitta pacifica, Sagitta helenae, S. hispida*

**Groups that are primarily oceanic**

Medusae – *Atolla wyvillei, Deepstaria enigmata, Nausithoe punctata*

Pteropods – *Cavolina inflexa, Diacria quadridentata, Hyalocylis striata, Limacina inflata*

Copepods – *Centropages violaceus, Corycaeus clausii, Lubbockia aculeata, Oithona setigera, Pontellina plumata*

Chaetognaths – *Pterosagitta draco, Sagitta bipunctata, S. decipiens, S. hexaptera, S. serratodentata*

**Groups that occur in both neritic and oceanic areas**

Medusae – *Aurelia aurita, Pelagia noctiluca, Stomolophus meleagris*

Copepods – *Centropages velificatus, Clausocalanus furcatus, Corycaeus americanus, Farranula gracilis, Oncaea venusta*

Chaetognaths – *Sagitta enflata, S. tenuis*

---

averaged 10.9 mL/100 m³, which is less than two-thirds of the density in continental shelf waters (17.1 mL/100 m³). The concentration of fish eggs in the open Gulf amounted to 18.7 percent and that of fish larvae only 2.6 percent of concentrations over the continental shelves.

In the 1960s the Soviets collected large zooplankton samples throughout open waters of the Gulf, and in the laboratory they sorted out the portion of the plankton that they considered to be useful food in the production of fish populations. This edible portion, which they called "food plankton," was roughly equivalent to the microzooplankton and macrozooplankton minus some of the larger gelatinous forms. The resulting density distribution pattern

for August–October 1964 is shown in figure 9.9. This map reveals that in the late summer and fall throughout most of the open Gulf, the zooplankton wet biomass is quite low (< 0.2 g/m³), but values are elevated (> 0.5 g/m³) in coastal areas affected by upwelling or Mississippi River effluent. If the study had been conducted during spring and early summer, much higher biomass values would have been shown on the Louisiana–upper Texas coast and certain other neritic areas. Biggs and Ressler (2001) have pointed out that against this background of low zooplankton production in open waters of the Gulf, there are "hot spots" where production and standing crops are much higher, approaching values more characteristic of the continental shelves. As in the case of the phytoplankton, these occur in areas affected by upwelling, entrainment and offshore transport of freshwater, frontal boundaries of strong currents, and within the core of cyclonic rings and eddies.

From a number of recent studies it has been possible to glean data on the numerical abundance of various invertebrate groups in epipelagic waters of the open Gulf (table S25). All densities are expressed as numbers/1,000 m³ of water sampled. Although collecting methods and depth covered vary from one study to another, the combined data set does provide order of magnitude estimates for the various groups. Radiolarians are clearly the most abundant group, but pteropods, copepods, and chaetognaths are also fairly abundant. Polychaetes, amphipods, and the various gelatinous forms are relatively rare. In a series of special studies carried out in the rotating eddies off south Texas and northern Mexico, it was found that within the (upwelling) cyclonic rings the gelatinous forms tended to be larger, more abundant, and more rich in species than in the (downwelling) anticyclonic rings. On a seasonal basis the gelatinous forms were most abundant in the fall, least abundant in the summer, and of intermediate abundance during the spring months. For several of the invertebrate groups, densities appeared to be somewhat higher in the eastern than in the western Gulf.

VERTICAL MIGRATION AND DAY/NIGHT COMPARISONS

In oceanic waters many zooplankton species live at greater depths during the daytime than during the night. In general, they rise toward the surface at sunset and descend to deeper waters at sunrise. This permits them to feed on the rich phytoplankton stocks in the dark and sink to darker waters where they are less susceptible to predation during daylight hours. The amount of vertical

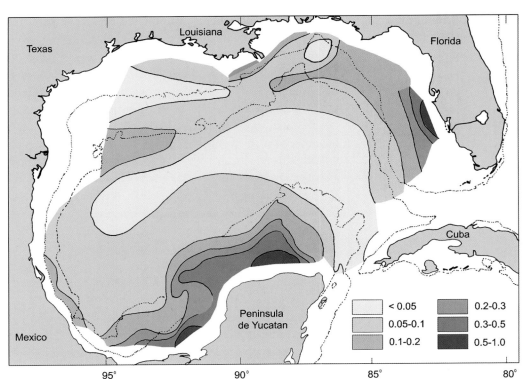

**Figure 9.9.**
Distribution of "food plankton" biomass in euphotic waters of the Gulf of Mexico, August–October 1964. Data are expressed as g/m³. (From Khromov 1965.)

migration varies from species to species, season to season, and also according to sex and developmental stage. The primary cue for vertical migration is the intensity and, perhaps, quality of light, but temperature may have a modifying effect. Some forms migrate many meters, whereas others move little or not at all. Lists of migrating and relatively nonmigratory species for three groups in the Gulf of Mexico are provided in table S26. It should be noted that migratory forms occur in most groups of invertebrates and in many fish families, but not all are well documented in the Gulf. However, in one interesting study, Berkowitz (1976) made a series of identical day and night collections at a single oceanic station off western Louisiana, permitting a clear comparison of day/night densities in the upper meter of water for several invertebrate groups (table S27). For all groups, nighttime densities exceeded those of the daytime, and the ratio of day to night densities ranged from 1:2.2 (in harpacticoid copepods) to 1:128 (in calanoid copepods). In most instances nighttime values were more than 15 times greater than those of the daytime. The implication is that forms found deeper in the water column during the day move up to the surface at night, even within the surface meter of water.

Supporting evidence comes from a study carried out by Bright, Ferrari, et al. (1972) during a total solar eclipse at noon on 7 March 1970. At a station in oceanic waters north of the Yucatán shelf, they found that as the light intensity decreased several species moved upward in the water column, the most prominent migrators being the copepods *Nannocalanus minor, Scolecithrix danae,* and *Undinula vulgaris* as well as the euphausiid *Stylocheiron carinatum.* Although other species did not show clear evidence of migration, the event lasted only about an hour and a half, and many species may have been too deep in the water column or too slow to respond to the event. The subject of vertical migration by the zooplankton is taken up in greater detail in chapter 14.

ACOUSTICAL MEASUREMENT

During recent years it has become possible to determine zooplankton density versus depth in the water column by acoustical means. In principle, sound signals emitted from a shipboard source are bounced off particles suspended in the water, and the echo is recorded back on the ship. Elapsed time between the sending and receipt of the signal is proportional to distance (in this case, the depth) of the target, and the intensity of the echo is proportional to the density of the reflecting particles. These quantities must be standardized by data from actual plankton samples taken from various depths in the water column.

Employing the Acoustic Doppler Current Profiler (ADCP), Biggs and Ressler (2001) provided day and night profiles of zooplankton density in the upper 200 m of the water column as they passed through a cyclonic ring and a Loop Current eddy (anticyclonic ring) and made a night profile through an area of convergence (frontal boundary with intense vertical shear). Near-surface zooplankton densities were far greater at night than during daylight hours and greater in the core of the cyclonic than in the anticyclonic eddy. Zooplankton density was also quite high in the area of convergence, even greater than in the center of the cyclonic ring. One of the great advantages of the acoustical method is that the data can be recorded while the ship is underway, thus permitting the coverage of large areas. A disadvantage is that not all zooplankton groups are equally good reflectors. Hence, the acoustical data must always be standardized against information from actual plankton samples.

## INDICATOR SPECIES

As noted earlier, many of the zooplankton species of the Gulf are largely limited to certain environments or particular water mass types, and their presence elsewhere provides information concerning water mass movement. The most obvious example is that many species of medusae, copepods, chaetognaths, and so forth are characteristic of neritic waters, and their presence in the open ocean is considered to be evidence of offshore transport. The appearance of oceanic species in coastal waters signifies onshore transport of oceanic water masses. In a similar vein, many zooplankton species have specific depth preferences, and the presence of deepwater species in near-surface waters provides strong evidence of upwelling. For instance, the finding of deep-living radiolarians in continental shelf waters off south Texas has been interpreted as the rising of subtropical underwater in the area (Casey et al. 1979).

Several groups have been suggested as indicators of Caribbean water in the Loop Current of the eastern Gulf. Austin (1971a) was able to differentiate water masses in the eastern Gulf by the occurrence, relative abundance, or species diversity of foraminiferans and pteropods. Species diversity, in particular, tended to be higher in the Loop Current than in nearby Gulf waters. He also found that upwelling areas were characterized by certain species of foraminiferans (*Globorotalia*) and pteropods (*Clio* and *Peraclis*). Cruise (1971) found that the sergestid shrimp *Lucifer typus*, a Caribbean species, is characteristic of the

Loop Current, whereas the related species *L. faxoni* is more indicative of Gulf waters. As the zooplankton of the Gulf becomes better known, other indicator species and groups will doubtless appear.

## ZOOGEOGRAPHY

The zooplankton of the Gulf of Mexico includes many species of worldwide occurrence in tropical and subtropical marine waters, and the fauna is quite similar to that of the subtropical west Atlantic and Caribbean Sea. However, the fauna does contain some unique elements, and even in those groups that have been well studied in the Gulf, some regional differentiation has been noted. For example, P. Phillips (1972) pointed out that the neritic cnidarian fauna of the northern Gulf (from south Texas to about the level of Tampa Bay) is clearly a northern fauna related to that of the south Atlantic coast of the United States (from Cape Canaveral to Cape Hatteras), but not continuous around the south Florida Peninsula. Apparently, during Pleistocene glacial times it was continuous, but now it exists as a relict fauna isolated in nearshore waters of the northern Gulf. Included are a number of species of hydrozoans, scyphozoans, and possibly other forms. This group is fairly distinct from the warm-water cnidarian fauna of the southern Gulf, which shows more affinity with that of the Caribbean and West Indies. This group includes a number of species of hydroids of the genera *Bythotiara, Calycopsis, Mitrocoma, Koellikerina,* and *Zancleopsis,* among others, with Indo-Pacific relations. These appear to be relicts from pre-Pleistocene times, when Atlantic Ocean waters flowed through the Straits of Panama into the Pacific Ocean. These species are no longer present in the eastern Pacific, but the relicts are now isolated in the southern Gulf. An interesting parallel exists in the radiolarian fauna (McMillen and Casey 1978). Within the Gulf of Mexico there persists a living assemblage of spongy radiolarians thought to have been extinct worldwide since the late Miocene. Like the warm-water cnidarians, these appear to be relicts isolated here since the closure of the Straits of Panama. The emergence of the Isthmus of Panama (ca. 3.5 million years ago) cut off ocean circulation with the Pacific, and the environment of the eastern Pacific changed considerably. However, the environment of the western Gulf remained relatively unchanged, and here the early fauna has persisted to the present day.

## NEUSTON

The surface layer of the sea is a variable and, in many ways, harsh environment much different from that of subsurface layers. Here solar radiation is very intense and rich in tissue-damaging ultraviolet light. Frequent and rapid changes in temperature occur in response to solar heating and atmospheric temperature shifts. Salinity may also change rapidly as a result of precipitation, on the one hand, and evaporation, on the other. There is much mechanical buffeting by water movement and the action of surface waves. The layer is rich in oil and other chemicals that tend to rise to the top or float on the water surface. Bodies and parts of bodies of dead organisms are frequently encountered at the surface.

In this strange marine environment there lives a tolerant group of bacteria as well as plant and animal species collectively called the *neuston*. Some are permanent residents, and others are temporary visitors or stragglers from deeper layers. A few live on or near the surface film, whereas others occur just below. The most common groups are certain phytoplankters, protozoans, and small crustaceans as well as fish eggs and larvae. In addition, since there is much dead organic matter, decomposing bacteria are quite abundant. Here also are encountered floating mats of seaweed (*Sargassum*) and jellyfishes, hydrozoans such as the Portuguese man-of-war, small snails (*Janthina*), salps, and other forms. Although no insects occur in the sea, one form of water strider, the so-called ocean skater (*Halobates*), scoots along above the surface film feeding on small organic particles. Investigators differ on the defined depth of this layer. Some limit it to the surface film itself, some to the top ten centimeters, and others include the top half meter. Since zooplankton makes up a significant fraction of the neuston, it is convenient to include this diverse group in the present chapter. Within the Gulf of Mexico the neuston has been studied by Finucane (1976, 1977), Finucane et al. (1977), L. Pequegnat and Wormuth (1977), and Wormuth, McEachran, and Pequegnat (1980). It should be mentioned that in recent years the surface layer of the sea in general, and of the Gulf of Mexico in particular, has been much modified by the addition of atmospheric pollutants, petroleum films, tar balls, pieces of Styrofoam, plastic bags, and other wastes of human civilization.

### South Texas neuston studies

The neuston data dicussed here were taken on a single transect across the continental shelf offshore from Corpus Christi, Texas, and represent the average of two years of seasonal collections. In this extreme environment population densities were quite low, but they still displayed the expected inshore/offshore pattern: inshore there were 211.0 individuals/m³, the midshelf had 163.4 individuals/m³, and the outer shelf had 71.7 individuals/m³. These numbers are all roughly a tenth of zooplankton abundances at the same stations (Transect II in fig. 9.8). Seasonal trends in taxonomic composition are provided in table S28. From a winter low the neuston rose in early spring and peaked in May and June before falling to the annual low in September and October. This seasonal pattern, exhibited by most groups, was particularly obvious in the copepods and in the "other" category. During the spring months this area is influenced by the downcoast flow of Mississippi River water as well as outflow from other streams of Louisiana and upper Texas. In addition to the invertebrate groups given in the table, the neuston contains the eggs, larvae, and juveniles of a great many types of fishes (table S29). From the collections 77 individual species and 45 families of fishes were recognized and many others could not be specifically identified. Despite its low density, the continental shelf neuston does include a fairly diverse group of organisms.

### Neuston of the open Gulf

The neuston of the open Gulf has not been thoroughly investigated, but a few studies do provide insight into the composition and community relations of organisms inhabiting the surface layer. As mentioned earlier, small surface-floating tar balls are ubiquitous, but effects on the flora and fauna are unknown. Berkowitz (1976) compared the density of neuston organisms (in the surface 9–19 cm) with that of the zooplankton slightly deeper (in the surface 1 m). During the daytime the neuston was only 46.1 percent and at night only 31.9 percent of the zooplankton catch, and he concluded that in terms of numerical density, neuston of the open Gulf is relatively impoverished. However, this is not always the case. Seasonally, especially in the spring and early summer, the Gulf is invaded by fleets of Portuguese man-of-war (*Physalia*), by-the-wind sailor (*Velella*), and the violet snail (*Janthina*), which are swept into the Gulf on the Yucatán Current. This current also brings in large rafts of the floating brown alga *Sargassum*, locally known as "gulfweed." All sargassum species have buoyant vesicles that enable them to float at the surface. Some species live attached to solid substrates, but they will float if dislodged. Two species, *Sargassum flui-*

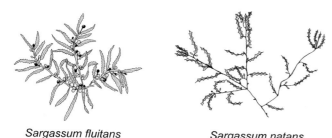

*Sargassum fluitans*          *Sargassum natans*

**Figure 9.10.** The two primary floating seaweeds of the Gulf of Mexico, *Sargassum fluitans* and *S. natans*. (After Earle 1972.)

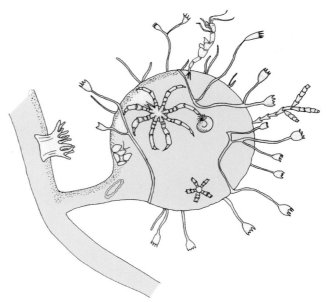

**Figure 9.11.** The fauna associated with a sargassum bladder from the Texas coast showing hydroids, an anemone, a flatworm, a bryozoan colony, polychaetes, a caprellid (amphipod), pycnogonids (sea spiders), and juvenile ophiuroids (brittle stars). (From Hedgpeth 1953.)

tans and *S. natans* (fig. 9.10), are obligate floaters, that is, they are never attached to substrates. Within the Gulf 80–90 percent of the sargassum is *S. natans*, somewhat less than 10 percent is *S. fluitans*, and the remainder is the attached species *S. filipendula*, the latter being most abundant in the northwestern Gulf.

Information on the composition of the sargassum community comes primarily from reports by J. Adams (1960) and Dooley (1972), who studied sargassum in the Florida Current off Miami. However, Berkowitz (1976) noted that the sargassum-associated fauna of the Gulf was quite similar to that described by Dooley. Additional references include Earle (1972), Hedgpeth (1953), Parr (1939), and W. Taylor (1954a, 1954b). Associated with the sargassum is a complex community of other organisms that find food or shelter among the floating rafts. Some species are firmly attached; others cling or climb among the "stems," "leaves," and floats; while others are more loosely associated and may be free floating or swimming around or beneath the algal mats. Worldwide, more than 100 species of invertebrates, belonging to most marine phyla, have been reported in the sargassum community. Shrimps and crabs appear to make up the bulk of the invertebrates, and these provide a major source of food for sargassum-associated fishes. More than 50 species of fishes, including many juveniles, have been taken around sargassum, and of these about 20 were closely associated. Ten species of invertebrates and two species of fishes (the sargassum pipefish, *Syngnathus pelagicus*, and the sargassumfish, *Histrio histrio*) are endemic to this community. A list of the commonly occurring sargassum-associated invertebrates and fishes is presented in table 9.4, and a sketch of the faunal association on a sargassum bladder from the Gulf is presented in figure 9.11. The sargassum community is discussed further in chapter 15.

Although no insects occur in the ocean, one group does live on the surface of the open sea, and this group includes the water striders or ocean skaters, sometimes called "grease bugs" because of their waxy cuticles (fig. 9.12). Properly these are included in the neuston (or epineuston). These tiny insects skate around on the surface and feed on organic matter upon or within the surface layer. Cheng and Wormuth (1992) examined several collections from the Gulf and concluded that there is probably an indigenous population of the species *Halobates micans* in the Gulf, which overwinters as eggs and which may complete two or three generations each year between March and November.

## ZOOPLANKTON IN RELATION TO FOOD CHAINS OF THE SEA

Zooplankton organisms are primarily particle feeders, and they display a wide variety of feeding methods and techniques. Some, such as the urochordates, pump water through filtering nets and remove all particles retained by the nets. Others produce large quantities of mucus in which small particles become entrapped. Still others employ cilia or spiny setae to glean small food particles from the water. Predaceous forms may stun their prey with stinging cells or capture prey through hunting and grasping techniques. Such mechanisms permit the zooplankton as a whole to utilize the diversity of available

**Table 9.4.** *Animals commonly associated with the sargassum community. (From J. Adams 1960; Dooley 1972; and other sources.)*

**Cnidaria**

    hydroids and relatives – *Aglaophenia, Clytia, Desmophyes, Gemmaria, Halecium, Monotheca, Obelia, Plumularia, Sertularia, Velella*

    siphonophores – *Physalia*

    anemone – *Anemonia*

**Platyhelminthes** – *Stylochus*

**Bryozoa** – *Flustra, Membranipora*

**Mollusca**

    gastropods

        pteropod – *Creseis*

        snail – *Janthina*

        nudibranchs – *Aeolidella, Cuthona, Doto, Fiona, Glaucus, Scyllaea, Spurilla*

        sea hare – *Tethys*

**Annelida** – *Alciopa, Spirorbis*

**Arthropoda**

    cladocera – *Evadne*

    barnacles – *Conchoderma, Lepas*

    copepod – *Copilia*

    amphipod – *Ampithoe*

    isopods – *Janira, Idotea*

    decapods

        shrimp – *Alpheus, Hippolyte, Latreutes, Leander, Sergestes, Virbius*

        crabs – *Neptunus, Planes*

    sea spiders – *Endeis, Tanystylum*

    insects – *Halobates*

**Chordata**

    tunicate – *Diplosoma*

    fishes – herrings (*Clupeidae*), frogfishes (*Antennariidae*), flying fishes and halfbeaks (*Exocoetidae*), needlefishes (*Belonidae*), pipefishes and seahorses (*Syngnathidae*), flying gurnards (*Dactylopteridae*), jacks (*Carangidae*), tripletails (*Lobotidae*), sea chubs (*Kyphosidae*), damselfishes (*Pomacentridae*), butterfishes (*Stromateidae*), leatherjackets (*Balistidae*), puffers (*Tetraodontidae*), and eel larvae (*leptocephali*)

**Figure 9.12.** The "ocean skater" *Halobates micans,* an insect highly modified for life on the surface of the open sea. (From Savilov 1967.)

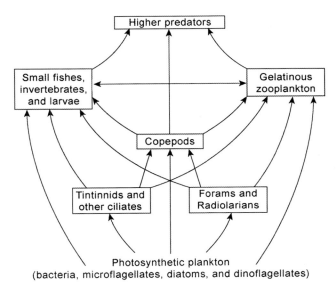

**Figure 9.13.** A simplified food web involving major elements of the zooplankton.

food types, and it is safe to say that within the range of particle sizes they can handle (from less than one micron to several millimeters in diameter), for every food resource of the water column there are many potential consumers. Practically everything gets used by one zooplankton consumer group or another. By such means the zooplankton plays a key role in consumer food chains of the sea. As illustrated in figure 9.13, the zooplankton depends heavily on a variety of small photosynthetic species. In addition, they take in various nonphotosynthetic bacteria and protozoans as well as decomposing particulate material. Likewise, the various zooplankton species feed on one another, larger forms generally consuming smaller ones. Because of their small size and great abundance, the copepods occupy a central position and play a major role in the transfer of energy and nutrients through marine food webs. They are large enough to consume most phytoplankton organisms, bacteria, and protozoa, but still small enough to be ingested by chaetognaths, larval fishes, and other larger zooplankton animals. Furthermore, they constitute a major food resource for gelatinous forms such as the cnidarians, ctenophores, and urochordates. All these organisms provide a food base for higher consumers of the water column.

Zooplankton organisms also contribute in a major way to the decomposer chains of the sea and supply food resources to deeper layers of the ocean and to the bottom communities. Through secretion and excretion of ammonia and organic nitrogen and phosphorus compounds, they facilitate remineralization and stimulate phytoplankton growth. Undigested food that passes through their digestive tracts is packaged into streamlined fecal pellets that sink rapidly from the warm surface layer into deeper, colder water where the rate of decomposition is retarded. Dead bodies and body parts, including shed exoskeletons, sink and decompose, providing food for other species. Films of mucus, produced by gelatinous and other species, are often referred to as "marine snow." This sticky material attracts or entraps bacteria, phytoplankton, and small zooplankton animals, forming complex organic aggregates that gradually sink and add to food resources of deeper layers. Because of its roles in supporting higher consumer chains on the one hand, and decomposer chains on the other, the zooplankton community is crucial to proper functioning of the biological economy of the sea.

The benthos includes all those organisms that live within, upon, or near the sea bottom. The *infauna* resides within the sediments, while the *epifauna* dwells upon the sediment surface. Some species are attached to the bottom but grow upright in the water column, whereas others exist in the water above the bottom but feed or otherwise closely affiliate with the benthic environment. Although most benthic organisms spend their entire lives in association with the bottom, others regularly or occasionally live elsewhere. Among the latter are those species that practice vertical migration, moving up into the water column during the nighttime hours. Many benthic species are attached and fixed in position, but others are mobile and move around. As mentioned earlier, many benthic forms produce planktonic eggs and/or larvae. As a group, the benthos embraces a very diverse array of organisms in which relations with the bottom are quite variable depending on life history details of the individual species. In practice, benthos is what is collected by bottom sampling devices.

From a taxonomic standpoint, benthos includes both plants and animals. Among the phytobenthos are the diatoms, dinoflagellates, and other unicellular and filamentous algae as well as thallic algae, the latter being generally larger and having more complex body structures. Rooted in the sediments are certain higher plants including marsh grasses and mangrove trees along the edges, and seagrasses in shallow protected marine waters. Although these higher plants are not strictly benthos, they are attached to the bottom, and their presence creates unique habitats for other benthic species. These higher plants and their associated communities will be discussed in a later chapter.

The zoobenthos includes representatives of all known animal phyla, some groups being much more prominent than others. The zoobenthos is often differentiated on the basis of feeding types. Predators capture and consume living prey; scavengers take dead bodies and body parts from the bottom surface; and suspension feeders selectively or unselectively remove particulate material from the water column by means of cilia, setae, straining nets, or mucus sheets. They retain what is digestible and discard the rest. Deposit feeders ingest the sediments, removing available living and dead organic matter and voiding the indigestible material. In the process of their daily activities benthic animals tunnel, plow through, or otherwise churn up surface layers of sediments, sometimes to a depth of several centimeters, and this constant reworking aerates the sediments, releases nutrients into the water column, and has important effects on the chemical and physical structure of the sea bottom.

Benthic organisms are most conveniently classified on the basis of size because such groups can readily be separated by sieves of appropriate mesh apertures, and also because sample size and collecting methodology vary depending on the size range of the organisms under study. Four size classes of benthic organisms are recognized. From smallest to largest, these include the *microbenthos*, *meiobenthos* (from the Greek *meion* = "smaller"), *macrobenthos*, and *megabenthos*. The size range and general taxonomic composition of each of these groups is given in table 10.1. The compositions are not precise because a given group may have small and large species, and as organisms grow they increase in size and move from one category to another.

Detailed studies have shown that in nearshore and continental shelf waters, the smaller benthic organisms, in particular, are distributed horizontally along gradients, often in patchy fashion, and generally in association with given substrate types. These gradients relate to grain size, organic content, chemical characteristics, and particularly to stability of the sediments. Few species do well in frequently shifting sands and muds. As discussed in the following section, characteristics of the overlying water and depth of the water column are also important in determining the distribution and abundance of benthic organisms. Most infaunal species are limited to the top few centimeters, but bacteria have been encountered at great depths. Examples of several types of animal-sediment relationships in soft bottoms of the continental shelf are illustrated in figure 10.1.

## The Benthic Environment

The benthic environment includes both the substrate, down to a depth of several centimeters, and the supraben-

**Table 10.1.** *Size range and general taxonomic composition of marine benthic faunal assemblages.*

### Microbenthos (< 62 µm)

Dominant forms: bacteria, single-celled algae, fungi, protozoans

### Meiobenthos (62 µm – 1.0 mm)

Dominant forms: nematodes, harpacticoid copepods, annelids, larger ciliated protozoans, foraminiferans

Others: flatworms, ribbon worms, kinorhynchs, tiny snails and bivalve mollusks, small polychaetes, small arthropods (pycnogonids, ostracods, cumaceans, tanaids, isopods, amphipods), invertebrate larvae

### Macrobenthos (1.0 mm – 25.4 mm)

Dominant forms: mollusks, polychaete worms, crustaceans, echinoderms

Others: worms and worm relatives, young of larger invertebrates, some fishes

### Megabenthos (> 25.4 mm)

Dominant forms: sponges, cnidarians (sea fans, sea pens, etc.), mollusks, larger polychaetes, crustaceans, echinoderms (starfishes, brittle stars, sea urchins, sea cucumbers, crinoids), acorn worms, sea squirts, lancets, some fishes

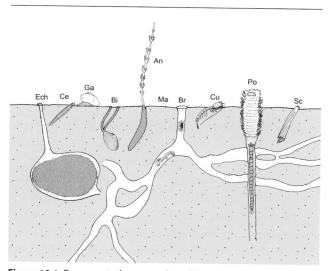

**Figure 10.1.** Representative examples of benthic animals and their relations with soft sediments, including tubes and burrows. Major groups and genera (in parentheses) are as follows: An = Anthozoa (*Virgularia*), Bi = Bivalvia (*Tellina*), Br = Brachyura (*Pinnixa*), Ce = Cephalochordata (*Branchiostoma*), Cu = Cumacea (*Cyclaspis*), Ech = Echinoidea (*Moira*), Ga = Gastropoda (*Polinices*), Ma = Macrura (*Callianassa*), Po = Polychaeta (*Diopatra*), and Sc = Scaphopoda (*Dentalium*). (Modified from Dörjes and Howard 1975.)

thic water column, to a height of half a meter to a meter above the bottom. Both habitats will be examined briefly.

### Sediment particle size composition

For soft substrates probably the most important factor determining species composition and abundance is the particle size composition of the sediments, that is, the relative percentages of sand, silt, and clay, and in some cases, coarser fractions such as shell hash. The typical patterns of sediment composition in terrigenous shelves of the Gulf of Mexico are shown in figure 10.2. Here it is seen that nearshore sediments are composed primarily of sand because wave action and alongshore currents remove and transport finer particles to other areas. Farther out the percentages of silt and clay are about equal, and both generally exceed that of sand. On the outer shelf (as well as on most of the slope and floor of the abyss), sand makes up less than 20 percent, and the percentage of clay often exceeds that of silt. Thus, there exists an across-shelf gradient from coarse particles near shore to fine-grained sediments on the outer shelf and in the deeper Gulf.

Off the mouths of estuaries and near river deltas this general pattern may be violated due to the deposition of silt and clay at all depth levels of the shelf. Toward the outer shelf away from rivers, sand sheets may be present, relics of a past time when sea level was rising and such areas were terrestrial or nearshore environments not yet covered with a veneer of silt and clay. In other cases, where swift currents sweep the outer shelf, coarser sand and/or shell hash remains as the dominant sediment type. On flat calcareous shelves with little river input, particularly

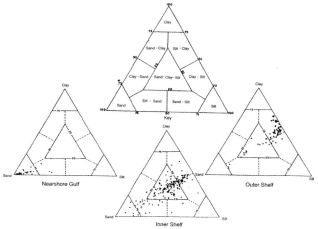

**Figure 10.2.** Triangle diagrams giving a key and showing typical sand-silt-clay characteristics of sediments across the continental shelf in the northern Gulf. (From Shepard and Moore 1955.)

**Table 10.2.** *Characteristics of oxic, intermediate, and anoxic layers of marine soft sediments.*
*Depths of the oxic and intermediate layers vary with locality and may change over time.*
*The intermediate layer displays a mixture of the characteristics of the other two layers.*

| | Sediment layer | | |
|---|---|---|---|
| Characteristic | oxic | intermediate | anoxic |
| Location | top | intermediate | bottom |
| Color | yellow | gray | black |
| Oxygen level | high | low | absent |
| Carbon forms | carbon dioxide and carbonates | mixture | reduced carbon compounds: alkanes (methane, etc.), alcohols, ketones, acids, etc. |
| Nitrogen forms | nitrates | nitrites | ammonium, organic compounds |
| Sulfur forms | sulfates | elemental sulfur | sulfides |
| Iron forms | ferric ion | both | ferrous ion |

off west Florida and northern Yucatán, the limestone platform is generally covered by a layer of coarse carbonate skeletal remains of marine organisms. Such materials may include foraminiferan tests, sponge spicules, and carbonate fragments of calcareous algae, corals, bryozoans, mollusk shells, polychaete tubes, and echinoderm skeletons. Therefore, instead of a simple nearshore-offshore gradient, many continental shelves show patchy patterns of erosional (coarse) and depositional (fine) sediment types.

**Sediment stratification**

It may be recalled from chapter 7 that soft sediments are stratified. The topmost layer, being in contact with the well-oxygenated water column, is an aerobic or *oxic* environment, which contains sufficient oxygen to support the respiration of higher animals. The deeper layer is totally devoid of free oxygen, that is, it is *anoxic* and is referred to as the reducing layer because elements and compounds here exist in the unoxidized or chemically reduced state. Between the oxic layer above and the anoxic layer below lies a thin zone of transition where the oxygen content is quite low but somewhat variable over time. The distinction between oxic and anoxic layers is critical to the survival of benthic organisms. Conditions in the anoxic layer are highly toxic to all higher forms of marine life. This strange environment is inhabited by only a few groups, including some bacteria, fungi, protozoans, small worms and worm relatives (flatworms, ribbon worms, kino-

rynchs, etc.), and a few others. A comparison of characteristics of the three layers is given in table 10.2. An excellent review of the nature and characteristic organisms of the anoxic sediments is provided by Fenchel and Riedl (1970), and a more technical discussion of the environment is given by Ponnamperuma (1972). Higher organisms that live buried in the sediments have developed tubes or burrows that open to the surface and permit the animals to respire in the water column above. Many of these species circulate water through the burrows, aerating the adjacent sediments in the process. So, even though buried, these organisms live in an oxygenated environment.

The thickness of the surface oxidized layer varies depending on sediment type and location. In the intertidal zone and just off the beach the pounding surf drives oxygenated water into the sand, and here the oxygenated layer may be up to a meter thick. Moving away from shore the layer becomes thinner, and throughout continental shelves, slopes, and oceanic basins the oxygenated layer is often only a centimeter thick or less. The same considerations apply in various estuarine and other coastal environments. On the back sides of barrier islands and in other protected areas the sediments are generally fine and have higher organic content. Here the thickness of the oxidized layer may be only a few millimeters. In the channels and other habitats subject to greater water movement, the layer is generally a centimeter thick or more.

## The suprabenthic environment

Characteristics of the suprabenthic water column, in part, define the environment of the benthos. Both the average levels and ranges of variation of bottom water temperature and salinity are important, as are the oxygen concentration and suspended sediment load of the overlying water. Latitudinal gradients in shallow-water species composition as well as seasonal changes in the fauna of the northern Gulf suggest that temperature is an important controlling factor, at least in shallow-water environments. Such differences are not so apparent in the benthic fauna of the outer shelf where seasonal temperatures remain more constant. Off west Florida the presence of estuarine fauna and flora on the middle continental shelf around freshwater springs suggests the importance of bottom salinity. On most continental shelves of the Gulf, oxygen levels of the water column range from about 4 to 8 mg $O_2$/L, and when levels drop below 2 mg $O_2$/L many benthic animal species become stressed. If, as regularly occurs off the Louisiana coast, levels drop to zero, the oxidized surface layer of the sediments disappears, and the anoxic layer extends up to the bottom surface and into the water column. Some of the fishes and more mobile invertebrates are able to escape to other areas, but most sedentary and less mobile forms perish. However, following hypoxic and anoxic events repopulation is often swift.

Where surface sediments are quite coarse, particularly on the limestone shelves off west Florida and Yucatán, but also where there are extensive sand sheets, the lower portion of the water column remains clear and is seldom turbid, even during stormy weather. Whatever sediment is raised quickly settles after the storm passes. By contrast, on shallow continental shelves with silt and clay sediments, bottom currents and storms raise the finer particles, and the resulting cloudy benthic *nepheloid* layer may persist for weeks. Small epibenthic and suprabenthic invertebrates such as crabs and shrimps tend to be found in greater abundance in such turbid bottom waters, presumably because they are less visible to predatory species.

Although the present chapter deals with soft rather than hard bottoms, in nature this distinction is not so clear cut, and there is much overlap in both habitat type and biota. The presence of dead shells and small rocky outcrops allows hard-bottom species to gain a foothold in otherwise soft-bottom areas. The veneer of soft sediments over rocky areas permits soft-bottom species to invade and persist where they would otherwise be precluded.

Many benthic species, especially those that are larger and more mobile, roam freely, independent of bottom types.

### BENTHIC LITERATURE

Most of the technical literature dealing with benthic organisms of the Gulf of Mexico concentrates on individual species or taxonomically related groups, and references important to each group are provided in the next section. However, numerous reports have appeared treating multispecies groups such as the meiofauna or macrofauna. Some of the most informative of these are listed below. For the northwestern Gulf (west of the Mississippi River Delta) the following are particularly significant: Bedinger (1981), Boesch and Rabalais (1991), Darnell, Defenbaugh, and Moore (1983), Darnell and Schmidly (1988), Flint (1981), Flint and Rabalais (1981), Gettleson (1976), Harper (1970, 1991), Harper, McKinney, and Nance (1985), Harper et al. (1981, 1991), H. Hildebrand (1954), Hooks, Heck, and Livingston (1976), Keith and Hulings (1965), McKinney, Nance, and Harper (1985), W. Pequegnat and Venn (1980), N. Rabalais (1990), and Shelton and Robertson (1981). References to benthic assemblages of the eastern and northeastern Gulf include Christmas and Langley (1973), Collard and D'Asaro (1973), Darnell and Kleypas (1987), Environmental Science and Engineering, Inc., LGL Ecological Research Associates, Inc., and Continental Shelf Associates, Inc. (1987), Franks et al. (1972), Harper (1991), Heard (1979b), Hoese (1972), Lyons et al. (1971), Lyons and Collard (1974), McNulty, Work, and Moore (1962), R. Parker (1956, 1960), N. Phillips, Gettleson, and Spring (1990), Stanton and Evans (1971), and Vittor and Associates (1985). Defenbaugh (1976) provided detailed information on distribution of the macroinvertebrate fauna of the continental shelves of the northern Gulf, with somewhat sketchier coverage of the western and southern Gulf. Macroinvertebrate assemblages of the southwestern part of the Yucatán shelf have been reported by H. Hildebrand (1955). Bullis and Thompson (1965) and S. Springer and Bullis (1956) provided distributional records of benthic invertebrates from continental shelves throughout much of the Gulf of Mexico. Hedgpeth (1954b) presented an overview of benthic communities of the Gulf of Mexico.

### COMPOSITION OF THE BIOTA

As mentioned above, many plant phyla and all animal phyla are represented in the soft-bottom benthos. As a guide to understanding this diversity, most of the impor-

**Table 10.3.** *Simplified classification of the bacteria, lower plants, and animal kingdom, including those groups important in relation to soft bottoms of the Gulf of Mexico.*

| Phylogenetic groups, subgroups, and common names | Phylogenetic groups, subgroups, and common names |
|---|---|
| **Bacteria and Lower Plants** | Arthropoda (continued) |
| Schizophyta (bacteria) | Mandibulata |
| Cyanophyta (blue-green algae) | Crustacea |
| Chlorophyta (green algae) | Ostracoda (ostracods) |
| Phaeophyta (brown algae) | Copepoda (copepods) |
| Rhodophyta (red algae) | Malacostraca |
| Mycophyta (fungi and yeasts) | Stomatopoda (mantis shrimps) |
| **Animal Kingdom** | Cumacea (cumaceans) |
| Protozoa (protozoans) | Tanaidacea (tanaids) |
| Ciliophora (tintinnids and aloricate ciliates) | Isopoda (isopods) |
| Sarcodina (foraminiferans and radiolarians) | Amphipoda (amphipods) |
| Porifera (sponges) | Decapoda |
| Cnidaria (coelenterates) | Natantia (swimming shrimps) |
| Anthozoa (sea anemones, corals, and relatives) | Macrura (lobsters) |
| Platyhelminthes (flatworms) | Anomura (hermit crabs and relatives) |
| Nemertea (ribbon worms) | Brachyura (true crabs) |
| Nematoda (roundworms) | Echinodermata (spiny-skinned animals) |
| Kinorhyncha (kinorynchs) | Asteroidea (starfishes) |
| Mollusca (mollusks) | Echinoidea (sea urchins, heart urchins, sand dollars) |
| Gastropoda (snails) | Ophiuroidea (brittle stars) |
| Bivalvia (bivalve mollusks) | Holothuroidea (sea cucumbers) |
| Scaphopoda (tusk shells) | Crinoidea (sea lilies) |
| Annelida (segmented worms) | Hemichordata (acorn worms) |
| Polychaeta (polychaetes) | Chordata (chordates) |
| Arthropoda (joint-legged animals) | Cephalochordata (lancets) |
| Chelicerata | Vertebrata (vertebrates) |
| Xiphosura (horseshoe crabs) | Chondrichthyes (cartilaginous fishes) |
| Pycnogonida (sea spiders) | Osteichthyes (bony fishes) |

tant groups are listed along with their common names in table 10.3. The following sections provide descriptions of each of the groups as well as some of the important literature references to their taxonomy and distribution within the Gulf of Mexico.

### Schizophyta

Bacteria include single-celled and colonial forms without discrete cell nuclei. Most species are heterotrophic, that is, they derive energy from the breakdown of preexisting organic compounds including petroleum hydrocarbons and natural gas. Some are photosynthetic, and others are chemoautotrophic, deriving energy from inorganic compounds such as ammonia, nitrites, or sulfides. Spherical, rodlike, and spiral forms have been recorded from the Gulf. Important genera reported from Gulf sediments include *Achromobacter, Flavobacterium, Micrococcus,* and *Pseudomonas.*

### Benthic algae

Marine benthic algae are primarily components of four phyla: Cyanophyta (blue-green algae), Chlorophyta (green algae), Phaeophyta (brown algae), and Rhodophyta (red algae). As the names imply, these groups are distinguished on the basis of their photopigments, although morphological differences are also important.

Many of the species are filamentous, while others have more complex thallic body forms, and all require light for photosynthesis.

Earle (1972) reported 647 species of marine algae from the Gulf of Mexico. About half of these are restricted to depths of 10 m or less, and the remainder range into deeper water, some to a depth of 100 m or more. A few species are limited to deeper waters of the continental shelf. In areas of high turbidity and low light penetration, such as occur along most of the northern Gulf, few species are found deeper than 10 m except on rocky outcrops of the outer shelf that project well above the bottom nepheloid layer. Records of deeper-living algae are from clearer waters of the south Florida and Yucatán shelves.

Most filamentous and thallic marine algae grow attached to hard substrates such as limestone bottoms, coral reefs, mangrove roots, and the blades of seagrasses, or to artificial surfaces such as pilings and breakwaters. However, some species with well-developed holdfasts grow anchored in soft substrates, especially in protected environments such as seagrass beds, but few species can become established in shifting sands or muds. Among the soft-bottom forms are species of the genera *Acetabularia, Agardhiella, Avrainvillea, Caulerpa, Codium, Dictyota, Laurencia, Lobophora, Padina, Penicillus,* and *Rhipocephalus.* Factors limiting the distribution and abundance of marine algae, in addition to light and substrate type, include temperature, salinity, water currents, wave action, nutrient availability, and grazing pressure by fishes and other marine animals.

Based largely on apparent temperature tolerances, the benthic algal flora of the Gulf can be divided into three categories: temperate, tropical, and cosmopolitan species. The temperate forms, which are limited largely to the northern Gulf during the winter season, have affinities with the New England coastal flora. Tropical species, which make up the largest group, occur largely in the southern Gulf and are related to the Caribbean algal flora. Some of these that have wide depth distributions in the southern Gulf also occur on deepwater outcrops of the northern Gulf shelf. The cosmopolitan species have wide distribution patterns in the Gulf and wide temperature tolerances. Many of the algal species are distinctly seasonal in occurrence, even in the southern Gulf. Common genera are listed in table S30, and several are shown in figure 10.3. Important references include Dawes and Van Breedveld (1969), Earle (1969, 1972), Humm (1973), Humm and Darnell (1959), Humm and Hildebrand (1962),

Humm and Taylor (1961), Schneider and Searles (1991), Steidinger and Van Breedveld (1971), and W. Taylor (1960).

## Mycophyta

Marine fungi and yeasts are colorless, nonphotosynthetic forms that break down preformed organic matter. Shallow-water species include terrestrial and freshwater groups that are able to survive and grow in salt water as well as some strictly marine forms. Spores of the nonmarine species are brought in by winds and rivers. Most of the spores are quite small (2–30 µm), and the species are identified by growing out colonies on agar plates. Counts are made by determining colony forming units (CFUs). Studies reported by P. Powell and Szaniszlo (1980) from sediment samples off south Texas showed densities ranging from 5 to 1,600 CFU/mL, with an average of 236 CFU/mL. The greatest density and diversity occurred during the fall, and abundance was directly correlated with the organic content of the sediments. The lowest density was found in the winter months. About half of the isolates tested were able to degrade crude oil, and these fungi were most abundant in coarse nearshore sediments. Important genera are listed in table S30.

## Tracheophyta

Tracheophytes (higher plants) of the Gulf include marsh grasses, seagrasses, and mangroves. These will be covered in chapter 12.

## Protozoa

The free-living benthic protozoans of the Gulf include the pigmented forms (euglenoids, chrysomonads, cryptomonads, etc.) and the Ciliophora and Sarcodina. Another group common in the Gulf is the Sporozoa, of which all members are parasitic. The pigmented forms are not well known, and only the ciliates and sarcodines are considered here. Sprague (1954) provided a review of the species reported from the Gulf in all these groups.

### CILIOPHORA

Both loricate forms (tintinnids) and aloricate ciliates are known from the Gulf. Many of the tintinnid species are also planktonic, and the literature does not relate to benthic species per se. However, a list of some of the more important genera is given in table 10.4. A list of the aloricate ciliates is given in Borror (1962). Most of the species are of rather cosmopolitan distribution, and their presence in the Gulf appears to be related more to the avail-

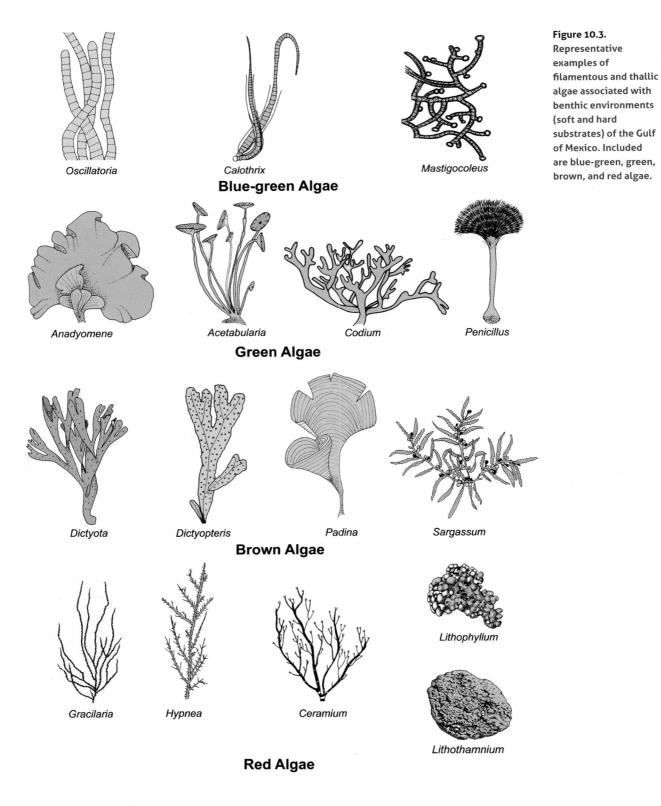

**Figure 10.3.**
Representative examples of filamentous and thallic algae associated with benthic environments (soft and hard substrates) of the Gulf of Mexico. Included are blue-green, green, brown, and red algae.

*Oscillatoria*

*Calothrix*

*Mastigocoleus*

**Blue-green Algae**

*Anadyomene*

*Acetabularia*

*Codium*

*Penicillus*

**Green Algae**

*Dictyota*

*Dictyopteris*

*Padina*

*Sargassum*

**Brown Algae**

*Gracilaria*

*Hypnea*

*Ceramium*

*Lithophyllum*

*Lithothamnium*

**Red Algae**

**Table 10.4.** *Important genera of lower invertebrates recorded from soft bottoms of the Gulf of Mexico. Some of the sponges may be primarily inhabitants of hard bottoms.*

**Protozoa**

Ciliophora

Tintinnoida

| | | |
|---|---|---|
| *Codonellopis* | *Helicostomella* | *Tintinnopsis* |
| *Favella* | *Salpingella* | *Tintinnus* |

Aloricate ciliates

| | | |
|---|---|---|
| *Kentrophorus* | *Myelostoma* | *Plagiopyla* |
| *Metopus* | *Parablepharisma* | *Sonderia* |

Sarcodina

Foraminifera

| | | |
|---|---|---|
| *Bolivina* | *Elphidium* | *Nouria* |
| *Buliminella* | *Nonionella* | *Streblus* |

**Porifera**

| | | |
|---|---|---|
| *Axinella* | *Haliclona* | *Spheciospongia* |
| *Geodia* | *Ircinia* | *Spongia* |

**Cnidaria**

Anthozoa

| | | |
|---|---|---|
| *Leptogorgia* | *Renilla* | *Virgularia* |

**Nemertea**

| | | |
|---|---|---|
| *Amphiporus* | *Cerebratulus* | *Micrura* |
| *Carinoma* | *Lineus* | *Zygeupolla* |

**Nematoda**

| | | |
|---|---|---|
| *Dorylaimopsis* | *Neotonchus* | *Terschellingia* |
| *Halalaimus* | *Sabatiera* | *Theristus* |

**Kinorhyncha**

| | | |
|---|---|---|
| *Centroderes* | *Pycnophyes* | *Trachydemus* |
| *Echinoderes* | *Semnoderes* | |

**Mollusca**

Amphineura

*Ischnochiton*

Scaphopoda

| | | |
|---|---|---|
| *Cadulus* | *Dentalium* | *Fustiaria* |

Gastropoda

| | | |
|---|---|---|
| *Busycon* | *Fasciolaria* | *Polinices* |
| *Conus* | *Murex* | *Strombus* |

Bivalvia

| | | |
|---|---|---|
| *Aequipecten* | *Chione* | *Mercenaria* |
| *Anadara* | *Laevicardium* | *Tellina* |

**Annelida**

Polychaeta

| | | |
|---|---|---|
| *Aedicira* | *Mediomastis* | *Protodorvillea* |
| *Cossura* | *Paraonis* | *Tharyx* |

ability of suitable habitats than to barriers to dispersal. A list of some of the aloricate ciliates reported from the Gulf is provided in table 10.4, and examples are shown in figure 10.4. Free-living ciliates are important in both the oxic and anoxic layers of marine sediments, and they are probably widespread, if poorly known, in benthic environments throughout the Gulf.

SARCODINA

Amoeboid forms include the Foraminifera and Radiolaria, which possess either external tests or internal skeletal elements. Among the foraminiferans both planktonic and benthonic forms are abundant, and their calcareous tests contribute much material to sediments throughout the Gulf. Because of their importance in defining benthic environments, foraminiferans are of interest to geologists, and they have been fairly well studied in the Gulf. In a series of papers, Phleger and coworkers (Greiner 1970; Phleger 1951, 1960; Phleger and Parker 1954a, 1954b) and others (Albers 1966; Bock 1979; Casey 1979a, 1979b; Culver and Buzas 1981) have demonstrated interesting patterns of foraminiferal distribution on the northern Gulf shelf. Seasonal patterns of species dominance are evident. The highest standing crops occur in the spring and the lowest in the winter. There are alongcoast variations in species dominance associated with sediment characteristics and other factors. Most striking are the depth distribution patterns. Recognizable species assemblages (the *biofacies* of the geologists) show faunal boundaries at the approximate depths of 100 m and 910 m. The most striking boundary appears at about 110 m, which marks the transition between continental shelf and continental slope environments. On the continental shelf itself, three subassemblages have been recognized, each with its own distinct grouping of foraminiferal species. These include the nearshore turbulent zone (0–18 m) with species of *Ammobaculites, Buliminella, Elphidium, Gaudryina, Quinqueloculina,* and *Streblus;* the inner shelf (0–55 m) with species of *Bolivina, Eggerella, Elphidium, Nonionella,* and *Nouria;* and the outer shelf (55–100 m) with species of *Bigenerina, Cancris, Cibicides, Eponides, Lenticulina, Nonion, Siphonina, Uvigerina,* and *Valvulineria.* Culver and Buzas (1981) provide a list of all known species and varieties of planktonic and benthic forams of the Gulf. A few of the continental shelf genera are listed in table 10.4, and examples are shown in figure 10.4. Benthic radiolarians of the Gulf are not as well known, but a number of species

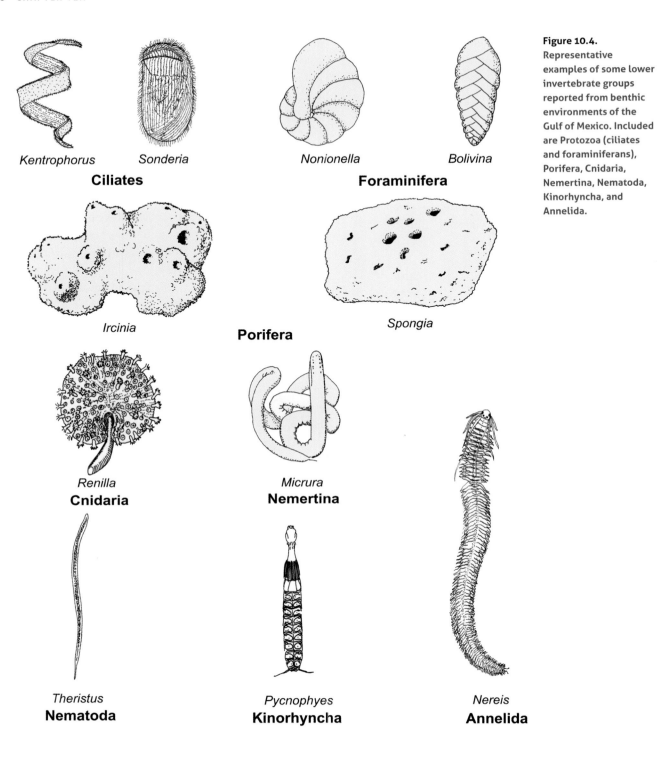

Kentrophorus          Sonderia
**Ciliates**

Nonionella          Bolivina
**Foraminifera**

Ircinia

**Porifera**

Spongia

Renilla
**Cnidaria**

Micrura
**Nemertina**

Theristus
**Nematoda**

Pycnophyes
**Kinorhyncha**

Nereis
**Annelida**

**Figure 10.4.**
Representative examples of some lower invertebrate groups reported from benthic environments of the Gulf of Mexico. Included are Protozoa (ciliates and foraminiferans), Porifera, Cnidaria, Nemertina, Nematoda, Kinorhyncha, and Annelida.

are given in Casey (1979a), McMillen and Casey (1978), and Sprague (1954).

### Porifera
Sponges are colonies of relatively undifferentiated cells that live together and function like a single individual. Lacking true organs or tissues, they are the most primitive of the multicellular animals. A common feature of all sponges is the presence of flagellated collar cells that line canals running through the body and opening to the outside through pores in the body wall. The beating flagella maintain water currents through the canals, bring-

ing in plankton and other particles on which the sponge cells feed. The simplest sponges are vase shaped and contain a large central cavity, open at the top, through which the water currents exit. In more advanced sponges the body wall is folded, the central cavity is greatly reduced, and the canal system is quite complex. The sponge body is supported by skeletal material made up of needlelike spicules of calcium carbonate or silicate or of thread-like proteinaceous fibers. Although a few species live in fresh and brackish waters, most sponges dwell in the sea, where they are found at all depths. Most species attach to hard substrates such as rocks, corals, and shells, but a few species do occur on soft bottoms, where they are anchored in the substrate. Within the Gulf the greatest abundance and diversity are encountered on the west Florida and Yucatán shelves, where hard substrates abound. A particularly large species, the loggerhead sponge (*Spheciospongia vesparium*), which may reach a diameter of two to three feet, is abundant on the west Florida shelf, and trawl hauls containing over a ton of these sponges have been recorded. The more complex marine sponges are virtual "apartment houses" harboring a great variety of small marine invertebrates and some fishes, which dwell in the canals and feed on passing particles. Some genera of Gulf sponges are given in table 10.4, and examples are shown in figure 10.4. Important references include de Laubenfels (1953), G. Green (1977), W. Hartman (1955), F. Smith (1954a), and Tierney (1954).

## Cnidaria

The Cnidaria, or coelenterates, are characterized by a two-layered body with a large digestive cavity, and a mouth at one end surrounded by tentacles. The tentacles are armed with stinging cells used to stun small prey. As noted in the previous chapter, there are two body forms: the medusa (in which the mouth is on the bottom, and the tentacles hang down) and the hydroid (in which the mouth and tentacles are on the top side). In the normal life history these two forms alternate, the medusa reproducing sexually and the hydroid reproducing asexually by budding. Three classes are recognized, the Hydrozoa and Anthozoa, both of which are abundant in benthic environments of the Gulf, and the Scyphozoa, which are entirely planktonic in the Gulf.

### HYDROZOA

In this group the hydroid form is generally dominant, although a planktonic medusa stage is usually present in the life history. Most are colonial, that is, they form bush-like structures containing many individual hydroids on branches. Among the sessile hydroids, most live attached to solid substrates, and these will be discussed in the next chapter.

### ANTHOZOA

In the Anthozoa the hydroid stage is dominant, and the medusa stage is eliminated entirely. Although some species are colonial, many are solitary. Included among the anthozoans are the corals, sea anemones, sea pens, sea fans, sea pansies, whip corals, and their relatives. Most species are attached to hard substrates, but a few live on soft bottoms. These include the stalked forms *Leptogorgia* and *Virgularia* and the sea pansy *Renilla*. Common genera are given in table 10.4, and an example is provided in figure 10.4. Important literature references include Bayer (1954), Carlgren and Hedgpeth (1952), Hedgpeth (1954a), and F. Smith (1954c).

## Lower worms and worm relatives

A number of minor phyla are represented in soft sediments of the Gulf. Most of these are small and rather inconspicuous organisms known only to specialists. However, they do appear in the literature, and some are of considerable importance in marine benthic food chains. Several of these groups are described here.

### PLATYHELMINTHES

Flatworms typically have three body layers and a digestive tract with a single opening. True organs and tissues are present. The body is bilaterally symmetrical, and there are definite front and hind ends. Two of the three classes, Trematoda (flukes) and Cestoda (tapeworms), are entirely parasitic and are not considered here. The third class, Turbellaria, includes all the free-living forms. In the marine environment most turbellarian species live on or in the sediments of shallow waters, although some dwell among algae (including floating rafts of sargassum) and various sedentary animals. A few are pelagic. Hyman (1954) reviewed the literature concerning free-living flatworms of the Gulf and listed, among others, the following genera: *Coronadena, Euplana, Gnesioceros, Hoploplana, Latocestus,* and *Stylochus.*

### NEMERTEA

Ribbon worms, sometimes called proboscis worms, are close relatives of the free-living flatworms, but all possess a flexible proboscis at the front end that is used for capturing prey. Most species are small and inconspicuous, but some reach a length of two meters. In marine waters most are bottom dwellers, but like the turbellarians, some are found in association with algae and larger sessile invertebrates. Some of the genera recorded from the Gulf are listed in table 10.4, and an example is given in figure 10.4. The primary reference to Gulf species is Coe (1954).

### NEMATODA

The nematodes, or roundworms, are more advanced than the previous phyla in the possession of a body cavity (pseudocoel) and a complete digestive tract with mouth and anus. Some species are parasitic, but many are free living in marine sediments, where they are often extremely abundant. Others are commensal and live in association with algae, sponges, colonial hydroids, bryozoans, mollusks, decapods, tunicates, and other marine organisms. Most free-living nematodes are quite small, and being poor swimmers, they are true planktonic forms, and not nektonic. However, nematodes are quite abundant in both the oxic and anoxic layers of soft bottoms, where they may reach densities of more than 500 organisms/mm³ of sediment. Some species are carnivorous, whereas others feed on bacteria, diatoms, filamentous and thallic algae, sponges, or decomposing organic matter. Primary references to free-living nematodes of the Gulf area include Chitwood (1951), Chitwood and Timm (1954), Hopper (1961a, 1961b; 1963), and King (1962). Important genera are listed in table 10.4, and an example is shown in figure 10.4.

### KINORHYNCHA

The kinorhynchs, or spiny-skinned worms, are distant relatives of the nematodes in which the body is surrounded by a segmented cuticle. Each segment bears several backward-pointing spines, and there are two or more large spines at the rear of the body. The protrusible and retractable mouth is surrounded by spines. These tiny animals, less than one millimeter in length, live in muddy bottoms of coastal marine waters, where they feed on diatoms and organic detritus. Most species are of rather cosmopolitan distribution. Genera recorded from the Gulf are listed in table 10.4, and an example is provided in figure 10.4. Although no major reference has been found for kinorhynchs of the Gulf, individual species have been recorded in various meiofaunal studies. Related groups not discussed here include the Tardigrada (Chit-

wood 1954), Brachiopoda (Cooper 1954, 1973), Phoronida (Hedgpeth 1954e), Echiuroida (Hedgpeth 1954e), and Sipunculida (Hedgpeth 1954e).

## Mollusca

As pointed out in the previous chapter, mollusks are unsegmented, soft-bodied animals with complete digestive tracts and definite front and hind ends. The body is generally surrounded by a thin tissue called the mantle, and there may be external or internal carbonate shells. Primitive mollusks do show some evidence of body segmentation, and this plus the presence of a true coelom (body cavity lined by mesoderm) and the trochophore larva indicate a relationship with the segmented worms. Six classes of living mollusks exist. One, the Monoplacophora, is limited to the deep Pacific and is not found in the Gulf of Mexico. The Amphineura is associated exclusively with hard substrates. Another, the Cephalopoda, while abundant in the Gulf, is primarily nektonic and is discussed in later chapters. The remaining three classes (Scaphopoda, Gastropoda, and Bivalvia) are considered below. Literature relating to the molluscan fauna of the Gulf is fairly extensive, and the following references are especially important: Bahr and Lanier (1981), Bishof (1980), Blake (1979), Butler (1954), Cruz-Ábrego, Flores-Andolais, and Solis-Weiss (1991), D. Farrell (1979), Kennedy (1959), Ladd (1951), Lyons (1979), Marcus and Marcus (1959), D. Moore (1961), Ode (1973), Opresko, Thomas, and Bayer (1976), W. A. Price (1954), Pulley (1952), Rehder (1954), and Rice and Kornicker (1962, 1965).

### SCAPHOPODA

The tusk or tooth shells are a small group of marine mollusks that have long, thin, often curved conical shells open at both ends. They burrow into soft bottoms and are found at all depths of the sea. Within the Gulf the tusk shells are widely distributed but never abundant. Several of the Gulf genera are listed in table 10.4, and one example is given in figure 10.5.

### GASTROPODA

Snails and their relatives make up a large group of mollusks typically having a single external shell with a conical spire. In some groups the shell is internal, while in others it is absent entirely. Some of the marine forms (heteropods and pteropods) are planktonic, but most are associated with the benthos. Here they occupy a variety of habitats including hard and soft bottoms from the shore-

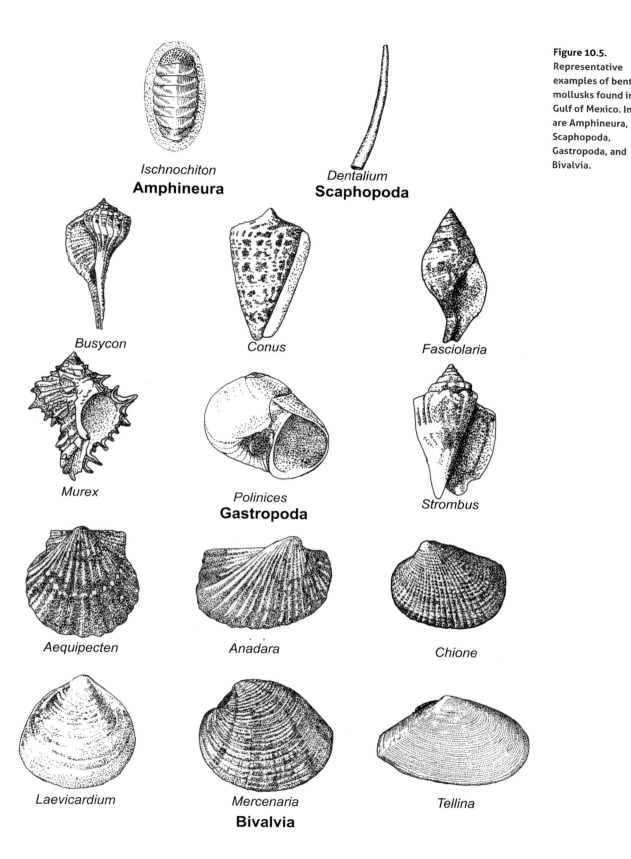

Figure 10.5. Representative examples of benthic mollusks found in the Gulf of Mexico. Included are Amphineura, Scaphopoda, Gastropoda, and Bivalvia.

*Ischnochiton*
**Amphineura**

*Dentalium*
**Scaphopoda**

*Busycon*

*Conus*

*Fasciolaria*

*Murex*

*Polinices*
**Gastropoda**

*Strombus*

*Aequipecten*

*Anadara*

*Chione*

*Laevicardium*

*Mercenaria*

*Tellina*

**Bivalvia**

line to great depths. Several genera are listed in table 10.4, and a few examples are shown in figure 10.5.

BIVALVIA

Clams, mussels, and their relatives are enclosed in a shell composed of two calcareous valves of about equal size and shape. Most of the marine species inhabit soft bottoms, but some are found on hard substrates, and members of one group, the shipworms, burrow into driftwood and wooden pilings. Since the soft-bottom bivalves live largely within the sediments and have limited powers of locomotion, they are among the best indicators of long-term prevailing environmental conditions in an area. Hence, any effort to determine depth-related or latitudinal biogeographic zonation in the Gulf should be based first on the distribution of infaunal bivalves. Although founded on limited information, the study of Pulley (1952) is an important beginning. Several Gulf genera are listed in table 10.4, and some are shown in figure 10.5.

## Annelida

Among the segmented worms, three primary groups are recognized. The Oligochaeta (earthworm relatives) and Hirudinea (leeches), although present, are relatively unimportant in Gulf sediments and will not be considered. The Polychaeta (polychaetes) are abundant and diverse, and they constitute one of the most important groups of marine animals. These are segmented worms in which each segment possesses a pair of lateral appendages (parapodia) used in locomotion and respiration. The anterior portion of the body has a relatively well-developed head that bears eyes, antennae, and a pair of fleshy palps. These worms range in size from a few millimeters to more than a meter in length. Most polychaetes are associated with soft bottoms, where they live within or upon the sediments, but some species occur on hard bottoms, and others are pelagic. Certain species build calcareous tubes, and large colonies of these create reefs or hard banks. Due to their great abundance and constant burrowing activities, benthic polychaetes stir and aerate bottom sediments and aid in releasing nutrients back into the water column. Important references include Fitzhugh (1984), Flint and Rabalais (1980), O. Hartman (1954, 1957), T. Perkins and Savage (1975), Uebelacker and Johnson (1984), and Vittor (1979). Representative genera are listed in table 10.4, and an example is shown in figure 10.4.

## Arthropoda

The phylum Arthropoda, or joint-legged animals, includes such recognizable forms as spiders, scorpions, insects, shrimps, and crabs. Primitively the body was made up of a long series of more or less similar segments, each bearing a pair of lateral appendages. These and other characteristics demonstrate the close relationship between the arthropods and the polychaetes (phylum Annelida). However, in the arthropods many of the segments have become fused, often into three body regions (head, thorax, and abdomen) or even two body regions (cephalothorax and abdomen). The paired appendages have become jointed and modified for a variety of functions, and the entire body is now encased in a tough outer cuticle, or exoskeleton, which must be shed for the animal to grow. Two main subdivisions of the Arthropoda are recognized, the Chelicerata and the Mandibulata. The former group possesses chelicerae, small appendages in front of the mouth generally used in feeding. Among the marine forms the Xiphosura (horseshoe crabs) and Pycnogonida (sea spiders) are present in the Gulf. The Mandibulata lack chelicerae but possess mandibles, appendages lateral or posterior to the mouth, which also aid in feeding. Included here are several groups, of which the Crustacea are most important in the marine environment.

XIPHOSURA

In the horseshoe crabs the body is divided into three regions: cephalothorax, abdomen, and long tail. Only a few species of this ancient lineage survive today, one of which (*Limulus polyphemus*) occurs in the Gulf of Mexico. In coastal waters these animals forage on the bottoms for algae, worms, mollusks, and other small invertebrates. The genus is listed in table 10.5 and illustrated in figure 10.6. The primary reference is Hedgpeth (1954h).

PYCNOGONIDA

Sea spiders are a small aberrant group of chelicerates not closely related to spiders or to any other group. They are characterized by short segmented bodies and long, thin legs that give them a somewhat spiderlike appearance. From each body segment a short process projects laterally, and to these the eight-segmented walking legs are attached. Pycnogonids feed on soft-bodied invertebrates such as sponges, colonial hydroids, soft corals, sea anemones, and bryozoans. They are found at all depths of the sea. Several genera are listed in table 10.5, and an example is shown in figure 10.6. The main reference is Hedgpeth (1954f).

CRUSTACEA

The crustaceans are a large and diverse group of mandibulate arthropods, all of which possess five pairs of head appendages including two pairs of antennae. Most species are aquatic, and the large majority are marine, where they occupy almost every conceivable ecological niche. Many of the benthic groups have planktonic representatives, which were discussed in the previous chapter. Many produce planktonic larvae.

*Ostracoda*

The ostracods are small primitive crustaceans in which the body is encased in a bivalve carapace hinged along the dorsal line. Some species are planktonic, but most live on or in the bottom as predators, scavengers, or detritus feeders. Some marine species produce bioluminescent secretions. Important genera are listed in table 10.5, and one form is shown in figure 10.6. References include Curtis (1960), Hulings (1967), Keith and Hulings (1965), Kornicker (1983, 1984, 1986), Machain-Castillo (1989), and Tressler (1954).

*Copepoda*

In the copepods the body is not encased in valves, and the antennae are often enlarged for locomotion. Most species are planktonic, but the group is well represented in the benthos. Although calanoids and cyclopoids do occur on the bottom, harpacticoids are the most abundant and diverse benthic copepod group. Common genera are given in table 10.5, and an example is provided in figure 10.6. An important reference is Venn (1980).

*Stomatopoda*

In the mantis shrimp the body is elongate and flattened, a carapace covers the anterior segments, and the large eyes are stalked. The second pair of thoracic appendages is much enlarged and modified for grasping prey. These shrimp live in crevices among rocks and shells as well as in burrows in soft sediments. Many are nocturnal and emerge to crawl or swim about at night. They are found at all depths in the sea. All are carnivorous and feed on crustaceans and other invertebrates as well as fishes. Some of the Gulf genera are given in table 10.5, and one is illustrated in figure 10.6. Important references include Camp (1973), Chace (1954), and Manning (1959).

**Table 10.5.** *Important genera of arthropods recorded from soft bottoms of the Gulf of Mexico.*

**Arthropoda**

| | | |
|---|---|---|
| <u>Chelicerata</u> | | |
|   Xiphosura | | |
|     *Limulus* | | |
|   Pycnogonida | | |
|     *Achelia* | *Anoplodactylus* | *Ascorhynchus* |
| <u>Mandibulata – Crustacea</u> | | |
|   Ostracoda | | |
|     *Cushmanidea* | *Eusarsiella* | *Krithe* |
|     *Cytheropteron* | *Hemicythere* | *Paracypris* |
|   Copepoda (Harpacticoids) | | |
|     *Ameira* | *Enhydrosoma* | *Pseudobradya* |
|     *Ectinosoma* | *Haloschizopera* | *Typhlamphiascus* |
|   Stomatopoda | | |
|     *Gonodactylus* | *Meiosquilla* | *Platysquilla* |
|     *Lysiosquilla* | *Nannosquilla* | *Squilla* |
|   Cumacea | | |
|     *Campylaspis* | *Cyclaspis* | *Glyphocuma* |
|     *Cumella* | *Eudorella* | *Leptostylus* |
|   Tanaidacea | | |
|     *Apseudes* | *Kalliapseudes* | *Paratanais* |
|     *Calozodion* | *Leptochelia* | *Tanais* |
|   Isopoda | | |
|     *Eurydice* | *Natatolana* | *Serolis* |
|     *Gnathia* | *Politolana* | *Sphaeroma* |
|   Amphipoda | | |
|     *Ampelisca* | *Heterophoxus* | *Photis* |
|     *Byblis* | *Monoculodes* | *Unciola* |
|   Decapoda | | |
|     Natantia | | |
|       *Acetes* | *Penaeus* | *Solenocera* |
|       *Parapenaeus* | *Sicyonia* | *Trachypenaeus* |
|     Macrura | | |
|       *Callianassa* | *Panulirus* | *Scyllarus* |
|       *Calocaris* | *Scyllarides* | *Upogebia* |
|     Anomura | | |
|       *Dardanus* | *Paguristes* | *Polyonyx* |
|       *Munida* | *Pagurus* | *Porcellana* |
|     Brachyura | | |
|       *Calappa* | *Hepatus* | *Ranilia* |
|       *Callinectes* | *Portunus* | *Raninoides* |

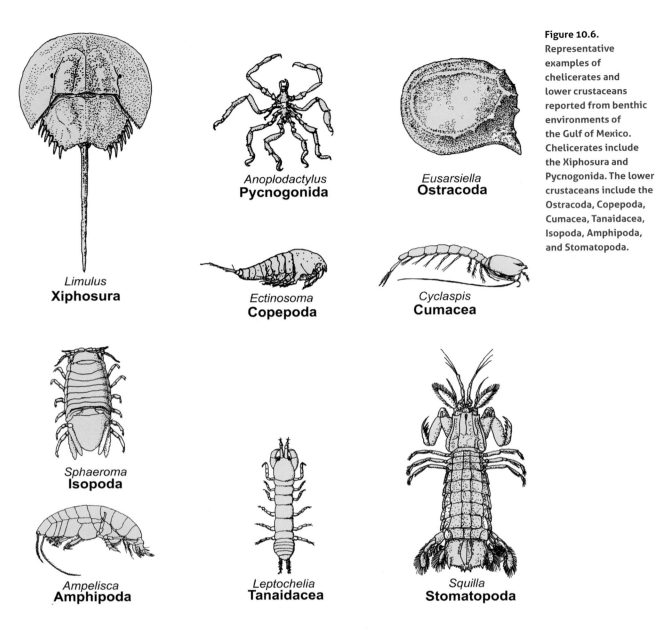

Figure 10.6. Representative examples of chelicerates and lower crustaceans reported from benthic environments of the Gulf of Mexico. Chelicerates include the Xiphosura and Pycnogonida. The lower crustaceans include the Ostracoda, Copepoda, Cumacea, Tanaidacea, Isopoda, Amphipoda, and Stomatopoda.

*Anoplodactylus*
**Pycnogonida**

*Eusarsiella*
**Ostracoda**

*Limulus*
**Xiphosura**

*Ectinosoma*
**Copepoda**

*Cyclaspis*
**Cumacea**

*Sphaeroma*
**Isopoda**

*Ampelisca*
**Amphipoda**

*Leptochelia*
**Tanaidacea**

*Squilla*
**Stomatopoda**

### Cumacea

The cumaceans are a small group of tiny benthic marine crustaceans related to the mysid shrimps. The head and thorax are greatly enlarged, and the large carapace has a pair of anterior hornlike extensions near the midline. Most species are encountered on continental shelves, where they live half buried in the sediments with the front and hind ends projecting above the sediment surface. They are capable of swimming and sometimes swarm at the surface of the water. Several genera are listed in table 10.5, and one is illustrated in figure 10.6.

### Tanaidacea

The tanaids are tiny marine crustaceans, related to both cumaceans and isopods, in which the second pair of walking legs is developed into a chela, or pincer. They live buried in soft sediments or in tubes that they construct. Both carnivory and filter feeding occur. Genera are listed in table 10.5, and one is shown in figure 10.6.

### Isopoda

The isopod body is dorsoventrally flattened, and the thoracic legs, none of which are chelate, are adapted for crawling. Most species are marine, where they occur at all depths. Most are fairly small, being an inch or less in length, but a giant species (*Bathynomus giganteus*), found in the deep Gulf and elsewhere, achieves a length of 14 inches and a width of 5 inches. Isopods occur on both soft and hard bottoms, and some (gribbles) are wood borers that attack driftwood, pilings, and even mangrove roots. Many species are parasitic. Several genera are given in table 10.5, and one is illustrated in figure 10.6. Major references for the Gulf are S. Clark and Robertson (1982) and Menzies and Kruczynski (1983).

### Amphipoda

Amphipods are small crustaceans with laterally flattened bodies and five pairs of thoracic walking legs. Among the different groups, these are adapted for a variety of functions such as walking, crawling, jumping, clinging, climbing, and, when modified or chelate, for grasping food. Although most species are minute, some deep-sea species exceed eight inches in length. In the marine environment amphipods occur on hard and soft bottoms, while others climb among algae, hydroids, bryozoans, and other invertebrates and plants. Some species bore into wood, and others live in tubes of their own construction. Several Gulf genera are listed in table 10.5, and an example is shown in figure 10.6. Important references include Lecroy (1995), McKinney (1977), Myers (1981), and Ortiz (1991).

### Decapoda

This group contains the largest and most conspicuous marine crustaceans. Included here are the Natantia (swimming shrimps), Macrura (lobsters), Anomura (hermit crabs), and Brachyura (true crabs). All have five pairs of thoracic legs, adapted primarily for walking, but some may be modified into chelae. Most species are found in the sea, where they occupy a variety of habitats at all depths. Important references to decapods of the Gulf include Behre (1954), Hulings (1961), Soto (1972, 1980), and Wass (1955). Also quite useful for identification of Gulf species are the works of Felder (1973) and A. Williams (1965).

### Natantia

In the swimming shrimps the body is somewhat flattened laterally, the abdomen is large, and the abdominal appendages are often oar-like and modified for swimming. These shrimps are abundant on hard and soft bottoms, and they occupy all depths of the sea. They may burrow in the sediments, crawl about on the surface, or swim in the water column. Included here are all the shrimp species of commercial importance. Several genera are listed in table 10.5, and examples are shown in figure 10.7. Literature references include Cobb, Futch, and Camp (1973), Dardeau (1984), Dardeau and Heard (1983), Huff and Cobb (1979), Lindner and Anderson (1954), Sánchez and Soto (1987), Soto and Garcia (1987), and Wood (1974).

### Macrura

The macrurans include lobsters and lobsterlike forms as well as some burrowing species. In this group the abdomen is large and fully extended, the thoracic walking legs are generally well developed, often with powerful chelae, and there is a large tail fan. Lobsterlike forms occur on both hard and soft bottoms of the Gulf. Several genera are given in table 10.5, and examples are illustrated in figure 10.7. References include Felder (1973), Lyons (1970), and F. Smith (1954b).

### Anomura

In the hermit crabs and their relatives the abdomen is asymmetrical and twisted to one side or flexed beneath the thorax. The tail fan is much reduced or absent. Most of the species possess large chelae. Although the group is primarily marine, some tropical species are terrestrial and climb trees. Some species burrow in the soft sediments, but most hermit crabs house themselves in discarded snail shells and crawl about on the sediment surface. In certain groups the abdomen is tucked beneath the thorax, giving them the appearance of true crabs. Several genera are listed in table 10.5, and two are illustrated in figure 10.7. Important references include Felder (1973) and Provenzano (1959).

### Brachyura

In the true crabs the abdomen is reduced and flexed tightly beneath the thorax. The body is generally flattened dorsoventrally and is wider than long. The first walking legs have become well-developed chelae. In the ocean they occur on all substrates and at all depths. Although most associate with the benthos, a few are pelagic. Genera are listed in table 10.5, and two are shown in figure 10.7. Important Gulf references include the works of Bender (1971), Felder (1973), García-Montes, Soto, and Garcia (1988), Gore and Scotto (1979), McRae (1950), Powers (1977), Soto (1972, 1979, 1980, 1985, 1986, 1991), and J. Sullivan (1979).

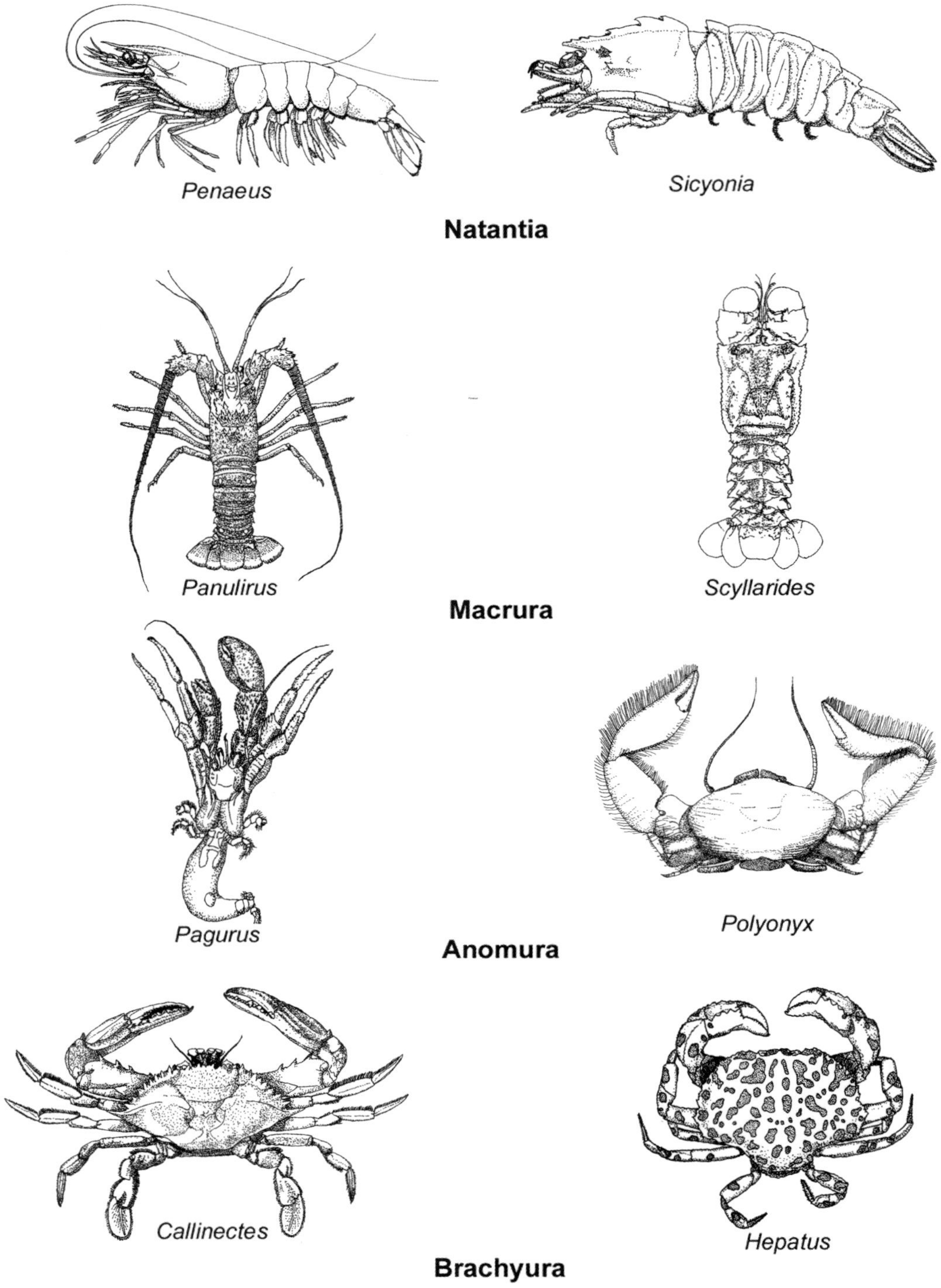

*Penaeus*

*Sicyonia*

**Natantia**

Figure 10.7.
Representative
examples of the
higher crustaceans
reported from benthic
environments of
the Gulf of Mexico.
Included are Natantia,
Macrura, Anomura, and
Brachyura.

*Panulirus*

*Scyllarides*

**Macrura**

*Pagurus*

*Polyonyx*

**Anomura**

*Callinectes*

*Hepatus*

**Brachyura**

### Echinodermata

In the basic plan of the echinoderms (starfishes and their relatives), there are five body divisions (often arms) radiating out from a central point. Thus, although derived from bilaterally symmetrical ancestors, echinoderms have secondarily developed a body based on radial sym-metry. Embedded in the leathery skin are calcareous plates with projecting spines, short in some groups and long in others. These are responsible for the name of the phylum, which means "spiny skinned." Echinoderms are exclusively marine, and they are found at all depths of the sea. Five classes are recognized, all of which occur in

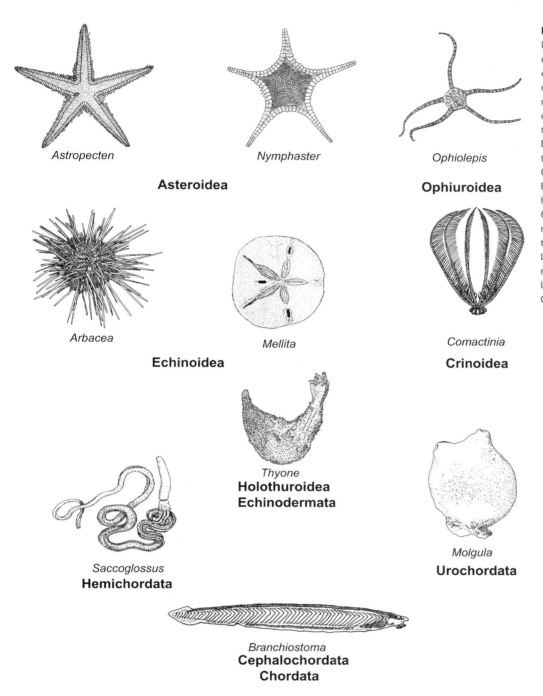

*Astropecten*

**Asteroidea**

*Nymphaster*

*Ophiolepis*

**Ophiuroidea**

*Arbacea*

**Echinoidea**

*Mellita*

*Comactinia*

**Crinoidea**

*Thyone*
**Holothuroidea**
**Echinodermata**

*Molgula*
**Urochordata**

*Saccoglossus*
**Hemichordata**

*Branchiostoma*
**Cephalochordata**
**Chordata**

**Figure 10.8.** Representative examples of the echinoderms, lower chordates, and chordate relatives from benthic environments of the Gulf of Mexico. Echinoderms include the Asteroidea, Ophiuroidea, Echinoidea, Holothuroidea, and Crinoidea. Chordate relatives include the Hemichordata. Lower chordates are represented by the Urochordata and Cephalochordata.

the Gulf. These include the Asteroidea (starfishes), Echinoidea (sand dollars and relatives), Ophiuroidea (brittle stars), Holothuroidea (sea cucumbers), and Crinoidea (sea lilies). An important general reference is A. Clark (1954).

ASTEROIDEA

In the starfishes there is no central disk, and the arms taper as they leave the body. Most species have five arms, but in some groups the number is larger. Many are carnivorous, preying on sponges, coral polyps, worms, mollusks, crustaceans, other echinoderms, and fishes. Others are plankton or deposit feeders. A number of common genera are given in table S31, and two forms are shown in figure 10.8. Key references are Downey (1973) and Hendler et al. (1995).

ECHINOIDEA

This group includes the sand dollars, sea urchins, and heart urchins. Here the skeletal plates have become fused into a solid case surrounding the body. The shape may be flat, ovoid, or spherical, and there are no arms. Spines, which project from the body surface, may be short or quite long, and in some species the spines contain toxins. Echinoids feed on a variety of living and dead organisms as well as on particles of organic detritus that they glean from the sediments. Important genera are listed in table S31, and examples are provided in figure 10.8. A major reference is Serafy (1979).

OPHIUROIDEA

In the brittle stars and their relatives the basket stars and serpent stars, the body is set off by a central disk, and the long arms are thin and fairly solid, not fleshy as in the starfishes. Most species are small and inconspicuous, but in some environments they are quite numerous. They generally feed on the bottom, taking in small living organisms and detritus particles. Several genera are listed in table S31, and an example is shown in figure 10.8. The primary reference is L. Thomas (1962).

HOLOTHUROIDEA

The sea cucumbers differ from all other echinoderms in having a long leathery body with the mouth at one end, surrounded by branched tentacles, and the anus at the other. Most species are several inches long, but some are almost a meter in length. Some species live on the sediment surface, some burrow into the bottom, and others swim in the water column. For the most part, sea cucumbers are deposit feeders, consuming sediments, but others filter plankton and detritus particles from the water. Several genera are listed in table S31, and one is illustrated in figure 10.8. Major references for the Gulf include Deichmann (1954) and Miller and Pawson (1984).

CRINOIDEA

The crinoids are unique among the echinoderms in that the mouth faces upward rather than downward. The mouth is surrounded by feathery tentacles that may be branched. Most species are sessile and anchored to the bottom on either hard or soft substrates. Many of the sessile forms, the sea lilies, are stalked. Others, the feather stars, are free living in deep water, where they swim about in the water column. Crinoids feed primarily on plankton and other suspended particulate matter, which they re-

move by means of their feathery tentacles. Several genera are listed in table S31, and a common unstalked sessile form is shown in figure 10.8. The primary reference is A. Clark (1954).

**Hemichordata**

The acorn worms, sometimes called proboscis worms, are long, thin marine invertebrates that may reach up to a half meter in length. The body is divided into three regions, the short anterior proboscis, collar section, and long trunk. Longitudinal rows of gill pores line the anterior portion of the trunk. On the basis of larval development and anatomy of the adults, the hemichordates are considered to be related to the echinoderms, on the one hand, and the chordates, on the other. Most species live in shallow coastal areas, where they burrow in soft sediments and feed largely on small invertebrates and detritus particles. One group, however, is limited to deep ocean bottoms. Gulf genera are listed in table S31, and an example is shown in figure 10.8. The primary reference is Hedgpeth (1954c).

**Chordata**

This phylum is characterized by the presence of a dorsal tubular nerve cord, a longitudinal strengthening rod (called a *notochord*), and paired pharyngeal gill slits. Although present in adults of some forms, the notochord and gill slits are embryonic features in the higher chordates. Three subphyla are recognized: Urochordata (sea squirts and relatives), Cephalochordata (lancets), and Vertebrata (vertebrates), but only the cephalochordates are discussed in detail here.

CEPHALOCHORDATA

The lancets are slim, streamlined animals, one or two inches in length, which, as adults, display all the chordate characteristics. The body is somewhat fish shaped, with dorsal and ventral fins at the posterior end, but no lateral fins. The anterior end includes a mouth or buccal cavity surrounded by a series of fingerlike tentacles called *cirri*. Lancets are found in shallow coastal areas, where they partially bury themselves in sandy or muddy sediments, leaving the anterior part projecting above the bottom into the water column. Through the mouth they take in water, which passes out through the numerous gill slits, and in the process the animals filter out fine living and nonliving particles for food. Adults occasionally appear in the nearshore plankton. So far, only one genus (*Branchiostoma*)

**Table 10.6.** *Average bacterial density and biomass values reported from continental shelves of the northern and southwestern Gulf of Mexico.*

| Area | Depth range (m) | Density (cells × $10^9$/cm³) | Biomass (g C/m²) | Reference |
|---|---|---|---|---|
| Central LA (hypoxic) | 17–22 | 3.53 | 15.63 | Cruz-Kaegi 1992 |
| Upper TX | < 100 | 1.92 | 8.45 | Cruz-Kaegi 1992 |
| | 185–800 | 2.20 | 9.97 | |
| Upper TX (base of Flower Garden Banks) | 90–223 | 16.00* | 2.62 | Yingst and Rhoads 1985 |
| South TX | 18–27 | 7.90** | — | Schwartz et al. 1980 |
| | 42–65 | 4.30** | — | |
| | 91–134 | 2.20** | — | |
| SW Gulf | 0–50 | — | 1.30 | Soto and Escobar-Briones 1995 |
| | 51–100 | — | 0.70 | |
| | > 100 | — | 0.30 | |

* Reported as number of cells × $10^9$/cm² of sediments
** Reported as number of cells × $10^5$/mL of wet sediments

and a few species are known from the Gulf. The genus is listed in table S31 and shown in figure 10.8. References include Boschung and Gunter (1966), Boschung and Shaw (1988), Hedgpeth (1954d), and Pierce (1965). The paper of Futch and Dwinell (1977), although not about the Gulf, is also of interest.

VERTEBRATA

The vertebrate animals posses an internal skeleton of cartilage or bone, the head is encased in a hard cranium, and in adults the notochord is replaced by a segmented vertebral column. Most groups have two sets of paired appendages. Included here are the fishes and higher vertebrates. Although some fish species are truly benthic, most are nektonic, and this group, as well as sea turtles and marine mammals, will be taken up in detail in later chapters.

**Distribution and Abundance**

The distribution and abundance of the benthic algal flora were taken up in the previous section, and the present discussion will focus on the bacteria and fauna of soft bottoms. Unfortunately, collecting techniques have varied from one study to another, and for the meiofauna and, to some extent, the macrofauna many species have been encountered that are new to science. For many of these species, little information is available concerning the life history and food habits. These circumstances pose limi-

tations on the ability to compare and generalize from one study to another. In general, the most thoroughly investigated area is the Texas-Louisiana continental shelf, and the least known area is the Yucatán shelf.

**Microbenthos**

As noted earlier, the microbenthos is composed primarily of bacteria, unicellular algae, fungi, and protozoans. Since bacteria greatly dominate in numbers and biomass, this group will be presented as representative of the total microbenthos. ZoBell (1954) provided a general discussion of bacteria in Gulf of Mexico environments but gave no numerical data. Since then a number of studies have appeared, of which the following are most relevant: Alexander, Schropp, and Schwartz (1982), Cruz-Kaegi (1992), Schwartz et al. (1980), Soto and Escobar-Briones (1995), and Yingst and Rhoads (1985). These studies cover the Texas-Louisiana shelf and that of the southwestern Gulf from below Veracruz to the base of the Yucatán Peninsula.

A summary of the primary density and biomass data for bacterial populations of these areas is presented in table 10.6. Here it is seen that bacterial populations tend to be high near shore and to decrease with depth across the continental shelf, but at least off upper Texas, the populations increase in density on the upper slope. Numerical density values off Louisiana and upper Texas range above $10^9$ cells/cm³, but the south Texas values reported by Schwartz et al. (1980) are on the order of only $10^5$ cells/cm³. More recent staining and counting techniques would undoubtedly reveal values for south Texas that are considerably higher but still well below those off Louisiana and upper Texas. Biomass values are much higher off Louisiana and upper Texas than in the southwestern Gulf.

Bacterial abundance in marine sediments of the Gulf, as elsewhere, is positively correlated with the organic content of the sediments, and this includes petroleum hydrocarbons. Therefore, the higher density and biomass values off Louisiana and upper Texas must reflect the greater availability of organic carbon, which may be derived from river-borne particulate matter, phytoplankton stimulated by river-borne nutrients, hydrocarbons from natural seeps and petrochemical operations, or a combination of these. Schwartz et al. (1980) found bacterial abundance to be highly correlated with levels of total alkanes (chain-like hydrocarbons) in the sediment. The lower levels of bacteria reported off south Texas and in the southwestern Gulf likely reflect the relative paucity of organic substrates in sediments of these areas. Cruz-Kaegi (1992) reported that about 70 percent of the bacteria occurred in the upper 15 cm of the sediment column, and that bacterial biomass accounted for between 73 and 99.5 percent of the total living biomass, the remainder being meiofauna and macrofauna. Some seasonal changes in bacterial abundance have been reported. Off south Texas, the populations were somewhat higher in the spring, probably in response to river-borne particulate organic matter. In the southwestern Gulf, biomass during the rainy season was about twice that of the dry season, again reflecting riverine input.

Because of their abundance in marine sediments and their ability to degrade a great variety of organic chemical compounds and release nutrients back into the water column, bacteria play a crucial role in recycling processes. Furthermore, as a food resource for many different types of animals, bacteria constitute an important link in marine benthic food chains.

**Meiobenthos**

The meiobenthic fauna has been investigated on continental shelves off Louisiana and Texas (Bedinger 1981; Gettleson 1976; Harper, Potts, et al. 1981; Montagna and Harper 1996; Murrell and Fleeger 1989; W. Pequegnat and Sikora 1977; W. Pequegnat and Venn 1980; E. N. Powell et al. 1983; Street and Montagna 1996; Venn 1980; and Yingst and Rhoads 1985), peninsular Florida and Alabama

**Table 10.7.** *Taxonomic groups reported in meiofaunal studies on the Texas-Louisiana continental shelf.*

| | | |
|---|---|---|
| Protozoa | Annelida | Crustacea |
| Foraminifera | Oligochaeta | Ostracoda |
| Other | Polychaeta | Copepoda |
| | | Calanoida |
| Porifera | Mollusca | Cyclopoida |
| | Amphineura | Harpacticoida |
| Cnidaria | Scaphopoda | Cirripedia |
| Hydrozoa | Gastropoda | Cumacea |
| Scyphozoa | Cephalopoda | Tanaidacea |
| Anthozoa | | Isopoda |
| | Arthropoda | Amphipoda |
| Platyhelminthes | Acarina | |
| | Pycnogonida | Tardigrada |
| Nemertea | | |
| | | Echinodermata |
| Minor Phyla | | Asteroidea |
| Gastrotricha | | Ophiuroidea |
| Kinorhyncha | | |
| Rotifera | | Hemichordata |
| Priapulida | | |
| Bryozoa | | Chordata |
| Phoronida | | Urochordata |
| Sipunculida | | |

(Ivester 1979), and the east coast of Mexico (Escobar-Briones and Soto 1997; and Soto and Escobar-Briones 1995). In all studies nematodes were the numerically dominant group, followed by lower percentages of harpacticoid copepods, kinorhynchs, polychaete annelids, and a variety of other invertebrate groups (table 10.7). In most studies the nematodes averaged 70–93 percent of the meiofauna, but on the outer shelf of upper Texas the percentage was much lower. Here harpacticoids made up 3–18 percent, and annelids accounted for up to 21 percent of the fauna (table 10.8). Overall densities were much lower off south Texas than elsewhere. In most studies densities were highest at the nearshore stations and decreased across the shelf with increasing depth of the water column (table 10.9), but occasionally meiofaunal density peaked at midshelf. In the eastern Gulf the highest densities occurred in fine to medium sands with moderate to high carbonate levels lying in less than 40 m of water. Seasonal changes in meiofaunal densities were

**Table 10.8.** *Densities and compositions of the meiofauna reported from the Louisiana, Texas, and west Florida continental shelves.*

| Area | Water depth (m) | Mean density (no./10 cm²) | Composition (%) | | | |
|---|---|---|---|---|---|---|
| | | | Nematoda | Harpact. | Annelida | Others |
| Louisiana | 8–13 | 1,810 | 92 | 3 | — | 5 |
| | 2–92 | — | 90 | 3 | 3 | 4 |
| Upper Texas | 19–20 | 941 | 72 | 15 | — | 13 |
| | 90–223 | 525* | 32 | 18 | 21 | 29 |
| South Texas | 0–140 | 133 | 93 | 4 | 4 | — |
| West Florida | 0–>100 | 723 | 70 | 14 | 5 | 11 |

* Includes foraminiferans, which made up 23% of the total.

**Table 10.9.** *Comparison of meiofaunal depth-density relationships off south Texas and west Florida.*

| South Texas | | West Florida | |
|---|---|---|---|
| Water depth (m) | Density (no./10 cm²) | Water depth (m) | Density (no./10 cm²) |
| 0–30 | 430 | 0–20 | 1,727 |
| 30–60 | 94 | 20–40 | 953 |
| 60–90 | 64 | 40–60 | 699 |
| 90–120 | 49 | 60–80 | 385 |
| 120–140 | 30 | 80–100 | 311 |
| | | > 100 | 264 |
| Average | 133 | | 723 |

**Table 10.10.** *Average meiofaunal biomass values reported from the western and southern Gulf in relation to depth of the water column. (From Escobar-Briones and Soto 1997; Soto and Escobar-Briones 1995.)*

| | Area | |
|---|---|---|
| Depth (m) | Rio Grande to Cabo Rojo (g C/m²) | Río Papaloapan to Yucatán (g C/m²) |
| 0–50 | 1.12 | 0.63 |
| 51–100 | 0.34 | 0.20 |
| 101 – ca. 200 | 0.12 | 0.10 |

never great, but populations were often somewhat higher during the summer months. An exception was recorded at a depth of 8–13 m off central Louisiana. Here populations were low in winter, reached a peak in the spring, and then dropped dramatically during the summer period of hypoxia. Although all taxonomic groups declined, harpacticoid copepods were the most severely affected. Not much is known about food habits of many of the species, but both carnivores and detritus feeders were present.

Two important studies have been done on the meiobenthos of the Mexican shelf, covering most of the area from the Rio Grande to the base of the Yucatán Peninsula. The most northerly consisted of a series of seasonal across-shelf transects from the Rio Grande to Cabo Rojo, just below Tampico (Escobar-Briones and Soto 1997).

The other covered the area from the Río Papaloapan, just below Veracruz, to the Yucatán Peninsula (Soto and Escobar-Briones 1995). All bottoms consisted of terrigenous gravels, sands, and silts except at the base of Yucatán, where carbonate sediments prevailed. Meiofauna of the terrigenous sediments was dominated by nematodes and harpacticoid copepods, but on the carbonate bottoms these gave way to foraminiferans. All meiofaunal biomass values were expressed in terms of their organic carbon content, that is, g C/m². In both studies the meiofauna was generally more abundant on the inner shelf and decreased seaward with increasing depth of the water column (table 10.10), although locally or seasonally the peak sometimes occurred at midshelf. Average biomass was somewhat higher on the northern transects. These values were always higher near the mouths of major rivers, indi-

**Table 10.11.** *Meiofaunal densities in stressed environments of the Texas-Louisiana continental shelf. (Data from Montagna and Harper 1996; E. N. Powell et al. 1983.)*

| Area | Depth (m) | Presumed stress agents | Location details | Mean densities (no./10 cm²) |
|---|---|---|---|---|
| Brine seep, East Flower Garden Bank | 72 – ca. 150 | brine, low oxygen, high sulfides | lake stream outfall | 30 67 108 |
| Oil platforms, Texas shelf | ca. 25–75 | trace metals, toxic organics, disturbance | nearfield farfield | 1,164 2,269 |

cating the importance of riverine contributions to benthic production, either directly through the input of terrestrial particulate organic carbon or indirectly through nutrient-stimulated marine phytoplankton production. Seasonal differences in meiofaunal biomass were noted. Biomass values were highest during the summer, when river flows were greatest and the water column was stratified. They were low during the winter, when the river outflow was reduced and the water column was mixed, especially by severe storms. These studies are particularly informative since they demonstrate the coupling of benthic biomass values with riverine outflow and seasonal hydrographic and meteorological conditions.

Three studies conducted on the Texas shelf reveal the responses of meiofaunal populations to stressed environments. E. N. Powell et al. (1983) reported on the meiofauna of a brine lake, stream, and outfall on the side of the East Flower Garden Bank on the outer shelf southeast of Galveston. The brine seep itself enters the lake, which lies near the apex of the bank at a depth of 72 m (see fig. 4.15). Outflow from the lake courses through a 96 m long submarine stream before reaching the outfall lower on the bankside. Within the lake the brine is highly concentrated (200–220 ppt) and charged with sulfides (up to 2,173 μg-at/L). During its course through the stream the brine and its sulfides become somewhat reduced, but the dissolved oxygen content is often low. As seen in table 10.11, meiofaunal densities were quite low throughout the system but tended to increase downstream as the brine becomes diluted. Many invertebrate groups were represented, but rarely were nematodes the dominant forms. In portions of the stream turbellarians and their relatives (gnathosto-

mulids) dominated in what was referred to as a "thiobios" (sulfur-dependent community).

Montagna and Harper (1996) studied the effects of active oil production platforms on meiofaunal density and composition (table 10.11). At stations 50 m from the platforms (nearfield) meiofaunal densities were about half those 500 or more meters away (farfield). Most severely affected were the harpacticoid copepods, which suffered great reduction in density, species diversity, and reproductive success. Street and Montagna (1996) investigated genetic diversity in mitochondrial DNA in harpacticoid copepods around the same platforms. Populations in the nearfield (< 50 m from the platforms) showed significant loss of genetic diversity as compared with those of the farfield (> 3 km away). These authors concluded that the pattern on the continental shelf seems to consist of a uniform level of genetic diversity punctuated by islands of lower diversity around oil and gas platforms. Selective pressures leading to gene loss may be chemical contaminants (heavy metals and/or toxic organics) or physical factors associated with frequent disturbance of the sediments.

In summary, the meiofauna of terrigenous bottoms on continental shelves of the Gulf of Mexico is composed largely of nematodes, with lower percentages of harpacticoid copepods, kinorhynchs, polychaetes, and other small invertebrates. Densities and biomass are elevated in areas of high sediment organic content, particularly off Louisiana and upper Texas, and to some extent off west Florida and the mouths of rivers along the Mexican coast. Densities and biomass are reduced off south Texas and on much of the Mexican shelf. Seasonal increases in density and biomass are associated with high temperatures (summer) and higher input of river-borne or plankton-derived organic material. Lower levels accompany low temperatures (winter) and reduced input of organic matter. Lower densities are also associated with stress from natural causes (hypoxia, concentrated brine, excess sulfides, etc.) and anthropogenic factors (heavy metals, toxic organics, etc.). Response to stress may include a general reduction in density, selective reduction of certain species or groups, shifts in meiofaunal composition, and loss of reproductive capacity, as well as a decrease in genetic diversity of certain groups. Harpacticoid copepods are particularly sensitive to stress agents.

Most foraminiferan species fall in the size range of meiofauna, some being smaller or larger, but often they are excluded from routine meiofaunal counts, probably because special techniques are required to distinguish living from dead organisms. Although there is a sizable literature on Gulf foraminiferans, most of this relates to habitats of individual species or communities (biofacies), indicator species, and relations with the geological record. Little information has been found on densities and biomass, and in reports on species diversity it is not always clear whether the author is dealing with living forms or a combination of living and dead animals. However, several points can be made. Foraminiferans are abundant in both terrigenous and biogenous (calcareous) bottoms. Species diversity is lower in coarse than in fine sediments. On the west Florida shelf, at least, species diversity increases across the shelf with depth of the water column. A summary of foraminiferan studies from the northern and eastern Gulf is presented by Bock (1979).

## Macrobenthos

It was noted earlier that the macrobenthos includes those organisms in the size range of 1.0–25.4 mm and that the megabenthos embraces all forms larger than 25.4 mm. In practice, these two groups are often lumped together as macrofauna, and a division is made based on habitat and method of capture. Thus, the macroinfauna includes those species that live within the sediments and are obtained by taking sediment samples with dredges, grabs, or coring devices and then sieving the top 10–15 cm of sediments through screens of 1.0 mm mesh size or finer. For this group very accurate data are obtained for the small areas sampled, but many samples are required to provide a clear picture of regional distribution patterns. The macroepifauna, on the other hand, includes those

organisms that live upon the sediment surface or in the near-bottom water column and are captured by means of trawl nets dragged along the bottom surface. Trawls are not as efficient in capturing all organisms, but they have the advantage of sweeping large areas, thus minimizing the effects of local patterns. These two groups will be discussed separately.

### MACROINFAUNA

Quantitative information is available concerning macroinfaunal distribution and abundance patterns for all continental shelf areas of the Gulf except the Yucatán shelf, but even here some information is available on the infaunal mollusks. The various data sets include information on species composition and numerical abundance in relation to depth, sediment type, and season. Abundance may be expressed as number of individuals/m² of bottom area or as biomass, that is, either wet weight or grams of organic carbon/m² of bottom area. Although not all data sets are complete, some generalizations are possible.

Throughout the Gulf shelves, polychaetes are generally the most diverse and abundant group, with crustaceans and mollusks next in rank. These three groups often account for over 90 percent of the shelf macroinfauna, the remainder including a variety of other invertebrate groups. The actual density varies from one region to another, but in most areas the highest densities occur on the inner shelf, and densities decrease with increasing depth of the water column. Seasonal changes in species composition are often obvious, especially on the inner shelf, but seasonal changes in density are generally not very great. Various feeding types occur, including deposit feeders, filter feeders, carnivores, and scavengers. Regional features of the macroinfauna are considered below.

The macroinfauna of the continental shelf west of the Florida Peninsula has been considered in reports by Blake (1979), Environmental Science and Engineering, Inc., LGL Ecological Research Associates, Inc., and Continental Shelf Associates, Inc. (1987), Heard (1979b), McNulty, Work, and Moore (1962), N. Phillips, Gettleson, and Spring (1990), and Vittor (1979), among others. Here the fauna is temperate, with an admixture of tropical (Caribbean) species becoming more prominent toward the south. Vittor (1979) reported that for the polychaete species 17 percent are of cosmopolitan distribution, 25 percent are Carolinian (found along the south Atlantic and northern Gulf), 28 percent are Caribbean, and about 30 percent are endemic to the Florida shelf. Species composition of the macroinfauna varies somewhat from season to season and from year to year, with the greatest variability at the nearshore (< 30 m) stations where the water shows the greatest annual temperature range. Species distribution patterns appear to be controlled primarily by sediment type and secondarily by salinity and food availability. Latitude and depth are also important. For polychaetes both diversity and abundance are greatest in coarse sediments.

For the southern half of the shelf (Tampa Bay to the Everglades), N. Phillips, Gettleson, and Spring (1990) reported that of the 1,121 species identified, 40.3 percent were crustaceans, 36.8 percent were polychaetes, 20.6 percent were mollusks, and the remainder included a variety of other invertebrate groups. Across the inner and middle shelf macroinfaunal densities varied from 5,700 to 7,240 individuals/m$^2$, dropping to 3,270/m$^2$ in the 80–90 m range, and 2,650/m$^2$ in deeper water (ca. 180 m). Although all groups declined in abundance with depth, the percentage of polychaetes increased with depth. Many of the polychaete species appear to be opportunists capable of responding to sudden increases in food supply resulting from pulses of organic matter swept out from rivers, bays, and estuaries and from phytoplankton blooms, which may occur seasonally along edges of Loop Current intrusions. However, the standing crop of macroinfauna appears to be too high to be supported by these food resources alone, and it is suspected that across the shelf in these clear waters there must be steady production by benthic microalgae that has never been measured. It has been noted that standing crops of macroinfauna are locally higher in the vicinity of rocky outcrops that support flourishing hard-bottom communities (N. Phillips, Gettleson, and Spring 1990).

The macroinfauna of the continental shelf off the Florida panhandle, Alabama, and Mississippi has been reported in a number of works, of which the following are most important: Blake (1979), Darnell (1991a, 1991b), Franks et al. (1972), Harper (1991), Heard (1979b), Hoese (1972), R. Parker (1956, 1960), Rowe, Polloni, and Horner (1974), J. Shaw et al. (1982), Stanton and Evans (1971), Vittor (1979), and Vittor and Associates (1985). This area, which extends from DeSoto Canyon in the east to the Mississippi River Delta in the west, is swept by strong currents and is subject to Loop Current intrusions. A comprehensive macroinfauna data set for three transects across this section of shelf (table S32) reveals distribution and density patterns in relation to depth, sediment type, and season. A sand sheet covered the bottom at the shallowest stations of all transects, and it extended southeastward across the middle shelf of the two eastern transects. Along the middle shelf of the eastern transect, near DeSoto Canyon, it was mixed with coarser materials (shell hash and carbonate gravel). Finer sediments prevailed near the Mississippi River Delta and at all deeper-water stations. Macroinfaunal densities here were much lower than on the Florida shelf, and densities tended to be highest near shore, decreasing with depth. This applied to total invertebrates as well as to individual groups. However, at the deepest station of the western transect, average polychaete values were high during the winter, reflecting the temporary outbreak of a single polychaete species in the soft sediments near the Mississippi River Delta. Densities of all groups tended to be higher in the coarse sediments. Seasonal differences were not great, but mollusks were about twice as abundant during summer as during winter. Overall, polychaetes made up about 60 percent of the infauna, mollusks and crustaceans about 15 percent each, and the remainder included a dozen different phyla. There is evidence that infaunal populations of this shelf area are supported in winter largely by phytoplankton and in the summer by outwelling of organic material from the extensive rivers, bays, estuaries, and marshlands bordering the area (Darnell 1991a). Since this shelf is bathed by very active currents, the organic matter probably has only a short residence time on the shelf before it is swept out to deeper water.

It should be pointed out that a major faunal break occurs in the vicinity of DeSoto Canyon. Although there is some overlap, by and large, the species that inhabit the carbonate sediments off peninsular Florida are not those encountered along the northern Gulf west of DeSoto Canyon. Likewise, R. Parker (1956, 1960) has shown that in the fine sediments around the Mississippi River Delta the infaunal community is somewhat distinct from that found in coarser sediments farther east and west of the Delta. He referred to this special group as the "Pro-Delta fauna."

The macroinfauna of the Louisiana shelf has been reported by Bedinger (1981), Gettleson (1976), McKinney, Nance, and Harper (1985), R. Parker (1960), and N. Rabalais and Harper (1992), among others. Most of the data relate to specific localities, and no across-shelf transect studies are available. Furthermore, the general picture is complicated by summer hypoxic events, proximity of oil and gas platforms, and general sediment contamination by hydrocarbons, heavy metals, and other chemicals. From widely scattered stations, Bedinger (1981) reported that polychaetes made up 28.9 percent of the taxa collected, followed by crustaceans (25.3 percent) and mollusks (22.1 percent). McKinney, Nance, and Harper (1985) provided data on seasonal densities at a location about 10 miles off the coast of central Louisiana. From a high of 5,262 individuals/m² in February, densities dropped to 2,127/m² in June and to 1,510 and 1,665/m² in July and August, respectively (during hypoxic conditions), and they recovered to more than 2,000/m² in September and November. Even with the low summer values included, the annual average was 2,599, and it is likely that without hypoxia the average would have been in the 5,000–7,000 range for this fertile shelf area. N. Rabalais and Harper (1992) reported even lower infaunal values during summer hypoxia.

Numerous reports have appeared concerning the macroinfauna of the Texas continental shelf. Most notably, these include Cruz-Kaegi and Rowe (1992), Flint (1981), Flint and Rabalais (1980, 1981), Harper (1970), Harper, McKinney, and Nance (1985), Harper, Potts, et al. (1981), Harper, McKinney, et al. (1991), Holland et al. (1980), Keith and Hulings (1965), N. Rabalais (1990), Rowe, Polloni, and Horner (1974), and Shelton and Robertson (1981). Studies carried out in shallow water (19–20 m) off Galveston Island on the upper Texas coast revealed high densities with a distinct seasonal pattern. The lowest levels (4,400 individuals/m²) occurred in January and the highest levels (7,300/m²) were observed in July, with an annual average of 5,800/m². This pattern of high densities peaking in the summer likely would characterize much of the Louisiana shelf in the absence of summer hypoxia. In the upper Texas study polychaetes, crustaceans, and mollusks made up 71.8, 18.8, and 5.0 percent of the infauna, respectively (Harper, Potts, et al. 1981).

Long-term studies of the macroinfauna of the south Texas shelf were conducted along four transects extend-ing offshore from near Matagorda Bay to the level of the Rio Grande (fig. 9.8). Off south Texas seasonal differences were slight. As shown in table 10.12, seasonally averaged densities were generally much higher on the inner shelf than at the two deeper stations, with an exception on Transect II off Corpus Christi. Average densities decreased from 2,901 individuals/m² on the inner shelf to 394/m² on the outer shelf. However, on the two southern transects densities at the outer shelf stations exceeded those at midshelf, suggesting an influence from the Rio Grande or from the offshore eddy often located in this region. Overall, polychaetes constituted about 61 percent of the infauna, and crustaceans and mollusks made up between 14 and 15 percent each (table 10.13). The percentage of polychaetes was highest near shore, whereas the percentage of the other groups increased with depth.

For the terrigenous bottoms of the Mexican coast from the Rio Grande to the Yucatán Peninsula, little information is available on the density and composition of the macroinfauna. Cruz-Ábrego, Flores-Andolais, and Solis-

**Table 10.12.** *Distribution of macroinfaunal densities along four transects across the south Texas continental shelf, all seasons averaged. Transect I extended seaward from Matagorda Bay, and Transect IV was near the mouth of the Rio Grande. (Data from Flint and Rabalais 1981.)*

| | Transect | | | | |
|---|---|---|---|---|---|
| Depth (m) | I | II | III | IV | Average |
| 0–39 | 2,940 | 767 | 4,671 | 3,426 | 2,901 |
| 40–89 | 837 | 350 | 223 | 486 | 479 |
| 90–140 | 334 | 235 | 258 | 752 | 394 |
| Average | 1,370 | 451 | 1,717 | 1,555 | 1,258 |

**Table 10.13.** *Distribution of major macroinfaunal groups in relation to depth across the south Texas continental shelf, all transects and seasons averaged. (Data from Flint and Rabalais 1981.)*

| Invertebrate groups | Percent composition by depth | | | | | |
|---|---|---|---|---|---|---|
| | 0–30 m | 30–60 m | 60–90 m | 90–120 m | 120–140 m | Average |
| Polychaetes | 81.6 | 67.6 | 49.6 | 53.2 | 52.6 | 60.9 |
| Crustaceans | 10.6 | 17.0 | 8.4 | 16.0 | 21.5 | 14.7 |
| Mollusks | 2.2 | 7.9 | 32.6 | 18.0 | 10.9 | 14.3 |
| Others | 5.6 | 7.5 | 9.4 | 12.8 | 15.0 | 10.1 |

Weiss (1991) studied the distribution of mollusks off the mouths of major rivers south of Tampico. Among the identified species were 12 gastropods, 61 bivalves, and 2 scaphopods. Although providing no quantitative data, they divided this shelf into four faunal zones based on species composition as follows: Zone A (off the Río Tuxpan), Zone B (Río Papaloapan to Río Grijalva), Zone C (Río San Pedro and Río San Pablo), and Zone D (off Laguna de Términos). The greatest species diversity was encountered in Zone B.

A number of investigators have reported on the molluscan fauna of the Yucatán shelf. The shallow-water mollusks were discussed by Vokes and Vokes (1983). Rice and Kornicker (1962) described the molluscan fauna on and around Alacrán Reef. In this work 90 gastropod, 40 bivalve, 2 scaphopod, and 1 amphineuran species were identified. The fauna was found to be similar to that of the Florida Keys and West Indies and somewhat similar to that of Blanquilla Reef (60 miles south of Tampico), but it showed little similarity to molluscan fauna of the northern Gulf shelf. Rice and Kornicker (1965) also reported on the deepwater molluscan fauna of the northwestern Yucatán shelf. Samples were taken in the depth range of 15–120 fm (28–227 m). Included here were 129 species of gastropods and 46 of bivalves. Both living species and dead shells were included, and some of the latter may have been relicts of earlier sea level stands. Depth and substrate relations were given. Treece (1980) reported on mollusk shells taken from carbonate bottoms of the continental shelf and upper slope off northeastern Yucatán in the depth range of 15–260 fm (28–476 m). Among the 339 species identified, 185 were gastropods, 143 were bivalves, and 11 were scaphopods. Sixty percent of the species occurred only on coarse-grained sediment, 15 percent only on fine-grained sediment, and 25 percent were found on both types. Rowe and Menzel (1971) reported that the average density of macroinfauna at two stations on the outer Yucatán shelf (185–295 m) was 680 individuals/m², somewhat higher than the average value given earlier for the south Texas outer continental shelf.

Macroinfaunal biomass values expressed as wet weight and/or organic carbon content have been published for several locations around the Gulf, and it is possible to provide estimates for additional areas based on reported density values and appropriate conversion factors. Studies have revealed that the wet weight of individual macroinfaunal animals averages 4.0 mg (Polloni et al. 1979) and that organic carbon averages 4.3 percent of the wet weight

(Rowe, Theroux, et al. 1988). Combining published values and the best information on density (given earlier), it has been possible to develop a record of the standing crop biomass for most of the terrigenous bottom shelf areas of the Gulf (table S33). The results are surprising. Based on values reported for other marine areas, the standing crop of organic carbon for the shallow-water macroinfauna should exceed 1,000 mg C/m² generally, and it would be expected to be well above this value for the fertile continental shelf areas on either side of the Mississippi River Delta. Anticipated values do occur along the Mexican coast, but levels in the northern Gulf are well below expectations. Several explanations are possible, and all are probably important in one area or another. The Alabama/Mississippi shelf, lying near the apex of the Gulf Loop Current, is bathed by strong and frequent bottom currents (Kelly 1991), and much of the organic matter produced locally by phytoplankton or brought in by streams probably has a very short residence time before being swept away (Darnell 1991a). Here the macroinfauna may be food limited. West of the Delta, overfertilization by nutrients from the Mississippi-Atchafalaya system results in seasonal hypoxia and reduced bottom fauna along the Louisiana and part of the upper Texas coast. The nepheloid layer, well developed in this area, reduces photosynthesis and transfer of organic material to the bottom at least seasonally. Rapid sedimentation buries some of the organic material, and seasonally strong bottom currents sweep organic matter off the shelf.

The continental shelf of south Texas receives little nutrient subsidy from the Mississippi-Atchafalaya system, and local streams bring in only modest amounts of freshwater and nutrients because they drain arid lands and much of the flow is diverted upstream for agricultural, municipal, and industrial use. The presence of a nepheloid layer may reduce the effects of local photosynthesis. In addition to these factors, it should be mentioned that the northern Gulf shelf is populated by very dense fish populations, particularly in the area from Mobile Bay to the Galveston area of upper Texas (see chapter 13, on nekton). Most of these grow up in the extensive marshlands and estuaries and feed on the shelf as adults. Since many are bottom feeders, heavy predation certainly plays a major role in reducing the standing crop of macroinfauna. Sorting out these details should be a major research thrust in the coming years.

Seasonal values for macroinfaunal carbon are available for shelves of the western and southwestern Gulf.

Above Tampico average values at inshore stations are as follows: April, 2,672; September, 1,524; and November, 802 mg C/m² (Escobar-Briones and Soto 1997). In both April and September, heavy outflow from rivers brings dissolved nutrients and particulate organic matter to the shelf and creates a stratified water column, but November is a period of low outflow and mixed water column. For the southwestern Gulf, high levels of infaunal carbon occur during spring and summer, which are both dry and wet seasons in this area, respectively, and lower values are seen in December, which is characterized by strong storms and intense mixing of the water column (Soto and Escobar-Briones 1995).

For the carbonate nonterrigenous bottoms off Florida and Yucatán, the macroinfaunal data are not sufficient to provide good biomass estimates, and the conversion factors employed for terrigenous bottom fauna may not apply here. It was pointed out earlier that standing crops for the south Florida shelf appear to be too high to be supported by the phytoplankton populations, and this also applies to the available biomass values from the Yucatán outer shelf. These carbonate systems appear to operate in a different pelagic/benthic relationship, which should be investigated in detail.

MACROEPIFAUNA

This group embraces the larger invertebrates that dwell on the surface of the bottom sediments. Generally speaking, they include the animals taken in bottom trawl nets, but in some studies dredges are also used, and in clear water these methods may be supplemented with still photographs, underwater video, and observations by divers. Demersal fishes and some of the larger, more mobile invertebrates are usually included in the nekton.

For macroepifauna of the continental shelf off the Florida Peninsula, the most important references are Environmental Science and Engineering, Inc., LGL Ecological Research Associates, Inc., and Continental Shelf Associates, Inc. (1987), T. S. Hopkins (1979), Lyons and Collard (1974), Lyons et al. (1971), and N. Phillips, Gettleson, and Spring (1990). Much of this shelf is a mosaic of soft- and hard-bottom types, which does not lend itself to easy characterization of one bottom type or the other. The broad sand sheet is punctuated with rocky outcrops of variable extent and with vertical relief up to about a meter. Below the sand the limestone bottom may be deeply buried or lie only a few centimeters beneath the sand surface, where it can be alternately exposed or covered. On the outer shelf, carbonate algal nodules and pavements may litter or cover the bottom surface. In such an environment, bottom trawls sample both soft and hard bottoms in a single haul, confounding efforts to sort out distribution patterns by bottom type. For these reasons, quantitative data on soft-bottom standing crops of the area are essentially nonexistent. As a practical solution, the workers have separated the sessile fauna (essentially, attached fauna anchored to a firm substrate) from the motile forms (which are nonattached and free to move around). The former require hard substrate, whereas the latter may be found on soft bottoms. Even this distinction does not help much in the definition of soft-bottom fauna because the density of soft-bottom species is often considerably higher in areas surrounding the patches of "live bottom" (hard substrate).

However, from the wealth of qualitative information available, some generalizations may be drawn concerning the soft-bottom macroepifauna. The broad west Florida shelf contains many species of West Indian affinity, and the percentage of these increases toward the south. Many species, particularly inshore and toward the north, are related to the Carolinian fauna of the south Atlantic states and northern Gulf. Quite a few species appear to be endemic to the Florida shelf, but this may change as the Yucatán fauna becomes better known. The primary determinant of species and bottom community distribution patterns is the depth of the water column, but within a given depth range different community types occur depending on sediment type. The mobile epifauna is composed largely of gastropod mollusks, decapod crustaceans, and echinoderms. More than 200 species of gastropods have been recorded. Among the decapods, the crabs are particularly diverse, with around 300 species known, but only a few species of lobsters are present. Echinoderm diversity is likewise quite high, especially in comparison with that of terrigenous bottoms, where only a few species occur. Included among the soft-bottom echinoderms are many species of asteroids, echinoids, ophiuroids, and holothurians. Although crinoids are quite abundant in deeper waters, these are generally sessile forms that inhabit mainly hard bottoms. A list of representative species of the depth-related macrofaunal assemblages of the west Florida shelf is provided in table S34. This is based on Defenbaugh (1976) and modified by information from T. S. Hopkins (1979).

A great deal of information is available concerning the macroepifauna of the Mississippi-Alabama continental shelf. Much of the literature dealing with the Mississippi Sound, shoreward of the barrier islands, has been summarized in Christmas and Langley (1973). Hoese (1972) and R. Parker (1960) discuss the fauna around the Chandeleur Islands and near the Mississippi River Delta. For the open shelf the following references are important: Darnell (1991a, 1991b), Defenbaugh (1976), Franks et al. (1972), Harper (1991), Soto (1972), and Vittor and Associates (1985). Since the most quantitative information is provided by Darnell (1991b) and Harper (1991), these data sets will be discussed below. As in the case of the macroinfauna, macroepifaunal collections were made seasonally at four depths (20, 50, 100, and 200 m) along three transects (Chandeleur, Mobile, and DeSoto Canyon). A total of 310 species was recognized, of which decapods made up 43.2, mollusks 30.3, and echinoderms 18.1 percent. The trawl data were normalized to provide information on the number of individuals captured per hectare (= 10,000 m²) (table S35). On the basis of the normalized data, decapods constituted 71.6, echinoderms 16.3, and mollusks 10.3 percent of the individuals taken. The average annual density was 465 individuals/ha, with the summer density being somewhat higher than that of the winter. Highest densities occurred on the western (Chandeleur) transect. In the summer, the highest overall densities were observed at midshelf and the lowest inshore. During the winter months densities of all the major groups increased with depth, and this was especially striking in the decapods, as vast numbers of small shrimps (particularly *Parapenaeus politus*) were encountered in deep water near the Mississippi River Delta and below Mobile. In the summer decapods were most abundant at midshelf, echinoderms in deep water, and mollusks at both shallow and deepwater stations. Cluster analysis revealed that the various faunal assemblages are associated with particular depth ranges, but because of the mobility of many of these forms, details change with the seasons and, to some extent, from year to year.

Species making up the macroepifauna of the Mississippi-Alabama continental shelf are essentially the same as those found at similar depths off Louisiana and Texas, but this northern Gulf epifaunal community is fairly distinct from that encountered on the west Florida shelf. The major faunal break occurs around DeSoto Canyon, although some of the Florida species also occur along the western rim of the canyon at midshelf and in deeper water. R. Parker (1960) and Defenbaugh (1976) recognized a specialized fauna on the fine-sediment bottoms in shallow water (4–20 m) close to the Mississippi River Delta. Characteristic species of this "Pro-Delta" fan assemblage are listed in table S36. Since the faunal composition of the remainder of this shelf is similar to that off Louisiana and Texas, details will be considered below in the discussion of that section.

Several major trawling surveys have been carried out on the Louisiana-Texas continental shelf, and as a result the epifauna of this area is well known. Important references to this work include Bedinger (1981), Darnell and Schmidly (1988), Darnell, Defenbaugh, and Moore (1983), Defenbaugh (1976), Flint and Rabalais (1980, 1981), H. Hildebrand (1954), and Holland et al. (1980). Although they provide a wealth of information concerning species distribution patterns, most of the databases have not been normalized to reveal catch per unit area or species density information. The chief exception is the large database analyzed by Darnell, Defenbaugh, and Moore (1983) for penaeid shrimps and fishes, and this will be discussed in the chapter on nekton. In the absence of quantitative data, detailed analysis of the macroepifauna of this area is not possible. However, some general conclusions can be drawn from the qualitative information available.

General macroepifaunal densities appear to be higher near shore and around midshelf than they are in deeper water. Details vary in relation to season, primarily because of the movements of the more mobile species such as squids, shrimps, and crabs. Many of these species grow up in brackish habitats of the extensive marginal marshes, bays, and estuaries and return to the shelf in vast numbers during the summer and fall months. Once on the shelf, most remain in nearshore to midshelf waters, but one species, the brown shrimp (*Penaeus aztecus*), overwinters on the outer shelf.

It also seems clear that despite the occurrence of summer hypoxic events, macroepifaunal densities tend to be higher off Louisiana and upper Texas than off south Texas. At least, this is true for the more mobile estuary-related species that find more suitable and extensive brackish-water habitat along the upper coast. A more subtle downcoast gradient is manifest in the general faunal composition. As one passes from the Mississippi to the Rio Grande there are changes in the densities of individual species, and some drop out while others come in. A number of tropical species extend northward above the Rio Grande to the lower Texas shelf. Thus, the numerically dominant species gradually change along this coastline. Even within a given depth range individual species

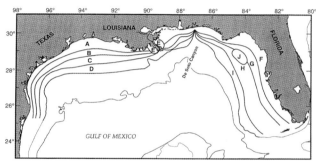

Figure 10.9. Distribution of depth-related macrofaunal assemblages on continental shelves of the northern and eastern Gulf of Mexico. A, B, C, and D represent the inner, intermediate, outer shelf, and upper slope assemblages, respectively, extending from the Rio Grande to DeSoto Canyon. E is the shallow-water Pro-Delta fan assemblage on both sides of the Mississippi River Delta. F, G, H, and I represent the inner, intermediate, outer shelf, and upper slope assemblages, respectively, on the east Florida shelf. J is the Florida Middle Ground assemblage, characterized by much hard bottom. Bathymetric contours are approximate only. (Modified from Defenbaugh 1976.)

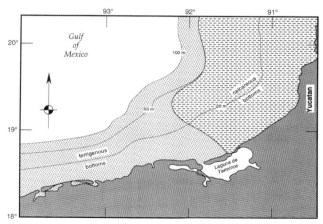

Figure 10.10. The dividing line between terrigenous and calcareous bottoms at the base of the Yucatán Peninsula. (From Yáñez-Arancibia and Sanchez-Gil 1986.)

may associate more with one sediment type or another. These downcoast gradients have not been well studied, although specific examples are known.

The depth-related macroepifaunal invertebrate assemblages along the entire northern Gulf of Mexico have been documented by Defenbaugh (1976) (table S37). He concluded that the Mississippi-Alabama shelf fauna is basically an eastward extension of that found at similar depths west of the Mississippi River. In addition to the Pro-Delta fan assemblage, which extends on the west as well as the east side of the Delta, he listed the composition of the faunal assemblages in each depth range essentially from the shoreline across the shelf to the 200 m depth contour. Sophisticated statistical analysis has confirmed the existence of depth-related faunal bands off the Mississippi-Alabama and south Texas coasts (fig. 10.9). Although there is much overlap, each assemblage does have its own recognizable suite of species.

As noted earlier, the Mexican continental shelf is divisible into two sedimentary provinces. The long stretch from the Rio Grande to the base of the Yucatán Peninsula is floored by terrigenous sediments with low calcium carbonate content, generally less than 20 percent. The Yucatán Platform, on the other hand, is carpeted by biogenic materials with high calcium carbonate content, generally over 70 percent. The dividing line between these two provinces is fairly sharp and occurs near the base of the Yucatán Peninsula (fig. 10.10). Although there is some

overlap, the macroepifaunal invertebrate communities of the two areas are fairly distinct and will be considered separately below.

For the terrigenous portion of the Mexican shelf, the only quantitative studies of the macroepifauna appear in the reports of Escobar-Briones and Soto (1997) and of Soto and Escobar-Briones (1995). For the northern portion, from the Rio Grande to Tampico, these workers reported average organic carbon values of 9, 8, and 2 mg $C/m^2$ for the inner, middle, and outer continental shelf zones, respectively. Corresponding values for the southern Gulf shelf, from Veracruz to the base of the Yucatán Peninsula, were somewhat higher, that is, 20, 10, and 10 mg $C/m^2$. Values tended to be lower during the dry season than in the rainy season. Composition of the macroepifaunal invertebrate communities of the terrigenous bottoms of the Mexican shelf have been reported by Defenbaugh (1976) and H. Hildebrand (1954), but the databases do not permit subdivision into depth-related zones (table S38). In composition this fauna is quite similar to that of the northern Gulf, but it contains more tropical species.

For the calcareous bottoms off Yucatán, no quantitative data are available, but a general species list has been compiled from information provided by Defenbaugh (1976) and H. Hildebrand (1955) (table S39). Again, depth-related faunal zones could not be constructed from the available data. Comparing data from tables S38 and S39, it is clear that although there is some overlap, the two faunal assemblages are fairly distinct. Of the 98 species on the calcareous-bottom list, 62 percent are not represented on the terrigenous-bottom list, and if the more mobile shrimps and crabs are left out, the distinctness of the calcareous-bottom assemblage rises to 70 percent. In overview, although much work remains to be done, the general picture of epifaunal distribution on continental shelves of the Gulf is becoming clear. Major features include the faunal distinction between terrigenous and calcareous bottoms, depth-related faunal zonation, and density and biomass decreases with increasing depth.

## Total benthos

It is of interest to know how the various compartments of the benthos relate to one another and to the total benthic biomass of the continental shelf. The only database that can provide such information is that of Soto and Escobar-Briones (1995) for the southern Gulf, although that of Escobar-Briones and Soto (1997) provides information for all compartments except the bacteria. These two databases are summarized in table S40. For the southern Gulf it is seen that for each category the biomass is highest on the inner shelf and decreases with depth, and the same applies to total benthos, which decreases from 3.72 to 0.61 g C/m². That is, the total benthos standing crop of the inner shelf is about six times that of the outer shelf. The ratios of the different compartments are given, and it is clear that at all depths most of the benthic biomass is tied up in the bacteria and the macroinfauna. On the other hand, the macroepifauna makes up less than 1 percent of the total benthos. In this study, bacteria ranged from 33.4 to 63.2 percent and averaged 44.2 percent of the total benthic biomass.

A similar picture is derived from data on the western Gulf shelf, although the ratios of meiofauna and macroinfauna to macroepifauna are generally higher than in the southern Gulf. Since these data are based heavily on the use of conversion factors derived from other areas, they are really order of magnitude estimates, and the conclusions should be accepted with some caution. Nor should it be assumed that the benthos of other continental shelves around the Gulf will conform to the patterns observed in the southern and western Gulf. For example, Cruz-Kaegi (1992) reported that off Louisiana and upper Texas the bacteria make up from 73 to 99.5 percent of the combined benthic biomass. Each section of the continental shelf represents a unique set of physical and sedimentary conditions, and the benthos of each must be analyzed by comparable quantitative techniques before a clear picture emerges. Nevertheless, the data presented in table S40 provide insight into the types of information that should eventually emerge from detailed studies of the total benthos of continental shelves around the Gulf.

Most of the information presented above relates to the composition, distribution, and standing crop of the various groups that make up the benthos. However, elsewhere, particularly along the Atlantic coast, many studies have been carried out on the internal dynamics of the benthos. Included are investigations of biochemical, physiological, and behavioral aspects of individual species and groups, species interactions, and organism-sediment relations. Although a limited amount of such work has also been conducted in the Gulf area, much remains to be done here before the internal workings of the benthos are well understood.

It is at least theoretically possible to assess the metabolic activity of the entire living biomass of the benthic community by measuring the rate of oxygen consumption by the sediments. Studies by Rowe (1998) in the hypoxic zone off the Louisiana coast have provided some data on sediment oxygen uptake. In the spring (April) uptake values were low (20.6 mg $O_2$/m² per hour), but by late summer (August) they had risen to 35 mg $O_2$/m² per hour. The limited data showed essentially a doubling of oxygen uptake for a temperature rise of 10°C if values were deleted from stations where oxygen available in the water column was quite low. Part of the oxygen demand was undoubtedly due to aerobic respiration of bacteria, which, as noted earlier, make up the bulk of the benthic biomass in this area and which, gram for gram, are metabolically much more active than larger organisms such as polychaetes or clams. Some of the uptake was the result of chemical oxygen demand by sulfides, ammonia, and nitrites, which themselves largely reflect anaerobic metabolism of the microorganisms. Sulfide oxidation alone could account for much of this. An additional very small percentage could have resulted from the oxidation of heavy metals such as iron and manganese. Since most of the oxygen requirement reflects biologically mediated processes, with due caution benthic oxygen uptake could provide a useful comparative index of biological metabolic activity in the sediments that might be indirectly related to the total standing crop of benthic biomass.

# 11 : BIOTA OF HARD SUBSTRATES
## BIOFOULING COMMUNITIES AND CORAL/ALGAL REEFS

Many species of marine plants and animals require hard substrates for attachment since these provide firm anchorage against the forces of waves and water currents. Thus situated, marine algae may project into the water column, where they have better access to light and nutrients. Firmly anchored animals also grow into the water, where they can more readily harvest plankton swept along by the currents. As these key species thrust upward and outward from the substrate, they create sheltered, organic-rich habitats and opportunities for colonization by a wide variety of other, more mobile forms, in this way building up the characteristic hard-substrate communities. Within the marine realm many of the hard substrates become covered with layers of fine sediments that permit invasion by soft-bottom species. On the other hand, soft-bottom areas frequently contain dead shells and fragments of other hard materials that permit occasional hard-substrate species to colonize soft-sediment areas. Despite such exceptions, in physical appearance and species composition the communities of hard substrates are quite distinct from those of soft sediments.

The dispersal stages of marine algae (sporelings) and the larvae of many marine animals are carried in the plankton to potential areas for settlement. Today the adults of many attached species are also transported around the world oceans on the hulls of ships or in the bilgewater, and most of these foreign species also produce planktonic larvae. Arriving in a new area, a larva must decide whether or not a given substrate is suitable for attachment. Larvae may spend some time apparently examining a substrate and rejecting it if certain specifications are not met. Although we cannot know all the factors involved in the larva's decision to settle or not to settle, fresh substrate placed in the marine environment is generally not as acceptable as one that has been allowed to age for a few days or weeks. During this period the substrate develops a coating of microorganisms called a "slime community." Bacteria are the first invaders, and within a few days these are followed by microscopic algae and then protozoa (primarily ciliates and flagellates). These early colonizers coat the substrate with mucus-like material that seems to be favored by the settling larvae. Although some species may

prefer natural substrates, most readily colonize artificial substrates when these are appropriately aged.

Species of algae and marine animals that colonize artificial substrates are referred to as biofouling organisms because their presence fouls up the hulls of ships and the surfaces of buoys and other human-made structures placed in the sea, and because keeping the surfaces relatively free of organisms requires frequent maintenance and considerable cost. Because of the economic significance of biofouling communities, they have been the subject of much scientific investigation. Studies have been carried out to determine the composition and dynamics of biofouling communities and methods for their prevention and removal. Such matters have been of particular concern for the various navies. Worldwide around 2,000 species, representing all groups of algae and most animal phyla, have been recorded among the biofoulers, although only about 100 species are of real significance. The literature contains much information on such topics as species composition, vertical and horizontal zonation (that is, seaward from the shoreline), seasonal abundance, latitudinal distribution (poleward from the equator), physical and biological factors associated with settlement and growth, and development and control of attached multispecies communities. Such studies on artificial substrates greatly add to our understanding of the biota of natural substrates, since essentially the same organisms and processes are involved in both instances.

Natural hard-surface substrates associated with marine environments include many types of igneous, metamorphic, and sedimentary rocks exposed above or below the surface of the sea. Limestone rocks abound in tropical areas. Some of the rocks are quite hard, whereas others are soft or friable. In highly turbid waters, such as near the mouths of large rivers, rocky outcrops are often covered with a layer of silt and the attached biota may be limited to a few tolerant species. In clear waters photosynthetic forms thrive, and these include various groups of algae as well as certain lower invertebrates that harbor photosynthetic algae within their tissues. In clear tropical waters, where winter temperatures seldom drop below 18°C, elaborate reefs made up of corals, algae, foraminife-

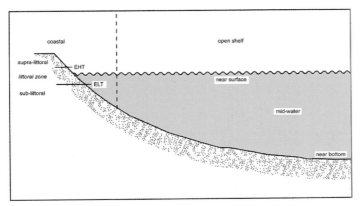

**Figure 11.1.** Habitat zonation in coastal and open-shelf environments of the sea. EHT = extreme high tide, ELT = extreme low tide. See text for explanation. Diagram is not to scale.

rans, and sometimes tube-building polychates are often widespread. As in the case of the biofouling communities, the biota of natural hard substrates has been the subject of much research worldwide. Vertical and horizontal distribution patterns are often quite obvious, and latitudinal patterns parallel those of artificial substrates. Experimental studies of competitive interaction often involve removal of one species and documentation of recovery patterns. Both natural and artificial hard substrates are abundant in the Gulf of Mexico, and they are present in a variety of environmental settings. Their study has provided much information on the composition and dynamics of marine biological communities in general.

## THE HARD-SUBSTRATE ENVIRONMENT

The suitability of a given substrate for biological colonization and development depends on two essential factors: the nature of the substrate itself and its location with respect to the various factors of the physical environment. Substrates that are porous or somewhat crumbly are preferred by certain sponges, polychaetes, mollusks, and other forms that bore or tunnel into the material and live inside where they are protected from predators and the physical environment, but such substrates do not provide firm anchorage for the external biota. Limestone and other hard calcareous rocks allow for firm anchorage, but only certain boring sponges, mollusks, and sea urchins are able to penetrate and live inside. Granite and igneous rocks are more difficult to penetrate, and metal structures such as the support columns of oil platforms totally preclude the boring species while providing excellent substrate for anchorage of the external forms.

Because of the effects of tides, waves, and (in high lati-

tudes) sea ice, the edge of the sea, or coastline, is a physically stressed environment. Since such areas are highly accessible to researchers, they have been the subject of much investigation around the world. Three primary zones (often with subzones) are recognized: the *supralittoral* (above high tide), *littoral* (intertidal), and *sublittoral* (below low tide) zones. These are illustrated in figure 11.1 and discussed below.

### Supralittoral zone

Sometimes called the "splash" or "spray" zone, the supralittoral, being above the high-tide level, receives seawater primarily in the form of droplets. Exposed to wind and sunlight, these evaporate and leave deposits of pure sea salt. Washed off by rainfall, these deposits reappear in short order. This zone is characterized by harsh extremes of solar radiation, temperature, moisture, and salinity. Surfaces exposed to direct sunlight are more extreme than those that are shaded. The lower portion of this zone is subject to physical buffeting by storm waves that reach higher than the usual high-tide levels. Sea birds deposit dung rich in phosphates and nitrates, which provide nutrients for tolerant plant species such as algae and lichens. In areas of the world with pounding surf, the supralittoral zone may extend upward for over 10 m, but around the Gulf it seldom exceeds a few feet. This zone is generally inhabited by a very sparse biota consisting of a few hardy blue-green algae, lichens, and specialized animals such as mobile isopods, snails, and barnacles. The chlorophyll content is generally low, seldom exceeding 0.50 mg/m², and the biological standing crop is also much reduced.

### Littoral zone

The intertidal zone is characterized by alternating periods of exposure and submersion, with details depending on the tidal range and period of the area. Alternating exposure subjects the habitat to frequent stretches of high light intensity, temperature extremes, and evaporation. Pounding by waves and scouring by sand may be severe. Here substrate stability is an important factor. The width of this zone depends on the tide range of the area, and around margins of the Gulf this is only about one meter. More species find homes here, and standing crops are higher than in the preceding zone. Included are several types of algae, mostly brown algae, as well as snails, some bivalves, and barnacles. The chlorophyll concentration averages around 0.87 mg/m², and the standing crop is higher than in the supralittoral.

## Sublittoral zone

Depending on the slope of the bottom and the wave characteristics, the sublittoral zone extends from the low-water line to a depth of 10 to 20 m. Here the factors of temperature, light, moisture, and salinity are more stable, and buffeting from waves and water currents is more subdued. In this habitat are encountered complex communities involving many species of green, brown, and red algae as well as hydroids, annelids, snails (including nudibranchs), bivalves, barnacles, hermit crabs, true crabs, echinoderms, and a host of smaller invertebrates. Both the chlorophyll content (ca. 1.07 mg/m²) and standing crop are much higher than in the preceding zones.

## Open continental shelf

Farther out on the continental shelf, beyond the sublittoral zone, temperature and salinity extremes and the effects of waves become even less pronounced, and the primary limiting factor here is the suspended sediment load of the water column. Near large rivers, especially in areas of terrigenous bottoms, proceeding seaward one passes from the turbid brown coastal zone, through the green water of midshelf, to the clear turquoise-blue water of the outer shelf. By reducing light penetration in the water column and settling out and coating surfaces, suspended sediments can inhibit or eliminate photosynthetic species. High levels of suspended inorganic particles also discourage most filter-feeding invertebrates that depend on planktonic food. For these reasons, rocky outcrops in silty areas harbor much reduced populations of a few silt-tolerant species. These effects are more pronounced in near-bottom than in near-surface waters. By contrast, in continental shelf areas where the water is clear and relatively free of suspended inorganic matter, algae and other photosynthetic forms as well as attached filter-feeding species compete for space on every bit of hard substrate available. Because of the obvious abundance and diversity of the large living forms that they support, such areas are often referred to as "live bottoms." Where the water is sufficiently warm throughout the year the live-bottom biota may include reef-building corals.

### COMPOSITION OF THE BIOTA

Although most major algal and animal groups have been reported to occur at one time or another among the biota of hard substrates of the Gulf, the present section will focus on those groups that require hard substrates for attachment and those mobile organisms that are found primarily in association with the attached species. For reference, table 11.1 provides a list of the groups to be discussed. Some of these, of course, are also represented on soft substrates, and for those groups treated in the previous chapter, descriptions here will be brief. Representative genera are listed, some are illustrated, and key literature references are provided. Although mention is made of some fish species, primary treatment of the fishes will be undertaken in chapter 13, on nekton.

## Algae

All four of the phyla containing filamentous or thallic algae occur in the Gulf in association with hard substrates such as rock, shell, mangrove roots, and seagrass blades, as well as artificial substrates of wood, concrete, or metal. The literature dealing with hard-substrate algae is fairly extensive and includes reports by Baca, Sorensen, and Cox (1979), Cheney and Dyer (1974), P. Edwards (1969, 1976), P. Edwards and Kapraun (1973), Eiseman and Blair (1982), Humm (1952, 1964), Humm and Caylor (1957), Humm and Darnell (1959), Humm and Hildebrand (1962), Kapraun (1980), Kim (1964), Lowe and Cox (1978), Minnery (1984, 1990), Minnery, Rezak, and Bright (1985), and W. Taylor (1954a, 1954b, 1960). Blue-green algae are represented by species of the genera *Anacystis, Calothrix, Lyngbya, Oscillatoria, Schizothrix,* and *Symploca,* among others. Prominent genera of green algae include *Anadyomene, Chaetomorpha, Enteromorpha, Halimeda, Udotea,* and *Ulva.* Among the hard-substrate brown algae are the genera *Colpomenia, Dictyota, Ectocarpus, Giffordia, Padina, Sargassum,* and *Sporochnus.* The two *Sargassum* species, *S. filipendula* and *S. vulgare,* normally grow attached to hard substrates. Although they may occasionally break off and float around, they are not truly pelagic like *S. fluitans* and *S. natans,* which were considered in a previous chapter. Prominent genera of red algae include *Bangia, Ceramium, Digenea, Fosliella, Gracilaria, Hypnea,* and *Polysiphonia.* Red coralline algae such as *Lithophyllum* and *Lithothamnium* often occur in deeper waters as concretions in the form of balls or a flat pavement. Several genera of green, brown, and red algae associated with hard substrates are illustrated in figure 11.2.

## Protozoa

Although several groups of protozoans are found in the Gulf in association with hard substrates (Sprague 1954), the foraminiferans have been most extensively studied. Prominent among the genera are *Amphistegina, Cancris,*

**Table 11.1.** *Simplified classification of the plant and animal kingdoms of the Gulf of Mexico including those groups associated with hard substrates.*

**Phylogenetic groups, subgroups, and common names**

**Plant Kingdom**
Cyanophyta (blue-green algae)
Chlorophyta (green algae)
Phaeophyta (brown algae)
Rhodophyta (red algae)

**Animal Kingdom**
Protozoa (protozoans)
    Sarcodina (foraminiferans)
Porifera (sponges)
Cnidaria (coelenterates)
    Hydrozoa
        Hydroida (hydroids)
    Anthozoa
        Octocorallia (soft corals)
        Zoantharia
            Actiniaria (sea anemones)
            Scleractinia (stony corals)
Mollusca (mollusks)
    Amphineura (chitons)
    Gastropoda (snails)
    Bivalvia (bivalve mollusks)
Annelids (segmented worms)
    Polychaeta (polychaetes)
Arthropoda (joint-legged animals)
    Isopoda (isopods)
    Cirripedia (barnacles)
    Decapoda
        Natantia (swimming shrimps)
        Anomura (hermit crabs and relatives)
        Brachyura (true crabs)
Bryozoa (Ectoprocta; moss animals)
Echinodermata (spiny-skinned animals)
    Asteroidea (starfishes)
    Echinoidea (sea urchins and relatives)
    Ophiuroidea (brittle stars)
    Holothuroidea (sea cucumbers)
    Crinoidea (sea lilies)
Chordata (chordates)
    Urochordata (tunicates)

*Carterina, Glabratella, Gypsina,* and *Sorites.* Examples are shown in figure 11.3. Important references include Bock (1979), R. A. Davis (1964), L. Lidz and B. Lidz (1966), Poag and Sweet (1972), Shifflett (1961), and Tresslar (1974b).

## Porifera

Sponges are important among the biological inhabitants of hard substrates, especially in areas that do not have high levels of suspended sediments that could clog up the pores. Major genera include *Axinella, Callyspongia, Homaxinella, Ircinia, Verongia,* and *Xenospongia.* In addition, the boring sponge, *Cliona,* is often found burrowed into soft rocks and old corals. Representative examples are provided in figure 11.3. References include C. C. Adams (1996), de Laubenfels (1953), G. Green (1977), Macias (1968), Storr (1964), and Teerling (1975). More than 70 species of sponges were reported from the southwest Florida shelf by Environmental Science and Engineering, Inc., LGL Ecological Research Associates, Inc., and Continental Shelf Associates, Inc. (1987).

## Cnidaria

Among the cnidarians four groups are of particular concern as inhabitants of hard substrates: the hydroids, soft corals, sea anemones, and stony corals.

### HYDROIDS

The life history of the hydroid typically entails two generations: a sessile asexually reproducing polyp stage followed by a free-floating sexually reproducing medusa stage. Medusae were discussed in the zooplankton chapter, and the polyp stage is considered here. In some species the attached hydroid is solitary, but in most it is colonial, where it is anchored to the substrate by a tough rootlike structure from which grow the tubular stalks supporting individual polyps like leaves on a twig. These stalks are often branched so that the entire colony assumes a bush-like appearance. Despite their complexity, most species reach a length of only a few centimeters. Along the stalks and branches two types of polyps develop, a feeding form with mouth and stinging tentacles and a reproductive form that buds off small medusae. The reproductive polyp receives nourishment from the nearest feeding polyps through the central tube of the branch. Within the Gulf, hydroids are widespread and are found at all salinities and attached to all types of substrates. Common genera include *Bimeria, Bougainvillia, Campanularia, Obelia, Plumularia,* and *Sertularia.* Several are shown in figure 11.3. Major references

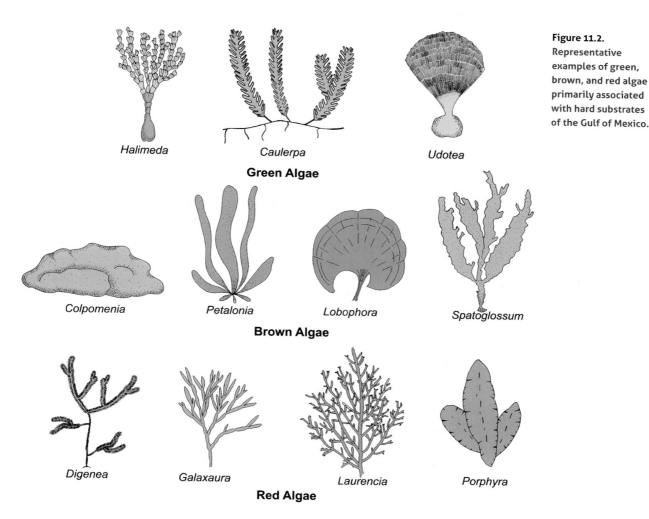

**Figure 11.2.** Representative examples of green, brown, and red algae primarily associated with hard substrates of the Gulf of Mexico.

*Halimeda*          *Caulerpa*          *Udotea*

**Green Algae**

*Colpomenia*          *Petalonia*          *Lobophora*          *Spatoglossum*

**Brown Algae**

*Digenea*          *Galaxaura*          *Laurencia*          *Porphyra*

**Red Algae**

are Deevey (1950, 1954), Defenbaugh (1974), Defenbaugh and Hopkins (1973), and C. Shier (1965).

Soft corals, sea anemones, and stony corals all belong to the class Anthozoa, in which the medusa stage is completely absent. Furthermore, the anthozoan polyp is much more complex than that of the hydroid. In the anthozoan polyp the mouth is connected with the digestive cavity by a tubular pharynx, and the digestive cavity itself is divided into compartments by fleshy septa that bear stinging cells.

### SOFT CORALS

In soft corals, sometimes called octocorals, the polyps always bear eight tentacles that possess side branches like a feather, and internally the digestive cavity is divided by eight septa. The skeleton of the colony is composed of horny material or fused calcareous spicules, and along the axis or network of the skeleton grow the small polyps. The shapes of the colonies are quite varied and may be rodlike, coiled and whiplike, feathery, bushy, or fan-like. Included are the sea pens, sea whips, sea fans, and their relatives. Examples are shown in figure 11.4. Representative genera include *Ellisella, Gorgonia, Lophogorgia, Muricea, Plexaura,* and *Telesto.* Important references are Chamberlain (1966), Giammona (1978), and Grimm (1978).

### SEA ANEMONES

Sea anemones are large, mostly solitary polyps, but a few are colonial. Although most are only a few centimeters in width and height, the largest may be almost a meter in width. The body of the sea anemone is a heavy column with a flat disk on the bottom for attachment. The top end contains a centrally located mouth surrounded by numerous tentacles armed with stinging cells. Pairs of septa in the pharynx usually occur in multiples of six. When disturbed, the polyp can contract, pulling in the mouth and tentacles and nearly or completely enclosing these with

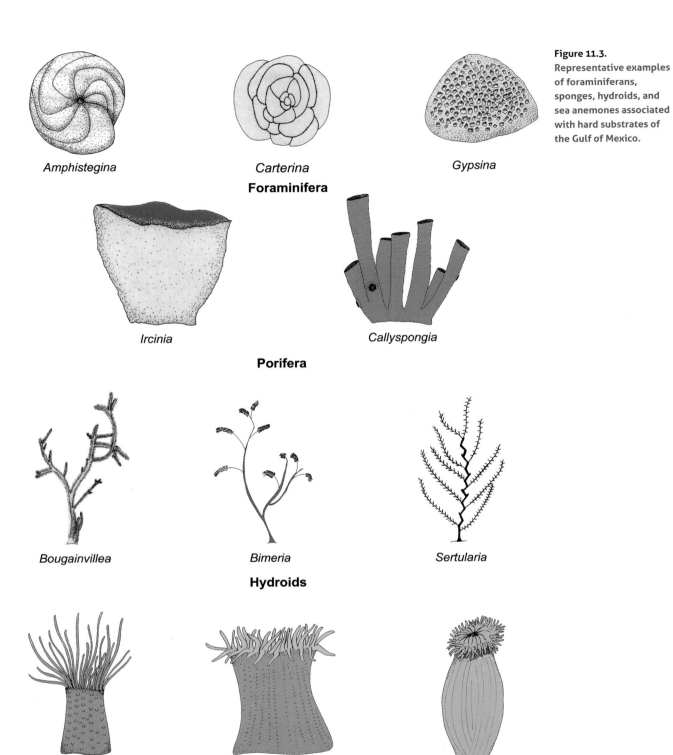

*Amphistegina*  *Carterina*  *Gypsina*

**Foraminifera**

*Ircinia*  *Callyspongia*

**Porifera**

*Bougainvillea*  *Bimeria*  *Sertularia*

**Hydroids**

*Botryon*  *Bunodosoma*  *Paranthus*

**Sea anemones**

**Figure 11.3.**
Representative examples of foraminiferans, sponges, hydroids, and sea anemones associated with hard substrates of the Gulf of Mexico.

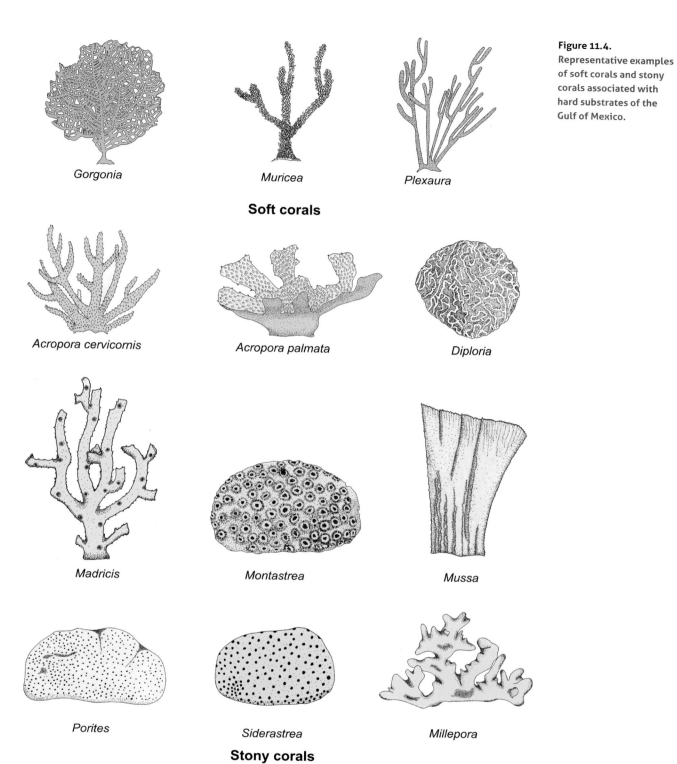

Gorgonia

Muricea

Plexaura

**Soft corals**

Acropora cervicornis

Acropora palmata

Diploria

Madricis

Montastrea

Mussa

Porites

Siderastrea

Millepora

**Stony corals**

**Figure 11.4.**
Representative examples of soft corals and stony corals associated with hard substrates of the Gulf of Mexico.

part of the lateral body wall. Many species are brightly colored. Most sea anemones dwell in coastal waters of the tropics, where they are attached to various types of hard substrates including the shells of living hermit crabs. Sea anemones consume zooplankton and other small invertebrates, which they stun with the stinging cells of their tentacles, but larger species also capture small fishes. Important genera in the Gulf include *Aiptasia, Botryon, Bunodactis, Bunodosoma, Calliactis,* and *Paranthus.* Several anemones are illustrated in figure 11.3. Significant references are Carlgren and Hedgpeth (1952) and Hedgpeth (1954a).

### STONY CORALS

Stony corals are closely related to sea anemones, but unlike the anemones they produce basal skeletons of calcium carbonate. There are also minor structural differences. For example, corals lack the ciliated groove of the pharynx (*siphonoglyph*) present in anemones. Although some species of stony corals are solitary, most are colonial and contain many tiny polyps, each only one or a few millimeters in diameter. Within a colony all the polyps are interconnected by a horizontal sheet of tissue. The skeleton is secreted by cells in the base and lower sides of each polyp and by cells in the bottom layer of the connecting sheet. The characteristic form of the stony skeleton depends on the size and arrangement of the polyps and on the growth pattern of the colony.

Reef-building, or *hermatypic,* corals occur only in warmer marine waters of the world where the water temperature seldom drops below 18°C. Since carbon dioxide is relatively insoluble in warm water, under this condition skeletal construction requires less energy than would be the case in colder water. Reef-building corals are generally limited to clear water and depths of 100 m or less. Symbiotic algae called *zooxanthellae* live within the tissues of all reef-building and some nonreef species. These require sunlight for photosynthesis, and they apparently supply oxygen, carbohydrates, and other useful chemical products to the host coral. These algae are actually dinoflagellates capable of escaping the coral polyps and living free in the water column as phytoplankton cells. In addition to receiving nutrition from the endobiotic algae, corals actively feed on zooplankton organisms and other small invertebrates at night, using both the stinging cells of the tentacles and the cilia to capture and transport food to the mouth. Among the solitary corals, some lack zooxanthellae, and some occur in deep water as well as colder water of high latitudes. About 200 species of hermatypic corals are found in the Great Barrier Reef off Australia, but only around 55 species are known from the West Indies, and most of these also occur in the Gulf of Mexico. Representative genera of reef corals of the Gulf include *Acropora, Diploria, Madracis, Montastrea, Mussa, Porites,* and *Siderastrea.* The fire coral *Millepora* is a stony coral also associated with reefs, but it is actually an aberrant member of the Hydrozoa and related to the hydroids. Several stony corals are shown in figure 11.4. There is a fairly large body of literature on stony corals of the Gulf, and some of the more important references include Busby (1965), Cairns (1977, 1978) G. Davis (1982), Jaap (1984), Kraemer (1982), Logan (1962), Rannefeld (1972), F. Smith (1954c, 1971), Tresslar (1974a), and Viada (1980).

## Mollusca

Although most species of marine mollusks are mobile rather than attached, many are characteristic residents of hard substrates. A few of the bivalves do anchor themselves firmly, and several bore into the substrate. Gastropods and bivalves are the most abundant and diverse groups of mollusks in such habitats, particularly around coral reefs, but amphineurans and cephalopods are also represented. Literature references include Ekdale (1974), T. S. Hopkins, Blizzard, and Gilbert (1977), Lipka (1974), Rehder (1954), Treece (1980), Tunnell (1974), Tunnell and Chaney (1970), Vokes and Vokes (1983), and Wiley, Circe, and Tunnell (1982).

### AMPHINEURA

The chitons are primitive mollusks with flattened bodies containing eight overlapping calcareous plates on the dorsal surface. Most species are only a few inches in length. Members of this group are fairly widely distributed in shallow tropical and subtropical seas, where they are generally found on hard substrates. Only a few genera have been recorded from the Gulf including *Acanthochitona, Chiton, Ischnochiton,* and *Tonicia.* One is illustrated in figure 11.5.

### GASTROPODA

The Gulf of Mexico supports a large benthic gastropod fauna, and many of the species occasionally or regularly occur on hard substrates. Individual species are frequently limited to particular habitats. Hard substrates

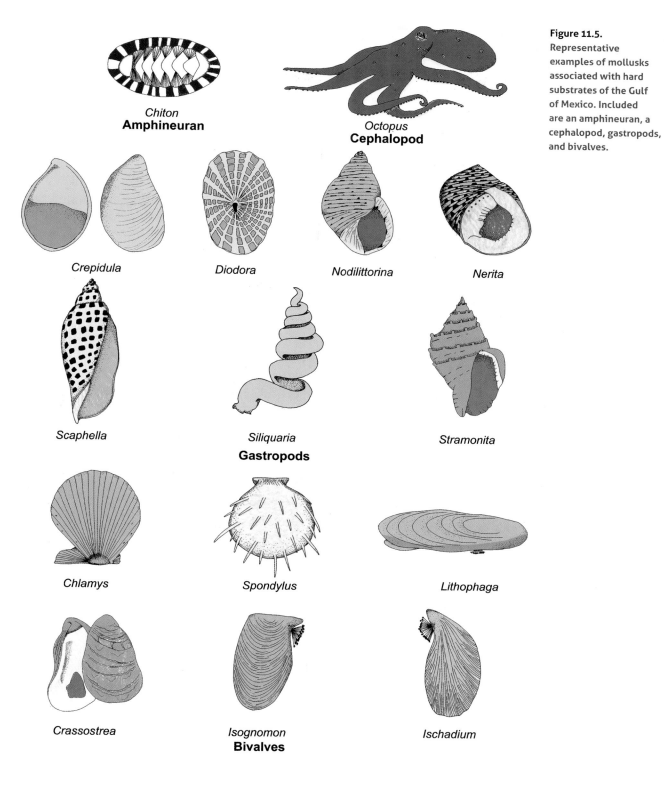

**Amphineuran**
*Chiton*

**Cephalopod**
*Octopus*

*Crepidula*

*Diodora*

*Nodilittorina*

*Nerita*

*Scaphella*

*Siliquaria*

*Stramonita*

**Gastropods**

*Chlamys*

*Spondylus*

*Lithophaga*

*Crassostrea*

*Isognomon*

*Ischadium*

**Bivalves**

**Figure 11.5.**
Representative examples of mollusks associated with hard substrates of the Gulf of Mexico. Included are an amphineuran, a cephalopod, gastropods, and bivalves.

along the shoreline (supralittoral, littoral, and sublittoral zones) are favored habitats for snails of the families Fissurellidae (keyhole limpets), Littorinidae (littorinids), and Phasianellidae (nerites), while subtidal hard banks and coral reefs often support those of the families Turbinidae (star shells), Cerithidae (ceriths), Coralliophyllidae (coral shells), and Turritellidae (worm shells). Moving around the periphery of the Gulf one encounters species substitutions and changes in the hard-substrate gastropod fauna. Even in essentially the same habitat type, the fauna of the northern Gulf is somewhat distinct from that of the southern waters, and that of the east is different from that of the west. Prominent hard-substrate gastropod genera include *Caecum, Crepidula, Diodora, Nerita, Nodilittorina, Scaphella, Siliquaria,* and *Stramonita,* among others. Several examples are shown in figure 11.5.

BIVALVIA

Hard-substrate bivalves are also well represented in the Gulf of Mexico, and like the gastropods, the bivalves are often characteristic of particular habitats. Some, such as the genera *Brachidontes* and *Crassostrea,* are limited to waters of less than full marine salinity, while others occur in fully marine coastal habitats, and still others are found in open shelf waters on hard banks and/or coral reefs. Characteristic genera include *Chlamys, Crassostrea, Ischadium, Isognomon, Lima, Ostrea,* and *Spondylus.* Although most species are free to move around, several are anchored to the substrate by strong threads or calcareous growths. In addition, a number of bivalves bore into the substrate and live in protected chambers with only the tips of tubelike siphons projecting into the open water. Genera of boring clams include *Botula, Gastrochaena, Gregariella, Lithophaga,* and *Spengleria.* Representative examples of hard-substrate bivalves are given in figure 11.5.

CEPHALOPODA

In the cephalopods the body is often elongate, and it possesses long flexible arms (ten in squids and eight in octopuses) generally equipped with disklike suckers. The shell is either internal or absent, except in the chambered nautilus and its relatives, where it is external. Although most species are nektonic, some live near or on the bottom, and one, the common octopus (*Octopus vulgaris*), is of interest here. This species is widespread on continental shelves of the Gulf, but it flourishes only where it can find protection. In the southern Gulf, large shells of the horse conch

(*Pleuroploca gigantea*) provide adequate housing, and in other areas it hides in rocky crevices around jetties and similar habitats. The octopus is shown in figure 11.5.

## Annelida

POLYCHAETA

The Gulf of Mexico is home to hundreds of species of polychaete worms, many of which find suitable living areas on or around hard-substrate environments. Here they occupy a variety of habitats and exhibit many different lifestyles. Most species are small and inconspicuous, living by day in nooks and crevices and becoming active at night. Some species bore or tunnel into soft or spongy carbonate rocks, coral heads, and algal concretions where they are relatively safe from predators. Others are found in dense growths of algae or sponges or among tangles of hydroids and other bushy life forms. Several groups, including some members of the families Sabellidae and Serpulidae, construct tubes of parchment, cemented sand grains, or calcium carbonate and live attached to the substrate. However, one species, the fire worm (*Hermodice carunculata*), is large and conspicuous and moves around reefs and outcrops at dawn and dusk. This animal escapes predation not by hiding, but by means of a powerful defense mechanism. Throughout the length of the body on the leglike parapodia there is a double row of tiny, highly toxic spines or setae that are normally retracted into pouches. When disturbed, the animal extends these spines, which break off and inflict painful wounds to any animal touching the worm. Polychaetes of hard substrates occupy all types of feeding niches. Some, such as the large fan worm (*Spirobranchus giganteus*), collect plankton by means of highly branched tentacles extending up into the water column. Some browse on algae, sponges, coral polyps, hydroids, or other attached forms. Others, including the fire worm, are predators on small mobile invertebrates, or they scavenge dead animal remains and other organic detritus found on the bottom. Important genera include *Amphitrite, Armandia, Dorvillea, Eunice, Harmothoe, Hermodice, Sabellaria,* and *Spirobranchus.* Two of these are shown in figure 11.6. References to hard-substrate polychaetes are Ebbs (1966), Hubbard (1977), McCarty (1974), and Wills and Bright (1974).

## Arthropoda

Although virtually all major types of marine arthropods have been reported around hard-substrate environments

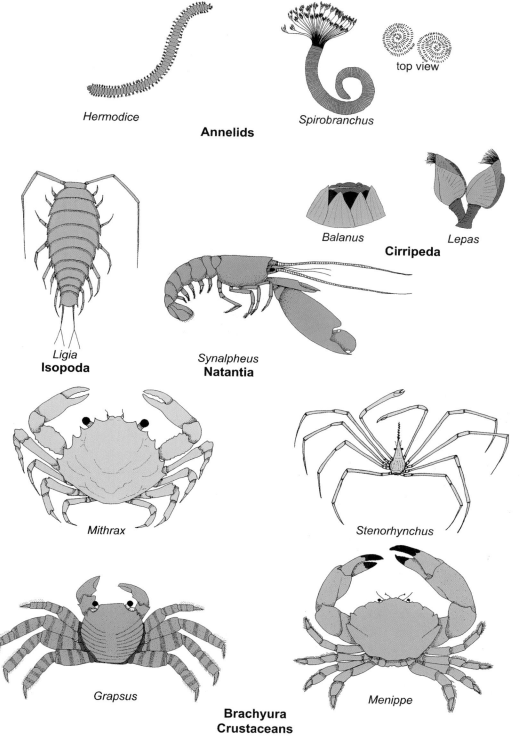

**Annelids**

*Hermodice*

*Spirobranchus*

top view

**Cirripeda**

*Balanus*

*Lepas*

*Ligia*
**Isopoda**

*Synalpheus*
**Natantia**

*Mithrax*

*Stenorhynchus*

*Grapsus*

*Menippe*

**Brachyura**
**Crustaceans**

**Figure 11.6.**
Representative examples of annelids and crustaceans associated with hard substrates of the Gulf of Mexico. Crustaceans include isopods, cirripedes, natantians, and brachyurans.

of the Gulf, many are small and rather inconspicuous. The larger or more prominent forms include mantis shrimps, an isopod, barnacles, and several groups of decapods (swimming shrimps, hermit crabs and their relatives, lobsters, and true crabs). Only these groups are discussed below.

### STOMATOPODA

Although most species of mantis shrimps of the Gulf inhabit soft bottoms, a few occur in the nooks and crevices of coral and algal reefs, most notably *Gonodactylus bredini,* for which the reef is its primary habitat. L. Pequegnat and Ray (1974) recorded both *G. bredini* and *G. lacunatus* from the West Flower Garden Bank of the northwestern Gulf.

### ISOPODA

Among the many species of isopods that occur around hard-substrate areas, only the sea roach (*Ligia*) is considered here because it is so conspicuous. This animal, a distant relative of the terrestrial sow bug, is found all around the periphery of the Gulf on docks, seawalls, jetties, and other hard substrates that project from salt water. Often mistaken for insects, these small crustaceans scurry around the supralittoral zone, particularly at night and during cooler parts of the day. They are never found far from water, to which they must return periodically to moisten their gills. Several species are represented, only one of which (*L. exotica*) is widespread. This is illustrated in figure 11.6. Certain isopods, such as species of *Limnoria* and *Sphaeroma,* bore into the wood of piers, pilings, and even the roots of mangrove trees and are responsible for considerable damage in some areas.

### CIRRIPEDIA

Barnacles are fairly primitive crustaceans distantly related to ostracods. The anterior end possesses cement glands that produce a powerful adhesive by which the animal attaches itself to the substrate. Compared to that of the ostracod, the body of the barnacle is upside down, and on the outside it typically has a series of calcareous plates, the actual number varying from one group to another. Surrounding the body proper is a carapace or mantle, open at the top, and through this opening the elongated feathery thoracic appendages can project to sweep plankton particles from the surrounding water. Except for some parasitic species, barnacles are the only sedentary crustaceans. All species are marine, although some do live in brackish water. Barnacles are divided into two major groups, the stalked forms (goose barnacles) and the unstalked or sessile forms (acorn barnacles). Most species live attached to rocks, pilings, driftwood, and other non-living surfaces (including artificial substrates), but many also attach to fishes, sea turtles, whales, and even some invertebrates. Important genera in the Gulf include *Balanus, Chthamalus, Conchoderma, Lepas, Megabalanus,* and *Verruca.* In addition, there is the rock-boring form, *Lithotrya.* Representative examples are given in figure 11.6. Key references are Bierbaum and Zischke (1979), Gittings (1985), Gittings, Dennis, and Harry (1986), D. Henry (1954), Spivey (1981), and Wells (1966).

### DECAPODA

As noted in the previous chapter, the decapods all have five pairs of walking legs, and some are modified into pincers or chelae. This group includes the Natantia, Macrura, Anomura, and Brachyura. References dealing with hard-substrate species include Felder (1971, 1973), Felder and Chaney (1979), Kropp and Manning (1987), L. Pequegnat and Ray (1974), E. H. Powell and Gunter (1968), Powers (1977), and J. Ray (1974).

#### Natantia

The swimming shrimps have large abdomens with the abdominal appendages modified for swimming. All are mobile, and around coral reefs and other rocky areas many species move about the bottom or hide in crevices. Of particular note are the pistol shrimps (*Alpheus, Synalpheus,* etc.), which stun small prey with explosive snapping sounds produced by an enlarged chela. Hydrophones lowered into the water around reefs pick up sounds like sizzling bacon produced by the incessant chorus of snapping shrimps. One example is shown in figure 11.6. Another interesting group is the cleaning shrimps (*Periclimenes*), which, instead of hiding, advertise their presence from prominent locations on top of sponges, coral heads, and other high places. Fish species recognize them as helpful cleaners, rather than prey, and solicit their services for removing external parasites from the skin and mouth cavity. Important genera of swimming shrimps include *Alpheus, Latreutes, Leander, Periclimenes, Processa, Synalpheus,* and *Thor.*

#### Macrura

This group includes the lobsterlike forms with well-developed walking legs and often with powerful chelae.

Several species occur in crevices around reefs and rocky areas and are active primarily at night. Prominent are the spiny lobster (*Panulirus*) and shovel-nose lobster (*Scyllarides*), shown in figure 10.7 of the previous chapter.

### Anomura

In hermit crabs and their relatives, the abdomen is either twisted to one side or flexed beneath the thorax, and most species possess enlarged chelae. Hermit crabs house themselves in discarded snail shells and crawl around the bottom, often around reefs and rocky outcrops. These include the genera *Dardanus, Paguristes, Pagurus,* and *Polypagurus,* among others. Related to hermit crabs are the galatheids, which do not utilize snail shells. Most are deepwater forms, but several species of the genus *Munida* have been recorded around reefs and outcrops of the outer shelf.

### Brachyura

True crabs have wide, flattened bodies, and the abdomen is flexed tightly against the thorax. The first pair of walking legs is developed into strong chelae. Around shallow rocky areas, particularly in the southern Gulf, one encounters the agile Sally Lightfoot crab (*Grapsus*), which dwells in the supralittoral, littoral, and sublittoral zones where it consumes algae and organic detritus picked from rock surfaces and crevices. The stone crab (*Menippe*) inhabits crevices of jetties and other rocky areas of the sublittoral zone. Its powerful pincers are used to crack or chip away at the shells of the bivalves it consumes. Deeper reefs and outcrops are home to a number of crabs including species of *Domecia, Mithrax, Pilumnus,* and the strange spider-like arrow crab, *Stenorhynchus.* Several hard-substrate crabs are shown in figure 11.6.

## Bryozoa

The moss animals, often placed in the phylum Ectoprocta, are fairly primitive organisms related to ancestors of the echinoderms. In common with two smaller phyla, the Phoronida and Brachiopoda (not discussed here), they possess a special food-gathering organ called a *lophophore.* This circular or horseshoe-shaped structure bears ciliated tentacles used for capturing plankton. An individual bryozoan animal is quite small, generally less than one millimeter in length, but most species live in upright or flat encrusting colonies that may be tens of centimeters in height or diameter. Although a few species inhabit freshwater, most live in the ocean, where they are widespread on hard substrates of coastal waters. A few have been reported from the deep sea. Some, such as *Bugula,* grow in upright bushy colonies that superficially resemble seaweeds or colonies of hydroids. Others, such as *Membranipora,* develop as flat, lacy, encrusting sheets that grow across and cover large surfaces of rocks, wooden pilings, algal fronds, seagrass blades, and other solid substrates. In many species the body wall is calcified. Because of their widespread distribution, tendency to overgrow on surfaces and other species, and difficulty of removal, this is a fairly important group of biofouling organisms. Prominent genera in the Gulf of Mexico include *Bugula, Cupuladria, Membranipora, Micropora, Microporella,* and *Schizoporella.* Examples of erect and encrusting colonial forms are given in figure 11.7. References to Gulf species include Cropper (1973), Lagaaij (1963), Leuterman (1979) Osburn (1954), and D. Shier (1964).

## Echinodermata

Although many species of echinoderms have been reported from coastal and offshore hard substrates of the Gulf, this fauna has not been thoroughly examined. Many of the species are cryptic, and most are active primarily at night. The really conspicuous species are generally the sea urchins, *Arbacia* and the rock-boring *Echinometra* in coastal waters, and *Diadema* and *Eucidaris* on offshore reefs and banks. Even these are primarily nocturnal. However, *Diadema,* the long-spined black sea urchin, is especially noted by divers seeking to avoid its long toxic spines. Probably the most numerous and diverse group is the brittle stars. Important genera within the five classes are the following:

Asteroidea (starfishes and relatives): *Asterinopsis, Astropecten, Linckia, Ophidiaster,* and *Narcissia*
Echinoidea (sand dollars, sea urchins, and relatives): *Arbacia, Clypeaster, Diadema, Eucidaris, Meoma,* and *Stylocidaris*
Ophiuroidea (brittle stars): *Astrocyclus, Astrophyton, Ophiactis, Ophiocoma, Ophioderma,* and *Ophionereis*
Holothuroidea (sea cucumbers): *Isostichopus*
Crinoidea (sea lilies): *Comactinia, Crinometra,* and *Ctenantedon*

Examples are shown in figure 11.7. References to hardsubstrate echinoderms include Burke (1974a, 1974b), Dubois (1975), Fairchild and Sorensen (1985), and Shirley (1974).

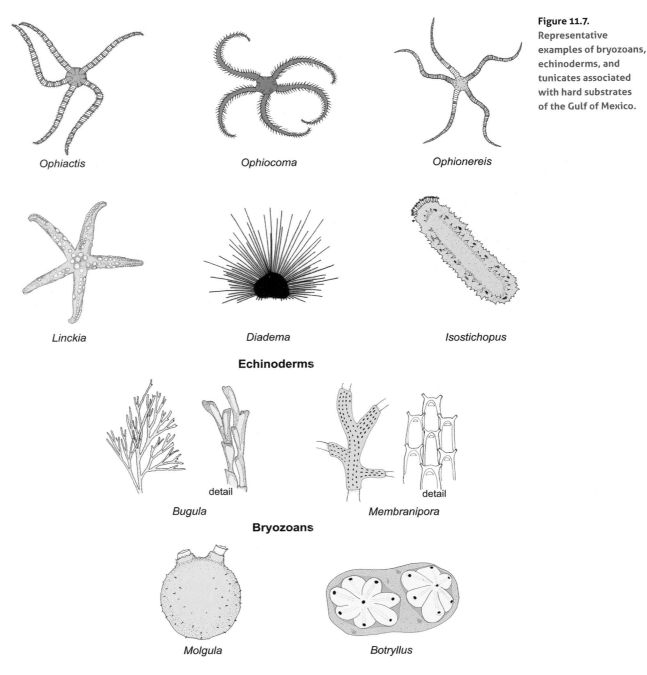

**Figure 11.7.**
Representative examples of bryozoans, echinoderms, and tunicates associated with hard substrates of the Gulf of Mexico.

*Ophiactis*          *Ophiocoma*          *Ophionereis*

*Linckia*          *Diadema*          *Isostichopus*

**Echinoderms**

detail                    detail

*Bugula*          *Membranipora*

**Bryozoans**

*Molgula*          *Botryllus*

**Tunicates**

## Chordata

Three subphyla of chordates are recognized, but only two, the Urochordata and Vertebrata, are considered here.

### UROCHORDATA

The larval urochordate possesses all the typical chordate characteristics (notochord, dorsal tubular nerve cord, and pharyngeal gill slits), but in the highly modified adults some of these characters are lost. All urochordates are marine, and most are planktonic. However, one group, the ascidians or sea squirts, is sessile, and most of these live attached to hard substrates. They may be solitary or colonial. The body of the adult sea squirt is covered by a tough leathery sac or tunic with two openings, one for water to enter and the other for water to exit. Thus, the animal is able to pump a constant stream of seawater

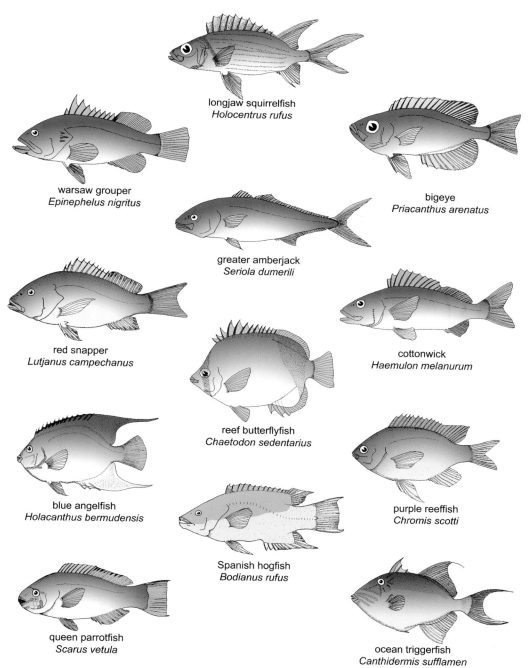

**Figure 11.8.** Representative examples of fishes often associated with reefs and other hard substrates of the Gulf of Mexico.

longjaw squirrelfish
*Holocentrus rufus*

warsaw grouper
*Epinephelus nigritus*

bigeye
*Priacanthus arenatus*

greater amberjack
*Seriola dumerili*

red snapper
*Lutjanus campechanus*

cottonwick
*Haemulon melanurum*

reef butterflyfish
*Chaetodon sedentarius*

blue angelfish
*Holacanthus bermudensis*

purple reeffish
*Chromis scotti*

Spanish hogfish
*Bodianus rufus*

queen parrotfish
*Scarus vetula*

ocean triggerfish
*Canthidermis sufflamen*

and filter out plankton for food. The group is successful only in clear water that is relatively free of suspended inorganic particles. Genera of sea squirts recorded from hard substrates of the Gulf include *Amaroucium, Botryllus, Clavelina, Didemnum, Eudistoma, Molgula,* and *Styela,* among others. Examples of solitary and colonial forms are depicted in figure 11.7. This group has not been well studied in the Gulf, but the following reports are useful:

Flores-Coto (1974), Plough (1978), N. Rabalais (1982), and Van Name (1954).

### VERTEBRATA

All vertebrate animals are mobile, and although many species associate with hard substrates, none are sedentary. A variety of fish species (fig. 11.8) find food or shelter around coastal piers, jetties, and breakwaters as well as

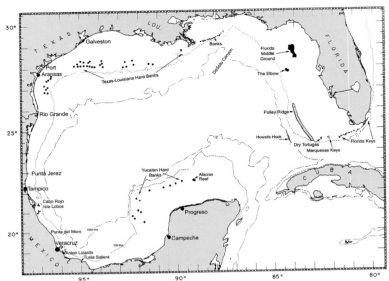

**Figure 11.9.** Map of the Gulf of Mexico showing the location of prominent islands, hard banks, and reefs of the continental shelves, as well as shoreline features mentioned in the text.

around offshore outcrops, natural and artificial reefs, and petroleum platforms. Sea turtles also occasionally find food or refuge around offshore structures, but marine mammals are not involved. These groups will be considered in detail in chapter 13, on nekton.

## DISTRIBUTION OF HARD-SUBSTRATE HABITAT

Before discussing biological development on hard substrates, it is desirable to review the distribution and nature of such substrates along the coastline and on continental shelves of the Gulf. The important features are shown in figure 11.9, which will also provide a useful reference for the biological discussions. Such features include rocky shoreline exposures, islands, and major submerged banks, outcrops, and reefs of the continental shelves. Not shown in the figure are the various human-made structures or the myriad smaller underwater outcrops.

### Coastal hard substrates

Prior to European discovery and settlement, few coastal areas of the Gulf were bordered by rocky outcrops. Broad beaches of siliceous sand, mixed with various amounts of silt and clay minerals and carbonate shell fragments, lined the shores of most mainland areas and barrier islands. On the low-lying coast of Florida north of Tampa Bay where wave energy is reduced, the Gulf was bordered largely by salt marshes and some mangroves and fronted by seagrass meadows. This was also true for much of the Louisiana coast, although here the seagrasses were absent. Around the Mississippi River Delta extensive mudflats and dense stands of marsh grasses prevailed. In

lowlands of southwestern and southern Florida, the Everglades met the Gulf with tangled swamps of mangroves. A few limestone outcrops occurred along the Mexican coastline near Punta Jerez, north of Tampico, and at local sites on the western and northern coasts of the Yucatán Peninsula near Campeche and Progreso. Igneous outcrops resulting from ancient lava flows protruded into the southern Gulf near Veracruz, that is, at several points near Punta del Moro (north of the city) and at the face of the San Andrés Tuxtla mountains, the Tuxtla Salient (well south of the city).

During the past century numerous human-made structures of concrete, limestone, or granite have been added around the periphery of the Gulf. In some locations seawalls and breakwaters provide protection again the erosive force of waves. The mouths of most navigable passes and rivers are now protected by pairs of parallel jetties extending into the Gulf a mile or more to prevent shoaling and closure of the channels by sediments transported by alongshore currents. These new structures now provide hard-substrate habitat all around the shoreline, habitat that hardly existed in the Gulf two centuries ago.

### Islands and emergent reefs

Limestone exposures as well as intertidal and subtidal reefs occur along the south Florida Keys and their extension, the nearby Marquesas Islands and Dry Tortugas. Off the Mexican coast below Tampico and Cabo Rojo are situated a series of small islands and reefs of the inner shelf, notably Isla Lobos and Blanquilla, Tanguijo, and Enmedio Reefs. Similar series are located off Veracruz (Blanquilla and other reefs) and off Antón Lizardo to the south (Isla de Media, Enmedio Reef, and others). On the middle and outer shelf off the western and northern coasts of the Yucatán Peninsula rises a series of carbonate islands and hard banks (Logan 1969). The most prominent are the island groups and reef tracts known as Arcos, Triangulos, Nuevo, Arenas, and Alacrán. Among the deeper submerged banks are the Obispo Shoals, Banco Nuevo, Banco Ingleses, and Nuevo Reef. Barrier reefs have also been reported along the narrow shelf of the northern coast of Cuba (W. A. Price 1954).

### Deeper banks and reefs

On the continental shelf of the northwestern Gulf (Mississippi River Delta to the Rio Grande), more than three dozen hard banks are known (Rezak, Bright, and McGrail 1985). Although a few low-relief features, such as Seven and One-Half Fathom and Sebree Banks off south Texas and Heald Bank off Galveston, arise from the inner shelf, most are located on the middle and outer shelf areas. Some, particularly toward the south, are remnants of old consolidated beach dunes, lake-bottom sediments, or drowned coral/algal reefs that developed during lower stands of the sea. Most of the deeper banks off the north Texas and Louisiana coasts represent the stony tops of salt domes or diapirs. These may rise 75 m or more from the surrounding bottom.

East of the Mississippi Delta on the middle and outer shelf are located a series of topographic features described as ridges, "patch reefs," flat-topped reefs, and pinnacles with elevations of up to 20 m (Rezak et al. 1989). Most appear to be remnants of consolidated beach dunes and other features developed during periods of lower sea level stands, and their surface is sufficiently durable to support hard-substrate flora and fauna. Large rectangular slabs of exposed limestone have been reported around the rim of DeSoto Canyon (Shipp and Hopkins 1978). In addition to these more prominent features, there are many largely unmapped smaller rocky outcrops along the continental shelf of the northern Gulf. These are discovered primarily by shrimpers when their nets become snagged. Similar outcrops occur along the terrigenous bottoms of the Mexican shelf, particularly between Tampico and Veracruz, where much of the area is considered to be too rocky for effective bottom trawling.

On the limestone platform off west Florida, low-relief (≤ 1 m) outcrops, terraces, and ledges as well as massive living and dead coral heads are scattered on the middle and outer shelf, but three areas are of particular note. The Florida Middle Ground, lying in the depth range of 40–50 m northwest of Tampa Bay, occupies an area of around 1,500 km² and displays a variety of carbonate exposures, terraces, ridges, and pinnacles up to 15 m in vertical relief. Like the features off Mississippi and Alabama, the taller structures are probably relict beach dunes from lower sea level stands. On the middle shelf west of Tampa Bay a smaller outcrop area is known as the Elbow. Off southwest Florida between the 70 and 100 m isobaths lies a large carbonate system known as Pulley Ridge. This 10 km wide feature extending from the level of the Florida Keys northward to the level of Charlotte Harbor consists of a series of parallel ridges, some now covered by sediments, representing ancient beach formations or coral/algal reefs. Finally, as noted earlier, a series of carbonate hard banks line the middle to outer shelf off the Yucatán

Peninsula. From this brief survey it is clear that the continental shelves of the Gulf display a remarkable array of natural hard-substrate exposures, and these occur in a variety of environmental settings.

## Artificial substrates of the continental shelf

During the past half century, humans have added two classes of artificial substrates to the continental shelf environment, emergent and submerged. Emergent structures include the various oil and gas platforms placed primarily on the north-central Gulf shelf, but these are also scattered elsewhere, with a secondary cluster in the Bay of Campeche of the southern Gulf. More than 1,200 are located on the continental shelf and slope off Louisiana alone, and another 300 or so occur off east Texas. The steel legs of these structures provide near-surface hard-substrate habitat for biofouling organisms and their ecological associates that stretches from the shoreline into the deep Gulf. This type of habitat was never before available anywhere in the Gulf of Mexico.

Artificial bottom habitats are of several types. On the sea floor around the bases of the steel legs that support the platforms lie piles of rocky drill cuttings (composed primarily of limestone chips). Oil and gas pipelines connecting platforms with each other and with the shore are initially buried in the bottom sediments, but over time many become exposed, adding additional hard substrate. Until the early 1970s, concrete rubble and other materials were freely dumped onto the continental shelf. Since 1972, federal law has prohibited unregulated dumping, and all dumping is now regulated by the US Army Corps of Engineers and the Environmental Protection Agency, from whom permits must be obtained. During the Second World War, German U-boats sank 55 merchant ships in the northern Gulf, mostly near the mouth of the Mississippi River. These and other wrecks are scattered around continental shelves of the Gulf, but they are most concentrated in the north-central sector. The military has established dump sites for unexploded ordnance at several locations, mostly in the deep Gulf, but one site near DeSoto Canyon off Pensacola, Florida, also includes part of the continental shelf. Although such dumping has been discontinued, this old material remains on the bottom. Finally, the various states bordering the Gulf have adopted a policy of establishing artificial reefs as attractants for fishes and other marine life. The reef material varies from clusters of old automobile tires, concrete rubble, and car bodies to old ship hulls. In addition, the

US Department of the Interior, in its "Rigs to Reefs" program, now permits the dumping of disused oil and gas rigs in designated areas of the northern Gulf shelf. These various human activities have vastly increased the extent of hard-substrate habitat on continental shelves, particularly in the north-central sector, but also in scattered locations elsewhere around the Gulf. While this material does provide habitat for many attached hard-substrate species, and it is definitely an attractant for a variety of fishes, the full biological impact on ecosystems of the Gulf has not yet been thoroughly assessed.

## HARD-SUBSTRATE COMMUNITIES OF THE GULF
### Flora and fauna of shoreline habitats

Of all the hard-substrate habitats around coastlines of the Gulf of Mexico, the stone jetties at Port Aransas, Texas, have been the most thoroughly studied. The biota of this area will be described in some detail, and this description will serve as a model against which other shoreline assemblages may be compared. Primary references to the Port Aransas jetty flora and fauna include Baca, Sorensen, and Cox (1979), Britton and Morton (1989), G. Edwards (1969), P. Edwards and Kapraun (1973), Hedgpeth (1953), Kapraun (1970, 1974, 1980), N. Rabalais (1982), and Whitten, Rosene, and Hedgpeth (1950). References to related shoreline areas of the northern Gulf are Fairchild and Sorensen (1985), Gunter and Geyer (1955), and Lowe and Cox (1978).

SOUTH TEXAS (PORT ARANSAS) JETTIES

The jetties themselves are trapezoidal in cross section, with a broad base (ca. 50 m wide), sloping sides, and narrow crest (ca. 4 m). Over a core of smaller stones, the jetties are faced with huge rectangular slabs of hard granite weighing up to 10 tons each. The flat crest may be capped with a concrete ramp 2 to 4 m wide, although some areas contain remnants of an asphalt roadway now in a state of disrepair. Cracks and crevices between facing stones of the side walls are sheltered from heavy wave action and intense radiation, and these provide habitat for more sensitive or secretive species. Channels separating the jetties vary in width from 800 m to almost 2 km. The south jetty at Port Aransas, which has been most thoroughly studied, extends from the shoreline into the Gulf for a distance of about 1.4 km. Tidal range in the area varies with the season and is lowest in January and February and highest in the summer months.

The general nature of floral and faunal zonation at

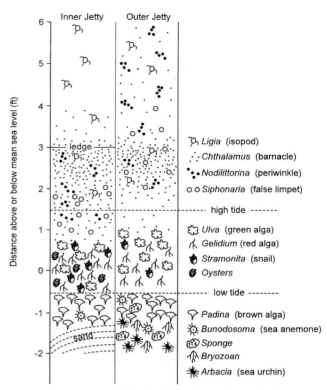

**Figure 11.10.** Patterns of floral and faunal zonation on the inner (nearshore) and outer (offshore) walls of the jetties at Port Aransas, Texas, during late May. (Modified from Hedgpeth 1953.)

two stations (nearshore and offshore) on the south jetty at Port Aransas is shown in figure 11.10. At the nearshore station, where wave heights are low and salinity is often less than that of the open Gulf, the supralittoral zone has no large algae. Microscopic filamentous blue-green algae (*Lyngbya,* etc.) are present primarily in the lower portion just above and below high-tide level. A protective gelatinous matrix provides this alga some protection against desiccation. Other species of microscopic algae are *endolithic,* that is, they reside within the rocks or concrete as a thin layer just beneath the surface. Animals characteristic of the supralittoral include two species of algae-eating snails (*Nodilittorina* and *Siphonaria*), sedentary barnacles (*Chthamalus*), and the isopod *Ligia,* which was inadvertently introduced around the world on wooden hulls of old sailing ships. The snails and barnacles are most concentrated in the lower portion of this zone, but the isopods stray well above the water's edge feeding on algae, carrion, and detritus particles.

The littoral or intertidal zone is habitat for several species of leafy algae (the sea lettuce *Ulva,* red alga *Gelidium,* etc.), snails (the oyster drill *Stramonita* and others), and young oysters (*Crassostrea* and *Ostrea*). Small worms and crustaceans (isopods and amphipods) abound. In general, life is much more abundant here than in the zone above, and this is largely because of the increased amount of time in contact with the water. Gulls, terns, and other seabirds often forage here. In the sublittoral zone of the inner jetty are found the brown alga *Padina* and the sea anemone *Bunodosoma.* The lower limit of growth is restricted not only by the proximity of the sandy bottom, but also by a zone of sand scour unfavorable for settlement and growth of attached species.

Biological zonation on the outer jetty follows the same basic pattern as that observed at the nearshore station, but there are differences. At the offshore station, where waves are higher and more intense and where sea spray hits higher on the jetty wall, the supralittoral zone is much wider and may extend up to 1.5 m above the high tide line. Dominant species of the supralittoral and littoral zones differ little from those at the nearshore station except that young oysters are absent from the more saline waters of the outer jetty. However, the biota of the sublittoral zone of the outer jetty is far richer in species composition. Although *Padina* remains the dominant algal genus, it shares the habitat with many others. The attached fauna includes several species of sponges, hydroids (*Bougainvillia, Obelia,* and others), sea whips (*Leptogorgia*), sea anemones (*Aiptasiomorpha, Anthopleura, Bunodosoma*), bryozoans, and tunicates. In addition, two types of stony corals (*Astrangia* and *Oculina*) form small colonies on rocks of the lower jetty wall. Crawling among the bush-like growths are a variety of worms and small crustaceans.

Among the more mobile forms of the outer jetty are several types of snails (*Siphonaria* and *Stramonita,* mentioned earlier, plus *Anachis, Caecum, Cantharus,* and others, and the large sea hare *Aplysia,* whose shell is internal). Hermit crabs (*Clibanarius, Pagurus,* and *Petrochirus*) scavenge the subtidal walls, as do the related porcelain crab (*Petrolisthes*) and a true crab (*Pachygrapsus*). The black sea urchin (*Arbacia*) is found in crevices and on outer surfaces of the submerged wall, and the large stone crab (*Menippe*) and common octopus (*Octopus vulgaris*) live among crevices of the lower jetty wall. At least 15 or 20 different kinds of fishes forage around the outer jetty. It is remarkable that such a thriving and diverse community should develop in a habitat that hardly existed a hundred years ago. The species are derived from nearby bays and estuaries as well as the open shelf, and some have likely arrived as sporelings or larvae from sources in the southern Gulf or more distant Caribbean.

MARINE ALGAE OF NORTHWESTERN GULF JETTIES
Since the algal flora is a very diverse group and has been studied in some detail, it can aid in placing the total jetty biota in clearer perspective. Around 90 different species of marine algae are now known from shoreline structures of the northwestern Gulf, and based on temperature tolerances and geographical affinities they fall into three general groups: cool temperate, tropical, and cosmopolitan. In nearshore waters of the northwestern Gulf, winter temperatures often fall below 20°C, especially off Louisiana and upper Texas, but summertime temperatures all along the coast reach about 30°C or a bit higher. The seasonal progression of temperature changes (perhaps augmented by differences in day length) is of great importance in determining the seasonal and geographic patterns of algal distribution. Cool-temperate species (of the genera *Bangia, Cladophora, Petalonia, Porphyra,* etc.) are related to those of the mid-Atlantic coast, where they extend as far north as Cape Cod and beyond. Within the Gulf these forms are winter residents, common in the north-central Gulf but diminishing toward south Texas. Tropical species of *Botryocladia, Ceramium, Dictyota, Padina,* and others are prevalent in the Caribbean, and many of these reach their northern distributional limits along the northern Gulf coast, where they are summer residents. Most of the red algae belong to this group. Cosmopolitan species (of the genera *Acrochaetium, Gelidium, Gracilaria, Grateloupia, Hypnea, Pterocladia,* etc.) range from Cape Hatteras to Brazil and beyond. These species have broad temperature tolerances and may be resident in the northern Gulf throughout the year. Thus, there is a definite seasonal shift in species composition, with cool-temperate and cosmopolitan species dominating the winter flora and tropical plus cosmopolitan species present in the summer months. This shift is most pronounced in the north-central Gulf. Most of the species recorded from Galveston jetties are cool temperate, and at least 25 of these are not found farther south. At Port Aransas tropical species are well represented, and at Port Isabel near the Mexican border they are the dominant flora. Clearly, the Texas coast represents a zone of transition between the temperate and tropical floras, and this change continues along the Mexican coast. A similar transition is seen along the west coast of Florida. This general pattern of change from temperate to tropical species, dem-

onstrated so clearly by the algae, also applies to the various invertebrate groups and even the fish fauna. Although temperature appears to be the primary factor controlling algal distribution, other factors are also important. Differences between the coastal flora of Louisiana and that of Galveston may reflect, in part, the more turbid and less saline nearshore waters of Louisiana. At the local level algal distribution may be affected by such factors as light intensity, wave exposure, tolerance to desiccation, and tidal fluctuations. Biological factors of interspecies competition and grazing by herbivores also come into play.

BIOTA OF MEXICAN SHORELINES

The flora and fauna of hard substrates along the Mexican coast from the Rio Grande to Cabo Catoche on the Yucatán Channel have received only a little study, the most important references being Britton and Morton (1989), Humm (1952), W. Taylor (1954a, 1954b; 1960), Vokes and Vokes (1983), and Wiley, Circe, and Tunnell (1982). However, enough is known to place the biota of this long stretch of shoreline in perspective. Proceeding southward from the Rio Grande, cool-temperate species become quite scarce, and in the southern Gulf they are replaced by tropical Caribbean species. The most luxuriant development of tropical shoreline biota of the southwestern Gulf occurs on the several igneous outcrops of the Punta del Morro coastline north of Veracruz. The flora and fauna of this area have been described in some detail by Britton and Morton (1989) following the work of Wiley, Circe, and Tunnell (1982). The shoreline community of the Punta del Morro area will serve as a model for the southern Gulf.

As on the Texas jetties, the supralittoral is populated by the ever-present isopod (*Ligia*) and the lined periwinkle (*Nodilittorina*). Additional snails include other species of this genus as well as representatives of other genera (*Littorina*, *Nerita*, and *Planaxis*). The tropical Sally Lightfoot crab (*Grapsus*) clings to emergent walls and scampers around rocks of the supralittoral and littoral zones. The littoral comprises three recognizable subzones. The upper littoral is marked by the densest concentration of barnacles (mostly *Chthamalus*) and several snails (*Nerita*, *Siphonaria*, and *Stramonita*). The midlittoral contains clusters of tube-dwelling polychaete worms (*Sabellaria*) and a number of keyhole limpets and other tropical snails (*Acmaea*, *Diodora*, *Fissurella*, *Planaxis*, *Purpura*, and others). Barnacles also extend into this subzone. The lower littoral is characterized by dense growths of marine algae (dominated by *Caulerpa*, *Padina*, *Sargassum*, and

*Ulva*, but including many others). Among these algal growths crawl several types of small snails (*Caecum*, *Fossarilittorina*, *Fossarus*, *Nitidella*, *Planaxis*, and *Tricholia*). Additional snails (*Petaloconchus*, *Purpura*, *Stramonita*, etc.) are found on the rocky substrate. Also occurring in the lower littoral are a number of bivalves (including *Brachidontes* and *Isognomon*) and at least three species of amphineurans (*Tonicia* and others).

The sublittoral zone has not been as thoroughly studied. Here are located large numbers of the colonial sea anemone (*Palythoa*), two sea hares (*Aplysia*), and several species of bivalves (*Arca*, *Arcopsis*, *Barbatia*, *Brachidontes*, and *Isognomon*). The black sea urchin of the northern Gulf (*Arbacia*) is absent and is replaced by the rock-boring red sea urchin (*Echinometra*). Cephalopods (*Octopus* and *Spirula*) also occur in the sublittoral. Altogether, 121 species of mollusks have been recorded from the Punta del Morro area, including 80 gastropods, 36 bivalves, 3 amphineurans, and 2 cephalopods (Wiley, Circe, and Tunnell 1982). Some of the species are quite abundant and undoubtedly constitute local breeding populations, whereas many others are scarce and probably represent survival of individuals from chance recruitment of planktonic larvae from distant breeding populations. When the sublittoral has been more thoroughly investigated, the list of polychaetes, crustaceans, and other groups will certainly increase.

The flora and fauna of limestone outcrops and human-made structures around the Yucatán coast are characterized by fewer species, and the composition of the communities varies from place to place depending on substrate type, wave exposure, nearby pollution sources, and other factors. However, all represent variations of the richer Punta del Morro community, and all are made up largely of tropical and cosmopolitan species. Some of the plants and animals of shoreline habitats in the southwestern and southern Gulf maintain endemic populations, whereas others appear to be periodic colonizers from the Caribbean or perhaps from islands of the Yucatán shelf (Alacrán Reef, Cayo Arenas, etc.). Whatever their sources and despite their diversity, the Punta del Morro and related biotas represent only a fraction of the species potentially available from the far richer flora and fauna of the Caribbean.

BIOTA OF SOUTH FLORIDA SHORELINES

Few studies have been carried out on the hard-substrate flora and fauna of the northeastern and eastern shores of the Gulf, the most thorough being the early work of Stephenson and Stephenson (1950) on the Florida Keys. More recently marine algae of this area have been investigated by Croley and Dawes (1970), Dawes, Earle, and Croley (1967), and Mathieson and Dawes (1975), and the intertidal barnacles have been examined by Bierbaum and Zischke (1979). The work of Stephenson and Stephenson (1950) will serve as the model for the south Florida shoreline biota, and this will be supplemented with information from the more recent investigations.

The Florida Keys represent low, more or less flat limestone platforms rising only a few meters above sea level. The more elevated upper platform of the interior is well vegetated with trees, shrubs, and herbaceous plants, and the strand vegetation often extends down to the lower platform near the water. The width, slope, and configuration of the rocky shoreline vary from place to place. In some areas it is more than 20 m wide and descends gradually or in steps or terraces to the water's edge. In other areas it may be only a few meters wide or end abruptly at the edge of the sea as a wall or undercut ledge. Sometimes the rock is flat, and in other cases it is broken or dissected. In places the shoreline is muddy or sandy and fringed by mangrove thickets. Whatever its horizontal distance, configuration, and composition, the height of the supralittoral is only about 1.5 m above sea level. Although biological development varies somewhat with the nature of the rocky shoreline, the basic pattern of zonation remains rather constant. The white limestone of the upper shoreline gives way to bands of gray and black near the water's edge, and to yellowish in the upper littoral zone. The darker bands of the supralittoral are apparently due to the presence of microscopic endolithic blue-green algae. Following Stephenson and Stephenson (1950), the shoreline zonation is classified as follows:

supralittoral—upper (white), middle (gray), and lower (black)
littoral—upper (yellow), lower ("lower platform")
sublittoral—"reef flat"

The zonal distribution of the marine algae is shown in table S41, and the distribution of marine animals is given in table S42. No effort has been made to update the taxonomy except in the case of some of the snails (*Dendropoma, Echinella,* and *Nodilittorina*). As seen in table S41, only two genera of macroscopic algae occur in the exposed supralittoral environment. *Bostrychia* extends through the middle and lower supralittoral, but *Polysiphonia* is found only in the lower subzone. By contrast, 15 genera are encountered in the less arid littoral zone, 11 in the upper and 10 in the lower littoral. Seventeen occur in the sublittoral, although a number of these grow on the sandy bottoms rather than hard substrates. In this subtropical locality red algae are numerically dominant, but green algae are well represented, and a few browns are present. Among the animals, the isopod *Ligia* occurs throughout the supralittoral and is the only marine animal found in the upper part of this zone. Eight genera of snails (including *Littorina, Nerita,* and *Nodilittorina,* among others) inhabit the middle supralittoral, and the same number occurs in the lower subzone together with the small crab *Pachygrapsus.*

Faunal diversity is much greater in the littoral zone, with 22 genera represented. Here are found several sea anemones, a chiton, and various snails, as well as a few bivalves, barnacles (*Chthamalus* and *Tetraclita*), a crab, and the red sea urchin *Echinometra.* For reef flats of the sublittoral zone, 33 kinds of animals are listed. These include sponges, gorgonians, sea anemones, stony corals, a chiton, several snails and bivalves, a crab, and 7 genera of echinoderms. As in the case of the algae, some are soft-bottom species, but most are associated only or primarily with hard substrates. Here, for the first time, we encounter the beginnings of the true tropical undersea "gardens" of sponges, gorgonians, stony corals, and their associates that characterize true coral reefs, to be examined later in the chapter.

Croley and Dawes (1970) carried out a detailed study of algal zonation of the middle Florida Keys (Content Keys) during a period of two and a half years. Of the 158 taxa identified, 58.1 percent were red, 30.6 percent green, and 11.2 percent brown algae. The zonation pattern described by Stephenson and Stephenson (1950) was found to be roughly applicable to the Content Keys area, but here there was also a well developed "littoral fringe" zone at the upper edge of the littoral characterized by dense growths of the algae *Bostrychia, Catenella, Caulerpa,* and *Gardneriella.* Croley and Dawes (1970) also divided the sublittoral into two subzones. The upper sublittoral, from mean low sea level to a depth of 5 m, contained perennials (*Dictyota, Eucheuma, Laurencia, Padina,* and *Sargassum*) and a number of ephemeral or sporadic forms. The lower sublittoral, between 5 and 8 or 10 m depth, had a number of soft-bottom algae as well as those attached to hard substrates (*Caulerpa, Chamaedoris, Gloiophloea, Halimeda, Halymenia,* and *Trichogloea*). The sublittoral flora was found to be primarily perennial, and most species showed strongest growth and reproductive activities during the winter and spring months.

Dawes, Earle, and Croley (1967) and Mathieson and Dawes (1975) provided information on the distribution of sublittoral and deeper-water marine algae on a variety of substrate types from the middle Florida Keys to the Tampa Bay area. They found that algal diversity was much greater in the Keys (characterized by higher winter temperatures, higher and less variable salinity, high light transmission, and an abundance of hard substrates), and the diversity decreased toward the north where environmental conditions were less favorable.

In overview, the shoreline flora and fauna of the Florida Keys, like that of Punta del Morro, is predominantly of Caribbean and cosmopolitan affinity, but the south Florida biota is far more diverse. On the broad, flat Florida shelf, with its many miles of hard-substrate shorelines, there is a much larger local stock of species to draw from, and many of the algal species are perennial and not dependent on constant renewal from external sources. The shoreline hard-substrate biota of the northern Gulf includes a mixture of temperate and tropical species, the former being more prominent in the winter months and the latter during the summer.

### Flora and fauna of offshore emergent structures

Prior to the 1940s, fixed structures of the northern Gulf coast consisted of shoreline construction such as docks, jetties, seawalls, and pilings, as well as a number of offshore shipwrecks. During the past half century the available hard surface has increased dramatically largely because of the construction of oil- and gas-related structures on the continental shelf and slope. A single platform in 30 m of water has been estimated to provide about 8,000 m² of hard substrate. In 1985 the National Research Council reported that more than 4,000 oil and gas platforms had been installed in the Gulf of Mexico, with most being on the Texas-Louisiana continental shelf. In addition, as of 1987 there were over 17,000 miles of pipelines on the continental shelf, many of which are exposed. This immense growth in offshore anthropogenic hard surfaces has added a large amount of artificial reef habitat where little existed before. This habitat has been populated by a diverse flora and fauna derived from species of previously existing hard banks and shoreline structures, planktonic young from more distant sources such as the southern Gulf and Caribbean, and fouling organisms from the hulls of ships that ply the area. Seasonally transient mobile species of fishes and other forms have now become temporary or permanent residents.

The organisms normally found in association with the anthropogenic structures fall into three categories: attached species, nonattached species that crawl around or live among the attached species, and larger mobile species that live in the water column or on the bottom but are loosely associated with the structures. The attached species include marine algae and a variety of invertebrate groups. These are the biofouling organisms that form the epibiotal mat on the surfaces of the structures. The nonattached forms living on the mat include many species of worms, snails, crustaceans, echinoderms, and other small invertebrates and fishes. These may be cryptic, or they may wander about the surface of the mat. Found here are a few blenny fishes that are sedentary and take shelter in old barnacle shells and other niches. The more mobile species of the water column and sea bottom include some of the larger crabs and lobsters, most of the fishes, and occasional sea turtles. Some of these are present because they feed on the biofouling mat or its inhabitants, whereas others appear to be attracted to the structures themselves. Some are residents, while others are seasonal transients.

The literature dealing with the biota of these artificial structures is not extensive, but it is adequate to provide a generalized picture of the living systems of at least the inner half of the continental shelf. Little is known about the biota of deeper-water structures except by extrapolation from other areas. The marine algae have been studied primarily by Bert and Humm (1979). The invertebrates and general population and community dynamics have been addressed by C. Adams (1996), Bright, Gittings, and Zingula (1991), Darnell and Schmidly (1988), Fotheringham (1981), Gallaway (1980), Gallaway and Lewbel (1982), Gallaway et al. (1981a, 1981b), George and Thomas (1979), Gunter and Geyer (1955), Howard et al. (1980), W. Pequegnat (1964), W. Pequegnat and L. Pequegnat (1968), S. Rabalais (1978), and P. Thomas (1975). The ichthyofauna has been reported by Gallaway and Lewbel (1982), Gallaway et al. (1981a, 1981b), R. Hastings, Ogren, and Mabry (1976), Klima and Wickham (1971), Rooker et al. (1997), Sonnier, Teerling, and Hoese (1976), and Wickham, Watson, and Ogren (1973). Most of these studies have been carried out on platforms of the north-central Gulf, and this area will provide the basis for the following discussions. Other work in the northeastern Gulf will be brought in as a supplement. Little is known about the biota of platforms elsewhere in the Gulf.

Altogether, several hundred species have been reported from around the platforms, and a partial listing of the algae and invertebrates is provided in table 11.2. Only the more common forms are given in the table, but a complete listing may be obtained from the references provided. Bert and Humm (1979) list 120 species of marine algae from shallow-water platforms off central Louisiana, and these are divided as follows: blue-green 18 percent; green 35 percent; brown 16 percent; and red algae 31 percent. All are limited to the euphotic zone, in this case, the upper few meters. Many invertebrate groups are represented, and the hydroids, polychaetes, mollusks, and crustaceans are particularly diverse. The biofouling mat itself is made up of those species that are attached directly to legs or bracings of the platform or that grow upon these attached forms. These include all the algae, sponges, hydroids, soft and stony corals, polychaete tubes, bivalve mollusks, barnacles, bryozoans, and tunicates. Around and through this mat crawl the various worms, snails, small crustaceans, echinoderms, and so forth. Lobsters, larger crabs, and echinoderms are found on the bottom around the base of the platform legs. Seventy-two species of fishes are also known from the platforms. These include several blennies (*Blennius, Hypsoblennius,* etc.), various reef-related species, and larger predatory forms. The fishes will be covered in some detail in chapter 13, on nekton.

In the north-central Gulf two basic biofouling community types are recognized, a nearshore coastal water community and an offshore "blue water" community. The environment of the nearshore community is variable and often extreme. The waters tend toward lower salinity, greater turbidity, higher nutrients, and seasonal temperature extremes. Oceanic water bathing the offshore community is of full marine salinity and clearer, has low nutrient levels, and is warm and seasonally less variable. Between the nearshore and oceanic areas, that is, in the approximate depth range of 20 to 60 m, the environment is variable and transitional and is inhabited by a transitional community. The basic characteristics of these community types are displayed in tables S43 and S44. The biomass of the fouling mat, which in surface water of the nearshore zone ranges up to 15.5 kg/m², becomes reduced to about 1 to 5 kg/m² in the offshore oceanic waters. Inshore, the fully developed mat may be up to 12 cm thick, but in blue water it is estimated to be

**Table 11.2.** *Representative genera of marine algae and invertebrates reported from drilling rigs and platforms of the Texas-Louisiana continental shelf. (Compiled from several sources.)*

**Algae**
Blue-green – *Microcoleus, Oscillatoria, Schizothrix*
Green – *Bryopsis, Chaetomorpha, Cladophora, Enteromorpha*
Brown – *Ectocarpus, Giffordia, Sargassum*
Red – *Acrochaetium, Callithamnion, Ceramium, Polysiphonia*

**Invertebrates**
Porifera – *Cliona, Halichondria, Haliclona, Verongia*
Cnidaria
    Hydroids – *Aglaophenia, Bougainvillia, Obelia, Tubularia*
    Soft corals – *Leptogorgia, Telesto*
    Stony corals – *Astrangia, Oculina*
Annelida
    Polychaetes – *Eunice, Haplosyllus, Neanthes, Nereis*
Mollusca
    Gastropods – *Cantharus, Crepidula, Murex, Stramonita*
    Bivalves – *Arca, Crassostrea, Isognomon, Ostrea*
Arthropoda
    Pycnogonids – *Tanystylum*
    Barnacles – *Balanus, Lepas, Megabalanus*
    Isopods – *Limnoria, Sphaeroma*
    Amphipods – *Caprella, Corophium, Jassa, Stenothoe*
    Decapods
        Shrimps – *Synalpheus*
        Lobsters – *Panulirus*
        Hermit crabs, etc. – *Pagurus, Petrochirus, Porcellana*
        True crabs – *Callinectes, Dromia, Eurypanopeus, Hexapanopeus, Menippe, Neopanope, Pachygrapsus, Panopeus, Pilumnus, Portunus, Stenorhynchus*
Echinodermata – *Arbacia, Ophiactis, Ophiothrix*
Bryozoa – *Acanthodesmia, Bugula, Membranipora*
Chordata
    Tunicates – *Enterogona*

2 to 4 cm thick. Brown and red algae, which are reduced inshore, are much more abundant in the clearer oceanic waters. Inshore, sessile barnacles (mostly *Balanus* and *Megabalanus*) dominate, and in oceanic waters these are replaced by stalked barnacles (*Conchoderma* and *Lepas*).

Bivalves (*Crassostrea* and *Ostrea*) make up only about 3 percent of the assemblage in the inshore waters, but bivalves (*Isognomon* and others) become the dominant fauna of the transitional and blue-water zones.

Biomass tends to decrease with depth in the water column, and the species composition also changes with depth. Marine algae are of course limited to the upper few meters. Hydroids, anemones, bryozoans, and tunicates become relatively more important below the photosynthetic zone. Depth zonation of sponge species on legs of a platform on the outer continental shelf is shown in table S45. Some forms appear limited to the near-surface waters, others to deeper layers, and several species occur throughout the depth range sampled. Each species survives in the environment most appropriate to its needs. Local deviations from these trends have been reported. The highest biofouling biomass recorded in the area (27 kg/m²) was observed at a transitional zone platform, but this particular community might have been enriched by sewage, petroleum hydrocarbons, and other effluents from the platform. In specific instances the biomass at 10 m was greater than at the surface. Thus, within the general context described above, actual development of the biofouling mat depends on locally prevailing environmental conditions.

Time is also an important factor. The development of a mature climax community has been estimated to take about 180 days under optimal conditions, or longer in some cases, depending on the time of year and other factors. The colonization and early development of the biofouling community on new bare surfaces depend on species availability, which in turn depends on the season of the year and uncertainties of dispersal. For example, barnacle larvae are not available during colder months of the year. In nearshore waters substrates placed near the surface and allowed to be colonized for about 180 days showed average daily gains in net weight of 100 g/m² when the colonization began during the summer but only 5 g/m² when growth began during the winter. Studies from other areas suggest that in proceeding from an initially bare area to the local climax community, alternate developmental pathways may be followed.

Some attempts have been made to understand nutrient relations of the biofouling community and associated organisms of the benthos and water column. The algae, of course, carry out photosynthesis. Most of the attached invertebrates are plankton feeders. These include particularly the sponges, cnidarians, some polychaetes, bivalves,

barnacles, bryozoans, and tunicates. The various worms, snails, crustaceans, and echinoderms that wander around the mat graze on the algae and attached species, and some prey on other mat wanderers. Some of the mat material falls to the sea bottom. This enriches the benthos and aids in supporting the bottom community. Some members of the fish community of the water column feed on organisms making up the biofouling mat, and the larger carnivorous species prey on larger invertebrates and the grazing fishes. The main support for the entire system is the plankton that drifts by, but this is supplemented in the euphotic zone by the attached algae, which depend on sunlight and dissolved nutrients.

W. Pequegnat and L. Pequegnat (1968) carried out detailed studies on the biofouling communities that developed on arrays of plastic floats set out on platforms off Panama City, Florida, just east of the head of DeSoto Canyon, an area bathed by several different water masses and influenced by the Gulf Loop Current. The stations were set at distances of 2 mi, 11 mi, and 25 mi from shore in 18 m, 30 m, and 46 m of water, respectively. Depths of the floats at the 2 mi station were 4, 10, and 17 m; at the 11 mi station they were 4, 10, 17, and 29 m; and at the 25 mi station they were 4, 10, 17, 29, and 44 m. Individual floats were removed, preserved, and taken to the laboratory at intervals of 2 weeks, 1 month, 2 months, 3 months, 6 months, and 12 months. The variables studied were taxonomic composition and organic biomass accumulation in relation to distance from shore, depth, length of exposure, and season of the year, but most of the results given related to the 2 mi and 11 mi stations. Detailed information was also provided for many of the biofouling species.

During the study 146 individual species were recognized, and another 43 were identified only to the level of genus or family. Several were undescribed species new to science. From a geographical standpoint, 14.3 percent were of northern affinity, 50.7 percent tropical, and 35.0 percent were cosmopolitan species. Taxonomically, the species were distributed as follows: 1 sea anemone, 35 hydroids, 25 polychaetes, 16 gastropods, 17 bivalves, 9 barnacles, 13 amphipods, 15 decapods, and 15 bryozoans. The accumulation of biomass on the floats in relation to four major variables is given in table S46. In practically all cases the quantity of organic material increased steadily with length of exposure. During the summer months the offshore station accumulated much more than did the inshore station, but during the winter the differences were not great. There was no clear pattern of accumulation in relation to depth. During the summer at the inshore station accumulation decreased with depth. During winter at the offshore station the reverse was true. At the inshore station accumulation was greatest during the winter, and at the offshore station it was greatest during the summer. The seasonal temperature and salinity regimes at the two stations were quite distinct, leading the authors to conclude that they were bathed by different water masses, and this conclusion was reinforced by the distinct nature of the fauna settling on the two sets of floats and by differences in biomass accumulations.

## Flora and fauna of offshore banks and hard bottoms

As noted earlier, offshore banks and hard bottoms occur on continental shelves throughout the Gulf of Mexico. Except in areas of high turbidity and heavy siltation, they support luxuriant and diverse communities of marine organisms, including (in warmer waters) well-developed coral reefs. Most of these communities have received sufficient study to permit general description of the biota, and for a few (such as the south Florida reefs and the Flower Garden Banks of the north-central Gulf), enough is known to allow insight into the dynamic functioning of the systems.

REEFS AND BANKS OF THE NORTHWESTERN GULF
Reefs and banks of the northwestern Gulf fall into three general categories: nearshore, midshelf, and outer shelf prominences. The nearshore structures include Seven and One-Half Fathom and Sebree Banks off Padre Island (of south Texas), West Bank off Freeport, and Heald Bank off Galveston. These consist largely of consolidated beachrock or old lake bottom sediments and have a vertical relief of 5 m or less. Among these features Seven and One-Half Fathom Bank has received the most scientific study, and this information is given in reports by B. Causey (1969), Felder (1971), Felder and Chaney (1979), McCarty (1974), Shirley (1974), Tunnell (1973), and Tunnell and Chaney (1970). This work has been reviewed briefly by Rezak, Bright, and McGrail (1985).

These shallow-water banks exist in rather harsh environments characterized by high turbidity, seasonally low temperature (sometimes below 15° and occasionally below 10°C off Louisiana), and variable salinity (periodically below 30 ppt). The attached and associated fauna consists primarily of sponges, hydroids, tube-dwelling polychaetes, mollusks, crustaceans, and bryozoans tolerant of turbid conditions. Many of the species also occur on

the coastal jetties. Although temperate forms are prominent on the northernmost banks, tropical elements become more important toward south Texas, and the fauna of Seven and One-Half Fathom Bank is considered to be transitional between temperate and tropical.

Around three dozen prominent hard banks are known from the middle and outer continental shelf of the northwestern Gulf (fig. 4.7). Rising from depths of 50–300 m, some approach within 15 or 20 m of the sea surface. Most have been surveyed by scuba divers and/or research submersibles, and information concerning their origins, geological structures, and biological characteristics has been presented by Rezak, Bright, and McGrail (1983, 1985), and Rezak, Gittings, and Bright (1990). Two of these prominences that support well-developed tropical coral reefs (the East and West Flower Garden Banks) have been extensively investigated, and these will be discussed in detail in a later section of the present chapter. Our concern here is to examine the general distribution and nature of the banks, the zonation of the biological communities growing on them, and the factors governing this zonation.

From a geological standpoint the deepwater banks off east Texas and Louisiana represent surface expressions of salt diapirs, that is, salt domes capped with hard rock (limestone, sandstone, siltstone, claystone, or basalt). These occur on both the middle and outer shelf. Midshelf banks of the area rise from depths of 50 to 80 m and have a relief of 4–50 m. The outer shelf banks rise from depths of 80 to 300 m and crest at depths of 15–100 m below the sea surface. The banks off south Texas below the level of Matagorda Bay (fig. 4.7) are entirely different in origin. These nondiapiric structures represent drowned coral/algal reefs that grew upon a relict carbonate shelf 18,000–10,500 years ago during a period of lower sea level stand. They occur only on the middle shelf in the depth range of 50–80 m and have relief up to 20 m.

Careful study of the various deep banks off Texas and Louisiana has revealed seven more or less distinct zones of biological development (table 11.3), which are correlated with such factors as substrate type, depth, distance from shore, and various hydrological factors (minimum winter temperature, salinity, turbidity, sedimentation, etc.). Although no bank has all the zones, all but one occur on two of the banks. The most vigorous and diverse coral reef development takes place in the uppermost (*Diploria-Montastrea-Porites*) zone, but hermatypic or reef-building corals are important community elements in the first four

**Table 11.3.** *Zonation of biological communities on hard banks of the middle and outer continental shelf of the northwestern Gulf of Mexico. Several of the zones are named for prominent genera of corals found there. (Modified from Rezak, Gittings, and Bright 1990.)*

A. Zones of major reef-building activity and primary production.

   These zones receive sufficient light to support photosynthesis by many algal species and to promote vigorous growth of most reef-building corals.

   1. Diploria-Montastrea-Porites zone. This includes the living, high-diversity coral reefs. Hermatypic corals predominant. Coralline algae are abundant, but leafy algae are limited.

   2. Madracis and leafy algae zone. Here the branching coral *Madracis mirabilis* produces large amounts of carbonate sediment, and leafy algae are conspicuous.

   3. Stephanocoenia-Millepora zone. This includes living but low-diversity coral reefs. Hermatypic corals still dominant. Coralline algae are abundant, but leafy algae are limited.

   4. Algal-sponge zone. This zone is dominated by coralline algae that produce large quantities of carbonate substrate, including algal modules. Here leafy algae are quite abundant, but hermatypic corals are scattered and include only a few species

B. Zone of minor reef-building activity. Here light limitation restricts the growth of corals and algae.

   5. Millepora-sponge zone. Sponges, the hydrozoan coral (*Millepora*), and other epifauna dominate this system. Coralline algae and scleractinian corals are rare.

C. Transition zone. Here reef-building corals are minor to negligible.

   6. Antipatharian zone. This zone has limited crusts of coralline algae and a few species of soft corals, the most conspicuous being the coiled or bedspring-shaped species of the genus *Antipathes*.

D. Zone of no reef-building activity. Limitation due to low light and other factors precludes the growth of algae or hermatypic corals.

   7. Nepheloid zone. This zone lies within the bottom nepheloid layer and is characterized by high turbidity and sedimentation. Rocks and ancient drowned reefs are covered with a carpet of fine sediments. Epifauna is depauperate and variable. Certain deepwater octocorals and solitary corals may be conspicuous. This zone occurs on all the banks below the transitional or antipatharian zone.

---

zones. Reef-building activity is severely restricted in the next three and is totally absent from the last or lowermost zone, which is shrouded in the turbid nepheloid layer. More information on the biological composition of these zones will be presented in the coral reef section of the present chapter.

The distribution and depth relations (where known) of the seven zones and soft bottom along 35 of the banks are given in table S47. The midshelf banks of the northwestern Gulf have only the *Millepora*-sponge or antipatharian zones above the nepheloid layer. Being in relatively shallow water and closer to shore, these low banks are all exposed to high levels of turbidity, and several also experience low winter temperature and seasonally low salinity, all of which are unfavorable for the development of hermatypic coral communities. On the other hand, most of the outer shelf banks support the algal-sponge zone, and on the few most favor-ably situated banks, one or more of the zones of reef coral development are represented. Taking into account all the banks, the general depth relations of the several zones are as follows: *Diploria-Montastrea-Porites* (15–36 m), *Madracis* (28–46 m), *Stephanocoenia-Millepora* (36–52 m), algal-sponge (45–98 m), *Millepora*-sponge (18–52 m), and antipatharian (52–123 m). Near shore, the top of the nepheloid layer can occur as shallow as 45 m, but on the outer shelf it is generally over 80 m and sometimes as deep as 123 m. Perspective drawings of two midshelf banks are shown in fig. 11.11. Sonnier Bank, in about 60 m of water off Louisiana, has the *Millepora*-sponge zone above the nepheloid layer. Southern Bank, in about 80 m off south Texas, has the antipatharian zone, although most of the animals here appear to be soft corals rather than antipatharians. Various invertebrates and fishes associated with these banks are illustrated in the figure.

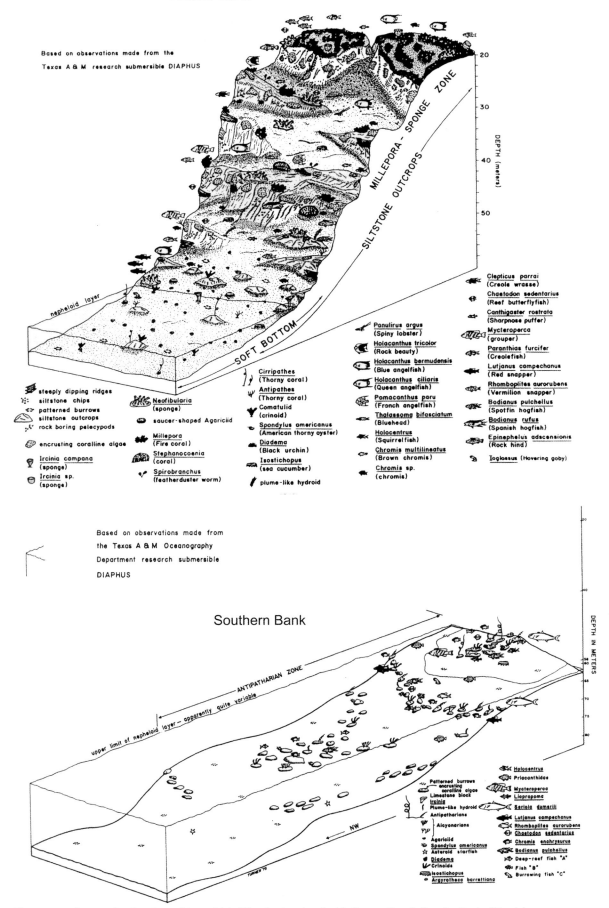

**Figure 11.11.** Perspective drawings of two midshelf banks showing the biotic zonation. A: Sonnier Bank off Louisiana; B: Southern Bank off south Texas. (From Rezak, Gittings, and Bright 1990.)

HARD BANKS OF THE NORTHEASTERN GULF

Between the Mississippi River Delta and DeSoto Canyon in the depth range of 53–110 m lies a series of hard banks that range from low mounds to ridges, flat-topped mesa-like features ("flat-topped reefs"), and sharp jagged pinnacles with vertical relief up to 18 m. The biotic assemblages growing on these features have been surveyed by Gittings, Bright, and Schroeder (1991) and Gittings, Bright, et al. (1992) using remotely operated video cameras supplemented by dredges and other collecting gear. In the often turbid water, identification was sometimes difficult, but over 93 distinct taxa were recognized, some to the level of species and others to genus or higher taxonomic category. The composition, diversity, and abundance of the biota were found to vary with location, depth, substrate type, and certain hydrographic factors, chief of which was potential for sedimentation. On the low features (< 6 m) and on the lower flanks of taller structures, species abundance and diversity were quite low, apparently due to suspension and resettlement of particulate matter in the turbid bottom layer. However, on higher structures biological growth was often luxuriant, particularly on horizontal surfaces of the flat-topped reefs. Here were located dense fields of sponges, soft corals (sea whips and sea fans), thorny corals (antipatharians), and crinoids, as well as many smaller forms including snails, bivalves, crustaceans, bryozoans, and echinoderms. The near-vertical sides of the banks and pinnacles were populated largely by ahermatypic (non-reef-building) corals. A list of some of the more important invertebrates is provided in table 11.4. Some calcareous algae were observed on taller structures, but never below a water depth of 78 m, beyond which they are probably light-limited. Although a few stony corals were observed, there was no evidence of tropical coral reef development. The influence of turbid plumes of Mississippi River water extends eastward from the Delta about 70 km (42 mi). In general, the biotic assemblages of the northeast Gulf banks resemble those of the midshelf banks of the northwestern Gulf, especially the algal-sponge, antipatharian, and nepheloid zones.

HARD SUBSTRATES OF THE WEST FLORIDA SHELF

The most prominent topographic features of the west Florida shelf include the complex series of carbonate terraces, ridges, and pinnacles lying in 40–50 m of water northwest of Tampa Bay, collectively known as the Florida Middle Ground. Although the taxonomic composition of the biota has been studied in some detail (T. S. Hopkins

**Table 11.4.** *Representative invertebrates reported from hard banks, particularly the surface of flat-topped reefs, of the middle and outer continental shelf of the northeastern Gulf of Mexico. (From Gittings, Bright, and Schroeder 1991, 1992.)*

**Porifera** – encrusting and upright sponges
**Cnidaria**
    Soft corals (antipatharians, sea fans, etc.) – *Antipathes, Cirripathes, Ellisella, Muricea, Nicella, Scleracis, Siphonogorgia, Telesto*
    Stony corals – *Agaricia, Madrepora, Oculina, Oxysmilia, Paracyathus, Rhizopsammia, Stephanocoenia*
**Mollusca**
    Gastropods – *Phalium, Polystira, Scaphella, Terebra*
    Bivalves – *Amygdalum, Eucrassatella, Lyropecten, Plicatula, Ventricolaria*
**Annelida** – *Hermodice*
**Crustacea**
    Anomurans – *Dardanus*
    Macrurans – *Scyllarides*
    Brachyurans – *Rochinia, Stenorhynchus*
**Echinodermata**
    Asteroids – *Astropecten, Linckia, Narcissia*
    Ophiuroids – *Astroporpa*
    Echinoids – *Clypeaster, Diadema, Eucidaris, Stylocidaris,* heart urchins
    Holothuroids – sea cucumbers
    Crinoids – *Antedon, Comactinia*

1974a, 1974b, 1976, 1979; T. S. Hopkins, Blizzard, et al. 1977; T. S. Hopkins, Blizzard, and Gilbert 1977; Rezak, Bright, and McGrail 1983), the authors have seldom made clear distinctions between the biota of soft and hard substrates. However, the general nature of the hard-substrate flora and fauna is clear. For example, 111 species of marine algae, mostly attached forms, have been reported including 29 species of green, 18 of brown, and 64 of red algae. Two transects across largely hard-bottom terraces are shown in figure 11.12. These assemblages are dominated by a fairly abundant and diverse group of algae, sponges, and soft corals. The sponges and corals have been found to harbor a great many commensal species of polychaetes, mollusks, crustaceans, and echinoderms, as well as fishes. Scattered patches of stony corals are present including species of *Agaricia, Dichocoenia, Madra-*

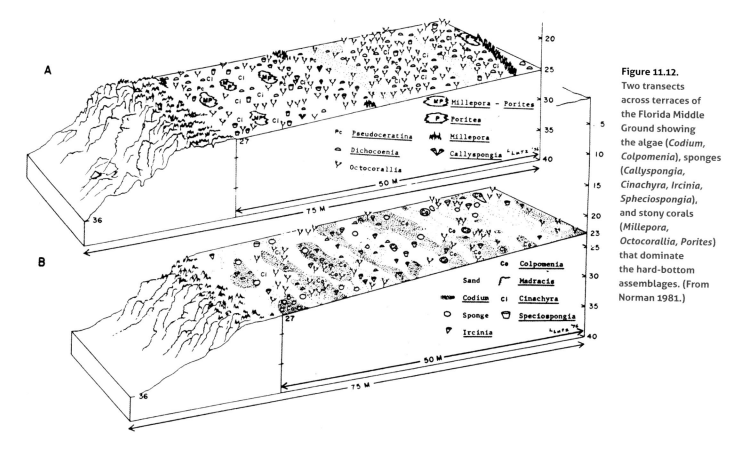

**Figure 11.12.**
Two transects across terraces of the Florida Middle Ground showing the algae (*Codium, Colpomenia*), sponges (*Callyspongia, Cinachyra, Ircinia, Spheciospongia*), and stony corals (*Millepora, Octocorallia, Porites*) that dominate the hard-bottom assemblages. (From Norman 1981.)

*cis, Manicina, Meandrina, Oculina, Porites, Scolymia,* and the fire coral *Millepora.* Some of these are known to die out periodically when bottom water temperatures plunge as low as 12°–13°C, well below the 18°C normally considered minimal for healthy growth of hermatypic corals. Although the Middle Ground assemblages exhibit moderate stony coral diversity and include an abundance of soft corals, tropical invertebrates, and fishes, the area does not support active coral reef development such as that found off the Florida Keys. Winter low temperatures prevent such development upon otherwise suitable substrates of the area.

The southwestern Florida shelf, from the level of Charlotte Harbor to just above the Florida Keys, has been studied fairly intensively (see N. Phillips, Gettleson, and Spring 1990, and references therein). Much of this limestone platform is covered by a veneer of sand, largely siliceous inshore and becoming progressively carbonaceous beyond a depth of 10–20 m. The sand veneer, which in many places is quite thin, averaging only 1–5 cm thick over the flat limestone base, appears to be quite mobile, at least on the inner and middle shelf. Rock outcrops are rare and mostly of low relief (< 1 m). Where they do occur they generally support dense growths of sessile epiflora and epifauna, but such growths are also found on thinly

covered carbonate substratum. Presumably, the sessile species can gain a foothold on algal nodules, larger shell fragments, or on the underlying limestone when it is exposed, and they continue to flourish after a thin sand layer has moved back in (fig. 11.13). Thus, the general aspect of this shelf is a mosaic of live-bottom patches surrounded by bare carbonate sand.

Although this live bottom occurs all across the shelf, it is most common in three depth ranges where the veneer is generally thin; 10–20 m (average coverage = 41 percent), 70–90 m (average coverage = 48 percent), and 120–160 m (average coverage = 65 percent). In the 60–90 m depth range coralline algal nodules occur along the crest of the buried reef tract known as Pulley's Ridge (fig. 4.5), increasing southward where it grades into a fused algal carbonate pavement. In the 120–160 m depth range another partially buried reef tract (Howell Hook) forms steep-walled outcrops or pinnacles protruding up to 2 m above the sea floor and providing substrate for epibiotal attachment.

The sessile flora and fauna of live bottoms typically include algae, sponges, hydroids, gorgonians, stony corals, bryozoans, and tunicates, but species composition varies with depth of the water column. On the inner shelf (10–20 m) the most conspicuous sessile forms are dense populations of large sponges (*Ircinia* and *Spheciospongia*) and

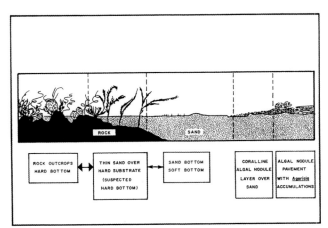

**Figure 11.13.** Schematic illustration of biological development on several substrate types across the southwest Florida shelf. The exposed and thinly covered rocky outcrops support attached algae, sponges, several types of soft corals, and crinoids. Deepwater algal pavement supports the green alga *Anadyomene* and the stony coral *Agaricia*. (From N. Phillips, Gettleson, and Spring 1990.)

gorgonians (*Eunicea, Muricea,* and *Pseudopterogorgia*). Beyond a depth of 20 m the gorgonians rapidly decline in abundance, giving way to sandy-bottom species of algae (*Caulerpa, Halimeda, Penicillus,* and *Udotea*). In the 60–90 m depth range associated with the coralline algal nodules are calcareous algae (*Halimeda* and *Peyssonnelia*) and populations of sponges, nonphotosynthetic gorgonians, stony corals, and crinoids. The fused algal pavement in the 60–80 m range is colonized by dense growths of green algae (*Anadyomene*), crustose red algae (*Peyssonnelia*), and stony corals (especially *Agaricia* and *Madracis*). On the outer shelf (100–200 m) algae are generally absent, and the most common epifauna are glass sponges, antipatharians (*Antipathes, Aphanipathes*), nonphotosynthetic gorgonians (*Ellisella, Nicella, Placogorgia*), stony corals (*Caryophyllia, Javania, Madrepora, Paranthus*), and crinoids (*Comactinia, Leptonemaster, Neocomatella*).

Table S48 summarizes depth distribution patterns of three numerically dominant groups. The sponges, with a total of 74 species, show 40–50 species within each depth range except the outer shelf, where only 8 species were recorded. Soft corals (octocorals) are represented by 71 species, many of which are of limited depth distribution. They are most diverse at the shallowest stations (< 20 m) and least diverse in the 20–60 m range. The stony corals are somewhat more diverse in the 20–32 m range. Most of these are sturdy, tolerant species characteristic of fringing environments of the Caribbean and Florida reef tract systems. Major hermatypic (reef-building) coral species

such as *Acropora cervicornis, A. palmata, Colpophyllia natans, Diploria labyrinthiformis, D. strigosa,* and *Montastrea annularis* have not been collected on this part of the southwest Florida shelf, and there is no evidence of major reef-building activity here.

The algae, of course, are photosynthetic, and many of the sessile animals harbor photosynthetic algae as intracellular symbionts. However, all the sessile animals also harvest plankton from the water column. The combination of autotrophic and heterotrophic nutrition permits high rates of production by these animals even in low-nutrient waters. Wet weight biomass estimates varied from 600 to 4,300 g/m², with sponges accounting for 40–70 percent of the total. Both biomass and percent cover are highest where there is exposed rock or only a thin veneer of sand. The sessile forms of these live-bottom areas provide habitat cover and food resources for a wide variety of mobile species such as polychaetes, mollusks, crustaceans, echinoderms, and fishes. Sea turtles also feed here on sponges, soft corals, and crustaceans. These live-bottom areas are thus of great importance in the biological economy of the area. A diagrammatic cross section of the southwest Florida shelf showing the general pattern of zonation is provided in fig. 11.14. Dominant organisms of each zone are indicated, but no distinction is made between the soft- and hard-bottom biota.

## Major coral reefs of the Gulf

Coral reefs are diverse and highly complex biological communities that develop and thrive in shallow (100 m or less) tropical and subtropical waters of full marine salinity. They grow on hard, stable substrates (often of their own manufacture) where water temperatures are seldom below 18°C and there is high and stable salinity, high light intensity, and very low levels of sedimentation. The corals, algae, foraminiferans, bryozoans, and other carbonate-producing organisms develop hard carbonate banks, which, over the millennia, grow upward toward the water surface. Annual carbonate accumulation rates of up to 20 kg/m² and vertical growth rates of 2.2 m/1,000 years have been reported for some south Florida reefs.

### DEVELOPMENT OF THE REEF BANK

Development of the reef bank involves a complex interaction of physical, chemical, geological, and biological processes. In the initial building phase the primary framework builders set down the first structure, and later colonizers add to the height and volume of the growing

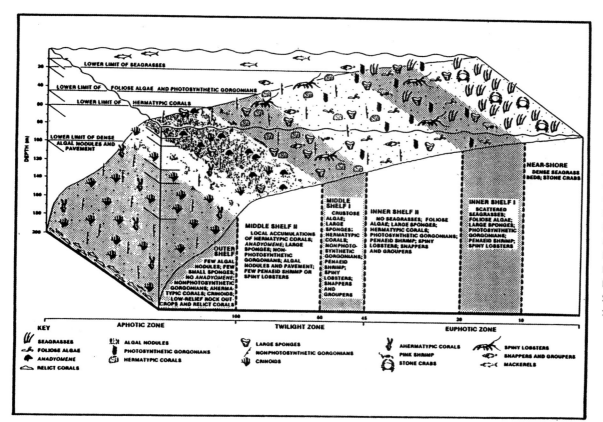

**Figure 11.14.** Diagrammatic section of the southwest Florida continental shelf just above the Florida Keys showing patterns of biotic zonation and general distribution of the dominant plant and animal groups. (From Environmental Science and Engineering, Inc., LGL Ecological Research Associates, Inc., and Continental Shelf Associates, Inc. 1987.)

carbonate bank. Soon after the first colonizers settle and start to grow, the breakdown of coral skeletons by boring organisms, chemical dissolution, and mechanical processes begins. Carbonate rubble and finer sediments created by these processes become part of the reef. Coarse fractions occupy the interstitial spaces of the reef framework, and fine sediments fill in the borings and voids. The primary carbonate sediment producers are the corals themselves and green algae, including *Halimeda* and its relatives. This material becomes incorporated into the reef framework largely by the binding action of crustose coralline algae and by in situ solidification of high-magnesium cements. The resulting structure is an extremely hard and durable limestone mass, and the reef wall is able to resist currents and wave energy characteristic of the marine environment.

As seen in figure 11.15, dominant organisms of the typical coral/algal reef of the Gulf of Mexico are distributed in definite depth-related zones. The zones shown in the figure refer to the Yucatán reefs, and although the actual depths may vary somewhat from one sector of the Gulf to another, the pattern of zonation is rather constant since it relates to the ability of the organisms to thrive under physical conditions characteristic of each depth zone. The lowest levels are dominated by calcareous red algae (*Lithophyllum* and *Lithoporella*). Next above is the

foraminifera–red algae (*Gypsina-Lithothamnium*) zone. Above this lies the lower zone of massive corals (*Agaricia-Montastrea*), and this is followed by the main zone of massive corals (*Diploria-Montastrea-Porites*). The uppermost zone, which develops up to the low-tide level of the sea surface, is dominated by the branching elkhorn coral (*Acropora palmata*), which is adapted to absorb and dissipate the high energy of surface waves and near-surface water currents. In the lagoonal habitat behind the emergent reef is a group of algae and corals that require a more protected environment.

This pattern of zonation helps explain the various types of reefs found in the Gulf. For example, banks that crest at depths greater than 36 m (118 ft) lack the massive reef corals, those that crest in the 9–36 m (30–118 ft) range are capped by massive coral heads, and those that approach the surface are dominated by elkhorn corals. The zonation pattern also aids in understanding the stages of reef development. A reef that begins in deep water is first colonized by deepwater species, and as the reef develops and builds to a shallower level it is colonized by species of shallower water affinity. Because the sea level reached its lowest stand during the last Ice Age, it is unlikely that any of the Gulf reefs are more than 18,000 years old, and most are probably less than half that age. From its lowest stand the sea rose rapidly during the first period of around

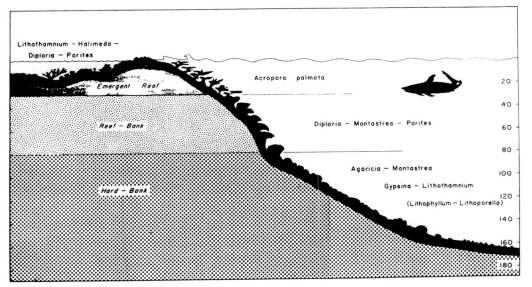

**Figure 11.15.** Depth-related zonation (in feet) of biological communities on a typical emergent coral reef in the Gulf of Mexico. The actual depths of the zones vary somewhat with location. (From Rezak and Edwards 1972.)

10,000 years. Reefs that began earlier could not grow fast enough to keep up and became drowned reefs in deeper water. However, during the past 7,000 years or so the sea level has risen more slowly, and reefs begun about that time or more recently have been able to continue growing without interruption. Detailed geological studies of coral reefs in the Bahamas reveal that these new reefs are often growing upon the hard substrate laid down by the earlier drowned reefs, and this is also true for many of the Gulf reefs that have been thoroughly studied (table S49).

## SOUTH FLORIDA REEF TRACT

From the level of Miami and arcing southwestward toward the Dry Tortugas, on the seaward (Atlantic) side of the Florida Keys, lies a discontinuous series of coral reefs collectively known as the Florida Reef Tract (fig. 11.16). Landward of the Keys (between the Keys and the Everglades) is situated a large shallow body of water known as Florida Bay. Bottomed largely by soft sediments and subject to periodic incursions of freshwater from the Everglades, the bay experiences occasional low temperatures during the winter and periods of excessively high temperatures during late summer. Following storms the turbidity may remain high for several days. Although many tropical species inhabit the bay, it is a poor environment for long-term growth of reef-building corals. However, the limestone platform on the seaward side of the Keys is bathed by warm waters of the Florida Current flowing up from the Caribbean. This platform, which slopes toward the shelf break, is terraced and dissected by deeper chan-

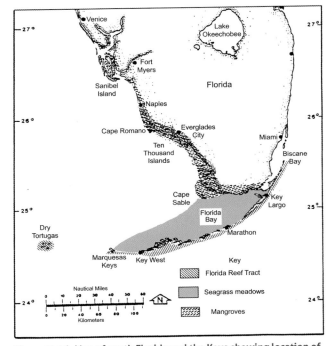

**Figure 11.16.** Map of south Florida and the Keys showing location of the Florida Reef Tract (cross-hatched) and associated features.

nels, resulting in "spur and groove" topography. The spurs and other rocky exposures of this area support the development of major coral reefs. These have been under study for over a hundred years, and much of the literature has been summarized by Jaap (1984) and Lidz et al. (2008).

Four basic community types are recognized. Proceeding seaward, these include the live-bottom and patch reefs of shallow water and the transitional and bank reefs of deeper water. Vertical relief and dominance by stony corals increase in this same progression. Each community type will be discussed briefly.

The *live-bottom* community occurs closest to shore, a short distance seaward of the intertidal zone, and may extend down to at least 30 m. This is dominated by sponges, soft corals, and algae, and it includes a few hardy stony coral species. These include *Manicina, Porites, Siderastrea,* and *Solenastrea,* and in deeper waters to these are added *Dichocoenia, Diploria,* and *Millepora.* This community does not accumulate carbonate and hence does not build reefs. Species composition varies greatly from the shallow to deeper water areas. In size, live-bottom areas vary from tens to hundreds of square meters, and in shallow water they may be surrounded by soft-bottom communities of seagrasses, sponges, and so forth. Many mobile invertebrates and fishes reside in the live-bottom habitat, and this community serves as an important nursery area for species whose adults are found elsewhere.

The *patch reef* is the most conspicuous element in the Florida Reef Tract. Characteristically, this is a dome-shaped structure of up to 3 m in vertical relief. Most patch reefs occur in 2–9 m of water, and the upper surface may be nearly exposed at low tide. They are roughly circular in outline and 30–700 m in diameter. The patch reef is a carbonate-accumulating community with a somewhat regular pattern of development. Primary invaders (probably *Favia, Manicina, Porites,* and others) become established and form hard substrate upon which further stages develop. These are followed by the primary framework builders (*Colpophyllia, Diploria, Montastrea,* and *Siderastrea*), which build the massive three-dimensional structures. Reef development proceeds with the formation of carbonate rubble, cementation, and arrival of new colonizers. By these means the structure builds upward toward the surface and outward along the edges, increasing the available niches and providing habitat for an elaborate array of associated flora and fauna. As the reef matures and approaches the sea surface, boring and tunneling organisms have been busy destroying the undersurfaces and interior of the massive coral heads, thereby weakening the structures. As a result the upper surface may collapse inward, forming a pile of coral rubble, rocks, and boulders. This irregular flat surface is then colonized by soft corals, smaller stony corals, and other invading species. Thus, the patch reef may decay into a live-bottom community. Patch reefs are often surrounded by a halo of bare sand and reef rubble largely devoid of algae and seagrasses. This halo appears to be a result of the grazing activities of sea urchins and some fishes that are resident on the reef and forage in surrounding areas at night. Patch reefs may be decimated by waves and surges associated with strong storms.

The *transitional reefs* represent early stages in the development of bank reefs, and they may result from direct growth on hard substrates or from coalesced patch reefs. They are characterized by some spur development of *Acropora palmata* (elkhorn coral) growths on the seaward fringe and well-developed reef flats behind. Characteristic stony corals of the reef flats include *Acropora cervicornis* (staghorn coral), *Diploria, Millepora, Montastrea,* and *Porites.* As with patch reefs, transitional reefs are subject to destruction by severe storms.

Major *bank reef* communities develop on exposed hard surfaces near the sharp change in bottom slope marking the seaward edge of the Florida Plateau. This occurs mostly in the depth range of 5–10 m. On the seaward flanks they are dominated by *Acropora palmata.* Downslope this community grades into a barren area that ultimately gives way to the buttress zone characterized by colonies of massive corals, such as *Montastrea,* and fields of staghorn corals (*A. cervicornis*). This reef platform extends down to a depth of about 27 m, but outlying communities continue to a depth of around 41 m. On the deep reefs *Madracis, Montastrea,* and *Siderastrea* are common on the horizontal surfaces, but *Agaricia* and *Helioceris* dominate vertical walls. Spur and groove topography occurs in both shallower and deeper zones of these major bank reef areas, and details of species composition and degree of development vary from one reef to another. The depth zonation of topography and reef communities off one of the Keys in the Florida Reef Tract is shown in figure 11.17.

YUCATÁN REEFS

Yucatán banks and reefs have been the subject of a fair amount of study. Important references to the geology and general environments include Busby (1965, 1966), Chávez, Hidalgo, and Izaguirre (1985), Farrell et al. (1983), Folk and Robles (1964), Fosberg (1962), Hoskins (1963), Kornicker and Boyd (1962), and Logan (1969). Although emphasizing geological aspects, most of these papers also treat the corals and other reef-associated biota. More specific biological references include Kim (1964) on marine algae, Rice and Kornicker (1962) on mollusks, and H. Hilde-

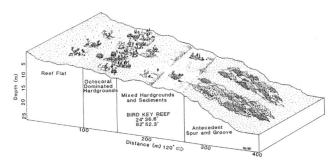

**Figure 11.17.** Depth-related zonation of topography and reef communities off Bird Key in the Florida Reef Tract. (From Jaap 1984.)

brand, Chávez, and Compton (1964) on fishes. The most comprehensive reference is Logan (1969), which provides information on the distribution, depth zonation, and biological composition of important reef communities of the various Yucatán banks.

As noted earlier, the reef masses of the northern and western portions of the Yucatán shelf are quite varied in size and configuration. Some are of low relief (one or a few meters), others of high relief (60 meters or more). Some are solitary knobs, others aggregated or of linear or crescent shape. Most do not reach the water's surface, but a few are emergent. The larger reef-wall structures are surrounded by fields of carbonate skeletal fragments derived from the reef above.

The general zonation of biotic communities of the Yucatán banks and reefs has already been described and illustrated in figure 11.15. From deeper to shallower water the five zones include the calcareous red algae, foraminifera–red algae, lower massive coral, main massive coral, and near-surface zones. In addition, on the crests of emergent reefs in protected waters behind the *Acropora palmata* rampart lies a reef-flat community consisting of live-bottom assemblages. The largest of the flat-topped reefs (Alacrán) features a broad, shallow lagoonal area. A perspective drawing of a typical solitary knoll (Nuevo Reef) showing the morphology of the wall and surrounding shelf and general community zonation is given in figure 11.18. A brief discussion of each of the community types, proceeding from deeper to shallower water, follows.

The calcareous–red algae (*Lithophyllum-Lithoporella*) zone normally occurs in the depth range of 30–60 m, but on some reefs it may extend as shallow as 18 m. It consists of extensive veneers of algal nodules on rocky surfaces of the inorganic banks. Also found here are fora-

miniferans (*Homotrema*) and scattered deepwater corals (*Diploria, Manicina,* and *Mussa*), as well as mollusks and bryozoans. This is a dead community composed of what were apparently early postglacial colonizers that came in when the water was shallower but that could not flourish as the water level rose. In many areas this relict community is now overgrown by foraminifera and red algae of the next zone above.

The foraminifera–red algae (*Gypsina-Lithothamnium*) zone may be found as shallow as 18 m, but it is best developed in depths of 30–60 m where it has overgrown the previous zone. Minor components of this community include the green alga *Halimeda,* sponges, calcareous tube-building polychaete worms, mollusks, and bryozoans.

The lower massive coral (*Agaricia-Montastrea*) zone occurs in the broad depth range of 15–60 m, where it is interspersed with areas dominated by the *Gypsina-Lithothamnium* community.

The main massive coral (*Diploria-Montastrea-Porites*) zone lies at depths of 9–24 m. On the windward side of emergent reefs it appears at deeper levels (below 9 m), but on the leeward side it may be found as shallow as 3 m. This community grows directly up to the base of the next higher zone without an intervening bare area such as occurs on the south Florida reefs. In this zone the massive interlocking corals achieve diameters of almost 2 m. Although coralline algae are present, they are relatively unimportant because of the coverage of the large corals themselves. The foraminiferan *Gypsina* aids in binding sediments of this zone.

The near-surface elkhorn coral (*Acropora palmata*) zone is found at depths of 0–9 m. On the windward side of an emergent reef it may extend down to 9 m, but on

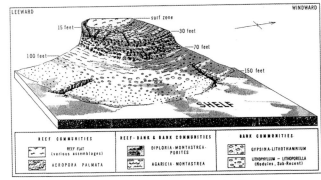

**Figure 11.18.** Perspective drawing of a solitary knoll structure (Nuevo Reef) on the Yucatán shelf showing the distribution of community zones and morphology of the wall from the reef flat to the floor of the surrounding shelf. (From Logan 1969.)

the protected leeward side it seldom occurs below 4.5 m. This community is especially adapted for the high-energy surface and near-surface environment subject to waves and storm surges. Encrusting coralline algae (*Porolithon*) and foraminiferans (*Homotrema*) are important sediment binders here. Also present are various sponges, soft corals, the fire coral *Millepora*, polychaetes, mollusks, crustaceans, echinoderms, and various species of fishes. Other stony corals (*Diploria, Montastrea, Porites,* etc.) found in this zone on the windward side tend to be stunted and reach diameters of only 1–2 ft.

Behind the low protective rim on the crest of the emergent reef lies the reef flat. Although this relatively flat area at the sea surface is subject to the surge of waves, it is not pounded by direct surf. In most cases, at least, development of the reef flat follows a regular series of steps. The initial topping is accomplished by the *Acropora palmata* community, which produces a ragged surface at the low-water level. Subsequently, this framework is welded into a loose meshwork by encrusting frame-binding organisms, chiefly the coralline alga *Porolithon* and small stony corals. Depressions in the structure are gradually filled by marginal growth and accumulations of fragmented skeletal material. Reef-flat assemblages continue to grow and accrete carbonate material in any remaining depressions. Finally, blankets of skeletal clastics and sands cover the older substrate, ramparts of coral boulders piled up by storms rim the windward margin, and islands of calcareous sand build up on the lee sides of the boulder ramparts. At the present time the various emergent reefs are in different stages of development. Reef-flat assemblages that grow in shallow water behind the windward reef rim include colonies of *Acropora cervicornis, A. palmata, Agaricia, Favia,* and *Porites,* as well as the algae *Halimeda* and *Penicillus.* Also present are crusts of the coralline alga *Amphiroa,* soft corals, and mats of colonial anemones. The various species assemblages of the reef flat are not clearly defined.

The largest of the emergent reefs (Alacrán), which has a crest about 23 km (14 mi) long and 12 km (7 mi) wide, features an *Acropora palmata* reef on the windward (eastern) side that protects a large shallow lagoonal area (fig. 11.19). About 60 percent of the lagoon floor is at or near low-water level, but deeper areas may achieve depths of around 15 m. The lagoon bottom is carpeted with carbonate skeletal material interspersed with clumps of large coral heads (especially *Montastrea* and *Porites*). Also present are large patches of *Halimeda* as well as coralline

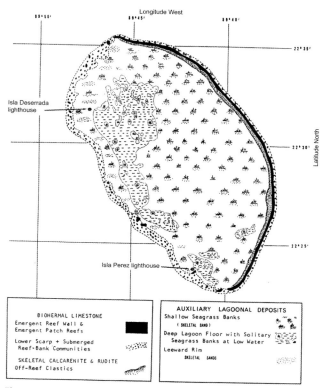

**Figure 11.19.** Alacrán Reef prominence showing reef wall, submerged reef bank communities, patch reefs, islands, and lagoonal features. (From Logan 1969.)

algae, soft corals, other stony corals (*Diploria, Manicina,* etc.), mollusks, and echinoids (sea urchins). Dense stands of seagrasses (*Thalassia*) cover large mounds rising from the floor to low-water level, and these form an interlocking network of rounded lagoonal banks. The dense seagrass beds trap, bind, and stabilize the loose sediments, gradually elevating them into banks. On the lee side of the lagoon lies a series of patch reefs and small islands, in places separated by deeper drainage channels.

REEFS OF THE WESTERN GULF

On the continental shelf of the western Gulf of Mexico coral reefs occur on nearshore prominences in two general localities (fig. 11.20). The northernmost are situated just below Cabo Rojo, about 120 km south of Tampico. These include Isla de Lobos and five other banks only 5–12 km from the mainland. The southern group includes several reefs in front of the city of Veracruz and another set about 11 km farther south off the coastal village of Antón Lizardo. These two farther sets, comprising about 23 reefs (some with small cays), are situated 1–10 km offshore. The various banks of the western Gulf are some-

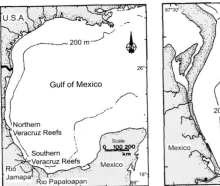

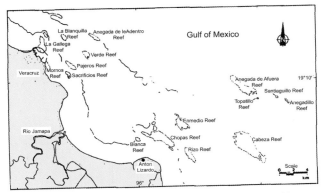

**Figure 11.20.** Location of major coral reefs of the western Gulf of Mexico. A: general location of the northern (Cabo Rojo) and southern (Veracruz) reef groups. B: detail of Cabo Rojo reef group. C: detail of Veracruz reef group showing the group off Veracruz city and the group off Antón Lizardo. (Redrawn from Tunnell 1988.)

between the two sets of reefs, and the larger Río Papaloapan debouches about 40 km south of the Antón Lizardo reefs. Devastation of the surface and near-surface zones in both sets of southern reefs suggests that outflow from one or both of these rivers has severely affected the reef-top and related communities. Both the Cabo Rojo and the Veracruz–Antón Lizardo reef groups show the effects of past hurricanes that have uprooted massive deepwater coral heads and strewn them in various locations from reef crest to deeper waters.

Technical literature dealing with the western gulf reefs is scattered, and many of the references are difficult to locate, but together they do provide a fairly clear picture of the reef communities. Referring to the northern or Cabo Rojo group, Rigby and McIntyre (1966) provided detailed descriptions of the geology and biological community zonation, particularly for Isla de Lobos. E. Chavez (1973) and E. Chavez, Sevilla, and Hidalgo (1970) also discussed the benthic communities. Specific biological groups have been addressed by Huerta and Barrientos (1965) on algae, Tunnell and Nelson (1989) on octocorals, D. Moore (1958) on stony corals, Tunnell (1974) on mollusks, and J. Ray (1974) on crustaceans. For the southern reefs off Veracruz and Antón Lizardo, Emery (1963) provided a general description, and P. Edwards (1969) discussed the surrounding sediments. Works dealing with particular biological groups include Huerta, Chávez, and Sanchez (1977) and Lehman and Tunnell (1992) on algae; L. Lidz and B. Lidz (1966) on foraminiferans; Aladro-Lubel (1984) on ciliate protozoans; G. Green (1977) and Macias (1968) on sponges; T. Nelson, Stinnett, and Tunnell (1988) and Tunnell and Nelson (1989) on octocorals; Heilprin (1890), Rannefeld (1972), and Santiago (1977) on stony corals; Tunnell (1974) on mollusks; Krutak and Rickles (1979) and Krutak, Rickles, and Argaez (1980) on ostracods; Henkel (1982) on echinoderms; and Flores-Coto (1974) on planktonic appendicularians (urochordates). In addition, Kuhlmann (1975) discussed the coral reefs in general, Villalobos (1971) described biotic communities of the reefs, and Tunnell (1988, 1992) elaborated on natural and human-induced stresses on coral reefs of the southwestern Gulf.

Community zonation on the Cabo Rojo reefs generally follows that observed earlier for the Yucatán reefs except that, rising from shallower water, they lack the lower coralline algal and foraminifera-algal zones, and the lower massive coral (*Agaricia-Montastrea*) zone, although present, is not clearly defined. However, the main massive

times referred to as "table reefs" since they rise abruptly from the shelf floor and are rather flat on top. The individual reefs trend from southeast to northwest and may represent organic development on ancient beach dunes, although the southern group could be growing on a base of igneous rock. Unlike the Yucatán reefs, those of the western Gulf are located near the continental shoreline, surrounded by sandy sediments of terrestrial origin, and influenced by river effluents.

The Cabo Rojo reefs rise from depths of 6–37 m, and most approach sea level. They are known to be bathed occasionally by fresher, sediment-laden water from the Río Pánuco, which enters the Gulf at Tampico some distance away. They are also influenced by winter storms, or "nortes," which chill nearshore waters of the western Gulf. However, considering their general health, neither river water nor low temperature seems to have had any permanent effect on this group of reefs. The Veracruz and Antón Lizardo banks rise from depths of 5–37 m, and many crest at sea level. The Río Jamapa enters the Gulf coral and *Acropora palmata* zones are quite well represented. On the windward reef the *A. palmata* community extends from the surface to a depth of about 6 m, and the main massive coral community occurs in the depth range of 6–24 m. Rigby and McIntyre (1966) subdivided the latter into three subzones: *Montastrea annularis–Porites* (6–9 m), *Diploria strigosa* (8–17 m), and *Montastrea cavernosa* (17–24 m). The lower slope of the bank is covered with a sand apron that precludes development of deeper hard-substrate communities. There are minor differences in depth zonation and species composition between the windward and leeward reefs.

The *Acropora palmata* community of both windward and leeward reefs is superimposed upon a base of massive hemispherical corals, and *Lithothamnium* crusts are evident. Also present in this community are various algae (*Caulerpa, Halimeda, Penicillus*), sponges, sea anemones, several other species of stony corals (*Diploria, Millepora, Montastrea, Porites,* and *Siderastrea*), sea urchins, and tunicates. In the massive coral zone are found algae (*Caulerpa*), soft corals (*Eunicea*), crinoids, and encrusting bryozoans. A number of stony coral species are present including, in deeper areas, *Agaricia* and *Oculina*.

In some places along the windward edge of the reef flat, ridges of coral boulders (largely heads of *Diploria* and *Montastrea,* some two and a half meters in diameter) have been wrenched from the massive coral zone of the lower reef bank and piled along the reef crest by hurricanes, and these ridges rest on a *Lithothamnium* pavement. This pavement also largely surrounds the inner lagoon, which averages about 1.5 m in depth but in places is over 2 m deep. On the sandy bottom grow several species of algae (*Caulerpa, Padina, Penicillus,* etc.), but most of the lagoon bottom is covered by seagrass (*Thalassia*) meadows. Inhabitants of these meadows include algae (*Halimeda*) and some isolated stony corals (primarily *Porites,* but also including *Diploria* and *Siderastrea*) as well as mollusks, sea urchins, and other invertebrates. On all sides the reef bank is incised by grooves extending from the crest into deeper water, giving the entire structure a typical spur and groove topography.

The southern reefs off Veracruz and Antón Lizardo follow the same general pattern of zonation except that on most reefs the *Acropora palmata* zone is barely represented or absent. In some cases a few clumps are present, and in others only dead stumps of *A. palmata* remain, but along the upper windward slopes of most of the banks there is no remaining evidence of this community even

though dead fragments of *A. palmata* may be found on the reef flat and in deeper water. In the few localities where this species is still present, it may be accompanied by other coral species (*A. cervicornis, Agaricia, Madracis, Scolymia, Solenastrea,* and *Stephanocoenia*) as well as coralline algae, foraminiferans, sponges, and colonial sea anemones (*Palythoa*).

On the windward slope of the bank, below the devastated *A. palmata* zone, the massive coral zone in the 9–17 m depth range remains largely intact. This community is dominated by the large corals (*Diploria, Montastrea,* and *Porites*) as well as *Colpophyllia* and *Stephanocoenia*. The deeper community, found below 17 m and at least as deep as 25 m (dominated by *Agaricia, Montastrea,* and *Stephanocoenia*), also appears to be intact. In the deeper water many of the corals are flattened like shingles and reduced to encrustations, apparently in response to reduced light.

The highest part of the reef is the reef flat, which crests on the windward side. This area, which is emergent at low tide and barely awash at high tide, slopes leeward toward the central lagoon. The reef flat itself is strewn with dead coral boulders and other coral fragments. Here are found encrusting coralline algae, leafy algae (*Caulerpa*), colonial sea anemones (*Palythoa*), a few encrustations of stony corals (*Diploria, Millepora,* and *Siderastrea*), and sea urchins (*Echinometra*). The lagoon, which is only 1–1.5 m deep, is floored with bare sand flats, and as in the Cabo Rojo banks, it supports extensive stands of seagrass (*Thalassia*). In this environment are encountered patches of filamentous algae, calcareous algae (*Halimeda*), small sponges, a few scattered corals (*Diploria, Millepora, Porites,* and *Siderastrea*), polychaetes, and sea urchins. On the lee side of the bank the bottom consists of fine carbonate sands and muds, and dense heads of the coral *Diploria* grow in the 0–6 m depth range. Outlying deeper reefs do not reach the surface, and some of these support quite dense growths of a few species of octocorals (*Erythropodium, Plexaura,* and others). Although there is much variation between individual reefs, the pattern given above generally prevails.

Tunnell (1985) noted that these southern reefs have deteriorated during the past few decades in which they have been regularly studied. He also documented that the water bathing these reefs is often fresher and brownish (presumably because of the presence of fulvic and humic acids from land drainage), and it contains much suspended particulate material that reduces light penetration. This layer, which is darker in color near shore, may extend down to a depth of 3 m. The water is most likely outflow from the Río Jamapa and perhaps also from the Río Papaloapan. It is particularly common around the southern reefs during the later summer months (July and August), the height of the rainy season. Although these reefs have undoubtedly been subjected to some river run-off for millennia, the effects appear to have been much more severe in recent years, and this in turn likely reflects human land practices in the drainage basins. In addition to increased runoff rates and heavier loads of suspended particulates resulting from deforestation and agricultural practices, the rivers may now carry loads of various agricultural and industrial chemical pollutants as well as municipal wastes injurious to *A. palmata, A. cervicornis,* and other biological forms that normally inhabit upper levels of the reef.

## REEFS OF THE NORTHWESTERN GULF (FLOWER GARDEN BANKS)

Although many hard banks and rocky outcrops occur on the northwestern Gulf continental shelf and some are known to harbor a few species of stony corals, hardy communities of tropical reef-building organisms occur only on the summits of the East and West Flower Garden Banks, about 190 km south-southeast of Galveston, Texas (fig. 11.21). These banks, about 12 km apart, rise from

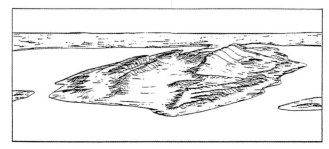

**Figure 11.22.** Sketch of the West Flower Garden Bank as it might have looked during the last Ice Age, when the level of the Gulf of Mexico stood 121–134 m lower than at present. (From G. Edwards 1971.)

100–150 m of water and crest at depths of 15–20 m. The East Flower Garden Bank is pear shaped and covers an area of about 67 km², whereas the West Flower Garden Bank is more oval and occupies around 137 km². These prominences represent surface expressions of salt diapirs capped with fractured bedrock pushed up from below and exposed on the ocean bottom. Their hard surfaces provided substrates for the initial growth of reef-building organisms. Environmental conditions were favorable, and complex reef communities developed on both banks, changing the stony covering of the crests and modifying the surrounding sediments. Since these banks were probably entirely exposed during the lowest sea level stand of the last Ice Age (fig. 11.22) and the summits were still emergent 6,000–8,000 years ago, the reefs are of fairly recent origin.

The continental shelf surrounding the banks is composed of soft sediments (sands, silts, and clays transported to the Gulf by streams entering the north shore), but terrigenous sediments do not occur on the bank slopes at depths shallower than 75–80 m. Above this level the sediments are all coarse carbonate skeletal sands and gravels as well as hard limestone structures built by the corals and other reef-building organisms themselves. Natural gas from subsurface deposits escapes from a number of sites around the banks, and brine seeps (resulting from dissolution of the underlying salt) occur along the sides of the banks.

Most banks of the northern Gulf are precluded from supporting active coral reefs by low winter temperatures, low salinity, high turbidity, heavy sedimentation, or a combination of these factors. Because they are distant from the coastline, well away from the influence of Mississippi River outflow, and somewhat below the surface layer

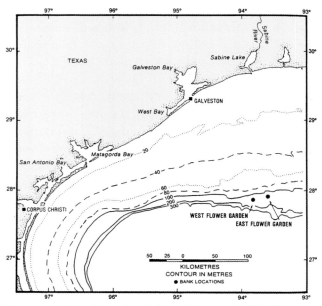

**Figure 11.21.** Map of a section of the northwestern Gulf shelf showing the location of the East and West Flower Garden Banks.

of the ocean, the Flower Garden Banks are uniquely situated. The summits are almost continuously exposed to clear subtropical oceanic water. The annual temperature at the bank crests ranges between 18° and 32°C, and the salinity remains close to 35 ppt. Water bathing the reefs has little suspended sediment, and light penetration is high. Studies have also shown that bottom water currents flow around the reefs and do not rise up from below and flow over them. Thus, the sediment-laden bottom nepheloid layer rarely rises above a depth of 75 m.

A substantial literature has developed concerning the geology and biology of the Flower Garden Banks. General references include Bright and Pequegnat (1974), Bright, Kraemer, et al. (1984), Bright, McGrail, et al. (1985), Rezak, Bright, and McGrail (1983, 1985), and Rezak, Gittings, and Bright (1990). G. Edwards (1971) described the geology. Specific biological groups have been addressed as follows: Eiseman and Blair (1982) on algae; Minnery (1984, 1990) and Minnery, Rezak, and Bright (1985) on coralline algae; Tresslar (1974b) on foraminiferans; Defenbaugh (1974) on hydroids; Giammona (1978) on octocorals; Gittings, Boland, et al. (1992), Gittings, Bright, et al. (1992), Tresslar (1974a), Viada (1980), and others on stony corals; Lipka (1974) on mollusks; Wills and Bright (1974) on polychaetes; Maddocks (1974), L. Pequegnat and Ray (1974), and J. Ray (1974) on crustaceans; Burke (1974a, 1974b) and Dubois (1975) on echinoderms; and Cropper (1973) and Leuterman (1979) on bryozoans. Ecological communities have been examined by Abbott (1975), Abbott and Bright (1975), Gittings (1983), Gittings, Bright, and Powell (1984), and Putt, Gettleson, and Phillips (1986). Other references of interest include Abbott (1979) on ecological processes; Gittings, Boland, et al. (1992) on coral spawning; Gittings, Bright, et al. (1992) and Kraemer (1982) on coral growth; Hagman and Gittings (1992) on coral bleaching; Gittings and Bright (1987) on sea urchin mortality; and E. N. Powell et al. (1983) on brine seep communities. The most complete summary report is Rezak, Bright, and McGrail (1985). Although no complete listing of all the flora and fauna of the Flower Garden Banks is available (a near-complete listing is given in Rezak, Bright, and McGrail 1985), a compilation of algae and invertebrate lists from the above sources includes about 100 species of filamentous, leafy, and coralline algae and nearly 400 types of invertebrates already known to occur there (table 11.5). In addition, well over 100 species of fishes have been recorded from the banks. Many of the smaller forms and cryptic species have not been studied thoroughly, if at all. Certainly, well over 1,000 different species of organisms inhabit the banks.

In an earlier section of the present chapter it was noted that the hard banks of the northwestern Gulf continental shelf support biological communities that are arranged in depth-related zones, and these are listed in table 11.3. Six of the seven zones occur on the Flower Garden Banks, and these will now be examined in more detail. Since the biota and zonation of the two banks differ only slightly, a single description will serve for both. To aid in visualizing the biota and zonation patterns, a perspective drawing of the East Flower Garden Bank is provided (fig. 11.23).

The zones and depth ranges of the six community types found on the Flower Garden Banks include the following: *Diploria-Montastrea-Porites* zone (15–36 m), *Madracis* zone (28–46 m), *Stephanocoenia-Millepora* zone (36–52 m), algal-sponge zone (46–88 m), antipatharian zone (82–89 m), and nepheloid zone (86+ m). Until recently, the near-surface *Acropora palmata* zone, so conspicuous on the reefs of south Florida, Yucatán, and the western Gulf, was considered to be totally absent from the Flower Garden Banks. No specimens of *A. palmata*, *A. cervicornis*, sea fans, sea whips, and other associates of this shallow-water community had ever been encountered here. This was considered to be in part because the banks crest at 15–20 m, below the depth where the shallow-water forms are most abundant. However, *A. palmata* has recently been discovered on the banks (Zimmer et al. 2006), and efforts are being made to locate other members of this community.

The topmost assemblage at the Flower Garden Banks is the main massive coral zone (high-diversity coral reefs) dominated by *Diploria, Montastrea,* and *Porites.* This reef consists largely of closely spaced coral heads up to 3 m in height and diameter. Sixteen of the 18 species of hermatypic corals known to occur on the banks are found here. Coralline algae are abundant, but leafy algae are scarce, possibly as a result of heavy grazing. Most of the invertebrates listed in table 11.5 were recorded from this zone. Particularly abundant are mollusks, polychaetes, crustaceans, bryozoans, and the sea urchin *Diadema,* a notorious algae grazer. On some large exposed knolls where massive coral heads and *Diadema* are rare, leafy algae (including *Caulerpa, Chaetomorpha, Codium, Dictyota, Lobophora, Rhodymenia, Stypopodium, Valonia,* and others) are quite abundant. At the lower margins of the main massive

**Table 11.5.** *Representative genera of organisms reported from the Flower Garden Banks of the northwestern Gulf. The approximate number of forms known for each group is given after the group name. In most cases these represent species, but some forms have been identified only to genus or a higher taxonomic level. (Data compiled from many sources.)*

| Group | No. of forms | Representative genera |
|---|---|---|
| Green algae | 27 | *Caulerpa, Cladophora, Halimeda, Udotea* |
| Brown algae | 8 | *Dictyopteris, Dictyota, Sargassum* |
| Red algae (leafy) | 55 | *Botryocladia, Chrysymenia, Galaxaura, Halymenia, Peyssonnelia, Rhodymenia* |
| Coralline algae | 10 | *Hydrolithon, Lithophyllum, Lithoporella, Lithothamnium, Tenarea* |
| Foraminifera | 34 | *Amphistigina, Carterina, Gypsina, Siphonina* |
| Sponges | 27 | *Ircinia, Neofibularia, Placospongia, Spongia, Xenospongia* |
| Hydroids | 17 | *Aglaophenia, Obelia, Plumularia, Sertularia* |
| Anemones | 2 | *Condylactis, Lebrunia* |
| Soft corals | 11 | *Ellisella, Leptogorgia, Scleracis* |
| Stony corals | 29 | *Agaricia, Colpophyllia, Diploria, Helioseris, Madracis, Millepora, Montastrea, Mussa, Oculina, Porites, Scolymia, Siderastrea, Stephanocoenia* |
| Antipatharians | 2+ | *Antipathes, Cirripathes* |
| Mollusks | | |
|     Chitons | 1 | *Acanthochitona* |
|     Gastropods | 41 | *Astraea, Cerithium, Conus, Coralliophila, Cypraea, Pisania, Siliquaria* |
|     Bivalves | 21 | *Arca, Chama, Isognomon, Lithophaga, Spondylus* |
|     Tusk shells | 1 | *Dentalium* |
|     Cephalopods | 2 | *Loligo, Octopus* |
| Polychaetes | 20 | *Dorvillea, Eunice, Harmothoe, Hermodice, Spirobranchus* |
| Arthropods | | |
|     Pycnogonids | 1 | *Eurycyde* |
|     Ostracods | 34 | *Cytherura, Orionina, Polycope* |
|     Amphipods | 7 | *Colomastix, Corophium, Melita* |
|     Barnacles | 6 | *Acasta, Balanus, Lepas, Scalpellum* |
|     Stomatopods | 2 | *Gonodactylus* |
|     Mysids | 1 | *Heteromysis* |
|     Shrimps | 28 | *Alpheus, Leander, Periclimenaeus, Periclimenes, Processa, Synalpheus* |
|     Anomurans | 12 | *Dardanus, Munida, Paguristes, Pagurus, Polypagurus* |
|     Lobsters | 2 | *Panulirus, Scyllarides* |
|     True crabs | 16 | *Carpilius, Domecia, Micropanope, Mithrax, Pilumnus, Stenorhynchus* |
| Bryozoans | 42 | *Antropora, Celleporaria, Schizoporella, Smittina* |
| Echinoderms | | |
|     Asteroids | 11 | *Asterinopsis, Astropecten, Linckia, Narcissia, Ophidiaster* |
|     Echinoids | 9 | *Arbacia, Clypeaster, Diadema, Eucidaris, Meoma* |
|     Ophiuroids | 13 | *Ophiactis, Ophiocoma, Ophioderma, Ophionereis, Ophiothrix* |
|     Holothuroids | 2 | *Isostichopus, Thyone* |
|     Crinoids | 5 | *Comactinia, Crinometra, Ctenantedon* |

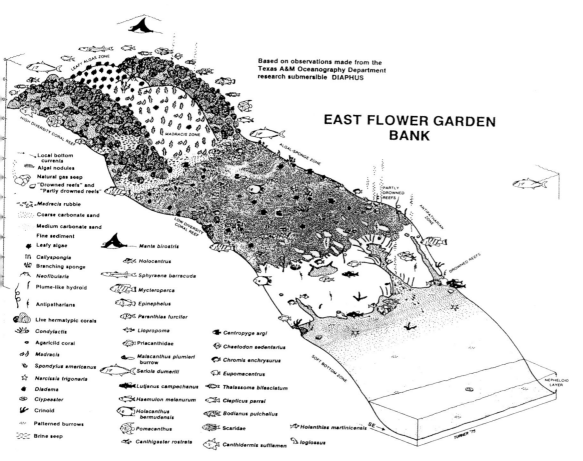

Based on observations made from the
Texas A&M Oceanography Department
research submersible DIAPHUS

**EAST FLOWER GARDEN
BANK**

Local bottom currents
Algal nodules
Natural gas seep
"Drowned reefs" and "Partly drowned reefs"
Madracis rubble
Coarse carbonate sand
Medium carbonate sand
Fine sediment
Leafy algae
Callyspongia
Branching sponge
Neofibularia
Plume-like hydroid
Antipatharians
Live hermatypic corals
Condylactis
Agaricild coral
Madracis
Spondylus americanus
Narcissia trigonaria
Diadema
Clypeaster
Crinoid
Patterned burrows
Brine seep

Manta birostris
Holocentrus
Sphyraena barracuda
Mycteroperca
Epinephelus
Paranthias furcifer
Liopropoma
Priacanthidae
Malacanthus plumieri burrow
Seriola dumerili
Lutjanus campechanus
Haemulon melanurum
Holacanthus bermudensis
Pomacanthus
Canthigaster rostrata

Centropyge argi
Chaetodon sedentarius
Chromis enchrysurus
Eupomacentrus
Thalassoma bifasciatum
Clepticus parrai
Bodianus pulchellus
Scaridae
Canthidermis sufflamen
Holanthias martinicensis
Ioglossus

**Figure 11.23.**
Perspective drawing of the southwest flank of the East Flower Garden Bank showing biotic zones and characteristic organisms. High-diversity coral reefs = *Diploria-Montastrea-Porites* zone; low-diversity reefs = *Stephanocoenia-Madracis* zone. Note bubbles of natural gas (methane) seeping from several places on the bank. Brine sweeps feed a small lake at the base of the carbonate bank (detailed in fig. 4.15). (From Rezak, Bright, and McGrail 1983.)

coral zone and on lower portions of the leafy algae knolls are found thickets of the coral *Madracis*. These are often accompanied by leafy algae and sponges.

Below the main massive coral and *Madracis* assemblages lies the lower massive coral zone (low-diversity coral reefs) dominated by *Stephanocoenia* and *Millepora*, but containing 12 of the 18 stony coral species. Here the coral cover is much less dense and more variable. The most abundant corals, in addition to *Stephanocoenia* and *Millepora*, are *Colpophyllia, Diploria, Montastrea, Mussa,* and *Scolymia*. Encrusting coralline algae are conspicuous, and large numbers of the bivalve *Spondylus* and the sea urchin *Diadema* are present.

The latter assemblage grades into the algal-sponge zone, the largest zone of the banks, which extends down to a depth of 88 m. Dominant organisms here are the foraminiferan *Gypsina* and the coralline alga *Lithothamnium*, whose nodules (1–10 cm or more) cover 50–80 percent of the bottom. Some leafy algae (*Halimeda* and *Udotea*) rise between the nodules. The few corals (primarily *Agaricia, Helioceris, Madracis,* and *Montastrea*) in this area grow as flat, platelike forms. Also present are sponges (*Chelotropella, Neofibularia*), thorny corals or antipatharians (*Antipathes, Cirripathes*), gastropods (*Siliquaria*), bivalves

(*Spondylus*), various crustaceans, starfishes (*Linckia*), sea urchins (*Arbacia, Pseudoboletia*), and abundant crinoids. A major component of the algal-sponge zone is the assemblage of the partially drowned reefs, old reef structures now covered with crusts of coralline algae. Also present here are some of the other organisms typical of the algal-sponge community. Conspicuous are large sea anemones (*Condylactis, Lebrunia*), small colonies of *Millepora* and other stony corals, and occasional basket stars and crinoids. Together, organisms of the above zones account for almost all of the primary production and reef-building activity of the banks.

The algal-sponge zone grades into the transition or antipatharian zone, which forms a small band at 82–89 m. Leafy algae, coralline algae, and hermatypic corals are quite rare or absent. Although some drowned reef outcrops occur, most of the bottom is composed of carbonate sands made up of dead tests of the foraminiferan *Amphistegina*. Conspicuous are the coiled antipatharians (*Antipathes, Cirripathes*), numerous species of octocorals, and several species of crinoids, as well as starfishes (*Chaetaster, Narcissia*) and sea urchins (*Clypeaster*). Unlike in the above areas, the fish fauna here is scarce and limited to a few deep-reef species.

The antipatharian zone marks the end of primary production and reef-building activity. Below this lies the nepheloid zone of turbid, sediment-laden water. The bottom here consists of soft, muddy sediments, and although patches of drowned reef occur, these are covered with a layer of silt. Light penetration is low. Living organisms are rare and restricted to a few tolerant species such as certain deepwater octocorals, solitary ahermatypic corals, antipatharians, and crinoids.

## OVERVIEW OF CORAL REEFS

On coral reefs photosynthesis is carried out by several groups of plants including the various free-living algae (leafy algae and coralline algae growing as crusts or nodules), endolithic blue-green algae (which live near the surface inside the skeletons of stony corals), and symbiotic zooxanthellae (dinoflagellates living within the cells and tissues of various marine invertebrates such as certain sponges, sea anemones, soft corals, and stony corals). Seagrasses associated with some reefs also contribute to the photosynthetic activity. As a result, coral reefs are highly productive ecosystems, and some of the highest primary production rates ever recorded on earth, daily production values of 2–10 g C/m$^2$, have been achieved by tropical Pacific reefs. Although no such studies have been carried out in the Gulf, it is safe to assume that these reefs are likewise very productive even when bathed by waters with low nutrient levels and low phytoplankton production. In addition to the organic matter derived by local production, reefs support many plankton feeders such as sponges, soft and stony corals, some polychaetes, barnacles, bryozoans, crinoids, and others. By capturing drifting plankton from the water column these species further add to the organic matter available to the reef community.

Larger reef species, such as sponges and the soft and stony corals, create complex vertical structures that provide protection from the force of water currents and shelter from predators, thus affording habitat for the smaller and cryptic fauna of the reef. The massive corals themselves are composed of hard but spongy aragonite, a form of carbonate that is hard but easily penetrated by the various boring and tunneling organisms of the reef. The abundance and diversity of such organisms is illustrated by a single large coral head from south Florida that, upon dissection, yielded 8,267 individual infaunal boring organisms (mostly polychaetes) representing 220 different species (Jaap 1984). All marine plant and animal phyla are represented in the reef biota. Ecologically, these include the primary producers and plankton feeders, mentioned above, as well as herbivores, several levels of carnivores, detritus feeders, parasites, and commensals. Some of the species interactions will be examined in chapter 15. Wherever it is located, the coral reef community is characterized by high species diversity, vertical zonation patterns, high production and standing crop biomass, rapid nutrient turnover, and close recycling of nutrients with little loss of nutrients to the outside water column. In these respects the coral reef community resembles that of the tropical rain forest.

The primary coral reefs of the Western Hemisphere are located around islands of the West Indies and mainland areas bordering the Caribbean Sea. Here, under ideal environmental conditions, the reefs achieve their most luxuriant growth and harbor the most diverse associated flora and fauna. Within the Gulf of Mexico the four areas of coral reef development, all ultimately derived from the Caribbean, are somewhat less diverse, and most are probably subject to at least occasional periods of environmental stress. The number of species of hermatypic corals known from each area is as follows: Caribbean (ca. 55), south Florida (ca. 51), Yucatán (ca. 34), western Gulf (33), and Flower Garden Banks (18). When the Yucatán banks are more thoroughly studied, the number of coral species there will certainly increase. If the shallow-water *Acropora palmata* community were represented, the Flower Garden Banks would also have more species. Nevertheless, there is a definite reduction in the number of coral species in the reefs farther from the Caribbean, and this progressive loss of tropical species also shows up in the associated biota. Particularly noticeable is the reduction in the number of species of gorgonians, or sea fans. More than 40 species occur in the Caribbean, 39 in south Florida, and around 20 on the Yucatán reefs, but none have been found on reefs of the western Gulf or at the Flower Gardens. Other tropical groups show similar patterns. This progressive loss of tropical reef-associated species relates, in part, to problems of larval recruitment at a distance through the available water currents, but it also relates to conditions of survival once they do arrive. Low winter or high summer temperatures, periodic low salinity, high turbidity, and heavy sedimentation are often suggested as limiting factors, but biological interactions such as competition and predation may also play a role in larval settlement and survival.

The Flower Garden Banks harbor the northernmost

thriving coral reef communities of the North American continent, and these might be expected to show some signs of stress, but those species that are present appear to be generally healthy despite their northerly location. When compared on a species-by-species basis, coral growth rates are comparable to those of reefs farther south. For example, *Montastrea annularis* at the Flower Garden Banks shows an accretionary growth rate of about 8 mm/yr compared to 11.2 mm/yr for south Florida and 6.5 mm/yr in the Caribbean. In all areas growth rates vary somewhat from one species to another and with depth in the water column, faster growth being associated with shallower depths. Recent coral reef studies have focused on such topics as spawning, recruitment, effects of pollution, and management of coral reefs of the Gulf. These complex but highly sensitive communities are important resources, and much yet remains to be learned concerning their composition, internal community dynamics, and relations with other living systems of the sea.

In 1992, the Flower Garden Banks were designated as a National Marine Sanctuary. Updates on the status and condition of the Banks have been conducted by Pattengill-Semmens and Gittings (2003) and the Office of National Marine Sanctuaries (2008).

# 12 : PLANT-DOMINATED COMMUNITIES
## SALT MARSHES, MANGROVE SWAMPS, AND SEAGRASS MEADOWS

Vascular plants are not prominent features of the world oceans, but along low-energy coastlines and in protected areas such as bays and saline lagoons, they are generally the dominant vegetational types. For the most part these plants are associated with soft, muddy or sandy substrates that provide anchorage for their roots and other subterranean structures, but in some instances they grow on hard surfaces that have sufficient soft sediment coverage or crevices for firm root attachment. The primary community types are salt marshes and mangrove swamps, which grow along the edge of the sea, and seagrass meadows, which are submerged in shallow marine waters. Cypress-gum swamps, associated with lowland streams, ponds, freshwater swamps, and other such habitats, do not grow beside the sea because their species cannot tolerate salt water.

Less than a quarter of the Gulf of Mexico is directly bordered by vascular plant communities, but their importance to the biological economy of the Gulf is far out of proportion to their actual physical coverage. Whether growing in bays and lagoons or bordering the Gulf, they provide feeding grounds for many resident species of invertebrates and fishes, as well as sea turtles and some marine mammals. They also serve as nursery areas (protected habitat and food resources) for the young of many species whose adults live elsewhere. In addition, these communities export organic material across the continental shelf and, to some extent, to the deep Gulf, thereby providing food subsidies to areas well beyond where they actually grow. They also serve as a buffer against wave action, thus retarding erosional loss of coastal shorelines.

### HIGHER PLANTS IN THE MARINE ENVIRONMENT

Why are there no trees or daffodils in the sea? This classical question, often posed to candidates on doctoral oral exams, has a complex answer, and some of the desired information is still poorly known. Since higher plants require light for photosynthesis, they are restricted to shallow portions of the continental shelf. Water, of course, is a very dense medium, and the force of waves and water currents in shallow-water environments can be very great, especially during stormy weather. Thus, there would be a need for very firm anchorage in the soft and shifting substrates, and the stems and branches would have to have some means of withstanding the force of the water. This could be accomplished by very rigid structures (such as the steel legs of an oil platform) or by very supple structures (such as the long, flexible, stemlike *stipes* of kelp plants that connect the bottom-anchoring holdfasts to the surface floats and fronds).

Besides these basic morphological adaptations, such plants would have to include adaptations to the anoxic conditions of submerged sediments surrounding the root systems, some physiological mechanisms for contending with the high salinity of the substrate and water column, and special adaptations for reproduction, especially fertilization, in marine waters. Actually, some higher plants have made appropriate adjustments and are highly successful as emergent or submerged forms around the margins of estuaries and lagoons and as submerged forms in the sea itself. A few remarks about these species, their environments, and their adaptations are appropriate.

Plants that regularly grow in saline environments are called *halophytes* (salt plants). Information concerning the species and their worldwide distribution patterns is given by Chapman (1960). Ponnamperuma (1972) has discussed the complex chemistry of submerged soils, and Waisel (1972) has described the physiological adaptations of plants to saline conditions. True halophytes are very specialized plants belonging to a limited number of plant families, and these tolerant species often have broad geographic distributions around the world. Where they are present, they often grow in great abundance in stands of only one or a few species because of the lack of competition from other species in the stressful habitats. An ecological classification of the marine halophytes of the Gulf of Mexico and the major genera is given as follows:

A. Low coast halophytes
   1. Salt and brackish-water marshes (*Distichlis, Juncus, Phragmites, Scirpus, Spartina,* etc.).
   2. Mangrove swamps (*Avicennia, Conocarpus, Laguncularia, Rhizophora*)
B. Submerged marine halophytes

1. Seagrass meadows (*Halodule, Halophila, Syringodium, Thalassia*)

Submerged soils are generally anoxic, necessitating special mechanisms for bringing oxygen to the root systems. Some halophytes (such as *Spartina,* mangroves, and others) have large air ducts in their tissues that transport oxygen from the leaves and stems down to the buried roots. In addition, certain mangroves have developed stilt roots, or long vertical projections from the roots (*pneumatophores*) with air ducts to supply the root system with oxygen. The roots, in turn, release some of the oxygen into the surrounding soil so they are not in direct contact with the anoxic sediments. In those species that do not have air ducts, the roots must contend with high concentrations of methane and other reduced organic compounds, hydrogen sulfide, and various metals in the chemically reduced state.

All halophytes must deal with high salinity in the soil, the water column, or both. This boils down to three basic problems. The first is the high external concentration of salts (primarily sodium and chloride ions). The second is the need to take in freshwater against a strong concentration gradient (the water inside the tissues having a tendency to leave the plant and move to the outside). The third problem is the need to retain the proper osmotic pressure, or *turgor,* of the plant cells (which tend to shrink as water is lost to the environment). Not only do the plants have to contend with the salinity problem in general, but they must do so in an environment in which the salinity sometimes changes on a daily or seasonal basis in response to the tidal cycle, droughts, floods, rainfall, and storm-induced rises in sea level. Such salinity shifts may be rapid and extreme. Although some plants tolerate high internal salt concentrations, most do not, and all must regulate the internal concentrations of specific ions (especially sodium and chloride, which can be lethal in high concentrations). Halophytes meet these challenges in several ways. Most limit salt intake. Some accumulate excess salts in specialized cells. Others eliminate excess salts through well-developed salt-secretion glands in the leaves or elsewhere. Detailed discussion of adaptive mechanisms is beyond the scope of this book, but a general appreciation of the problems faced by halophytes is important to an understanding of their distribution and ecology.

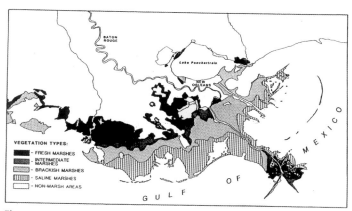

**Figure 12.1.** Distribution of plant communities in the lowlands of southeastern Louisiana showing the extensive areas of salt and brackish-water marshes bordering the Gulf sounds and bays. (From Gosselink 1984, after Chabreck and Linscombe 1978.)

## SALT MARSH COMMUNITIES

Salt and brackish-water marsh communities dominate low-energy tidal shorelines of the northeastern Gulf between Tampa Bay and Cedar Key, Florida, as well as the complex coastline of southeastern Louisiana, especially around the Mississippi River Delta (fig. 12.1). They are also abundant in low-energy saline environments on the landward side of barrier islands and in lowland areas bordering sounds, bays, and estuaries along the northern and northwestern edge of the Gulf. However, below about Tampa Bay in the east and Tampico in the west, they are largely replaced by the more tropical mangrove swamps. Altogether, the marginal marshes of the Gulf cover about 2.3 million hectares (5.7 million acres) (Eleuterius 1976), of which nearly two-thirds are found in coastal Louisiana. Literature dealing with these marshlands is fairly extensive. Among the most important references are those that provide thorough community descriptions (Britton and Morton 1989; Gosselink 1984; and Stout 1984) and those treating more specific topics, including distribution and species composition (Eleuterius 1973, 1976; Thorne 1954), environmental relations (Gagliano and van Beek 1970; Reimold and Queen 1974), community zonation (Chabreck 1970, 1972; Eleuterius and Eleuterius 1979; Penfound and Hathaway 1938), production (Day, Smith, et al. 1973; de la Cruz 1973; Gabriel and de la Cruz 1974; Keefe 1972; E. Odum 1974; R. Turner 1976; R. Turner and Gosselink 1975), algal associates (Sage and Sullivan 1978), decomposition (Post and de la Cruz 1977), and economic value (Gosselink, Odum, and Pope 1973).

Salt and brackish-water marshes thrive primarily in the intertidal area, although they are found a quarter to a half meter above and below this zone. In their normal habitats salinity may range up to about 28 ppt, but some species tolerate higher salinities such as those encountered in the hypersaline lagoons of south Texas. Substrates range from fine silt to coarse sand, and many have a high organic content, in part because of the decomposing roots and rhizomes of the marsh plants themselves. Salinity of the interstitial water within the soil varies and may exceed 24 ppt depending on the overlying water, elevation, and other factors, and within a given area the soil salinity may change in response to the tidal cycle, flooding, and drought.

## Vascular plants: Species composition and zonation

Coastal marshland communities of the Gulf are characterized by low species diversity. Within the more saline areas there are often vast acreages of homogeneous stands consisting of only one or a few vascular plant species. However, toward the brackish and fresher water areas the number of resident and occasional species increases significantly. Since these communities are typically dominated by just a few species, it is convenient to characterize them in terms of these dominant forms, a few of which are shown in figure 12.2. Some of the plant species exhibit different growth forms, with a tall type occupying lower elevations and a short form occurring some meters away toward the upland. Another characteristic of coastal marshes is their patterns of zonation, that is, the tendency of species to be distributed in bands with respect to environmental gradients. This is particularly pronounced in relation to the salinity gradient, where different groups of species occupy the salt, brackish, intermediate, and fresher water marshes. Species zonation also occurs along the elevation gradient from lower to upper intertidal and then to the high meadow above the normal high-tide line. Examples of such zonations are shown in figure 12.3.

Within the salt marsh, smooth cordgrass (*Spartina alterniflora*) is generally the dominant species fronting the open water, and some distance back from the open water this is joined by the black needlerush (*Juncus roemerianus*). Extensive acreages of one or both of these species plus others fringe many of the coastlines of bays and estuaries of the northern Gulf. In somewhat less saline water the smooth cordgrass becomes greatly reduced, and the community is joined by other species such as salt grass (*Distichlis spicata*), saltmeadow cordgrass (*Spartina patens*), three-square (*Scirpus olneyi*), and several other

halophytes. The brackish marsh itself may grade into a lower-salinity (intermediate) marsh dominated by giant cordgrass (*Spartina cynosuroides*), Roseau cane (*Phragmites australis*), arrowhead (*Sagittaria latifolia*), bulrushes (*Scirpus* spp.), and other forms. This community, in turn, may grade into freshwater marshes or swamps, each with its own characteristic suite of dominant vascular plant species. Dominant species of salt, brackish, and lower-salinity marshes are given in table 12.1.

If the salt or brackish marsh grades into higher ground, it may be bounded by a fairly open, wet, saline area inhabited by saltwort (*Batis maritima*), glassworts (*Salicornia* spp.), dwarf specimens of black mangrove (*Avicennia germinans*), and other salt-tolerant species. On slightly higher ground these give way to the high meadow dominated by needlerush (*Juncus*) and saltmeadow cordgrass. In this zone are found many different plant species including shrubs such as marsh elder (*Iva frutescens*) and groundsel (*Baccharis halimifolia*) and a variety of herbaceous species including asters (*Aster* spp.), goldenrods (*Solidago* spp.), and others. These zonation patterns are observed all around the periphery of the northern Gulf.

In addition to the tall, emergent, grasslike plants, the marshes support large communities of benthic diatoms and filamentous, mostly blue-green, algae, prominent among which are *Anacystis, Microcoleus, Oscillatoria,* and *Schizothrix*. The diatoms, which make up a more or less continuous cover on the marsh sediments, are quite diverse, and more than 100 species have been identified. Many algal species also grow as epiphytes on the submerged stalks and leaves of the dominant marsh plants. Among the most frequently occurring are the filamentous green algae *Cladophora, Enteromorpha,* and *Rhizoclonium* and the red alga *Bostrychia*, often encountered on hard substrates around the edge of the Gulf. The coastal marshes also support many species of fungi, which aid in the decomposition processes taking place in the marshland.

## Animals of salt and brackish marshes

Salt and brackish-water marshes of the Gulf support an interesting assortment of resident and transient animal species. Residents are those forms that are specifically adapted to intermediate salinities (around 5–30 ppt) and that tend to be found in the marshlands throughout the year. Only a few species belong in this category. The transients are derived from fresher or more saline waters or from the uplands. Some of the transient species regu-

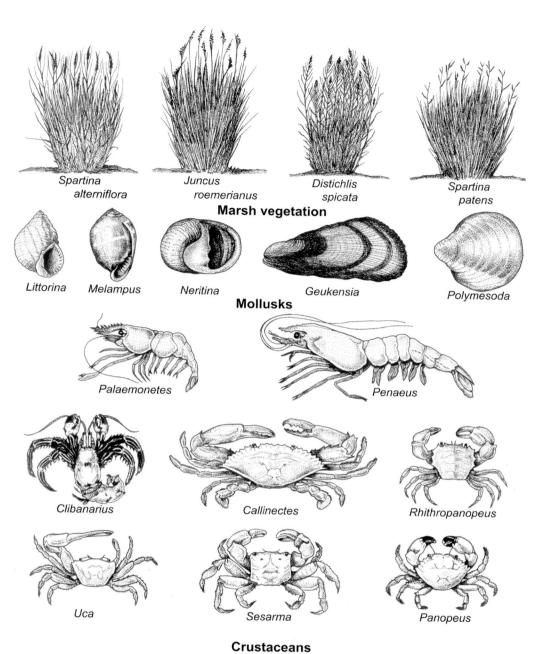

**Figure 12.2.**
Representative
examples of the
dominant vascular
plants and common
invertebrates of
brackish and saline
marshes bordering
the Gulf of Mexico.

| Spartina alterniflora | Juncus roemerianus | Distichlis spicata | Spartina patens |

**Marsh vegetation**

| Littorina | Melampus | Neritina | Geukensia | Polymesoda |

**Mollusks**

| Palaemonetes | Penaeus |

| Clibanarius | Callinectes | Rhithropanopeus |

| Uca | Sesarma | Panopeus |

**Crustaceans**

larly inhabit the marshes at certain seasons of the year and then migrate out to other areas. Others are irregular inhabitants that forage in the productive marshlands on an opportunistic basis. Particularly important among the seasonal transients are those derived from nearby marine areas. Many of these arrive as planktonic larvae or juveniles. After growing up in the marshes, they move out to the marine waters for spawning as subadults or adults.

ZOOPLANKTON

Zooplankton populations of coastal marshes of the northeastern Gulf have been studied by Cuzon du Rest (1963) and L. Shipp (1977). Plankton samples taken in the marshlands generally contain large amounts of decomposing fibrous plant material, which clogs up meshes of the collecting nets. Hence, it is very difficult to obtain quantitative samples and exact counts of marsh zooplankton. Available information shows that the populations vary in relation to season and hydrographic regimes, especially

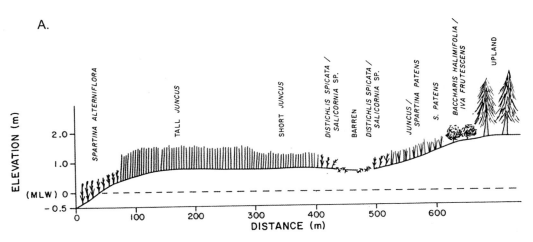

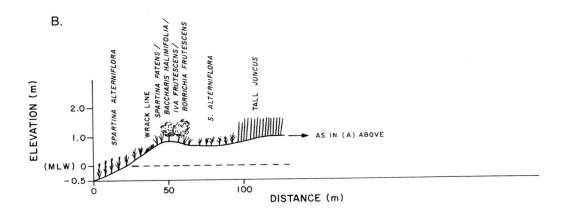

**Figure 12.3.**
Generalized diagrams of
salt marsh zonation on
protected low-energy
(A) and open moderate-
energy (B) shorelines
of the Gulf of Mexico.
(From Stout 1984.)

salinity. The very successful estuarine copepod *Acartia tonsa* is numerically dominant in most areas and throughout the year, its abundance often far exceeding that of all other species combined. All life history stages of this species are present. From time to time other estuarine copepods such as *Eurytemora* and *Oithona* are also quite abundant. Among the diverse forms that appear regularly or occasionally in marsh plankton samples are foraminiferans, medusae, ctenophores (*Mnemiopsis* and *Beroë*), polychaetes (*Tomopteris*), leeches, cladocerans, ostracods, free-living and parasitic copepods (*Argulus, Centropages, Cyclops, Ergasilus, Eucyclops, Labidocera, Macrocyclops, Paracalanus, Pseudodiaptomus, Tortanus,* etc.), mysids, isopods, amphipods, small shrimps (*Palaemonetes, Penaeus*), small crabs, and urochordates. In addition, the plankton is often rich with larval stages of polychaetes, mollusks, copepods, barnacles, shrimps, crabs, and echinoderms. Fish eggs, larvae, and juveniles are seasonally abundant. Larval stages of over a dozen species of estuarine and marine decapod crustaceans have been identi-

fied in the marsh zooplankton, and those of fiddler crabs (*Uca* spp.) are generally present in the greatest numbers. Larvae of the several species peak at different seasons of the year, and larvae of marine species are most frequent in the more saline marshes near open Gulf waters. Vertical migration is a common phenomenon in decapod larvae and other zooplankton groups. Many, particularly the more advanced larval stages, apparently remain on the bottom during the daytime and move up into the water column at night, where they are more available to collecting nets. Compared to that in open waters of the continental shelf, the zooplankton of coastal marshes is generally less diverse, but it may be more abundant in terms of numbers of individuals and biomass.

### MEIOFAUNA

The meiofauna of tidal marshes of the northeastern Gulf has been discussed by Harp (1980), Humphrey (1979), and Stout (1984). The fauna is depauperate in terms of both abundance and species diversity. Most populations

**Table 12.1.** *Dominant species of vascular plants in salt, brackish, and low-salinity marshes of the northeastern Gulf of Mexico. (Modified from Stout 1984.)*

| Scientific name | Common name |
|---|---|
| **Salt marsh** | |
| *Distichlis spicata* | salt grass |
| *Juncus roemerianus* | black needlerush |
| *Salicornia* spp. | saltworts |
| *Scirpus olneyi* | three-square |
| *Scirpus robustus* | leafy sedge |
| *Spartina alterniflora* | smooth cordgrass |
| *Spartina cynosuroides* | giant cordgrass |
| *Spartina patens* | saltmeadow cordgrass |
| **Brackish marsh** | |
| *Distichlis spicata* | salt grass |
| *Juncus roemerianus* | black needlerush |
| *Limonium carolinianum* | sea lavender |
| *Sagittaria latifolia* | arrowhead |
| *Scirpus olneyi* | three-square |
| *Spartina alterniflora* | smooth cordgrass |
| *Spartina cynosuroides* | giant cordgrass |
| *Spartina patens* | saltmeadow cordgrass |
| **Low-salinity (intermediate) marsh** | |
| *Cladium jamaicense* | saw grass |
| *Iris virginica* | blue flag |
| *Juncus roemerianus* | black needlerush |
| *Phragmites australis* | Roseau cane |
| *Scirpus validus* | bullwhip |

are dominated by nematodes, followed by harpacticoid copepods, although ostracods are sometimes well represented. Other forms include foraminiferans, polychaete and oligochaete annelids, tiny mollusks, and larvae of larger invertebrates including insects. Seasonal differences in biomass occur, with a peak in the spring followed by a summer low, probably because of predation by fishes and larger invertebrates. Although there is some local differentiation, meiofaunal populations of different plant associations show a high degree of similarity. The generally low biomass and diversity values for the marsh populations could reflect adverse physical conditions of the soil or water column, but these are probably reinforced by biological factors such as food availability, competition, and predation.

MACROINVERTEBRATES

The larger invertebrate fauna of the coastal marshes and associated tidal creeks and ponds has been discussed by a number of authors including Britton and Morton (1989), Heard (1979a), Stout (1984), Subrahmanyam and Coultas (1980), Subrahmanyam and Drake (1975), and Subrahmanyam, Kruczynski, and Drake (1976). Remarkably, there are only a few truly resident species, which consist chiefly of annelids, mollusks, crustaceans, and insects (table 12.2). Some of the more prominent forms are illustrated in figure 12.2. However, some of these resident groups are abundant and may be very important in the breakdown of organic material and aeration of the soils. Particularly significant in this regard are the large populations of fiddler crabs (*Uca* spp.), which feed on algae, small invertebrates, and organic detritus and which construct burrows in both submerged and slightly emergent soils. Seasonally, the marshes are invaded by waves of crustacean larvae and young, especially those of the brown and white shrimps (*Penaeus aztecus* and *P. setiferus*) and the blue crab (*Callinectes sapidus*). These arrive during the spring and summer, and after growing up in the marshes, most are gone by late fall. Some species of insects are year-round residents, but many more utilize the marshes seasonally or stray in from the neighboring uplands. Most of the major insect groups are represented. Anyone who has worked in coastal marshes is familiar with the hordes of salt marsh mosquitoes (*Culex* spp.), but ants, grasshoppers, and other forms may also be temporarily abundant.

FISHES

Coastal marsh fishes of the northeastern Gulf have been studied by Hackney and de la Cruz (1981), Kilby (1955), Subrahmanyam and Coultas (1980), and Subrahmanyam and Drake (1975). The fish fauna of northern Florida marshes was divided by Subrahmanyam and Drake (1975) into five categories that are probably typical of all coastal marshes. These include (a) permanent residents, (b) juveniles of nonresident species, (c) adult migrants, (d) individuals foraging from outside the marshes, and (e) rare sporadic visitors. As seen in table 12.2, the true residents consist of only a few small species of the families Cyprinodontidae (killifishes) and Poeciliidae (live-bearers), whereas the transients represent nearly 100 species from more than 30 families. Major families of transient species include the Clupeidae (herrings), Mugilidae (mullets), Carangidae (jacks), Gerreidae (mojarras), Sparidae

**Table 12.2.** *Common animal inhabitants of salt and brackish marshes of the northern Gulf of Mexico, including both resident and transient forms.*

**Residents**

Meiofauna

    Foraminifera, nematodes, polychaete and oligochaete annelids, mollusks, ostracods, harpacticoid copepods, invertebrate larvae

Macroinvertebrates

    Polychaetes – *Heteromastis, Lycastopsis, Nereis*

    Snails – *Littorina, Melampus, Neritina*

    Bivalves – *Geukensia, Modiolus, Polymesoda*

    Crustacea

        Isopods – *Cyathura*

        Amphipods – *Gammarus, Orchestia*

        Shrimps – *Palaemonetes*

        Hermit crabs – *Clibanarius*

        True crabs – *Panopeus, Rhithropanopeus, Sesarma, Uca*

    Insects – Many, primarily of the order Diptera, Hemiptera, and Hymenoptera

Vertebrates

    Fishes – several residents, primarily cyprinodonts (*Adinia, Cyprinodon, Fundulus, Lucania*) and poeciliids (*Poecilia*)

    Reptiles – alligator (*Alligator*), snakes (*Nerodia*), and turtles (*Malaclemys, Pseudemys*)

    Birds – 19 species listed as year-round residents (see table 12.3)

    Mammals – nutria (*Myocastor*), muskrat (*Ondatra*), and marsh rabbit (*Sylvilagus*)

**Transients**

Macroinvertebrates

    Crustaceans

        Shrimps – brown shrimp (*Penaeus aztecus*) and white shrimp (*Penaeus setiferus*)

        True crabs – blue crab (*Callinectes sapidus*)

    Insects – more than 200 species representative of many orders

Vertebrates

    Fishes – nearly 100 species, mostly juveniles of marine groups, but also including adults of some freshwater and estuarine forms

    Birds – more than 100 species recorded including many casual visitors. Most frequently associated species listed in table 12.3.

    Mammals – rats, rabbits, minks, otters, weasels, raccoons, foxes, bobcats, deer

(porgies), Sciaenidae (drums), Gobiidae (gobies), and Bothidae (lefteye flounders), most of which are represented by juveniles. Although a few transients (bass, sunfish, etc.) are derived from fresher areas, most are marine. In fact, the outstanding feature of the coastal marsh ichthyofauna is the seasonal appearance of waves of young of marine species that somehow find their way into the marshlands, where they feed and grow until large enough to survive outside the marshes.

Although the dominant transients vary somewhat with habitat, the following seasonal pattern is typical. Winter and spring dominants include the Gulf menhaden (*Brevoortia patronus*), bay anchovy (*Anchoa mitchilli*), striped mullet (*Mugil cephalus*), and rough silverside (*Membras martinica*). Other spring dominants include the spotfin mojarra (*Eucinostomus argenteus*), pinfish (*Lagodon rhomboides*), and spot (*Leiostomus xanthurus*). Bay anchovies are abundant through the summer months, and the spotfin mojarra and pinfish dominate in the fall. Adult migrants of many species utilize the marshes, especially during the summer and fall months. Foraging species, derived largely from estuarine and marine waters, enter the marshes primarily in warm weather during periods of high tide. These include the silverstripe halfbeak (*Hyporhamphus unifasciatus*), silver perch (*Bairdiella chrysoura*), and sand seatrout (*Cynoscion arenarius*), among others. Prominent among the sporadic visitors are the American eel (*Anguilla rostrata*), hardhead catfish (*Arius felis*), sheepshead (*Archosargus probatocephalus*), spotted seatrout (*Cynoscion nebulosus*), great barracuda (*Sphyraena barracuda*), and Atlantic threadfin (*Polydactylus octonemus*). The seasonal appearance of individual species in the marshlands largely reflects, of course, the seasonal breeding patterns of the individual species. Local biomass is related not only to these seasonal patterns, but also to the tidal cycle. During periods of high tide more of the marshland is flooded, and the fishes are more widely dispersed, whereas at low tide they are concentrated in the smaller portion of the marsh that is still flooded. Also, at high tide the foraging and incidental fishes move into the marshes, but these retreat to the deeper water of bays and estuaries as the tide goes down and the water of the marshes becomes shallower. Because of these daily and seasonal changes and the difficulty of operating collecting nets, it is hard to obtain reliable quantitative data on the biomass of fishes utilizing the marshlands.

## AMPHIBIANS AND REPTILES

Amphibians are generally absent from salt and brackish marshes of the Gulf. However, the American alligator (*Alligator mississippiensis*), a water snake (*Nerodia fasciata*), a few species of turtles (*Pseudemys* spp.), and the diamondback terrapin (*Malaclemys terrapin*) do occur in the marshlands.

## BIRDS

The coastal avifauna of the Gulf of Mexico has been the subject of much study. Important and useful references include Clapp, Banks, et al. (1982), Clapp, Morgan-Jacobs, and Banks (1982), Clapp (1983), Gosselink (1984), Imhoff (1976), Keller, Spendelow, and Greer (1984), Lowery (1974), Lowery and Newman (1954), Oberholser and Kincaid (1974), and G. Sutton (1951). In dealing with the subject of coastal birds of the Gulf, Lowery and Newman (1954) remarked, "Fully nine-tenths of the species of birds in eastern North America have been recorded at one time or another in the counties bordering the Gulf coasts of Florida, Alabama, Mississippi, Louisiana, and Texas, and it is likely that all but a few of these species have occurred at certain times and places within a hundred yards of the surf." The potential avifauna of the coastal marshlands is very great if one considers the resident species and regular visitors as well as all the seasonal transients and incidental forms that enter, approach, or migrate across the coastal wetlands. However, from the literature it is often difficult to determine which species actually occur in the marshes and to what extent each makes use of the marsh habitat. Table 12.3 lists those species considered to be most closely associated with marshes of the northeastern Gulf. These include various swimming birds (ducks and their relatives), wading birds (egrets, herons, ibises), shorebirds (gulls, terns, plovers, rails, sandpipers), and some upland birds (blackbirds, crows, larks, sparrows, and wrens).

In dealing with marsh birds of the northeastern Gulf, Stout (1984) listed only three species as being confined to the marshes: the seaside sparrow (*Ammodramus maritimus*), marsh wren (*Cistothorus palustris*), and Louisiana clapper rail (*Rallus longirostris*). However, a total of 19 species are considered to be year-round residents, all of which breed locally, and another 11 species are summer residents, 7 of which are local breeders. Twenty-six species are winter visitors to the marshes, and 10 additional species are migrants. Beyond these, many bird species from other habitats, especially the local uplands,

**Table 12.3.** *Bird species associated with salt and brackish marshes of the northeastern Gulf of Mexico. B = breeding population; M = migrant species. (Modified from Stout 1984.)*

**Year-round residents**

| | | | |
|---|---|---|---|
| Turkey vulture | B | Black-crowned night heron | B |
| King rail | B | Mottled duck | B |
| Clapper rail | B | Fish crow | B |
| Forster's tern | B | Marsh wren | B |
| American oystercatcher | B | Eastern meadowlark | B |
| Snowy plover | B | Red-winged blackbird | B |
| Killdeer | B | Boat-tailed grackle | B |
| Great blue heron | B | Sharp-tailed sparrow | B |
| Great egret | B | Seaside sparrow | B |
| Snowy egret | B | | |

**Summer residents**

| | | | |
|---|---|---|---|
| Purple gallinule | B | Green-backed heron | B |
| Common moorhen | B | Little blue heron | B |
| Wilson's plover | B | Tricolored heron | B |
| Whimbrel | | Least bittern | B |
| Short-billed dowitcher | | White ibis | |
| Great white heron | | | |

**Winter visitors**

| | | | |
|---|---|---|---|
| Northern harrier | | Fulvous whistling duck | |
| Virginia rail | M | Mallard | |
| Sora | M | American black duck | |
| Yellow rail | | Gadwall | |
| Caspian tern | | Northern pintail | |
| Semipalmated plover | | Green-winged teal | |
| Piping plover | | American widgeon | |
| Black-bellied plover | M | Northern shoveler | |
| Ruddy turnstone | M | Redhead | M |
| Long-billed curlew | M | Lesser scaup | M |
| Least sandpiper | M | Surf scoter | |
| Dunlin | M | Sedge wren | |
| Western sandpiper | M | Western meadowlark | |

**Additional migrants**

| | |
|---|---|
| Gull-billed tern | |
| Willet | Semipalmated sandpiper |
| Greater yellowlegs | American avocet |
| Redknot | American bittern |
| Stilt sandpiper | Blue-winged teal |
| | Tree swallow |

occur incidentally in the coastal marshlands from time to time. The marsh habitat provides nesting areas relatively safe from predation, extensive feeding grounds, and places of concealment for resting birds. Food habits vary from one species to another, but the marshes provide ample supplies of small fishes, crustaceans, snails, and insects, as well as seeds and other vegetation parts, all of which are utilized by one species or another. Fishes and fiddler crabs are the favored foods of many. As a bird habitat, coastal marshes are especially important.

MAMMALS

Only a few mammal species utilize the marshes (table 12.2). The three resident species, nutria (*Myocastor*), muskrat (*Ondatra*), and swamp rabbit (*Sylvilagus*) are all herbivores, and when abundant they can destroy large areas of marsh vegetation. Other herbivores, including rats (particularly the rice rat, *Oryzomys*), cotton rat (*Sigmodon*), and white-tailed deer (*Odocoileus*), sometimes descend from higher ground and feed on marsh vegetation. Several species of carnivorous mammals including minks and weasels (*Mustela*), otters (*Lontra*), red foxes (*Vulpes*), and even bobcats (*Felis*) forage in the marshes, taking rats, eggs, and young birds, as well as crustaceans and fishes. The omnivorous raccoons (*Procyon*) seek out live food as well as carrion. As a group, except for the resident species, mammals are not important inhabitants of the coastal marshes.

## Standing crop and primary production of marsh vegetation

Studies of standing crop and primary production in coastal marshes of the northern Gulf have been summarized by a number of authors including Gosselink (1984), Keefe (1972), Stout (1984), and R. Turner (1976). The vascular plants themselves consist of aboveground structures (primarily leaves and stems) and those that lie belowground (roots and rhizomes). Although all photosynthesis is carried out by the aboveground parts, plant growth or accumulation of organic material (net production) is apportioned between the aboveground and belowground structures. In marsh plants of the northern Gulf, measurements of aboveground standing crop have given values in the range of 650–2,500 g dry weight/m², with an average of about 1,250 g/m². Estimates of aboveground net production vary widely, in part because of differences in methods of assessment. However, it is clear that there are real dif-

ferences in production rates of the different species, and for a given species at different seasons and in different environments. Growth is highest during spring and summer and least in fall and winter. The more favorable environments are those that are better flushed by moving water and where the soil is partially aerated. Less favorable environments include more stagnant areas away from the open coast and tidal creeks where soils are anoxic. For *Spartina alterniflora* and *Juncus* marshes, aboveground net production values range from a few hundred to over 9,000 g dry weight/m²/yr. A conservative estimate of the average net production rate would be around 2,000 g dry weight/m²/yr (= ca. 17,840 lb/A/yr).

The biomass and net production of the roots and rhizomes exceed those of the stems and leaves. In favorable environments the biomass of the belowground portions may be only two or three times that of the aboveground structures, but in unfavorable soil conditions, such as in stagnant water or certain upland areas, the ratio may be 20 or 40 to 1. In poor environments more root surface is required to support a given amount of aboveground material. Nitrogen has been found to be the primary limiting nutrient for marsh vegetation, and this element, which is acquired by the root system, is more readily available in well-flushed environments.

The periphyton, consisting mainly of filamentous algae and diatoms, grows on submerged portions of the vascular plant stems and leaves. Although it has a much lower standing crop biomass than the macrophytes, it is metabolically far more active, and in certain environments it has been found to be responsible for 4–11 percent of the total photosynthesis. Because it is smaller and less fibrous than vascular plants, and hence more easily consumed and richer in proteins, the periphyton is doubtless of considerable importance to the consumer species of the marsh ecosystem.

## Decomposition and export of organic matter from coastal marshes

For the marsh community as a whole, the most important part of the vascular plant is the aboveground portion. In the fall and winter over half the standing crop of stems and leaves dies back and falls into the water or onto the soil surface. Processes of biological decomposition and mechanical breakdown quickly release any fluid cell components, and the remaining fibrous material is gradually reduced to fine particulate matter, a process that may take a year or more to complete. Early on, the detritus is colonized by various species of bacteria, fungi, algae (including diatoms), protozoa, and small invertebrates, many of which play active roles in the decomposition process. These detritus particles, now enriched with small organisms, may continue decomposing on the marsh floor or they may be eaten by various consumer species or exported from the marshes by the flushing action of water currents associated with tides and periodic floods. The amount of marsh-derived organic detritus exported to neighboring bays, estuaries, and continental shelf waters appears to vary with local circumstances, but only a few quantitative estimates are available. Day, Smith, et al. (1973) concluded that about half the annual net production was exported from the marsh, and Gosselink (1984) cited potential export values of 300 and 1,120 g/m²/yr. Happ, Gosselink, and Day (1977) estimated that a Louisiana bay exported to the Gulf about 150 g of organic carbon/m²/yr, most of which could be accounted for by marsh export to the bay. Kennicutt (1991), also discussed by Darnell (1991a), provided direct evidence (from ¹³C measurements) of terrestrial detritus on the continental shelf off eastern Louisiana, Mississippi, and Alabama. Thus, organic detritus originally swept from marshes into bays and estuaries winds up, at least in some cases, on the continental shelf as a subsidy for marine systems.

The marshes also export organic matter to coastal waters through the migratory activities of many types of marine fishes and invertebrates that feed and grow up in the marshes before migrating back to the continental shelf to spawn. R. Turner (1977) clearly demonstrated the importance of coastal marshes for shrimp production, and Copeland (1965) calculated the annual biomass of fishes and invertebrates passing seaward through a tidal pass in south Texas. For the entire bay system, including marshes and open water, the annual export was about 576 kg/ha (513 lb/A), of which about 3.3 percent was penaeid shrimp (see Darnell and Soniat 1979). Through the export of organic detritus and the emigration of fishes and invertebrates, the coastal marshes clearly make important contributions to ecosystems of the continental shelf, even if many of the details remain to be worked out.

## MANGROVE SWAMP COMMUNITIES

Mangroves are trees and shrubs that grow in lowland coastal habitats in tropical and subtropical regions of the world. All are adapted to live in saline habitats on loose, wet soil subject to periodic tidal submergence. As in the case of coastal marshes, mangrove systems are limited by factors associated with the climate, salinity, tidal fluctuation, and substrate. Since they are basically tropical species, mangroves do not develop well where the average annual temperature is below about 19°C (66°F), and most do not survive freezing temperatures for any length of time. Although they tolerate the salinity of full seawater, mangroves do not require it, and most species also grow well in freshwater. Tidal action, involving alternate wetting and drying, is also not required by mangroves, but such action serves to bring in nutrients, flush out accumulated organic material and reduced sulfur compounds, aid in the transport of seedlings, and reduce competition with nontolerant plant species. Since the tides often carry salt water well inland, in many low-lying areas they allow mangroves to penetrate up water courses well away from the coast. Mangroves grow best in low-energy depositional environments. High wave energy prevents the accumulation of fine sediments, destroys the shallow mangrove root systems, and inhibits the establishment of new plants. For these reasons, mangrove swamps and forests are best developed along deltaic coasts, behind barrier islands, and in protected bays and estuaries where the substrate is composed of silt or clay and has a high percentage of organic matter. Anoxic sediments, which restrict many potential competitive species, pose no major problem for mangroves because of their special mechanisms for transporting oxygen to the root systems.

Most mangrove species are viviparous; that is, the seeds develop while still attached to the parent tree. Following fertilization, the zygote develops continuously through the embryo stage to the seedling while still attached to the parent plant. There is no resistant resting stage or seed, and the new developing seedling is referred to as a *propagule*. The degree of development and morphological appearance of propagules vary greatly from one species to another. Most cast-off seedlings develop in substrates near the parent plant, but some float to distant shores. Since all mangrove propagules are dispersed by water, they must be able to survive floating in seawater for extended periods of time, in some species for up to a year.

Because they are found on tropical and subtropical coasts around the globe, mangrove swamps and forests have been the subject of considerable study. General references to mangroves and mangrove systems include Chapman (1976a, 1976b), Lugo and Snedaker (1974), Macnae (1968), Waisel (1972), G. Walsh (1974), and G. Walsh, Snedaker, and Teas (1974). Worldwide more than 50 species of mangroves belonging to 12 different plant families are recognized, but only 3 or 4 species occur within the Gulf of Mexico, and these are found primarily in southern Florida from the Cedar Key area, north of Tampa Bay, south through the Florida Keys. They also occur in protected bays and lagoons of Mexico from above the level of Tampico south and eastward along the coastal lagoons of Yucatán. A few scattered thickets of one species (the black mangrove) occur sporadically on barrier islands of the northern Gulf. The most extensive mangrove swamps and forests of the Gulf are those of southern Florida, which cover an estimated 180,000 ha (ca. 450,000 A) including the Ten Thousand Islands area and lower portions of the Everglades.

General accounts of the mangrove systems of the Gulf are provided by Pool, Snedaker, and Lugo (1977) and Thorne (1954). For the south Florida area more specific information is given by a number of authors as follows: vegetation identification (Carlton 1975), general ecology (Odum, McIvor, and Smith 1982), plant succession (J. Davis 1940), productivity (Lugo et al. 1975), trophic relations (Odum and Heald 1972), macrofauna (P. T. Sheridan 1992), hurricane damage (McCoy et al. 1996), and relations with other communities (Schomer and Drew 1982). Mangroves of the northern Gulf have been discussed by Sherrod and McMillan (1985), and studies on mangrove systems of Mexico have been published by Day, Conner, et al. (1987), Ley-Lou (1985), Lot, Vasquez-Yáñez, and Menendez (1975), R. Sánchez and Elana (1963), Vargas Maldonado, Yáñez-Arancibia, and Amezcuna Linares (1981), and Yáñez and Day (1982).

### Vascular plants: Species composition and zonation

Four treelike species (the red, black, and white mangrove and the buttonwood) make up most of the woody vegetation found in mangrove forests of the Gulf coast. These are illustrated in figure 12.4, and each is briefly described below.

The red mangrove (*Rhizophora mangle*) is easily recognized by the presence of prop roots, which stem directly from the main trunk or drop from the upper branches. This mangrove may grow as a low bush or, in more favorable locations, as a tree up to 25 m (80 ft) in height. It is generally found in pure stands at the seaward edge of the swamp forest, but in many areas it also occurs in mixed stands with other mangrove species. The propagules can reach a length of 30 cm (12 in) while still attached to the parent plant, and once detached, they may survive for at least a year floating in seawater.

The black mangrove (*Avicennia germinans*) is recognized by the absence of prop roots and the presence of many short, pencil-like pneumatophores projecting vertically from the shallow horizontal root system. The shiny green leaves are often coated with salt crystals, and the dark green bean-shaped propagules are only about an inch long while still attached to the parent plant. Black mangroves often grow in relatively pure stands just inland from the red mangrove fringe, or they may occur in mixed stands. Of all the mangrove species occurring in the Gulf region, this species is the most tolerant of low temperatures, and it is the only species found around the northern Gulf, where it occurs sporadically as stunted thickets about a meter in height. Although freezing weather kills off upper portions of the plant, regeneration takes place from the surviving root system.

The white mangrove (*Laguncularia racemosa*) lacks prop roots and exhibits only a few short vertical pneumatophores. Trees reach a height of 15 m (49 ft). The leaves are broad and flat, and the propagules are small, only about half an inch long. A pair of salt-secreting glands is present at the base of the leaf. White mangroves normally grow on higher ground behind the red and black mangroves, but mixed stands do occur.

The buttonwood (*Conocarpus erectus*) is not considered to be a true mangrove since it produces thin-shelled seeds rather than propagules. It also lacks prop roots and pneumatophores. As in its near relative, the white mangrove, it possesses salt glands at the base of the leaves. Trees reach a height of 20 m (64 ft). The buttonwood generally occurs on higher ground, not flooded by normal tides, behind the true mangrove species. However, it is an important component of most mangrove forests of the southern Gulf.

In addition to the four woody treelike species, a number of other plants are also found in mangrove swamps and forests. These have been described by Carlton (1975) and others, and a few are worth brief mention. Two vine-like shrubs, the nicker bean (*Caesalpinia crista*) and coin vine (*Dalbergia ecastaphyllum*), often sprawl over mangrove thickets on the landward side. Another vine, marine ivy (*Cissus incisa*), climbs through the crowns of mangroves and sends down to the ground long cord-like aerial roots. Smaller herbaceous plants that may grow as forest

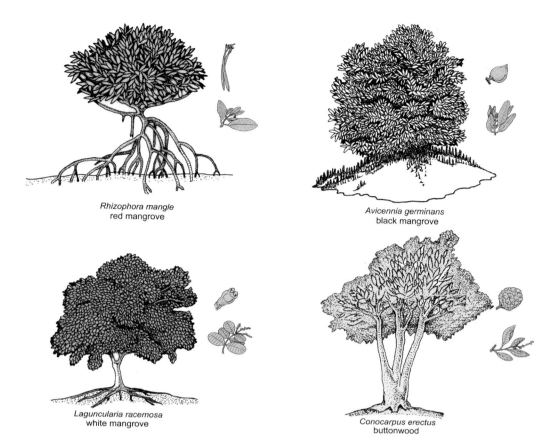

*Rhizophora mangle*
red mangrove

*Avicennia germinans*
black mangrove

*Laguncularia racemosa*
white mangrove

*Conocarpus erectus*
buttonwood

**Figure 12.4.**
The four primary species of woody plants (red mangrove, black mangrove, white mangrove, and buttonwood) making up mangrove forests of the Gulf of Mexico, showing the adult plant as well as flowers, leaves, and propagules (or seeds). (Mangrove species from Odum, McIvor, and Smith 1982.)

understory or along the landward edge include saltwort (*Batis*), glasswort (*Salicornia*), sea purslane (*Sesuvium*), and sea blite (*Suaeda*). On higher ground, associated with buttonwood thickets, there are many additional species including grasses, asters, goldenrods, and others.

Since mangrove forests thrive in a variety of lowland environments, the size and shape of individual plants and the overall aspect of the forest vary considerably from one habitat to another. This has led scientists to classify the mangrove forests into six different categories based on such environmental factors as location, soil characteristics, and flushing rates. These categories are as follows: (1) *overwash forests*—somewhat isolated plants subject to frequent tidal washing and high flushing rates; (2) *fringing mangrove forests*—mangroves growing as a thin fringe along the slightly elevated natural levees lining waterways; (3) *riverine mangrove forests*—mangrove forests growing in lowland floodplains along the flowing waters of rivers and tidal creeks; (4) *basin mangrove forests*—forests occurring on inland depressions that channel terrestrial runoff toward the coast; (5) *hammock forests*—forests similar to the previous type but growing on slightly elevated ground; and (6) *scrub dwarf forests*—

forests limited to flat coastal plains in poor soil where nutrients appear to be limiting (Lugo and Snedaker 1974). These different forest types vary in species composition, tree height and growth form, soil organic accumulation, and other factors. Among the different mangrove forest types, the overwash and fringing forests are tide dominated and most directly related to the marine environment, but the riverine forest, greatly influenced by the outflow of freshwater, may also export large quantities of organic material to the nearby continental shelf. The basin and hammock forests are farther removed from the marine environment, and the scrub forest produces only a small amount of leaf litter.

In many areas the mangroves grow in distinct, almost monospecific zones along the coast (fig. 12.5). The most seaward zone is dominated by the red mangrove. More landward zones are dominated respectively by the black mangrove, white mangrove, and buttonwood. Earlier it was thought that this pattern of zonation represented a classical case of plant succession involving invasion of the sea, building up of soil, and gradual elevation of the land; all beginning with the pioneer species, the red mangrove, and leading to the regional climax community, the ter-

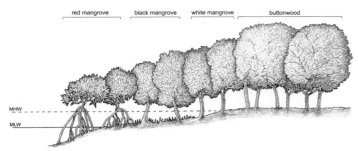

**Figure 12.5.** Generalized diagram of zonation in a fringing mangrove forest. Note that most of the trees grow in the intertidal zone between the mean high and mean low tide levels.

restrial tropical hardwood forest. Most investigators now believe that the coastal mangrove forests act primarily as shoreline stabilizers rather than as land builders and that the zonal pattern simply reflects the differential response of each competing plant species to such factors as wave action, water depth, length of exposure to tidal inundation, salinity, and soil characteristics.

In other areas two or more mangrove species may grow together in mixed stands, and it is not entirely clear why in some cases the stands are pure and arranged zonally while in other cases the species are mixed together. It has been suggested that the zonal pattern is the normal condition resulting from competition between the different species in an environmental gradient. However, from time to time outside disturbance factors such as storms and hurricanes, lightning strikes, fires, herbivory, and disease kill or injure some of the dominant plants and cause breaks in the forest canopy, and these breaks permit the invasion of competing species, resulting in mixed-species stands. Depending on their locality, mangrove swamps and forests grade into various freshwater and upland community types. On the seaward or lagoonal side they may be bounded by seagrass meadows as in south Florida and in the Términos Lagoon of the southern Gulf (fig. 12.6).

## Nonvascular plants

Odum, McIvor, and Smith (1982) have summarized literature dealing with the algae, fungi, and bacteria of the mangrove swamps of south Florida. More than 70 species of filamentous algae have been recorded as epiphytes from prop roots and pneumatophores of the mangroves. Red algae are the most numerous, but blue-green, green, and brown algae are also well represented. These are often arranged in vertical zones on the roots, but zonation also occurs in relation to salinity and other factors. These algae are most abundant along edges of the forest and where there are breaks in the forest canopy, but in the deep shade of the forest they are greatly reduced or absent. Diatoms may also be abundant on prop roots and pneumatophores as well as on the mud surface, but

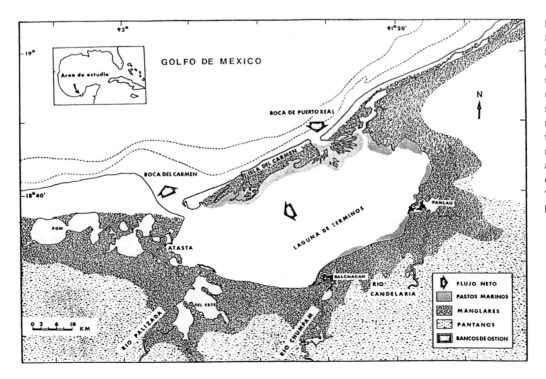

**Figure 12.6.** Map of Términos Lagoon bordering the southern Gulf of Mexico showing the directions of the net flow (flujo neto), seagrass beds (pastos marinos), mangrove forests (manglares), marshes (pantanos), and oyster reefs (bancos de ostiones). (From Yáñez-Arancibia and Day 1982.)

they also are reduced under low light conditions. Bacteria and fungi, important decomposing agents in the abundant leaf litter of the forest floor, are well represented and quite diverse. Although little work has been done on the phytoplankton of mangrove swamps, apparently regular marine and freshwater species occur, depending on the prevailing flow patterns, and these communities can change with the tidal cycle and runoff regimes. In these shallow waters benthic diatoms and dinoflagellates, stirred from the bottom by wind and water currents, frequently appear in the plankton samples.

## Animals of salt and brackish mangrove swamps

### ZOOPLANKTON

Virtually no work has been done on the zooplankton of waters within mangrove swamps of the Gulf of Mexico. However, from studies conducted in marine waters just outside the swamps, it appears likely that swamp tidal waters contain numerous polychaetes, calanoid copepods (especially the abundant *Acartia tonsa*), crustacean eggs and larvae, and small fishes. At night such small benthic crustaceans as mysids, isopods, and amphipods may migrate vertically and feed in the water column.

### MEIOFAUNA

Apparently no meiofaunal studies have been carried out in mangrove swamps of the Gulf area, although work on the decomposition of mangrove leaves has revealed diverse populations of nematode worms.

### MACROINVERTEBRATES

Odum, McIvor, and Smith (1982) have reviewed studies of macroinvertebrates in the mangroves of south Florida. This fauna includes inhabitants of several habitat types (arboreal, prop roots, mud flats, and water column). Although some species are limited to a given habitat, many of the more mobile species move readily from one habitat to another. Lists of representative animals from each of the habitats are given in table 12.4.

The arboreal fauna of mangrove forests is dominated by insects, and Simberloff (1976) and Simberloff and Wilson (1969) list more than 200 species associated with mangroves of overwash islands in the Florida Keys. In the forest canopy these are joined by the curious little mangrove crab (*Aratus pisonii*), which may reach densities of 1–4/m². This crab and a number of the insects feed on mangrove leaves, and in some areas up to a quarter of the leaves are destroyed. Together, these herbivores cause

**Table 12.4.** *Representative macroinvertebrates of the mangrove forests of south Florida. (From Odum, McIvor, and Smith 1982.)*

**Arboreal**

Gastropods – *Cerithidea, Littorina, Melampus*

Isopod – *Ligia*

Crab – *Aratus*

Insects – more than 200 species

**On prop roots**

Polychaete – *Nereis*

Gastropods – *Anachis, Bulla, Crepidula, Diodora, Littorina, Urosalpinx*

Bivalves – *Arca, Brachiodontes, Crassostrea*

Isopod – *Sphaeroma*

Barnacle – *Balanus*

Shrimps – *Palaemon, Periclimenes, Synalpheus, Thor*

Anomuran – *Petrolisthes*

Crabs – 8 species

Tunicate – *Ascidia*

**On mud flats**

Within forest

    Gastropods – *Cerithidea, Melampus*

    Crabs – *Aratus, Rhithropanopeus, Sesarma*

    Insects – 16 species

Adjacent to forest

    Crabs – *Eurytium, Uca*

**In water column**

Shrimps – *Alpheus, Palaemonetes, Penaeus, Periclimenes*

Crab – *Callinectes*

Insects – mosquito larvae

the dropping of many leaves and much fecal material to the forest floor. Additional macroinvertebrates found in the trees include several snails and the semiterrestrial isopod *Ligia exotica*.

On prop roots are found polychaetes, gastropods, bivalves, isopods, barnacles, several small shrimps and crabs, an anomuran, and a tunicate. These feed mainly on the epiphytic algae and organic detritus, but the barnacles, bivalves, and tunicate filter particles from the water column. One of the bivalves, *Crassostrea virginica*, is the common edible oyster. The prop root fauna is distributed in zones with respect to water level and length of submergence. Above mean high water level are found a snail (*Littorina*) and an isopod (*Ligia*). Within the intertidal zone occur polychaetes, a gastropod (*Bulla*), several bivalves, barnacles, and a tunicate. Two isopods of the genus *Sphaeroma* are found below the level of mean low water. One of these, *S. terebrans,* is a wood borer, and this species tunnels into prop roots, permitting the invasion of bacteria and fungi that decompose and weaken the affected roots. In areas of heavy infestation where a number of roots of a given plant are weakened or destroyed, the entire mangrove plant can be toppled by strong storms and hurricanes. These pests occur mostly in high salinities, and they do not appear to constitute a major problem for the mangrove forests as a whole.

On mud flats within the mangrove forest are found several gastropods, mud and marsh crabs, and at least 16 species of insects. These are primarily detritus feeders, nourished by the decomposing leaf litter, but several are omnivorous and consume both living and dead organic material. On mud flats adjacent to the mangrove forest are found several species of fiddler crabs (*Uca*) and a small mud crab (*Eurytium*), all of which burrow into the substrate and consume detritus particles.

Larger invertebrates of the water column consist of several shrimps (including the commercially important pink shrimp, *Penaeus duorarum*), crabs (including the edible blue crab, *Callinectes sapidus*), and the larvae of a number of mosquitoes (principally species of *Aedes* and *Culex*).

The lists given above are clearly incomplete because the invertebrate fauna of most Gulf coast mangrove swamps and forests has not been thoroughly investigated. However, they do provide information on the general nature of the macroinvertebrate fauna, and they point to the ecological importance of mangrove leaf litter in food chains of the benthos and water column of such areas.

FISHES

More than 200 different fish species have been recorded from the various mangrove-associated habitats of south Florida, and the relevant literature has been summarized by Odum, McIvor, and Smith (1982). Most of the species are normal inhabitants of nearby estuaries and the shallow continental shelf, but they run the gamut from purely freshwater forms, such as sunfishes and largemouth bass, to fully marine species, such as parrotfishes and mackerels. Representative examples are given in table 12.5. Although the list is far from complete, it illustrates the wide range of types occurring in the various habitats. About half the species have been found in all the habitat types. Species tolerant of lower salinities are limited to the tidal streams, whereas the more marine forms are confined to the fringes of oceanic bays.

Only a few fish species are true residents of the deeply shaded swamps, where waters are characterized by low oxygen, high acidity, and often high sulfide levels. These include several small freshwater and brackish-water species such as killifishes (*Cyprinodon, Floridichthys, Fundulus,* and *Rivulus*) and live-bearers (*Gambusia* and *Poecilia*). These feed mainly on algae, detritus, and small invertebrates, including mosquito larvae. During the winter dry season these fishes are largely isolated in pools, but spring and summer floods wash many into creeks and streams, where they become prey for larger predatory fishes and birds.

One of the more striking aspects of the mangrove ichthyofauna is the seasonal invasion of the swamps by waves of juvenile, mostly marine, fishes following peak periods of offshore spawning. After moving inshore these young fishes take up residence in protected waters among the trunks, prop roots, and pneumatophores of their mangrove haven. Most arrive in late spring and early summer. As they grow and mature they move into deeper and more open water, and by late fall most have migrated back to the offshore waters where they overwinter. Species following this general pattern include the ladyfish, tarpon, menhaden, snook, snappers (especially the gray snapper), sheepshead, pinfish, silver perch, spotted seatrout, red drum, and striped mullet, among others. In addition to the annual invasion by juveniles, the adults of many estuarine and marine species also move into tidal rivers and creeks during warmer weather. These often accompany the incoming tide, prey on the smaller fishes and invertebrates, and exit with the falling tide. For the most part, these also pass the winter in warmer offshore waters. Included in this group are such forms as adult ladyfish, tarpons, needlefishes, jacks, mangrove snappers, jewfish, and barracudas. In estuarine bays and lagoons additional summer residents include various species of snappers, sea robins, toadfishes, grunts, mojarras, porgies, drums, mullets, soles, and flounders. If seagrasses are nearby, the mangroves may also harbor pipefishes, gobies, scorpionfishes, and puffers. In areas where there are neighboring reefs the mangroves may provide shelter and feeding grounds for such reef-associated species as snappers, sea basses, grunts, porgies, parrotfishes, wrasses, damselfishes, surgeonfishes, and triggerfishes.

**Table 12.5.** *Representative examples of fish species characteristic of three habitats associated with mangrove systems of south Florida. TS = tidal streams, EB = estuarine bays, and OB = oceanic bays. In general, the tidal streams are the least saline, and the oceanic bays are the most saline environments. (From Odum, McIvor, and Smith 1982.)*

| Common name | Scientific name | Habitat TS | EB | OB |
|---|---|:---:|:---:|:---:|
| Bull shark | *Carcharhinus leucas* | + | | |
| Blacktip shark | *Carcharhinus limbatus* | | | + |
| Bonnethead | *Sphyrna tiburo* | | + | + |
| Southern stingray | *Dasyatis americana* | + | + | + |
| Ladyfish | *Elops saurus* | + | + | + |
| Tarpon | *Megalops atlanticus* | + | + | + |
| Bonefish | *Albula vulpes* | | | + |
| Yellowfin menhaden | *Brevoortia smithi* | + | + | |
| Scaled sardine | *Harengula jaguana* | + | + | + |
| Atlantic thread herring | *Opisthonema oglinum* | + | + | + |
| Bay anchovy | *Anchoa mitchilli* | + | + | + |
| Inshore lizardfish | *Synodus foetens* | + | + | + |
| Sea catfish | *Arius felis* | + | + | + |
| Gulf toadfish | *Opsanus beta* | + | + | + |
| Key brotula | *Ogilbia cayorum* | | + | + |
| Halfbeak | *Hyporhamphus unifasciatus* | | + | + |
| Atlantic needlefish | *Strongylura marina* | + | + | |
| Diamond killifish | *Adinia xenica* | + | | |
| Sheepshead minnow | *Cyprinodon variegatus* | + | + | + |
| Gulf killifish | *Fundulus grandis* | + | | |
| Mangrove gambusia | *Gambusia rhizophorae* | + | | |
| Sailfin molly | *Poecilia latipinna* | + | | + |
| Tidewater silverside | *Menidia beryllina* | + | + | + |
| Dwarf seahorse | *Hippocampus zosterae* | | + | + |
| Gulf pipefish | *Syngnathus scovelli* | + | + | + |
| Snook | *Centropomus undecimalis* | + | + | + |
| Black sea bass | *Centropristis striata* | | | + |
| Jewfish | *Epinephelus itajara* | + | + | + |
| Crevalle jack | *Caranx hippos* | + | + | + |
| Florida pompano | *Trachinotus carolinus* | | | + |
| Schoolmaster | *Lutjanus apodus* | + | | + |
| Gray snapper | *Lutjanus griseus* | + | + | + |
| Dog snapper | *Lutjanus jocu* | | | + |
| Lane snapper | *Lutjanus synagris* | + | + | + |
| Spotfin mojarra | *Eucinostomus argenteus* | + | + | + |
| Porkfish | *Anisotremus virginicus* | | | + |
| Pigfish | *Orthopristis chrysoptera* | | + | + |
| Sheepshead | *Archosargus probatocephalus* | + | + | + |
| Pinfish | *Lagodon rhomboides* | + | + | + |
| Silver perch | *Bairdiella chrysoura* | + | + | + |

**Table 12.5.** *Continued*

| Common name | Scientific name | Habitat | | |
| --- | --- | --- | --- | --- |
| | | TS | EB | OB |
| Spotted sea trout | *Cynoscion nebulosus* | + | + | + |
| Atlantic croaker | *Micropogonias undulatus* | + | | + |
| Red drum | *Sciaenops ocellatus* | + | + | + |
| Sergeant major | *Abudefduf saxatilis* | | | + |
| Blue parrotfish | *Scarus coeruleus* | | | + |
| Striped mullet | *Mugil cephalus* | + | + | + |
| Great barracuda | *Sphyraena barracuda* | + | + | + |
| Marbled blenny | *Paraclinus marmoratus* | | | + |
| Florida blenny | *Chasmodes saburrae* | | | + |
| Fat sleeper | *Dormitator maculatus* | + | | |
| Emerald goby | *Gobionellus smaragdus* | + | + | + |
| Clown goby | *Microgobius gulosus* | + | + | |
| Spanish mackerel | *Scomberomorus maculatus* | + | + | + |
| Bighead sea robin | *Prionotus tribulus* | + | + | + |
| Bay whiff | *Citharichthys spilopterus* | | + | + |
| Gulf flounder | *Paralichthys albigutta* | + | + | + |
| Dusky flounder | *Syacium papillosum* | | | + |
| Lined sole | *Achirus lineatus* | + | + | + |
| Hogchoker | *Trinectes maculatus* | + | + | + |
| Planehead filefish | *Monacanthus hispidus* | | + | + |
| Scrawled cowfish | *Acanthostracion quadricornis* | + | + | + |
| Southern puffer | *Sphoeroides nephelus* | + | + | + |
| Striped burrfish | *Chilomycterus schoepfi* | | + | + |

Thus, the fish fauna of southern Gulf mangrove forests varies from place to place and from one season to another. To a large extent, this fauna depends on habitats available in neighboring waters. Some fish species exhibit considerable movement between different habitats, one area serving as a daytime refuge and another serving as a nighttime foraging area. Furthermore, the details vary along the salinity gradient and also along the coastline, more tropical species being found in south Florida and on the Mexican coast, and more temperate species occurring around Tampa Bay and the Cedar Keys.

AMPHIBIANS AND REPTILES

The amphibians and reptiles known or presumed to utilize mangrove swamps and forests of southern Florida are listed in table 12.6. These species include 3 frogs, 10 turtles, 3 lizards, 6 snakes, and 2 crocodilians. Two species (the squirrel treefrog and mud turtle) are abundant, and 10 species (the giant toad, Cuban treefrog, striped mud turtle, Florida softshell turtle, loggerhead turtle, green anole, Cuban brown anole, green water snake, mangrove water snake, and American alligator) are listed as common. The remaining 12 species are uncommon or rare. The list includes species representative of fresh, brackish, and salt water portions of the mangroves.

Because of difficulties in osmoregulation (regulation of salts and water in the internal body fluids), it is unlikely that any of the amphibians regularly occur in brackish or saltwater areas. Among the turtles, several are probably limited to freshwater, but the mud turtle and diamondback terrapin are often found in brackish water. The mud turtle occurs in salt marshes as well as brackish mangrove swamps, but mangroves are the principal habitat for the diamondback terrapin, and this species is distributed

**Table 12.6.** *Amphibians and reptiles recorded from mangrove areas of south Florida. A = abundant, C = common, R = rare. (From Odum, McIvor, and Smith 1982.)*

| Common name | Scientific name | A | C | R |
|---|---|---|---|---|
| **Amphibians** | | | | |
| Giant toad | *Bufo marinus* | | X | |
| Cuban treefrog | *Hyla septentrio-nalis* | | X | |
| Squirrel treefrog | *Hyla squirella* | X | | |
| **Reptiles** | | | | |
| Florida red-bellied turtle | *Chrysemys nelsoni* | | | X |
| Chicken turtle | *Deirochelys reticularia* | | | X |
| Striped mud turtle | *Kinosternon bauri* | | X | |
| Mud turtle | *Kinosternon subrubrum* | X | | |
| Diamondback terrapin | *Malaclemys terrapin* | | | X |
| Florida softshell | *Trionyx ferox* | | X | |
| Loggerhead | *Caretta caretta* | | X | |
| Green turtle | *Chelonia mydas* | | | X |
| Hawksbill | *Eretmochelys imbricata* | | | X |
| Atlantic ridley | *Lepidochelys kempi* | | | X |
| Green anole | *Anolis carolinensis* | | X | |
| Bahaman bank anole | *Anolis distichus* | | | X |
| Cuban brown anole | *Anolis sagrei* | | X | |
| Eastern cottonmouth | *Agkistrodon piscivorus* | | | X |
| Eastern indigo snake | *Drymarchon corais* | | | X |
| Rat snake | *Elaphe obsoleta* | | | X |
| Striped swamp snake | *Liodytes alleni* | | | X |
| Green water snake | *Nerodia cyclopion* | | X | |
| Mangrove water snake | *Nerodia fasciata* | | X | |
| American alligator | *Alligator mississippiensis* | | X | |
| American crocodile | *Crocodylus acutus* | | | X |

in mangrove areas of the mainland and throughout the Florida Keys. Four of the turtles (the loggerhead, green, hawksbill, and Atlantic ridley) are marine species that spend their lives in estuaries or the open sea. All of these marine turtles are associated with mangrove vegetation at some stage of their lives, and green turtles and loggerheads may use mangrove estuaries as nursery areas. All four species will feed on mangrove vegetation, and the hawksbill relies heavily on mangrove leaves, fruit, roots, wood, and bark.

The three lizards of the genus *Anolis* are arboreal and feed on insects, and all have been reported from the mangroves of south Florida. Of the six species of snakes, the mangrove water snake is the most dependent on mangrove swamp habitats, where it feeds on fishes and invertebrates. Alligators occur mainly in lower-salinity sectors, but the American crocodile inhabits brackish and fully marine waters and is dependent on the mangrove habitat. Historically, the distribution of this species was centered in mangrove-dominated areas of the Florida Keys (particularly Key Largo) and nearby mangrove-lined shores and mud flats along the northern edge of Florida and Whitewater Bays of the mainland. During the past half century, largely because of human encroachment and habitat modification, the range of the species has decreased greatly, and it is now virtually eliminated from most of the Keys south of Key Largo. Mangrove thickets along creeks and canal banks of red and black mangrove swamps are critical areas for nest building and are thus important for the reproductive success of this endangered species. The crocodile feeds on larger invertebrates, fishes, and occasional waterbirds.

BIRDS

The various habitats of south Florida support an abundant and diverse avifauna, and important references include Howell (1952), Kushlan (1974), Kushlan and White (1977), Odum, McIvor, and Smith (1982), Robertson (1955), and Robertson and Kushlan (1974). This fauna comprises species that live in south Florida throughout the year and breed there. Included also are migrants from northern areas that overwinter in the warmer climate of south Florida, as well as those that spend the winter farther south and pass through during the spring and fall migratory flights. A few species of tropical birds are present only during the summer months. Of the hundreds of bird species known from south Florida, nearly 200 appear to make use of the mangrove forests and associated habitats, and for some species the mangroves provide critical habitat.

The common and abundant birds of the mangrove forests are listed in table S50, which also provides information on the season(s) of occurrence and abundance status of each species. For convenience, the birds are divided into six groups based on similarities in habitat, behavior, and methods of obtaining food. These include (1) wading birds, (2) probing shorebirds, (3) floating and diving waterbirds, (4) aerially searching birds, (5) birds of prey, and (6) arboreal birds, the latter a catchall group of species that feed and nest mainly in the mangroves.

The wading birds (herons, egrets, storks, ibises, and their long-legged relatives) are all year-round residents, and all nest in the mangroves. Only the wood ibis is considered to be abundant. Two-thirds of the species feed almost exclusively on fishes, but some also take in mollusks, crustaceans, frogs, and even mice. The sand-hill crane, one of the rarer species and not listed in the table, feeds on roots and rhizomes of aquatic vegetation. All these food resources are abundantly available in the waters and lands associated with the mangrove forests. The Louisiana heron, snowy egret, and cattle egret are the most numerous breeders of the south Florida mangroves, and these and other large wading bird species often aggregate in sizable breeding colonies that may include several thousand breeding pairs. Spring is the season of greatest breeding activity, but during every month of the year at least one species is breeding. These wading birds provide large amounts of guano to the soil and water under the mangroves, and beneath rookeries this amounts to at least 1 g/m$^2$/day. The nitrogen and phosphorus from these droppings enhance growth of the mangroves and stimulate growth of the phytoplankton, and in turn the aquatic animal life, but these effects are quite localized.

Probing birds and shorebirds (rails, plovers, sandpipers, and relatives) are associated with intertidal and shallow-water habitats. Twenty-five species are included in this category, of which 2 (spotted sandpiper and semipalmated plover) are abundant, and 16 species are common. Two species, the king rail and willet, are year-round residents that nest in the mangroves, and most of the remainder are winter residents and/or spring and fall migrants. The black-necked stilt is a summer resident only. Most of the probing birds and shorebirds feed on small worms, mollusks, crustaceans, and insects found around shorelines and in the intertidal areas within and adjacent

to the mangrove forests, but a few also consume small fishes such as killifishes (*Fundulus* spp.).

Floating and diving waterbirds associated with the south Florida mangroves include 29 species of ducks, grebes, loons, pelicans, cormorants, and gallinules. Five of the species (pintail, blue-winged teal, ring-necked duck, lesser scaup, and American coot) are abundant, and 11 species are common. Six of the species (pied-billed grebe, brown pelican, double-crested cormorant, anhinga, blue-winged teal, and common gallinule) are present throughout the year, and the rest are winter residents and/or seasonal transients. Six species nest in the mangroves. Many of these birds feed primarily on fishes, while others take in various benthic invertebrates and plant material. Two species, the brown pelican and double-crested cormorant, both of which nest in the mangroves, appear to be highly dependent on the mangrove habitat. The mangrove-fringed estuaries and waterways are also important for several species of ducks, coots, and related species.

Among the aerially searching birds, 14 species hunt in the ponds, creeks, and other waterways lined by mangroves, and such forms as gulls, terns, skimmers, kingfishers, and fish crows feed on invertebrates and fishes associated with the mangrove food webs. Among the common species, the laughing gull, black skimmer, kingfisher, and fish crow are year-round residents. Three species occur only in the winter, and one, the lesser tern, utilizes the mangroves only during the summer months.

Birds of prey inhabiting mangrove forests of south Florida include 20 species of hawks, falcons, eagles, vultures, owls, and frigatebirds. Only 9 are common. All except the American kestrel nest here. Most live in the area throughout the year, but the peregrine falcon and American kestrel are present only during the winter, and the swallow-tailed kite lives in the area only during the summer. Prey taken by these birds includes a wide range of fishes, frogs, snakes, lizards, small birds, waterfowl, rodents, and carrion. Use of the mangrove forests varies greatly from one species to another, but the bald eagle, peregrine falcon, and osprey are dependent on these forests for their continued existence in south Florida, since they nest and roost in the trees and feed extensively on mangrove-associated fishes. The mangroves are also particularly important for other nesting species including the magnificent frigatebird, both vultures, and the swallow-tailed kite. Peregrine falcons and related hawks feed extensively on waterfowl and other birds associated with the mangroves.

The largest group, the arboreal birds, constitutes a diverse assemblage of 71 species that roost in the mangrove canopy. Included here are cuckoos, woodpeckers, flycatchers, thrushes, vireos, warblers, blackbirds, pigeons, and sparrows. Five species (mockingbird, American robin, yellow-rumped warbler, palm warbler, and northern waterthrush) are considered abundant, and 25 species are common. Seven live in the mangrove areas throughout the year, but most are winter residents or seasonal transients. Three species (yellow-billed cuckoo, gray kingbird, and eastern wood pewee) occur in the area only in the summer. All the year-round residents nest here. Many of the birds in this group feed on invertebrates, chiefly insects, while others consume seeds, berries, and fruits. A number of species are recent immigrants or occasional visitors from the West Indies, and these seem to prefer the mangrove habitat.

As mentioned earlier, the mangroves of south Florida represent a diversity of habitat types, and these in turn are used by a variety of bird species for resting, roosting, nesting, and feeding. Included are permanent and seasonal residents as well as migratory transients. The fringing and riverine mangrove forests with their tall trees and well-developed canopies are particularly favored for roosting and nesting sites, and proximity to shorelines and aquatic feeding areas is also an important consideration for the shorebirds and fully aquatic species. For many of the bird species the mangroves provide critical habitat without which they would be greatly reduced in abundance or absent from the area.

MAMMALS

Twenty species of mammals are known to make use of the mangrove swamps, forests, and waterways of south Florida. Important references include Hamilton and Whittaker (1979), Layne (1974), Odum, McIvor, and Smith (1982), and Shemitz (1974). A list of these species and abundance estimates are provided in table 12.7. Four of the species (Virginia opossum, marsh rabbit, cotton rat, and raccoon) are considered to be abundant, and four more (striped skunk, bobcat, white-tailed deer and its subspecies the Key deer, and black rat) are common. The remaining species are uncommon, rare, or occasional residents of the mangroves. The tiny Key deer is absent from the mainland but common on Big Pine Key and

**Table 12.7.** *Mammal species reported from mangrove areas of south Florida. A = abundant, C = common, R = rare. (From Odum, McIvor, and Smith 1982.)*

| Common name | Scientific name | A | C | R |
|---|---|---|---|---|
| Virginia opossum | *Didelphis virginiana* | X | | |
| Short-tailed shrew | *Blarina brevicauda* | | | X |
| Marsh rabbit | *Sylvilagus palustris* | X | | |
| Gray squirrel | *Sciurus carolinensis* | | | X |
| Fox squirrel | *Sciurus niger* | | | X |
| Marsh rice rat | *Oryzomys palustris* | | | X |
| Cudjoe Key rice rat | *Oryzomys argentatus* | | | X |
| Cotton rat | *Sigmodon hispidus* | X | | |
| Gray fox | *Urocyon cinereo-argenteus* | | | X |
| Black bear | *Ursus americanus* | | | X |
| Raccoon | *Procyon lotor* | X | | |
| Mink | *Mustela vison* | | | X |
| Striped skunk | *Mephitis mephitis* | | X | |
| River otter | *Lontra canadensis* | | | X |
| Panther | *Felis concolor* | | | X |
| Bobcat | *Felis rufus* | | X | |
| White-tailed deer | *Odocoileus virginianus* | | X | |
| Key deer | *Odocoileus v. clavium* | | X | |
| Black rat | *Rattus rattus* | | X | |
| Bottle-nosed dolphin | *Tursiops truncatus* | | | X |
| West Indian manatee | *Trichechus manatus* | | | X |

adjacent islands. Among the large and medium-sized carnivores, the raccoon is abundant, and the striped skunk and bobcat are common. Three of the rarer carnivores (mink, otter, and panther) appear to be highly dependent on the mangrove habitat for their continued existence in the area. The small mammal fauna is primarily arboreal and terrestrial but adapted to periodic flooding. Included here are the opossum, shrew, and marsh rabbit as well as several rodents. Two fully marine mammals make use of mangrove-lined waterways. The bottlenose dolphin feeds here on fish populations, and the West Indian manatee, which feeds exclusively on aquatic vegetation, also frequents rivers, canals, and embayments close to the mangroves. The mink and otter consume fishes, and several of the mammals catch crabs and other crustaceans. These and the two marine mammals appear to be the only mammal species of the mangroves that have any real relations with organisms of the marine environment.

## Standing crop and primary production of mangrove vegetation

For the south Florida area few estimates of mangrove biomass are available. Lugo, Sell, and Snedaker (1976) determined that in an overwash forest the aboveground biomass averaged nearly 12,500 g/m², and this was partitioned as follows: leaves 5.7 percent, propagules 0.1 percent, wood 56.5 percent, and prop roots 36.7 percent. Studies elsewhere suggest that the biomass of the belowground root system is somewhat greater than that of the aboveground portion of the plant. Odum and Heald (1975) found that even in very different forest types red mangrove leaf biomass averaged between 700 and 800 g/m², and mangrove litter fall appeared to average about 2–3 g/m²/day. On the floor of the overwash forest, mentioned above, the leaf litter detritus averaged about 1,565 g/m², but this is likely to vary widely with different flushing regimes and from one forest type to another.

Estimates of net primary production for mangrove forests of southern Florida have been brought together by Odum, McIvor, and Smith (1982) (table S51). Expressed on a dry weight basis, the average values for net annual primary production are as follows: red mangroves 6,248, black mangroves 4,952, and mixed forests 9,785 g dry weight/m²/yr. Equivalent values in other units are also provided in the table. All calculations are based on the assumptions that 62 percent of the dry weight is organic matter and that organic carbon represents 50 percent of the organic matter (Darnell and Soniat 1979). Not included in the figures are the very low production values for the red mangrove scrub forests, which, as mentioned earlier, are clearly growing in unfavorable environmental conditions. Their net primary production values are less than a fifth those of red mangroves growing under optimal conditions. Within their zones of optimal growth discussed earlier, red mangroves tend to show the highest, black mangroves intermediate, and white mangroves the lowest production values, but production in mixed forests is the highest of all. Among the factors associated with high mangrove production are high nutrient levels, high flushing rates, and lack of stress (from such factors as ex-

treme salinity, disease, heavy grazing by herbivores, and competition with other species).

Studies on mangrove forests of the Mancha Lagoon of eastern Mexico near Veracruz by Barriero-Gilemes and Balderas-Cortes (1991) and Rico-Gray and Lot-Helgueras (1983) provide average annual leaf primary production values of 1.36 and 2.28 g dry weight/m²/yr, respectively, which are surprisingly low, and only about one-tenth of the values reported from south Florida. Along the southwestern Gulf, mangrove production is lower in the winter dry season, when both temperature and light levels are reduced, than during the summer wet season with its higher temperatures and increased light. However, even with these considerations, the Mexican mangroves studied must be subject to some sort of stress and not typical of the Mexican forests in general.

## Decomposition and export of organic matter from coastal swamps

All the material that drops from mangrove trees to the forest floor contributes to the litter. This includes leaves, twigs, flowers and flower parts, propagules, and excrement from insects, crabs, birds, and others. Although mangrove leaves are shed continuously throughout the year, minor peaks occur in the early part of the summer wet season and following periods of stress due to cold snaps, high soil salinities, and pollution events. Typically, leaves represent 68–86 percent, twigs 3–15 percent, and miscellaneous material 8–21 percent of the total litter fall. This material is very important since it returns nutrients to the soil and provides organic carbon and energy for detritus-based food webs of the mangrove swamps and related coastal waters. Estimates of the daily and annual rates of litter fall in several mangrove forest types are provided in table S52. The average litter fall from healthy red mangroves averages around 2.8 g/m²/day, or 1,022 g/m²/yr. Mixed forests produce a little less, black mangroves about half, and the small scattered trees of scrub forests about a seventh of the litter fall of healthy red mangroves. The greatest values occur in the highly productive fringing, overwash, and riverine red mangrove forests.

A number of important studies have been conducted on the biological decomposition of red mangrove leaf litter, and these have been summarized by Odum, McIvor, and Smith (1982). It has been found that decomposition proceeds most rapidly under marine conditions, more slowly in freshwater, and slowest of all on the dry forest floor. Experiments have shown that after four months under marine conditions 91 percent of the litter had disappeared, whereas in brackish water and freshwater 61 and 46 percent were gone, respectively, and in dry conditions only 35 percent had been removed. During early stages of decomposition most of the weight loss is a result of the release of dissolved organic matter. In the process of decomposition the particulate material is invaded by bacteria and fungi. Their presence greatly increases the nitrogen content of the particles, which can more than double over the first six months. Black mangrove leaves appear to decompose more rapidly than those of red mangroves. Since the decomposing litter is swept out more rapidly from frequently flooded fringing, riverine, and overwash forests, these show little accumulation of litter on the forest floor, but other forest types have lower flushing rates and greater litter accumulation.

Studies on the export of organic matter from mangrove communities have been reviewed by Odum, Fisher, and Pickral (1979) and Odum, McIvor, and Smith (1982). Although it is clear that the mangrove forests of south Florida do export organic matter to neighboring aquatic systems, the actual amounts have not been firmly established. Present estimates range from 0.2 to 1.8 g/m²/day or 73 to 657 g/m²/yr, but these values are probably underestimates since they do not take into account amounts exported by certain mechanisms (bed-load transport, dissolved organic matter of the water column, and offshore transport associated with major floods, storms, and hurricanes). Odum, McIvor, and Smith (1982) suggested that mangrove-based food webs are dominant in mangrove-lined creeks, rivers, and small bays but that in larger, more open bodies of water mangrove detritus may be an important but not the dominant source of organic carbon. It has been estimated that mangroves supply about a third of the organic carbon available to aquatic systems of Rookery Bay, south of the Everglades. Although mangrove-derived carbon is definitely exported to the open continental shelf, the amounts have not been quantified, and this represents just one of several sources of organic material available to the ecosystems of the shelf. It should not be overlooked that many fully marine species (fishes, sea turtles, dolphins, etc.) regularly feed in the creeks, rivers, and bays of mangrove-dominated systems and that the young of many more marine species (shrimps, lobsters, crabs, fishes, etc.) use such systems as nursery areas. Their subsequent emigration back to the conti-

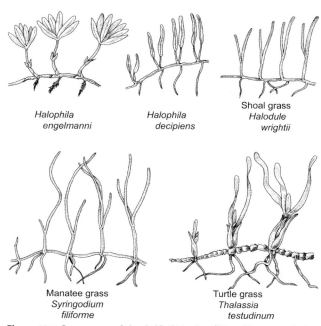

**Figure 12.7.** Seagrasses of the Gulf of Mexico. (After Zieman 1982.)

nental shelf also constitutes a substantial movement of mangrove-derived carbon to the marine environment.

## SEAGRASS MEADOW COMMUNITIES

Seagrasses are flowering plants that grow in shallow coastal marine waters, and although derived from terrestrial ancestors, they pass their entire life histories fully submerged. Some species superficially resemble the grasses of land, but seagrasses are not closely related to the true grasses. The various seagrass species are distributed in bays and coastal lagoons and on shallow continental shelves from the tropics to boreal regions throughout the world. About 45 species are known worldwide, but only 4 genera and 5 species occur in the Gulf, and all of these are essentially tropical or subtropical in distribution. These include turtle grass (*Thalassia*), manatee grass (*Syringodium,* formerly called *Cymodocea*), shoal grass (*Halodule,* formerly known as *Diplanthera*), and two species of *Halophila* (*H. decipiens* and *H. engelmannii*) (fig. 12.7). Of all these species, turtle grass is the most common, widespread, and abundant in the Gulf, where it often forms vast undersea meadows.

The typical seagrass plant consists of stems, stalks that bear leaves, and roots. The stems, called *rhizomes,* grow horizontally through the sediments. They produce nodes from which spring the roots and short vertical stalks that grow up into the water column and give rise to new leaves as well as buds and tiny flowers. In turtle grass new leaves are produced throughout the year, averaging about one new leaf per shoot every 14–16 days, or longer during cool weather. Leaf shape and size vary among the different species, and the leaves are most pronounced in turtle grass, where they grow to 35 cm (14 in) or more in length. In this species the older portion of the leaf (farthest removed from the rhizome) becomes brittle and coated with epiphytic plant and animal growth, until it finally breaks off and contributes to the pool of particulate organic detritus of the sediments. The underground rhizomes, which contain conductive and storage tissues, generally grow only a few centimeters deep in the sediments, but they can be found as deep as 25 cm (10 in). In the denser stands the rhizomes and root systems form complex interlacing mats that provide firm anchorage against the force of waves and water currents. All seagrasses exhibit the following five characteristics: the ability to (1) live in the marine environment, (2) function while fully submerged, (3) remain firmly attached in the sediments, (4) complete their reproductive cycle while fully submerged, and (5) compete with other organisms in the marine environment. Important general references to seagrasses include den Hartog (1970), McRoy and Helfferich (1980), R. Phillips (1978), R. Phillips and McRoy (1979), and Waisel (1972).

All five species of seagrasses are found, at least sporadically, around the entire periphery of the Gulf of Mexico. The two areas of greatest coverage and concentration are off south Florida (around the Keys and in Florida Bay) and in the Big Bend area north of Tampa Bay. In the northeastern Gulf, east of the Mississippi River Delta, significant beds occur in Mississippi Sound and Chandeleur Sound behind the barrier islands. However, west of the delta, off Louisiana and the upper Texas coast, they are absent, apparently because of such factors as high turbidity, reduced salinity, and low winter temperatures. Turtle grass, manatee grass, and shoal grass occur in saline bays of south Texas, and all five species are found in protected waters of eastern Mexico from around Tampico south and along the coast of Yucatán. Turtle grass beds are also prominent in coral reef lagoons at Isla de Lobos (south of Tampico), off Veracruz, and on the Yucatán shelf, particularly at Alacrán Reef. All five species have also been reported from the shallow shelf of northwestern Cuba. Zieman (1982) has estimated that within the Gulf of Mexico as a whole, seagrasses cover an estimated 10,000 km² (3,860 mi²) of sea bottom, about 85 percent of

which is in Florida waters. Around 5,500 km² (2,120 mi²) are located in shallow waters of the Florida Reef Tract and adjacent Florida Bay (below the Everglades), and another 3,000 km² (1,158 mi²) lie in the Big Bend area, essentially from Tampa Bay north through Apalachee Bay. The most important references to seagrasses of the Gulf of Mexico are Zieman (1982) and Zieman and Zieman (1989), which provide community profiles of the extensive seagrass meadows off southern and western Florida, respectively. Major additional references dealing with the distribution, temperature and salinity tolerances, productivity, and community relations include Esperanzo-Avalos (1996), Humm (1956), Humm and Caylor (1957), Lot-Helgueras (1971), D. Moore (1963), K. Moore and Wetzel (1988), Onuf (1996), Thorne (1954), and Zimmerman and Livingston (1976, 1979).

## Habitat and environmental relations

In general, seagrasses flourish best in shallow marine waters with low turbidity, high light penetration, and moderate wave action. In shallow bays and other nearshore areas where turbidity is high, they are often limited to depths of only 1 or 2 m, but in the clear waters of some continental shelves they may be found at depths of at least 42 m (138 ft). Moderate currents appear to enhance growth and production of seagrasses by bringing in nutrients and removing wastes, and the densest beds are often found in channels that carry bottom currents. However, heavy wave action accompanying storms and hurricanes may expose the rhizomes and uproot and remove some of the seagrasses from the area.

Since the seagrasses of the Gulf are all essentially tropical species, they have limited tolerance for low temperatures. The optimal temperature range for seagrasses in shallow water off south Florida appears to be about 20°–30°C (68°–86°F), and growth rates decrease sharply below and above that range. However, seagrasses of the northern Gulf are adapted for somewhat lower temperatures, and even off south Florida seagrasses of shallow water are more cold tolerant than those growing in deeper water. When the plants are exposed to extreme temperatures the shoots and leaves may die off, but buried in the sediments the rhizomes are buffered against temperature extremes, and they send up new shoots when conditions are more favorable. In general, shoal grass is more cold tolerant than turtle grass, and both are more cold tolerant than manatee grass or the species of *Halophila*.

The optimum salinity for all seagrasses is near that

of full seawater, but all have fairly wide ranges of tolerance and can withstand salinities as low as 3.5 ppt for short periods. Photosynthetic efficiency is highest at full marine salinity and decreases with reduced salinity, being only about a third at half seawater salinity. Shoal grass is more tolerant of reduced salinity, followed by turtle grass, and then manatee grass and *Halophila*. All seagrasses have low tolerance for exposure to air and desiccation, but again, shoal grass is the most tolerant. Following exposure at extreme low tides, some leaves die and are lost. However, no permanent damage results, since the plant quickly develops new leaves.

Although seagrasses grow in a variety of sediment types from fine muds to coarse sands, they do best when fine particles make up at least 15 percent of the substrate. Actually, the seagrass beds themselves tend to retain fine particles as well as organic detritus particles derived from outside and inside the beds, and they build up their own fine-particle organic substrate, which eventually becomes elevated a few centimeters above the level of the surrounding sea bottom. These high-organic substrates, rich in bacteria and other microscopic forms, become anoxic and develop chemically reducing conditions. In this environment sulfur becomes reduced to hydrogen sulfide and nitrogen compounds are converted to the soluble ammonium ion, which may be utilized as a nutrient by the local vegetation or lost to the water column.

Seagrasses reproduce both vegetatively and sexually. Vegetative reproduction involves continual growth of the rhizome to produce new roots, shoots, and leaves in a linear chain of essentially new plants. Sexual reproduction entails the release of pollen grains by the stamens of male flowers and the transport of this pollen by water currents to the stigmas of female flowers. Following fertilization, tiny seeds and fruits are produced, and these fruits, in turn, are carried by water currents to new areas where they germinate in the sediments to form new colonies. Sexual reproduction by seagrasses of the Gulf takes place primarily during the spring and summer months.

## Zonation

Seagrasses may grow in monospecific or in mixed-species stands, and these are often distributed in depth-related zones (fig. 12.8). Closest to shore and to a depth of about one meter lies a band of shoal grass (*Halodule*), the species most tolerant of environmental extremes such as low temperature, low salinity, and periodic desiccation. Because of its wide range of tolerance and its shoreward

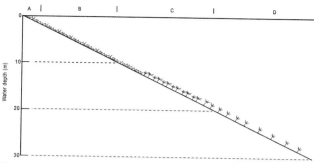

**Figure 12.8.** Diagrammatic depiction of zonation in seagrass beds off the west coast of Florida. A = shoal grass, B = turtle grass (and manatee grass), C = shoal grass (and manatee grass), and D = *Halophila* spp.

position, this is often referred to as the pioneer species. However, shoal grass is not limited to the shallow nearshore zone, and in some areas it extends beyond a depth of 30 m (ca. 100 ft). Beyond the nearshore band of shoal grass lie the dense beds of turtle grass, in pure stands or mixed with manatee grass, which in favorable locations can stretch along the coast for tens of kilometers. In some localities manatee grass in pure stands replaces turtle grass. This zone extends to a depth of about 10–12 m (33–39 ft). Beyond this sometimes lies a band of shoal grass and/or manatee grass to about 15 m (50 ft), and this gives way to the outer band consisting of diffuse stands of one or both species of *Halophila,* which in some areas may grow out to a depth of 40 m (130 ft) or more. Some *Halophila* plants may also be found in the dense beds of turtle grass. This general zonation pattern, observed off west Florida, is subject to much local variation, depending on the prevailing environmental conditions.

## Algae and animals of seagrass communities

Seagrass meadows are a distinct ecosystem that produces and accumulates organic material, develops a bed of fine organic-rich sediments, and protects the bottom and near-bottom environment from the force of waves and water currents. Hence, these meadows are inhabited by a distinct group of organisms that grow as epiphytes on the seagrass plants, dwell near or on the bottom surface, or live within the substrate. Seagrass beds also provide shelter and/or food resources for many fishes and other mobile species of the water column. So far as is known, there is no distinct phytoplankton or zooplankton associated with the beds, but planktonic organisms are gleaned from the water column, trapped in the slower-moving water that circulates among the leaf blades, and ultimately added to the bottom, thus enriching the sediments. Also, certain algal sporelings and animal larvae find suitable conditions here, and they settle and grow within the seagrass beds. Some of these become permanent residents, while others use the beds as nursery areas and move out to other habitats as they become adults.

### EPIPHYTIC PLANTS AND ANIMALS

Leaf blades of the seagrasses, particularly the long, broad leaves of turtle grass, provide firm substrate for attachment of epiphytic algae, and Humm (1964) recorded 113 species, three-fifths of which were red algae, epiphytic on turtle grass leaves in Florida waters. These included seasonal as well as year-round residents, a number of which were calcareous species. During the winter and spring months the epiphytes sometimes become so abundant that they shade the leaves, reducing photosynthesis by the seagrass, but the epiphytic algae themselves are important photosynthesizers of the community. Also listed among the species living on seagrass leaves are a variety of attached animal species including certain sessile protozoans, sponges, hydroids, polychaetes, bryozoans, and others. As the terminal portions of the seagrass leaves break off, these fragments and their associated epiphytes drop to the bottom surface and contribute to the sediments, both organic material and inorganic carbonates. Zieman (1975) estimated that the lifetime of a single turtle grass leaf is only 30–60 days, so the turnover of epiphytic species is quite rapid.

### MACROALGAE

Benthic macroalgae are often present within and around seagrass beds, but because of the limitations of soft substrates and problems with shading, only a few species are able to do well here. The most important genera of seagrass-associated macroalgae are *Acetabularia, Caulerpa, Halimeda, Penicillus, Rhipocephalus,* and *Udotea,* which have been mentioned in previous chapters and which all possess some sort of anchoring devices for maintaining position in soft sediments. Where rocky outcrops are present, many additional species of algae may find suitable habitat in the seagrass meadows. Most of the soft-substrate macroalgae produce large quantities of calcareous material, which eventually adds to the carbonates of the sediments. In addition to these larger algae, seagrass meadows may contain a carpet of benthic microalgae, which in some cases are also quite productive of organic material.

**Table 12.8.** *An abbreviated list of common invertebrate genera encountered in seagrass beds of south Florida. (Modified from Zieman 1982.)*

Sponges – *Ircinia, Spongia, Tethya*
Corals – *Manicina, Porites*
Polychaetes – *Arenicola, Eunice, Onuphis, Spirorbis, Terebelloides*
Mollusks
  Gastropods – *Aplysia, Fasciolaria, Pleuroploca, Strombus*
  Bivalves – *Atrina, Chione, Codakia, Laevicardium, Lucina, Tellina*
  Cephalopods – *Octopus*
Crustaceans
  Amphipods – *Cymadusa, Gammarus, Grandidierella, Melita*
  Shrimps – *Alpheus, Hippolyte, Palaemonetes, Penaeus, Periclimenes, Thor, Tozeuma*
  Hermit crabs – *Pagurus*
  Lobsters – *Panulirus*
  True crabs – *Menippe*
Echinoderms
  Starfish – *Oreaster*
  Sea urchins – *Diadema, Lytechinus, Tripneustes*
  Sea cucumbers – *Actinopyga, Holothuria*

### INVERTEBRATES

Many species of larger or mobile invertebrates inhabit the seagrass meadows, and although the composition varies somewhat in relation to location and plant composition, the general nature of the invertebrate fauna is relatively constant (table 12.8). Large sessile forms include several sponges and a few types of corals. A number of species of mollusks and small crustaceans, particularly amphipods and small shrimps, graze on epiphytes of the seagrass leaves. Larger snails such as the tulip shell (*Fasciolaria tulipa*), horse conch (*Pleuroploca gigantea*), and queen conch (*Strombus gigas*) move along the bottom grazing the epiphytes or feeding on infauna and bottom detritus. Also residing and foraging among the seagrasses are larger forms such as penaeid shrimps, hermit crabs, tropical lobsters, stone crabs, various echinoderms, and a small predatory octopus (*Octopus briareus*). The infauna consists of meiofaunal species as well as a number of polychaetes, bivalves, and small crustaceans. Some of the invertebrates, such as the long-spined sea urchin and

**Table 12.9.** *Fish species commonly associated with seagrass meadows of south and west Florida. (From R. Livingston 1984; Reid 1954; Zieman 1982.)*

| Family and scientific name | Common name | Family and scientific name | Common name |
|---|---|---|---|
| Muraenidae | | Sparidae | |
| *Gymnothorax nigromarginatus* | black edge moray | *Archosargus probatocephalus* | sheepshead |
| Ophichthidae | | *Calamus arctifrons* | grass porgy |
| *Myrichthys breviceps* | sharptail eel | *Diplodus holbrooki* | spottail pinfish |
| *Myrichthys ocellatus* | goldspotted eel | *Lagodon rhomboides* | pinfish |
| Synodontidae | | Sciaenidae | |
| *Synodus foetens* | inshore lizardfish | *Bairdiella chrysoura* | silver perch |
| Batrachoididae | | *Cynoscion nebulosus* | spotted seatrout |
| *Opsanus beta* | gulf toadfish | *Leiostomus xanthurus* | spot |
| Gobiesocidae | | *Sciaenops ocellatus* | red drum |
| *Acyrtops beryllinus* | emerald clingfish | Scaridae | |
| Syngnathidae | | *Sparisoma chrysopterum* | redtail parrotfish |
| *Anarchopterus criniger* | fringed pipefish | *Sparisoma radians* | bucktooth parrotfish |
| *Hippocampus erectus* | lined seahorse | *Sparisoma rubripinne* | redfin parrotfish |
| *Hippocampus zosterae* | dwarf seahorse | Clinidae | |
| *Syngnathus floridae* | dusky pipefish | *Paraclinus fasciatus* | banded blenny |
| *Syngnathus louisianae* | bay pipefish | *Paraclinus marmoratus* | marbled blenny |
| *Syngnathus scovelli* | gulf pipefish | Blenniidae | |
| Serranidae | | *Chasmodes saburrae* | Florida blenny |
| *Centropristis striata* | black sea bass | Gobiidae | |
| *Mycteroperca microlepis* | gag | *Gobionellus boleosoma* | darter goby |
| Lutjanidae | | *Gobiosoma bosc* | naked goby |
| *Lutjanus analis* | mutton snapper | *Gobiosoma robustum* | code goby |
| *Lutjanus griseus* | gray snapper | *Microgobius gulosus* | clown goby |
| *Lutjanus jocu* | dog snapper | Bothidae | |
| *Lutjanus synagris* | lane snapper | *Paralichthys albigutta* | gulf flounder |
| *Ocyurus chrysurus* | yellowtail snapper | Balistidae | |
| Gerreidae | | *Monacanthus ciliatus* | fringed filefish |
| *Eucinostomus argenteus* | spotfin mojarra | *Monacanthus hispidus* | planehead filefish |
| *Eucinostomus gula* | silver jenny | Ostraciidae | |
| Haemulidae | | *Acanthostracion quadricornis* | scrawled cowfish |
| *Haemulon plumieri* | white grunt | Tetraodontidae | |
| *Orthopristis chrysoptera* | pigfish | *Chilomycterus schoepfi* | striped burrfish |
| | | *Sphoeroides nephelus* | southern puffer |

several others, use the beds as young and move out into other habitats as they mature. Other species make use of the seagrass beds only seasonally or at night.

**FISHES**

Many types of fishes occur within and around seagrass meadows of the Gulf, and these include the year-round residents, seasonal residents, regular migrants, and occasional visitors. Of the more than 200 fish species recorded from shallow marine waters of south Florida, most, if not all, have been taken at one time or another from the grass flats of the area. Some of the more common forms are listed in table 12.9. The permanent residents are generally small, less mobile, fairly cryptic species that can hide among the grass blades. These include such forms as the emerald clingfish (*Acyrtops*), several small seahorses (*Hippocampus*) and pipefishes (*Micrognathus, Syngnathus*), as well as blennies (*Chasmodes, Paraclinus*), and gobies (*Gobionellus, Gobiosoma, Microgobius*). A few small eels (*Myrichthys*) burrow in the bottom or slither through the grass blades. Most of these permanent residents feed on plankton, epiphytes, or small mobile invertebrates such as polychaete worms, amphipods, and small shrimp.

The seagrass meadows also provide nursery habitat for the young of many species whose adults are found elsewhere. These juveniles occupy the beds seasonally, and as they mature they move out to other habitats, primarily in deeper water. Prominent among the seasonal residents is the pinfish (*Lagodon*), which is probably the most abundant fish species occupying the beds during the spring and early summer months. Also prominent are the planehead and fringed filefishes (*Monacanthus* spp.), the former being most abundant during the summer, and the latter during the fall and winter. Other species whose young utilize the seagrass beds seasonally include the black edge moray (*Gymnothorax*), some snappers (*Lutjanus, Ocyurus*), mojarras (*Eucinostomus*), grunts (*Haemulon, Orthopristis*), porgies (*Archosargus, Calamus, Diplodus*), drums (*Bairdiella, Cynoscion, Leiostomus, Sciaenops*), flounders (*Paralichthys*), cowfish (*Lactophrys*), and puffers (*Chilomycterus, Sphoeroides*). Most of these feed largely on invertebrates and small fishes. The predatory lizardfish (*Synodus*) and several flatfishes occur in the more open areas of the grass flats, and in deeper channels and shell banks are found the toadfish (*Opsanus*) and a few other species. Most of the permanent and seasonal residents are small fishes measuring about eight inches or less in length, that is, small enough to remain concealed among the seagrass blades. Larger individuals are more vulnerable to predation and must find haven elsewhere.

Other nonresident species visit the grass beds regularly or occasionally. Among the regular visitors are several species of parrotfishes (*Sparisoma*) that come to feed directly on blades of the seagrasses. Off south Florida a number of predatory fish species such as squirrelfishes, grunts, and snappers, which during the daytime find shelter around coral reefs, at night migrate to the nearby grass beds to forage for fishes, shrimps, and other invertebrates. Less common predators include several sharks, tarpons, ladyfishes, bonefishes, sea basses, snappers, groupers, jacks, snooks, barracudas, and mackerels. Although locally common, these large predators also forage in other habitats.

Thus, the seagrass meadows of the Gulf provide important habitat for a variety of marine fishes. For some this is a permanent home. For others it is an important nursery area for the young. For yet others the beds provide critical grazing or foraging grounds. The density and diversity of fish species in the seagrass meadows are far higher than on the barren stretches of the shallow continental shelf. Without the beds numerous fish species would be greatly reduced in abundance or absent altogether.

## REPTILES

The only reptile known to make much use of the seagrass meadows of the Gulf is the green sea turtle (*Chelonia mydas*). Adults are herbivorous and feed directly on the leaves of seagrasses, especially turtle grass. All grazing takes place during the daylight hours, and the turtles spend the night in deeper water around offshore reefs or buried in the bottom. When feeding, the turtles discard the older outer sections of the leaves that are heavily coated with epiphytes, and they consume only the younger portions of the leaves, which they crop to within a few centimeters of the rhizomes. Day after day a turtle will return to the same grazing spot from which the older leaf sections have already been removed. Studies have shown that the younger portion of the leaf is much richer in nitrogen, so the grazing turtle consumes the most nutritious portion of the leaf. It has been estimated that a medium-sized turtle consumes about 500 g dry weight of turtle grass per day. In an average turtle grass stand, this would translate to about half a square meter, but several square meters of cropped vegetation would be required to supply this much seagrass. Because green sea turtles are not very abundant in the Gulf, their overall effect on seagrass beds is small. However, during pre-Columbian times these turtles were much more abundant in the Gulf of Mexico, perhaps a hundred or a thousand times more numerous than at present. Heavy harvesting by sailors, settlers, and then commercial hunters has decimated the Gulf populations. The impact of turtle grazing on seagrass beds in pre-Columbian times can only be guessed, but it must have been far greater than at present.

Another sea turtle species, the loggerhead (*Caretta caretta*), is omnivorous and has been known to consume some seagrass, but this turtle is also fairly rare and certainly has little effect on the seagrass meadows of the

Gulf. The American crocodile (*Crocodylus acutus*), which lives and nests in sandy and mangrove-covered shorelines of south Florida, feeds at night in the local bays and sounds. It undoubtedly forages for fishes and larger invertebrates over the seagrass beds of Florida Bay, but its effect on the grass beds themselves is likely minimal.

BIRDS

The bird life associated with seagrass meadows of south Florida has been discussed by Zieman (1982), and the most common species are listed in table 12.10. Based on their modes of feeding, these birds are divided into three groups: waders, swimmers, and flyers-plungers. All are primarily fish eaters, but the waders also consume some invertebrates. The shallowest seagrass meadows are heavily used as feeding grounds by herons, egrets, and other wading bird species, primarily during periods of low tide when the grasses are more exposed. Swimming species such as cormorants, mergansers, and the white pelican seek their prey underwater and hence require deeper water for swimming and feeding. The flying species seek their prey while on the wing. Ospreys and

eagles grasp near-surface fishes in their talons, whereas the brown pelican plunges into the water with its bill open and engulfs small fishes schooling near the surface. Since the flyers and plungers operate less effectively in shallow water, they often fish over the deeper seagrass beds. Prey species vary somewhat among the three feeding groups.

MAMMALS

Two species of marine mammals make use of the seagrass meadows of the Gulf, the West Indian manatee (*Trichechus manatus*) and the bottlenose dolphin (*Tursiops truncatus*). Their relations with seagrass meadows have been addressed by Gunter (1954), Zieman (1982), and others. The manatee is a tropical species native to shallow waters of the Caribbean and the southern half of the Gulf of Mexico. Nowadays the species is far less abundant than it was in pre-Columbian times, and although formerly present along much of the eastern, southern, and southwestern Gulf, it is now apparently limited to a few hundred individuals living in the shallow coastal and inland waters of southern and southwestern Florida and the Bay of Campeche. During the summer it occasionally strays into the northern Gulf, but because it is extremely sensitive to cold, it apparently has never been a year-round resident there. The manatee is strictly herbivorous and feeds directly on the seagrasses. Weighing up to 500 kg (1,012 lb), the manatee can consume up to 20 percent of its body weight in vegetation per day. Unlike the green sea turtle, the manatee ingests both the leaves and rhizomes, which it removes in large bites, each denuding an area in the soft sediments of about 30 × 60 cm (12 × 24 in). Such areas are quickly overgrown by new vegetation. Considering the small size of the existing manatee population, the effect on seagrass meadows is quite small, but little can be said about the impacts of the larger, pre-Columbian populations.

The bottlenose dolphin is a wide-ranging species found in coastal waters around the entire Gulf. It feeds on fishes, which it often catches in shallow water over the seagrass beds. However, since this dolphin is not restricted to the grass beds and forages widely elsewhere, its effect on the local fish populations is probably quite limited.

**Table 12.10.** *Birds that commonly make use of the seagrass meadows of south Florida. (Modified from Zieman 1982.)*

| Common name | Scientific name |
| --- | --- |
| **Wading birds** | |
| Great blue heron | *Ardea herodias* |
| Great white heron | *Ardea occidentalis* |
| Great egret | *Casmerodius albus* |
| Reddish egret | *Egretta rufescens* |
| Louisiana heron | *Egretta tricolor* |
| Little blue heron | *Florida caerulea* |
| Roseate spoonbill | *Ajaia ajaja* |
| Willet | *Catoptrophorus semipalmatus* |
| **Swimming birds** | |
| Double-crested cormorant | *Phalacrocorax auritus* |
| White pelican | *Pelecanus erythrorhynchos* |
| Red-breasted merganser | *Mergus serrator* |
| **Flying and plunging birds** | |
| Osprey | *Pandion haliaetus* |
| Bald eagle | *Haliaeetus leucocephalus* |
| Brown pelican | *Pelecanus occidentalis* |

## Standing crop and primary production of seagrass meadows

Data concerning the biomass and primary production of seagrasses of the Gulf have been summarized by Zieman

(1982). The biomass of the seagrass plant consists of two portions: aboveground (leaves) and belowground (rhizomes and roots). Of all the Gulf seagrass species, turtle grass has by far the best-developed system of rhizomes and roots. Hence, the aboveground portion of this plant represents only 15–20 percent of the total plant biomass. By contrast, in the other seagrasses the aboveground portion often represents 50–60 percent of the weight of the plant. Within a given species the ratio varies somewhat in relation to substrate type, and plants growing on fine sediments have a higher proportion of leaves than those on coarse substrates. Turtle grass has the highest total plant biomass values, with averages in the range of 150–3,100 g dry weight/m² and maximum values of 8,100 g dry weight/m² for south Florida stands. Average values for manatee grass are in the range of 45–300 g dry weight/m², and for shoal grass they are 50–250 g dry weight/m².

Primary production of seagrasses has been measured by several techniques, the most reliable of which consist of either measuring $^{14}C$ uptake or simply marking new leaves and later measuring the amount of growth. For most studies these two methods give comparable results. As might be expected, turtle grass has the highest production rates, with daily production rates in the range of 0.6–16.0 g C/m². For manatee grass comparable values are 0.6–9.0 g C/m², and for shoal grass they are 0.5–2.0 g C/m² (the shoal grass values are from North Carolina since Gulf values are unavailable). Translated to annual production rates, these would be as follows: turtle grass 219–5,840, manatee grass 219–3,285, and shoal grass 183–730 g C/m² (= about 390–18,705 g dry weight/m²). These are extremely high production rates, and they illustrate the very productive nature of seagrass ecosystems, especially those where turtle grass is the dominant species.

## Decomposition and export of organic matter from seagrass meadows

Seagrasses and their associated epiphytes provide food for higher consumer levels in three ways: direct grazing by herbivores, detrital food webs within the grass beds, and export of plant material and organic detritus to other areas outside the grass beds. Although no investigations have been carried out in the Gulf on the percentage of seagrass production passing through each route, studies elsewhere suggest that about 70 percent is available through the detrital system, and it seems clear that organic detritus serves as the major pathway of energy flow in seagrass ecosystems (Zieman 1982). Direct grazing on living sea-

grasses and their epiphytes occurs in a variety of invertebrates (particularly snails, crustaceans, and sea urchins) and vertebrates (fishes, sea turtles, and manatees), but most of the organic matter produced through photosynthesis is lost as dissolved organic matter or passes through decomposition as particulate organic detritus.

Living seagrass leaves and their epiphytes release into the water column a great deal of organic carbon as dissolved compounds, in some cases up to 10 percent of the recently photosynthesized carbohydrates. Organic compounds including carbohydrates, proteins, and other substances are also released into the water column by decomposing seagrass leaves, and this may amount to 20 percent of their carbon content. Additional amounts are given off by the microbes and other decomposing organisms themselves, so the quantity released by the community as a whole is substantial. Most of this released material is in chemical forms readily utilizable by plants and other microbes, resulting in a great deal of internal recycling of organic material within the community. However, much of the dissolved organic material escapes and is exported for use by ecosystems outside the seagrass beds. Regardless of its source, once released into the water column, this dissolved material functions as organic detritus, even though it is not in particulate form.

Over half of the particulate matter making up fresh organic detritus from seagrass leaves is in the form of structural compounds, particularly cellulose, but only a few of the higher consumer species have enzyme systems capable of digesting this material. However, the bacteria, fungi, and certain other small forms that rapidly invade the remains of the dead particulate plant chemically attack the fibrous material, breaking it down into finer and finer particles and, in the process, enriching it with nitrogen taken from the water column and the sediments. As this breakdown proceeds, the particles are invaded by a host of small invertebrates that feed on the microbes. In this form the detritus substrate, together with the microbes and small invertebrates, is consumed by larger animals, which digest the nutrient-rich smaller organisms and pass the indigestible fibrous matter in their fecal material. In turn, this is reinvaded by the smaller organisms, and the cycle repeats until the fiber is completely broken down and the energy content is exhausted (Darnell 1967a).

Comparative studies have shown that turtle grass undergoes very rapid decomposition, much faster than that of mangroves, which in turn is faster than that of

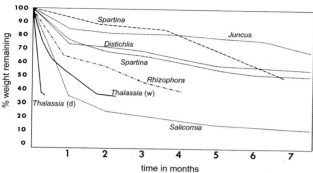

**Figure 12.9.** Comparative decay rates showing the rapid decomposition of turtle grass (*Thalassia*) compared with the red mangrove (*Rhizophora*), several marsh grasses (including *Spartina*, *Juncus*, and *Distichlis*), and glasswort (*Salicornia*). (From Zieman 1982.)

various marsh grasses (fig. 12.9). Only a few weeks or a month, at most, is required for half the turtle grass leaf blades to be reduced to very tiny particles, thus facilitating the very rapid turnover of organic compounds and nutrients in seagrass ecosystems. Within the seagrass beds many deposit-feeding invertebrates (notably polychaete worms, certain gastropods, isopods, amphipods, and echinoid and ophiuroid echinoderms) as well as a number of fish species feed on and derive their nutrition from the microbes and smaller invertebrates associated with the detritus particles. Some of these higher consumers have been shown to assimilate the chemical compounds with high efficiencies (50 to nearly 100 percent), whereas their assimilation of undecayed plant material is of very low order (less than 5 percent efficiency).

Material produced in the seagrass beds is exported to other ecosystems by several mechanisms. Reference has already been made to the loss of dissolved organic matter. Many of the fine detritus particles are resuspended and transported away by water currents, and the finer the particles, the more readily available they are for resuspension and export. Sometimes portions of leaves and even entire mats of seagrass are removed and transported elsewhere by storms. Seagrass leaves clipped off by fishes and sea turtles also float away to other areas, where they eventually sink to the bottom or are deposited along the shoreline as beach wrack. Seagrass remains not only enrich other areas of the continental shelf, but they have even been reported from the bottom of the deep Gulf off the Florida west coast and north of Yucatán (W. Pequegnat 1983). Many of the fishes as well as the sea turtles and manatees that feed on seagrasses and their associated organisms move to other areas, where they deposit fecal material containing partially digested organic matter, thus effectively transporting seagrass production to other ecosystems. Finally, it is recalled that seagrass meadows provide nursery grounds for spiny lobsters and a variety of fish species that pass juvenile stages here, and as they mature they move to other habitats, taking with them organic matter originally produced in the seagrass beds. Most enriched are the nearby mangrove swamps and coral reefs, where these are present, but much of the seagrass contribution goes to the open continental shelf, and ultimately some makes it to the deep Gulf.

# 13 : NEKTON

The nekton embraces larger and stronger animals of the sea that are capable of swimming horizontally as well as vertically. Unlike the tiny zooplankton organisms, which are small and basically at the mercy of the water currents, nektonic animals can control where they go despite the currents. This distinction between zooplankton and nekton is conceptually useful in our discussions, but in nature the two groups broadly overlap. The small animals in this area of overlap are often classified as micronekton, and on a practical basis, these are the animals collected in nets having mesh openings of about 1–4 mm. Included in the nekton of the Gulf are various cephalopods, some decapods (shrimps and crabs), fishes, sea turtles, and marine mammals (manatees, dolphins, and whales). Many species that are capable of swimming do not do so all the time, and these may rest on or roam about the bottom in search of food or breeding partners, or they may live a cryptic life in holes and crevices. For present purposes all these groups are included as nekton. Those associated with the bottom or near-bottom environment are called *demersal* species, whereas those that remain largely within the water column are referred to as *pelagic.* Species that occur over the continental shelf are *neritic,* and those associated with off-shelf waters are *oceanic* species. Again, there is much overlap between these general categories, and individual animals may freely move from one habitat to another.

Our knowledge of the nektonic species of the Gulf is incomplete and biased toward those forms that are most easily studied. The primary method of collecting many species is by means of bottom trawl nets (for demersal species) and midwater trawls (for pelagic species) pulled behind ships, but such nets generally capture only the slower-swimming forms. Smaller and cryptic species around reefs are often taken by poisoning (with rotenone). Additional information on the larger carnivorous species derives from hook-and-line catches, especially those of commercial long-line fisheries, sport tournaments, and other recreational catches monitored at dockside. These catch data provide information concerning such forms as sharks, snappers, groupers, jacks, mackerels, tunas, and

billfishes. Direct observations by scuba divers have added greatly to our knowledge of nektonic species inhabiting shallow waters of the continental shelf, especially around coral reefs, oil rigs, and artificial reefs. Supplemental information derives from observations from research submersibles and cameras aboard remotely operated vehicles (ROVs). During recent years populations of sea turtles and marine mammals have been censused by experienced observers aboard ships and low-flying aircraft. Information on migration patterns stems from programs in which animals are individually tagged, released, and later recaptured.

As a result of the many types of studies carried out on the nekton of the Gulf, there is now a substantial body of knowledge on such topics as species distribution and abundance, habitat utilization, seasonal occurrence, migration, life history patterns, behavior, food habits, spawning, and reproduction. In general, the coastal forms are fairly well known, but considerably less information is available about species inhabiting the off-shelf oceanic waters, and for those species that live in deep water well below the surface, little is known except their presence, and in some instances, their food habits.

As in the case of most other groups of organisms in the Gulf, one can distinguish between the temperate and tropical neritic nektonic fauna. Distinctions can also be made between those species that spend a portion of their lives (generally as juveniles) in the lower-salinity waters of estuaries and those that are more strictly marine. Of all the nektonic forms in the Gulf, the fishes are by far the most diverse group, and because of their commercial importance and recreational interest, they have been the most thoroughly studied. Most of the Gulf nekton literature relates to specific groups such as cephalopods, fishes, and so forth, but there are a few regional reports on the nekton as a whole. Although nektonic animals occur at all depths of the sea, the present chapter will deal primarily with nekton of the continental shelves and epipelagic waters of the open Gulf. Nektonic animals of the deeper Gulf (mesopelagic and bathypelagic realms) will be taken up in the next chapter.

## MODIFICATIONS FOR NEKTONIC LIFE

Before discussing the individual nektonic groups it is important to say a few words about how large swimming animals are adapted for life in the water column. In order to live and move effectively in the relatively dense medium of water, the nektonic animals have had to develop a number of special anatomical, physiological, and behavioral adaptations, and as we shall see, different groups have sometimes come up with different solutions to common problems. A few of the adaptations have been mentioned in the chapter on zooplankton, but because they are larger, denser, and more mobile, the nektonic forms often require more well-developed mechanisms.

### Buoyancy

Although it varies somewhat in relation to temperature, salinity, and pressure, the density of normal seawater averages about $1.02–1.03$ g/cm³. However, the density of such materials as muscle, bone, shell, and scales averages about $1.05–2.00$ g/cm³. Since their bodies are significantly denser than seawater, nektonic animals tend to sink, and without special mechanisms to reduce body density, much energy would be used up simply maintaining the proper depth in the water column. One method of reducing density is to get rid of as much dense body mass as possible. The ancestors of modern squids lived on the bottom and had large external shells, but in modern squids the shell, which is now internal, is much reduced or absent, being replaced in many forms by a small, stiff, chitinous structure called a *pen.* In a few fish species that inhabit great depths of the sea, the amount of bone and body muscle has been much reduced so that they are now little more than a large mouth attached to a small flabby body.

A more general approach involves reducing body density by the inclusion of less dense oils, fats, and waxes in the tissues or gases in special bladders. Some squid species have substituted the very light ammonium ion for the much heavier calcium and magnesium ions in their body fluids. Many species of fishes and marine mammals store in their tissues large quantities of fats and oils (density $0.8–0.9$ g/cm³). For example, the blubber of the blue whale, which accounts for about a fifth of its body weight, is about 80 percent fat. This blubber also aids in streamlining the body and provides insulation against low water temperatures. Large amounts of fats and oils are stored in the liver or muscle tissues of many sharks and bony

fishes, and in some sharks the liver may account for a quarter of the body weight.

In surface water, the density of common gases is only about a tenth that of seawater. This density increases with depth, and at 7,000 m the gas is so compressed that its density is about $0.7$ g/cm³, which is close to that of fat. A great many fish species possess a gas-filled bladder, known as the *swim bladder,* which develops embryologically as a pouch from the esophagus. Some surface-dwelling fishes still have an open duct connecting the swim bladder with the esophagus, and they can fill the bladder by gulping air. However, the usual method is for the nitrogen and oxygen gases to be added to or removed from the bladder by a special highly vascularized gland at the rear of the bladder. Hence, the rate at which such a fish can change its buoyancy is controlled by the rate of glandular gas exchange, and this in turn controls the rate at which the fish can rise or sink in the water column without using swimming movements. Swim bladders are present in most species of nektonic bony fishes, but they are often absent in those forms that live primarily on the bottom and those that are strong, active swimmers. Some mesopelagic and deep-sea species have developed fat-filled swim bladders.

### Locomotion

#### BODY FORM

Most nektonic species have developed more or less streamlined bodies, and the faster and more active the species, the more streamlined they have become, culminating in the near-perfect streamline shape of the tuna fishes. The problem is to reduce drag when progressing through the relatively dense aqueous medium. Actually, three components of drag must be overcome. *Form drag* stems from the necessity of displacing an amount of water equal to the cross-sectional area of the body as it moves forward. *Frictional drag* is caused by the interaction between the outside of the animal's body and the liquid medium. *Turbulence* refers to the drag associated with nonlaminar flow or swirling of water, particularly around posterior portions of the animal's body. Studies have shown that the ideal shape for a body moving through a fluid is for the body to be rounded in cross section and to have a small, bluntly rounded front end that achieves maximum cross-sectional area about a quarter of the way back and tapers to a point at the rear (i.e., teardrop shaped). Most actively swimming animals, be they squids, fishes, or whales, tend to conform to this general body shape. However, for those nektonic species that live on or near the bottom or around coral reefs, seagrasses, or other structures, streamlining is less important. Here the ability to dart, turn, hide, or maintain a fixed position places a premium on other body forms.

PROPULSION

To propel the body through the water, each nektonic species must have some mechanism for providing thrust. In cephalopods this is accomplished through jet propulsion. Taking water in through the mantle cavity, they can squirt it out at rapid speed through a "nozzle" at the end of the tubular siphon. Since the siphon can be aimed in any direction, the squid is capable of very rapid swimming movements in any direction.

The primary locomotion of most species of fishes is accomplished by undulating body movements. Bending the head from side to side sends waves of undulations back along the body, and the major thrust comes from the backward push of the posterior part of the body and tail (or caudal) fin against the water. Alternating pushes from one side and then the other keep the fish moving in a forward direction. The caudal fin itself may take one of several shapes. Rounded caudal fins are useful for slow locomotion and for making small adjustments in body position in the water. Caudal fins that are squared off or forked provide for moderate or fairly fast swimming speeds. However, in very rapid and continuous swimmers such as mackerels, tunas, and billfishes the caudal fin is tall, stiff, and arched or lunate in shape. The tail of a shark is *heterocercal,* that is, it is not symmetrical, the top portion being longer than the bottom section. As a result, when the shark swims its back end tends to rise. This effect is counterbalanced by the paired fins, particularly the broad, flat pectoral fins near the front that act as planes to lift the front end of the body, thus keeping the shark level. In open water the shark must keep swimming because, being denser than seawater, it would otherwise sink to the bottom. Whereas fishes swim by moving their posterior body parts from side to side, marine mammals achieve thrust by moving the posterior part of their body and lateral tail fins or flukes up and down. Although the direction of motion is different, the effect is the same. In both cases the locomotion must be powered by very strong sets of muscles fed by efficient blood supply systems.

POSITIONING

Since they are generally not required for major propulsion, the paired pectoral and pelvic fins of bony fishes are available to aid in balancing, braking, hovering, and other fine body movements. In some bottom and near-bottom fishes, enlarged fanlike pectoral fins are used for jerky swimming movements. In many skates and rays the outer edges of the flattened pectoral fins aid in locomotion, and in large pelagic rays, such as the manta and eagle rays, the pectoral fins flap up and down, and these rays actually "fly" through the water. In flying fishes the greatly enlarged pectoral fins are used for gliding through the air above the wave tops. The fins do not flap, and all propulsion occurs underwater. Here the fish builds up speed and then breaks the water surface and glides for distances of up to several hundred yards. This is thought to be a mechanism for escaping predators. In a number of bottom-dwelling fishes the pectoral fins serve as primitive legs to crawl about the sediment surface. In a few fish species the unpaired dorsal and anal fins are used for locomotion, but in most species they act primarily as stabilizers. In nektonic shrimps locomotion is provided by the beating of the oar-like paired abdominal appendages and tail fans. In the swimming crabs (Portunidae) the last pair of walking legs has been modified into effective paddles that provide for slow locomotion but remarkable maneuvering capabilities.

SPEED

Some porpoises have been clocked at 40 km/hr (24 mi/hr), and tunas and their near relatives, wahoos, have been shown to be capable of speeds of 75–77 km/hr (43–46 mi/hr) in short bursts. Claims of 110 km/hr (66 mi/hr) for some of the larger tunas do not seem unreasonable, but so far they are unverified. Very fast swimming speeds result from strong streamlining of the body, exceptional development of muscular and caudal propulsion systems, and development of internal oxygen distribution and other circulatory and physiological mechanisms.

## Schooling

In the three-dimensional environment of the water column two major feeding strategies are prominent, filter feeding and predation. Filter feeders, which are usually smaller species but include some of the larger sharks and (baleen) whales, strain zooplankton and micronekton from the water. Since phytoplankton tends to have a patchy distribution, so does the zooplankton that feeds on it. Thus, the filter-feeding nektonic species such as herrings, anchovies, silversides, and others that consume the zooplankton find it advantageous to run in schools that can thoroughly exploit the clustered food supply. Schooling apparently also provides some protection from predatory species. Whereas a long, slow-swimming, filter-feeding fish would be quite vulnerable to attack by a fast-swimming predator, in schools the abundance of darting prey makes it difficult for the predator to home in on a single individual. However, even if successful, the predator cannot eat more than will fill its belly, and the remaining prey individuals are thus rendered safe after the sacrifice of a few.

In its most highly developed form, schooling behavior in prey species involves a series of complex instinctive behavior patterns in which the individual animals swim parallel to one another while maintaining a constant distance from their nearest neighbors. Animals of the school swim and turn almost in unison. This behavior is based largely on visual cues, but in bony fishes, at least, the lateral line system, which is remarkably sensitive to vibrations and slight pressure changes, may also be involved.

In some groups of fishes and marine mammals the predator species may hunt in packs, but they do not exhibit true schooling behavior since their alignment is not so precise, their movements are not so coordinated, and they are capable of taking independent action. Predators that hunt in packs may make more efficient use of the food supply, and some are known to herd the prey schools, keeping them available for later meals. Some pack-hunting predators are also able to make use of prey larger than themselves, as in the case of a pod of killer whales attacking and feeding on much larger individual whales.

## Special sensory and behavioral adaptations

Nektonic species possess the same senses (sight, hearing, smell, taste, and touch) as we do, and some of these, particularly the chemical senses, may be especially acute. Nektonic species are also quite sensitive to changes in water temperature, and all have a good sense of equilibrium. The lateral line system, as mentioned, is capable of detecting minute vibrations and changes in water pressure that could aid in schooling, but its primary function is likely to be to alert an animal to the approach of predators and prey. Sharks and some rays and bony fishes

are able to sense electrical charges given off by their own bodies and those of other individuals, including other species. Such charges may aid in the detection of objects in the water including members of their own species as well as prey organisms. There is also evidence that sea turtles and possibly tunas and other species that migrate long distances are capable of sensing the earth's magnetic field. The built-in compass presumably aids the animal in locating itself on the earth's surface and is an important mechanism underlying the adult sea turtle's ability to home in on the beach where it was born. The presence of numerous receptor-like cells and organs in the skin and elsewhere that are closely associated with the nervous system suggests that many nektonic animals possess senses of which we are totally unaware.

In addition to their specialized senses, many nektonic species are capable of producing effects on the environment. Some fish species generate electrical currents or electrical fields around their body that may be useful for sensing objects in the water, communicating among members of the same species, and/or for stunning prey and discouraging predators. Various groups of nektonic organisms including squids, shrimps, and fishes produce light, primarily in the deep sea. This serves a variety of functions, and the topic of bioluminescence will be taken up in the next chapter. One of the most remarkable adaptations of nektonic animals is the ability of certain marine mammals to produce and receive sonar signals and to use echolocation to sense features of the environment and locate prey organisms. Killer whales may also use sudden loud explosive sound bursts to stun prey. Life in the open water of the ocean is quite different from what we are accustomed to on land, and we are just beginning to understand some of the remarkable ways in which nektonic animals sense, modify, and ultimately survive in this open-water environment.

### COMPOSITION OF THE NEKTON

As pointed out earlier, the nekton of the Gulf of Mexico consists of cephalopods, some decapods, fishes, sea turtles, and marine mammals. Each of these groups will be considered in the present section, and a table showing the higher classification (table 13.1) provides a useful guide for the discussions that follow. Actually, other groups are sometimes included in the nekton. Large heteropods and euphausiid shrimp, which are abundant in the Gulf, are often listed among the micronekton, but

**Table 13.1.** *Higher classification categories for the major groups of nektonic animals inhabiting the Gulf of Mexico. Some of the less important and primarily deep-sea groups have been omitted, particularly among the bony fishes.*

| | |
|---|---|
| Phylum | Mollusca – mollusks |
| Class | Cephalopoda – cephalopods |
| Order | Sepioidea – cuttlefishes |
| Order | Teuthoidea – squids |
| Order | Vampyromorpha – vampire squid |
| Order | Octopoda – octopuses, paper nautilus |
| Phylum | Arthropoda – arthropods |
| Subphylum | Mandibulata – mandibulates |
| Class | Crustacea – crustaceans |
| Order | Decapoda – decapods |
| Suborder | Natantia – swimming shrimps |
| Family | Penaeidae – penaeid shrimps |
| Family | Sergestidae – sergestid shrimps |
| Suborder | Brachyura – true crabs |
| Family | Portunidae – swimming crabs |
| Phylum | Chordata – chordates |
| Subphylum | Vertebrata – vertebrates |
| Class | Chondrichthyes – cartilaginous fishes |
| Order | Lamniformes – thresher sharks, etc. |
| Order | Squaliformes – dogfish sharks |
| Order | Rajiformes – skates, rays |
| Class | Osteichthyes – bony fishes |
| Order | Elopiformes – tarpons |
| Order | Albuliformes – bonefish |
| Order | Anguilliformes – eels |
| Order | Clupeiformes – herrings |
| Order | Siluriformes – catfishes |
| Order | Aulopiformes – lizardfishes, etc. |
| Order | Gadiformes – cods, etc. |
| Order | Batrachoidiformes – toadfishes |
| Order | Lophiiformes – frogfishes, batfishes |
| Order | Atheriniformes – silversides, flying fishes, etc. |
| Order | Beryciformes – squirrelfishes |
| Order | Gasterosteiformes – pipefishes, seahorses |
| Order | Scorpaeniformes – scorpionfishes, etc. |
| Order | Perciformes – sea basses, snappers, etc. |
| Order | Pleuronectiformes – flounders, soles |
| Order | Tetraodontiformes – leatherjackets, puffers |
| Class | Reptilia – reptiles |
| Order | Testudinata – turtles |
| Family | Chelonidae – shelled sea turtles |
| Family | Dermochelyidae – leatherback sea turtle |

**Table 13.1.** *Continued*

| Phylum | Mollusca – mollusks |
|---|---|
| Class | Mammalia – mammals |
| Order | Sirenia – sirenians |
| Family | Trichechidae – manatees |
| Order | Cetacea – whales, dolphins, porpoises |
| Suborder | Mysticeti – baleen whales |
| Family | Balaenidae – right whale |
| Family | Balaenopteridae – rorquals |
| Suborder | Odontoceti – toothed whales |
| Family | Physeteridae – sperm whales |
| Family | Ziphiidae – beaked whales |
| Family | Delphinidae – dolphins |
| Family | Phocoenidae – porpoises |
| Order | Pinnipedia – seals, sea lions |
| Family | Phocidae – hair seals |

these have already been treated in the zooplankton chapter. Stomatopods are capable of swimming and sometimes enter the nekton, but most live in association with the bottom, and they have also been considered earlier. A few species of echinoderms (sea cucumbers and crinoids) swim in deeper waters, but these are not known to be important in the Gulf. Sea snakes and saltwater crocodiles, present in the Indo-Pacific, are absent from Gulf waters. Diving and swimming birds, so important at higher latitudes, are rare in the Gulf of Mexico and are not included here. Important literature references are presented in the individual group discussions.

**Cephalopods**

The cephalopods constitute a very specialized and highly organized group of mollusks. The head region is well developed, and the various ganglia have become fused to form a brain. Some species have very complex eyes that are in many respects structurally similar to those of vertebrate animals. In most species, eight prehensile arms, generally armed with suckers, project from the head region. These are used for seizing prey, but among the octopods they also serve for locomotion. The bases of these arms surround the mouth, which contains a horny beak for cutting off chunks of food from prey organisms.

Present-day cephalopods are derived from a large and diverse group of mollusks, now extinct, called ammonites. These had well-developed straight, arched, or coiled shells divided into chambers by transverse calcareous septa. Although most species were probably bottom dwellers, the gas-filled chambers may have created neutral buoyancy, permitting some species to float or move around in the water column. Modern cephalopods, however, are all modified for a nektonic existence, and although some such as the octopuses have secondarily moved back to the bottom, all are capable of swimming by jet propulsion. One group of modern cephalopods, the nautiloids (including the Indo-Pacific chambered nautilus) retains an external coiled shell, but this group is not represented in the Gulf. In all other cephalopods the shell is internal and reduced in size or absent altogether. In some groups it is replaced by a long, flat, chitinous strengthening structure called a pen. Evolution of the shells and pens among the different cephalopod groups is illustrated in figure 13.1. All living cephalopods except the nautiloids produce a black or brown fluid called "ink" that is stored in a special sac. When the animal is alarmed, the fluid can be ejected into the water, creating a cloud to confuse predators and facilitate escape. The fluid may be noxious or anesthetic to the chemical sense organs of some predator species, and in one group of deep-sea squids the ink is luminescent. Outside the nautiloids, modern cephalopods are divided into four orders: Sepioidea (cuttlefishes), Teuthoidea (squids), Vampyromorpha (vampire squid), and Octopoda (octo-

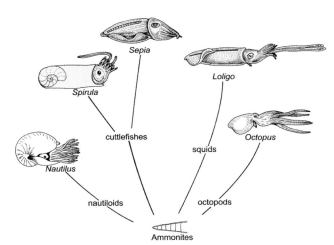

**Figure 13.1.** Major evolutionary lines leading to modern cephalopods, with emphasis on the body support structures. In modern nautiloids the chambered shell remains external, in cuttlefishes it is internal and spiral shaped or straight, in squids it is replaced by the chitinous pen, and in octopods it is absent.

puses), all of which are represented in the Gulf. Worldwide more than 700 species of cephalopods are known to science (Sweeney and Roper 1998). Most living species of cephalopods are nektonic and oceanic and are found at all depths of the sea. Many mesopelagic species appear to undergo diel vertical migrations, appearing in the upper layers at night, but little is known of the habits and life histories of most species found in deeper waters. All pelagic species are carnivorous and feed on zooplankton and nektonic animals, including other cephalopods. Most of the pelagic species also have pelagic juvenile stages as components of the micronekton, and these are often very difficult to identify to the species level.

Major references to the cephalopod fauna of the Gulf of Mexico include Berry (1934), Cairns (1976), Hanlon (1985), Lipka (1970, 1975), Nessis (1987), Passarella (1990), Salcedo-Vargas (1991), and G. Voss (1954, 1956). An updated list of species known to occur in the Gulf has been provided by Salcedo-Vargas (1991), but several additional species have been reported by Passarella and Hopkins (1991). Further information concerning some Gulf species and their non-Gulf relatives appears in N. Voss et al. (1998). Because of the unevenness of collecting effort, the cephalopod fauna of the northern and eastern sectors of the Gulf is much better known than that of the western and southwestern sectors, and this is particularly true for the oceanic and deepwater fauna. So far, more than 80 species of cephalopods have been reported from the Gulf of Mexico, and the number will likely increase

as the deeper waters of the Gulf are more thoroughly explored. Although several species appear to be endemic, the cephalopod fauna of the Gulf as a whole is broadly representative of that of the western Atlantic and Caribbean regions. The neritic and endemic as well as a few common epipelagic species of the Gulf are listed in table 13.2, and examples are shown in figure 13.2. These forms are discussed briefly below.

NERITIC SPECIES

Eight cephalopod species, including five squids and three octopods, are found primarily or exclusively in neritic waters of the Gulf. *Lolliguncula brevis,* a small squid generally occurring in large schools, is the common coastal species all around the Gulf. During warmer months it enters bays and estuaries (the only Gulf cephalopod to do so), and on the continental shelf it is largely confined to waters of 30 m (102 ft) or less. *Loligo pealei,* a much larger species, is common over the continental shelf all around the Gulf, and it extends from near shore to the outer edge of the shelf. Its young are often collected together with adults of the previous species. *Loligo plei* (formerly of the genus *Doryteuthis*) is often called the arrow squid because of its long, slender body and pointed tail fin. Its distribution is similar to that of its near relative, *L. pealei,* although it is less abundant. *Sepioteuthis sepioidea* has been taken only around coral reefs of south Florida and Yucatán, and it is not known to occur at the Flower Garden Banks of the northwestern Gulf. *Pickfordiateuthis pulchella* is a very small squid, generally less than 85 mm (3.5 in) long. It appears to be limited to the extensive seagrass meadows and adjacent habitats of south Florida. A number of upper-slope species such as *Semirossia equalis* and *S. tenera* also occur from time to time on the outer continental shelf of the eastern and northern Gulf.

Among the three species of primarily neritic octopods, the most common and widely distributed is *Octopus vulgaris.* It occurs across the shelf all around the Gulf wherever suitable hiding places are available. Along the shoreline it inhabits crevices among the stones at the base of jetties, and on the carbonate shelves of Florida and Yucatán it occupies the discarded shells of large gastropods such as those of the horse conch (*Pleuroploca gigantea*). The other shallow-water octopus species appear to have more limited distribution patterns. These include *O. briareus* (shallow shelf off western Florida) and *O. maya* (shallow waters, including seagrass flats and rocky areas off Yucatán). The latter species is easily recognized by the presence of a prominent ocellus, or pigmented "eye spot" marking, on the side of the body (G. Voss and Solis Ramirez 1966). Several species characteristic of the upper slope, such as *Octopus burryi* and *O. joubini,* also occur on the continental shelf.

OCEANIC SPECIES

The remaining Gulf cephalopods are oceanic, and most are nektonic, although for some demersal species of the slope and deep Gulf it is not clear whether they live on the bottom or hover just above it. Most cephalopod species of the open Gulf occur in discrete depth ranges, but some are widely distributed in both surface and deeper layers. According to Lipka (1975), species of the following genera are probably epipelagic (0–200 m depth): *Argonauta, Cranchia, Chtenopteryx, Heteroteuthis, Illex, Leachia, Liocranchia, Ommastrephes, Onychoteuthis, Onykia, Ornithoteuthis,* and *Tremoctopus.* In one of the epipelagic species, *Argonauta hians* (the paper nautilus), the female produces a delicate and beautifully sculpted spiral shell that floats at the surface and serves as an egg case (fig. 13.2). Passarella and Hopkins (1991) discussed the epipelagic micronektonic cephalopod fauna of the eastern Gulf. Most species are vertical migrators, being found in the daytime in the depth range of 100–400 m and at night above 200 m. Most are small species that feed on crustaceans, such as copepods, as juveniles and rely more on larger prey, such as fishes and other squids, as they mature. These cephalopods, in turn, are important food resources for larger predatory fishes and marine mammals.

Among the endemic cephalopods of the Gulf is the shallow-water form *Octopus maya* as well as several species characteristic of the upper continental slope of the northern Gulf around the Mississippi River Delta. These include *Lycoteuthis springeri, Pholidoteuthis adami, Rossia bullisi,* and *Selenoteuthis scintillans.* The small squid *Pickfordiateuthis pulchella* is endemic to south Florida, but its range extends to the Atlantic side of the peninsula and into Biscayne Bay near Miami.

## Crustaceans

Although most species of crustaceans are capable of swimming to some extent, three groups are sufficiently large and well enough adapted for swimming to be classified as nekton. These include members of the shrimp families Penaeidae and Sergestidae and the crab family Portunidae, all inhabitants mainly of neritic or epipelagic waters and all producers of planktonic larvae. Also

**Table 13.2.** *Distribution of cephalopods of the continental shelf and epipelagic waters of the Gulf of Mexico. Endemic species of the Gulf are marked with an asterisk.*

| Class | Cephalopoda |
|---|---|
| Order | Sepioidea |
| Family | Sepiolidae |
| | *Heteroteuthis hawaiiensis* – epipelagic |
| | *Rossia bullisi\** – upper slope, northern Gulf |
| | *Rossia tortugaensis* – outer shelf, upper slope |
| | *Semirossia equalis* – outer shelf, upper slope |
| | *Semirossia tenera* – outer shelf, upper slope |
| Order | Teuthoidea |
| Family | Pickfordiateuthidae |
| | *Pickfordiateuthis pulchella* – shallow shelf, south Florida |
| Family | Loliginidae |
| | *Loligo (Doryteuthis) pealei* – across shelf, entire Gulf |
| | *Loligo (Doryteuthis) plei* – across shelf, entire Gulf |
| | *Lolliguneula brevis* – shallow shelf, entire Gulf |
| | *Sepioteuthis sepioidea* – coral reefs, Florida and Yucatán |
| Family | Lycoteuthidae |
| | *Lycoteuthis springeri\** – upper slope, northern Gulf |
| | *Selenoteuthis scintillans\** – upper slope, northern Gulf |
| Family | Onychoteuthidae |
| | *Onychoteuthis banksi* – epipelagic |
| | *Onykia carriboea* – epipelagic |
| Family | Chtenopterygidae |
| | *Chtenopteryx sicula* – epipelagic |
| Family | Ommastrephidae |
| | *Illex coindetii* – epipelagic |
| | *Ommastrephes bartramii* – epipelagic |
| | *Ornithoteuthis antillarum* – epipelagic |
| Family | Lepidoteuthidae |
| | *Pholidoteuthis adami\** – upper slope, northern Gulf |
| Family | Cranchidae |
| | *Cranchia scabra* – epipelagic |
| Order | Octopoda |
| Family | Octopodidae |
| | *Octopus briareus* – shallow shelf, Florida |
| | *Octopus burryi* – outer shelf, upper slope |
| | *Octopus joubini* – continental shelf, upper slope |
| | *Octopus maya\** – shallow shelf, Yucatán |
| | *Octopus vulgaris* – across shelf, entire Gulf |
| Family | Tremoctopodidae |
| | *Tremoctopus violaceus* – epipelagic |
| Family | Argonautidae |
| | *Argonauta argo* – epipelagic |

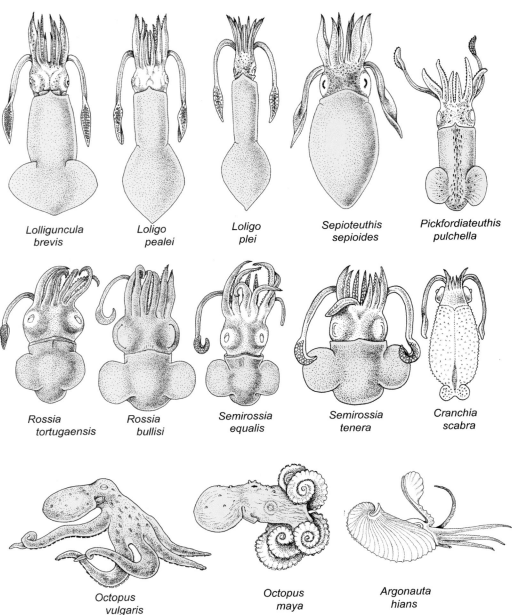

**Figure 13.2.**
Examples of neritic and epipelagic cephalopods recorded from the Gulf of Mexico.

*Lolliguncula brevis*

*Loligo pealei*

*Loligo plei*

*Sepioteuthis sepioides*

*Pickfordiateuthis pulchella*

*Rossia tortugaensis*

*Rossia bullisi*

*Semirossia equalis*

*Semirossia tenera*

*Cranchia scabra*

*Octopus vulgaris*

*Octopus maya*

*Argonauta hians*

included among the nekton are a number of deepwater shrimps representative of the groups Mysidacea, Euphausiacea, Penaeidea, and Caridea, which will be taken up in the next chapter.

PENAEID SHRIMPS

Shrimps of the family Penaeidae generally have laterally compressed bodies, thin external shells, or *carapaces,* and long muscular abdomens equipped with five pairs of well-developed *pleopods,* or swimmerets. These oar-shaped ap-

pendages provide propulsion for swimming. An exception is found in species of the genus *Sicyonia* (rock shrimps), in which the carapace is quite heavy and the pleopods are reduced in size, and this group is strictly benthic. All penaeid shrimps of the neritic zone feed on small benthic invertebrates, carrion, and detritus particles that they glean from the bottom. When not feeding, they may rest upon or lie buried within the surface sediments. However, many, if not most, also spend some time in the water column, and some species undergo migrations of tens of kilometers on their way to breeding grounds. Most are active primarily at night, but a few species are active in the daytime. Juveniles of several species, primarily of the genus *Penaeus,* regularly use estuaries as nursery grounds, but all spawning takes place on the continental shelf. Although penaeid shrimps occur on continental shelves throughout the Gulf, distribution patterns of individual species appear to be controlled chiefly by the availability of suitable sediment types and, for some species, the availability of estuarine nursery areas. Females tend to grow to larger sizes than males, and they often outnumber males in trawl catches. Species of the genus *Penaeus,* which make up the bulk of the commercial catch of edible shrimps of the Gulf, reach maximum body lengths of around 200 mm (8 in), but most of the neritic species average only a third to a half this length. Primary references to the taxonomy, distribution, and life history patterns of penaeid shrimps of the Gulf of Mexico include Burkenroad (1934), Costello and Allen (1966), Darnell, Defenbaugh, and Moore (1983), Darnell and Kleypas (1987), Defenbaugh (1976), H. Hildebrand (1954, 1955), Huff and Cobb (1979), Pérez-Farfante (1969, 1971, 1977), Pérez-Farfante and Bullis (1973), Sánchez and Soto (1987), and Williams (1965). Upper-slope penaeids such as *Hadropenaeus* (= *Hymenopenaeus*) *affinis* and *Pleoticus* (= *Hymenopenaeus*) *robustus,* which sometimes appear on the outer continental shelf, have been reported in the several papers by Pérez-Farfante (see above) and by T. Roberts and Pequegnat (1970). Thumbnail sketches of the individual species of neritic penaeids of the Gulf are given in table 13.3, and representative species are shown in figure 13.3.

### SERGESTID SHRIMPS

Shrimps of the family Sergestidae have compressed bodies, thin carapaces, and elongated abdomens with five well-developed pairs of pleopods, but unlike in their relatives of the family Penaeidae, in the sergestids the last two pairs of legs are reduced in size or lost altogether, and the gills at the bases of the legs have also been reduced or lost. The sergestids of the Gulf are quite small, averaging less than 30 mm (1.2 in) in length, and they belong to the micronekton. They exist largely in the water column, although the neritic species may also spend some time on the bottom. All apparently exhibit vertical migration and are more concentrated in near-surface waters at night. References to Gulf species include Berkowitz (1976), Burkenroad (1934), Cruise (1971), Harper (1968), Huff and

Cobb (1979), and Woodmansee (1966). Three species are known from the Gulf, the neritic forms *Acetes americanus* and *Lucifer faxoni,* and the oceanic form *L. typus. Acetes americanus* has apparently been recorded only from the northeastern Gulf in the depth range of 57–60 m (Franks et al. 1972; S. Springer and Bullis 1956). *Lucifer faxoni* is more widespread and has been reported from both the eastern and northern Gulf (Florida to Texas). Although found primarily over the continental shelf, this species sometimes enters estuaries, and it also occurs in epipelagic waters of the deep Gulf. *Lucifer typus* is an oceanic species that occurs occasionally over the outer shelf (Harper 1968). Examples of sergestid shrimps are shown in figure 13.3. Deepwater pelagic mysid, euphausiid, penaeid, and caridean shrimps are taken up in the next chapter.

### PORTUNID CRABS

In the portunid or swimming crabs, the body is flattened, more or less oval in outline, and generally equipped with a pair of lateral spines. There are usually five to nine points, or "teeth," between the lateral spine and the eye and two to four pairs between the eyes. On the last pair of legs the terminal section has been modified into a flat, rounded or oval plate, and together these plates serve as oars or propellers for locomotion and maneuvering within the water column. Many of the swimming crabs tolerate low salinities and even freshwater, and they are often found in brackish estuaries, bays, and mangrove swamps, and some even enter coastal streams. However, the primary habitat is the continental shelf, and here they occupy a variety of habitats including seagrass meadows, soft and hard bottoms, and coral reefs. Several species swim in surface waters, especially at night, and one species is pelagic and associated with sargassum and other materials floating on the surface of the open ocean. Most portunid crabs are omnivorous and consume a variety of plants and small animals as well as carrion and organic detritus.

There is a large body of literature dealing with the portunid crab fauna of the Gulf of Mexico, and much of this has been summarized by Powers (1977). Some of the more important references treating the taxonomy and distribution of Gulf species include Abele (1970), Chace (1940), Contreras (1930), Felder (1973), Felder and Chaney (1979), Franks et al. (1972), García-Montes, Soto, and Garcia (1988), H. Hildebrand (1954, 1955), Leary (1967), Lemaitre (1991), Powers (1977), Soto (1979, 1980),

**Table 13.3.** *Brief characterization of the individual penaeid shrimp species (exclusive of the genus* Sicyonia*) of neritic waters of the Gulf of Mexico.*

*Mesopenaeus tropicalis.* Florida and northeast Yucatán shelves. Middle to outer shelf to a depth of 915 m. Rare. Maximum length ca. 92 mm (3.6 in).

*Metapenaeopsis goodei.* Florida and Yucatán shelves. Most common in 18–55 m depth range, but extends to 329 m depth. Maximum length 157 mm (6.2 in).

*Parapenaeus longirostris.* Around periphery of entire Gulf. Mostly in depth range of 50–183 m, but recorded to depth of 732 m. Maximum length 104 mm (4.1 in).

*Parapenaeus politus.* Southern Gulf at base of Yucatán Peninsula. Depth range of 75–170 m. Maximum length ca. 100 mm (3.9 in).

*Penaeus aztecus* (brown shrimp). Continental shelves all around the Gulf from Alabama to base of Yucatán Peninsula. Absent from carbonate platforms of west Florida and Yucatán, but present off south Florida (Florida Bay and the Keys). Young utilize estuaries as nursery grounds. Most abundant in northern Gulf from Alabama to northeastern Mexico and in Bay of Campeche, where it forms the basis for large commercial fisheries. Adults most concentrated on sandy and silty bottoms in depth range of 27–55 m, but recorded to depth of 183 m. Maximum length 211 mm (8.3 in).

*Penaeus brasiliensis.* Restricted to southern tip of Florida (Florida Bay to the Tortugas) and along northern coast of Cuba. Rare. Most common in depth range of 37–73 m, but recorded to depth of 275 m. Maximum length 214 mm (8.5 in).

*Penaeus duorarum* (pink shrimp). Found in low densities around entire periphery of Gulf. High concentrations off southwest Florida (Tortugas and Sanibel grounds) and in southwest Gulf at base of Yucatán Peninsula, where it is basis for major shrimp fisheries. Young use high-salinity bays, seagrass meadows, and mangrove areas as nursery grounds. Adults most concentrated in depth range of 11–37 m, but recorded to depth of 275 m. Maximum length 210 mm (8.3 in).

*Penaeus schmitti.* Restricted to a few localities on northern coast of Cuba. Mostly from shore to depth of 22 m, but occasionally to 48 m. Maximum length 235 mm (9.3 in).

*Penaeus setiferus* (white shrimp). From Apalachee Bay in northern Florida along most of coast of northern, western, and southern Gulf to base of Yucatán Peninsula. Absent from carbonate shelves of western and southern Florida and Yucatán. Young use low-salinity estuaries as nursery grounds. Heaviest concentrations in northern Gulf from Mobile Bay to southern Texas and in southern Gulf. This is a major commercial species. Adults concentrated near shore to a depth of 37 m, but occasionally to 82 m. Maximum length 197 mm (7.8 in).

*Solenocera atlantidis.* Found on continental shelf and upper slope throughout Gulf. Most abundant in depth range of 9–75 m, but recorded to depth of 329 m. Generally rare, but locally abundant on sandy substrates. Maximum length 56 mm (2.2 in).

*Solenocera necopina.* Reported from eastern flank of DeSoto Canyon in northeastern Gulf. An upper slope species that rarely strays onto the continental shelf. Has been taken as shallow as 39 m, but normal depth range is 200–250 m. Maximum size not reported.

*Solenocera vioscai.* Found on soft bottoms of northern, western, and southern Gulf across shelf and onto upper slope. Most common in depth range of 37–68 m, but recorded in range of 16–366 m. Fairly common in some areas. Maximum length 100 mm (4.0 in).

*Trachypenaeus constrictus.* Found all around periphery of Gulf, primarily on coarse (sandy) substrates. Depth range of 4–91 m, but most common around 18 m depth. Locally abundant. Maximum length 92 mm (3.6 in).

*Trachypenaeus similis.* Present throughout northern, western, and southern Gulf and in south Florida around Tortugas. Absent from carbonate shelves of west Florida and Yucatán. Most abundant on soft, silty sediments, especially off Louisiana and Texas, where it is quite common. Found from 6 to 108 m; most common in depth range of 11–72 m. Maximum length 90 mm (3.5 in).

*Xiphopenaeus kroyeri* (seabob). Found on sandy and muddy bottoms from Apalachee Bay through northern, western, and southern Gulf to base of Yucatán Peninsula. Apparently absent from carbonate shelves of west Florida and Yucatán. Habitat is estuaries and nearshore zone of continental shelf, especially in depth range of 5–37 m, rarely to 46 m. Locally abundant. Maximum length 127 mm (5.0 in).

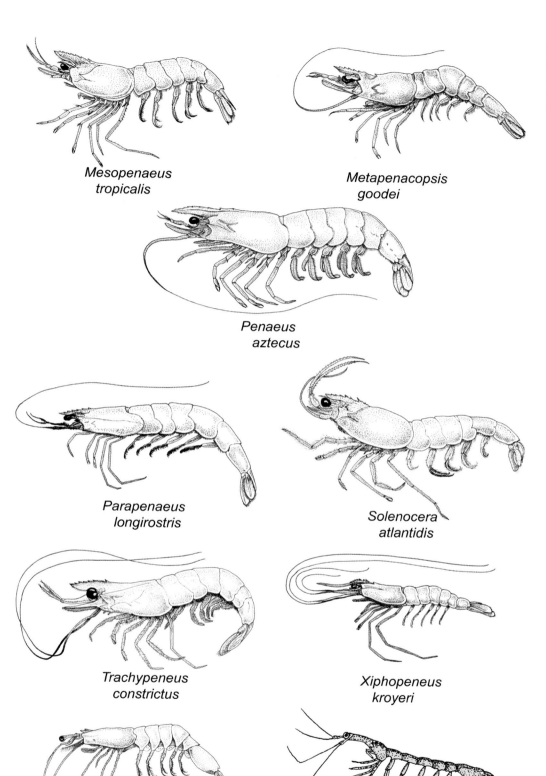

Mesopenaeus
tropicalis

Metapenacopsis
goodei

Penaeus
aztecus

Parapenaeus
longirostris

Solenocera
atlantidis

Trachypeneus
constrictus

Xiphopeneus
kroyeri

Acetes
americanus

Lucifer
sp.

**Figure 13.3.**
Examples of littoral and epipelagic penaeid and sergestid shrimps of the Gulf of Mexico.

S. Springer and Bullis (1956), and A. Williams (1965, 1966, 1974). Twenty-six species are known to inhabit neritic and oceanic waters of the Gulf. Each of these is characterized briefly in table 13.4, and examples are illustrated in figure 13.4. Distributional patterns given in the table are approximate because of uncertainties stemming from early taxonomic confusion of several species and lack of collecting along some portions of the Mexican shelf. Most of the Gulf of Mexico species are also widely distributed along the Caribbean coast of Central America and/or the West Indies, and the three endemic species of the Gulf (*Callinectes rathbunae, Portunus floridanus,* and *P. vossi*) are derived from tropical relatives. Detailed life history information is incomplete or lacking for most species, but the blue crab (*Callinectes sapidus*) of commercial importance has been fairly well studied. In this species, mating takes place in low-salinity estuaries and bays, and thereafter the female migrates back out to the continental shelf, where she carries the eggs on her swimmerets until they hatch into planktonic larvae (Darnell 1959). Along the west coast of Florida, mated females leave the estuaries and migrate northward, sometimes for considerable distances, to special nursery areas in Apalachee and Apalachicola Bays, where they remain until egg release (Oesterling and Evink 1977). The larvae are then carried southward on oceanic currents, and eventually some of them settle to the bottom or make their way as planktonic larvae into the Florida estuaries where they metamorphose, feed, and grow into adults. Exactly how the larvae move through passes into the estuaries is a complex subject that is not fully understood (Lochmann, Darnell, and McEachran 1995). Although the adult females of all species of portunid crabs carry the eggs on swimmerets, details of the various life histories have yet to be worked out. As seen in the case of the blue crab, it is possible that local populations have developed different strategies for completing their life histories in response to different prevailing conditions of their physical environment, particularly the water currents that transport planktonic larvae.

## Fishes

Fishes are vertebrates possessing a brain, dorsal tubular nerve cord, pharyngeal gill slits, and generally fins. The brain and spinal cord are encased in a protective cranium and vertebral column made up of cartilage or bone. All fishes are aquatic or semiaquatic, and they obtain oxygen by passing water through the mouth, across the gill membranes where gas exchange takes place, and out through the gill slits. Fishes make up the most diverse group of vertebrates. A recent estimate puts the number of living fish species in the world at around 25,000, of which about 15,000, or 60 percent, are marine (J. Nelson 1994). The vast oceans provide a great diversity of habitats, and fishes have adapted to all depths and are found even at the bottom of the deepest trenches. Many species are bottom dwellers, and some live around reefs and other structures, but a great many are pelagic. In the open ocean some are limited to particular depth ranges, whereas others are vertical migrators, remaining in deep waters in the day and moving upward to feed in the more productive epipelagic waters at night. Fishes vary in size from less than an inch to about 18.3 m (60 ft) and over 20 tons, in the case of the great whale shark, but most fall in the size range of 5–150 cm (ca. 2–60 in). The shapes of fishes vary greatly, some being streamlined, some round, some flat, and so forth, and as mentioned earlier, these shapes reflect the lifestyles of the individual species. Color patterns serve as camouflage, species recognition signs, or mating displays. Deepwater species often have light-producing organs that may serve as signals for species recognition or as lures to attract prey. In order to survive in the marine environment, fishes have developed a great variety of life history patterns. Most species feed on invertebrates and/or other fishes, and some are plankton feeders, but a few feed on vegetation, while others are detritus feeders.

Fishes of the modern world are descendants of several ancient groups dating back at least 300 million years, and the recognition of modern fish groups will be enhanced by some understanding of the anatomical changes associated with fish evolution. The most primitive forms lacked jaws or paired fins, the skeleton was cartilaginous, and numerous individual gill openings were present on the sides of the pharynx. Today these fishes are represented by a few eel-like species referred to as hagfishes and lampreys. Their ancestry goes back at least 250 million years to the ancient ostracoderms or closely related groups. More advanced fishes developed upper and lower jaws (derived from supporting elements of the first gill arch) and paired fins. The skeleton was still made up of cartilage, and individual gill slits were still present but reduced in number. The tail was of the heterocercal type, that is, asymmetrical, with the top section being longer than the bottom. In today's world this group is represented by the sharks, skates, and rays, collectively known as the cartilaginous fishes. They were derived from the placoderms or their relatives, and essentially modern

**Table 13.4.** *Brief characterization of the individual portunid crab species of neritic waters of the Gulf of Mexico.*

*Arenaeus cribrarius* (speckled crab). Found around eastern, northern, western, and southern Gulf from Florida Keys to base of Yucatán Peninsula. From shoreline to depth of 68 m.

*Callinectes bocourti.* A West Indian species recorded in Gulf only in Biloxi Bay, Mississippi, at a depth of 3 m.

*Callinectes danae.* Recorded from Florida Keys and along north coast of Cuba to a depth of 75 m. Found in brackish and marine waters. Erroneously listed from eastern and northern Gulf due to confusion with *C. similis* (A. Williams 1974).

*Callinectes exasperatus.* Occurs in Florida Keys and Dry Tortugas, in southwestern Gulf from Veracruz to Yucatán, and on north coast of Cuba. One questionable record from south Texas. Inhabits low-salinity estuaries and continental shelf to a depth of 7.5 m.

*Callinectes marginatus.* Occurs in southwest Florida, Keys, Dry Tortugas, southwestern Gulf from Veracruz to Yucatán, and north coast of Cuba. One record from Louisiana. Lives around mangrove swamps, seagrass meadows, tide pools, and oyster reefs. Tolerates low salinity. Found primarily from shore to depth of 15 m, rarely to 25 m.

*Callinectes ornatus.* Found around Florida Keys, Dry Tortugas, Florida shelf to level of Tampa Bay, Gulf coast of Yucatán, and on north coast of Cuba. Often in low to moderate salinity and on shelf to a depth of 75 m. Previously confused with *C. similis* (A. Williams 1974).

*Callinectes rathbunae.* Occurs on Mexican east coast from Tampico to southern Veracruz, rarely north to Rio Grande. Lives in brackish and nearshore marine waters. Endemic to Gulf of Mexico.

*Callinectes sapidus* (blue crab). Found along entire Gulf coasts of United States, Mexico, and Cuba. Present in low-salinity waters of bays and estuaries and on continental shelf to a depth of 35 m, rarely to 90 m. This is the common edible crab of the northern Gulf, and it supports commercial fisheries all around the Gulf.

*Callinectes similis.* Occurs along entire Gulf coastline of United States and Mexico. Found in higher salinity bays and estuaries and on shelf to a depth of 92 m. Often swims at surface at night. Many early records of *C. danae* and *C. ornatus* around eastern and northern Gulf are referable to this species (A. Williams 1974).

*Cronius ruber.* Recorded from south Florida, Dry Tortugas, Texas, Yucatán (Campeche), and north coast of Cuba. Occurs from shoreline to a depth of 105 m.

*Cronius tumidulus.* Found in Florida Keys, Dry Tortugas, west coast of Florida, and north coast of Cuba, Occurs in depth range of 5–73 m.

*Lupella forceps.* Known from north coast of Cuba in depth range of 13–15 m.

*Ovalipes floridanus* (lady crab). Found from southwest Florida along eastern and northern Gulf coast to south Texas. Occurs in bays and lagoons and on continental shelf to a depth of 31 m. Early records refer to this species as *O. guadalpensis* or *O. ocellatus* (A. Williams 1974).

*Portunus anceps.* Occurs in Florida Keys and on north coast of Cuba in brackish and marine waters to a depth of 103 m.

*Portunus binoculus.* Recorded from south Florida (Florida Straits) and on north coast of Cuba.

*Portunus depressifrons.* Found in south Florida, Keys, Dry Tortugas, Bay of Campeche, on Yucatán shelf, and north coast of Cuba. Occurs to a depth of 29 m.

*Portunus floridanus.* Recorded around coral reefs at Key West, Florida, at a depth of 82 m. Endemic to Gulf.

*Portunus gibbesii* (iridescent swimming crab). Occurs in south Florida, Keys, Dry Tortugas, and around eastern and northern Gulf to south Texas. Also found in Bay of Campeche to base of Yucatán Peninsula. Occurs from shore to a depth of 88 m.

*Portunus ordwayi.* Recorded from south, west, and northwest coast of Florida and north coast of Cuba to a depth of 106 m.

*Portunus sayi* (sargassum crab). Occurs from Florida Keys and Dry Tortugas along eastern and northern Gulf to south Texas and on north coast of Cuba. This is a pelagic species normally associated with sargassum and other floating objects, and it is sometimes observed swimming free at the surface of the open Gulf.

*Portunus sebae.* Found in Florida Keys and Dry Tortugas in depth range of 4–18 m.

*Portunus spinicarpus.* Occurs around periphery of Gulf from south Florida to base of Yucatán Peninsula and on north coast of Cuba. Found in depth range of 9–550 m.

*Portunus spinimanus* (blotched swimming crab). Found around periphery of Gulf from south Florida to base of Yucatán Peninsula. Occurs from shore to a depth of 91 m.

*Portunus ventralis.* Recorded from Dry Tortugas and north and west coasts of Cuba. Found from shore to a depth of 25 m.

*Portunus vocans.* Occurs on north coast of Cuba in depth range of 37–309 m.

*Portunus vossi.* Found in only a few localities along the west coast of Florida from the Everglades to south of Pensacola in the depth range of about 20–30 m. This is apparently a burrowing crab, not readily captured, and its distribution may eventually prove to be wider than presently indicated.

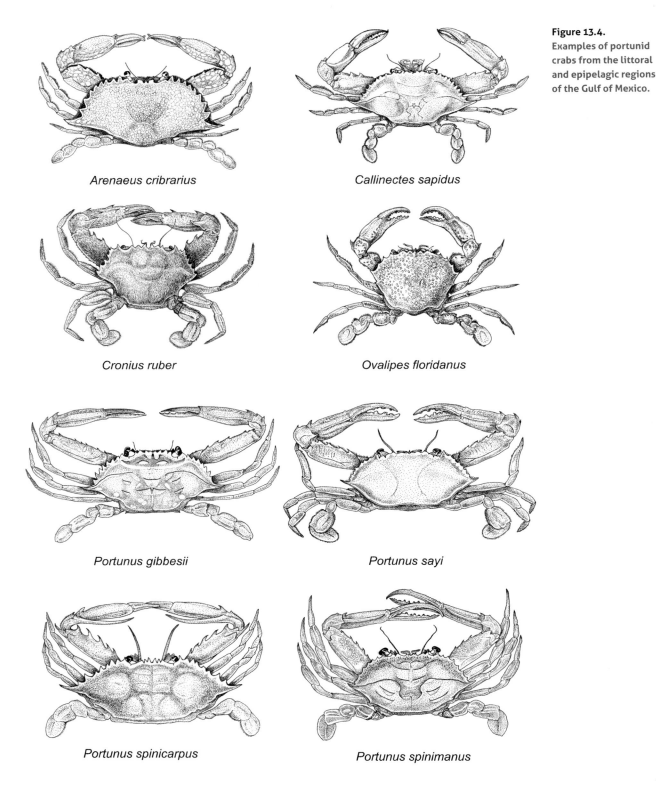

*Arenaeus cribrarius*

*Callinectes sapidus*

*Cronius ruber*

*Ovalipes floridanus*

*Portunus gibbesii*

*Portunus sayi*

*Portunus spinicarpus*

*Portunus spinimanus*

**Figure 13.4.**
Examples of portunid crabs from the littoral and epipelagic regions of the Gulf of Mexico.

types of sharks had appeared by the Jurassic period (± 160 million years ago).

In more advanced fishes the jaws and paired fins are present, but the cartilage has been replaced by true bone, and the tail is generally of the homocercal (symmetrical) type. In this group the gills are covered by a protective bony plate, or operculum. The bony fishes are believed to be derived from the palaeoniscids, another ancient group. Several of the surviving bony fish groups evolved independently, and such fishes as the sturgeons, gars, lungfishes, and some others are relicts of very ancient forms, and they do not have close living relatives. Among the modern bony fishes two general groups are recognized. The more primitive forms (sardines, herrings, and their relatives) are soft bodied, the fins are supported by soft rays only, and the body scales are generally of the *cycloid* type (more or less circular in outline and lacking comblike teeth). The maxillary bone forms part of the upper border of the mouth, and the swim bladder is still connected with the esophagus by an open duct. Both the pectoral and pelvic fins are generally abdominal, that is, positioned on or near the lower margin of the body. By contrast, in the more advanced bony fishes (typified by such forms as the snappers, groupers, perches, and basses) the fins are supported by spines as well as soft rays, and scales are normally of the *ctenoid* type (having comblike teeth on the posterior edge). The maxillary bones are excluded from the gape of the mouth due to elongation of the premaxillaries, and the swim bladder no longer opens to the esophagus. The pelvic fins, when present, have moved forward so that they are now near the throat (*jugular*) or below the pectoral fins (*thoracic*). Examples of fishes showing these basic body plans are illustrated in figure 13.5.

In marine waters of the Gulf of Mexico there are about 1,500 different species of fishes, or about one-tenth of all the living marine fish species of the world. This diversity reflects the size and variety of available habitats on the extensive continental shelves as well as the vast areas and volumes of shallow and deep waters in the Gulf that are connected with the Caribbean Sea and open Atlantic, facilitating colonization of neritic and oceanic species. In a monumental two-volume treatise, McEachran and Fechhelm (1998, 2005) have listed and described all the known fish species of the Gulf. Other, generally less technical works useful for the identification of shallow-water and epipelagic Gulf fish species include Böhlke and Chaplin (1993), Boschung (1992), Boschung et al. (1983), Hoese

and Moore (1998), Humann (1994), Longley and Hildebrand (1941), Murdy (1983), Nakamura (1976), Randall (1968), Robins, Ray, et al. (1986), and Walls (1975). Information on the distribution and ecology of continental shelf fishes is provided in the following works: northwestern Gulf (Darnell, Defenbaugh, and Moore 1983; D. Moore, Brusher, and Trent 1970), northeastern and eastern Gulf (Darcy and Gutherz 1984; Darnell and Kleypas 1987), and southwestern Gulf (Castro-Aguirre 1978; H. Hildebrand 1955; Sánchez-Gil, Yáñez-Arancibia, and Amezcua Linares 1981; Yáñez-Arancibia 1985a, 1985b; Yáñez-Arancibia and Sánchez-Gil 1986; and Yáñez-Arancibia, Linares, and Day 1980). Reef fishes of the Gulf are given in the following studies: northern Gulf (Bright and Cashman 1974; Cashman 1973; B. Causey 1969; Dennis 1985; Dennis and Bright 1988), eastern Gulf (Austin 1971b; Longley and Hildebrand 1941; G. Smith 1976; F. Smith et al. 1975), and southern Gulf (Castro-Aguirre and Márquez-Espinosa 1981; H. Chávez 1966; and H. Hildebrand, Chávez, and Compton 1964).

The following discussion is designed to introduce the reader to the major groups of fishes inhabiting the continental shelves and epipelagic waters of the Gulf, but it is, of necessity, a simplification. A number of the smaller and more obscure groups are omitted, and only those that are most numerous, ecologically important, or better known to the layman are included. Authorities disagree on the details of the higher classification categories of the world's fish groups, and herein we follow the familiar listing of orders, families, and common names as given in Robins, Bailey, et al. (1991), even though that of J. Nelson (1994) may be more technically correct. Not included here are the deepwater species of the Gulf, which are addressed in the next chapter. In the following discussion each of the orders will be taken up sequentially as given in table 13.5, and one or more representatives of each order is illustrated in figures 13.6 through 13.8.

All the neritic and epipelagic fishes of the Gulf belong to two classes: the cartilaginous fishes (Chondrichthyes), which lack true bone, and the bony fishes (Osteichthyes), in which bone is present. As considered here, the former embraces the first 3 orders discussed below, and the latter the next 17 orders.

LAMNIFORMES

The lamniforms include those sharks in which an anal fin is present. There are two dorsal fins, generally without spines. This large group includes most of the shark

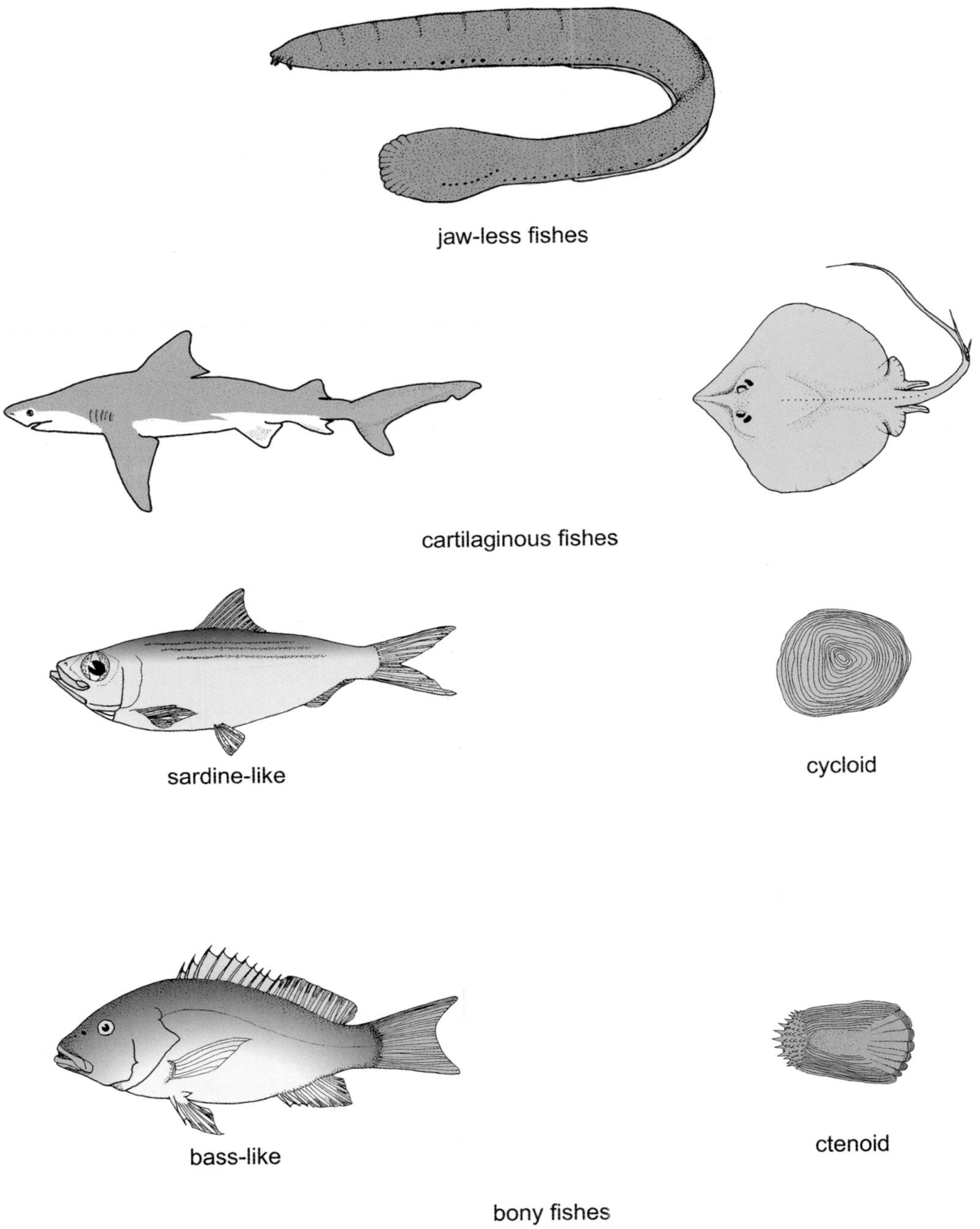

Figure 13.5.
Basic body plans and
body scales of different
groups of modern
fishes. See text for
detailed explanation.

jaw-less fishes

cartilaginous fishes

sardine-like

cycloid

bass-like

ctenoid

bony fishes

types inhabiting the Gulf. Most are predators and have jaws filled with sharp teeth, but the group also contains the great whale shark, which strains plankton from surface waters of the open sea. Common species include nurse sharks, makos, threshers, sand tigers, cat sharks, hammerheads, tigers, oceanic whitetips, and bull sharks, among others. They are widespread on all continental shelves as well as in the open oceanic waters. Common genera include *Alopias, Carcharhinus, Carcharodon, Mustelus, Rhincodon,* and *Sphyrna.*

SQUALIFORMES
This small group of sharks is characterized by the presence of two dorsal fins, generally with spines, and the absence of an anal fin. Although most have the normal sharklike body, in some forms the body is flattened, and

**Table 13.5.** *Orders and families of fishes important in neritic and pelagic waters of the Gulf of Mexico.*

**Lamniformes**
  Rhincodontidae – whale shark
  Alopiidae – thresher sharks
  Cetorhinidae – basking sharks
  Lamnidae – mackerel sharks
  Odontaspidae – sand tigers
  Scyliorhinidae – cat sharks
  Carcharhinidae – requiem sharks
  Sphyrnidae – hammerhead sharks

**Squaliformes**
  Squalidae – dogfish sharks
  Squatinidae – angel sharks

**Rajiformes**
  Pristidae – sawfishes
  Torpedinidae – electric rays
  Rhinobatidae – guitarfishes
  Rajidae – skates
  Dasyatidae – stingrays
  Urolophidae – round stingrays
  Myliobatidae – eagle rays
  Mobulidae – mantas

**Elopiformes**
  Elopidae – tarpons

**Albuliformes**
  Albulidae – bonefishes

**Anguilliformes**
  Muraenidae – moray eels
  Ophichthidae – snake eels
  Congridae – conger eels

**Clupeiformes**
  Clupeidae – herrings
  Engraulidae – anchovies

**Siluriformes**
  Ariidae – sea catfishes

**Aulopiformes**
  Synodontidae – lizardfishes

**Gadiformes**
  Gadidae – cods
  Ophidiidae – cusk-eels

**Batrachoidiformes**
  Batrachoididae – toadfishes

**Lophiifomes**
  Antennariidae – frogfishes
  Ogcocephalidae – batfishes

**Atheriniformes**
  Atherinidae – silversides
  Belonidae – needlefishes
  Exocoetidae – flying fishes

**Beryciformes**
  Holocentridae – squirrelfishes

**Gasterosteiformes**
  Syngnathidae – pipefishes

**Scorpaeniformes**
  Scorpaenidae – scorpionfishes
  Triglidae – sea robins

**Perciformes**
  Centropomidae – snooks
  Serranidae – sea basses

  Priacanthidae – bigeyes
  Apogonidae – cardinalfishes
  Echeneidae – remoras
  Carangidae – jacks
  Coryphaenidae – dolphinfishes
  Lutjanidae – snappers
  Gerreidae – mojarras
  Haemulidae – grunts
  Sparidae – porgies
  Sciaenidae – drums
  Mullidae – goatfishes
  Chaetodontidae – butterflyfishes
  Pomacanthidae – angelfishes
  Pomacentridae – damselfishes
  Mugilidae – mullets
  Sphyraenidae – barracudas
  Labridae – wrasses
  Scaridae – parrotfishes
  Clinidae – clinids
  Blenniidae – combtooth blennies
  Gobiidae – gobies
  Acanthuridae – surgeonfishes
  Scombridae – mackerels
  Xiphiidae – swordfish
  Istiophoridae – billfishes

**Pleuronectiformes**
  Bothidae – lefteye flounders
  Soleidae – soles

**Tetraodontiformes**
  Balistidae – leatherjackets
  Ostraciidae – boxfishes
  Tetraodontidae – puffers

---

the pectoral fins are much enlarged. Included here are several dogfish sharks. The common genus is *Squalus*.

RAJIFORMES

In all fishes of this order the body is flattened, the pectoral fins are greatly enlarged, the eyes and gill openings (*spiracles*) are situated on the dorsal surface, the mouth is on the ventral surface, and the anal fin is absent. Here

are included various species of skates and rays. The skates have short, stout tails not well differentiated from the body, but in the rays the tail is slender and sometimes equipped with a venomous barbed spine (stingrays). Most species are demersal and consume bottom invertebrates, but some, such as the large manta rays, are pelagic and feed on plankton. Genera include *Dasyatis, Gymnura, Manta,* and *Raja.*

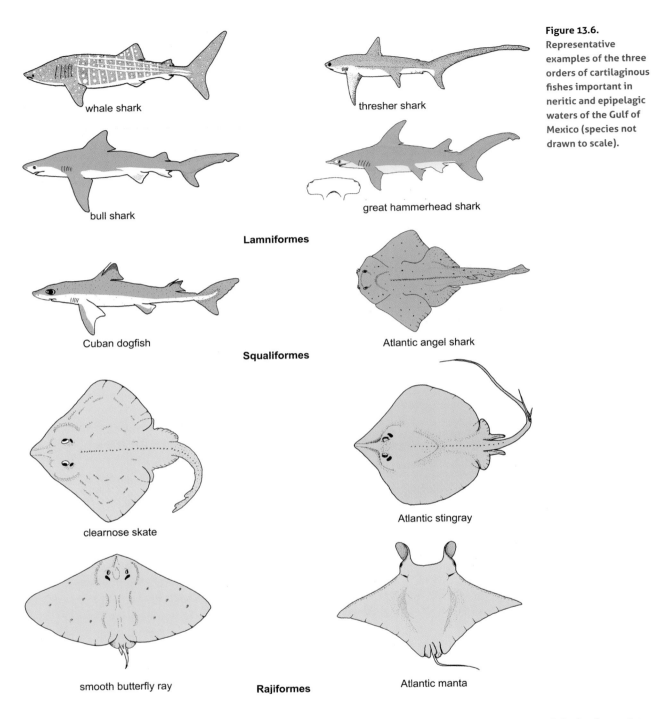

**Figure 13.6.**
Representative examples of the three orders of cartilaginous fishes important in neritic and epipelagic waters of the Gulf of Mexico (species not drawn to scale).

whale shark

thresher shark

bull shark

great hammerhead shark

**Lamniformes**

Cuban dogfish

Atlantic angel shark

**Squaliformes**

clearnose skate

Atlantic stingray

smooth butterfly ray          **Rajiformes**          Atlantic manta

ELOPIFORMES

In the elopiform fishes the body is elongate, the dorsal fin is situated behind the pelvic fins, and the mouth is at the tip of the snout and angled upward. A bony plate (*gular plate*) is present on the ventral surface of the head at the center of the lower jaw. In this group of primitive bony fishes the life history includes a planktonic larval stage (*leptocephalus* larva) that bears no resemblance to the adult fish. The tarpon and ladyfish, both predators, are popular shallow-water game fishes. Genera include *Elops* and *Megalops.*

ALBULIFORMES

In all members of this order the mouth is subterminal (located beneath the snout), and the sensory groove running the length of the lower jaw is not roofed over. The life

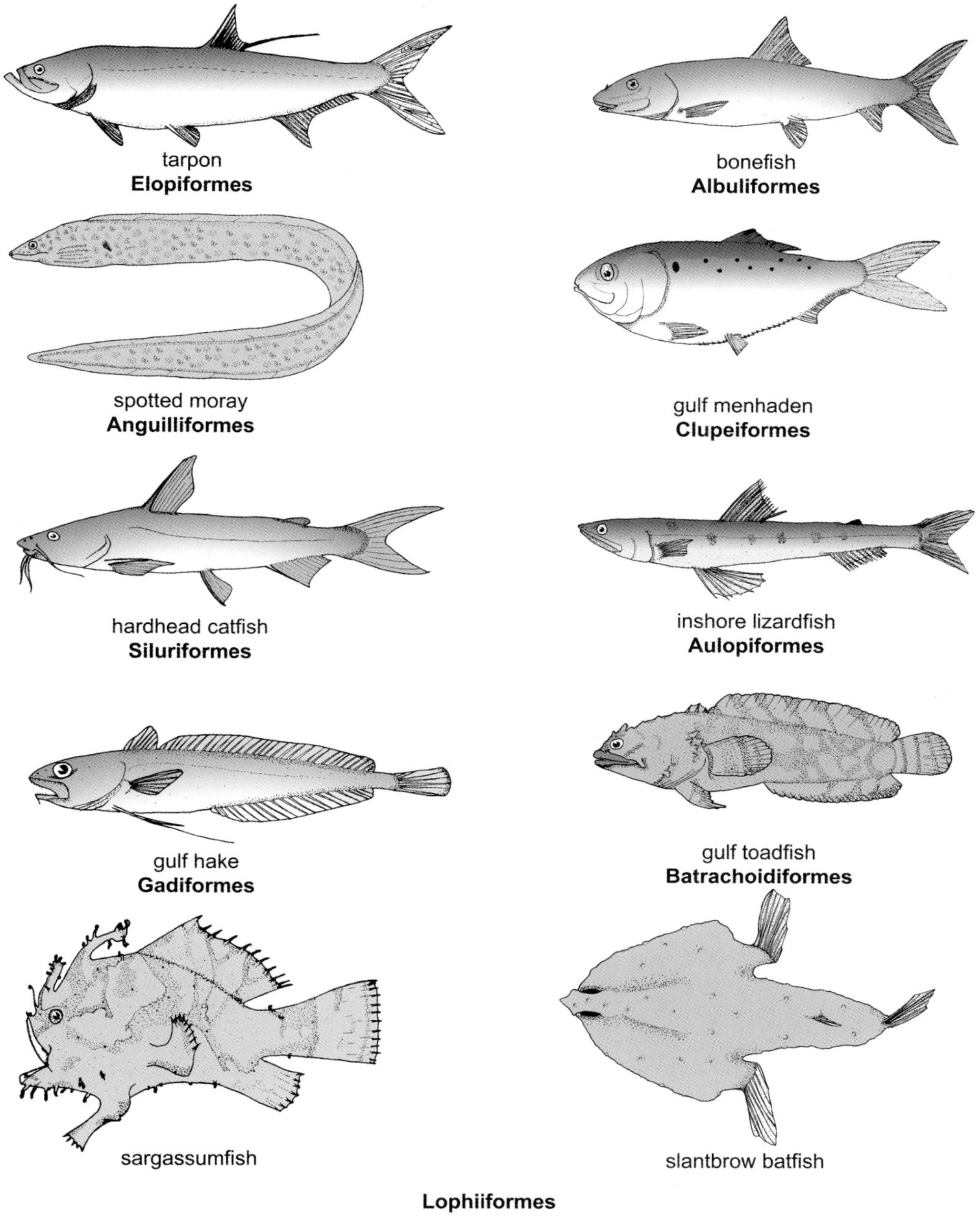

**Figure 13.7.**
Representative examples of bony fishes of the orders Elopiformes through Lophiiformes important in neritic and epipelagic waters of the Gulf of Mexico (species not drawn to scale).

tarpon
**Elopiformes**

bonefish
**Albuliformes**

spotted moray
**Anguilliformes**

gulf menhaden
**Clupeiformes**

hardhead catfish
**Siluriformes**

inshore lizardfish
**Aulopiformes**

gulf hake
**Gadiformes**

gulf toadfish
**Batrachoidiformes**

sargassumfish

slantbrow batfish

**Lophiiformes**

cycle of most, if not all, members of this order includes a leptocephalus-type larva (flattened and more or less transparent), which affirms their relationship with the eels. Three families are recognized: the Albulidae, which occurs in warm, tropical continental shelf waters, and two additional families found only in deeper waters of the continental slope and abyss. Within the Albulidae the bonefish is a prime sport species around seagrass meadows of south and southwest Florida. It feeds largely on benthic invertebrates. The single Gulf genus is *Albula*.

ANGUILLIFORMES

Eels, of course, have very elongated bodies. Gill openings are generally narrow, pelvic fins are absent, and body scales are usually absent or buried in the skin. The presence of a leptocephalus larva indicates their relationship with fishes of the previous order. Eels are well adapted for entering holes and crevices, and they are especially numerous around coral and rocky reefs and other submerged structures, but some species live in holes in the bottom sediments from which they emerge to feed at

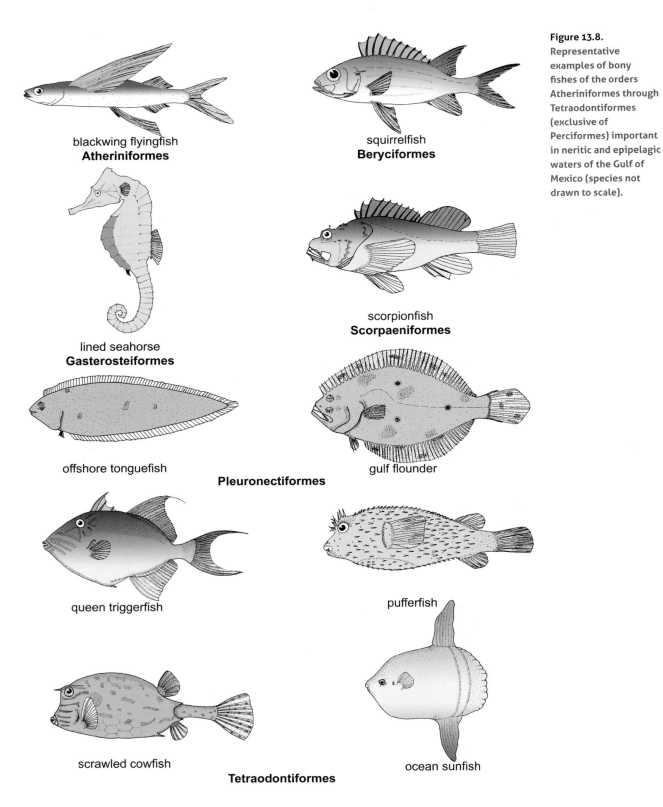

blackwing flyingfish
**Atheriniformes**

squirrelfish
**Beryciformes**

lined seahorse
**Gasterosteiformes**

scorpionfish
**Scorpaeniformes**

offshore tonguefish

gulf flounder

**Pleuronectiformes**

queen triggerfish

pufferfish

scrawled cowfish

ocean sunfish

**Tetraodontiformes**

**Figure 13.8.**
Representative examples of bony fishes of the orders Atheriniformes through Tetraodontiformes (exclusive of Perciformes) important in neritic and epipelagic waters of the Gulf of Mexico (species not drawn to scale).

night. All are carnivorous. Included are a large variety of eel-like types such as morays, false morays, congers, and cutthroat, duckbill, snake, and spaghetti eels. Common genera are *Uroconger, Gymnothorax, Hoplunnis, Myrophis,* and *Ophichthus.*

## CLUPEIFORMES

Clupeiform fishes lack a gular plate, and there is no leptocephalus larva. The head normally lacks scales, and the lateral line system is not developed. The single dorsal fin is composed of soft rays only, and the pectoral fins are placed low on the body. Most species have long and numerous gill rakers for straining their planktonic food. The clupeiforms are all schooling fishes that occur together in vast numbers. Worldwide they include many important commercial species (sardines, herrings, etc.), and in the Gulf the gulf menhaden is the most heavily harvested species, accounting for hundreds of millions of pounds per year. Anchovies are smaller relatives of the herrings that inhabit bays and nearshore coastal waters. They are important forage fishes linking the zooplankton, on the one hand, and higher consumers, on the other. Common genera of anchovies are *Anchoa* and *Anchoviella;* common herring genera are *Brevoortia, Etrumeus, Harengula,* and *Sardinella.*

## SILURIFORMES

Worldwide, catfishes are an extremely important and widely distributed group, but in the Gulf they are represented by only three species, the hardhead and gafftopsail, as well as a poorly known species of the southern Gulf. They have three pairs of barbels (fleshy, whisker-like projections) around the mouth and a small fatty (*adipose*) fin behind the dorsal. Both dorsal and pectoral fins have strong serrated spines. The skin lacks scales. The marine catfishes are mouth brooders (i.e., males carry the fertilized eggs in their mouths). Catfishes are both predators and scavengers, and they feed on small animals and carrion near the bottom and in the water column. Common genera are *Arius* and *Bagre.*

## AULOPIFORMES

Fishes of this group are long and cigar shaped. They have large mouths and adipose fins, and internally they possess very specialized gill structures not known in other fishes. Within the Gulf they are represented by the lizardfishes (Synodontidae), all of which are demersal. Although some live in deeper waters of the continental slope, sev-

eral are common on the continental shelf. All are predators that consume crustaceans and fishes. Genera include *Saurida, Synodus,* and *Trachinocephalus.*

## MYCTOPHIFORMES

These are small slender fishes with laterally compressed bodies, large eyes, and adipose fins. The large mouth is usually terminal. Most species bear *photophores* (light-producing organs) arranged in rows or groups on the head and body. Many are mesopelagic in the open sea and rise to the surface waters at night. In the Gulf they are represented by the lanternfishes (Myctophidae). Common genera include *Diaphus, Hygophum,* and *Myctophum.*

## GADIFORMES

Gadiform (codfish-like) fishes all have a unique type of inner ear bone (*otolith*). Most species display long dorsal and anal fins, the former often being subdivided into two or three sections. Pelvic fins, when present, are generally inserted in front of and below the pectoral fins. Most gadiforms are cold-water species living at high latitudes or in deeper layers of the sea. Two families are important on continental shelves of the Gulf, the Gadidae (hakes) and Ophidiidae (cusk-eels). Both consist of demersal predators. Common genera include *Brotula, Lepophidium, Ophidion,* and *Urophycis.*

## BATRACHOIDIFORMES

These fishes have large heads with the eyes placed near or on top of the head. The mouth is large, pelvic fins are situated far forward, and the body lacks scales. Internally there are no rib bones. These carnivorous fishes are all demersal. Several species, including the toadfishes and midshipman fish, occur on the continental shelf, but others are found in deeper waters of the continental slope. Common genera are *Opsanus* and *Porichthys.*

## LOPHIIFORMES

In this strange group of fishes the first ray of the dorsal fin has moved forward to a position above the mouth. It is movable and has a fleshy structure on the end, which, when wiggled, serves as a fishing lure. The body is covered with loose skin, sometimes containing tubercle-like scales. In one group (batfishes) the body is flattened. The pectoral and/or pelvic fins may be foot-like and used for creeping or crawling. Two families are represented, the frogfishes (Antennariidae) and batfishes (Ogcocephalidae). The frogfishes are predatory, and with their large mouths and elastic stomachs, some are capable of swallowing prey larger than themselves. Batfishes are demersal and feed on polychaetes and other small benthic animals. Common genera include *Antennarius, Dibranchus, Halieutichthys,* and *Ogcocephalus.*

### ATHERINIFORMES

The atheriniform fishes are characterized by unique skeletal features among which are protractile (extendible) jaws containing a special ball-and-socket mechanism. All are slender fishes with the pectoral fins placed high on the sides of the body. Three families are represented. The flying fishes (Exocoetidae) have very elongated pectoral fins that serve as "wings," allowing the fishes to execute long glides above the ocean surface. In the needlefishes (Belonidae) both upper and lower jaws are long and thin with many tiny pin-like teeth. In the silversides (Atherinidae) the jaws are small and placed at the front of the snout. All three families are made up of schooling fishes that live at or near the surface of the sea. The silversides and needlefishes tend to be coastal, but the flying fishes largely inhabit the open ocean. All feed on small invertebrates. Common genera include *Cypselurus, Exocoetus, Hirundichthys, Menidia,* and *Strongylura.*

### BERYCIFORMES

Beryciform fishes are characterized by the presence of specialized sensory canals in the head both above and below the eyes. All are marine, and most live in deep water. In the shallow Gulf they are represented by the family Holocentridae, or squirrelfishes. In most species the eyes are large, and the fins have both spines and soft rays. Squirrelfishes live around reefs of the outer shelf and upper slope, where they feed on small invertebrates. The common genus is *Holocentrus.*

### GASTEROSTEIFORMES

This group is represented in the shallow Gulf by the family Syngnathidae, which includes the pipefishes and seahorses. The body may be long and slender, as in the pipefishes, or somewhat more stocky with a curled tail, as in the seahorses. On the outside the body is covered by an armor of dermal plates. In this family the males carry the fertilized eggs in a brood pouch located on the belly. They have very small mouths at the end of long snouts, and they feed on tiny zooplankton animals. These fishes are specialized to live in vegetated areas, and they are quite common in the seagrass meadows off Florida. Common genera are *Hippocampus* and *Syngnathus.*

### SCORPAENIFORMES

This order is represented by the scorpionfishes (Scorpaenidae) and sea robins (Triglidae). All have spines around the head region, particularly on the preopercle and opercle bones of the gill cover. Some have poison glands associated with spines of the dorsal, anal, and pelvic fins. In the Scorpaenidae the pectoral fins are normal and undivided, but in the Triglidae the anterior rays of the pectoral fin are separated from the rest of the fin and act as fingers enabling the fishes to crawl along the bottom. Most scorpaeniform fishes are demersal, and many are found around rocks and reefs, but the sea robins are common on the open shelf. Both groups have deepwater relatives. All are carnivorous. Common genera include *Bellator, Peristedion, Pontinus, Prionotus,* and *Scorpaena.*

### PERCIFORMES

This is the largest and most diverse order of living fishes, and more than one-third of all known fish species of the world fall into this group. There are more than 50 different families in the neritic and epipelagic waters of the Gulf. As noted earlier, perciform fishes are distinguished from other fish groups by the following combination of characters: fin rays supported by spines, ctenoid scales, swim bladder without an open connection to the esophagus, upper jaw bordered only by the premaxilla, and a number of other skeletal and fin ray details. As further research is carried out, this large group will undoubtedly be broken up into a number of different orders. The perciforms are a relatively recent group of fishes. There are no certain fossils dating back to the Cretaceous period, but during the past 65 million years (i.e., during the Tertiary and Quaternary periods) they have become quite diverse and are now the dominant group of fishes in most freshwater, brackish, and marine environments of the world. Several of the perciform families (Serranidae, Apogonidae, Sciaenidae, Pomacentridae, Labridae, Blenniidae, and Gobiidae) constitute the world's largest marine fish families. Brief sketches of some of the most important of these families are given in table S53, and representatives of each family are shown in figures 13.9 and 13.10. These families were selected on the basis of their diversity, numerical abundance, general ecological significance, commercial importance, or their status as game fishes.

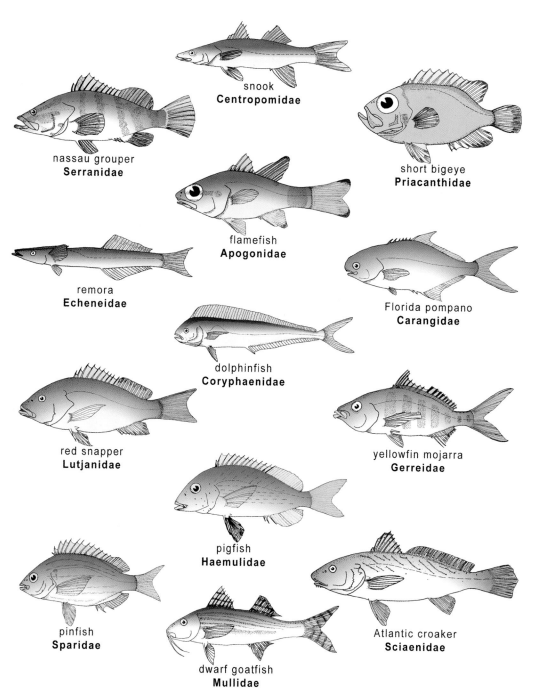

**Figure 13.9.**
Representative examples of perciform fishes of the families Centropomidae through Mullidae important in neritic and epipelagic waters of the Gulf of Mexico (species not drawn to scale).

snook
**Centropomidae**

nassau grouper
**Serranidae**

short bigeye
**Priacanthidae**

flamefish
**Apogonidae**

remora
**Echeneidae**

Florida pompano
**Carangidae**

dolphinfish
**Coryphaenidae**

red snapper
**Lutjanidae**

yellowfin mojarra
**Gerreidae**

pigfish
**Haemulidae**

pinfish
**Sparidae**

Atlantic croaker
**Sciaenidae**

dwarf goatfish
**Mullidae**

## PLEURONECTIFORMES

This group embraces the flatfishes of the world: the flounders, soles, tonguefishes, and their relatives. In the adults the bodies are flattened (flat on the bottom and somewhat rounded above), and both eyes are situated on the top side of the head, often projecting above the body surface. Very young flatfishes are bilaterally symmetrical and swim upright like most other fishes, but when they become about an inch long one eye migrates across the top of the skull to a position near the other eye, and the young fish begins swimming with the eyeless side down. Flatfishes live on the bottom and often bury themselves in the sediments with just their eyes projecting above. All are carnivorous and feed on bottom invertebrates and fishes.

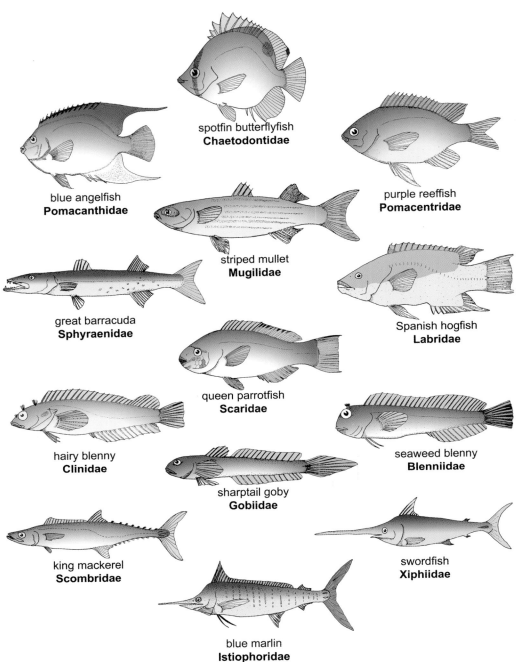

**Figure 13.10.**
Representative examples of perciform fishes of the families Chaetodontidae through Istiophoridae important in neritic and epipelagic waters of the Gulf of Mexico (species not drawn to scale).

Larger species such as the flounders are excellent food fishes. Representative genera include *Citharichthys, Paralichthys, Syacium,* and *Symphurus.*

TETRAODONTIFORMES
In this strange group of fishes (triggerfishes, cowfishes, puffers, ocean sunfishes, etc.) the skull bones are much modified, with some bones being fused together and others lost entirely. They have small gill openings, and the scales are usually modified into spines, shields, or plates. They exhibit many different body forms, and the flesh of some (puffers) carries powerful toxins and is poisonous to eat. Some species produce sounds by grinding the teeth or vibrating the swim bladder. Most feed on small benthic invertebrates, but the ocean sunfishes, which are pelagic, feed largely on jellyfishes, ctenophores, and salps. Representative genera include *Aluterus, Diodon, Lactophrys, Mola, Monacanthus,* and *Sphoeroides.*

## Sea turtles

Sea turtles occur in warm-temperate oceans, and throughout the world they are represented by fewer than a dozen species. In these marine reptiles the entire body is encased in a shell composed internally of bony plates and covered externally with a layer of horny material generally divided into geometrically arranged plates, or scutes. The shell, consisting of a top (*carapace*) and bottom (*plastron*) section, is open at the front and rear to accommodate the head, limbs, and tail, but unlike freshwater species, marine turtles cannot withdraw the head and limbs inside the shell. Horny plates may also be embedded in the skin of the head and limbs. Although sea turtles lack true teeth, they all possess a sharp horny bill that may be equipped with points or cusps, and some have tooth-like projections along the edge of the jaw. The limbs, or flippers, are paddle shaped and may possess one or two claws. Males are usually distinguished from females by their longer tails. Propulsion through the water is accomplished by simultaneous "flight" movements of the two front flippers, and the turtles are surprisingly fast swimmers, some being capable of sustained speeds of 35 km/hr (21 mi/hr).

Sea turtles reach maturity around the age of 6–13 years. Prior to nesting, copulation takes place in shallow nearshore waters off sandy beaches, and the females later crawl ashore, generally at night, and dig nests in the sand above the high-tide line. Here they deposit from 75 to 150 (rarely up to 200) eggs and cover the nest before returning to the sea. Several such clutches may be laid in one season, but a given female probably does not lay eggs every year. Nesting occurs during the warmer months, usually between April and September. After 6–9 weeks the young hatch, dig their way to the surface, and scamper down to the sea. The eggs and very young are subject to heavy predation by terrestrial mammals, seabirds, and fishes. During the first year of life the young turtles disperse, some associated with floating rafts of sargassum weed, but little is known about sea turtle life histories until they reach sexual maturity and appear offshore of their nesting areas. Some species apparently remain in coastal waters, where they make use of bays, estuaries, and the shallow continental shelf, often around (but not on) reefs and offshore islands. Others are highly migratory, making long voyages far out at sea, but all have a strong homing instinct that causes them to return to lay their eggs on the same beaches where they were born. Most sea turtles are carnivorous, but some subsist chiefly on vegetation. Efforts to census sea turtle populations by trained observers on ships or airplanes have met with limited success because of the difficulty of identifying individual species (based on shell size, shape, and color and on flipper size) and because sea turtles spend most of their lives underwater. Beach counts of nesting females are somewhat more reliable since the species can be identified with certainty, but again, only a portion of the population is represented, and little information is available during the nonbreeding season or for nonnesting areas. However, during the past few centuries and particularly during the past few decades, all evidence has pointed to a dramatic decline in sea turtle populations of all species around the world, and they are now endangered or threatened with extinction. Recent protective legislation and local enforcement, particularly of nesting beaches, and breeding programs have begun to reverse the worldwide decline.

Within the Gulf of Mexico the sea turtles are represented by five species: the loggerhead (*Caretta caretta*), green (*Chelonia mydas*), Kemp's ridley (*Lepidochelys kempii*), Atlantic hawksbill (*Eretmochelys imbricata*), and leatherback (*Dermochelys coriacea*). Originally present in some numbers throughout the Gulf, sea turtles are now quite rare, especially in the northern Gulf. This is apparently because of such factors as extensive harvesting of eggs and adults during the past few centuries, and more recently to the loss of shoreline and aquatic habitat in the bays and lagoons, dredging and siltation of habitat areas, pollution by toxic chemicals and solid trash (plastic bottles, bags, etc.), and extensive shrimp trawling in nearshore marine waters, which entraps and drowns sea turtles. Increasing boat and ship traffic disturbs the turtles and often results in propeller damage. All species of the Gulf are listed as endangered or threatened. Protective measures are now in place that should reduce the mortality and permit the several species to recover and repopulate much of their ancestral ranges. Literature dealing with sea turtles of the Gulf is not extensive, and much of this relates to populations along the Florida coastline and to protection and restocking efforts in the northwestern Gulf. Some of the more important references include Bjorndal (1982), A. Carr (1952, 1967), A. Carr and Caldwell (1956), H. Chávez (1969), H. Chávez and Kaufmann (1974), H. Chávez, Contreras G., and Hernandez D. (1968), Fritts, Hoffman, and McGehee (1983), Fritts et al. (1983), Gitschlag, Herczeg, and Barcak (1997), H. Hildebrand (1963), Plotkin, Wicksten, and Amos (1993), S. Rabalais and Rabalais (1980), Rebel (1974), F. Smith (1954d), and Yerger

(1965). Identification of the sea turtles of the Gulf is aided by keys and species descriptions in A. Carr (1952), C. Ernst and Barbour (1972), Márquez (1978), and F. Smith (1954d). Figure 13.11 shows general aspects of the five species of sea turtles of the Gulf.

### LOGGERHEAD TURTLE (*CARETTA CARETTA*)

The loggerhead has a heavy, rounded, and somewhat heart-shaped shell. On the carapace there are five pairs of lateral scutes, instead of the usual four, and the first pair is in contact with the precentral (middle marginal) scute. The head is large and broad, and the flippers each bear two claws. The shell of the adult is reddish brown above and pale yellow beneath. This is a large turtle, and individuals have been known to reach a length of 213 cm (84 in) and a weight of 409 kg (900 lb), although most are much smaller. Loggerheads travel long distances and have been encountered at sea hundreds of kilometers from land, but they do not seem to follow regular seasonal migratory pathways. The great majority are concentrated in continental shelf waters less than 50 m deep, and they are sometimes seen in bays and estuaries. These turtles are omnivorous and feed largely on sea pens, clams, oysters, crabs, sea urchins, jellyfishes, and other invertebrates, as well as fishes. Females nest only every two or three years. Egg laying takes place from May to August, and incubation of the eggs requires 46–62 days, depending on the temperature.

This is the most common sea turtle in the Gulf of Mexico. During the 1800s it was apparently abundant all around the Gulf coasts, and together with the green sea turtle, it supported a large fishery for turtle eggs and meat. Although now rare around much of the Gulf, during the warmer months it is still common off the west coast of Florida, which appears to be a major feeding area. It has recently been estimated that the population in Florida waters numbers around 40,000 individuals, but this estimate includes the east coast, which is a major nesting area for the species. Limited nesting also occurs on the west coast of the state. While at sea these turtles often bask at the surface, where they provide temporary perches for gulls, terns, and other seabirds. Loggerheads are not seen in the northern Gulf during the winter months, and it is assumed that they either migrate to warmer waters farther south or remain dormant in the bottom sediments, and there is evidence that they actually do pass the winter locally, buried in the mud.

### GREEN SEA TURTLE (*CHELONIA MYDAS*)

The shell of the green sea turtle is flattened and somewhat elongate. On the carapace there are only four pairs of lateral scutes, and the first pair is not in contact with the precentral. On top of the head between the eyes there is only a single pair of prefrontal plates, instead of two pairs as in other species. Strong ridges are present on the inner surface of the upper jaw, and the edge of the lower jaw is serrated or coarsely toothed. Each flipper bears a single claw. The upper surface of the shell is an olive brown with radiating yellow, brown, or black spots on each scute, and the bottom of the shell is pale yellow or creamy white. The common name of this turtle derives from the greenish color of its fat. The record size is about 152 cm (60 in) in length and 386 kg (850 lb) in weight. The green turtle is largely a coastal species, and the adults are closely associated with vegetation beds of bays, lagoons, and the shallow continental shelf, but occasionally they are seen far out at sea. The young are apparently carnivorous, but as discussed in the previous chapter, the adults are essentially herbivores that sometimes consume some animal flesh. Egg laying occurs from June to September, and incubation requires 45–60 days.

This is the second most common sea turtle in the Gulf. As in the case of the loggerhead, the green turtle is most abundant off the west coast of Florida, where the population in 1956 was estimated to be around 5,600 individuals. With its extensive seagrass meadows, this is a major feeding ground. Some nesting takes place on the west Florida coast and on other sandy beaches around the Gulf, but many more nest on Florida's Atlantic beaches. This turtle seems to disappear by late October, and its apparent absence could be because of migration southward or winter dormancy. There is evidence that at least some individuals overwinter buried in muddy bottom sediments. Formerly fairly abundant around much of the Gulf, the green turtle has been overharvested for its eggs and highly prized meat, and like the other sea turtles it has suffered from habitat loss, water pollution, and drowning in the nets of commercial fishermen.

### KEMP'S RIDLEY (*LEPIDOCHELYS KEMPII*)

In this, the smallest of the sea turtles, the shell is short, wide, and nearly circular in outline. There are five pairs of lateral scutes, of which the first pair is in contact with the precentral. The small head bears two pairs of prefrontal plates between the eyes. The horny beak may be serrated, and each flipper has only a single claw. Unique

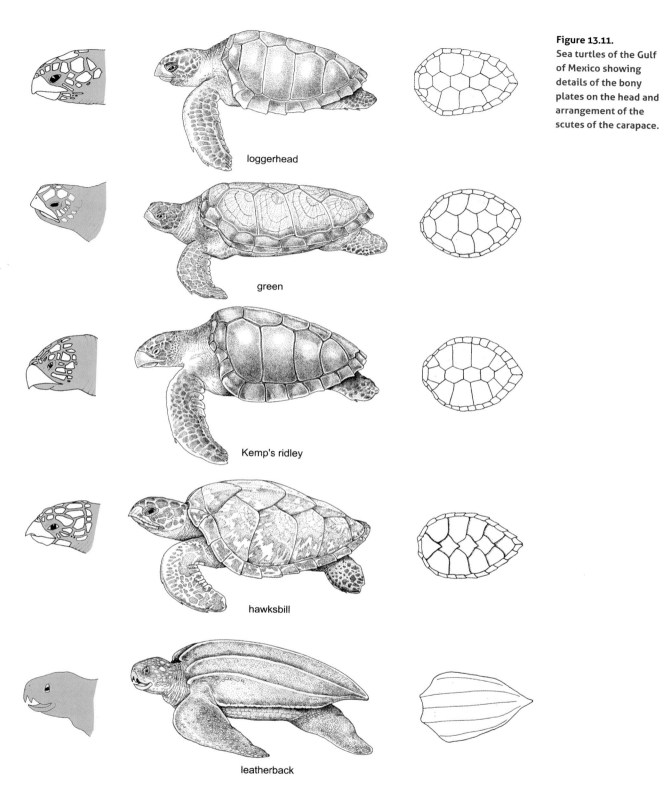

loggerhead

green

Kemp's ridley

hawksbill

leatherback

**Figure 13.11.**
Sea turtles of the Gulf of Mexico showing details of the bony plates on the head and arrangement of the scutes of the carapace.

to this species is the presence of a pore near the rear edge of the outside (inframarginal) scutes of the plastron. The carapace, which may have a ridge running down the center, is grayish, and the underside of the body is pale yellow or whitish. The maximum recorded size is only about 73 cm (29 in) in length and 49 kg (108 lb) in weight. Adults feed chiefly on mollusks and crustaceans, supplemented by some jellyfishes and fishes. The ridley nests from April to July, and the young hatch after 45–60 days.

Unlike other sea turtles that nest along many shores throughout their range, all the Kemp's ridley turtles of the Gulf, Atlantic, and Caribbean nest on a single shoreline of the northwestern Gulf of Mexico. The main nesting site is a strip of beach 15 km (9 mi) in length near the Mexican village of Rancho Nuevo and about 270 km (162 mi) south of the US border. Limited nesting takes place north and south of the main beach, and historically some nesting occurred on beaches of Padre Island in far south Texas. Young ridleys tagged at Rancho Nuevo have been recovered all around the Gulf, and individuals of this species have been found in the Caribbean, along the Atlantic seaboard, and as far away as Bermuda and Europe. Although primarily a coastal species, this turtle has been observed in the open sea. Shallow flats and waters off mangrove-lined shores are favored habitats, and the west Florida shelf appears to be a major feeding ground. Like the previous species, the ridley probably remains dormant during the winter. Once quite abundant around the Gulf, this species has now been reduced to fewer than 1,000 breeding females. It has been estimated that during the 1960s as many as 40,000 females would nest in a single day, but at the present time more than 150 per day is rare. Efforts are now being made to protect the nesting beaches, to rear the young and release them at a less vulnerable size, to protect them from the trawling nets of shrimpers, and to reestablish nesting populations along the south Texas shoreline.

### ATLANTIC HAWKSBILL (*ERETMOCHELYS IMBRICATA*)

The carapace of the hawksbill is heart shaped or oval in outline and has a ridge running down the center. Unlike those of other sea turtles, the dorsal scutes overlap like shingles (i.e., they are imbricated). There are four pairs of lateral scutes, the front pair not touching the precentral. The head is elongated, and the upper jaw has a prominent hook-like beak. Situated between the eyes are two pairs of prefrontal plates. Two claws are present on each flipper. The carapace is dark brown, richly mottled with yellow and reddish streaks, and the plastron is pale yellow. This is a fairly small turtle with a maximum recorded size of 91 cm (36 in) in length and 127 kg (280 lb) in weight. The beautiful marbled tortoiseshell pattern of the dorsal scutes appears only after the shell is polished. The hawksbill is omnivorous and has been reported to consume sponges, mollusks, sea urchins, crustaceans, fishes, algae, and seagrasses. Egg laying takes place from April to June, and the eggs hatch in about 45–55 days. This turtle has been recorded all around the Gulf in bays and lagoons lacking extensive vegetation and on the shallow shelf, especially around coral islands and rocky coasts. Once fairly common, it is now rare, particularly in the northern Gulf. The eggs and flesh are used for food locally, and the carapace yields valuable tortoiseshell that is used in the manufacture of jewelry and other decorative items.

### LEATHERBACK TURTLE (*DERMOCHELYS CORIACEA*)

The leatherback is quite distinct from all other sea turtles in possessing a smooth leathery skin rather than a hard shell with horny scutes. Actually, a mosaic of small inconspicuous plates is embedded in the skin. The carapace is shield shaped, pointed at the rear, and including the lateral pair, there are seven narrow ridges extending the length of the back. Five ridges run the length of the plastron. The head is relatively small, and the horny beak of the upper jaw has a well-defined point or cusp on each side. There is also a cusp in the center of the lower jaw. The flippers are large and without claws, and in adults the hind flippers are broadly connected with the tail by a web of skin. The body is dark brown to almost black, sometimes with white spots on the neck, and the bottom is white with black markings. This is the largest of the sea turtles, and individuals have been recorded up to 244 cm (96 in) long and weighing up to 717 kg (1,600 lb). Egg laying takes place from March to September, and the young hatch in about 50–70 days. This turtle feeds largely on jellyfishes, ctenophores, and other gelatinous invertebrates, but it also consumes other invertebrates and small fishes. It is a remarkably strong and rapid swimmer, and in the open ocean it is known to make dives of over a thousand meters.

The leatherback is a long-distance migrator. Turtles tagged in South America have been recaptured in the Gulf, and migrations of at least 5,000 km (3,000 mi) have been documented. This species was formerly thought to be limited to waters of the open ocean except during the nesting season, but recent records show that individuals may spend considerable time in coastal waters, often less than 30 m deep, and they sometimes enter estuaries. This turtle is known from all around the Gulf, and limited nesting takes place on west Florida beaches. Although generally solitary, it is sometimes found in groups, and an aggregation of about 100 individuals has been recorded in the surf off the Texas coast apparently feeding on a swarm of jellyfishes. Unlike other sea turtles, the leatherback is tolerant of fairly low temperatures, and it is unlikely that this species overwinters in a dormant state anywhere in the Gulf.

## Pelagic birds

Some species of marine birds typically live over the open sea far from shore and beyond the continental shelf. These marine birds are often referred to as "pelagic" even though they live above rather than within the water column. Thirty-one species of these oceanic birds representing eight families have been reported from offshore waters of the northern Gulf of Mexico (table 13.6), and about half of these occur in the area on a regular basis. In addition, they are joined seasonally by several species more characteristic of coastal areas and shallow marine waters such as the laughing, herring, and ring-billed gulls and the royal, sandwich, common, and black terns (Peak 1999).

## Marine mammals

Marine mammals of the world represent three distinct mammalian orders: Cetacea (whales and dolphins), Sirenia (manatees and dugongs), and Carnivora (seals, sea lions, walruses, and sea otters), and historically all three groups have been represented in the Gulf of Mexico. They are all derived from terrestrial ancestors and exhibit varying degrees of adaptation to the marine environment. All show a tendency toward development of a streamlined body shape with reduction or loss of hind limbs. The skin has become thickened and underlain by a layer of insulating fat, particularly in the cetaceans and sirenians, which have lost much of their body hair. All have had to cope with the salt problem. Unable to obtain a drink of freshwater, these animals must extract water from their salty food and remove excess salts by way of specialized kidneys, thus excreting highly saline urine. Seals and their relatives still must come ashore to mate, give birth, and nurse their young, but cetaceans and sirenians carry out these reproductive functions at sea and never come ashore. Each of the groups will be described and some of their special adaptations discussed below.

**Table 13.6.** *Oceanic seabirds reported from offshore waters of the northern Gulf of Mexico.*

**Diomedeidae** – albatrosses
Yellow-nosed albatross – *Diomedea chlororhynchos*

**Procellariidae** – gadfly petrels and shearwaters
Black-capped petrel – *Pterodroma hasitata*
White-chinned petrel – *Procellaria aequinoctialis*
Cory's shearwater – *Calonectris diomedea*
Greater shearwater – *Puffinus gravis*
Sooty shearwater – *Puffinus griseus*
Manx shearwater – *Puffinus puffinus*
Audubon's shearwater – *Puffinus lherminieri*

**Hydrobatidae** – storm-petrels
Wilson's storm-petrel – *Oceanites oceanicus*
Black-rumped storm-petrel – *Oceanodroma castro*
Leach's storm-petrel – *Oceanodroma leucorhoa*

**Phaethontidae** – tropicbirds
Red-billed tropicbird – *Phaethon aethereus*
White-tailed tropicbird – *Phaethon lepturus*

**Sulidae** – gannets and boobies
Northern gannet – *Sula bassanus*
Masked booby – *Sula dactylatra*
Brown booby – *Sula leucogaster*
Red-footed booby – *Sula sula*
Blue-footed booby – *Sula nebouxii*

**Fregatidae** – frigatebirds
Magnificent frigatebird – *Fregata magnificens*

**Scolopacidae** – sandpipers and relatives
Red phalarope – *Phalaropus fulicarius*
Red-necked phalarope – *Phalaropus lobatus*

**Laridae** – jaegers, gulls, terns, and skimmers
Pomerine jaeger – *Stercorarius pomarinus*
Parasitic jaeger – *Stercorarius parasiticus*
Long-tailed jaeger – *Stercorarius longicaudus*
Black-legged kittiwake – *Rissa tridactyla*
Sabine's gull – *Xema sabini*
Arctic tern – *Sterna paradisaea*
Roseate tern – *Sterna dougallii*
Bridled tern – *Sterna anaethetus*
Sooty tern – *Sterna fuscata*
Brown noddy – *Anous stolidus*

## CETACEA

The cetaceans are one of the oldest groups and the most highly modified for life in the sea. They appear to have made the transition from land to freshwater and then to the sea 50–60 million years ago around the shorelines of the Tethys Sea, which at that time separated the northern land masses (Asia, Europe, and North America) from those to the south (Indian subcontinent, Africa, and South America). Apparently they are derived from an early mammalian group, the mesonychids, which were carnivorous. During the course of their evolution they lost the hind limbs (bony vestiges remain) and most of the body hair. They developed dorsal fins, and the front limbs became modified into flippers used primarily for steering and balancing. Since propulsion was derived from vertical movements of the rear section of the body, they developed powerful posterior muscle masses and forked tails. Great elongation of bones of the upper and lower jaws (maxillae and mandibles) pushed the nasal bones toward the rear so that now the nostrils making up the blowhole are on the top of the head. Whales and dolphins vocalize a great deal and exhibit various types of social organization built around the mother-offspring bond. Of the 4,000+ species of mammals living today, around 80 species are cetaceans.

The whales early divided into two evolutionary lines. One group, known as the Mysticeti, specialized in feeding on swarms of small shrimps and fishes in the water column. To this end they developed very large mouths with long, stiff, fringed plates (baleen plates) hanging from the edges of the upper jaw. This system allows the feeding whale to engulf large volumes of water containing many prey animals. As the whale closes its jaws it expels the water while retaining the prey. This group now constitutes the baleen whales, which specialize in feeding on swarms of euphausiid shrimps (krill) or schools of herrings and other clupeid fishes. Teeth were present in the ancestral forms and may still be present in the fetus, but they are lacking in adults of all modern mysticetes. This group includes the giant blue whale and its relatives, the largest animals ever to live on the earth. Whales of this group are not deep divers, and although they produce a variety of sounds, they are not known to use sonar for echolocation. Most species inhabit colder waters of the Arctic and Antarctic, although some species overwinter in tropical waters where the calves are born.

The second group of whales is the Odontoceti, which includes the sperm whales, beaked whales, killer whales, and smaller dolphins. These all lack baleen plates, and their jaws are lined with small undifferentiated teeth. Whales of this group do not engulf swarms but apprehend and consume individual prey organisms, generally squids and fishes. Most of the smaller whales and dolphins do not make really deep dives, but sperm whales and some of the beaked whales are known to dive to depths of over a kilometer in search of squids and to remain submerged for an hour or more at a time. It is not clear how their bodies handle the immense pressure changes and store enough oxygen for such excursions, but various modifications of the circulatory systems and large quantities of hemoglobin in the blood and muscle tissues are certainly involved. Although all marine mammals are capable of sound production, this ability is most highly developed in the toothed whales, which produce pulses of sound at extremely high frequencies, a type of sonar used for echolocation. This allows the whale to "see" potential prey and other objects in waters that have poor visibility because of either poor lighting or the presence of much suspended material. Underwater, sound travels about 4.5 times faster than it does in air, and there is little loss of loudness with distance. Some whale sounds may be heard over tens and perhaps hundreds of kilometers, allowing whales to keep in touch with each other over long distances. Considering the variety of sounds produced, these undoubtedly serve other social functions in whale groups. Some whale species remain in a given area throughout life, but others such as the sperm and some of the baleen whales may migrate long distances, either following the coastline or moving across the open sea. Some of the smaller species exhibit seasonal onshore and offshore movements. How whales find their way and orient themselves in what appears to be featureless ocean water remains a mystery.

## SIRENIA

The sirenians represent another ancient group that arose along the shores of the Tethys Sea 50 million years ago or more, and anatomical characters link them with ancestors of the modern elephants. Sirenians have completely lost the hind limbs and developed very rotund bodies with thick layers of insulating fat and flat, rounded tails for locomotion. The forelimbs are reduced in length and used for steering and manipulating food. Body hair has largely disappeared, but bristles remain on the snout. Bones of the skeletal system are very dense and aid in counteracting the floating effect of the body fat. All sirenians are vegetarians that feed on rooted aquatic plants of coastal streams, lagoons, and the nearshore continental shelf. To handle this coarse, low-energy food, they have developed compartmentalized stomachs and extremely long intestines. These quiet, sluggish, nonaggressive creatures never come ashore and carry out all reproductive activities at sea. All present species are inhabitants of warm tropical and subtropical waters, but a recently extinct species, the Steller's sea cow, lived in cooler waters of the north Pacific.

## CARNIVORA

Several groups of the order Carnivora have independently invaded the sea. The forms of interest here are the Pinnipedia, which made the transition about 20 million years ago. These are the seals, sea lions, and walruses, apparently derived from bearlike ancestors. Since these are more recent arrivals than the cetaceans and sirenians, they are not as highly modified, but the same selective forces are at work to shape their bodies and behavior patterns. Adaptations to marine life include general streamlining of the body, reduction (but not loss) of the hind limbs, and modification of the front limbs into flippers that are effective for swimming and steering. Although body hair has not been lost, the layer of subcutaneous fat is well developed. Various internal modifications adapt the animals for diving and for obtaining freshwater from their salty diet of invertebrates and fishes. Unlike the cetaceans and sirenians, the pinnipeds still mostly come ashore for mating, giving birth, and nursing their young. Two families are of interest, the true seals (Phocidae), which lack external ears, and the eared seals (Otariidae). One phocid, the West Indian monk seal, was, until recent years, a resident of the Gulf of Mexico.

Thirty-four types of marine mammals are known from the Gulf of Mexico, including 33 native and 1 introduced species. Knowledge of these mammals has developed slowly. The early explorers and adventurers left only a few records concerning the marine mammals they encountered. The Spaniards Oviedo and de Landa wrote of harpooning manatees along the Mexican coast, particularly the Yucatán coast north of Campeche, because these animals were especially prized for their fat (Baughman 1952). In 1675 the British adventurer Daumpier, referring to the West Indian monk seal of Alacrán Reef, wrote, "The Spaniards do come hither to make oyl of their fat" (Gunter 1954). Records of old whaling ships testify that during the eighteenth and nineteenth centuries sperm whales were regularly taken from the Gulf. During the early part of the twentieth century newspaper accounts and technical articles noted occasional strandings, but a systematic effort to document marine mammal strandings did not begin until the mid-1970s, with the establishment of the Southeastern United States Marine Mammal Stranding Network.

During the past two decades, under sponsorship of federal agencies, a series of ship and airplane surveys has been carried out to assess marine mammal populations off the Gulf states. Most of the surveys have extended across the continental shelf into open waters of the deep Gulf where most of the species of marine mammals actually live, and these surveys have greatly expanded our knowledge of the composition, distribution, abundance, and behavior of the marine mammal species inhabiting the Gulf. Much of the existing knowledge of marine mammals of the Gulf has recently been summarized by Würsig, Jefferson, and Schmidly (2000). Other important references include Blalock et al. (1995), Caldwell and Caldwell (1973), Darling et al. (1995), R. W. Davis, Scott, et al. (1994), R. W. Davis, Fargion, et al. (1998), Fritts et al. (1983), Geraci and Lounsbury (1993), Gunter (1954), D. Hartman (1979), Jefferson (1995), Jefferson and Schiro (1997), Leatherwood, Caldwell, and Winn (1976), Mullin et al. (1994), O'Shea, Ackerman, and Percival (1995), Schmidly (1981), and Schmidly and Shane (1978). A list of all marine mammal species known from or presumed to occur in the Gulf is presented in table 13.7, and the common species are illustrated in figure 13.12.

Three of the 34 species of marine mammals listed in table 13.7 (long-finned pilot whale, short-beaked common dolphin, and long-beaked common dolphin) are mentioned as of *possible* occurrence in the Gulf. These are known to be present in nearby waters of the Atlantic or Caribbean and are likely to be found as the Gulf is more thoroughly investigated. Of the 31 species definitely known from the Gulf, one (West Indian monk seal) is now presumed to be *extinct,* and another (California sea lion), although not native to the Gulf, has definitely been identified and probably represents *introduced* individuals that escaped from captivity. Among the remaining forms, 3 (northern right whale, blue whale, and Sowerby's beaked whale) are considered to be *extralimital.* Although known from the Gulf on the basis of a few records, their presence probably results from unusual wanderings of a few individuals from their normal ranges in the northwestern Atlantic. Six species are listed as *rare.* These (fin whale, sei whale, minke whale, humpback whale, Cuvier's beaked

**Table 13.7.** *Marine mammals known or presumed to occur in the Gulf of Mexico. Definitions of the status of the various species are explained in the text. (From Würsig, Jefferson, and Schmidly 2000.)*

| Scientific and common names | Status |
| --- | --- |
| Cetacea – whales and dolphins | |
| Mysticeti – baleen whales | |
| Balaenidae – right whales | |
| *Eubalaena glacialis* – northern right whale | extralimital |
| Balaenopteridae – rorquals | |
| *Balaenoptera musculus* – blue whale | extralimital |
| *Balaenoptera physiculus* – fin whale | rare |
| *Balaenoptera borealis* – sei whale | rare |
| *Balaenoptera edeni* – Bryde's whale | uncommon |
| *Balaenoptera acutirostrata* – minke whale | rare |
| *Megaptera novaeangliae* – humpback whale | rare |
| Odontoceti – toothed whales and dolphins | |
| Physeteridae – sperm whale | |
| *Physeter macrocephalus* – sperm whale | common |
| Kogiidae – pygmy and dwarf sperm whales | |
| *Kogia breviceps* – pygmy sperm whale | uncommon |
| *Kogia simus* – dwarf sperm whale | uncommon |
| Ziphiidae – beaked whales | |
| *Ziphia cavirostris* – Cuvier's beaked whale | rare |
| *Mesoplodon bidens* – Sowerby's beaked whale | extralimital |
| *Mesoplodon densirostris* – Blainville's beaked whale | rare |
| *Mesoplodon europaeus* – Gervais' beaked whale | uncommon |
| Delphinidae – ocean dolphins | |
| *Orcinus orca* – killer whale | uncommon |
| *Globicephala macrorhynchus* – short-finned pilot whale | common |
| (*Globicephala melas* – long-finned pilot whale) | possible |
| *Pseudorca crassidens* – false killer whale | uncommon |
| *Feressa attenuata* – pygmy killer whale | uncommon |
| *Peponocephala electra* – melon-headed whale | common |
| *Steno bredanensis* – rough-toothed dolphin | common |
| *Grampus griseus* – Risso's dolphin | common |
| *Tursiops truncatus* – bottlenose dolphin | common |
| *Stenella attenuata* – pantropical spotted dolphin | common |
| *Stenella frontalis* – Atlantic spotted dolphin | common |
| *Stenella longirostris* – spinner dolphin | common |
| *Stenella clymene* – clymene dolphin | common |
| *Stenella coeruleoalba* – striped dolphin | common |
| (*Delphinus delphis* – short-beaked common dolphin) | possible |
| (*Delphinus capensis* – long-beaked common dolphin) | possible |
| *Lagenodelphis hosei* – Fraser's dolphin | common |
| Sirenia – manatees and dugongs | |
| Trichechidae – manatees | |
| *Trichechus manatus* – West Indian manatee | common |
| Carnivora – carnivores | |
| Otariidae – sea lions, fur seals | |
| *Zalophus californianus* – California sea lion | introduced |
| Pinnipedia – true seals | |
| Phocidae – fur seals | |
| *Monachus tropicalis* – West Indian monk seal | extinct |

whale, and Blainville's beaked whale) are also western Atlantic species that are present in such small numbers that they are seldom seen in the Gulf. Seven species are listed as *uncommon*. These may or may not be widely distributed in the Gulf, but they never occur in large numbers. Included in this group are Bryde's whale, pygmy sperm whale, dwarf sperm whale, Gervais' beaked whale, killer whale, false killer whale, and pygmy killer whale. The remaining 13 species are *common* in the Gulf, that is, widely distributed over the area and abundant where they do occur. These include 3 whales, 9 dolphins, and 1 sirenian. Each of these common forms as well as the extinct and introduced species will be discussed individually below.

## SPERM WHALE (*PHYSETER MACROCEPHALUS*)

Sperm whales are easily recognized by the blunt head extending a quarter or more of the length of the body, lower jaw shorter than the upper jaw, and dorsal fin reduced to a hump and followed by several smaller humps. This is the largest of the common whales in the Gulf, with males reaching a length of up to 15 m (50 ft) and a weight of 40 tons. Females are only about two-thirds as large. Sperm whales are capable of diving up to 3,200 m (10,500 ft), but most dives are only about 1,000 m or less. Such dives, which may last an hour or more, are followed by 10–20 minutes of breathing at the surface. A reservoir in the head contains up to 1,900 L (502 gal) of a fine waxy oil (*spermaceti*), which may serve to focus sound in echolocation. Sperm whales feed largely on deepwater squids, but they are known to take a variety of other deepwater organisms including sponges, jellyfishes, octopuses, lobsters, crabs, and numerous types of fishes. Typically, groups of females and young remain together throughout the year in tropical and subtropical waters. Adult males are solitary and roam widely into higher latitudes to feed during the summer, but they return to the tropics to breed in the winter. There appears to be a resident population of sperm whales in the Gulf of Mexico, and census estimates suggest that about 300–400 individuals are present in the northwestern Gulf alone. They are generally encountered over continental slope waters (100–1,000 m deep) and in the open Gulf.

## SHORT-FINNED PILOT WHALE (*GLOBICEPHALA MACRORHYNCHUS*)

The short-finned pilot whale, sometimes called "blackfish," has a stocky body with a globular head, large dorsal

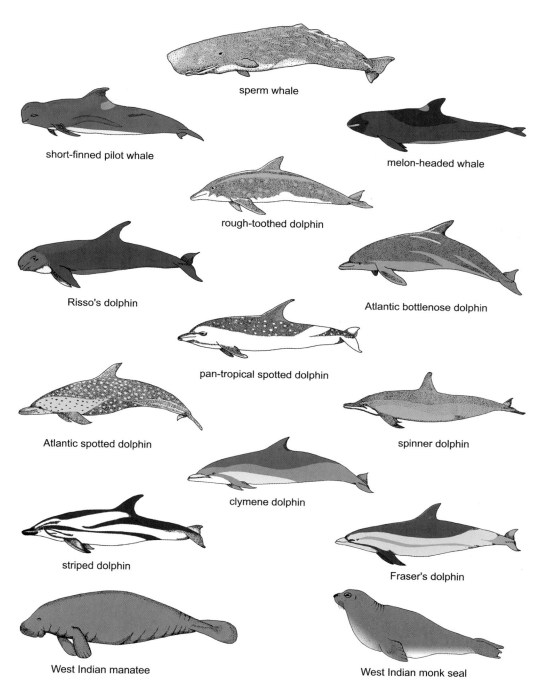

**Figure 13.12.**
Common marine mammals of the Gulf of Mexico including the recently extinct West Indian monk seal.

sperm whale

short-finned pilot whale

melon-headed whale

rough-toothed dolphin

Risso's dolphin

Atlantic bottlenose dolphin

pan-tropical spotted dolphin

Atlantic spotted dolphin

spinner dolphin

clymene dolphin

striped dolphin

Fraser's dolphin

West Indian manatee

West Indian monk seal

fin located forward of the center of the back, and long, slender, pointed flippers. The body is dark gray or black and marked with lighter streaks and patches, notably a light streak in advance of the dorsal fin and a white patch behind and beneath the dorsal fin. Males may attain a length of 6.1 m (20 ft) and a weight of four tons, but females are somewhat smaller. Short-finned pilot whales often occur in groups of up to 60 animals. They are found throughout the year over the continental slope, and esti-

mates suggest that up to 1,500 individuals inhabit the northern Gulf.

MELON-HEADED WHALE (*PEPONOCEPHALA ELECTRA*)
This is a small, slender whale with a globular head, rounded on top and sides and flat below. The dorsal fin is situated at the center of the back, and the flippers are long and pointed. There is a white patch near the mouth and a white stripe on top of the head. Adults may reach a

length of 2.7 m (9 ft) and a weight of 275 kg (605 lb), with the females being slightly smaller than the males. These whales often travel in large groups of more than 100 individuals. They live over the continental slope at depths of 200–2,000 m. Northern Gulf population estimates range from about 1,000 to 4,000 individuals.

### ROUGH-TOOTHED DOLPHIN (STENO BREDANENSIS)

In this small dolphin the head is somewhat conical, with the prominent beak sloping smoothly to the forehead. The dorsal fin is slightly curved and pointed, and the flippers are set far back on the body. The sides of the beak, throat, and chin are white, and light spots may be present on the side of the body. These dolphins attain a length of 2.3 m (7.5 ft) and a weight of 158 kg (350 lb). They feed on octopuses, squids, and fishes. They occur in small groups of 10–20 individuals. Rough-toothed dolphins inhabit waters over the continental slope and deep Gulf, and they are present at all seasons. Tentative estimates place the population size over the continental slope of the northern Gulf at around 89–351 individuals.

### RISSO'S DOLPHIN (GRAMPUS GRISEUS)

The grampus, as it is sometimes called, is a medium-sized robust dolphin with a blunt head and no beak. The dorsal fin is tall and sickle shaped (falcate), and the flippers are long and pointed. The bodies of older individuals are often heavily scarred. Adults range in size up to about 3 m (10 ft) in length and 300 kg (660 lb) in weight. They feed largely on squids but take in some crustaceans and fishes. These dolphins are normally seen in groups of 30 or fewer, but they sometimes occur in much larger groups. They live over the outer edge of the shelf and continental slope. Risso's dolphin is a year-round resident, and estimates suggest that as many as 3,000 individuals inhabit deeper waters of the northern Gulf.

### BOTTLENOSE DOLPHIN (TURSIOPS TRUNCATUS)

This medium-sized dolphin has a short snout in which the lower jaw protrudes slightly beyond the upper jaw. The tall dorsal fin is falcate and located at the middle of the back. The coloration is uniform dark gray above and lighter gray beneath, and lighter spots are rare. Gulf specimens average about 2.7 m (9 ft) or less in length and 160 kg (350 lb) in weight. Food consists of a variety of types of fishes (small sharks, rays, young tarpons, eels, menhadens, anchovies, catfishes, mullets, and seatrout) as well as shrimps. Individuals may feed alone, or groups of individuals may cooperate in herding schools of prey fishes. Bottlenose dolphins are very vocal, and they engage in a wide variety of social interactions. Groups of 2–15 individuals are common, but aggregations of several hundred have been observed. This is the common dolphin of inshore waters of the Gulf, and it enters rivers, estuaries, and lagoons and is especially common around passes. However, there are two distinct forms, an inshore and an offshore form, the latter being found over the outer shelf and upper continental slope, and there appears to be no interbreeding between the two groups. It has been estimated that 78,000 bottlenose dolphins inhabit the northern Gulf of Mexico, of which about 23,000 occur in bays and along the coast, and 55,000 make up the deepwater populations.

### PANTROPICAL SPOTTED DOLPHIN (STENELLA ATTENUATA)

This small dolphin has a robust body and long beak. The dorsal fin is tall, narrow, and curved toward the rear, and the flippers are long and pointed. The dark gray body is marked with white spots on the back. The beak is black with a white tip, and black lines run from the beak to the black eye patch and from the lower jaw to the flipper. Individuals average 1.6–2.5 m (5–8 ft) in length and about 100 kg (220 lb) in weight. This dolphin feeds at the surface on squids, fishes (flying fishes, mackerels, etc.), and shrimps. Groups of fewer than 30 are seen in coastal waters, but in oceanic areas groups of more than 1,000 individuals are not uncommon. This is one of the most common cetaceans in deep waters of the Gulf, and it has been estimated that the northern Gulf population consists of about 50,000 individuals, mostly beyond the 100 m depth contour. The species is particularly abundant in the northeastern Gulf.

### ATLANTIC SPOTTED DOLPHIN (STENELLA FRONTALIS)

This is a stocky little dolphin with many small spots on the sides of the body. These spots are pale above and darker on the belly, and there is often a pale area on the side below the dorsal fin. The dorsal fin, flippers, and flukes are all dark colored and unspotted. The upper jaw has a white tip, and the lower jaw is whitish. This dolphin averages about 2.1 m (7 ft) in length and 109 kg (240 lb) in weight. It feeds on squids and small fishes (herrings, anchovies, carangids, and flounders), and groups of individuals cooperate in herding the fish schools at the surface. Small groups of 10 individuals or fewer are common, but larger aggregations have been observed. The Atlantic spotted dolphin is common over the continental shelf, mostly in water depths of 20–200 m. In summer it moves inshore, and in the winter it migrates back offshore, possibly in response to temperature changes or food availability. The population in the northwestern Gulf has been estimated at 562–2,332 individuals, and the species is known to be abundant in the southwestern Gulf.

### SPINNER DOLPHIN (*STENELLA LONGIROSTRIS*)
These small dolphins average only 1.8 m (6 ft) in length and 75 kg (165 lb) in weight. Both the body and beak are slender. The triangular dorsal fin is slightly curved, and the flippers are long and pointed. The beak is black above and white below, and its tip is dark. There is also a dark stripe from the flipper to the eye. These dolphins feed at night on squids and mesopelagic fishes. When excited they often make spectacular leaps and spin around up to six times before falling back into the water. Group size varies from a few dozen to thousands. Spinner dolphins are found primarily over the continental slope and in the open Gulf, and the population in the northeastern Gulf alone has been estimated at 9,000 individuals.

### CLYMENE DOLPHIN (*STENELLA CLYMENE*)
This is another small dolphin averaging only 1.8 m (6 ft) in length and 75 kg (165 lb) in weight. It has a light gray band on top of the head. The beak is gray above and white below, and the lips are black. The dorsal fin is gray with black margins. This dolphin consumes squids and small fishes and appears to feed at night in mesopelagic waters of the open Gulf. Individuals sometimes leap and spin but not in as complex a fashion as the spinner dolphin. Groups usually consist of 2–20 individuals. Some are seen over the continental shelf, but most occur beyond the 100 m contour. Population estimates include the following: eastern Gulf near shore, 2,000; northwestern Gulf outer shelf (100–200 m), 827–3,474; and oceanic waters of the northern Gulf, as many as 10,000 individuals. (Note: *clymene* means "celebrated" or "famous.")

### STRIPED DOLPHIN (*STENELLA COERULEOALBA*)
In this long, slender dolphin the upper portion of the body is uniformly dark, and the chin and belly are white and marked with characteristic black stripes. A prominent stripe runs from the eye back to the anal region, a stripe extends from the eye to the flipper, and another passes from the side stripe to above the flipper. The dor-

sal fin is tall and curved. The average length of adults is about 2.4 m (8 ft), with a weight of 100 kg (220 lb). This is an oceanic species that inhabits deep waters beyond the edge of the continental shelf. It subsists on squids and mesopelagic fishes, especially lanternfishes (myctophids). This dolphin often occurs in large groups of hundreds or thousands of individuals. More than 4,858 animals are estimated to live in the northern Gulf, and a large concentration of about 2,000 inhabits the DeSoto Canyon region of the northeastern Gulf.

### FRASER'S DOLPHIN (*LAGENODELPHIS HOSEI*)
This dolphin has a robust body and relatively small dorsal fin, flippers, and tail flukes, and it exhibits a distinctive pattern of stripes on the side of the body. A dark line runs down the upper back, and a gray stripe passes from the facial area to the anus. Another dark line runs from the beak to the black eye patch and extends on back to the flipper. Adult males measure up to 2.7 m (9 ft) in length and may weigh 210 kg (460 lb), with the females being somewhat smaller. Fraser's dolphins occur in deep water beyond the shelf edge, where they feed on squids and midwater fishes. They are very rapid swimmers, and throughout their range they appear in large groups from several dozen to thousands of individuals. However, in the northern Gulf the total population probably numbers fewer than 1,000 animals.

### WEST INDIAN MANATEE (*TRICHECHUS MANATUS*)
The West Indian manatee has a plump, robust body ending in a flat, rounded tail. The squarish head has bristles on the snout, and the flippers are paddle shaped. The thick, wrinkled skin is often scarred and may bear growths of algae and barnacles. Adults reach a length of 4.6 m (13 ft) and a weight of 600 kg (1,320 lb). As pointed out in the previous chapter, these sluggish animals are strictly herbivorous and feed on various types of aquatic vegetation including turtle grass, manatee grass, and water hyacinths, among others. They inhabit bays, estuaries, and other shallow coastal waters and seldom stray far out at sea. In the Gulf they are found in greatest numbers on the Florida coast. There is also an indigenous population on the Mexican coast, and at one time they were probably quite abundant along the coast of Yucatán. During warmer months the manatee may stray northward, and it has been recorded from time to time all along the northern Gulf. However, the species is extremely sensitive to cold and cannot tolerate temperatures much below 18°C

(64°F), so it probably never did overwinter in the northern Gulf. In the fall the wanderers doubtless head back south to warmer waters. The Florida population has been estimated to include about 2,600 individuals.

WEST INDIAN MONK SEAL (*MONACHUS TROPICALIS*)

Historically, there were three monk seals in the world, the Mediterranean, West Indian, and Hawaiian. The West Indian species is now apparently extinct, and the other two are in danger of being lost. As a member of the true seals, the West Indian monk seal had no external ears, the front flippers were small, and at rest the hind flippers turned posteriorly. The fur was a grayish brown on the back, becoming lighter on the sides and yellowish on the belly. Adults averaged about 2.3 m (7.5 ft) in length. Little is known about its life history, but it probably ate squids and fishes. The species was formerly distributed on islands throughout the Caribbean, Bahamas, and Florida Keys, as well as islands of the Yucatán shelf. Archaeological evidence suggests that it was also once present in the northwestern Gulf. As an island species, it was quite unafraid of humans and was easily hunted and killed. It was hunted extensively for the valuable oil rendered from its fat and had already become rare by the mid-1800s. It was reported around Alacrán Reef as late as 1948, and a small colony was observed in the Caribbean in 1952. Despite extensive efforts to locate survivors, it has not been seen alive since. Its final demise was probably at the hands of local fishermen who regarded these seals as competitors for the marine fish stocks.

CALIFORNIA SEA LION (*ZALOPHUS CALIFORNIANUS*)

California sea lions have external ears and hind flippers that can be reversed and brought forward beneath the body for movement on land. Males grow up to 2.4 m (8 ft) in length and weigh up to 300 kg (660 lb), but females are much smaller, being only about two-thirds the length and one-third the weight of the males. These animals appear occasionally along the Atlantic and Gulf coasts, and since it is highly unlikely that they could make the journey from the Pacific coast on their own, they are assumed to be escapees from zoos or aquaria. Photographs verify that this species has appeared in the coastal waters of Alabama, and reports of its occurring off Florida and Louisiana are probably also correct. Lacking the rocky habitats of their homeland, sea lions in the Gulf haul out on buoys, barges, and other structures situated some distance off-shore. Sight records suggest that these feral animals survive for several months before disappearing.

## PATTERNS OF NEKTON DISTRIBUTION AND BEHAVIOR

For a number of reasons, the nekton cannot be described and treated as a whole as has been done for plankton and benthos. There is no single type of sampling gear that can collect all nekton groups as a unit, the various groups making up the nekton differ radically in their life history and behavior patterns, and for many species basic life history patterns are simply not known. Since the most substantial body of knowledge relates to the fishes of the Gulf, we will describe and illustrate patterns observable in the fish fauna and then mention how these patterns relate to the other nekton groups.

### Habitat associations

In previous chapters, reference has been made to a number of specific habitat types associated with waters of the Gulf of Mexico. In general terms, these include estuaries and coastal lagoons, plant-dominated communities (salt marshes, mangrove swamps, and seagrass meadows), soft- and hard-bottom areas of the continental shelf, natural and artificial reefs, and the water column above the shelf and in the epipelagic realm of the open Gulf. The lives of individual nektonic species revolve around their relationships with one or more of these habitat types. Being mobile, they can select their habitats, and many nektonic species utilize more than one habitat type during the course of their life histories.

ESTUARIES AND LAGOONS

Information concerning the distribution of fishes around the periphery of the Gulf can be obtained by comparing the fish fauna of three widely separated coastal lagoons: Galveston Bay in the north, Tampa Bay in the east, and Laguna de Términos (Terminos Lagoon) in the south, bordering the Bay of Campeche (table S54). The fish fauna of each of these areas has been fairly well studied, although collecting techniques and habitat sampling intensity varied somewhat between the different localities. The combined list for all three estuaries includes 283 species representing 84 families of fishes. Forty-four species (15.5 percent) are of freshwater affinity, and 239 (84.4 percent) are marine species. From the table it can be seen that Galveston Bay contains 161 species, of which 31 (19.3 percent) are freshwater, and 130 (80.7 percent) are marine. Comparable figures for Tampa Bay are 139 species total with 8 (5.8 percent) freshwater and 131 (94.2 percent) marine, and for Terminos Lagoon there are 158 species total with 12 (7.6 percent) freshwater and 146 (92.4 percent) marine. These figures illustrate the important fact that the fish fauna of estuaries and lagoons bordering the Gulf consists largely (often over 90 percent) of marine-derived species, and this same principle generally applies to the nektonic shrimps, crabs, turtles, and mammals. Many of the marine species use estuaries as nursery grounds, and hence they depend on such areas to complete their life histories.

However, the location and conditions within each estuary and their fish faunas are somewhat unique. For example, Galveston Bay, being close to the Mississippi River and the extensive coastal wetlands of Louisiana and being fed by a large freshwater stream (Trinity River), is inhabited by a large number of freshwater species mostly derived from the great Mississippi River fauna. Included are representatives of 14 different families (1 paddlefish, 4 gars, 3 herrings, 2 minnows, 2 suckers, 4 bullhead catfishes, 1 pirate perch, 2 killifishes, 2 live-bearers, 2 temperate sea basses, 5 sunfishes, 1 perch, 1 drum, and 1 sleeper). Tampa Bay, much farther removed from the Mississippi River, lacking much freshwater inflow and somewhat isolated from other bays and coastal lagoons, has only 8 freshwater species belonging to 6 families (1 gar, 1 herring, 1 bullhead catfish, 1 killifish, 2 live-bearers, and 2 sunfishes). Six of these species are in common with the fauna of Galveston Bay. On the other hand, Terminos Lagoon, which is far removed from the Mississippi River and other large bays and lagoons, has 12 species of freshwater fishes representing 3 families (1 bullhead catfish, 9 cichlids, and 2 sleepers). Except for the catfish, which is ultimately derived from the Mississippi River fauna, all are of southern origin. Only one of the species (a sleeper) is in common with the fauna of Galveston Bay, and none are present in Tampa Bay. Thus, the data show that Galveston Bay has the highest proportion of freshwater fish species and that Galveston and Tampa Bays share a number of freshwater elements, but that the freshwater fauna of Terminos Lagoon is largely distinct.

Turning to the marine ichthyofauna, the marine component of Galveston Bay makes up 80.7 percent, that of Tampa Bay 94.2 percent, and that of Terminos Lagoon 92.4 percent. Fifty-eight marine species are common to all three areas. This represents 20.5 percent, or about a fifth of all the marine species listed. Galveston Bay shares an additional 31 marine species with Tampa Bay and 12 species with Terminos Lagoon. Tampa Bay and Terminos Lagoon share an additional 16 species. Unique to Galveston Bay are 60 marine species, to Tampa Bay 34 species, and to Terminos Lagoon 72 species. Thus, from 12 to 31 species are common to only two bays, but from 34 to 72 species are unique to a given bay. Although Galveston and Tampa Bays share many species, Tampa Bay shows a slightly closer affinity with Terminos Lagoon than does Galveston Bay. In addition to the Gulf-wide coastal species more or less common to all the bays, there is a northern Gulf marine fish fauna that is best represented in Galveston Bay but that extends in diminished form to about Tampa Bay in the east and to near Tampico in the west. There is also a shallow-water marine coastal fauna largely derived from the Caribbean and extending to about Tampa Bay in the east and straggling up to northern Mexico or southern Texas in the west (Reid 1954; Darnell 1962b). Examples of these three faunal groups (northern, southern, and cosmopolitan) can be derived by examining the information in table S54.

OPEN CONTINENTAL SHELVES

In table S55 information is provided on fishes collected from several sectors of the continental shelves around the periphery of the Gulf of Mexico. English common names are given for most species, but they are not available for a few of the deepwater species and most of the Caribbean forms not yet reported from shelf waters of the United States. The four sectors include the following: northwestern Gulf (Rio Grande to the Mississippi River Delta), northeastern Gulf (Mississippi River Delta to DeSoto Canyon), eastern Gulf (DeSoto Canyon to the Florida Keys), and southwestern Gulf (from a few miles west of Terminos Lagoon through the western base of the Yucatán Peninsula to about the level of the city of Campeche). All the data are based on major across-shelf, multiseasonal trawl surveys of the bottom-dwelling invertebrates and fishes. It is recognized that bottom trawls (large bag-like nets dragged along the bottom) are selective devices that capture primarily the slower-moving species located on or just above the bottom. Their capture efficiency depends on such factors as the size, mesh, and configuration of the nets, as well as towing speed. Despite the obvious sources of bias, trawl capture data provide the best single means of assessing the composition of the local fish fauna and the best means of comparing the fish faunas of different areas.

The combined data set for all four sectors of the Gulf includes a total of 439 fish species belonging to 91 families. Representing the northwestern Gulf are 164 species (37.4 percent of the total); northeastern Gulf, 197 species (44.9 percent); eastern Gulf, 316 species (72.0 percent); and southwestern Gulf, 252 species (57.4 percent). Clearly, the fish fauna of the northwestern Gulf is by far the least and that of the eastern Gulf the most diverse, and the reasons are fairly obvious. Because of the outflow from the Mississippi and other large rivers, much of the northwestern Gulf is bathed in highly turbid, silt-laden water that precludes photosynthesis and vegetational development on the bottoms, and this restricts the variety of types of invertebrates and fishes that can tolerate such conditions. Furthermore, thick layers of silt cover most of the bottom including rocky outcrops. Toward middle and south Texas the water becomes clearer and the bottoms less silty, but here, in the relative absence of nutrients, plant production and hence animal production is low and less diverse. The prevalence of low winter temperatures, of course, restricts the distribution of many of the tropical species. Most of these considerations also apply to the northeastern Gulf, characterized for the most part by high turbidity, heavy sedimentation, and low winter temperatures. Habitats that are stressed or monotonous tend to support less diverse faunas.

By contrast, waters of the eastern Gulf display low turbidity and sedimentation, high light penetration, and generally higher winter temperatures, and many types of benthic habitats are available. Bottom sediments vary from quartz sand near shore to carbonate rubble offshore to finer ooze on the outer shelf. Widespread rocky outcrops, pavements, and ledges are populated by dense growths of algae and sessile and mobile invertebrates. Extensive seagrass meadows are present in the "big bend" area above Tampa Bay and inside the lower Keys. Mangrove swamps and forests line the southern Florida coast, and major coral reefs occur toward the south. These diverse habitat types provide food and shelter for a wide array of fish species. The Gulf Loop Current, which often bathes portions of the Florida shelf, brings in tropical fishes and their larvae, some of which find compatible living conditions on the west Florida shelf. As a result, the fish fauna of the eastern Gulf shelf is a composite of the faunas associated with the different habitat types, and it includes a fair number of tropical species. These different faunas have been discussed in some detail in Darnell and Kleypas (1987).

Like the eastern Gulf, the southwestern sector also provides generally clear water and a number of different habitat types including sandy, silty, and carbonate bottoms. Rocky bottoms and outcrops and coral reefs are present toward the northeast, and seagrass meadows occur along the base of the Yucatán Peninsula. Because of its southern location and warm waters, the area is populated by a fair number of Caribbean species. Some seasonal upwelling apparently occurs in the area, and as a result a number of deepwater species appear on the outer shelf.

The importance of local habitat diversity is illustrated by examination of the number of fish species endemic to each sector of the Gulf. Out of the total of 439 species, 191 species (43.5 percent) were found to be restricted to a single sector. Of these, 98 species (31.0 percent of the local fauna) were taken only in the eastern Gulf, 68 species (26.0 percent) in the southwestern, 14 species (8.5 percent) in the northwestern, and 11 species (5.6 percent) in the northeastern sector. These figures show the high levels of endemicity in the eastern and southwestern and low levels in the northwestern and northeastern sectors of the Gulf. Out of the total, 202 species (46.0 percent) were found to be limited to the southern Gulf (eastern plus southwestern sectors), but only 29 species (6.6 percent) were limited to the northern Gulf (northwestern plus northeastern sectors), a dramatic difference. For the most part, the differences between the ichthyofaunas of the different sectors represent occasional species from a wide variety of families, but the endemic species of the eastern Gulf include heavy representation (35 species) from six specific families: pipefishes, cardinalfishes, damselfishes, clinids, combtooth blennies, and gobies. These are all small, mostly cryptic species, common in weed beds or around rocky outcrop or coral reef areas.

REEFS AND BANKS

As noted earlier, continental shelves all around the Gulf are punctuated by reefs, hard banks, and rocky outcrops. These are inhabited by an array of fishes including some species for which this is primary habitat and some that stray from other environments. Primary reef fishes are defined as those species that are characteristically associated with natural reefs, that is, hard substrate with at least one meter of vertical relief, as well as rocky rubble around the base of such structures. Even though these fishes may sometimes be found elsewhere, the reef provides appropriate habitat during some stage of their life history. Secondary reef fishes are those that are sometimes encountered in reef environments but are more characteristic of flat bottoms, artificial reefs, or the open water column. This distinction is a very useful and ecologically meaningful one. Since reef and hard-bank environments generally cannot be effectively sampled with trawls, most of our information concerning reef fishes derives from scuba observations and collections, supplemented in deeper water by observations from submersibles or recordings from video cameras on remotely operated vehicles (ROVs).

In order to gain insight into the distribution of the reef fish fauna around the Gulf, we will compare the primary reef fishes of three areas that have been fairly well studied, the East and West Flower Garden Banks on the outer shelf of the northwestern Gulf, the hard banks of the middle and outer shelf off Mississippi and Alabama, and the reefs and outcrops of the eastern Gulf shelf including the Florida Middle Ground. Appropriate references to the ichthyofauna of these three areas include the following: Flower Garden Banks—Cashman (1973), Dennis (1985), and Dennis and Bright (1988); Mississippi-Alabama banks—Gittings, Bright, and Schroeder (1991); eastern Gulf shelf—T. S. Hopkins, Schroeder, et al. (1981), G. Smith (1976), and G. Smith et al. (1975). Although some studies have been carried out on the obviously rich reef fish fauna of the Mexican shelf (e.g., Castro-Aguirre and Márquez-Espinosa 1981; H. Chávez 1966; H. Hildebrand, Chávez, and Compton 1964), this fauna is still too poorly known to provide a basis for comparison. Since the habitats of the Flower Garden, Mississippi-Alabama, and eastern Gulf banks and reefs have already been described (chapter 11), this information will not be repeated here.

A tabulation of the primary reef fishes recorded from the three areas is provided in table S56. A total of 138 species is listed, with 71 species, or about half, having also been taken in trawl collections (see table S55). Among the trawl-collected species, 58 primary reef fishes occurred in the eastern and 43 in the southwestern Gulf, whereas only 12 were taken in the northwestern and 9 in the northeastern Gulf. Clearly, trawls were sampling some low-relief reef habitat in both the eastern and southwestern Gulf areas. The primary reef fishes listed in table S56 represent 33 different families, but most of these (91 species, or 65.9 percent) belong to only 9 families, including the squirrelfishes, sea basses, cardinalfishes, butterflyfishes, angelfishes, damselfishes, wrasses, parrotfishes, and leatherjackets. Sea basses are the numerically dominant family, with 26 species. Of the 138 species of primary reef fishes found on the three reef/bank areas, 96 (68.6 percent) occurred on the Flower Garden Banks, 38 (27.5 percent) on the Mississippi-Alabama banks, and 94 (68.1 percent) on the eastern Gulf reefs. These figures indicate the high diversity in the Flower Garden and eastern Gulf areas and the relatively low diversity in the Mississippi-Alabama area. Although this generalization apparently holds, the difference is not as great as it appears. The fish fauna of the Mississippi-Alabama banks could not be assessed as effectively using only video cameras in often turbid water. Here the primary fish fauna was more diverse than that listed but apparently still well below that of the other two areas, which were studied extensively by scuba divers. Only 19 species (13.8 percent) occurred in all three areas, and 67 species (48.6 percent) appeared in only a single area. Thirty-three species (23.9 percent) were unique to the Flower Gardens, 7 species (5.1 percent) to the Mississippi-Alabama banks, and 27 species (19.6 percent) to the eastern Gulf shelf. Among the species observed only at the Mississippi-Alabama banks were two species of sea bass, one bigeye, one cardinalfish, two wrasses, and one boxfish. Some of the Mississippi-Alabama primary reef fishes are characteristic of deeper water banks around the Gulf, which have been poorly studied.

The reefs and banks off Mississippi and Alabama and in the eastern Gulf are at least occasionally bathed by Caribbean waters of the Gulf Loop Current, which must repeatedly bring to these areas tropical reef species and their larvae, some of which find temporary or permanent homes in the northeastern and eastern Gulf. Furthermore, the eastern Gulf reefs can be populated by species spreading northward from the lower Keys and other environments of south Florida. But a question arises concerning the origins of the obviously diverse primary reef fish fauna of the Flower Garden Banks of the northwestern Gulf. Many of the colonizing species must come northward, hopping from one bank to another, along the continental shelf of the western Gulf. Some may arrive as planktonic larvae in eddies spun off by the Loop Current that head into the western Gulf. Perhaps some species are derived from the eastern Gulf via the Mississippi-Alabama banks. Interestingly, the Flower Garden Banks and eastern Gulf reefs share only 59 species (42.8 percent of the total primary reef fish fauna), whereas many species are present in only one area or the other. The same general question also applies to the origins of the reef corals and other tropical invertebrates found at the Flower Garden Banks, and at present there are no clear-cut answers.

OPEN WATER COLUMN

Many of the pelagic fish species of the Gulf are good swimmers and live much of their lives in the open water column well above the bottom and away from reefs and banks. Some dwell in coastal waters of the inner shelf, some live primarily over the outer shelf or in epipelagic waters of the deep Gulf, and others roam widely through both coastal and oceanic waters. During the fall a number of the better swimmers (some of the jacks, mackerels, tunas, and billfishes) abandon inshore waters of the northern Gulf and spend the winter offshore or in the southern Gulf, only to return in the spring when the water warms up. Records of the open-water fishes derive from a number of sources including midwater trawls, occasional individuals appearing in bottom trawls, the harvest of hook-and-line fisheries of commercial long-line operations, and the catch of sport fishermen often monitored at dockside by representatives of state and federal agencies. Midwater trawl collections in the deep Gulf capture primarily the smaller and slower-swimming species. To these are added observations by scuba divers and experienced observers aboard surface ships, submersibles, and airplanes. Because of the difficulties of sampling and study, the habitat relations and life history patterns of many pelagic species are still poorly known.

Representative examples of fishes of the open water column are presented in table S57. For brevity, a number of the groups, primarily epipelagic and mesopelagic fishes of the deep Gulf, have been omitted. (These are treated in the next chapter.) For each species the table also provides information on general habitat distribution. Listed are 31 species of sharks, 6 species of rays, and 141 species of bony fishes. A number of these (including some sharks, tarpons, herrings, anchovies, and jacks) also occurred in the bottom trawl collections (table S55), and a few were noted around reefs and banks (table S56), but the majority did not appear in the previous lists. Essentially unique to the present list are the lanternfishes, flying fishes, halfbeaks, needlefishes, mackerels, swordfishes, billfishes, butterfishes, and molas.

Some ecological information is available concerning the epipelagic species. Except for some herrings and mullets, which feed largely on phytoplankton or benthic vegetation and detritus, all are carnivorous. Some of the sharks, rays, and cobias feed largely on benthic animals. The young of most species and the adults of manta rays, many herrings, anchovies, lanternfishes, codlets, flying fishes, halfbeaks, needlefishes, silversides, some jacks, most butterfishes, and molas feed largely on zooplankton and small nektonic animals. Among the open-water fishes, top carnivores include most sharks, tarpons, and bluefishes, and some jacks, dolphins, barracudas, mackerels, swordfishes, and billfishes.

There are few hiding places in the open sea, and protection from predators is a major problem for most prey species. A few, such as the pilotfish (jacks) and remoras, accompany sharks and other large swimming species where they are relatively safe and can feed on scraps left by the feeding of their hosts. The juveniles of some jacks (pompanos) and butterfishes (man-of-war fish, harvestfish, butterfish, etc.) associate with larger medusae, Portuguese man-of-wars, or floating sargassum, and some consume the flesh of the medusae and hydrozoans. In deep waters of the Gulf beyond the continental shelf, many of the small pelagic zooplankton-feeding fishes live during the daytime in the dark mesopelagic zone and migrate up into surface waters at night to feed on zooplankton and small nektonic animals. By living in the deeper layers during the day and feeding in surface waters at night, they are able to obtain food while avoiding the heaviest predation.

## Spatial and seasonal patterns

During the course of their life histories many marine species undergo regular seasonal migrations, and some of their spatial and temporal patterns have been well documented. Using data from extensive bottom trawl surveys, Darnell, Defenbaugh, and Moore (1983) and Darnell and Kleypas (1987) published two bio-atlases showing distribution patterns of 13 penaeid shrimp and 372 fish species across the continental shelves of the northern and eastern Gulf of Mexico, from the Rio Grande to the Florida Keys. Where the numbers were sufficient, the maps depict the seasonal (winter, spring, summer, and fall) and areal patterns of numerical density distribution. Despite certain inherent weaknesses of trawl catch data, these maps represent the best available approximations of actual density distribution patterns, and they are particularly useful in portraying those species most vulnerable to capture. The discussions that follow are based on examples from the bio-atlases selected to illustrate particular distribution patterns of the shrimps and fishes.

Figures 13.13 and 13.14 represent the summer versus

winter distribution patterns of white shrimp (*Penaeus se-tiferus*) and brown shrimp (*P. aztecus*) on the continental shelf of the northwestern Gulf. In both species spawning occurs offshore, and the planktonic larvae make their way to the low-salinity bays and estuaries where they meta-morphose and grow into subadults. These in turn mi-grate back to the continental shelf, where they mature and reproduce. Few white shrimp overwinter in the bays and estuaries, but many brown shrimp do so. Therefore, whereas almost all the white shrimp and many of the brown shrimp move to the outside waters in late summer and fall, a portion of the brown shrimp population over-winters in the inside waters and exits in late spring and early summer. Once on the continental shelf the white shrimp remain close to shore, being largely confined in-side the 20 m depth contour. As seen in figure 13.13, dur-ing the summer white shrimp densities on the continen-tal shelf are generally quite low, and the highest densities occur off the mouths of passes. In the winter, nearshore densities are much higher, but the populations are still largely confined inside the 20 m contour, with a few shrimp straying into deeper water. By contrast, as seen in figure 13.14, the brown shrimp occurs all across the conti-nental shelf during both summer and winter. In the sum-mer the spring migrants added to the adults that overwin-tered on the shelf create high densities out to about the 40 m contour, and these shrimp are particularly dense off the coastline of Texas. However, during the winter most of the brown shrimp have abandoned the nearshore waters, and all along the coast their densities are generally greatest in warmer waters beyond the 40 m depth contour. Thus, al-though both shrimp species utilize the estuarine nursery areas as juveniles, on the continental shelf their distribu-tion and behavior patterns are quite distinct. The patterns for both species seen in the northwestern Gulf also apply to their populations in the northeastern Gulf off Missis-sippi and Alabama, but neither species is very abundant on the shelf of peninsular Florida.

The pink shrimp (*Penaeus duorarum*) occurs in very low density on the continental shelf of the northwestern Gulf, but it is abundant on portions of the Florida shelf. In this species spawning takes place offshore and the plank-tonic larvae make their way to inshore nursery areas that are generally of higher salinity then those used by white and brown shrimps. Off Florida the main nursery areas include seagrass meadows of the "big bend" area above Tampa Bay, the Everglades, and Florida Bay below the Everglades. Spring collections (fig. 13.15) show the pres-ence of overwintering adults on the continental shelf, mostly beyond the 20 m depth, and some individuals in the Big Bend seagrass meadows. By fall the new crop of subadults and adults moves to the spawning grounds offshore from the northern seagrass meadows, the Ever-glades, and Florida Bay. As seen in the figure, two distinct spawning areas exist off south Florida. Although all three penaeid shrimp species display roughly similar life his-tory patterns, these patterns differ significantly in detail, and the differences are clearly depicted in the seasonal density distribution maps.

Turning now to the fishes, figure 13.16 shows the sea-sonal distribution of numerical density for all the com-bined species of fishes captured on the northwestern Gulf shelf. In the spring months densities are fairly high all across the shelf, but the highest concentrations occur off the mouths of passes. These represent chiefly estuary-related species, of which some are preparing to enter the bays, some are exiting, and others are spawning near the passes. By fall many young of the year have left the bays and are now concentrated on the inner half of the con-tinental shelf. Particularly notable is the very high con-centration of mostly estuary-related species off the coast of Louisiana. In fact, because of the local availability of low-salinity coastal estuaries and marshlands and the nutrient-rich outflow of the Mississippi-Atchafalaya and related stream systems, within the nearshore zone (0–46 m) the poundage of fishes caught off Louisiana per hour of trawling effort is over three times that off Texas. Beyond this depth range the difference diminishes, and toward the outer shelf the catch rates for the two areas are about the same (Darnell and Schmidly 1988). The down-coast gradient in fish density is interrupted off far south Texas by a dense concentration of fishes derived in part from inside waters of the lower Laguna Madre.

In contrast with the gradient-like distribution of the total fish catch of the northwestern Gulf, that of the east-ern Gulf shelf (fig. 13.17) is spotty and more clearly related to the distribution of specific habitat types. In the spring high densities appear on the silty and sandy noncarbo-nate bottoms off Mississippi and Alabama where these chiefly represent concentrations of fishes associated with the Mississippi River deltaic marshes and the Pascagoula River–Mobile Bay nursery areas. Very high densities also appear off Apalachicola Bay and in the Big Bend seagrass meadows. Localized areas of high density also occur around the 50 m and 60 m depth contours, and these represent mixtures of many midshelf species associated

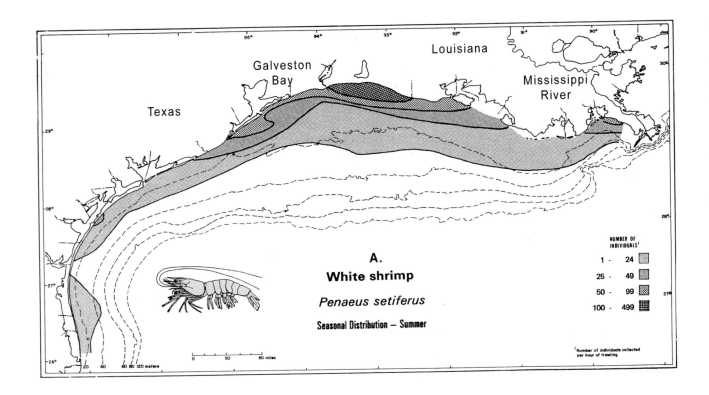

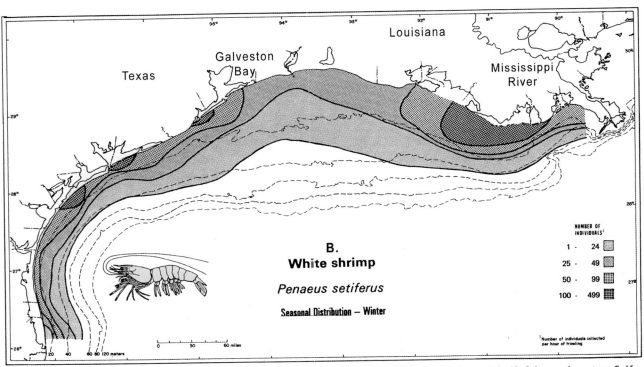

**Figure 13.13.** Seasonal and areal density distribution of white shrimp (*Penaeus setiferus*) on the continental shelf of the northwestern Gulf of Mexico. Density is expressed as the number of individuals captured during 60 minutes by a 45-foot flat trawl with 2.3-inch stretch mesh. A = summer, B = winter months. (From Darnell, Defenbaugh, and Moore 1983.)

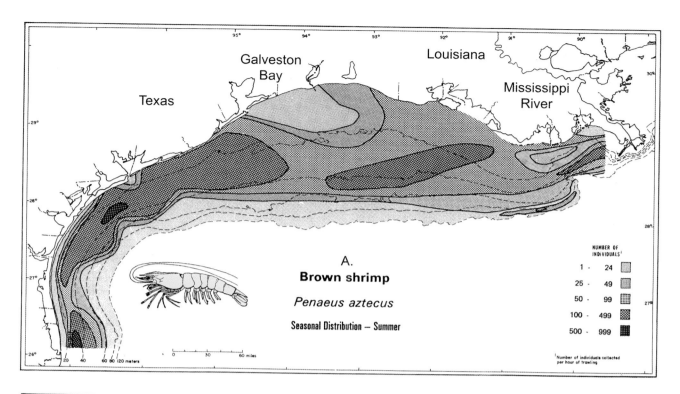

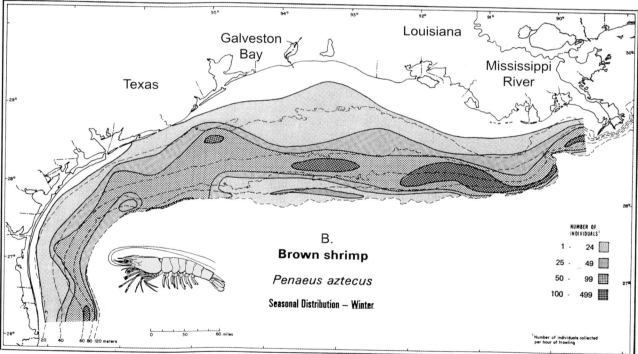

**Figure 13.14.** Seasonal and areal density distribution of brown shrimp (*Penaeus aztecus*) on the continental shelf of the northwestern Gulf of Mexico. Density is as given in figure 13.13. A = summer, B = winter months. (From Darnell, Defenbaugh, and Moore 1983.)

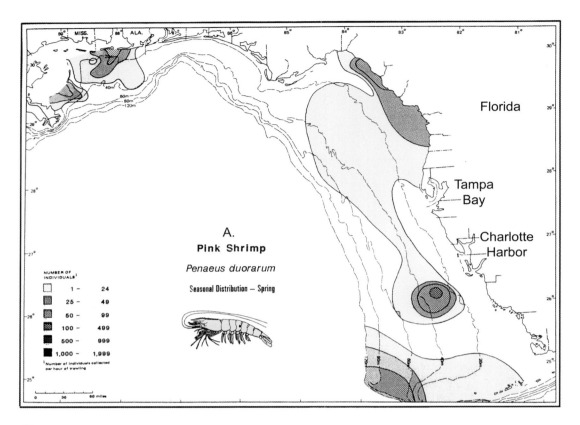

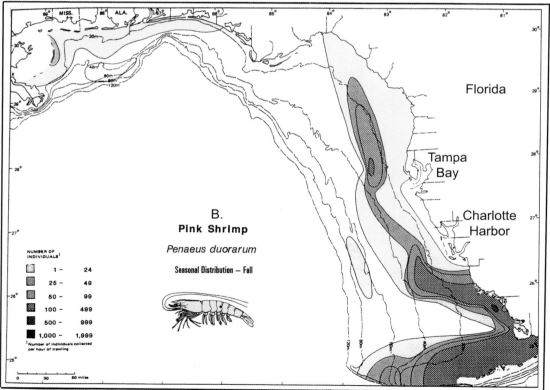

**Figure 13.15.** Seasonal and areal density distribution of pink shrimp (*Penaeus duorarum*) on the continental shelf of the eastern Gulf of Mexico. Density is as given in figure 13.13. A = spring, B = fall months. (From Darnell and Kleypas 1987.)

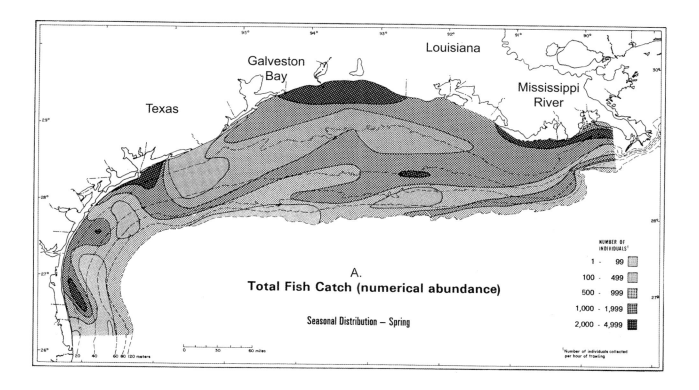

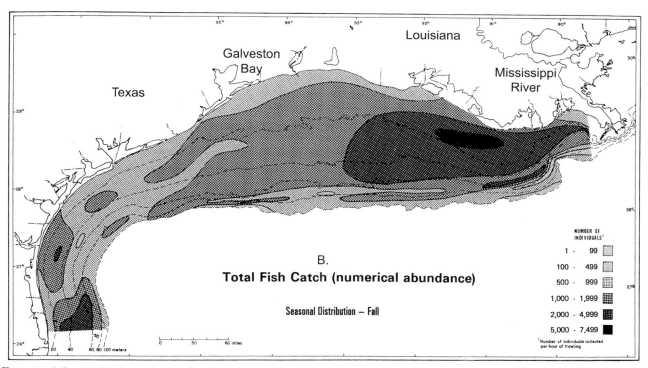

**Figure 13.16.** Seasonal and areal density distribution of the total fish catch (numbers of individuals) on the continental shelf of the northwestern Gulf of Mexico. Density is as given in figure 13.13. A = spring, B = fall months. (From Darnell, Defenbaugh, and Moore 1983.)

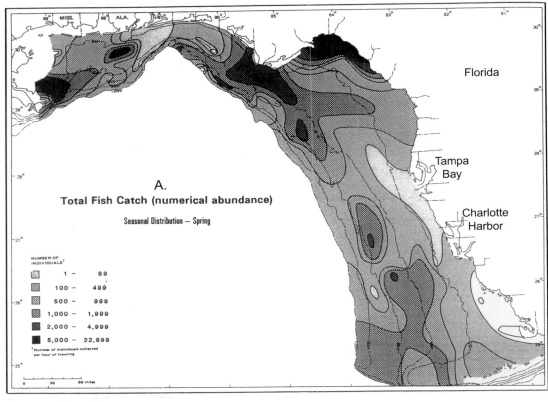

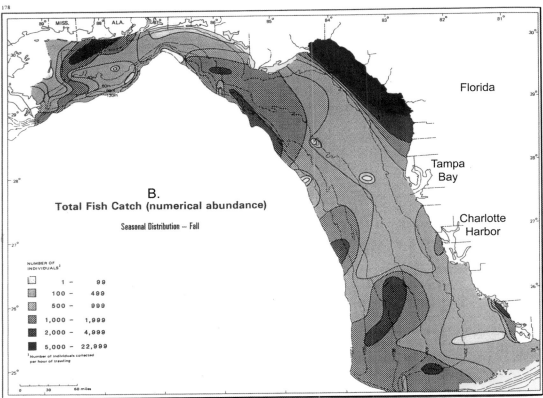

**Figure 13.17.** Seasonal and areal density distribution of the total fish catch (numbers of individuals) on the continental shelf of the eastern Gulf of Mexico. Density is as given in figure 13.13. A = spring, B = fall months. (From Darnell and Kleypas 1987.)

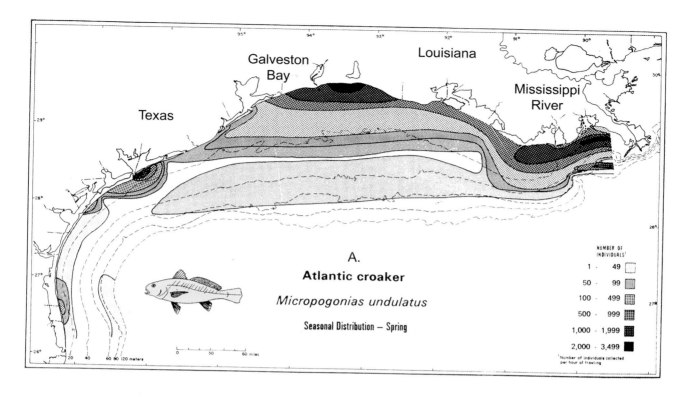

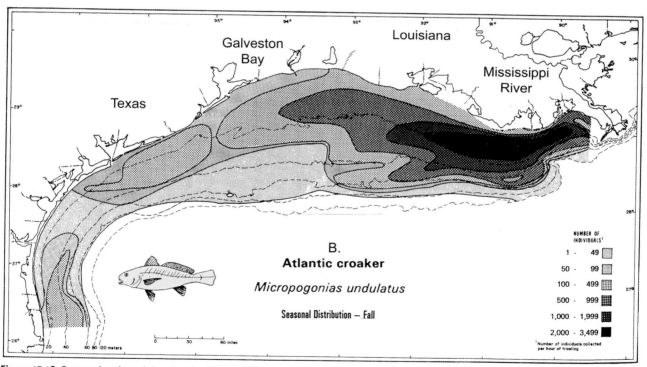

**Figure 13.18.** Seasonal and areal density distribution of the Atlantic croaker (*Micropogonias undulatus*) on the continental shelf of the northwestern Gulf of Mexico. Density is as given in figure 13.13. A = spring, B = fall months. (From Darnell, Defenbaugh, and Moore 1983.)

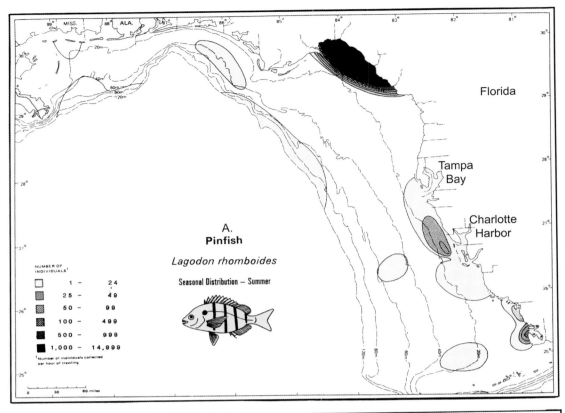

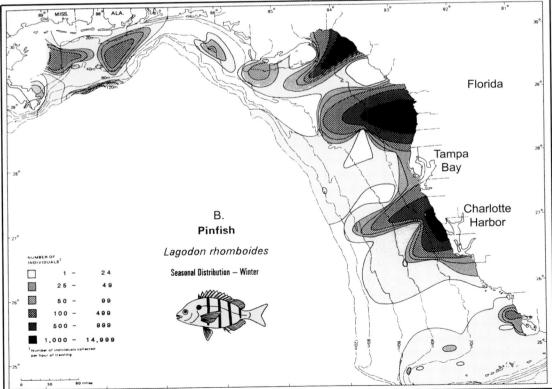

**Figure 13.19.** Seasonal and areal density distribution of the pinfish (*Lagodon rhomboides*) on the continental shelf of the eastern Gulf of Mexico. Density is as given in figure 13.13. A = spring, B = fall months. (From Darnell and Kleypas 1987.)

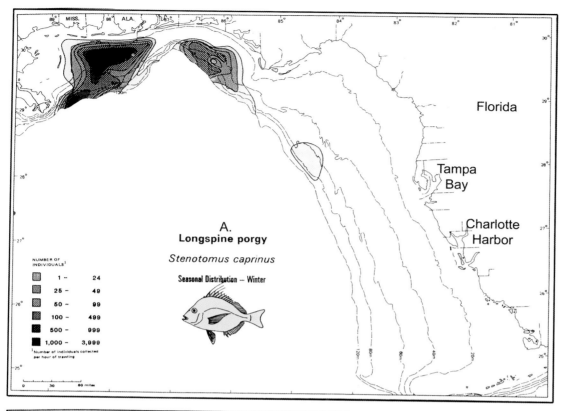

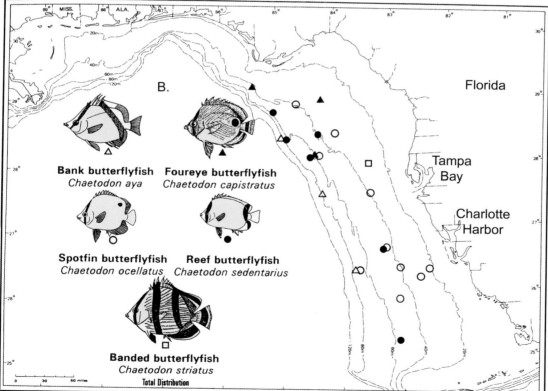

**Figure 13.20.** Winter density distribution of the longspine porgy (*Stenotomus caprinus*) (A) and total distribution of five species of butterflyfishes (B) on the continental shelf of the eastern Gulf of Mexico. (From Darnell and Kleypas 1987.)

with carbonate bottoms and outcrops together with some species whose young matured inshore and whose adults have now moved to deeper water to spawn. During fall very high densities are seen off Mississippi and Alabama where young are leaving the inshore nurseries and moving out onto the continental shelf (as off Louisiana), and in the big bend seagrass meadows where nursery young are now maturing. Areas of high density also occur off Apalachicola Bay and the Everglades where young are migrating onto the shelf. Additional areas of high density appear on the middle and outer shelf south of Apalachicola Bay and off Charlotte Harbor. These represent aggregations of many species of shelf fishes augmented by some coastal species that have moved into deep water for spawning.

These contrasting patterns from the northwestern and eastern Gulf and from spring to fall can be interpreted in greater detail when one examines the patterns species by species on the basis of distribution maps available in the two bio-atlases. For example, the density distribution of the numerically dominant Atlantic croaker (*Micropogonias undulatus*) in the northwestern Gulf (fig. 13.18) helps explain the total fish catch in the area, especially off Louisiana and upper Texas. Off lower Texas other species such as the sand seatrout (*Cynoscion arenarius*) and silver seatrout (*C. nothus*) assume greater importance. The croaker and seatrout distribution patterns also help explain the seasonal density distribution patterns off Mississippi and Alabama, but not those off peninsular Florida, where these particular species are quite rare. Here, for example, the pinfish (*Lagodon rhomboides*) (fig. 13.19) is far more important. For its nurseries this species utilizes the Big Bend seagrass meadows as well as the grass beds of Tampa Bay and Charlotte Harbor and to a lesser extent those of the Everglades and Florida Bay. Thus, in the summer on the shelf the pinfish is largely limited to shallow waters of the Big Bend area, but in the winter dense concentrations occur off the several nursery areas mentioned above.

To further comprehend the spotty nature of density concentrations of the eastern Gulf, one may examine the winter distribution of the longspine porgy (*Stenotomus caprinus*) (fig. 13.20-A) and the total distribution of five species of butterflyfishes (fig. 13.20-B). The porgy is largely limited to the noncarbonate bottoms off Mississippi and Alabama, where it is very abundant, and to the eastern flank of DeSoto Canyon, where fine sediments also prevail, but it is virtually absent from the sandy and carbonate sediments off peninsular Florida. The five butterflyfishes, including the bank butterflyfish (*Chaetodon aya*), foureye butterflyfish (*C. capistratus*), spotfin butterflyfish (*C. ocellatus*), reef butterflyfish (*C. sedentarius*), and banded butterflyfish (*C. striatus*), are all reef- and bank-related species associated primarily with outcrops of the middle and outer shelf, where they are occasionally taken by bottom trawls.

Space limitations preclude the provision of further examples of the distribution patterns of individual species, but the point has been made that each species displays a spatial and seasonal pattern that reflects its own ecological and life history needs. Some species do best on muddy bottoms in turbid water, while others prefer carbonate sediments in clear water. Some species are limited to particular depth ranges, whereas others occupy different depths during their life histories. Many use estuarine or nearshore nurseries, but others do not. In some cases, such as the estuary-related shrimp and fish species, the patterns are somewhat similar in general aspect, but as noted earlier, they are quite different in detail. The same principles apply to other nektonic animals of the Gulf. The sea turtles all return to the beach to lay their eggs, but at sea their lives appear to be quite distinct. Many species of marine mammals inhabit waters over the outer edge of the continental shelf and upper slope, and examination of their food suggests that they must have somewhat different life history patterns, even if these are not yet completely known to science.

About two-thirds of the surface of the Gulf of Mexico is underlain by water that is 200 m or more in depth, and this vast deepwater realm is the subject of the present chapter. Included are the water column bounded by the deep basin as well as the seafloor representing the surface of the continental slope and rise and the broad abyssal plain, much of which lies at a depth of about 3,400–3,750 m (11,152–12,300 ft). However, since life in the deep basin is intimately involved with and ultimately dependent on life in the overlying waters, it is necessary to consider the entire water column lying beyond the 200 m isobath, that is, everything seaward of the continental shelves, even though the surface waters have been addressed to some extent in previous chapters.

## VERTICAL ZONATION

Since we will be dealing with various depths of the water column, it is convenient to review the terminology associated with the various depth zones of the pelagic environment as they apply to the Gulf of Mexico. As seen in figure 14.1, the upper 200 m of the water column is referred to as the *epipelagic* zone. In effect, this corresponds to the euphotic zone where the water receives sufficient light to support photosynthesis by the phytoplankton organisms. Although the euphotic zone may not always reach 200 m off the mouths of the Mississippi and other major rivers because of local turbidity, for the clear open Gulf in general, the correspondence between euphotic and epipelagic is fairly reasonable. Below the epipelagic, and extending down to a depth of 1,000 m, lies the *mesopelagic* zone. This is often referred to as the "twilight" zone because here, even though the light level is too low to support photosynthesis, there is enough light to allow for some vision by marine animals with highly sensitive eyes, and apparently many marine animals actually do have such keen visual acuity. Lying beneath the mesopelagic and stretching to the floor of the Gulf is the *bathypelagic* zone, a realm of stygian darkness where the sun's radiation does not penetrate sufficiently to be visible to any known organism. Extending through the depth range of 1,000 m to the bottom at about 3,750 m, this is the broadest depth zone of the Gulf.

It should be noted that the Gulf is not one of the world's deepest oceans. In the Atlantic and Pacific, for example, the bathypelagic extends down to 4,000 m, and beyond this lies the *abyssal* zone, which continues down to 6,000 m. In deeper trenches lies the *hadal* zone, which reaches a maximum depth of about 10,900 m (35,750 ft). The Gulf of Mexico is only slightly deeper than the average depth of the world oceans (3,300 m = 10,824 ft) and only a little more than one-third of the depth of the deepest ocean trench. The vast floor of the major oceans lies within the abyssal zone and is referred to as the abyssal plain. That of the Gulf is situated in the bathyal zone and should properly be called the bathyal plain, but in conformity with other ocean bottoms, it will here be referred to as the abyssal plain. As will be shown subsequently, the upper three zones (epipelagic, mesopelagic, and bathypelagic) have real meaning in terms of their biological in-

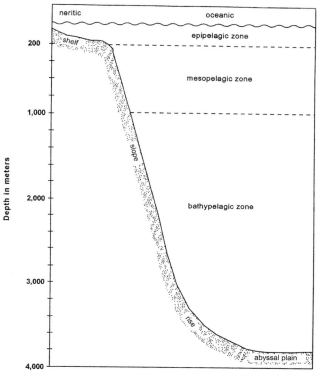

**Figure 14.1.** Vertical zonation of the open waters and benthic environments of the Gulf of Mexico.

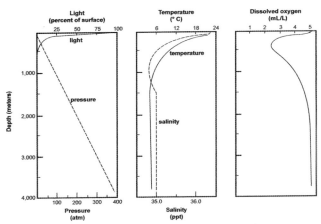

**Figure 14.2.** Vertical profiles of major physical factors (light, pressure, temperature, salinity, and dissolved oxygen) in the Gulf of Mexico. For full explanation see chapter 6.

habitants. As for the seafloor, one could apply the zonal terms developed for the pelagic realm. However, as we shall see later, analysis of the depth distribution patterns of the bottom fauna reveals a related but slightly different set of zonal depth boundaries.

## PHYSICAL ENVIRONMENT OF THE DEEP GULF

A brief review of salient features of the physical environment will serve as background for the biological discussions that follow. Seven factors will be considered: light, pressure, temperature, salinity, density, dissolved oxygen, and deepwater currents. Attention will also be given to major geological features and to the nature and distribution of the bottom sediments of the deep Gulf. Vertical distribution patterns for several of the physical factors are depicted in figure 14.2.

### Light

Light is derived from the sun; hence its intensity is greatest at the ocean surface. Within the water column the radiation is scattered by suspended particles, and the various wavelengths are differentially absorbed by the water molecules. Infrared and ultraviolet are removed in the upper few meters. Among rays of the visible spectrum, reds and yellows are absorbed preferentially, leaving greens and blues to penetrate most deeply. The intensity of the light diminishes logarithmically with depth. About half is gone by a depth of 20 m, and less than 1 percent of the surface illumination reaches a depth of 200 m. In the clearest ocean water the human eye can still detect light at about 500 m, and photographic plates exposed at various depths reveal that some light penetrates as deep

as 1,000 m. This diminishing supply of light with depth is one of the overriding factors controlling the depth distribution of life in the sea. Since light distribution limits photosynthesis, and hence plant production, to the epipelagic zone, it means that the deep ocean is nourished primarily from above. Animals of the deeper layers may move up to the epipelagic to graze (which they do primarily at night), feed on sinking organic materials (phytoplankton, fecal matter, dead bodies, etc.), or prey on the vertical migrators when they return to the deep.

### Pressure

Atmospheric pressure reflects the weight of a column of air pressing down from above, and at sea level this averages about 1.04 kg/cm$^2$ (14.7 lb/in$^2$, a standard pressure called one atmosphere, or 1 atm). Seawater is much heavier than air. Water pressure increases about 1 atm for every 10 m of depth, and this continues regularly to the bottom. Thus, the pressure at the bottom of the Gulf is about 380 atm, or around 2.8 tons/in$^2$. Human divers spending time at, say, 100 m may experience gas-exchange problems if they ascend to the surface too fast. Water pressure at this depth, about 10 atm, presses on the body and squeezes the lungs, which are filled with compressible gases (air or other breathing mixtures). The gases, in turn, are forced into the bloodstream and body cells under pressure. On the way back to the surface the diver must take care not to ascend faster than the blood and tissues can normally de-gas, or else the gas will come out as bubbles and cause serious systemic problems (known as the "bends"). This same general situation is faced by sea turtles and marine mammals that descend from the surface to great depths and by fishes that have gas-filled swim bladders and make significant vertical migrations. All these marine animals have had to develop special physiological and/or anatomical mechanisms that allow them to change depths without experiencing bodily harm. When deepwater fishes with swim bladders are brought rapidly to the surface in trawls, they are always dead and have their expanded swim bladders protruding through their mouths. Some fish species have gotten around this problem by developing fat-filled swim bladders, while others have eliminated the bladders entirely. For fishes without swim bladders and for invertebrates in general, the absence of gas-filled chambers eliminates much, but not all, of the pressure-change problem. In fact, some invertebrate species and even fishes can be brought slowly to the surface and remain alive in shipboard aquaria for days, weeks, or even longer, depending on the species and rate of ascent, if the aquarium water is kept cold.

## Temperature

In the open Gulf the surface water temperature varies seasonally from about 22° to 28°C (or even 30°C in late summer), and the depth of the surface mixed layer ranges from about 50 m in the summer to around 100 m in the winter. Aside from these variations in the near-surface waters, the vertical temperature profile remains relatively constant throughout the year. From the 20°C or higher surface values the temperature drops to 10°C by 400 m and to about 5°C by 1,000 m. Below this depth and all the way to the bottom it remains relatively constant. What role this temperature profile plays in the lives of the deep-Gulf species is not clear. Deep-diving mammals are well insulated and clearly can handle the temperature changes they encounter, as can the various fish and invertebrate species that regularly move into warm surface waters at night and spend daylight hours in the cooler depths. However, living specimens brought up from deeper waters are much more likely to survive if they are insulated against sudden temperature shock.

## Salinity

As in the case of temperature, salinity in the open Gulf is most variable in the surface layer, where it is affected by rainfall and evaporation and some mixing from riverine and shelf waters. Near the surface the salinity varies between about 35.5 and 36.5 ppt. Below the mixed layer it drops gradually, reaching a low of 34.9 ppt at 800 m and rising thereafter to about 35.0 ppt in the deeper water. So far as is known, this small range of salinity variation (only about 2.5 ppt) is of little consequence in the lives of deep-sea organisms.

## Density

The density of seawater is determined by temperature, salinity, and pressure acting in combination. Since water is relatively incompressible, even at great depths, the effect of pressure on seawater density is quite small. In vertical profile the range of salinity variation is also fairly small. Thus, temperature is the primary factor controlling the density of seawater. In the open Gulf the density increases with depth from a surface value of about 1.00225 g/cm³ (22.5 $\sigma_T$, or sigma-t) to a bottom value of around 1.00274 g/m³ (27.4 $\sigma_T$). Although the range of values appears small, the density of the medium is a fac-

tor that marine organisms must take into account. If marine animals are to remain at the proper depth level without great expenditure of energy, they must maintain neutral buoyancy, that is, the density of their body must equal that of the external medium. Furthermore, organisms that undertake major vertical migrations must be able to change their own body densities in order to rise or sink in the water column. The subject of buoyancy was discussed in some detail in the last chapter and need not be repeated here. However, it should be mentioned that most pelagic animals of the deep sea tend to be found at characteristic depth ranges, which means that they must adjust their body densities appropriately. How they sense the proper depth range and how they constantly fine-tune their body densities are not fully understood.

## Dissolved oxygen

Oxygen is added to seawater through surface gas exchange with the atmosphere and as a by-product of photosynthesis by the phytoplankton. Oxygen is removed through oxidation of organic matter by bacteria and higher organisms. The addition of oxygen predominates in the near-surface mixed layer, and the removal of oxygen is the dominant process below this layer. A vertical profile in the open Gulf shows the dissolved oxygen level in the surface mixed layer to be about 5.0 mL/L, although it may run somewhat higher during periods of maximum photosynthesis. Thereafter, it drops sharply, reaching 3.2 mL/L by a depth of 200 m and a minimum value of 2.6 mL/L between 350 and 600 m. Below this depth it gradually rises, achieving levels of 4.5–5.1 mL/L below about 1,700 m. The salient feature of this profile is the oxygen minimum layer between about 300 and 750 m, where the dissolved oxygen level is 3.0 mL/L or less, but this appears to be well within the tolerance range of most pelagic species. Certainly, the oxygen minimum layer is well populated with a diversity of marine animals. Where the oxygen minimum layer impinges on the continental slope of the northwestern Gulf the organic content of the sediments is somewhat higher than at adjacent depths, but how this may affect the benthic life is unknown. In general, it appears that at most, if not all, depths of the open Gulf there is sufficient oxygen for the normal activities of most marine organisms.

## Deepwater currents

Currents are known to be very active in surface waters of the Gulf of Mexico, and recent evidence suggests that they are also active, if at reduced speeds, throughout deeper waters of the Gulf. The Gulf Loop Current, which enters through the Yucatán Channel, penetrates northward for varying distances, and exits through the Straits of Florida, has surface velocities of 150–200 cm/sec. Decreasing with depth, the velocity at 700 m often exceeds 10 cm/sec, and the effects of this current are felt to around 1,000 m. (For reference, 44.7 cm/sec = 1 mi/hr.) Subsurface currents off the Mississippi River Delta and upcanyon currents in DeSoto Canyon have been associated with Loop Current forcing. Large eddies detached from the Loop Current travel west or southwest and eventually decay by interaction with continental shelves or slopes or with each other or by the shedding of smaller rings. Within such eddies, current speeds initially may be 150–200 cm/sec at the surface, 35 cm/sec at 450 m, and 10 cm/sec at 700 m. Satellite data reveal that at any given time surface waters throughout the Gulf are populated by numerous cyclonic and anticyclonic eddies and rings interacting with one another.

Wind forcing appears to be responsible for the permanent cyclonic gyre in the Bay of Campeche. However, forcing by energetic episodic wind events such as frontal passages, extratropical cyclones, and tropical storms and hurricanes can induce surface speeds in excess of 100–200 cm/sec, and their effects are felt to depths of 900 m or more. These episodic wind events can cause major currents in the deep water of the Gulf, and the resulting internal oscillations may last for up to 10 days.

Within the deeper water of the Gulf (1,000 m or deeper) the circulation is influenced by the Loop Current, which apparently generates horizontal Rossby waves that propagate westward at about 10 cm/sec. Subsurface intensified currents, called "jets," have been measured at various depths in the Gulf down to about 1,500 m. Although generally having speeds of 30–50 cm/sec, some in the near-surface waters have been found to exceed 150 cm/sec, and in the 1,200–1,500 m depth range short bursts of 100 cm/sec have been recorded. Some bottom currents have been measured with speeds of up to 40 cm/sec.

From the above information it is clear that at all depths, from surface to bottom, the water in the Gulf of Mexico is actively moving around. Within the water column most of the currents are associated with rotational spinning of eddies and rings, which themselves move around, so a given point in the water column may be bathed alternately by currents from different directions. However, this is not universally the case with boundary currents that flow along the deep slope or the Gulf bottom. Evidence from fields of long parallel furrows on the seafloor at the base of prominent features (Sigsbee Escarpment, Green Knoll, etc.) in the northwestern Gulf in the depth range of 2,000–3,000 m strongly suggests that currents here are directionally stable.

What effect these currents may have on the lives of individual deep-Gulf organisms is not known. What is clear, however, is that the currents allow for the transport and dispersal of planktonic and pelagic species throughout the Gulf. This is likely of particular importance in the dispersal of larvae. It also means that vertically migrating species will be feeding in different pastures each night. For sedentary or attached benthic organisms the presence of significant bottom currents promises the transport of food particles if the organisms can afford to wait for the scant suspended resources of the deep water. Finally, the existence of currents at all levels ensures that particulate material settling from above will become widely distributed throughout the water column and the bottom. This averages out the harvest so that it does not all fall directly beneath upwelling sites and other areas of high surface productivity.

## Bottom topography and surface sediments

The ocean bottom beyond the continental shelf is often thought of as a relatively flat surface grading gently downward across the continental slope and rise and out onto a broad, featureless abyssal plain. Although this model aptly describes some regions of the deep Gulf, it is totally inappropriate for others. Because of the Gulf's complex geological history, the dynamic nature of its subsurface salt deposits, and the deposition of enormous quantities of clastic sediments by the Mississippi River (especially during periods of continental glaciation), at all depths the floor of the deep Gulf is marked by prominent topographic features. Among these are steep escarpments, intraslope basins and knolls, ridge and valley systems, channels, and submarine canyons. The escarpments lie off the shelves of the Florida and Yucatán Peninsulas and along the lower slope of the northwestern Gulf (Sigsbee Escarpment). Knolls and basins characterize continental slopes of the northern and southern Gulf, where they are underlain by massive salt deposits. Ridge and valley systems roughly parallel the coastline along much of the continental slope of the western Gulf, surface expressions of underlying folded shales or salt deposits. Channels of various sizes and shapes are particularly prominent on the upper and lower portions of the Mississippi Fan. Large submarine canyons that traverse the upper slope include DeSoto Canyon off the Florida panhandle, the Mississippi Trough southwest of the Mississippi River Delta, and Campeche Canyon at the western base of the Yucatán Peninsula. In the northwestern Gulf, and possibly elsewhere, in the depth range of 2,000–3,000 m the otherwise flat bottom consists of fields of parallel furrows from 5 to 10 m deep and extending for tens of kilometers (created and kept open by bottom currents). The abyssal plain of the western Gulf is punctuated by a series of knolls (Sigsbee Knolls) that represent surface expressions of underlying salt diapirs. In the southeastern Gulf the slope and abyssal plain north of Cuba feature a pair of knolls (Catoche and Jordan Knolls) of probable igneous origin. In addition to these prominent features, many smaller topographic irregularities occur on the bottom surface of the deep Gulf.

As noted earlier, continental shelves of the Gulf fall into two general categories on the basis of surface sediment types: those dominated by coarse biological carbonate deposits and those in which clastic terrigenous materials predominate. Calcareous shelves occur off the Florida and Yucatán Peninsulas, and river-borne clastic sediments dominate elsewhere. On the outer continental shelves and throughout most of the deep Gulf the surface sediments are rich in pelagic ooze, that is, deposits of the calcareous shells of planktonic protozoans (coccolithophores and foraminiferans). These are mixed with varying amounts of siliceous diatom frustules and fine silt and clay particles. Off the northern Gulf, particularly in areas influenced by the present-day or ancestral Mississippi River, the slope and deeper-water sediments often contain higher percentages of sand and other clastic materials.

Outer continental shelf and slope areas subject to rapid sedimentation, such as off the Mississippi River Delta, build up unstable deposits that may be subject to periodic slumping. Such materials then charge downslope to be redeposited on the rise or across the abyssal plain. This slumping may be a result of the downward pressure of gravity alone, or it may be triggered by episodic strong currents such as those associated with storms and hurricanes. The former slump areas are marked by scars and downslope channels and the exposure of surfaces charac-

terized by coarse-grained clastic sediments. Topographic features serve to guide the various bottom currents, and locally they must play a role in determining the particle grain-size distribution of the surface sediments.

From a biological standpoint the various bottom types of the deep Gulf present a mosaic of substrates: rocky and hard in some areas and soft and particulate in others, coarse grained grading to fine grained, clastic grading to calcareous, and higher organic to lower organic. The organic content of bottom sediments ranges from about 2 percent on portions of the outer shelf to around 1 percent or less on the deep ocean floor. The complex topography ensures a variety of bottom current patterns and speeds. Thus, contrary to what might be expected, the deep Gulf provides a diversity of benthic habitat types, and habitat diversity is the key to biological species diversity. Although slumping must devastate the local benthic fauna caught up in a mudslide or downslope turbidity current or buried by the final deposition, it is probably a boon to the fauna of the deep Gulf. Turbidity currents are considered to be one of the major agents for the transference of organic matter from shallow to deeper water sediments.

Finally, it should be mentioned that in many areas of the continental slope of the northern Gulf, hydrocarbons (methane gas and petroleum) escape from near-surface deposits into the water column. Also, at the base of the Florida Escarpment hydrogen sulfide gas escapes through the sediments into the water. Both the hydrocarbon and hydrogen sulfide seeps provide energy to support specialized microbes and other organisms, many of which are unique to such areas.

## DISTRIBUTION OF LIFE AND NUTRIENTS IN THE DEEP GULF

The general distribution of life in the open Gulf of Mexico is illustrated in figure 14.3. The uppermost, or epipelagic, is the most important zone because here there is sufficient sunlight to support photosynthesis and the production of the phytoplankton on which most of the rest of the system depends. For this reason the epipelagic supports a great diversity of life forms and a relatively high density of living biomass. In addition to the planktonic algae, here are encountered many different types of bacteria, fungi, protozoa, medusae, siphonophores, polychaetes, pteropods, heteropods, cephalopods, crustaceans, chaetognaths, salps, and fishes, as well as sea turtles and marine mammals. Among the fishes are a number of types of

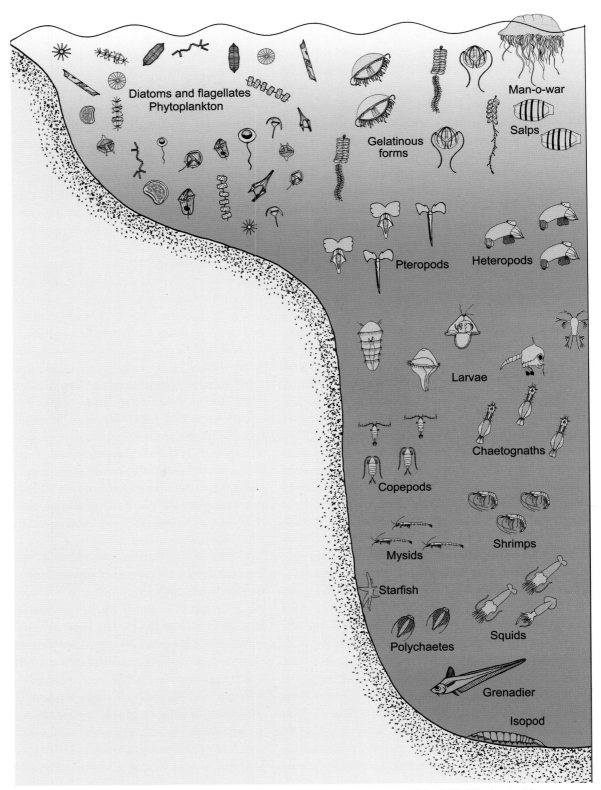

**Figure 14.3.** A generalized picture of the distribution of life in the several depth zones of the Gulf of Mexico. In this panoramic view the various organisms are not drawn to the same scale. The microscopic and other small forms including the phytoplankton and zooplankton (protozoans, polychaetes, pteropods, heteropods, copepods, mysid shrimps, chaetognaths, and larvae) as well as gelatinous types, although shown only in the left panel, are widespread and would be found at their

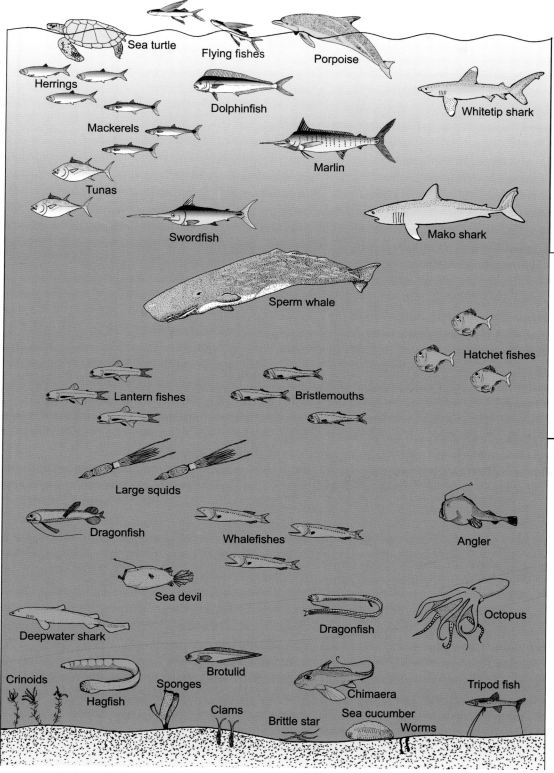

Epipelagic

Mesopelagic

Bathypelagic

Sea turtle
Flying fishes
Porpoise
Herrings
Dolphinfish
Whitetip shark
Mackerels
Marlin
Tunas
Swordfish
Mako shark
Sperm whale
Hatchet fishes
Lantern fishes
Bristlemouths
Large squids
Dragonfish
Whalefishes
Angler
Sea devil
Dragonfish
Octopus
Deepwater shark
Crinoids
Brotulid
Hagfish
Sponges
Chimaera
Tripod fish
Clams
Brittle star
Sea cucumber
Worms

appropriate depths on the right, as well. Most of the larger animals (squids, octopuses, fishes, sea turtles, and marine mammals) appear in the right panel. The phytoplankton and most of the sea turtles and marine mammals are more or less limited to the upper or euphotic zone. Invertebrates and fishes occur throughout all depths from the surface to the bottom of the deep Gulf.

sharks (whale, thresher, basking, white, mako, ocean whitetip, tiger, blue, hammerhead, etc.), manta and eagle rays, tarpons, herrings, needlefishes, flying fishes, halfbeaks, dolphinfishes, mackerels, tunas, billfishes, and many others with more or less familiar names. This dynamic pasture is also home to the eggs and larvae of many different invertebrates and fishes, not only of species that spend their entire lives in the epipelagic, but also of those in which the adults live in deeper layers of the Gulf. At night the epipelagic also harbors many types of zooplanktonic and nektonic animals that move up from deeper layers to feed.

Since phytoplankton production also depends on the availability of inorganic nutrients, not all areas of the open Gulf are equally productive. The most productive areas are generally those associated with the outer shelf and upper slope, where river outflow and upwelling are most common, and the least productive portions occur in the open Gulf more remote from the shelf. However, even in the open Gulf fairly productive areas resulting from local upwelling may be associated with the Gulf Loop Current or with rings, eddies, and filaments generated by the Loop Current. In response to the increased food supply, regions of high phytoplankton production are often areas of high zooplankton and nekton production, and the upper continental slope underlying such areas is often the site of high benthic production.

From the epipelagic there is a constant rain of organic material to deeper layers of the Gulf. Phytoplankton cells, carcasses of dead animals, bits and pieces of dead animals, quantities of fecal material, and veils of mucus sink from the epipelagic, providing nutritive support for organisms in the dark, nonphotosynthetic zones below. This constant loss of organic material drains nutrient elements from the epipelagic that must be restored through upwelling and other mechanisms, as noted above. Much of the sinking organic matter becomes oxidized in the layer just below the epipelagic, and this partially depletes the dissolved oxygen supply, resulting in the oxygen minimum layer.

The mesopelagic or twilight zone, where there is some faint light penetration, is populated by bacteria, countless tiny flagellates and other microalgae, and a variety of invertebrates and fishes. The latter include such forms as argentines, lanternfishes, hatchetfishes, bristlemouths, pearleyes, barracudinas, deep-sea smelts, snaggletooths, and many other groups less familiar to the layman, and some of these small fishes exist in the world oceans in enormous numbers, even if at low densities. Many of the animals pass their entire life histories in this zone, but others produce eggs and larvae that develop in the epipelagic, where small particulate food items are more plentiful. Likewise, many of the mesopelagic zooplankton organisms as well as cephalopods, shrimps, and fishes carry out regular vertical migrations that permit them to feed at night in the epipelagic. On the other hand, many of the marine mammals dive down at night to feed in the lower epipelagic and upper mesopelagic on cephalopods, shrimps, and fishes that are resident there or have moved up from deeper waters to feed. Here in the mesopelagic the density of living organisms is lower than it is in surface waters.

In the bathypelagic zone conditions are much the same as in the mesopelagic except that sunlight is for all practical purposes absent, and food particles falling from above are even less plentiful. Therefore, living organisms are much rarer, and in order to survive in this cold, lightless world the animals have had to develop many special adaptations. The same general groups of invertebrates and fishes found in the mesopelagic are encountered in the bathypelagic, but others are unique to the deeper water. Among the fishes are the smoothheads, lightfishes, viperfishes, dragonfishes, loosejaws, seadevils, anglers, and whalefishes, many with bizarre body shapes that are often adorned with strange appendages. Some of the bathypelagic species apparently migrate regularly or occasionally up into the mesopelagic or even the epipelagic to feed. In some species the eggs float, and larvae develop in epipelagic waters.

In the benthic and near-benthic environment conditions are different from those in the overlying waters, and this is true for all depths. Here food is somewhat more available. Large carcasses and smaller particles falling from above come to rest on the bottom surface, where they are attacked by a host of bacteria, protozoans, small and large invertebrates, and fishes. Largely because of the greater availability of food and the presence of the substrate itself, the density and variety of animal life increases in the benthic and near-benthic environment. Here animals may spend part or all of their time swimming or hovering in the water column, lying or crawling about the bottom surface, anchored and permanently attached to the substrate, or burrowing in the sediments. Some of the better swimmers may move well up into the water column to feed, but most apparently seldom venture far from the bottom surface.

Among the sedentary or nearly sedentary benthic

species are various sponges, coelenterates (hard and soft corals, sea pens, and hydroids), several types of worms (sipunculids, many kinds of polychaetes, etc.), brachiopods, shelled mollusks (gastropods, bivalves, and tusk shells), certain crustaceans (some shrimps, lobsters, hermit crabs and their relatives, and true crabs), and echinoderms (starfishes, brittle stars, sea urchins, sea cucumbers, and crinoids), as well as bryozoans and attached tunicates. Bottom currents apparently raise particulate organic matter in sufficient quantity to support attached filter feeders such as sponges, coelenterates, barnacles, bryozoans, and tunicates. Fishes prominent in the benthic environment of the deep Gulf include such forms as hagfishes, ratfishes, deepwater sharks, skates, halosaurid and notacanthid eels, synaphobranchid eels, cusk-eels, tripod fishes, brotulids, and grenadiers.

In overview, most (probably over 90 percent) of the organic matter supporting organisms of the deep Gulf is ultimately derived from phytoplankton that rains down from above as particles of various sizes, but there are other mechanisms for getting organic material to the deep Gulf. One potential method is through the chain of vertically migrating animals. In this scenario, organisms from a given layer move upward to feed in a shallower layer and then move back down to where they, in turn, are consumed by predators that have come up from even deeper layers to feed. Thus, organic matter moves downward from the epipelagic into deeper waters through a stepwise progression of living predators. This mechanism is sometimes referred to as a "food ladder" or "ladder of vertical migration." Water currents sweep organic material from the continental shelves, and much of this comes to rest on the upper continental slope, although some must settle out in deeper water. Slumping and turbidity currents remove organic material from the outer shelf and upper slope and deposit it in the deep Gulf.

Finally, in certain areas of the continental slope major hydrocarbon seeps support locally dense concentrations of specialized organisms, and in the deep Gulf a hydrogen sulfide seep also supports a local community. It seems likely that pockets of hydrocarbon and possibly hydrogen sulfide are much more widespread on the slope and in the deep Gulf than is currently appreciated.

## VERTICAL PATTERNS OF PRODUCTION AND STANDING CROP

In the open Gulf of Mexico, away from the continental shelves and upwelling areas, the standing crop of phytoplankton and the rate of primary production are quite low, even when compared with other areas of the world oceans. Data presented by El-Sayed (1972) showed that in offshore waters of the Gulf, phytoplankton standing crops (as measured by chlorophyll-a concentrations) averaged 0.13 mg/m$^3$ in surface waters, and the average depth-integrated value was 10.94 mg/m$^2$. Surface primary production averaged 0.21 mg C/m$^3$ per hour, and the average depth-integrated value was 5.45 mg C/m$^2$ per hour. Such values are typical of those encountered in nutrient-poor (*oligotrophic*) waters of central oceanic gyres, such as the Sargasso Sea, and are much lower than those of strong upwelling areas or those influenced by boundary currents. Fredericks (1972) showed that in the open Gulf the particulate organic carbon concentration in the upper 90 m averages 0.050 mg C/L (= 50 mg C/m$^3$), and from 90 m to the bottom of the deep Gulf it averages 0.028 mg C/L (= 28 mg C/m$^3$) with very little vertical change.

T. L. Hopkins (1982) studied vertical distribution of the zooplankton standing crop in the upper 1,000 m at a station in the open-water region of the eastern Gulf located west of Tampa Bay in an area influenced by the descending arm of the Gulf Loop Current. His values (day and night collections combined), expressed as zooplankton dry weight for a vertical column beneath one square meter of surface water, are as follows: 0–200 m = 737 mg, 0–500 m = 1,011 mg, and 0–1,000 m = 1,241 mg. These biomass values are typical of those encountered in boundary current areas elsewhere. They are about a tenth of those found in upwelling areas and somewhat greater than those of central oceanic gyres. In Hopkins's study the concentration (dry weight biomass) in the 200–500 m depth range was only about a quarter (24.25 percent) of the value in the upper 200 m, and in the 500–1,000 m range it was only about an eighth (12.4 percent) of that in the surface waters. Thus, over half of the zooplankton biomass is concentrated in the upper 200 m of the water column, and there is a logarithmic decrease in zooplankton biomass density with depth. The zooplankton maximum was found to correspond to the depth of the maximum phytoplankton production rate and not to the depth of the chlorophyll peak. Unfortunately, no comparable data are available on the vertical distribution of zooplankton numbers or biomass for the central Gulf.

The distribution of the benthic macrofauna of the Gulf from the continental shelf to the abyssal plain has been reported by Rowe and Menzel (1971) and Rowe, Polloni, and Horner (1974). Their data are expressed in terms of

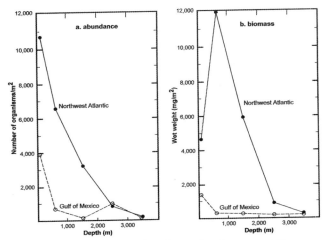

**Figure 14.4.** Vertical distribution of macrobenthos numerical abundance and biomass densities in the northwest Atlantic (off New England) and the Gulf of Mexico. (Data from Rowe and Menzel 1971; Rowe, Polloni, and Horner 1974.)

abundance (number of individuals/m²) and biomass (wet weight, dry weight, and organic carbon weight/m²), and values from the Gulf are compared with those from the northwest Atlantic. Abundance and wet weight biomass values for the Gulf and northwest Atlantic are plotted in figure 14.4. Within the Gulf the abundance values decreased from 3,834 individuals/m² on the continental shelf (0–200 m) to a low of 75/m² in depths greater than 3,000 m, and biomass values decreased from 1,289 mg/m² on the shelf to 80 mg/m² on the abyssal plain. By contrast, in the northwest Atlantic comparable figures for abundance were 10,507 and 272 individuals/m², and for biomass they were 4,534 and 198 mg/m². In both the Gulf and northwest Atlantic the concentrations decreased logarithmically with depth. However, on average, whether comparing abundance or biomass on the continental shelf or in the abyss, the concentration of benthic macrofauna in the Gulf was only about a third that of the northwest Atlantic. Both on the continental shelf and on the deepwater bottoms the Gulf values are quite low, and concentrations on the abyssal bottoms are comparable to those below the nutrient-poor Sargasso Sea. At any given depth the abundance of benthic life reflects the organically bound energy available in the food supply, and low macrobenthic density in the deep Gulf apparently reflects the low rates of primary production in surface waters of the open Gulf.

Parenthetically, it must be emphasized that all the above production, abundance, and biomass values are based on only a few studies, each with limited databases,

and the results could reflect, to some extent, the particular habitats and seasons sampled. This is particularly true for the continental shelf and upper slope, where benthic faunal density is known to show considerable variability. For example, in his detailed study of macrofaunal distribution on the continental shelf off Mississippi and Alabama, Harper (1991) repeatedly sampled 12 continental shelf stations at different times of the year and found considerable variability in both the abundance and biomass of the benthic macroinfauna. Abundance values ranged from 561 to 4,908 individuals/m² and biomass densities ranged from 150 to 24,660 mg/m², with most values falling in the range of 5,000 to 9,000 mg/m². Escobar-Briones and Soto (1997) showed that benthic faunal density varied in relation to the proximity of sampling sites to the mouths of large rivers. However, for the open Gulf, data from studies of the phytoplankton, zooplankton, and macrobenthos are all consistent in showing very low floral and faunal densities.

## ADAPTATIONS FOR LIFE IN THE DEEP GULF

During the course of its life history every animal must obtain food, avoid predators, locate a sex partner of the same species, and reproduce. In order to carry out these functions, animals of the deep Gulf have had to develop adaptations that allow them to cope with the special circumstances under which they live. Although the survival value of some of the adaptations is obvious, the significance of others is obscure, and we have only the vaguest clues about the functions and sensitivities of some of the sense organs and other structures found on the bodies of many deep-sea species. In the following sections we will discuss some of the more obvious morphological, physiological, and life history adaptations, but a great deal remains to be learned about how life really goes on in the deep undersea world. Much of the research on adaptations has been conducted elsewhere, but it is applicable to animals of the Gulf, and a brief review of this work will provide a better understanding of life in the deep Gulf. Some general references include Bright (1968), Gage and Tyler (1991), Goode and Bean (1896), and Marshall (1954, 1980).

## Morphological adaptations

The various anatomical adaptations of animals to conditions of the deep-sea environment are most clearly seen in the fishes. Hence, emphasis will be placed on the fishes, but information concerning adaptations of other groups will be brought in, as appropriate.

## COLORATION

In the open ocean there are few hiding places, and for prey and predators alike a premium has been placed on reducing their own visibility, particularly through the use of protective coloration. This is especially true for those animals that live in the well-lighted epipelagic zone, but it also applies to those of at least the upper mesopelagic, as well. In the epipelagic zone some species are nearly transparent, and for most of the others the less visible colors such as blue, silver, and light gray predominate. Furthermore, many of the animals are countershaded, that is, darker toward the top and lighter on the lower portions of the body. Viewed from the side the animal blends with the watery background, viewed from above it is less visible against the darker water below, and seen from beneath it tends to disappear in the bright sunlight coming from the surface. In the mesopelagic, especially the upper mesopelagic, the most common prey fishes (hatchetfishes, lanternfishes, bristlemouths, etc.) tend to be silvery, grayish, light brown, or pale colored, and they are generally countershaded. However, in this zone the countershading is frequently reinforced by the presence of rows of small luminous organs along the lower portions of the body that provide a ventral glow. As studies with photographic plates have shown, this greatly reduces the fish's visibility when viewed from below. On the other hand, many of the predatory fishes of the mesopelagic are dark colored, generally black or brown, and the skin may be somewhat iridescent with glints of gold, copper, or green. This suggests that the predators often attack from beneath, where they are less visible against the darker background. Below the mesopelagic most of the prey and predatory fishes are dark brown to jet black, although a few are silvery. Surprisingly, some of the invertebrates in the mesopelagic and many in the bathypelagic (nemertean worms, some cephalopods, copepods, numerous shrimps, and arrow worms) are scarlet or blood red. Since longer wavelengths of light are largely absorbed in the upper few meters of water, red animals must all appear to be black when viewed in their natural habitats in the deeper water.

## BODY SHAPE

As discussed in the previous chapter, many of the fish species of the epipelagic zone have strong bodies, prominent fins, and streamlined shapes and are built for speed. Many others, while not as fast, at least have the terete body shape of a typical fish. However, deeper in the unlighted waters of the ocean, body forms take on strange and different shapes (fig. 14.5). First of all, there is a general miniaturization of body size (partially contested by Polloni et al. 1979). Most deep-sea fishes, even the fierce-looking predators, often are only a few centimeters long, and rarely do they exceed a meter in length. Second, the body often appears flabby and weakly constructed and has small, often degenerate, fins. Actually, a few of the mesopelagic and bathypelagic fishes (some sharks, lancetfishes, snake mackerels, cutlassfishes, etc.) are larger and/or streamlined, and more powerfully built, but these are rare exceptions. Third, in several groups there is a tendency to develop large heads, very large mouths generally full of fang-like teeth, jaws that can become temporarily unhinged for swallowing very large prey, and stomachs and abdominal walls that are elastic and distensible so that they can accommodate large prey items (which may be as large as or larger than the predator itself). Fishes that live on or near the bottom may have elongate pointed snouts (most prominent in the grenadiers) that may be sensory for locating prey, structural for plowing through the sediments in search of food items, or both. The bodies of many near-bottom fishes (ratfishes, various eels, grenadiers, etc.) also have developed long tails that taper to a point. This adaptation apparently permits such fishes to swim smoothly through the water while greatly reducing turbulence and pressure signals that might be useful to would-be predators.

## BODY PARTS

Among the body parts modified for life in the deep sea, the eyes are particularly important (fig. 14.6). Many of the mesopelagic fishes have enlarged eyes. The pupils may be widened, the lens enlarged and rounded, and the retinas densely packed with rods. All these features serve to increase visual sensitivity under conditions of greatly reduced light. It has been speculated that with such extreme development the eyes of midwater fishes are among the most light sensitive in the animal kingdom and that they may be a hundred or more times as sensitive as those of humans. Above a depth of 500 m they should be capable of forming good images of other animals, but even down to at least 1,000 m they should be able to detect light and silhouettes. Below 1,000 m the eyes of most species are smaller and somewhat degenerate and are probably limited to perception of the light produced by bioluminescent organisms. In extreme cases the lenses have been lost. The eyes are especially reduced in the whalefishes, many deep-sea anglerfishes, some brotulids, and the ip-

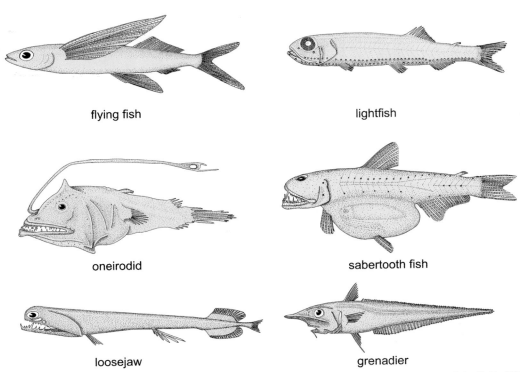

**Figure 14.5.** Some general anatomical adaptations of fishes for life in open-ocean environments of the Gulf of Mexico. Flying fish (*Cheilopogon*) showing typical countershading of epipelagic fishes. Lightfish (*Vinciguerria*) showing photophore-enhanced countershading seen in many mesopelagic species. Oneirodid angler fish (*Oneirodes*) with solid dark body coloration, large head and mouth, and weak body and fins characteristic of some bathypelagic species. Sabertooth fish (*Evermannella*), a deepwater species, with a large squid in its distensible stomach. Loosejaw (*Malacosteus*) showing very large jaws that can become unhinged to swallow oversized prey. Grenadier (*Caelorinchus*) with long, stiff snout, probably used for plowing through bottom sediments for food, and elongated tapering tail to reduce water turbulence while swimming.

nopids, and in the latter group the eyes of some species have been reduced to patches of light-sensitive tissue. However, in some of the deep-sea bottom fishes (notably the grenadiers) the eyes remain large and apparently functional, and since many of the grenadiers have luminous bellies, these fishes may be able to see bottom food organisms illuminated by their own bodies.

Another remarkable development, primarily among the mesopelagic fishes, is elongation of the eyeballs. In most of these tube-eyed fishes (certain hatchetfishes, lightfishes, spookfishes, sabertooth fishes, and pearleyes) the eyes point upward, but in a few, such as some of the deepwater smelts, microstomatids, and giganturids, they point forward. Since the optical axis of both eyes points in the same direction, it appears that tubular-eyed fishes have binocular vision. Furthermore, in these fishes there is often an accessory retina along the inside of the tube, and it is likely that these fishes also have bifocal vision, that is, they can focus on both distant and near fields. If true, these adaptations would be of great advantage in

locating prey at a distance and also viewing it at close range before attack.

The lateral line system of fishes is designed to pick up low-frequency vibrations or pressure waves traveling through the water at the speed of underwater sound (1,500 m/sec, about four and a half times as fast as sound travels in air). In effect, the system is an accessory auditory organ adapted for the reception of vibrations caused by swimming movements of invertebrates and fishes, although it undoubtedly has other functions such as locating stationary objects and the bottom surface for benthic fishes. The system extends along the trunk of the fish from behind the gill cover to the base of the tail fin, and at the front it has branches running around the head (over the eye, under the eye, across the cheek, and along the lower jaw). Typically, it consists of a canal buried beneath the skin and opening to the surface through a series of pores. Sensory hairs located along the canal pick up the vibrations. The system is variously elaborated by different groups of deep-sea fishes. In a few species the canal is not

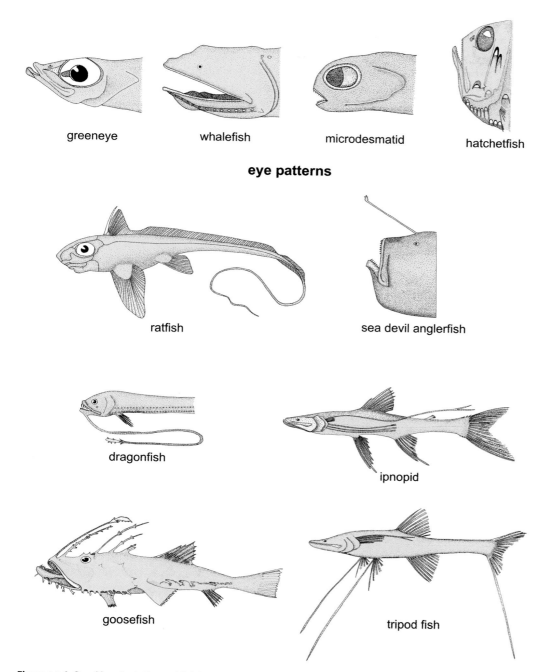

greeneye

whalefish

microdesmatid

hatchetfish

**eye patterns**

ratfish

sea devil anglerfish

dragonfish

ipnopid

goosefish

tripod fish

**Figure 14.6.** Specific adaptations of fish body parts for life in open-ocean environments of the Gulf of Mexico. The top line shows eye modification patterns. The greeneye (*Chlorophthalmus*) has the large eyes characteristic of many mesopelagic predators. The whalefish (*Cetomimus*) displays the degenerate eyes seen in many bathypelagic species. Both the microstomatid (*Xenophthalmichthys*) and hatchetfish (*Argyropelecus*) have tubular eyes, forward-directed in the former and upward-directed in the latter. The ratfish (*Hydrolagus*) externally displays the tract of the lateral line system. The seadevil anglerfish (*Cryptosaras*) displays a lighted lure at the end of a long stalk growing from the top of its head. The black dragonfish (*Idiacanthus*) carries a long whiplike chin barbel with a phosphorescent lure at its tip. In the bathypelagic ipnopid fish (*Bathypterois bigelowi*) some rays of the pectoral fin are greatly elongated and probably serve as feelers. The goosefish (*Lophiodes*) carries on its back several rods and lures derived from anterior rays of the dorsal fin. In the tripod fish (*Bathypterois grallator*) a single ray from each pelvic fin and the caudal fin is elongated and stiffened to provide tripod support and possibly to probe the sediments for food items.

roofed over by skin. In many the pore walls have become extended as papillae of various shapes containing sensory hairs exposed to the water. In a few groups, such as the lanternfishes and melamphaeids, the system is most highly developed in mesopelagic and bathypelagic fishes with poor vision or in places where it is too dark to rely on vision alone. However, the system is poorly developed in some groups, such as many of the anglerfishes, which employ luminous lures to attract prey animals. When any fish swims through the water it produces pressure waves, and somehow it must be able to distinguish between its own vibrations and those produced by other individuals. The best way is to stop swimming or to swim very slowly and listen, and many of the bathypelagic fishes (and squids) appear to hover motionless in the water column much of the time. The general reduction of the fins of many deep-sea fishes may be, in part, a reflection of this waiting type of lifestyle. When another organism is detected, the otherwise sluggish fish is capable of a sudden dash to capture prey or to avoid becoming prey. Exactly how the fish distinguishes food from foe is unclear, but considering its various anatomical modifications, the lateral line system likely provides critical information concerning the nature of the nearby animal.

As discussed in the previous chapter, the fins are used by epipelagic fishes primarily for propulsion and steering, and this continues to be the case for most mesopelagic and bathypelagic species. However, in deeper waters, where rapid swimming ability is not at a premium, some of the fins may be reduced (as in certain species of deep-water smelts, smoothheads, tubeshoulders, bristlemouths, and anglerfishes), or the fins may be put to other uses. In many different families the whole fin or certain fin rays have become greatly elongated and probably serve as feelers. Like the long antennae of shrimps, they must be sensitive to touch and other stimuli. Such modifications can be seen in the dorsal fin (certain viperfishes, ipnopids, grenadiers, goosefishes, and anglerfishes) as well as in the pectoral or pelvic fins (some ipnopids, cusk-eels, codlets, phycids, and grenadiers).

In another remarkable development, the first ray of the dorsal fin has become stiffened and detached from the remainder of the fin and moved forward to the top of the head, where it serves as a fishing rod (*illicium*) with a fleshy lure or bait (*esca*) attached on the end. This bait, which is generally luminous, may be simple, divided, or elaborately branched. As the rod is twitched around by a special set of muscles, the luminous lure dances back and forth emulating a small worm or shrimp, but always within reach of the fish's capacious mouth. Although a few species of anglerfishes inhabit shallow water, they are most diverse and highly modified in the mesopelagic and bathypelagic realms. Some, such as the remarkable goosefishes, live on the bottom.

In yet another development, certain rays of the pectoral and caudal fins have become stiffened and greatly elongated, permitting the fish to stand on the bottom on stilts, as it were. In this position the body of the tripod fish can remain motionless while elevated off the bottom away from benthic surface predators. At the same time the elongated fin rays placed on or within the substrate may be able to sense movements of small worms and crustaceans within the sediments that could represent potential food items. Whether or not the fish actually probes around with these fin rays is uncertain.

From the chin or lower jaw of many deep-sea fishes (certain species of snaggletooths, dragonfishes, beardfishes, cusk-eels, anglerfishes, grenadiers, etc.) there hang one or more fleshy appendages called barbels. In benthic species these are well supplied with sensory cells and are probably used for locating prey by touch and taste. However, barbels also occur on various midwater fishes, where they are often elongated, in extreme cases being several times the length of the body. Here they probably function like the antennae of shrimps and aid in the detection of nearby animals. In some species the chin barbels are equipped with luminescent bulbs and other structures, and these likely serve as lures for prey or as false lures to attract predators away from the body.

In this brief survey we have described some of the more obvious morphological adaptations of the fishes that inhabit the deep Gulf of Mexico, and we have attempted to describe some possible functions and uses of these adaptations. The bodies of some cephalopods and shrimps may also be highly modified, but the functions of the modifications are not so apparent.

## Bioluminescence

Bioluminescence refers to the production of light by living organisms. Although the phenomenon is observed in terrestrial and shallow marine environments and in the epipelagic zone of the sea, nowhere is it as prevalent as in the deeper mesopelagic and bathypelagic realms, where 90 percent or more of the species produce light. In this dark, monotonous environment, glows, sparkles, and flashes play important roles in the lives of most inhabitants. Chemically, the light-producing reaction is mediated by an enzyme (*luciferase*) in which a substrate (*luciferin*) is oxidized by molecular oxygen with the subsequent release of photons. The luciferins and luciferases of the various species may be quite different, so much so that the luciferin of one species may not work with the luciferase of another. Within the open ocean, bioluminescence is produced by various microorganisms and by representatives of most animal phyla. Among the planktonic forms it is seen in bacteria, dinoflagellates, medusae, copepods, pyrosomes, and salps, and in the nektonic species it is most widespread in squids, crustaceans, and fishes. It also occurs in benthic forms such as sponges, soft corals, sea pens, hydroids, starfishes, and sea cucumbers. There is a large literature dealing with marine bioluminescence, most of it relevant to the Gulf of Mexico, but only a limited amount of work has been done in the Gulf itself. Important references include Bennington (1979), Harvey (1952), J. Hastings (1983), Herring (1978), Morin (1983), Tett and Kelly (1973), and R. Young (1983).

Among marine animals light production is accomplished by three basic mechanisms: intracellular, extracellular, and symbiotic. Intracellular light production involves the emission of light by specialized cells (*photocytes*), which may be isolated or grouped together in organs called photophores. Photocyte light production is widespread among marine animals. Extracellular light production entails the manufacture and storage of light-producing chemicals in special sacs or glands. When this material is extruded into seawater it produces a luminous cloud. This method is rare. Among animals of the Gulf it appears to be limited to one species of squid (*Heteroteuthis dispar*), two species of deepwater caridean shrimps (*Oplophorus spinicauda* and *Systellaspis debilis*), and several fishes of the family Platytroctidae (tubeshoulders). In this group of fishes the gland is located on the side of the body behind the gill cover, and it opens to the outside by a prominent tube through which the fluid can be discharged. Symbiotic light production involves a mutualistic relationship between certain species of luminescent bacteria and a host species, generally a deepwater fish. In this arrangement the host provides a glandular sac that houses a colony of luminescent bacteria (primarily species of *Beneckea* or *Photobacterium*), and it also provides nutrients by which the bacteria are nourished. The bacteria in turn produce light, which is useful to the host. Luminescent bacteria are not confined to photophores. Many species are free living in the water column, some

are saprophytes that decompose dead organic matter, some are external parasites on shrimps and other small crustaceans, and still others live in the digestive tracts of fishes.

Whether composed of photocytes or colonies of symbiotic bacteria, the photophores may be very simple or highly complex structures. The luminous cells are generally surrounded by a cup of black-pigmented cells, and in many species the cup is lined by a layer of reflecting cells that intensify the beam of light. In more elaborate organs, there may be a lens to focus the light, an adjustable diaphragm to control the size of the opening through which the light passes, and even a color filter. Sometimes there is a shutter to permit on/off control of light emission. The light itself may be of low or high intensity, and it may be emitted as short flashes (< 2 sec) or as longer glows (> 5 sec duration). Colors of the transmitted light cover the visible spectrum, but the most favored colors are blue, green, yellow, and white. Within a given individual, different colors may be emitted by photophores on different parts of the body. Squids are the real masters of colored light production, and the body of a single individual may have, at the same time, photophores in different body regions glowing vividly with colors such as ultramarine and sky blue, pearly and snowy white, and ruby red. Deep-sea squids clearly communicate with one another by means of colored light patterns, but their conversations are beyond our present understanding.

Luminous organs may be located in almost any position, and in some species of squids they occur all over the body, including the fins and arms. In a few types of fishes (some dragonfishes and viperfishes), upon stimulation the entire body glows pinkish, apparently because of secretions of glands located in bulbs, or *caruncles*, situated in front of the dorsal fin. However, in squids, crustaceans, and fishes the most favored photophore locations are around the eyes and along lower portions of the body. Photophores on or near the head light up the visual field, permitting the animal to see prey or predators. Those located along the lower sides and belly serve primarily as countershading, especially for mesopelagic species. Animals that employ bioluminescent countershading generally possess photoreceptors capable of providing accurate readings of the ambient light to which the animals must constantly adjust their own luminous output. Other favored locations for photophores are on the escas (lures) of anglerfishes and on the barbels of the various seadevils and their relatives. Some of the locations and patterns of

**Table 14.1.** *Behavioral functions of bioluminescence in marine organisms.*
*(Modified from Morin 1983.)*

| Behavioral types and functions | Signal |
| --- | --- |
| **Predator evasion** | |
| Camouflage effect | |
| Counterillumination – uniform ventral glow | concealing glow |
| Disruptive illumination – mottled ventral glow | concealing glow |
| Flashbulb effect – temporary blinding of predator | contact flash |
| Decoy effect | |
| Deluding cloud – release of chemical secretions | attracting glow |
| Sacrificial lure – release of luminous structure | attracting glow |
| Danger signal – warning that prey is dangerous | contact flash |
| **Prey capture** | |
| Camouflage effect – to conceal predator | concealing glow |
| Flashlight effect – to illuminate prey | flash or glow |
| Lure effect – to attract prey | attracting glow |
| Stun effect – to stupefy prey | contact flash |
| **Intraspecific communication** | |
| Spacing | |
| Aggregation – attraction for groups | flash or glow |
| Territoriality – to maintain distance | flash or glow |
| Reproduction – for courtship and mating | flash or glow |
| Complex signals to language | complicated patterns |

luminescent structures in fishes are shown in figures 14.5 and 14.6.

The roles played by bioluminescent structures in the lives of deep-sea animals have been the subject of much study and speculation, and they boil down to three basic functions: avoidance of predation, location and apprehension of prey, and intraspecific communication (table 14.1). As discussed earlier, to avoid predation many mesopelagic species have developed patterns of countershading illumination. This is generally manifested by rows of tiny photophores along the lower sides of the body, and their luminous output can often be adjusted to match ambient light levels. Another method of predator avoidance is based on the "startle effect." When a predator gets too close, the prey animal suddenly emits a very bright flash or releases a luminous cloud, temporarily confusing or blinding the predator while the prey darts away into the darkness. Some prey species have luminescent structures on long fin rays or barbels that might serve as decoys to attract predators to the distantly placed light organ and away from the prey's body. It has also been speculated that in some cases bioluminescence can serve as a warning signal to notify predators that the potential prey animal is dangerous and is better left alone.

Mechanisms for the location and apprehension of prey are more straightforward. Luminous, sometimes flashing, lures that mimic the behavior of small worms and crustaceans are employed by anglerfishes and others to attract prey within striking distance. Some lanternfishes have small photophores within their mouth that may serve the same purpose, and in some viperfishes (*Thaumatichthys*) a fleshy, wormlike, luminous bait hangs from the roof of the large mouth. A widespread method of locating potential food involves lighting up the visual field with one or more photophores on the head. Often one of these provides a powerful beam and serves as a spotlight that not only aids in finding prey but, when suddenly turned on full blast, can also blind and stun the prey just before the predator strikes. Luminous glows on the chin or abdomen of bottom-dwelling fishes may serve to light the feeding area of surface sediments.

Finally, it is noted that for all deep-sea species the arrangement of photophores along the body is constant, and these unique patterns undoubtedly provide a means for members of a given species to recognize one another. Species recognition could aid in schooling or, as in the case of some crustaceans, in the formation of breeding swarms. If a prey animal flashes a light to startle a predator, the sudden flash could send a warning signal for other members of the species to scatter. Luminous organs must also aid in sex recognition. In lanternfishes the arrangement of platelike photophores near the base of the tail fin is distinct in the two sexes, and the cheek light of some species of dragonfishes is very large in males and small or absent in females. Thus, photophores probably transmit information for species recognition, sex recognition, and warning of predators. How much more information is communicated among members of the same species by the arrangement, flashing, and coloration of the lights can only be guessed.

## Physiological adaptations

In moving from shallow to deeper waters of the sea, animals have faced the necessity of making biochemical and physiological adjustments to the conditions that exist at great depths, most particularly to low temperature, high pressure, and greatly reduced food resources. The nature of these adjustments is discussed below, but since the relevant studies are still in their infancy the generalizations must be accepted with some caution. Good summaries of the existing information have been published by Heremans (1982), Hochachka and Somero (1973), Jaenicke (1981), A. MacDonald (1975), Somero (1982), and Somero, Siebenaller, and Hochachka (1983).

The lowest temperature encountered by animals of the deep Gulf is around 4.0°C, but shallow-water marine animals of polar regions must contend with even lower temperatures. It is a well-known physiological principle that for every rise of 10°C the rate of biochemical reactions (and physiological processes in general) increases by a factor of about two. Conversely, a drop of 10°C reduces the velocity of such reactions by about half. Enzyme systems capable of functioning fairly efficiently at low temperatures have been developed by many high-latitude marine organisms, and if these animals invaded the deep sea at

high latitudes their enzyme systems would already be adjusted to the temperatures encountered at great depths. Still, considering a subtropical marine environment such as the Gulf Mexico, where the temperature differential between surface and deep waters may be 15°C or more, the rates of enzyme-mediated reactions in deepwater animals should be less than half of those occurring in organisms near the surface. Within the animal body, chains of chemical reactions are linked in tandem, and to maintain the smooth functioning of the overall system these various reactions must be carefully coordinated. At low ambient temperatures the price of this regulatory control is a general reduction in metabolic rate.

High pressure poses different problems. As discussed earlier, the hydrostatic pressure of seawater rises about 1 atm for every 10 m increase in depth, and at the bottom of the Gulf the pressure stands at about 380 atm. Increased pressure can affect the structure of proteins, the binding properties of various chemicals (such as enzymes and substrates), and the volume characteristics of the reaction products. Therefore, many enzyme systems that are very efficient in shallow water lose this efficiency under high pressure. Even modest pressure increases of, say, 50–100 atm are sufficiently perturbing to favor the adoption of (a) enzymes specifically adapted to function under conditions of high pressure or (b) pressure-insensitive enzymes that work fairly well over a wide range of pressures. In either case, these enzyme systems also appear to be less efficient than those adapted to the low-pressure environment of surface waters.

Both temperature and pressure affect the viscosity of lipids. At low temperature and high pressure, lipids tend to become more solid and may approach a semicrystalline state. For deepwater animals this poses several problems. The phospholipids of cell membranes are of particular concern because if they are too greatly modified the enzyme-mediated transfer of materials across the membranes could be impaired. The central nervous system also could be affected by the breakdown of the electrochemical processes involved in the transmission of nerve impulses. Likewise, modification of the lipids could alter the buoyancy of marine animals, and hence their ability to remain at the proper depth without expending extra energy.

These problems are compounded for species that undergo regular vertical migrations of hundreds of meters. Such animals must be able to function under a wide range of pressures and a narrower, but still significant, range of temperatures. One means of handling this situation would be to maintain two sets of enzymes, one for the low pressure and high temperature of shallow waters and the other for the high pressure and low temperature of deep waters. Another method would be to rely on temperature- and pressure-insensitive enzymes that can function over the entire range of temperatures and pressures. In either case, there is a price to pay. The result is a more costly and less efficient system. The same problem is faced by a species in which the young live in shallow surface water and the adults live in deep water. In such species there must again be two sets of enzymes or one set of inefficient enzymes acting over a wide range of conditions.

Animals that live at great depths dwell in an environment where there are few food resources, and those that are present are, gram for gram, lower in average caloric content and less nutrient-rich than food resources of near-surface waters. As a result, animals of the deep sea have lowered body protein and lipid contents and elevated water contents, and so do their food resources. For these reasons they also have reduced levels of enzymatic activity and lower metabolic rates than do their counterparts in surface waters. For deep-sea predators there are two basic strategies: "float and wait" versus vigorous searching for food. The "float and wait" strategy is employed by anglerfishes and many others that possess lures and other devices for attracting prey. These species tend to have especially low levels of enzymatic activity and low metabolic rates. The more active predators such as grenadiers have higher muscle enzymatic activities and metabolic rates, but these rates are still well below those for shallow-water predators.

In this brief review we have touched on some of the physiological problems confronted by animals that live in the relatively inhospitable environment of the deep sea. Compared to the warm, low-pressure, high–food resource surface waters where life proceeds at a fast pace, life in the depths operates near the borderline where biochemical inefficiency and reduced metabolic rates are the rule. However, these are not insurmountable drawbacks. If the predator is inefficient by shallow-water standards, so is the prey, and life can still carry on at its own reduced pace.

## Life history adaptations

Life history patterns of oceanic animals have been discussed by many investigators, including W. Ernst and Morin (1982), Marshall (1954, 1980), and C. Young and Eckelbarger (1994), among others. As we have seen earlier, many marine species pass their entire lives in the epipelagic waters, but for many other planktonic and nektonic groups this region is habitat for the larval stages only, the adults being found at greater depths. Here in the epipelagic, the food supply is most concentrated, and much of it is present in the small particle sizes most suitable for ingestion by smaller animals such as the larval and juvenile stages of invertebrates and fishes. Among species of the mesopelagic zone, a large percentage undergo regular vertical migrations, moving up at dusk to feed in the epipelagic and back down in the early morning hours. Some inhabitants of the upper bathypelagic also carry out regular vertical migrations of hundreds of meters, moving up into the mesopelagic or epipelagic to feed, and even a few of the middle and lower bathypelagic species carry out vertical migrations. Although most of the benthic and near-benthic animals appear to be nonmigrators, some species of invertebrates and fishes do move up into the water column to feed.

With respect to reproduction and larval life, among open-ocean species there appear to be two basic life history patterns (table 14.2). In the first group, regardless of the habitat of the adults, the larval and juvenile stages are passed in the epipelagic zone, and in the second group the developmental stages are not epipelagic. Many species whose adults are resident in the mesopelagic and some from the bathypelagic, including both vertical migrators and nonmigrators, have larval stages that develop in the epipelagic. In some cases the females may move up and deposit their eggs in near-surface waters, but for most species of cephalopods, shrimps, and fishes the females appear to spawn at depth and release eggs containing oil globules. These eggs then rise to the shallower waters where they hatch. In the case of crustaceans, most species must pass through a series of molts and complex larval stages. After a period of feeding and growth the young graduate to the adolescent or preadult stage, in which there is a sudden and rapid descent to the depths occupied by the adults. Although some bathypelagic species have epipelagic larval stages, many do not, and the same may be said for benthic animals of the deep sea. Such long vertical migrations for tiny young and subadults at a time when they are especially vulnerable to predation seem to rule out this possibility for many species. In some deep-sea crustacean groups, such as the mysidacean and caridean shrimps and brachyuran crabs, the female carries

**Table 14.2.** *Life history patterns of representative deep-sea animals resident in the Gulf of Mexico. Note: The life histories of most deepwater species are largely unknown.*

**Pattern A** – Eggs produced (generally at depth), larvae develop in epipelagic waters.
 – Many deep-sea cephalopods including various squids and certain octopods (*Japetella*, etc.)
 – Many crustaceans (some copepods, many euphausiids, and sergestid shrimps)
 – Most midwater fishes (bristlemouths, hatchetfishes, lightfishes, viperfishes, dragonfishes, etc.)
 – Some bathypelagic and benthic fishes (some deepwater eels, dragonfishes, and anglerfishes)
**Pattern B** – Early life history passed in deep waters with adults.
 – Some cephalopods (*Vampyroteuthis*, etc.)
 – Caridean shrimps (*Systellaspis*, etc.), some euphausiid shrimps (*Bentheuphausia*, etc.), brachyuran crabs
 – Some deepwater fishes (brotulids, etc.)

the eggs, and the larval stages are abbreviated and passed while still within the egg capsule. When they hatch, they are already subadults. In other deepwater species the eggs are not carried by the female, but when released they contain a large amount of yolk, permitting them to bypass the free-living larval stages. However, some deepwater euphausiid shrimps and fishes (deepwater eels) produce eggs that develop into planktonic larvae that remain in deep water, where they somehow manage to find sufficient food resources.

The advantages of vertical migration for adult animals are clear enough. Since the epipelagic waters provide the most concentrated and nutrient-rich food resources of the sea, it is worth making the daily trip to feed in the best pastures even if energy is expended in getting there and back and even if predation rates are higher during the migration. However, there are additional reasons for having the larval and juvenile stages develop in the epipelagic zone. These smaller individuals require food in very tiny packages such as can be supplied by individual phytoplankton cells, protozoans, and other microplanktonic organisms of the epipelagic, and it helps if the young and adults are not competing for the same food resources. Also, it is of great advantage if the young can develop and mature at a depth where they are not subject to

intense predation pressure by adults of the same species. Frequently, but not always, species with planktonic larvae must produce very large numbers of eggs so that enough will survive predation in the plankton to parent the next generation. This can be accomplished either by releasing large numbers of eggs at infrequent spawnings or by producing smaller numbers of eggs per clutch at more frequent intervals. Both methods are employed.

In order to reproduce it is necessary to locate a partner of the same species but of the opposite sex. How this is accomplished by deep-sea animals generally is not known. It has already been noted that in some species of deep-sea fishes there are anatomical differences between the sexes, but in the darkness such differences must be hard to detect. Sexual differences in bioluminescence patterns or signals would certainly be useful in this regard. Although most of the senses may be involved in one way or another, it seems likely that in most species males are attracted to the appropriate partners by specific chemicals, or *pheromones*, released into the water by the adult females.

Since animals of the deep sea are widely dispersed, the chances of finding an appropriate breeding partner are greatly diminished, and some deep-sea animals have developed special adaptations for increasing the odds. In several families of mesopelagic and bathypelagic fishes the adults are *synchronous hermaphrodites*, that is, they possess both male and female functional sex organs at the same time. Although fertilization of one's own eggs is probably not widely practiced, the arrangement does mean that any encounter among adults of the same species could result in production of fertile eggs by one or both individuals. Synchronous hermaphroditism is encountered in a number of families of Gulf of Mexico fishes including bristlemouths, notosudids, pearleyes, sabertooth fishes, and barracudinas.

In several groups of deep-sea anglerfishes the female grows to a large size (half a meter or more) and has a well-developed rod and lure, which is often luminous. The tiny free-living males (generally an inch or less in length) lack the rod and lure and often lack a functional digestive tract, but they have large eyes to aid in locating a female. Once the encounter is made, the male attaches himself to her body by biting her skin. Eventually, the tissues fuse, and the male spends the rest of his life as a sexual parasite on the body of the female. The eyes are lost, and some other organs degenerate, but the testes enlarge (fig. 14.7). A given adult female may carry more than one parasitic male. Although this arrangement may seem strange, it

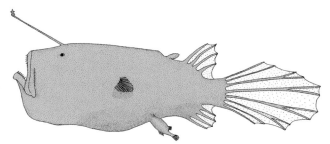

**Figure 14.7. Female seadevil anglerfish (*Cryptosaras*) with small parasitic adult male attached to the lower flank.**

means that during its entire life a male must encounter only one female of the same species, and thereafter he is available to fertilize the eggs, a helpful arrangement in a vast, dark world where encounters are rare.

### PLANKTON AND NEKTON OF THE OPEN GULF

In dealing with animals of the open water column, it is convenient to treat the plankton and nekton together. In nature these groups are not sharply divided. Rather, there is a continuous gradient of size classes from the microscopic forms all the way up to the great sharks and whales, and a large percentage of the animals fall right into the micronekton category, which straddles the fuzzy dividing line between plankton and nekton. Furthermore, in their life histories many animals pass from tiny planktonic larval and juvenile stages to nektonic adults, and the literature is not always clear about which sizes and stages an investigator has encountered.

Before discussing the fauna of the open Gulf, it is important to note some of the limitations of the existing data. Until recently most open-ocean collections were made with nonclosing nets, and samples were often taken by vertical or oblique tows over wide depth ranges. Even when such nets were towed at discrete depths, the catch was often contaminated by specimens from shallower levels that entered during the lowering or raising of the nets. Fortunately, most of the more recent studies on the Gulf have been carried out with nets that open and close and can sample discrete depths with little or no contamination from other depths. In many investigations samples were taken whenever a ship reached a predetermined collecting site, and comparative day and night samples were seldom taken at the same location. Several types of collecting gear and mesh sizes have been employed, thus confounding efforts to compare the results of different investigators. There is also the problem of possible net avoidance. Although small planktonic animals probably cannot escape the collecting nets, many of the faster-swimming nektonic species clearly do avoid capture by slow-moving towed nets, and this avoidance is likely greater during the daytime, when the nets can be seen, than at night. As a result, there is a definite, but often indeterminate, bias in samples of larger fast-swimming animals such as some cephalopods and fishes, and this is also likely the case for some of the intermediate-sized euphausiid, penaeid, and caridean shrimps. Finally, it is not always clear whether a given species lives on or near the bottom or well above it. This is particularly true for some of the deepwater cephalopods, shrimps, and fishes, which may, in fact, spend part of the time associated with the bottom and part up in the water column. For the reasons given above, much of our knowledge of the depth distribution and vertical migration patterns of zooplankton and nekton of open Gulf waters must be considered somewhat tentative.

### Composition of the zooplanktonic and nektonic invertebrates

Although much work remains to be done, we already have a fairly good idea of the species composition of most zooplankton and nekton groups inhabiting the deep Gulf. Clearly, the open-water fauna is closely related to that of the Caribbean and western Atlantic and, in a broader sense, to the subtropical open-water fauna of the world oceans in general. A few species of copepods and some other groups are known only from the Gulf, but it is too early to conclude that any of these open-water species are truly endemic.

In previous chapters dealing with zooplankton and nekton, some attention was given to the epipelagic species, but here our concern is the fauna of the total vertical water column. A general classification of this fauna is provided in table 14.3, and representatives of several groups are shown in figure 14.8. With the exception of some of the Protozoa, Ostracoda, and certain minor groups in which the pelagic forms appear not to have been well documented, we will address each of the groups in the following discussion. Species lists are provided for most, and where possible some information on depth distribution is presented. Much of the available information exists in sources difficult to obtain, and it has never been brought together in the open technical literature.

**Table 14.3.** *Higher classification categories for the major groups of zooplankton and micronekton (exclusive of larval forms) inhabiting open waters of the Gulf of Mexico.*

| | |
|---|---|
| Phylum | Protozoa – protozoans |
| Class | Ciliophora – ciliates |
| Class | Sarcodina – sarcodinids (foraminiferans) |
| Phylum | Cnidaria – coelenterates |
| Class | Hydrozoa – hydrozoans |
| Order | Hydroida |
| Suborder | Anthomedusae |
| Suborder | Hydromedusae |
| Order | Siphonophora – siphonophores |
| Class | Scyphozoa – scyphozoans |
| Phylum | Mollusca – mollusks |
| Class | Gastropoda – gastropods |
| Order | Pteropoda – pteropods |
| Order | Mesogastropoda – mesogastropods |
| Superfamily | Heteropoda (Atlantacea) – heteropods |
| Class | Cephalopoda – cephalopods |
| Order | Decapoda – decapods |
| Suborder | Sepioidea – cuttlefishes |
| Suborder | Teuthoidea – squids |
| Order | Vampyromorpha – vampire squid |
| Order | Octopoda – octopuses |
| Phylum | Annelida – annelid worms |
| Class | Polychaeta – polychaetes |
| Phylum | Arthropoda – arthropods |
| Subphylum | Mandibulata – mandibulates |
| Class | Crustacea – crustaceans |
| Subclass | Ostracoda – ostracods |
| Subclass | Copepoda – copepods |
| Subclass | Malacostraca – malacostracans |
| Order | Mysidacea – mysid shrimps |
| Order | Amphipoda – amphipods |
| Order | Euphausiacea – euphausiid shrimps |
| Order | Decapoda – decapods |
| Suborder | Natantia – swimming shrimps |
| Section | Penaeidea – penaeoid shrimps |
| Family | Penaeidae – penaeid shrimps |
| Family | Sergestidae – sergestid shrimps |
| Section | Caridea – caridean shrimps |
| Phylum | Chaetognatha – chaetognaths |
| Phylum | Chordata – chordates |
| Subphylum | Urochordata – urochordates (tunicates) |
| Subphylum | Vertebrata – vertebrates |
| Class | Chondrichthyes – cartilaginous fishes |
| Class | Osteichthyes – bony fishes |

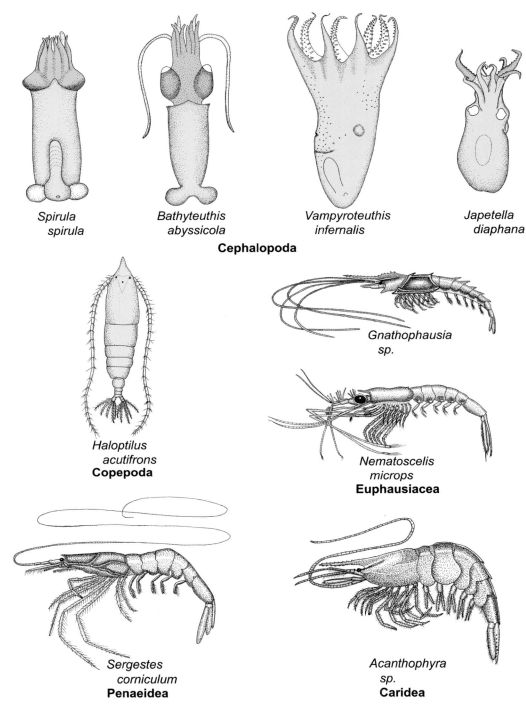

**Figure 14.8.**
Representative examples of zooplankton and smaller nektonic animals of the open Gulf. Included are four cephalopods (*Spirula spirula*, *Bathyteuthis abyssicola*, *Vampyroteuthis infernalis*, and *Japetella diaphana*), a copepod (*Haloptilus acutifrons*), a mysid shrimp (*Gnathophausia* sp.), a euphausiid shrimp (*Nematoscelis microps*), a penaeid shrimp (*Sergestes corniculum*), and a caridean shrimp (*Acanthophyra* sp.).

*Spirula spirula*

*Bathyteuthis abyssicola*

*Vampyroteuthis infernalis*

*Japetella diaphana*

**Cephalopoda**

*Haloptilus acutifrons*
**Copepoda**

*Gnathophausia sp.*

*Nematoscelis microps*
**Euphausiacea**

*Sergestes corniculum*
**Penaeidea**

*Acanthophyra sp.*
**Caridea**

FORAMINIFERA

Planktonic foraminifera of the southeastern Gulf were reported by Jones (1968), who identified 13 species representing 8 genera (table S58). Within the upper 650 m each species showed a special depth preference, and each was shown to be associated with a particular water mass.

Snyder (1978) identified 36 species of planktonic foraminifera in surface sediments of the deep Gulf. Most represented tropical or subtropical species, but a few temperate and cold-water species appeared in sediments of the northern Gulf.

## CNIDARIA

Pelagic cnidarians of open waters of the Gulf of Mexico have been reported by P. Phillips (1972), and a list of the species and their depth distributions, as reported by Phillips, is given in table S59. Among the 83 species recognized, 10 are scyphozoans, 29 are hydromedusae, and 44 are siphonophores. Of the total, 57 species are exclusively epipelagic, 13 occur only in deeper waters, and the remaining 13 species are found in both zones. Biggs, Bidigare, and Smith (1981) and Biggs, Smith, et al. (1984) provided population density estimates of gelatinous forms in surface waters of the open Gulf, and D. Smith (1982) discussed the abundance and foraging abilities of some siphonophores.

## GASTROPODA

The pelagic pteropods and heteropods of the open Gulf have been reported by Hughes (1968), Snider (1975), D. Taylor (1969), and D. Taylor and Berner (1970), and these are listed in table S60. Of the 45 species, 29 are pteropods, and 16 are heteropods. Although the collections were made throughout the epipelagic and mesopelagic zones, reliable depth distribution data are not available. The pelagic gastropods are not a very diverse group, but some of the individual species are fairly abundant.

## CEPHALOPODA

The cephalopod fauna of the open Gulf has been reported by a number of investigators, most notably Lipka (1970, 1975), Nessis (1975, 1987), Passarella (1990), Passarella and Hopkins (1991), Salcedo-Vargas (1991), G. Voss (1954, 1956), and N. Voss et al. (1998). In his 1991 report Salcedo-Vargas listed 71 identified species of cephalopods from the Gulf as well as unidentified species belonging to 14 genera. Although most are open-water species, some are neritic, while others are associated primarily with benthic environments of the continental slope and abyssal plain. Table S61 lists those species that appear to be associated primarily with epipelagic and mesopelagic waters of the open Gulf. Of the 57 species given, 4 are sepioids, 47 are teuthoids, 1 is a vampyromorph, and 5 are octopods. Depth distributions are given for most species based on the best available information, but since many of the records certainly or probably refer to juveniles, the list must be considered tentative. A total of 20 species is listed as epipelagic only, and 28 species are given as inhabiting both the epipelagic and mesopelagic zones.

Bathypelagic species include 7 teuthoids (*Bathyteuthis abyssicola, Mastigoteuthis glaucopsis, M. grimaldi, Joubiniteuthis portieri, Cycloteuthis sirventi, Grimalditeuthis bomplandi,* and *Grimpoteuthis* sp.) and 1 vampyromorph (*Vampyroteuthis infernalis*). The latter species is primarily bathypelagic, but it has also been reported from mesopelagic depths in the Gulf. *Grimalditeuthis bomplandi* may be largely benthic in the deep Gulf. As noted later, a number of the cephalopods are strong vertical migrators, and as a group the cephalopods play important roles in the oceanic food chains.

## POLYCHAETA

The pelagic polychaetes of the open Gulf have received little attention. In his studies of the neuston, Berkowitz (1976) identified 10 species and noted the presence of representatives of four additional genera (table S62).

## COPEPODA

Copepods are the most numerous and taxonomically diverse of all zooplankton groups inhabiting tropical and subtropical waters of the world oceans. At least 170 species are known from the open Gulf even though the surface waters have not been examined extensively, and only a few studies have addressed the deepwater fauna. Among the most important references are Bennett and Hopkins (1989), Berkowitz (1976), Cummings (1982), Ferrari (1973), Fleminger (1956), González (1957), Grice (1969), Park (1970, 1975a, 1975b), and Schmitt (1954). Most of the studies have dealt only with calanoid copepods, but Berkowitz (1976) reported 8 species of cyclopoids and 5 species of harpacticoids, as well as 13 species of calanoids. Fleminger (1956) identified 97 species of calanoids, and Cummings (1982) listed 88 species, mostly from epipelagic waters. In the most thorough investigation of vertical zonation, Park (1970) reported 97 species, 28 of which were new to science, and only 21 of which were also on the list of Cummings (1982). Table S63 lists the calanoid copepod species encountered by Park (1970) and provides information on the depth zones at which each species occurred. Note that the mesopelagic is divided into upper and lower subzones. Most copepod species tend to be rather depth specific. In this study the epipelagic was poorly represented by samples; nearly half the species (47.4 percent) were encountered in the bathypelagic, and nearly a third (32.9 percent) were found exclusively in this depth zone. The data of Cummings (1982) reveal that the average density of calanoid copepods in the epipelagic was around 1,619 individuals/100 m³ of water, but Park (1970) reported that in the lower mesopelagic and bathypelagic the density was only about 6–35 individuals/100 m³. Many of the copepod species are diurnal migrators, being found in the mesopelagic during the day and in the epipelagic at night. Because of their small size, great abundance, rapid reproduction, and widespread distribution, the copepods as a group are extremely important components of food chains at all depths of the sea.

## MYSIDACEA

The known mysid shrimp fauna of the open Gulf of Mexico, consisting of 17 species belonging to 4 families, is given in table S64. The list is based on reports by T. L. Hopkins and Lancraft (1984), Springer and Bullis (1956), Tattersall (1951), and an updated but unpublished list (L. Pequegnat 2000). This list includes identifications of specimens in the oceanographic collections at Texas A&M University. An additional species (*Chalaraspidium alatum*), not included in the table, was reported by Springer and Bullis (1956) with a question mark. It is not otherwise known from the Gulf. Since most of the specimens were taken by nonclosing nets, depth distribution patterns are not given. However, all species of the family Mysidae were collected at or near the surface. Species of the remaining families generally inhabit deep water, although some are known to undergo vertical migrations.

## AMPHIPODA

The pelagic amphipod fauna of the Gulf has received little attention. Table S65 lists 7 species and an additional 8 genera given in Berkowitz (1976) and Morée (1979). All specimens were collected in the epipelagic zone.

## EUPHAUSIACEA

The euphausiid shrimps of the Gulf have been treated by a number of authors, most thoroughly by Hopkins and Lancraft (1984), James (1966, 1970), W. Pequegnat, L. Pequegnat, et al. (1971), and Schroeder (1971). Of the 90-odd described species, 35 have been recorded from the Gulf of Mexico (table S66). Information concerning depth distribution patterns is incomplete, but 17 species are definitely known to inhabit the epipelagic, 24 have been taken in the mesopelagic, and 2 in the bathypelagic. Many species carry out vertical migrations, but some apparently do not. Mauchline and Fisher (1969) discussed the important roles played by euphausiid shrimps in marine food chains.

## PENAEIDEA

In the open Gulf shrimps of the section Penaeidea are represented by two families, the Penaeidae and Sergestidae. Although most members of the former family are benthic, several species are considered to live up in the water column. The known pelagic species of the two families inhabiting the Gulf of Mexico are listed in table S67, which is based on data given in Berkowitz (1976), Hopkins and Lancraft (1984), W. Pequegnat, L. Pequegnat, et al. (1971), T. Roberts (1970), and T. Roberts and W. Pequegnat (1970), a list of the deepwater members of the family Penaeidae (L. Pequegnat 2000), and specimens in the oceanographic collections of Texas A&M University. Included in the table are 6 species of the family Penaeidae and 14 species of the family Sergestidae. Depth distribution data are not available for most species, but some are known to make regular daily vertical migrations. Information on the vertical distribution and feeding ecology of some penaeid and sergestid shrimp species is provided by M. Flock (1989) and Hefferman and Hopkins (1981).

## CARIDEA

Species of caridean shrimps recorded from open waters of the Gulf of Mexico are given in table S68. This information derives from the reports of Hopkins and Lancraft (1984), T. L. Hopkins, Gartner, and Flock (1989), L. Pequegnat and Wicksten (2006), and W. Pequegnat, L. Pequegnat, et al. (1971), and a list provided by Dr. L. Pequegnat (2000) of specimens present in the oceanographic collections at Texas A&M University. Included are 24 species representing 4 families: Bresiliidae (1), Oplophoridae (18), Pandalidae (3), and Pasiphaeidae (2). Several of these families are also represented in the benthic fauna. Depth distribution patterns, based primarily on data provided in T. L. Hopkins, Gartner, and Flock (1989), are also given in the table. Here it is seen that 1 species is known from the epipelagic, 15 from the mesopelagic, and 23 from the bathypelagic. Three species (*Acanthephyra purpurea, Parapandalus richardi,* and *Systellaspis debilis*) together make up a large fraction of the catch. All three are strong vertical migrators, and at least a dozen other species show at least a weak tendency to migrate. Some, such as *S. debilis,* appear to breed throughout the year, whereas others, including *P. richardi,* are seasonal breeders. As in the case of the other shrimps, the pelagic carideans of the Gulf are part of the broader western Atlantic marine fauna (W. Pequegnat, L. Pequegnat, et al. 1971).

### CHAETOGNATHA

Pelagic chaetognaths of the deep Gulf have been reported by Every (1968), McLelland (1984, 1989), McLelland and Perry (1989), and Michel (1984), and the known species are listed in table S69. These include 24 species representing 11 genera. Twelve species inhabit the epipelagic, 8 the mesopelagic, and 7 the bathypelagic zone.

## Composition of the zooplanktonic and nektonic fishes

Our knowledge of the ichthyofauna of open waters of the Gulf stems largely from the catch by midwater trawls, the commercial long-line fishery, the sport fishery, and, for the larval and smaller fishes, the catch in towed plankton nets. As noted earlier, the faster-swimming fishes of the open waters easily avoid towed nets, and information on their occurrence is rare and based largely on juveniles and the hook-and-line catch of adults. Actually, hundreds of species of strange and often little-known fishes inhabit open waters of the deep Gulf of Mexico, and for thorough discussions of each of the known species the reader is referred to the two-volume treatise by McEachran and Fechhelm (1998, 2005) and references contained therein. Omitting some of the less important groups, we will deal here with 56 families representing 13 orders of open Gulf fishes (table S70). Brief discussions of each of these families are provided in table S71, and representatives of each family are shown in figure 14.9a–d. The 13 orders, some of which were discussed in the previous chapter, are taken up individually below.

### LAMNIFORMES

This order, which includes most of the common sharks, was discussed in the previous chapter. In the open Gulf it is represented by 10 families, of which 7 are listed here. Some, such as the whale and basking sharks, are plankton feeders. These and the hammerheads appear to be limited to the epipelagic. Most of the cat sharks are found in mesopelagic and bathypelagic waters. The thresher, mackerel, and requiem sharks are primarily epipelagic, but some species range down into the mesopelagic. Many also forage near the bottom. About 30 species are known from the Gulf. The most important families include the

Rhincodontidae, Alopiidae, Cetorhinidae, Lamnidae, Scyliorhinidae, Carcharhinidae, and Sphyrnidae.

### SQUALIFORMES

The squaliform sharks were taken up in the previous chapter. They are characterized by the absence of an anal fin. In the Gulf they are represented by two families, only one of which, the dogfish sharks (Squalidae), is considered here. They occur from the shallow to the deepest water, and some species frequent the bottom. About 20 species have been reported from the Gulf.

### RAJIFORMES

Most of the skates and rays are bottom dwellers, but there is one pelagic stingray (*Pteroplatytrygon violacea*). Two groups of rays, the eagle rays (Myliobatidae) and mantas (Mobulidae), although most frequently encountered over the continental shelf, also venture into epipelagic waters of the open Gulf.

### OSMERIFORMES

The osmeriform or smeltlike fishes were not considered in the previous chapter. In most species of this group, the maxillary bone is included in the gape of the mouth, and two of the skull bones (*basisphenoid* and *orbitosphenoid*) are missing. The pectoral fins are abdominal or set low on the sides of the body, and the pelvic fins are always abdominal and set far back along the belly. The dorsal adipose fin may be present or absent. These fishes inhabit mesopelagic or bathypelagic waters, and some species appear to be associated with the bottom. Six major families are represented in the Gulf: Argentinidae, Bathylagidae, Microstomatidae, Opisthoproctidae, Alepocephalidae, and Platytroctidae.

### STOMIIFORMES

The stomiiform fishes (bristlemouths, hatchetfishes, etc.) were not discussed in the previous chapter. In this group both the premaxilla and maxilla are included in the gape of the mouth, and both bones bear teeth. Photophores are present. In most species the mouth is large and extends back past the eyes. Chin barbels are present in some species. The pectoral, dorsal, or adipose fins are absent in some species, and a ventral adipose fin is present in a few. Most species are dark brown or black, but some are silvery. Fishes of this group may be found at all depths, and some, particularly the bristlemouths, hatchetfishes,

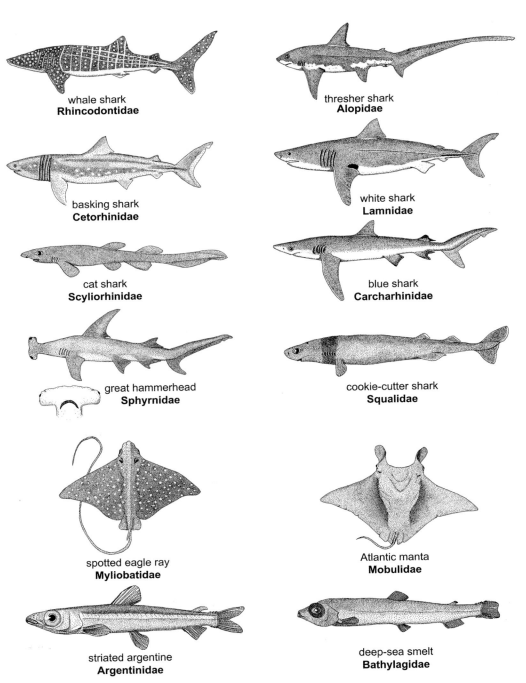

**Figure 14.9a–d.**
Representative examples of important families of fishes inhabiting the open water column of the deep Gulf of Mexico.

whale shark
**Rhincodontidae**

thresher shark
**Alopidae**

basking shark
**Cetorhinidae**

white shark
**Lamnidae**

cat shark
**Scyliorhinidae**

blue shark
**Carcharhinidae**

great hammerhead
**Sphyrnidae**

cookie-cutter shark
**Squalidae**

spotted eagle ray
**Myliobatidae**

Atlantic manta
**Mobulidae**

striated argentine
**Argentinidae**

deep-sea smelt
**Bathylagidae**

and snaggletooths, are extremely abundant in offshore waters. Most are vertical migrators. The order is represented in the open Gulf by 9 families: the Gonostomatidae, Sternoptychidae, Phosichthyidae, Astronesthidae, Chauliodontidae, Idiacanthidae, Malacosteidae, Melanostomiidae, and Stomiidae.

AULOPIFORMES

The aulopiform fishes were discussed in the previous chapter. They all exhibit unusual internal structures associated with the second and third gill arches. Tubular eyes and luminous tissues are present in some species. Only a single family has representatives on the continental shelf, but 8 families are represented in the open Gulf, including the Giganturidae, Chlorophthalmidae, Ipnopidae, Noto-

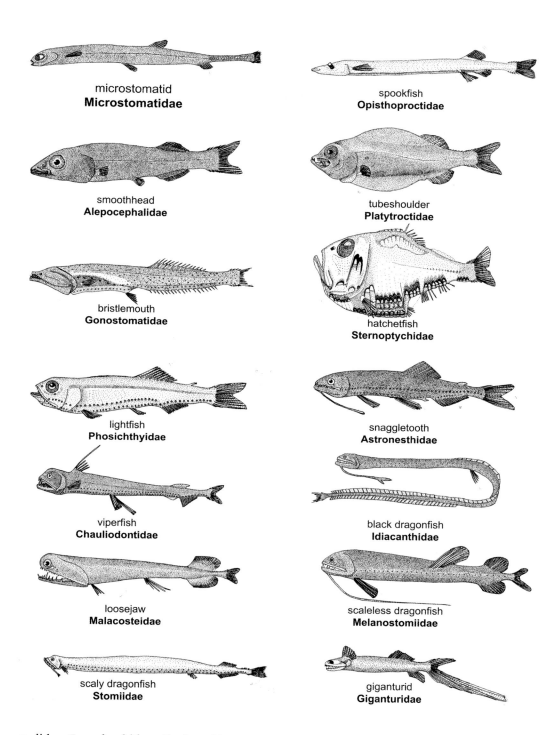

microstomatid
**Microstomatidae**

spookfish
**Opisthoproctidae**

smoothhead
**Alepocephalidae**

tubeshoulder
**Platytroctidae**

bristlemouth
**Gonostomatidae**

hatchetfish
**Sternoptychidae**

lightfish
**Phosichthyidae**

snaggletooth
**Astronesthidae**

viperfish
**Chauliodontidae**

black dragonfish
**Idiacanthidae**

loosejaw
**Malacosteidae**

scaleless dragonfish
**Melanostomiidae**

scaly dragonfish
**Stomiidae**

giganturid
**Giganturidae**

sudidae, Scopelarchidae, Alepisauridae, Evermanellidae, and Paralepididae. Most are pelagic, but a few are associated with bottom habitats.

MYCTOPHIFORMES

The mycotophiforms are slender fishes with laterally compressed bodies, large eyes, and dorsal adipose fins.

The large mouth is usually terminal. Most species bear photophores arranged in groups or lines on the head and body. Of the two families present in the Gulf, one, the lanternfishes (Myctophidae), is considered important, and this family is represented by 41 species. These fishes are found at all depths of the open Gulf, and members of many species migrate into surface waters at night.

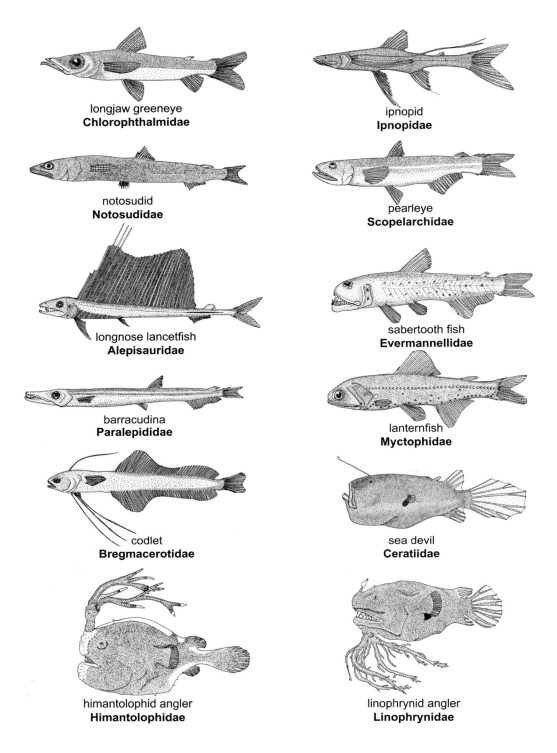

longjaw greeneye
**Chlorophthalmidae**

ipnopid
**Ipnopidae**

notosudid
**Notosudidae**

pearleye
**Scopelarchidae**

longnose lancetfish
**Alepisauridae**

sabertooth fish
**Evermannellidae**

barracudina
**Paralepididae**

lanternfish
**Myctophidae**

codlet
**Bregmacerotidae**

sea devil
**Ceratiidae**

himantolophid angler
**Himantolophidae**

linophrynid angler
**Linophrynidae**

GADIFORMES

The gadiform fishes were discussed in the previous chapter. Most of these cod-like fishes are found on or near the bottom, but in one family, the codlets (Bregmacerotidae), several species are at least semipelagic.

LOPHIIFORMES

The lophiiform fishes were addressed in the previous chapter. Some families are associated with the bottom, but several families of deep-sea anglerfishes are pelagic in the world oceans. These strange, often bizarrely shaped fish have very large heads, gaping mouths, and relatively

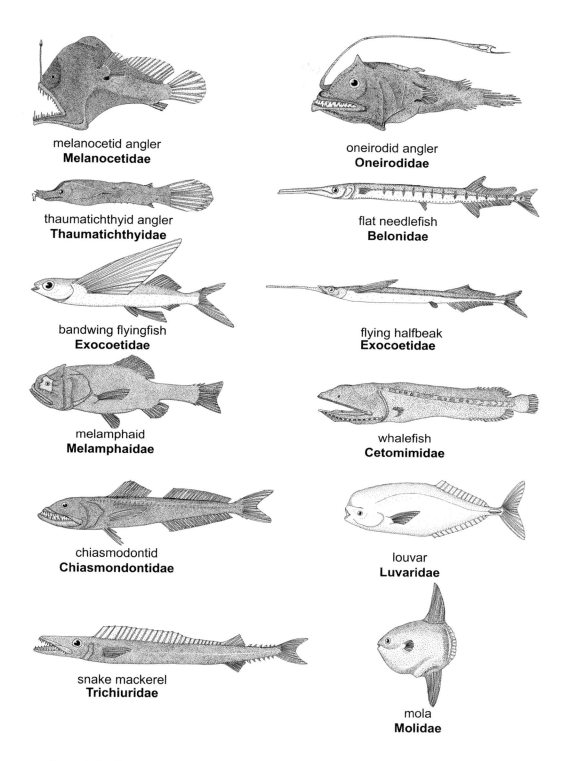

melanocetid angler
**Melanocetidae**

oneirodid angler
**Oneirodidae**

thaumatichthyid angler
**Thaumatichthyidae**

flat needlefish
**Belonidae**

bandwing flyingfish
**Exocoetidae**

flying halfbeak
**Exocoetidae**

melamphaid
**Melamphaidae**

whalefish
**Cetomimidae**

chiasmodontid
**Chiasmondontidae**

louvar
**Luvaridae**

snake mackerel
**Trichiuridae**

mola
**Molidae**

small, weak bodies. The dorsal and anal fins are set far back near the caudal fin. These fishes possess a rod and luminous lure (illicium and esca), the latter often being elaborated into branches, bulbs, and filaments. Some species have branched luminous chin barbels. These fierce-looking but sluggish fishes are ambush predators that float around in mesopelagic and bathypelagic waters and attract prey organisms by means of their luminous lures. Most species are sexually dimorphic, with the males being much smaller and anatomically different from the females, and in some species the males are sexual parasites on the adult females. Within the Gulf this order is represented by the following families: Ceratiidae, Himantolophidae, Linophrynidae, Melanocetidae, Oneirodidae, and Thaumatichthyidae.

### ATHERINIFORMES

The atheriniform fishes were described in the previous chapter. These are more or less slender fishes that live at the sea surface over the continental shelves or in the open Gulf. Some have enlarged pectoral fins that act as wings, permitting them to glide above the surface of the sea. In the open Gulf this group is represented by 3 families, the Belonidae, Exocoetidae, and Hemirhamphidae.

### STEPHANOBERYCIFORMES

The stephanoberyciform fishes (pricklefishes, whalefishes, dragonfishes, etc.), are characterized by a number of common internal features (thin skull bones, absence of an orbitosphenoid bone, etc.). Externally these fishes have somewhat rounded bodies, and the dorsal and anal fins are set far back near the caudal fin. Within the Gulf 7 families are represented, 2 of which (Melamphaidae and Cetomimidae) are considered here. Together they include about 18 species.

### PERCIFORMES

The perchlike fishes were discussed in the previous chapter. This group, which is so diverse and well represented in fresh waters and on the continental shelves, is represented in open oceanic waters by only a few important families. For the most part, these are fast-swimming predatory fishes of epipelagic waters, including jacks, dolphinfishes, mackerels, tunas, and billfishes. A few (chiasmodontids and snake mackerels) are mesopelagic. One group (remoras, or sharksuckers) is specialized for hitchhiking on sharks, manta rays, and other large marine animals, and one group of highly modified fishes (louvars) appears to feed on jellyfishes, comb jellies, and other gelatinous forms of the open ocean. Important families of perciform fishes of the open Gulf include the Echeneidae, Carangidae, Coryphaenidae, Chiasmodontidae, Luvaridae, Trichiuridae, Scombridae, Xiphiidae, and Istiophoridae.

### TETRAODONTIFORMES

The tetraodontiform fishes were addressed in the previous chapter. On the continental shelves they are represented by the puffers and their relatives, but in the open Gulf there is a single family, the molas or ocean sunfishes (Molidae), members of which are solitary and float or swim at the surface.

## Zooplankton and micronekton standing stocks

Much of our quantitative knowledge concerning the day and night vertical distribution patterns of zooplankton and micronekton of the Gulf stems from studies by T. L. Hopkins and coworkers (especially T. L. Hopkins 1982, and T. L. Hopkins and Lancraft 1984), who carried out a series of investigations over several years at a single benchmark station in the east-central Gulf of Mexico. Located at 27°N and 86°W, this station was situated in deep water (ca. 3,000 m) just west of the Florida shelf and somewhat south of the level of Tampa Bay. The location was within or adjacent to the southward-flowing arm of the Gulf Loop Current, or eastern boundary current. In this area during the summer months the mixed layer includes the upper 30–50 m, and the thermocline extends down to about 150 m. Here the highest primary production takes place in the upper 50 m of the water column (El-Sayed 1972). At this station, Hopkins and coworkers made quantitative collections with several types of gear so that they could determine the taxonomic composition and standing crop (in terms of numbers of individuals, wet weight, and dry weight) of the microzooplankton (< 1 mm), zooplankton (1–25 mm), and micronekton (> 25 mm). This was done day and night for several depths down to 1,000 m so that abundance values were obtained for day and night samples at various depths as well as for the entire water column from the surface to a depth of 1,000 m.

The results of these studies are most interesting. Table 14.4 shows the counts and biomass values for the various fractions calculated to lie beneath one square kilometer and from the surface to a depth of 1,000 m. The smallest organisms, the microzooplankton, were by far the most numerous, being about 35 times as abundant as the zooplankton animals, and these, in turn, were about 13.2 times as abundant as the micronekton animals. However, being very small, the microzooplankton made up only 4 percent of the biomass of the zooplankton. The micronekton, although larger, was only about 26 percent of the biomass of the zooplankton. These biomass estimates are rather typical of values obtained elsewhere for low-nutrient subtropical boundary currents. They are somewhat higher than values obtained from very low nutrient central gyres located away from coastal areas and much lower than values obtained for nutrient-rich upwelling areas.

**Table 14.4.** *Estimated standing crop (numbers of individuals and biomass) of microzooplankton, zooplankton, and micronekton in the depth range of 0–1,000 m beneath one square kilometer of sea surface in the central Gulf of Mexico. All data represent nighttime values. Microzooplankton includes all metazoan animals (protozoans excluded) less than 1 mm in size. Zooplankton includes all animals in the 1–25 mm size range. Micronekton includes those animals larger than 25 mm. Dry weight is assumed to be 10% of the wet weight. (Data from T. L. Hopkins 1982; T. L. Hopkins and Lancraft 1984.)*

| Category | No. of individuals (no./km²) | Wet weight (kg/km²) | Dry weight (kg/km²) |
|---|---|---|---|
| Microzooplankton | 1,386,000 × 10³ | 400 | 40.0 |
| Zooplankton | 39,600 × 10³ | 9,830 | 983.0 |
| Micronekton | 3,017 × 10³ | 2,558 | 255.8 |
| Totals | 1,428,617 × 10³ | 12,788 | 1,278.8 |

The microzooplankton fraction was dominated by immature copepods, which accounted for about 94 percent of the numbers and 88 percent of the biomass. Since protozoans are often lost or destroyed by the techniques employed, they were not included. In the zooplankton fraction adult copepods were dominant, consisting of over 80 percent of the numbers and over half of the biomass. Within the micronekton the principal groups, in terms of wet weight, were the medusae, fishes, and crustaceans, which made up 48.3, 34.7, and 12.6 percent, respectively. Diversity of the micronekton was high and included 148 species of crustaceans and fishes alone. Fishes of the gonostomatid genus *Cyclothone* were very abundant and constituted about 34 percent of the micronekton numbers.

The data reveal day and night patterns of vertical distribution. The microzooplankton was most concentrated in the upper 200 m of the water column. Here were encountered 94 percent of the numbers and 61 percent of the biomass, although day and night differences were slight. The average day and night zooplankton biomass values in the epipelagic and mesopelagic zones are given in table 14.5. The total zooplankton biomass was most concentrated near the surface and decreased logarithmically with depth. During the day the concentration ratio of epipelagic to mesopelagic zooplankton was 3.6:1, and at night this increased to 11.0:1. At night, largely because of vertical migration, the biomass concentration in the upper 200 m almost doubled (288.0:507.8 mg dry weight/100 m³). Copepods made up over half the epipe-

lagic zooplankton biomass and were relatively more important in the daytime (61.6 percent) than at night (45.1 percent). Euphausiids, which are individually much larger than copepods, contributed a much larger percentage to the biomass at night. All the major taxonomic groups displayed a nighttime increase in biomass in the epipelagic, and this was most notable in the euphausiids, where the day to night ratio was 1:8.7. However, this may have been due, in part, to daytime net avoidance and possibly to the patchy distribution of euphausiids, that is, to the chances of hitting or missing a swarm of these shrimps.

The day and night distribution patterns of zooplankton within the upper 200 m are shown in figure 14.10. Here the data are expressed as percentages of the total daytime or nighttime biomass. Even within the epipelagic the biomass of most groups was shallower at night than during the day. This was particularly true for the total zooplankton as well as the hydrozoans, polychaetes, copepods, and "other crustaceans," for which most of the biomass was distributed at 30 m or below during the daytime and in the upper 15 m at night. This is an important point, because by moving into the near-surface waters these zooplankton animals are apparently able to escape the heaviest predation by the larger vertically migrating carnivores, which by and large do not move into such shallow water at night.

Table 14.6 provides information on the composition of the micronekton within the upper 1,000 m expressed as a percentage of the invertebrate or fish catch. Within this size fraction the invertebrate fauna was dominated

**Table 14.5.** *Average biomass of zooplankton (mg dry weight/100 m³) of various taxonomic groups in the epipelagic (0–200 m) and mesopelagic (300–1000 m) zones of the eastern Gulf of Mexico in day and night collections. Day/night and mesopelagic/epipelagic ratios are also provided. (Data from T. L. Hopkins 1982.)*

| | Day | Night | Day | Night | Day | Night | Day | Night |
|---|---|---|---|---|---|---|---|---|
| | **Total** | | **Hydrozoans** | | **Gastropods** | | **Polychaetes** | |
| **Average biomass density (mg/100 m3)** | | | | | | | | |
| epipelagic | 288.0 | 507.8 | 19.3 | 26.9 | 5.8 | 12.8 | 3.9 | 9.1 |
| mesopelagic | 80.3 | 46.3 | 1.8 | 1.5 | 1.5 | 0.2 | 0.4 | 0.4 |
| **Ratios** | | | | | | | | |
| epipelagic day/night | 1:1.8 | | 1:1.4 | | 1:2.2 | | 1:2.3 | |
| meso/epi | 1:3.6 | 1:11.0 | 1:10.7 | 1:17.9 | 1:4.5 | 1:64.0 | 1:9.8 | 1:22.8 |
| | **Ostracods** | | **Copepods** | | **Amphipods** | | **Euphausiids** | |
| **Average biomass density (mg/100 m3)** | | | | | | | | |
| epipelagic | 6.5 | 10.4 | 177.3 | 229.1 | 6.8 | 14.2 | 10.8 | 93.8 |
| mesopelagic | 2.9 | 1.6 | 45.2 | 30.6 | 1.1 | 0.4 | 16.9 | 8.5 |
| **Ratios** | | | | | | | | |
| epipelagic day/night | 1:1.6 | | 1:1.3 | | 1:2.1 | | 1:8.7 | |
| meso/epi | 1:2.2 | 1:6.5 | 1:3.9 | 1:7.5 | 1:6.2 | 1:35.5 | 1:0.6 | 1:37.5 |
| | **Other crustaceans** | | **Chaetognaths** | | **Tunicates** | | **Other zooplankton** | |
| **Average biomass density (mg/100 m3)** | | | | | | | | |
| epipelagic | 5.6 | 26.5 | 30.5 | 54.8 | 8.0 | 12.6 | 12.7 | 17.8 |
| mesopelagic | 2.3 | 0.3 | 5.7 | 6.0 | 0.6 | 0.5 | 2.0 | 2.0 |
| **Ratios** | | | | | | | | |
| epipelagic day/night | 1:4.7 | | 1:1.8 | | 1:1.6 | | 1:1.4 | |
| meso/epi | 1:2.4 | 1:88.3 | 1:5.4 | 1:9.1 | 1:13.3 | 1:25.2 | 1:6.4 | 1:8.9 |

by shrimps, which together made up 86.6 percent of the total invertebrate catch. Penaeid shrimps alone constituted 46.4 percent, and the mysids, euphausiids, and carideans each made up between 12 and 14 percent of the invertebrates taken. Among the fishes the stomiiformes made up 73.2 and the myctophiformes 17.3 percent of the catch, the remaining fish groups together contributing less than 10 percent of the fish catch. The stomiiformes were dominated by the gonostomatids and particularly by species of the genus *Cyclothone.* However, the sternoptychids were also well represented. Among the myctophids the genera *Diaphus, Lampanyctus,* and *Lepidophanes* were most prominent.

The data presented in this section represent our best estimates of the distribution of zooplankton and micronekton in the upper 1,000 m of the water column based on state-of-the-art technology. However, the information cannot be accepted as fully representative of the actual situation in nature. Rather, it is an approximation. No attempt was made to assess the protozoans. Other groups such as the hydrozoans, ctenophores, and polychaetes are very delicate and are largely destroyed by the collecting devices. Net avoidance by some squids, shrimps, and fishes further biases the results, and chance encounters with concentrations such as swarms of euphausiids may skew the catch data. Despite such potential sources of

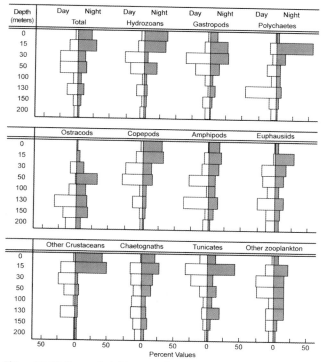

**Figure 14.10.** Day and night comparisons of the average biomass of zooplankton of various taxonomic groups at several depth levels in the epipelagic zone of the eastern Gulf of Mexico. The values shown represent percentages of the total biomass taken during the day or night samplings. (Data from T. L. Hopkins 1982.)

**Table 14.6.** *Micronekton organisms captured in oblique tows (0–1,000 m) in the east-central Gulf of Mexico. The numbers represent percentages of the total invertebrate or fish catch. (Data from T. L. Hopkins and Lancraft 1984.)*

| Taxonomic group | Catch (%) |
|---|---|
| Cnidarians | 2.3 |
| Ctenophores | 2.9 |
| Mollusks | |
|   Pteropods | 0.5 |
|   Heteropods | 2.6 |
|   Cephalopods | 1.5 |
| Annelids | |
|   Polychaetes | 0.4 |
| Arthropods | |
|   Amphipods | 1.4 |
|   Stomatopods | 1.4 |
|   Mysidaceans | 12.6 |
|   Euphausiids | 13.8 |
|   Penaeids | 46.4 |
|   Carideans | 13.8 |
| Tunicates | 0.3 |
| Total invertebrates | 99.9 |
| Fishes | |
|   Osmeriformes | 0.3 |
|   Stomiiformes | 73.2 |
|   Aulopiformes | 0.4 |
|   Myctophiformes | 17.3 |
|   Gadiformes | 0.4 |
|   Lophiiformes | 0.2 |
|   Stephanoberyciformes | 0.8 |
|   Perciformes | 0.2 |
|   Leptocephalus larvae | 4.3 |
|   Miscellaneous fishes | 2.8 |
| Total fishes | 99.9 |

error, the available information does provide considerable insight into the composition, abundance, and day and night dynamics of animal populations inhabiting the water column in the eastern Gulf of Mexico. The abundance values presented are probably somewhat higher than those of central gyres of the western Gulf and much lower than those of upwelling areas. Only further research can clarify these issues.

## BENTHOS AND DEMERSAL FISHES OF THE DEEP GULF

In this section we will discuss those animals that live on or near the bottom in waters seaward of the continental shelf, that is, in the depth range of about 200 to 3,750 m. Since virtually the entire area lies below the photic zone, photosynthetic plants are not significant inhabitants of this realm. The animal populations are often diverse and abundant on the upper slope, but in the world oceans in general, both diversity and abundance tend to decrease with depth (fig. 14.4). Our knowledge of the deepwater benthic fauna stems from the use of several types of sampling gear. Information on the meiofauna and macro-

fauna derives from sediment samples usually obtained from box cores. The megafauna is sampled with bottom dredges, skimmers, and trawls. Although the meiofauna and macrofauna are sampled quantitatively and without contamination, the gear used in collecting megafauna is nonquantitative, and the samples may be contaminated by midwater organisms captured as the gear is lowered or retrieved through the water column. Additional information on the larger animals derives from photographs of the sediment surface and near-bottom waters. The larger animals that live within the sediments (infauna) are poorly known, but some specimens have appeared in collections made by dredges and other gear, and indirect evidence of the infaunal animals (holes, burrows, trails, piles of fecal material, etc.) often appears in bottom photographs. References to the benthos of the deep ocean bottoms of the world include works such as Gage and Tyler (1991), Heezen and Hollister (1971), and Menzies, George, and Rowe (1973).

Most of our knowledge of the benthic fauna of the deep Gulf stems from collections made by W. Pequegnat and his students and coworkers. The primary references to the benthic fauna of the deep Gulf are Gallaway, Martin, and Howard (1988), W. Pequegnat (1983), W. Pequegnat and Chace (1970), W. Pequegnat, L. Pequegnat, et al. (1971), W. Pequegnat, Darnell, et al. (1976), and W. Pequegnat, Gallaway, and L. Pequegnat (1990). Additional references, primarily related to specific groups, include Abele and Martin (1989), Bright (1968, 1970), Carney (1971), Firth (1971a, 1971b), Halpern (1970), Holthuis (1974), Holthuis and Mikulka (1972), Hubbard (1995), James (1972), Kennedy (1976), Lockhart et al. (1990), D. M. Martin (1978, 1984), Monniot and Monniot (1987), Mukai (1974), D. Opresko (1972), L. Pequegnat (1970a, 1970b), L. Pequegnat and W. Pequegnat (1970), W. Pequegnat (1970), W. Pequegnat and L. Pequegnat (1971), W. Pequegnat, James, et al. (1972), Pérez-Farfante (1977), Pérez-Farfante and Bullis (1973), Perry, Waller, et al. (1995), Rayburn (1975), T. Roberts (1970, 1977), T. Roberts and Pequegnat (1970), Rowe (1966), Rowe and Menzel (1971), Rowe, Polloni, and Horner (1974), Soto (1985, 1986, 1991), Stock (1986), and Waller et al. (1995). Citations of earlier studies may be found in the bibliographies of the above references. Some records of benthic animals of the deep Gulf are provided in the lists of Bullis (1956), Bullis and Thompson (1965), and S. Springer and Bullis (1956). References to the benthic fauna of deepwater hydrocarbon and hydrogen sulfide seep areas are given later in connection with discussions of those topics. As in the case of the continental shelf, the benthic animals of the deep Gulf are studied in relation to size classes (micro-, meio-, macro-, and megafauna), and each of these groups will be reviewed briefly below.

## Microfauna

The microfauna, sometimes referred to as the nanofauna, includes those tiny animals that pass through screens with mesh openings of 63 microns (0.0063 mm). This group consists primarily of protozoans (amoebas, flagellates, ciliates, and sporozoans) and some yeast-like cells of unknown affinity. These organisms are quite numerous in the surface layer of deep-sea sediments, and they doubtless play important roles in the metabolism of benthic communities, feeding on bacteria and bits of organic detritus and providing a food resource for the larger sediment feeders. They are delicate and difficult to study, and hence little work has been carried out on the benthic microfauna of the world oceans in general, and this group apparently has not been examined in the deeper waters of the Gulf.

## Meiofauna

The meiofauna consists of those small animals that pass through mesh openings of 0.3 mm but are retained by mesh with 63-micron openings. Gage and Tyler (1991) noted that among the benthic macrofaunal animals there is a tendency toward miniaturization with depth so that many macrofaunal groups of shallow-water animals appear in deep-sea collections as meiobenthos. The meiobenthos of the continental slopes of the north-central and northeastern Gulf have been reported by Gallaway, Martin, and Howard (1988) and W. Pequegnat, Gallaway, and L. Pequegnat (1990). These investigators identified representatives of 43 major groups of invertebrates, but nearly 98 percent of the individuals belonged to only 6 groups. In descending order of abundance, these were nematodes, adult harpacticoid copepods, copepod nauplii, polychaetes, ostracods, and kinorynchs (nematode relatives).

## Macrofauna

The macrofauna includes all those benthic invertebrates taken from sediment samples that are retained on screens with a mesh size of 0.3 mm, and these animals range in size up to a centimeter or more. In the Gulf this fauna has been reported by Gallaway, Martin, and Howard (1988) and W. Pequegnat, Gallaway, and L. Pequegnat (1990). They identified 40 major taxonomic groups representing 8 phyla, but the 6 most abundant groups made up 85 percent of the total numbers. In descending order of abundance, these included polychaetes, ostracods, bivalves, tanaids (isopod relatives), bryozoans, and isopods. At the species level, many of the invertebrate groups were represented by very few individuals.

## Megafauna

The megafauna includes all the invertebrates and fishes taken by trawl, dredge, and other such gear and has a minimum size of about 1 cm. In their trawl studies of the continental slopes of the north-central and northeastern Gulf, Gallaway, Martin, and Howard (1988) and W. Pequegnat, Gallaway, and L. Pequegnat (1990) captured more than 40,000 specimens of invertebrates and fishes, and the percentage distribution of the catch among the major megafaunal groups is shown in table 14.7. Arthropods made up over half the catch, and echinoderms and fishes each made up about 15 percent. Six major groups together made up over 84 percent of the catch. These include decapods (45 percent), fishes (14 percent), ophiuroids (11 percent), sponges (7 percent), soft corals (4 percent), and barnacles (4 percent). The composition of the benthic fauna, taken up in the next sections, is based primarily on the megafauna. Depth distribution patterns for the various size fractions are addressed in a later section.

## Composition of the benthic invertebrate fauna of the deep Gulf

In approaching the composition of the benthic fauna of the deep Gulf, it is important to recognize two main points. In the first place, the taxonomic diversity is quite great on the upper and middle continental slope, although it drops off in deeper water. Practically every major group of the animal kingdom found in the sea is represented, exceptions being the hermatypic corals (which require light), ctenophores and chaetognaths (which are planktonic), and higher vertebrates (which must breathe air). The second point is that the same major groups of animals found in the benthos of shallow marine waters are those that inhabit the depths. Most differences occur at the level of genus and species, with a few at the level of family or higher taxonomic category. An exception is the glass sponges (Hexactinellida), which are rare in shallow waters but relatively abundant in the deep sea. By and large, the groups with which we will be dealing are already familiar to the reader (see chapter 10). In fact, it is remarkable that so many groups of shallow-water animals have independently colonized the deep oceans of the world.

A list of the groups to be discussed in the present section is presented in table 14.8. As mentioned above, these

**Table 14.7.** *Numerical abundance of the various benthic megafaunal groups captured in transects across the continental slopes in the north-central and northeastern Gulf of Mexico. The data are expressed as a percentage of the total catch. (Data from Gallaway, Martin, and Howard 1988.)*

| Group | Percentage |
|---|---|
| Porifera | 6.7 |
| Cnidaria | |
|   Hydrozoa (hydroids) | 0.1 |
|   Alcyonaria (soft corals) | 4.0 |
|   Actiniaria (sea anemones) | 1.7 |
|   Scleractinea (stony corals) | 0.5 |
|   Miscellaneous cnidarians | 0.1 |
| Mollusca | |
|   Gastropoda | 0.7 |
|   Bivalvia | 2.8 |
|   Cephalopoda | 0.1 |
| Annelida | |
|   Polychaeta | 1.3 |
| Arthropoda | |
|   Isopoda | 2.6 |
|   Amphipoda | 0.1 |
|   Cirripedia (barnacles) | 3.6 |
|   Decapoda | 45.2 |
| Echinodermata | |
|   Asteroidea | 2.5 |
|   Echinoidea | 0.4 |
|   Ophiuroidea | 10.6 |
|   Holothuroidea | 1.4 |
|   Crinoidea | 0.8 |
| Miscellaneous invertebrates | 0.4 |
| Total invertebrates | 85.6 |
| Chordata | |
|   Fishes | 14.3 |
| Total catch (percentage) | 99.9 |
| Total catch (number) | 40,439 |

are mostly the megafaunal groups taken in trawl collections. For the sake of brevity some of the smaller and less numerous groups are omitted. These include certain wormlike lower invertebrates (turbellarians, nemerteans, kinorynchs, priapulids, nematodes) and some mollusks (aplacophorans), annelids (oligochaetes, sipunculids, archiannelids), and arthropods (acarinids, copepods, cuma-

**Table 14.8.** *Higher classification categories for the major groups of animals inhabiting the bottom sediments and near-bottom waters of the deep Gulf of Mexico.*

| | |
|---|---|
| Phylum | Protozoa – protozoans |
| Class | Sarcodina – sarcodinids (foraminiferans) |
| Phylum | Porifera – sponges |
| Phylum | Cnidaria – coelenterates |
| Class | Hydrozoa – hydrozoans (hydroids) |
| Class | Anthozoa – anthozoans |
| Subclass | Alcyonaria – soft corals |
| Subclass | Zoantharia – zoantharians |
| Order | Actiniaria – sea anemones |
| Order | Scleractinia – stony corals |
| Phylum | Mollusca – mollusks |
| Class | Gastropoda – gastropods |
| Class | Bivalvia – bivalves |
| Class | Scaphopoda – tusk shells |
| Class | Cephalopoda – cephalopods |
| Phylum | Annelida – annelid worms |
| Class | Polychaeta – polychaetes |
| Phylum | Arthropoda – arthropods |
| Subphylum | Chelicerata – chelicerates |
| Class | Pycnogonida – pycnogonids |
| Subphylum | Mandibulata – mandibulates |
| Class | Crustacea – crustaceans |
| Subclass | Ostracoda – ostracods |
| Subclass | Cirripedia – barnacles |
| Subclass | Malacostraca |
| Order | Tanaidacea – tanaids |
| Order | Isopoda – isopods |
| Order | Amphipoda – amphipods |
| Order | Decapoda – decapods |
| Suborder | Natantia – swimming shrimps |
| Suborder | Reptantia – crawlers |
| Section | Macrura – lobsters |
| Section | Anomura – hermit crabs and relatives |
| Section | Brachyura – true crabs |
| Phylum | Bryozoa – bryozoans |
| Phylum | Echinodermata – spiny-skinned animals |
| Class | Asteroidea – starfishes |
| Class | Echinoidea – brittle stars |
| Class | Ophiuroidea – sea urchins |
| Class | Holothuroidea – sea cucumbers |
| Class | Crinoidea – crinoids |
| Phylum | Chordata – chordates |
| Subphylum | Urochordata – urochordates (tunicates) |
| Subphylum | Vertebrata – vertebrates |
| Class | Agnatha – jawless fishes |
| Class | Chondrichthyes – cartilaginous fishes |
| Class | Osteichthyes – bony fishes |

ceans), as well as the brachiopods and pogonophorans. For information on these groups, see Gallaway, Martin, and Howard (1988).

FORAMINIFERA

Benthic foraminiferans, or "forams," are tiny shelled protozoans. Although not taken in trawl collections in the Gulf, they have been the subject of a number of investigations since they are useful indicators of water depth and are of keen interest to those concerned with oil exploration. Their calcareous tests are important components of deepwater sediments. Consolidation of the calcareous ooze produces limestone whose depth of deposition can often be determined by the foraminiferal species content. The benthic foraminifera of the deep Gulf have been reported by a number of authors, most notably by Pflum and Frerichs (1976). Based on transects from the shore, across the shelf and slope, and onto the abyssal plain, mostly in the northern Gulf, these authors identified 328 species, of which 99 were considered to be useful as bathymetric indicators. Species diversity increased from the shoreline out onto the upper continental slope and decreased somewhat on the lower slope and abyssal plain. At least 20 species were found to be characteristic of clastic sediments of the northwestern Gulf, but only four species were associated primarily with carbonate sediments of the northeastern Gulf. Most species displayed regular zonation patterns, but in the sedimentary environments off the Mississippi River Delta these patterns were sometimes altered (presumably by sediment slumping, etc.). Important genera of deepwater foraminifera of the Gulf of Mexico include *Ammolagena, Cibicides,* and *Eponides.*

PORIFERA

Sponges are sessile animals that may grow attached to solid substrates, such as rocks or shells, or as solitary animals anchored in soft substrates. In deep waters of the world oceans they are best represented by the Hexactinellida, in which the supporting skeletal structure is composed of glass fibers or solid spicules. Many species grow as vase- or cup-shaped structures, and some are supported on long thin stems. Although the deepwater sponges of the Gulf apparently have not been well studied, Gallaway, Martin, and Howard (1988) and W. Pequegnat (1983) provide lists of species collected from various depths. Over 50 different types were collected, but only 7 were identified to species and 41 to the level of genus. Important genera include *Euplectella, Hyalonema,* and *Thenea* (fig. 14.11).

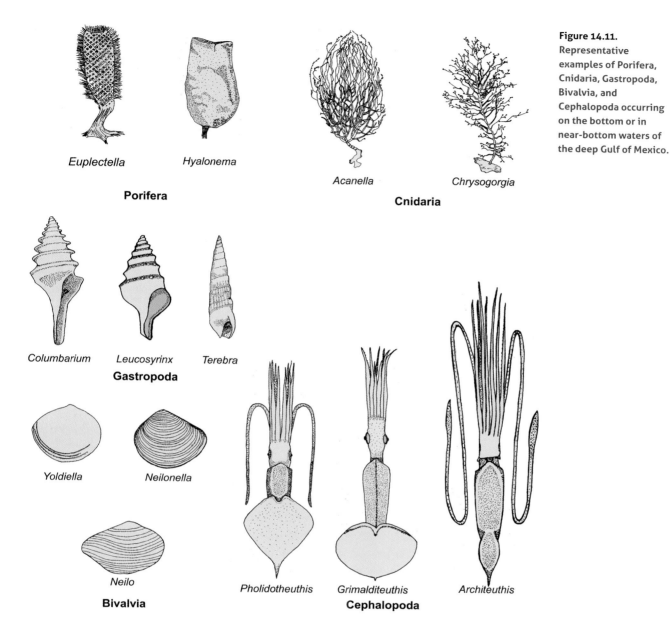

**Figure 14.11.**
Representative examples of Porifera, Cnidaria, Gastropoda, Bivalvia, and Cephalopoda occurring on the bottom or in near-bottom waters of the deep Gulf of Mexico.

*Euplectella*      *Hyalonema*

**Porifera**

*Acanella*      *Chrysogorgia*

**Cnidaria**

*Columbarium*      *Leucosyrinx*      *Terebra*

**Gastropoda**

*Yoldiella*      *Neilonella*

*Neilo*

**Bivalvia**

*Pholidotheuthis*      *Grimalditeuthis*      *Architeuthis*

**Cephalopoda**

## CNIDARIA

Cnidarians of the deep Gulf have been reported by Galla-way, Martin, and Howard (1988), Opresko (1972), and W. Pequegnat (1983). These forms are represented by at least 60 species of hydroids, sea fans, sea anemones, and both soft and solitary stony corals. Their depth distribution patterns were discussed by W. Pequegnat (1983). Representative genera of the various groups include the following: hydroids (*Acryptolaria, Cladocarpus, Opercularella*), sea fans (*Funiculina, Pennatula, Umbellula*), sea anemones (*Actinauge, Antholoba, Halcurias*), soft corals (*Acanella, Acanthogorgia, Chrysogorgia*), and stony corals (*Anomocora, Enallopsammia, Solenosmilia*) (fig. 14.11).

## GASTROPODA

Gastropod mollusks of the deep Gulf have been reported by Gallaway, Martin, and Howard (1988) and W. Pequegnat (1983). The latter author listed 193 species representing 40 families. Thirty-six percent of the species and 64 percent of the total individuals belong to the family Turridae. Although found across the slope, rise, and abyssal plain, the gastropods achieved their highest density in the depth range of 800–1,200 m. They were especially diverse in the DeSoto Canyon area. Representative genera include *Columbarium, Leucosyrinx,* and *Terebra* (fig. 14.11).

## BIVALVIA

Bivalve mollusks of the deep Gulf have been reported by Gallaway, Martin, and Howard (1988), James (1972), and W. Pequegnat (1983). Since most species live within rather than upon the sediments, the bivalves are classified as infauna, and they are collected most effectively by dredges rather than by trawls. The most abundant bivalves of the deep Gulf, below 500 m, belong to the subclass Palaeotaxodonta. W. Pequegnat (1983) listed 73 species of bivalves representing 18 families, but the list is incomplete because some specimens were lost during shipment. Bivalves were collected to a depth of 3,500 m. Representative genera include *Neilo, Neilonella,* and *Yoldiella* (fig. 14.11).

## SCAPHOPODA

The scaphopods, or tusk shells, of the deep Gulf have been reported by Gallaway, Martin, and Howard (1988) and W. Pequegnat (1983). Like the bivalves, scaphopods belong to the infauna. W. Pequegnat (1983) recorded the shells of 17 species from the deep Gulf, although only 10 species were collected alive. Some of the scaphopod species displayed broad depth ranges, up to 2,400 m. Representative genera are *Dentalium, Fissidentalium,* and *Laevidentalium.*

## CEPHALOPODA

Cephalopods of the deep Gulf have been reported by a number of authors including Gallaway, Martin, and Howard (1988), Lipka (1975), and W. Pequegnat (1983). Since the deepwater cephalopods are collected by trawls, the collections could be contaminated by midwater species. Actually, few cephalopods live directly on the bottom, and photographs often show them a few meters above the bottom. However, some forms, such as species of *Grimpoteuthis,* are known to feed on polychaetes and other benthic animals, so some of the species must at least visit the bottom for feeding. On the basis of available information a tentative list of cephalopods in which the adults probably live on or near the bottom of the deep Gulf is given in table S72, and examples are shown in figure 14.11. Several of these species are also listed among the nekton of the open Gulf (table S61), and a few are listed in the benthos of the continental shelf (table 13.2). The various benthic or benthopelagic cephalopods have been collected from the edge of the continental shelf to depths of over 3,700 m. Representative genera include *Architeuthis, Grimalditeuthis,* and *Pholidoteuthis.*

## POLYCHAETA

Polychaetes of the deep Gulf have been reported by Gallaway, Martin, and Howard (1988), Hubbard (1995), and W. Pequegnat (1983). As in the case of most meiofaunal and macrofaunal groups, the polychaete fauna of the deep Gulf has not been well investigated despite its obvious richness. A large percentage of the species reported in the above-referenced studies are undescribed and new to science. Hubbard (1995) recognized 450 individual species, but at least 150 others could be identified only to the level of genus, family, or higher taxonomic category. Existing data show that polychaetes are widespread, abundant, and diverse in the deep Gulf. Among the chief factors determining local abundance and diversity are particle size and carbonate content of the sediments. Common genera include *Aricidea, Lumbrinerides,* and *Prionospio.* Only 6 species of oligochaetes were recognized, 3 of which are new to science. Genera include *Bathyhydrilus, Limnodriloides,* and *Tubificoides.*

## PYCNOGONIDA

Pycnogonids of the deep Gulf have been reported by Gallaway, Martin, and Howard (1988), W. Pequegnat (1983), and Stock (1986). Twenty species representing 12 genera and 5 families are known from the area. They are fairly widespread on the upper slope and have been taken down to a depth of about 1,200 m, but nowhere do they appear to be abundant. Since many species are small and can pass through the meshes of the trawl, they may be more abundant than it appears at present. Representative genera include *Ascorhynchus, Nymphon,* and *Paranymphon.*

## OSTRACODA

Ostracods of the deep Gulf have been reported by Gallaway, Martin, and Howard (1988) and W. Pequegnat (1983). Twenty-two forms are listed, of which 18 have been identified to the level of species. Representative genera are *Angulorostrum, Euphilomedes,* and *Philomedes.*

## CIRRIPEDIA

Both Gallaway, Martin, and Howard (1988) and W. Pequegnat (1983) recorded barnacles from the deep Gulf, and these occurred in the depth range of 329–2,401 m. The known fauna includes 16 species in 12 genera. Specimens of two of the species (*Octolasmis geryonophilia* and *Trilamis kaempferi*) were found attached to the carapaces of deepwater crabs. Representative genera include *Amigdoscalpellum, Arcoscalpellum,* and *Verruca* (fig. 14.12).

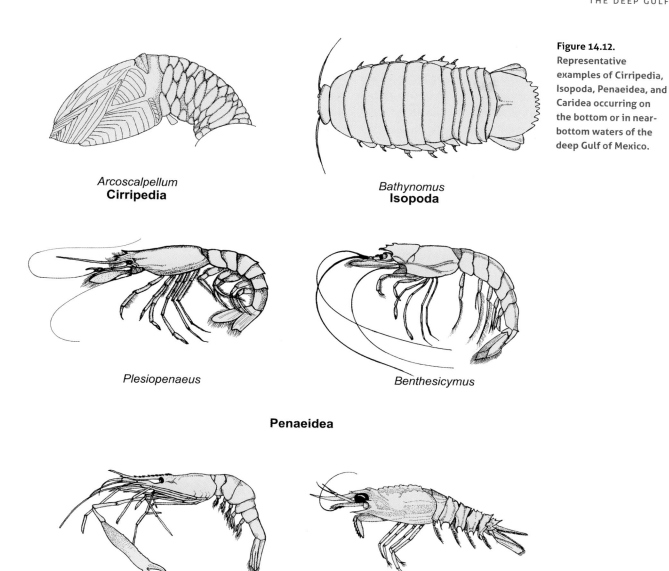

**Figure 14.12.**
Representative examples of Cirripedia, Isopoda, Penaeidea, and Caridea occurring on the bottom or in near-bottom waters of the deep Gulf of Mexico.

*Arcoscalpellum*
**Cirripedia**

*Bathynomus*
**Isopoda**

*Plesiopenaeus*

*Benthesicymus*

**Penaeidea**

*Bathypalaemonella*

*Glyphocrangon*

**Caridea**

TANAIDACEA

The tanaids are tiny crustaceans related to isopods, but unlike isopods they have chelae or pincers on the first pair of walking legs. In the deep Gulf 186 taxa have been recognized, of which 168 were identified to species and another 13 to the level of genus (Gallaway, Martin, and Howard 1988). In samples taken across the continental slope, densities ranged from 18 to $512/m^2$, with the highest densities occurring in the 600–1,000 m depth range. Representative genera include *Leptognatha, Mesotanais,* and *Pseudotanais.*

ISOPODA

Among the isopods of the deep Gulf, 133 taxonomic groups have been recognized, of which 119 have been identified to species and another 8 to the level of genus (Gallaway, Martin, and Howard 1988; W. Pequegnat 1983). Densities in the bottom sediments ranged from 18 to $580/m^2$, and highest densities occurred in the depth range of 1,000–1,500 m. Although most of the marine isopods are quite small, a centimeter or two at the most, on the continental slopes of the deep Gulf there lives a giant isopod (*Bathynomus giganteus*) (fig. 14.12), which reaches a maximum length of about 356 cm (14 in) (Holthuis and Mikulka

1972). This is presumed to be both a predator and scavenger. Commensalistic barnacles have been found attached to its abdomen. Specimens of the giant isopod have been collected in the depth range of about 400–2,250 m, and maximum population densities occur around 1,200 m. Another deepwater isopod (*Livoneca*) is parasitic on fishes. Representative genera include *Gnathia, Ischnomesus,* and *Prochelator.*

### AMPHIPODA

Gallaway, Martin, and Howard (1988) collected 79 taxa of amphipods from the deep Gulf, of which 50 were identified to the level of species. Their densities ranged from 0 to 232/m², and the greatest densities were encountered in the depth range of 500–900 m. Most amphipod species are quite small, averaging only a few millimeters in length, but recently a giant amphipod (*Eurythenes gryllus*) has been discovered in the deep Gulf (G. T. Rowe, personal communication). This amphipod, which reaches a body length of about 14 cm (5.6 in), occurs at depths of around 3,600 m, where it is quickly attracted to baits lowered to the ocean floor. Since it has not been taken in trawls, it is possible that this amphipod lies buried in the sediments or swims in the water column several meters above the bottom. As scavengers these amphipods are quickly attracted to carcasses of larger animals that sink to the ocean floor. Representative genera of amphipods include *Byblis, Melita,* and *Pandisynopia.*

### PENAEIDEA

Shrimps of the family Penaeidae that live on the bottom or in near-bottom waters of the deep Gulf have been reported by Abele (1970), Gallaway, Martin, and Howard (1988), W. Pequegnat (1983), W. Pequegnat, L. Pequegnat, et al. (1971), Pérez-Farfante (1977), Pérez-Farfante and Bullis (1973), T. Roberts (1970), T. Roberts and Pequegnat (1970), and Soto (1972).

Dr. L. Pequegnat has aided the author in determining which species are associated with the benthic habitat. A tentative list of such species is given in table S73, and representative examples are shown in figure 14.12. Twenty-nine species are listed. A few are basically continental shelf species whose range includes the upper portion of the continental slope. About a third are characteristic of the upper and middle slopes. Eleven species extend below a depth of 1,000 m, and four of the species occur on the abyssal plain. Representative genera include *Benthesicymus, Hymenopenaeus,* and *Plesiopenaeus.*

### CARIDEA

The caridean shrimps that live on or near the bottom of the deep Gulf have been reported by Abele (1970), Abele and Martin (1989), Gallaway, Martin, and Howard (1988), L. Pequegnat (1970a, 1970b), W. Pequegnat (1983), W. Pequegnat, L. Pequegnat, et al. (1971), and Soto (1972). Dr. L. Pequegnat has aided the author in determining which species are associated with the benthic habitat. These species are listed in table S74, and representative examples are shown in figure 14.12. Included are 43 species belonging to 10 families. It is striking that many of the species exhibit very broad depth ranges, often extending over 1,000 m, and in a few cases, over 2,000 m. Most are continental slope species, but three extend onto the abyssal plain. Representative genera include *Bathypalaemonella, Glyphocrangon,* and *Plesionika.*

### MACRURA

The lobsters of the deep Gulf have been documented by Firth (1971a, 1971b), Gallaway, Martin, and Howard (1988), Holthuis (1974), and W. Pequegnat (1983). These are listed in table S75, and examples are illustrated in figure 14.13. Only 15 species are known, representing 4 families. Since some of the species burrow into the substrate, it is likely that the list will grow as further studies are carried out in the deep Gulf. Several of the species appear to be limited to the upper slope, but at least 5 species penetrate to depths greater than 1,000 m, and 2 occur on the abyssal plain. Representative genera include *Nephropsis, Polycheles,* and *Scyllarus.*

### ANOMURA

The anomurans include the hermit crabs and their relatives. Species that inhabit the deep Gulf have been reported by Gallaway, Martin, and Howard (1988), W. Pequegnat (1983), L. Pequegnat and W. Pequegnat (1970), and W. Pequegnat and L. Pequegnat (1971). A list of the known species is given in table S76, and examples are illustrated in figure 14.13. Fifty-seven species are known, representing 5 families, by far the largest being the family Galatheidae with 35 species, followed by the Paguridae with 17 species. Twenty-two species are found as deep as 1,000 m, and 4 species occur as deep as 3,000 m. The genera *Munida* and *Munidopsis* are especially well represented, the former having 11 species, and the latter with 24 species. A number of these deepwater forms are known only from the Gulf of Mexico basin. Representative genera are *Munida, Munidopsis,* and *Paguristes.*

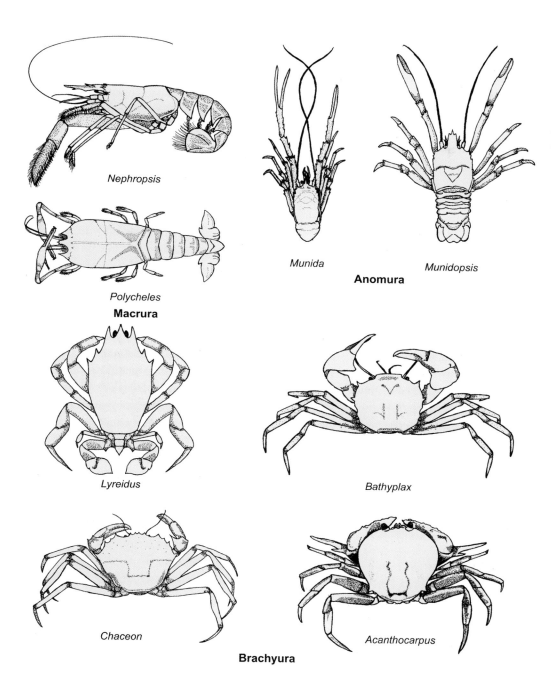

**Figure 14.13.**
Representative
examples of Macrura,
Anomura, and Brachyura
occurring on the bottom
or in near-bottom
waters of the deep
Gulf of Mexico.

*Nephropsis*

*Munida*

*Munidopsis*

**Anomura**

*Polycheles*

**Macrura**

*Lyreidus*

*Bathyplax*

*Chaceon*

*Acanthocarpus*

**Brachyura**

### BRACHYURA

Brachyuran crabs of the deep Gulf have been reported by Gallaway, Martin, and Howard (1988), Lockhart et al. (1990), W. Pequegnat (1970, 1983), W. Pequegnat, L. Pequegnat, et al. (1971), Perry, Eleuterius, et al. (1995), Soto (1972, 1985, 1986, 1991), and Waller et al. (1995). About 100 species are known from the deep Gulf. Representative species are listed in table S77, and several are shown in figure 14.13. As a group, the brachyurans are limited primarily to the upper continental slope. Most species reach their greatest population densities at 550 m or above, and both numbers and diversity decrease below a depth of 400 m. Soto (1986, 1991) provided information on the abundance and depth distribution patterns of brachyurans in the upper 1,000 m of the continental slope in the Straits of Florida. Only four species are known to extend to depths greater than 1,000 m, but one species, *Ethusina abyssicola,* has been taken at a depth of 3,750 m. Deepwater forms are generally small and delicate, but one species, *Chaceon quinquedens,* is large and robust. Representative genera include *Acanthocarpus, Bathyplax,* and *Chaceon.*

### BRYOZOA

Lagaaij (1963) and W. Pequegnat (1983) reported on bryozoans of the deep Gulf. To date, nearly 250 species have been recorded from Gulf waters (Leuterman 1979), but of these, only 23 are known to inhabit depths greater than 100 m, and these are listed in table S78. The paucity of deepwater species clearly reflects lack of collecting effort rather than habitat limitation. To locate specimens of the smaller species, special techniques are required. Of the known deepwater species, most are limited to the upper slope. Only 4 species have been taken deeper than 500 m, and only 2 have been collected below 1,000 m. Representative genera include *Cupuladria, Euginoma,* and *Nellia.*

### ASTEROIDEA

Starfishes of the deep Gulf have been reported by Downey (1973), Gallaway, Martin, and Howard (1988), Halpern (1970), Mukai (1974), and W. Pequegnat (1983). More than 60 species are now known to inhabit the deep Gulf (table S79), and representatives are illustrated in figure 14.14. Of this number, about two-thirds are found deeper than 500 m, and nearly half occur below 1,000 m. Four species have been taken below 3,000 m. Since some of the deepwater species bury themselves in the sediments, further collecting by dredges should increase the known number of deepwater species. Representative genera include *Astropecten, Litonaster,* and *Nymphaster.*

### ECHINOIDEA

Echinoids of the deep Gulf have been recorded by Booker (1971), Gallaway, Martin, and Howard (1988), and W. Pequegnat (1983). As seen in table S80, at least 29 species representing 15 families have been reported. Representatives are illustrated in figure 14.14. Ten species have been recorded from depths greater than 1,000 m, 6 species from depths greater than 2,000 m, and only a single species below 3,000 m. As in the case of the starfishes, some of the echinoids bury themselves in the bottom sediments. Representative genera are *Brissopsis, Phormosoma,* and *Plesiodiadema.*

### OPHIUROIDEA

Serpent stars of the deep Gulf have been reported by Gallaway, Martin, and Howard (1988) and W. Pequegnat (1983). Thirty-two species are listed in table S81, and an example is shown in figure 14.14. Half the species reach a depth of 1,000 m, 9 species reach 2,000 m, and 4 species occur as deep as 3,000 m. Representative genera include *Bathypectinura, Homalophiura,* and *Ophiomusium.*

### HOLOTHUROIDEA

Sea cucumbers of the deep Gulf have been reported by Carney (1971), Gallaway, Martin, and Howard (1988), and W. Pequegnat (1983). Identified species are listed in table S82, and representatives are shown in figure 14.14. So far, 23 species representing 7 families are known from the deep Gulf. Sixteen species occur below 1,000 m, 14 below 2,000 m, and 6 have been taken below 3,000 m. This high diversity of sea cucumbers on the lower slope and on the abyssal plain is in marked contrast with the other echinoderm groups, which are most diverse on the upper and middle portions of the continental slope. By processing large volumes of nutrient-poor sediments in their long digestive tracts, sea cucumbers are able to thrive where most other echinoderms appear to be food limited. Some holothurians of the deep Gulf, such as *Pelagothuria* and *Scotoanassa,* are benthopelagic and may swim above the bottom sediments. Representative genera are *Benthodytes, Molpadia,* and *Psychropotes.*

### UROCHORDATA

The tunicate fauna of the deep Gulf of Mexico has been discussed by Monniot and Monniot (1987) based on specimens collected by Gallaway, Martin, and Howard (1988). Eleven species were recognized, representing 6 families (table S83). Most of these species are also found in the South Atlantic, but one, *Bathystyeloides mexicanus,* is so far known only from the Gulf of Mexico. The Gulf tunicates have been taken down to a depth of 2,850 m. All of the species are representative of deepwater groups, and none are present on the continental shelf. Interestingly, the tunicate fauna of the northern Gulf slope achieves its greatest density and diversity in the depth range of 700–1,400 m, whereas in other oceanic areas the greatest diversity occurs at 4,000–5,000 m. Clearly, the deepwater tunicate fauna of the northern Gulf is derived from the South Atlantic rather than from the adjacent shelf, and the larvae are probably brought in by the Antarctic Intermediate Water at depths between about 500 and 1,000 m. This fauna resides largely in the deep Gulf water mass, which is characterized by very stable environmental conditions. Representative genera include *Araneum, Bathystyelloides,* and *Minipera.*

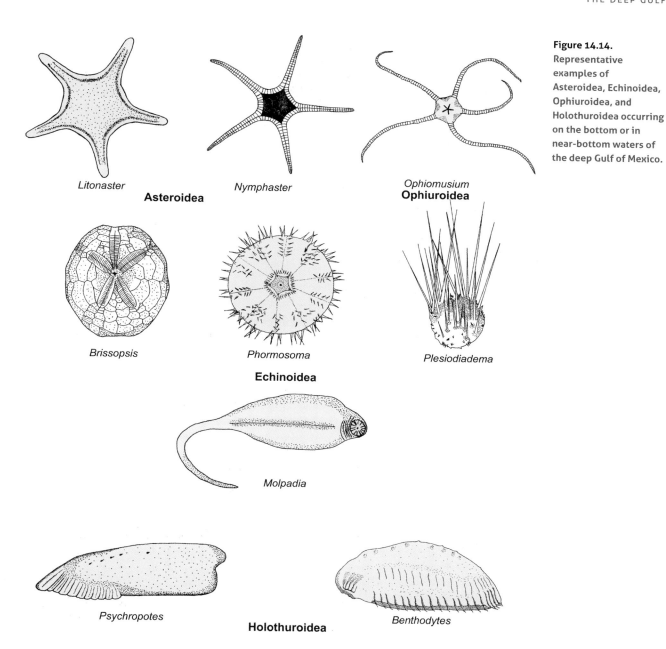

**Figure 14.14.** Representative examples of Asteroidea, Echinoidea, Ophiuroidea, and Holothuroidea occurring on the bottom or in near-bottom waters of the deep Gulf of Mexico.

*Litonaster*  *Nymphaster*

**Asteroidea**

*Ophiomusium*
**Ophiuroidea**

*Brissopsis*  *Phormosoma*

*Plesiodiadema*

**Echinoidea**

*Molpadia*

*Psychropotes*

**Holothuroidea**  *Benthodytes*

## Composition of the demersal fish fauna of the deep Gulf

Many of the bottom and near-bottom fish species of the outer continental shelf extend onto the upper continental slope, where their ranges overlap those of the true continental slope species. By a depth of about 500 m most of the shelf fishes have disappeared, having been replaced by upper- and middle-slope species. These, in turn, ultimately give way to the fish fauna of the lower slope and abyssal plain. In addition, food analysis has shown that some pelagic predators, such as sharks, often forage in bottom waters at least to a depth of 1,000 m. Our atten-

tion here is focused on those bottom-related fishes that inhabit the lightless world of the deep Gulf from the upper slope to the abyss, and on some of the pelagic predators that feed here. A list of the major groups of deepwater demersal fishes is provided in table S84. Included are 27 families representing 13 orders, some of which have already been discussed in the present or previous chapter. The primary reference to the deepwater demersal fishes of the Gulf is McEachran and Fechhelm (1998, 2005). Additional works of some interest include Anderson et al. (1985), Bekker, Shcherbachev, and Tchuvasov (1975), Bright (1968), Gallaway, Martin, and Howard (1988), D. M.

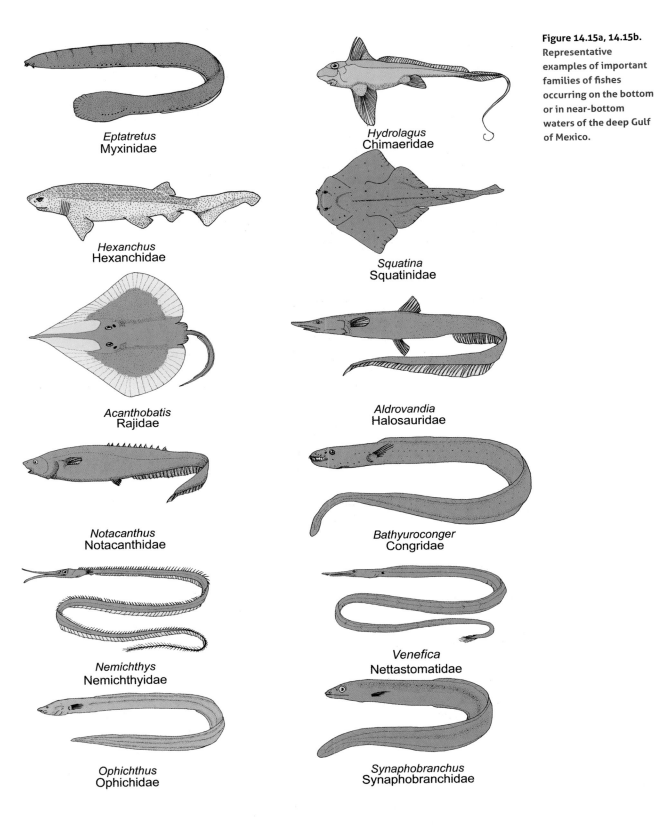

*Eptatretus*
Myxinidae

*Hydrolagus*
Chimaeridae

*Hexanchus*
Hexanchidae

*Squatina*
Squatinidae

*Acanthobatis*
Rajidae

*Aldrovandia*
Halosauridae

*Notacanthus*
Notacanthidae

*Bathyuroconger*
Congridae

*Nemichthys*
Nemichthyidae

*Venefica*
Nettastomatidae

*Ophichthus*
Ophichidae

*Synaphobranchus*
Synaphobranchidae

**Figure 14.15a, 14.15b.** Representative examples of important families of fishes occurring on the bottom or in near-bottom waters of the deep Gulf of Mexico.

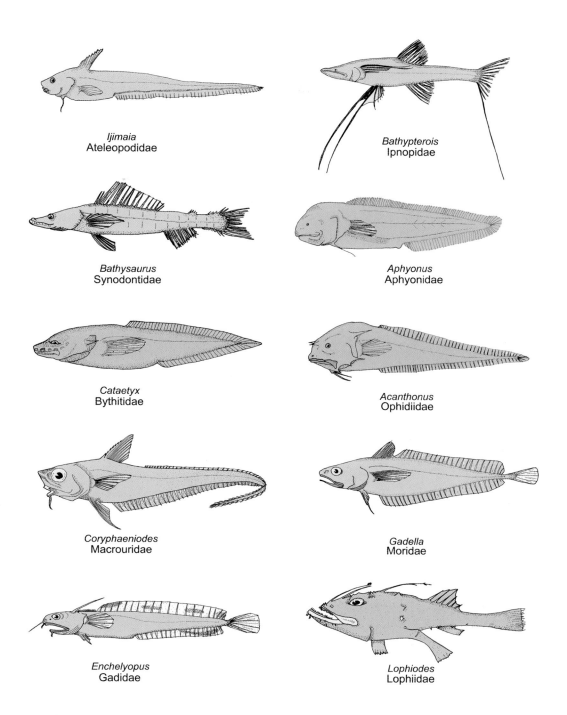

*Ijimaia*
Ateleopodidae

*Bathypterois*
Ipnopidae

*Bathysaurus*
Synodontidae

*Aphyonus*
Aphyonidae

*Cataetyx*
Bythitidae

*Acanthonus*
Ophidiidae

*Coryphaeniodes*
Macrouridae

*Gadella*
Moridae

*Enchelyopus*
Gadidae

*Lophiodes*
Lophiidae

Martin (1978), W. Nelson and Carpenter (1968), W. Pequegnat (1983), Potts and Ramsey (1987), Rass (1971), and Rayburn (1975).

Representative examples of many of the families are illustrated in figures 14.15a and 14.15b. Examination of these figures reveals that many of the deep-sea demersal fishes have eel-like bodies. Some, such as *Ijimaia, Aphyonus, Cataetyx, Acanthonus*, and *Lophiodes,* have large heads and reduced bodies. Most have reduced (*Gadella,*

*Enchelyopus*), pointed, or ratlike tails to reduce rear turbulence as they swim or slither along. Additional morphological adaptations include fleshy lips, long diverging jaws for snaring the antennae of shrimps (*Nemichthys*), chin barbels for sensing the sediments (*Coryphaenoides, Enchelyopus*), highly developed lateral line systems (*Hydrolagus*), stilt-like elongated fin rays for standing above or probing the sediments (*Bathypterois*), leg-like fins for creeping along the bottom (*Lophiodes*), and luminous lures for attracting prey (*Lophiodes*). In many of the deep-water benthic species the pelvic fins are greatly reduced or absent. Most of the species probably have highly developed senses of hearing (or feeling water pressure changes), touch, smell, and taste, which together aid in avoiding predators, locating food, and recognizing potential sex partners. Descriptions of the orders are given below, and individual families are taken up in table S85.

### MYXINIFORMES

The myxiniformes (hagfishes) are the most primitive group of living fishes. The long, scaleless body lacks jaws or paired fins. The skeleton contains no bone, and in adults the lateral line system is absent. Behind the mouth along both sides of the body lie a series of pore-like external gill openings. In the deep Gulf there is a single family, the Myxinidae.

### CHIMAERIFORMES

The chimaeras, or ratfishes, primitive cartilaginous fishes, have fairly large heads, enlarged pectoral fins, and tapering tails. The mouth, which is equipped with fleshy lips, is positioned on the bottom of the head (i.e., inferior), and the lateral line system, running around the head and down the length of the body, is much in evidence. There are two dorsal fins, which may be connected or slightly separated, and the tall anterior fin is preceded by a sharp spine, which may be hollow and connected to a venom gland. In the deep Gulf this order is represented by two families, Chimaeridae and Rhinochimaeridae, the latter possessing a long, pointed snout and relatively short tail.

### LAMNIFORMES

The lamniform sharks, discussed earlier in this chapter, are largely pelagic. Only a few species extend into the very deep waters, where they may forage on bottom animals. Deepwater species occur primarily in the families Lamnidae and Scyliorhinidae.

### HEXANCHIFORMES

These slender sharks (frill and cow sharks) are easily recognized by the presence of six or seven gill slits (instead of the usual five) and a single dorsal fin located far back near the tail. Only a single family, the Hexanchidae, is represented in the Gulf of Mexico.

### SQUALIFORMES

The squaliform sharks were taken up in the last chapter. Included are both pelagic and demersal species. Among the latter are the angel sharks, which are intermediate between the regular sharks and the ray-like fishes. In this group (Squatinidae) the body as well as the pectoral and pelvic fins are broad and flattened, but the paired fins are partially separated from the body and from each other. Although represented on the continental shelves, the squaliform fishes are widely distributed on the continental slope and have been recorded to a depth of 3,500 m. In the deep Gulf two families are represented, the Squalidae and the Squatinidae.

### RAJIFORMES

The rajiform fishes, including the skates and rays, were discussed in the previous chapter. Most species are limited to the continental shelf or upper slope, but a few members of the family Rajidae occur beyond a depth of 1,000 m.

### ALBULIFORMES

This order was discussed in the previous chapter, where the focus was on the shallow-water bonefish. However, two groups, the halosaurids and notacanthids, occur as demersal species in deeper waters of the Gulf. Their bodies tend to be long and slender, although a few are more deep bodied. The mouth is subterminal, and a spine projects backward and downward from the maxilla bone of the upper jaw. There may be a single dorsal fin or separated spines. The pectoral fins are located well up on the sides of the body, and the pelvic fins are abdominal and connected with each other at the base. The tail is tapered to a point, and the anal fin is continuous to the tail tip. These fishes are restricted to the continental slope, where they may extend below 2,500 m. Two families are represented in the Gulf, the Halosauridae and the Notacanthidae.

### ANGUILLIFORMES

The anguilliforms, containing the true eels, were discussed in the previous chapter. They are restricted to benthic environments. Although best represented on the continental shelf, some species extend onto or across the continental slope, and they have been recorded to depths of over 4,000 m. Important families with deepwater representatives include the Congridae, Nemichthyidae, Nettastomatidae, Ophichthidae, and Synaphobranchidae.

## OSMERIFORMES

The osmeriform fishes were addressed earlier in the present chapter. Most of those found in the Gulf are deepwater pelagic species, but some appear to inhabit near-bottom waters. This is particularly true for species of the family Alepocephalidae, some of which are known from depths of over 4,000 m.

## ATELEOPODIFORMES

These strange deepwater fishes (jellynose fishes) are of uncertain affinity, but they may be related to the Stomiiformes, discussed earlier. They have large bulbous heads and long tapering tails. The mouth is subterminal, and the skeleton is largely cartilaginous. There is a single high dorsal fin located behind the head. The pelvic fins of adults consist of a single long ray followed by several shorter rays. The long anal fin stretches back to the end of the tail. This order is represented by a single family, the Ateleopodidae.

## AULOPIFORMES

The aulopiform fishes were discussed in the previous chapter and earlier in the present chapter. Some species live on the bottom of the continental shelf, others are pelagic in the open Gulf, and yet others are encountered on or near the bottom on the continental slope or the abyssal plain. Families with demersal representatives in the deep Gulf include the Chlorophthalmidae, Ipnopidae, and Synodontidae.

## GADIFORMES

Gadiform fishes have been discussed in the previous chapter and earlier in the present chapter. Among the deepwater forms, many have short, eel-like bodies and tapering tails. Most possess chin barbels, and when present the pelvic fins are inserted far forward on the throat. All species are carnivorous, and most live on or near the bottom. In the deep Gulf members of this group are common on the continental slope, and some extend to a depth of over 3,500 m. Important deepwater families include the Aphyonidae, Bythitidae, Gadidae, Ophidiidae, Macrouridae, Moridae, and Phycidae.

## LOPHIIFORMES

The lophiiform fishes were taken up in the previous chapter and earlier in the present chapter. In benthic waters of the deep Gulf they are represented by a single family, the Lophiidae. These strange fishes are fairly rare, and they

extend across the continental slope to depths in excess of 800 m.

## Depth-related changes in the benthic meiofauna and macrofauna

### MEIOFAUNA

The reports by Gallaway, Martin, and Howard (1988) and W. Pequegnat, Gallaway, and L. Pequegnat (1990) provide our best information on the composition, numerical abundance, and biomass of both the meiofauna and macrofauna of the continental slope of the Gulf in relation to depth, season, and environmental factors. Within the meiofaunal collections 43 major invertebrate groups were recognized. Of these, 2 groups (nematodes and harpacticoid copepods) made up nearly 70 percent of the specimens, and these together with 4 additional groups (copepod nauplii, polychaetes, ostracods, and kinorhynchs) constituted 98 percent of the individuals collected. Most of the species were represented by just a few individuals. Across geographical areas, seasons, and depths the relative abundances of the major groups remained quite constant.

Meiofaunal densities ranged from 125,000 to 1,141,000 individuals/m². As seen in figure 14.16, the average densities decreased with depth from 575,700/m² (in the 200–1,000 m depth range) to 258,000/m² (in the 2,000–3000 m range), a decrease of 55.2 percent. Meiofaunal biomass dropped from 1,905 mg/m² (200–1,000 m range) to 794 mg/m² (2,000–3000 m range), a decrease of 58.3 percent.

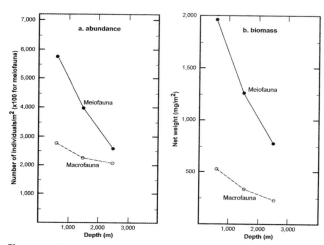

**Figure 14.16.** Numerical abundance and biomass density of meiofauna and macrofauna in relation to depth on the continental slope of the north-central and northeastern Gulf of Mexico. (Data from Gallaway, Martin, and Howard 1988; W. Pequegnat, Gallaway, and L. Pequegnat 1990.)

**Table 14.9.** *Average densities (no. of individuals/m²) of the most important meiofaunal and macrofaunal invertebrate groups in relation to depth based on transects in the north-central and northeastern Gulf of Mexico. (Data from Gallaway, Martin, and Howard 1988.)*

| Taxonomic group | Depth range (m) | | | | Percentage of total |
|---|---|---|---|---|---|
| | 290–492 | 500–900 | 1,000–1,500 | 2,000–2,945 | |
| Porifera | 16 | 28 | 22 | 50 | 0.7 |
| Nemertea | 51 | 44 | 30 | 13 | 0.7 |
| Nematoda | 1,025 | 1,198 | 1,156 | 719 | 25.7 |
| Polychaeta | 1,982 | 1,787 | 1,441 | 482 | 35.7 |
| Oligochaeta | 11 | 21 | 6 | 19 | 0.4 |
| Gastropoda | 20 | 20 | 15 | 7 | 0.4 |
| Aplacophora | 73 | 65 | 50 | 7 | 1.2 |
| Bivalvia | 271 | 227 | 260 | 124 | 5.5 |
| Scaphopoda | 36 | 20 | 21 | 17 | 0.6 |
| Ostracoda | 131 | 446 | 273 | 135 | 6.2 |
| Copepoda | 154 | 262 | 348 | 228 | 6.2 |
| Cumacea | 36 | 38 | 28 | 15 | 0.7 |
| Tanaidacea | 114 | 330 | 201 | 63 | 4.4 |
| Isopoda | 117 | 152 | 178 | 66 | 3.2 |
| Amphipoda | 66 | 102 | 78 | 12 | 1.6 |
| Sipunculida | 78 | 55 | 15 | 4 | 1.0 |
| Bryozoa | 213 | 256 | 91 | 121 | 4.3 |
| Brachiopoda | 1 | 2 | 2 | 26 | 0.2 |
| Ophiuroidea | 40 | 45 | 37 | 18 | 0.9 |
| Holothuroidea | 5 | 27 | 11 | 5 | 0.3 |
| Totals | 4,440 | 5,125 | 4,263 | 2,131 | 100.1 |

Although the trend of logarithmic decrease with depth is clear, locally high densities were observed near hydrocarbon seeps. The abundance of both nematodes and harpacticoids was highest in fine-grained sediments and decreased as grain size increased.

MACROFAUNA

In the continental slope collections referred to above, Gallaway, Martin, and Howard (1988) and W. Pequegnat, Gallaway, and L. Pequegnat (1990) obtained nearly 50,000 individual macrofaunal organisms representing 1,569 separate taxa. Although 40 major invertebrate groups made up the collections, they were dominated by polychaetes followed by nematodes, and these 2 groups accounted for 64 percent of the collections (table 14.9). These plus the next 6 groups (ostracods, copepods, bivalves, tanaids, isopods, and bryozoans) made up about 90 percent of the individuals taken. The various invertebrate groups reached peak densities in different depth ranges. Among the 20 most abundant invertebrate groups, the depth ranges at which each peaked are given as follows:

290–492 m: seven groups (nemerteans, polychaetes, gastropods, aplacophorans, bivalves, scaphopods, sipunculids)

500–900 m: ten groups (nematodes, oligochaetes, gastropods, ostracods, cumaceans, tanaids, amphipods, bryozoans, ophiuroids, holothuroids)

1,000–1,500 m: two groups (copepods, isopods)

2,000–2,945 m: two groups (poriferans, brachiopods)

Note that the gastropods were equally abundant in the two shallowest depth ranges. Some of the individual macrofaunal species exhibited very wide depth ranges. For example, among the polychaetes, the most thoroughly studied group (Hubbard 1995), it was found that some of the more abundant species extended throughout the entire depth range sampled (ca. 300–3,000 m). The results of these studies suggest that the macrofauna of the north-central and northeastern Gulf is composed of a great many rare species and that within given samples there is little tendency toward numerical dominance by any one species.

Macrofaunal abundance values (fig. 14.16) ranged from 2,100 to 2,800 individuals/m². Average values decreased from 2,848/m² (200–1,000 m depth) to 2,112/m² (2,000–3,000 m), a decrease of 25.8 percent. Macrofaunal densities decreased from 554.4 mg/m² (200–1,000 m) to 205.3 mg/m² (2,000–3000 m), a drop of 63.0 percent. Local anomalies did occur where higher than expected densities were observed on the slope, and in several instances densities at a given station were higher during the spring than in the fall. On the upper slope the meiofauna was 202.1 times as numerous as the macrofauna, but on the lower slope it was only 122.2 times as numerous. In terms of biomass, on the upper slope the meiofauna was 3.44 times that of the macrofauna, and on the lower slope it was 3.87 times as abundant. Thus, despite being 100 to 200 times as numerous, the meiofauna represented only about 3.5 times the biomass, and this was because of the smaller size of the individual meiofaunal animals.

The slope macrofaunal abundance and biomass data given above can be compared with those of Rowe and Menzel (1971) and Rowe, Polloni, and Horner (1974), discussed earlier (see fig. 14.4). The databases are not exactly comparable since Rowe and coworkers used sieves of 0.42 mm mesh and Gallaway and coworkers employed sieves of 0.30 mm mesh, the smaller-mesh sieves being expected to retain more individuals than the larger-mesh screens. However, mesh size alone cannot account for the differences in the results. The extensive data of Gallaway and coworkers show that in the depth range of 2,000–3,000 m the slope macrofauna of the Gulf is almost ten times (9.6 times) as abundant and has a biomass over twice (2.3 times) as great as that reported by Rowe and co-workers. There appears to be considerable local variation in macrofaunal density on the continental slope, as on the shelf, and this in turn probably reflects local variations in food availability, competition, and predation pressure.

## Benthic megafaunal zonation in the deep Gulf

Just as there is vertical zonation in species composition and abundance in the water column (fig. 14.1), so there is zonation in the fauna associated with the bottom, and this shows up most clearly in the benthic megafauna (Carney, Haedrich, and Rowe 1983; Haedrich, Rowe, and Polloni 1975, 1980; Menzies, George, and Rowe 1973). However, since the ecosystems of the water column and the benthos are controlled in part by different factors, the faunal boundaries of the two realms do not necessarily coincide. Conceptually, at least, benthic faunal zones exist as broad depth bands within which there is little change in faunal composition, and these are separated by narrow bands where the rate of species replacement is high. These benthic bands or zones are defined in part by depth and characteristics of the water column and bottom sediments, but mostly by biological factors such as food availability and competition. Although the benthic zones seem to be real, their apparent limits may vary somewhat depending on where the zonation is studied, the taxonomic groups being considered, the sampling patterns, and the statistical techniques used to define the zones.

Based on studies throughout the Gulf and particularly in the north-central and northeastern Gulf, Gallaway, Martin, and Howard (1988) and W. Pequegnat, Gallaway, and L. Pequegnat (1990), using cluster analysis and chi-square tests, defined the benthic zones as follows:

Shelf/Slope Transition Zone (118–475 m)
Archibenthal Zone (500–975 m)
Horizon A (500–775 m)
Horizon B (800–975 m)
Upper Abyssal Zone (1,000–2,275 m)
Meso-Abyssal Zone (2,300–3,225 m)
Lower Abyssal Zone (3,250–3,750 m)

This classification differs slightly from that proposed by W. Pequegnat (1983). However, these classification systems have not been borne out by the work of Hubbard (1995) and ongoing studies of the deep Gulf by Rowe (personal communication). Until more definitive data are available it seems appropriate to divide the benthic megafaunal zones of the Gulf conservatively as follows:

Upper Slope (200–1,000 m)
Mid-Lower Slope (1,000–2,000 m)
Deep Slope and Abyssal Plain (2,000–3,750 m)

The benthic megafauna of each of these zones is described briefly below, and some common or characteristic species of each zone are listed in table 14.10.

**Table 14.10.** *Megafaunal species that are common or characteristic of the various depth zones of the continental slope, continental rise, and abyssal plain of the deep Gulf of Mexico. (Data from Gallaway, Martin, and Howard 1988; W. Pequegnat 1983; W. Pequegnat, Gallaway, and L. Pequegnat 1990.)*

### Upper Slope (200–1,000 m)

| | |
|---|---|
| Penaeids | *Penaeopsis serrata* |
| Carideans | *Parapandalus willisi, Plesionika holthuisi* |
| Galatheids | *Munida forceps, M. longipes, M. valida, Munidopsis robusta* |
| Brachyurans | *Acanthocarpus alexandri, Bathyplax typhla, Benthochascon schmitti, Chaceon quinquedens, Lyreidus bairdii, Rochinia crassa* |
| Asteroids | *Astropecten nitidus, Luidia elegans, Midgardia xandaros* |
| Echinoids | *Brissopsis alta, B. atlantica, Phormosoma placenta* |
| Fishes | *Bathygadus macrops, B. melanobranchus, Bembrops gobioides, Caelorhinchus caribbaeus, C. caelorhinchus, Dibranchus atlanticus, Nezumia aequalis, Poecillopsetta beanii* |

### Mid-Lower Slope (1,000–2,000 m)

| | |
|---|---|
| Carideans | *Glyphocrangon aculeata, Nematocarcinus rotundus* |
| Macrurans | *Stereomastis sculpta* |
| Holothuroids | *Benthodytes sanguinolenta, Mesothuria lactea* |
| Fishes | *Gadomus longifilis* |

### Deep Slope and Abyssal Plain (2,000–3,750 m)

| | |
|---|---|
| Carideans | *Nematocarcinus acanthitelsonis, N. ensifer* |
| Galatheids | *Munidopsis bermudezi, M. geyeri* |
| Asteroids | *Ampheraster alaminos, Dytaster insignis* |
| Holothuroids | *Protankyra brychia, Psychropotes depressa* |
| Fishes | *Barathronus bicolor, Bassozetus normalis, Bathytroctes macrolepis, Dicrolene kanazawi* |

## UPPER SLOPE (200–1,000 M)

This zone extends from the edge of the continental shelf to a depth of 1,000 m. W. Pequegnat, Gallaway, and L. Pequegnat (1990) placed the upper limit at 118 m (more or less appropriate for the Texas-Louisiana slope), but as noted by Martin and Bouma (1978), in the Gulf of Mexico the outer edge of the continental shelf may occur at any depth from 60 to 270 m. Many outer-shelf species extend beyond the shelf break where their ranges overlap those of upper-slope species, but the details have not been worked out. This transitional area is a very productive zone containing a great abundance of megafaunal animals. Biodiversity is very high, and the area is particularly rich in species of crabs, starfishes, and fishes. By the middle portion of the Upper Slope (500–750 m) there is some reduction in the number of species, and fewer species reach peak densities here. The continental shelf species have largely disappeared, but abundance and diversity are still high, especially among the fishes. There is also a great increase in the number of species of caridean shrimps in this depth range. The lower portion of the Upper Slope (750–1,000 m) is characterized by a great reduction in the abundance and diversity of several invertebrate groups such as the brachyuran crabs. There is some diminution in the number of fish species and a great reduction in the number that reach peak population densities here.

## MID-LOWER SLOPE (1,000–2,000 M)

In moving from the Upper to the Mid-Lower Slope there is a significant change among the galatheid crabs (fig. 14.17), with the genus *Munidopsis* (now with 11 species) largely replacing the shallower-water *Munida* (now with only 3 species). The diversity of brachyuran crabs is greatly reduced (only 4 species here as contrasted with 35 species in upper portions of the Upper Slope). There is a major increase in the abundance of large sea cucumbers (Holothuroidea). The number of fish species is greatly reduced, there being only half as many as in the previous zone, but many of those present reach peak population densities here.

## DEEP SLOPE AND ABYSSAL PLAIN (2,000–3,750 M)

The upper portion of this zone is characterized by a sharp faunal break that is due, in part, to the presence of steep escarpments at this depth. Compared with the previous zone, the fauna here is quite depauperate. For most invertebrate and fish groups there is a reduction in both abun-

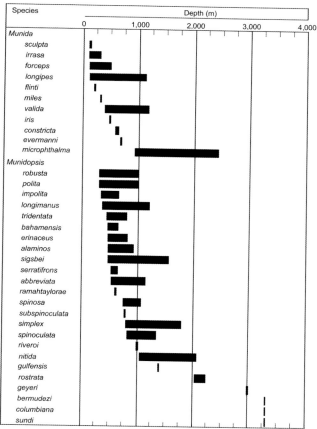

**Figure 14.17.** Comparison of the known vertical distribution ranges of species of the anomuran crab genera *Munida* and *Munidopsis* found in the Gulf of Mexico. Only 3 of the 11 species (27.3 percent) of *Munida* occur as deep as 1,000 m, but 16 of the 24 species (66.7 percent) of *Munidopsis* achieve this depth, and 4 species (16.7 percent) occur as deep as 3,000 m. (Data from Gallaway, Martin, and Howard 1988; W. Pequegnat 1983.)

dance and diversity, and many of the species encountered are unique to this zone.

Further perspective on the depth zonation of the megafauna can be achieved by examination of data derived largely from W. Pequegnat (1983) and based on collections made throughout the Gulf. Table 14.11 presents lists of the numbers of species in 18 megafaunal invertebrate groups and fishes collected within different depth intervals. These intervals essentially coincide with the zones and horizons proposed in Pequegnat's zonal classification mentioned above. Here it is seen that all groups are fairly well represented down to a depth of 2,250 m. However, below that depth the diversity drops off sharply, and the fauna of the two deepest intervals contains far fewer species (162 and 126) than any of the previous inter-

vals (all over 400 species). Only the sponges, polychaetes, gastropods, bivalves, sea cucumbers, and fishes are represented by more than 10 species. Obviously, more megafaunal species inhabit the deep continental slope and abyssal plain than are indicated in the table, but when population densities are so low, many collections are required to provide an adequate picture of the real diversity. Furthermore, many deep Gulf species live buried in the sediments, and the true nature of the density and diversity will emerge only after techniques have been worked out to provide adequate samples of this hidden fauna. Although estimates of megafaunal biomass have been attempted, trawl samples are only semiquantitative at best, and such estimates must be considered tentative.

## Seep communities

From deep reservoirs underlying the northern Gulf continental slope, and possibly elsewhere, petroleum and natural gas migrate upward through faults, ultimately passing through layers of unconsolidated surface sediments and seeping into the water column. Some of this material is retained in the sediments, where it tends to spread out over a larger area than the fault through which it migrated. In the upper meter or so of the sediments, microbial degradation of the labile hydrocarbons depletes the available oxygen and reduces the sulfates of the seawater to hydrogen sulfide. Microbial activity also increases the alkalinity, resulting in extensive local precipitation of carbonate as layers and rocks within and upon the sediments. The buildup of carbonates and the accumulation of mud and fluids ultimately produce low mounds 10–500 m in diameter on the seafloor. From and around such mounds, methane gas and liquid hydrocarbons may bubble from the seafloor, often accompanied by hydrogen sulfide and high-salinity brine dissolved from subsurface salt deposits. Where gas becomes trapped beneath layers of rock or other shallow obstructions it may form gas hydrates, methane-rich ice-like deposits. This strange environment (rich in methane, liquid petroleum, hydrogen sulfide, brine, mud, and carbonate precipitates) provides habitat for a specialized group of seafloor organisms collectively referred to as hydrocarbon seep chemosynthetic communities.

Chemosynthesis is a process whereby specialized bacteria derive energy by oxidizing chemically reduced substrates, especially methane ($CH_4$) or hydrogen sulfide ($H_2S$). To accomplish this, the bacteria require not only

**Table 14.11.** *Numbers of species in various megafaunal groups collected at different depth intervals in benthic and near-benthic environments of the deep Gulf of Mexico. (Data from Gallaway, Martin, and Howard 1988; W. Pequegnat 1983.)*

| Taxonomic group | Depth range (m) | | | | |
|---|---|---|---|---|---|
| | 150–450 | 475–950 | 975–2,250 | 2,275–3,200 | 3,225–3,750 |
| Porifera | 16 | 19 | 18 | 12 | 14 |
| Cnidaria | 16 | 34 | 27 | 4 | 1 |
| Polychaeta | 63 | 83 | 29 | 23 | 17 |
| Mollusca | | | | | |
|   Gastropoda | 59 | 122 | 117 | 21 | 15 |
|   Bivalvia | 21 | 14 | 34 | 20 | 12 |
|   Scaphopoda | 9 | 10 | 12 | 5 | 5 |
|   Cephalopoda | 6 | 11 | 19 | 3 | 4 |
| Arthropoda | | | | | |
|   Natantia | | | | | |
|     Penaeidea | 7 | 7 | 14 | 7 | 6 |
|     Caridea | 9 | 19 | 14 | 4 | 5 |
|   Macrura | 2 | 6 | 5 | 2 | 2 |
|   Anomura | | | | | |
|     Galatheids | 13 | 23 | 13 | 2 | 3 |
|     Pagurids | 16 | 11 | 4 | 1 | 1 |
|   Brachyura | 34 | 13 | 4 | 1 | 1 |
| Echinodermata | | | | | |
|   Asteroidea | 18 | 25 | 31 | 10 | 8 |
|   Echinoidea | 20 | 9 | 6 | 2 | 1 |
|   Ophiuroidea | 10 | 25 | 17 | 10 | 6 |
|   Holothuroidea | 4 | 20 | 23 | 17 | 10 |
|   Crinoidea | 9 | 4 | 4 | 0 | 0 |
| Total invert. spp. | 332 | 455 | 395 | 144 | 111 |
| Fishes | 90 | 108 | 72 | 18 | 15 |
| Total species | 422 | 563 | 467 | 162 | 126 |

the substrate but also an electron acceptor such as oxygen or nitrate. The chemosynthetic bacteria may be either free living or symbiotic, that is, living within the cells of certain specialized invertebrate hosts. The primary species that make up the chemosynthetic communities of the northern Gulf continental slope are listed in table 14.12 and described below.

FREE-LIVING CHEMOSYNTHETIC BACTERIA
The free-living chemosynthetic bacteria (*Beggiatoa* sp.) consist of long filaments that, in aggregate, constitute large colonies or mats growing over the surface of hydrocarbon-

rich sediments. Within the individual cells large vacuoles store nitrates, useful as electron acceptors as the bacteria oxidize hydrogen sulfide. In order to obtain the hydrogen sulfide the bacterial filaments extend from the surface as deep as 10 cm into the sediments. At the hydrocarbon seeps the bacterial mats may extend laterally for several meters. Some portions of a mat may be white while others are orange, and the two color phases may represent bacteria utilizing different modes of nutrition (autotrophic and heterotrophic). Beneath the mats there is often a high density of living foraminifera, and in the near-bottom waters there is often a high concentration of bacterioplankton.

**Table 14.12.** *Species composition of seep communities of the northern and eastern Gulf of Mexico and associated fauna. (From I. MacDonald 2000.)*

**Primary seep species**

Bacteria – *Beggiatoa* sp.

Annelids

    Vestimentiferan tube worms – *Escarpia* sp., *Lamellibrachia* sp.

    "Ice worms" – *Hesiocaeca methanicola*

Bivalves

    Vesicomyid clams (Vesicomyidae) – *Calyptogena ponderosa, Vesicomya cordata*

    Lucinid clams (Lucinidae) – *Lucinoma atlantis, Thiasira oleophila*

    Seep mussels (Mytilidae) – *Bathymodiolus brooksi, B. childressi, B. heckerae, Idas macdonaldi, Tamu fisheri*

**Associated fauna**

Sponges and hydroids – epifaunal on both species of tube worms

Bivalves – *Acesta bullisi* – on tubes of *Lamellibrachia*

Gastropods – *Cataegis* sp. – epifaunal on mussels

Isopods – *Bathynomus giganteus*

Decapods

    Caridean shrimps – *Alvinocaris* sp. – on mussels

    Galatheid crabs – *Munidopsis* sp. – on escarpiid tube worms

    Brachyuran crabs – *Bathyplax typhla, Benthochascon schmitti, Rochinia crassa*

Fishes – *Chaunax pictus, Eptatretus* sp., *Hoplostethus* sp., *Peristedion greyae, Urophycis cirrata*

---

VESTIMENTIFERAN TUBE WORMS

The vestimentiferans are highly specialized polychaete worms that lack digestive tracts (mouth, gut, and anus) and secrete long tubes of ß-chitin within which they live. The tube is firmly anchored within the sediments (for up to 30 percent of its length), and the free end extends well up into the water column. At its anterior end the worm has a highly vascularized red plume, which normally projects into the water column but which can be withdrawn into the tube for defense. At the posterior end a complex rootlike structure projects below the tube into the sulfide-rich sediments. Inside the tube the worm contains a saclike organ (*trophosome*) that houses a colony of symbiotic chemosynthetic bacteria that live within spe-

cialized host cells. The hemoglobin of the worm's blood is capable of simultaneously and reversibly binding oxygen and hydrogen sulfide. In practice, the worm absorbs hydrogen sulfide from the sediments and transfers it to the trophosome, where it is oxidized by the bacteria. The bacteria supply the host with energy-rich chemicals useful in its metabolism, and the worm ultimately excretes the waste sulfur compounds. In the northern Gulf there are two species of tube worms (*Lamellibrachia luymesi* and *Escarpia laminata*), both new to science. The former are larger and grow up to two meters in length, of which a meter or more projects into the water column, and they live in clusters of several hundred around the hydrocarbon seeps. Worms of the latter species are smaller and project only 10–15 cm above the surface of the sediments. These generally live around the base of the *Lamellibrachia* colonies. The tube worms have slow growth rates of about one centimeter per year, and considering the length of the adults, their life span must exceed 200 years.

"ICE WORM"

The so-called ice worms (*Hesiocaeca methanicola*) inhabit shallow burrows in the surface of gas hydrate deposits—an incredible habitat. Although these recently discovered polychaetes apparently do not harbor internal symbiotic bacteria, their nutrition seems somehow to be derived from the hydrocarbons.

SEEP MUSSELS

Associated with the northern Gulf hydrocarbon seeps are several species of the mussel family Mytilidae. Some species contain methane-oxidizing bacteria in their gill tissues, some harbor sulfide-oxidizing bacteria, and at least one species has both types. Some also possess complete guts and appear to be capable of ingesting particulate food. Seep mussels attach to one another by byssal threads, creating mats that may extend across the seafloor for tens of meters. It has been estimated that these mussels live up to 40 years.

VESICOMYID CLAMS

Two species of giant clams of the family Vesicomyidae have been encountered in association with seep communities of the northern Gulf, *Calyptogena ponderosa* and *Vesicomya cordata*. Adults reach lengths of 75–90 cm and have deep, heavy-bodied shells. Aggregations of these surface-dwelling clams are found in petroleum flow fields where expulsion of oil-rich mud generates shallow anoxic layers charged with sulfides. In feeding, the clam pushes its foot deep into the sediments, where it absorbs hydrogen sulfide. This is transferred by specialized hemoglobin molecules to the gills, where it is oxidized by intracellular symbiotic bacteria.

## LUCINID CLAMS

Two species of these small clams (*Lucinoma atlantis* and *Thyasira oleophila*) have been encountered around northern Gulf hydrocarbon seeps. These small round clams live in deep U-shaped burrows, and like the vesicomyids they harbor sulfur-oxidizing bacteria in their enlarged gills.

## ASSOCIATED FAUNA

Loosely associated with the methane- and sulfide-dependent species of the seep community proper are a variety of invertebrate and fish species derived from the general slope fauna. As noted in table 14.12, a number of species (sponges, hydroids, gastropods, bivalves, caridean shrimps, and galatheid crabs) are epifaunal on tube worms or mussels. Bathypelagic species of the water column (squids, tunicates, and trichiurid fishes) swim or hover nearby. Benthic species (giant isopods and several species of brachyuran crabs and fishes) are also attracted. Although not strictly dependent on the seep fauna, these associated species find here opportunities for attachment and/or feeding. Primary references to the northern Gulf hydrocarbon seeps and seep communities include Kennicutt, Brooks, et al. (1985), Kennicutt, Brooks, Bidigare, et al. (1988), I. MacDonald (2000), I. MacDonald, Boland, et al. (1989), and I. MacDonald, Callender, et al. (1990). The seeps and their inhabitants are also well illustrated in an article in *National Geographic Magazine* (I. MacDonald and Fisher 1996).

## *TYPES AND DISTRIBUTION OF NORTHERN GULF SEEP COMMUNITIES.*

The development of seep communities is controlled by the intensity and composition of the seep discharges. Rapid fluid flux produces "mud volcanoes." Here, although hydrocarbons are abundant, the slow-growing seep fauna may be subject to frequent burial by mud flows. Active fluid flow often results in quantities of brine being discharged, and this may form dense pools in depressions or channels from brine runoff. Around the edges of such pools and channels live colonies of seep mussels. Decreased rates of venting permit full development of the seep communities and result in the accumulation of mounds or domes on the bottom that may be capped with lithified carbonate layers. These represent the terminal stages of seeps. Lithification reduces the porosity and limits seepage to faults and fissures in the crust. Layers of bivalve shells may remain long after chemosynthetic production has ceased.

To date, more than 50 seep community sites have been discovered on the northern Gulf slope between 88°00′ and 94°47.5′ W longitude (Mobile Bay to Galveston Bay) and in the depth range of about 500–3,300 m. At all depths vestimentiferan tube worms, mussels, and bacterial mats dominate, with occasional areas of vesicomyid clams and pogonophoran tube worms, but the species composition of the worms and mollusks tends to change with depth (H. Roberts et al. 2007). There are also changes in the associated fauna, with gastropod mollusks being more abundant in shallower water (above 900 m) and echinoderms (a brittle star, a sea cucumber, and heart urchins) being more prominent in deeper water (below 1,400 m).

## SEEP COMMUNITIES OFF THE FLORIDA ESCARPMENT

At the base of the Florida Escarpment at 26°02′ N and 84°55′ W (west of the Everglades) at a depth of about 3,200 m there is a continental margin brine seep formed by the dissolution of materials (probably halite and gypsum) deep beneath the Florida Plateau. The emerging brine is enriched with hydrogen sulfide and possibly methane. The seep area is inhabited by vestimentiferan tube worms and *Bathymodiolus*-like mussels (*Bathymodiolus heckerae*). The tube worms are dependent on the hydrogen sulfide, and the mussels apparently rely on methane. However, because of its depth, the site can be visited only by submersible, and it has not been well studied. References pertinent to this site include Hecker (1985), Paull and Neumann (1987), Paull, Hecker, et al. (1984), and Paull, Jull, et al. (1985).

The hydrocarbon and hydrogen sulfide seeps of the Gulf of Mexico are referred to as "cold seeps" because their temperature is not elevated above the background levels. In this respect, they stand in contrast with the more widely known hydrothermal vents of scalding water located in areas of seafloor spreading off the Galápagos and in other rift valleys of the Pacific and elsewhere. However, the hydrothermal vents supply sulfides that support chemosynthetic communities made up, in part, of large colonies of vestimentiferan tube worms and vesicomyid clams related to those of the Gulf of Mexico, and in both the cold and thermal seeps the communities are supported, in part, by symbiotic chemosynthetic bacteria harbored by their invertebrate hosts.

## Photographic views of the deep Gulf bottom

Since pictures are often more eloquent than verbal descriptions, we shall conclude this chapter with a few photographs of the surface of the deep Gulf (figs. 14.18 and 14.19). These have been selected to illustrate sediment types, topography, biological species in their natural habitats, and evidence of biological activity. A few preliminary remarks will enhance the viewer's ability to interpret the photos. The sediments themselves are all composed of mixtures of particles of different sizes. The predominance of silts and clays produces a coherent, often compact, material that would popularly be characterized as "mud," and this is the dominant sediment type on the northern Gulf shelf, slope, and fan, especially where they receive outflow particulates from the Mississippi River. This material is good for burrowing by a variety of invertebrates and some fishes. Coarse material would be recognized as "sand," and this occurs in both shallow and deep areas where water currents sweep away the finer particles. This does not make good burrows, but some species of bivalves, echinoderms, and others do bury themselves in coarse sediments.

The material making up the surface sediments includes both river-borne clastic inorganic particles and biologically derived carbonates and other substances including planktonic skeletal material (primarily from coccolithophores, foraminifera, and pteropods) as well as shell and skeletal fragments of benthic animals (chiefly snails, bivalves, tusk shells, and echinoderms). Upon the surface of the sediments may lie fragments of land-derived woody materials and marine vegetation, turtle grass (*Thalassia*) from shallow-water beds in the West Indies and floating seaweed (*Sargassum*). Turtle grass is limited largely to the deep-bottom areas of the eastern Gulf, but *Sargassum* is found on bottoms throughout the Gulf. Both are transported northward by the Yucatán Current.

A salient feature of the surface of deepwater sediments is the evidence they provide of disturbance by various types of biological activity (*bioturbation*). This manifests itself in tracks and trails; mounds and depressions; burrows, holes, and groups of holes; feeding marks; and fecal deposits. Some are distinctive and can be attributed to known species, but the majority cannot be identified with known animals. In many areas, especially in fine sediments, the subsurface material is lighter in color than the material on the surface. Therefore, patches and deposits of lighter color are generally indicative of recent biological activity. Many photographs of the deep ocean sediments may be found in works such as Heezen and Hollister (1971) and Menzies, George, and Rowe (1973). Photographs of the bottom of the deep Gulf may be seen in W. Pequegnat (1983) and W. Pequegnat, James, et al. (1972).

**Figure 14.18.**
Photographs of the bottom surface of the Gulf of Mexico in the depth range of 500–1,000 m. (The photos in this and the following figure were taken by W. E. Pequegnat and made available by L. H. Pequegnat.)

a. Western Gulf, southeast of Corpus Christi, Texas (96°10′ W), 714 m. Close-up of well-consolidated fine particulate sediments showing much evidence of biological activity (holes of various sizes, trails, and fecal deposits).

b. Western Gulf, southeast of Corpus Christi, Texas (96°10′ W), 715 m. Close-up of well-consolidated sediments showing various types of bioturbation and a brittle star (*Bathypectinura heros*) moving from the lower left toward the upper right. Note depressions made in the sediment surface by the brittle star's arms as it moves.

c. Western Gulf, southeast of Corpus Christi, Texas (96°10′ W), 715 m. Wider view of surface sediments showing various types of bioturbation, a few white shell fragments, and a penaeid shrimp (*Plesiopenaeus edwardsianus*), in the upper left, swimming about a meter above the bottom and casting a shadow on the bottom (lower left).

d. Eastern Gulf, DeSoto Canyon area (87°04′ W), 770 m, near the east wall of the DeSoto Canyon trough. Here the sediments are composed of silty carbonate sand smoothed by bottom currents. Some shell fragments and bioturbation are evident. The fish on the bottom is probably a macrourid.

**Figure 14.19.**
Photographs of the
bottom surface of the
Gulf of Mexico in the
depth range of 1,000–
2,000 m. Photographs
by Ian McDonald.

a. Eastern Gulf, near east wall of DeSoto Canyon (87°29.5' W), 1,100 m. The sediment is very fine clayey mud. The brachyuran is the giant red crab *Chaceon quinquedens,* with a leg span of about 30 cm (12 in.). The white objects on its legs are barnacles of the genus *Scalpellum.* Note the numerous slash marks made in the sediment surface by walking movements of the crab's legs..

b. Eastern Gulf, below DeSoto Canyon (87°20' W), 1,564 m. The bottom is silty mud with some shell hash. The fish swimming just above the bottom is a synaphobranchid eel. The various holes and mounds were created by unknown animals, but the ring of holes toward the left is similar to that made by the caridean shrimp *Glyphocrangon.*

# Part IV

ECOLOGICAL PROCESSES

# 15 : ECOLOGICAL PROCESSES I
## ENVIRONMENTAL FACTORS AND SPECIES ADAPTATIONS

In the previous chapters we have examined the Gulf of Mexico from the perspective of classical oceanography. However, in order to gain a deeper understanding of the processes at work and how the system really functions, we also need to examine the Gulf and its inhabitants from the standpoint of classical ecology. Although some of the material will already be familiar to the reader, this information will be integrated with new knowledge and examined from a different perspective. Through the years classical oceanography has built up a body of knowledge about marine systems in terms of how and by whom the data were collected, that is, by meteorologists, geologists, and physical and chemical oceanographers, on the one hand, or by biologists studying the phytoplankton or zooplankton, nekton, or benthos, on the other. Classical ecology cuts across these disciplines and focuses more on the individual organism and its adaptations to environmental factors, life history patterns, and interspecies relationships. It is concerned with how small species aggregations form and how they build up in hierarchical fashion into larger integrated systems and ultimately into the total functioning ecosystem.

The discussion of ecological processes will occupy two chapters. The present chapter deals with various environmental factors that, taken together, constitute the habitats of the marine species. Attention will focus on the physical and biological factors, how these affect the organisms, and how the organisms cope with the environmental constraints. The subsequent chapter will be concerned with various aspects of life histories and with species aggregations of various size and complexity. Although real understanding of marine ecosystems appears to lie some years in the future, this double approach, through oceanography and ecology, should place us a step closer to this long-term goal.

### RANGES OF TOLERANCE AND LIMITING FACTORS
Earlier we examined the various physical and chemical conditions occurring in different areas of the Gulf of Mexico, but for the individual plant or animal species these boil down to whether a given environment is or is not a suitable place to settle down, live, and reproduce.

Hence, it is important to explore what really constitutes a suitable habitat for an organism. Laboratory studies have shown that for each physical factor (such as temperature, salinity, etc.) there is a range of values in which the body functions best, the various enzyme systems work well together, and the body is not under stress (fig. 15.1). Outside this range, say at higher or lower temperatures, there is an upper and a lower suboptimal range in which regular functions can take place, but the body is under some stress, and the functions are not being carried out with the greatest efficiency. At the extremes the subopti-

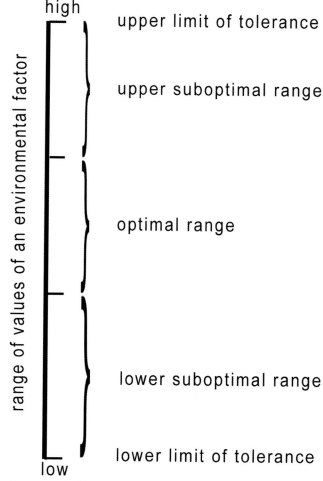

**Figure 15.1.** An idealized diagram of an organism's vital range and limits of tolerance with respect to a given environmental factor.

mal ranges are bounded by conditions (high and low temperature, for example) that represent the limits of tolerance beyond which the body cannot function, and death ensues. The vital ranges and limits of tolerance may vary with different life history stages of a given species (seeds or planktonic larvae versus adults), and they can be shifted somewhat by gradual acclimation rather than sudden exposure, but even here absolute limits are soon achieved. This general physiological pattern of optimal and suboptimal ranges and limits of tolerance applies to all macroscopic organisms with respect to such factors as temperature, salinity, pressure, oxygen content of the water, and so on. For photosynthetic plants it also applies to the factor of light. If there is insufficient light, photosynthesis cannot take place, but if the radiation is too intense, photoinhibition results.

The ranges and limits with respect to one factor are affected by other environmental factors acting in concert. Thus, the tolerance for low or high salinity may be affected by prevailing conditions of temperature, dissolved oxygen, and so forth. In nature, organisms must face and adjust to a number of factors acting in combination, and the ability to function and survive is determined, in part, by how well the individual can handle ambient conditions in a multifactor and often changing environment.

Most oceanic and coastal environments are inhabited by a great many different species whose vital ranges overlap and whose limits of tolerance may be nearly identical. There is a great redundancy of species that could inhabit most marine environments. Hence, the success of a given species in an area often depends on its ability to outcompete its neighbors while adjusting to the physical conditions. Slight differences in tolerances and competitive abilities can determine the outcome. Variable and extreme environmental conditions are most often encountered in shallow coastal areas, and it is here that the physical factors are most likely to play the dominant role. In waters of the open Gulf and on bottoms of the outer shelf and deep Gulf, physical factors tend to be more stable and less extreme. Here biological factors such as competition, food supply, and predation appear to be the most critical.

## PHYSICAL FACTORS OF THE ENVIRONMENT

It is instructive to review the primary physical factors in terms of their effects on species inhabiting the Gulf and in light of the physiological ranges and limits discussed above.

## Temperature

Temperature is sometimes referred to as the "master factor." Since it controls the rates of all chemical reactions, including those mediated by enzymes, it operates within the organism's body. Most species inhabiting the Gulf are *poikilothermic* (i.e., they are unable to regulate their own body temperatures), and thus, their internal body temperatures are set by those of the external environment. To a considerable extent, temperature also affects the body's ability to deal with changes and extreme conditions of other factors.

The Gulf of Mexico occupies an area extending from about 18° to 30°N latitude and is bisected by the Tropic of Cancer, which runs from east to west at around 23.5°N latitude. Its northern portion lies in the temperate zone, which, during the winter months, is affected by continental cold fronts pushing down from the north. During this period air temperatures along much of the northern Gulf coast may fall below freezing, and shallow waters of the bays, estuaries, and coastal lagoons may dip below 10°C (50°F). By contrast, the southern Gulf lies in the subtropical zone, where freezing weather never occurs, and water temperatures, even in the shallowest lagoons, seldom fall below 18°C (64°F). Summer temperatures in the shallowest coastal waters of both the northern and southern Gulf often exceed 30°C (86°F).

These temperature conditions have only recently been established. During the Ice Ages both the air and shallow-water temperatures around the margins of the entire Gulf were lower than they are at present, and most tropical species, being intolerant of low temperatures, were probably restricted to refugia in the Caribbean, especially considering that the continental shelves were at that time quite restricted. During cold periods associated with glacial advances, the shallow-water flora and fauna of much of the Gulf was undoubtedly limited to temperate and cold-temperate species, with possibly some subtropical algae and other forms surviving in or invading the southern Gulf during the summer months. With the retreat of the last continental glaciers and the gradual warming and deepening of continental-shelf waters, the tropical species recolonized the southern Gulf and began moving northward along two fronts, the Florida peninsular coast in the east and the Mexican coast in the west (fig. 15.2).

Today the coastal flora and fauna of the southern Gulf is made up primarily of tropical species year-round, and the biota of much of the eastern and western Gulf coasts, although mixed, is also largely tropical. Along the north-

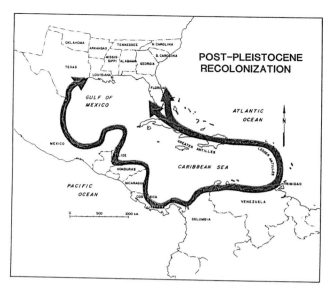

**Figure 15.2.** Post-Pleistocene recolonization of mangroves into the Gulf of Mexico from refugia in the Caribbean. (From Sherrod and McMillan 1985.)

ern Gulf coast tropical species are prominent only during the summer months. Humm and Darnell (1959) reported a variety of tropical algal species in summer collections from a protected channel shoreward of the Chandeleur Islands (east of the Mississippi River Delta), demonstrating that these algae are capable of colonizing the northern Gulf and strongly suggesting that their absence during the winter is most likely because of low temperature limitation. Four species of tropical mangroves occur in suitable habitats bordering the southern half of the Gulf (up to Cedar Keys in the eastern Gulf and the mouth of the Río Soto la Marina in the western Gulf) (Sherrod and McMillan 1985). However, only one species, the black mangrove (*Avicennia germinans*), extends along portions of the northern Gulf, where it grows as a low stunted shrub. Here it is subject to dieback during severe winters. For this species, low temperature limitation has been demonstrated experimentally.

The role of temperature in limiting attached tropical fauna in the northern Gulf is most clearly demonstrated by the distribution of hermatypic, or reef-building, corals. It is well documented that these animals cannot tolerate prolonged exposure to water temperatures below about 18°C (64°F). Many coral species inhabit reefs of the southern Gulf, but despite the presence of suitable rocky habitats, especially off the south Texas coast, they are largely absent from the northern Gulf continental shelf. The only real coral reef development in the northern Gulf occurs

at the East and West Flower Garden Banks, located about 100 nautical miles south of the Sabine River mouth (bordering Louisiana and Texas). The banks crest at depths of 15–20 m from the surface. Here the water bathing the banks seldom falls below the minimum limiting temperature. Not only the corals themselves, but the associated tropical flora and fauna of the banks are probably subject to low temperature limitation in other potential habitats of the northern Gulf.

Prominent among the shallow-water mobile fauna of the northern Gulf are species of penaeid shrimps, portunid crabs, and fishes of the drum family (Sciaenidae). These and other forms, all derivatives of tropical families, have through the years become more or less adapted to conditions in the temperate zone. Although some pass their entire lives on the continental shelf, many utilize the coastal estuaries and lagoons as summer nursery areas. Most leave the shallow nurseries in the fall and spend the winter in warmer waters of the continental shelf, thereby avoiding the coldest water temperatures, but some of the shrimp species apparently burrow into the bottom and overwinter in the estuaries. Early or severe cold snaps sometimes catch the migrators within the shallow coastal waters, and in a chilled condition near the lower limit of tolerance they enter a state referred to as "cold torpor" where they are still alive but too cold to move. If such low temperatures persist or if the water temperature falls further, they expire. Mass mortalities of fishes and invertebrates as a result of cold events have been recorded periodically around the northern Gulf from south Texas to Key West, Florida (Gunter 1941; Gunter and Hildebrand 1951; R. Moore 1976; Overstreet 1974; Rinckey and Saloman 1964). Historically, sea turtles and some marine mammals (manatees and monk seals) also visited shallow coastal waters of the northern Gulf during the summer months.

In addition to the tropical and subtropical forms, the northern Gulf coast harbors a number of temperate and cold-temperate species. During the winter months the shallow-water attached algal flora of the northern Gulf is dominated by species characteristic of the mid-Atlantic and New England coasts. A number of species of hydroids and medusae of the shallow northern Gulf are also cold-temperate forms. These and some other groups are considered to be relicts left over from the old Pleistocene Gulf flora and fauna of the last Ice Age that are still able to survive at least seasonally in the northern Gulf (Deevey 1950, 1954; Defenbaugh and Hopkins 1973; P. Phillips 1972).

Today the inhabitants of the southern Gulf are essentially tropical and subtropical, whereas those of the northern Gulf represent a mixture of tropical and temperate species. Here the picture is one of advancing warm-adapted and dwindling cold-adapted species, each limited by prevailing temperatures and temporarily waxing or waning in response to annual weather conditions. This also applies to the inhabitants of the eastern and western Gulf, where the northerly advance of tropical species is periodically interrupted by cold snaps (Darnell 1962a; Reid 1954). This general scenario, focusing on warm and cold-adapted species, presents a somewhat oversimplified picture. Found throughout the Gulf and superimposed on both groups is a suite of rather cosmopolitan species with wide temperature tolerances.

The situation in the open Gulf is somewhat different from that in the coastal waters. In the open Gulf seasonal temperature changes are much less extreme, but as much as a 15°C temperature differential occurs between the shallow surface waters and the deep Gulf. The depth-related temperature zones may limit some species. This is attested by the fact that individuals raised up from fairly deep, cold water often expire if they are not temperature insulated, but many survive if maintained at their normal habitat temperature. However, it has been noted that some benthic species have very wide depth ranges (hundreds of meters), and hence, temperature distributions, and many species of pelagic invertebrates and fishes regularly make vertical migrations traversing hundreds of meters of depth and a wide range of temperatures. Somehow, these species are able to adjust to the large temperature changes involved.

## Salinity

The salinity of the open Gulf remains nearly constant at around 34–36 ppt, and the proportions of the various ions making up the sea salts are also fairly constant. Sodium and chloride are by far the most abundant ions, together accounting for about 86 percent of the total by weight. Coastal waters, on the other hand, are diluted by rivers, and in estuaries the salinity may range from that of freshwater (0–1.0 ppt) at the head of the estuary all the way up to near full marine salinity where the estuary meets the Gulf. Waters of the continental shelves, particularly around the mouth of the Mississippi River and neighboring streams, may also fall below full marine salinity, and in reality these are extensions of the estuary onto the continental shelf. Salinities of coastal waters tend to vary

seasonally depending on river outflow, with flooding particularly common during spring in the northern Gulf and during the summer rainy season in the southern Gulf. Ion concentrations in coastal waters may vary somewhat from those in the open Gulf as a result of salts brought in by the rivers. The calcium ion is particularly abundant in shallow waters of the eastern and southern Gulf where streams drain limestone basins. In the coastal lagoons of south Texas and northeastern Mexico that normally receive little freshwater inflow, hypersaline conditions often develop as a result of high summer evaporation and poor circulation. Here salinity values twice that of seawater are not uncommon, and values of nearly 300 ppt have been recorded in the Laguna Madre of south Texas (Copeland 1967). In the present section we will first address the osmotic and ionic problems faced by plants and animals, especially in coastal waters of the Gulf, and their general means of coping with the problems, and then we will consider salinity-related adaptations of specific groups.

Organisms must maintain the concentrations of their internal body fluids (cell sap, blood, lymph, etc.) within appropriate physiological limits. In the face of varying and sometimes extreme environmental salinity conditions, they must be concerned with the internal concentrations of water, total salts, and specific ions. Two inherent problems are involved: (1) maintenance of low, or at least tolerable, osmotic pressure on cell membranes exposed to the environment and (2) regulation of internal ionic concentrations. The osmotic problem relates to the total concentrations of salts (and other osmotically active compounds) on either side of the exposed membranes. If the internal and external concentrations are equal, there is no pressure on the membranes, and the body is in balance with the environment, or iso-osmotic. If the total concentration inside the body is much higher than that outside, water tends to move in, the body swells, and pressure is exerted on the membranes, rupturing them in extreme cases. If the total concentration outside is greater, water tends to leave the body, shrinking the cells and raising the internal salt concentrations to dangerous, or possibly lethal, levels. To cope with the osmotic problems many marine organisms, particularly those that reside in coastal waters where the salinity is most variable, must have mechanisms that can obtain and retain or remove excess water and/or salts.

Marine organisms deal with osmotic and ionic problems in three basic ways: tolerance, avoidance, and regulation. Some marine algae and invertebrates have the ability to withstand a wide range of salinity conditions in their internal fluids, and since their internal salt concentrations tend to match those of the environment, there is no major osmotic problem. However, even these species must regulate the internal concentrations of specific ions. Most species do not have such wide ranges of tolerance. Clams and oysters, for example, although adapted to live in the brackish waters of bays and estuaries and often faced with periods of fresher or more saline conditions, cannot tolerate wide fluctuations in the concentrations of their body fluids. Being unable to move rapidly to more favorable locations, these bivalves escape unfavorable salinities by avoidance, that is, by nearly or completely closing their shells for many minutes or a few hours until conditions ameliorate. However, since they must eventually open their shells to respire, they cannot withstand prolonged exposure to extreme salinity conditions. Mobile invertebrates and fishes can avoid unfavorable salinities by simply moving away.

Regulation is the most important and widespread means by which plants and animals handle the water and salt problems. Many marine forms are largely covered by integuments that are nearly or completely impervious to the passage of water and salts, thereby limiting the flow to specific areas of the body such as roots or gills. All higher animals have excretory or secretory organs (including kidneys and salt-secreting glands) capable of eliminating excess water and/or salts. Some have "ion pumps" for moving specific ions into or out of the body, sometimes against considerable concentration gradients. The seawater concentrations of sodium and chloride, in particular, are generally far in excess of those tolerated by most organisms, and most species are able to substitute one ion for another, for example, ammonium ($NH^+$) or potassium ($K^+$) for sodium ($Na^+$), and bicarbonate ($HCO_3^0$) or phosphate ($HPO_4^{00}$) for chloride ($Cl^0$). In so doing, they can maintain their total salt concentrations at desired levels while avoiding unfavorable internal sodium and chloride concentrations.

Many marine algae maintain internal salt concentrations somewhat higher than that of seawater. Thus, water moves into the cells and aids in maintaining the necessary internal pressure, or cell turgor, to keep the membranes taut, and any excess water must be pumped out. Some diatoms, dinoflagellates, and filamentous blue-green, green, brown, and red algae are very successful in estuarine waters, but few species of marine macroalgae live in low-salinity brackish waters, probably because

they cannot handle all the excess water that would enter their cells.

All higher plants that live in estuarine and shallow marine environments (including marsh grasses, mangroves, and seagrasses) are derived from freshwater or terrestrial ancestors, and they have had to develop ways of dealing with the high and often variable salinities. In these plants the internal osmotic concentrations are kept somewhat higher than those of the environment. This is because not only salts but also osmotically active organic compounds (organic acids, etc.) are present in their body fluids. Even in quite saline habitats this allows them to take up water and transport it from the roots up through the conductive tissues to the leaves, where it can be lost through transpiration. Some also have mechanisms for storing or eliminating the excess salts that enter. Most of these plants are able to adjust their internal osmotic concentrations rapidly in response to changing external conditions. As a result of this regulatory ability, some are able to tolerate a wide range of salinity conditions. For example, turtle grass (*Thalassia testudinum*) has been shown to tolerate salinities ranging from 10 to 48 ppt.

Among the animals, most of the invertebrates that normally live in the sea have internal salt concentrations more or less equal to that of seawater, but of course they do control the internal concentrations of specific ions. Lacking sophisticated osmoregulatory mechanisms, most cannot survive for long in the reduced salinities of estuaries. In fact, many of the marine invertebrates that do spend part of their lives in estuaries must return to the open Gulf to spawn because their eggs cannot tolerate lower salinities. Among the most successful invertebrate invaders of estuaries of the Gulf are certain shrimps, crabs, and other crustaceans that have developed the ability to pump out excess water by means of a special gland (the "green gland") located at the base of the antennae.

In elasmobranch fishes (sharks, skates, and rays) the blood is more or less iso-osmotic with seawater. The internal salt concentration is only one-third to one-half that of seawater, and certain osmotically active organic compounds, especially urea and trimethylamine, make up the remaining concentrations. Within the Gulf only a few elasmobranchs frequent lower-salinity bays and estuaries, notably the bull shark (*Carcharhinus leucas*) and a few rays, and these must have special osmoregulatory abilities.

In marine bony fishes the internal salt concentrations are roughly similar to those of elasmobranchs, that is, only one-third to one-half that of seawater, but they lack effective osmotically active organic compounds. Thus, their body fluids are hypo-osmotic to seawater, and through simple osmosis they tend to lose freshwater to the more saline marine environment. Their problem is obtaining enough freshwater to overcome this loss. This they accomplish by removing water from their food (which may also be quite salty) and drinking seawater. Some of the excess salts are removed by the kidneys, but most are pumped out by special cells in the gills. Freshwater remains behind to be used in the body's metabolism. Those marine species that successfully invade and live in bays and estuaries have slightly lower internal salt concentrations than their cousins that remain in the sea. In high-salinity bays they remove excess salts as they do in seawater, but in lower-salinity areas they must get rid of excess water. This they do by producing copious quantities of dilute urine.

Higher vertebrates associated with the Gulf include sea turtles, marine birds, and marine mammals. All have internal salt concentrations about one-third to one-half that of seawater, and since their food comes from the sea, all are faced with the difficulties of obtaining enough freshwater and eliminating excess salts. This is less of a problem if the food is fishes (which have low-salinity body fluids), but a much greater problem if the food is marine vegetation or marine invertebrates. Sea turtles and marine birds obtain water from their food, and possibly from drinking seawater. They remove excess salts partially by the kidneys but mostly by special salt-secreting glands located near the eyes (in sea turtles) or in the nasal cavities (in marine birds). Marine mammals, however, have developed very powerful kidneys that are capable of removing the excess salts, even if their food consists largely of invertebrates such as squids and crustaceans. Some of the higher vertebrates are known to frequent lower-salinity coastal waters. Among these are some sea turtles (green and loggerhead), various shorebirds (gulls, terns, pelicans, etc.), and marine mammals (bottlenose dolphins and West Indian manatees). Here the salt excretion problem is much reduced, and excess water is eliminated by the kidneys.

## Oxygen

With the exception of some bacteria, fungi, and lower invertebrates, all plants and animals of the Gulf require a constant supply of oxygen for respiration. Oxygen is used in the body to oxidize certain organic compounds, thereby releasing a continuous supply of energy to fuel the body's various metabolic functions. Most, if not all, marine animals appear to be comfortable in dissolved oxygen concentrations of 3.0 mg $O_2$/L and above, but below 2.0 mg/L many are clearly under stress and cannot long survive. Between 2.0 and 3.0 mg/L the response depends on the tolerances of the individual species and the length of time of exposure. Throughout the open Gulf of Mexico, oxygen levels are mostly above 3.0 mg/L, but within the oxygen minimum layer (roughly between 250 and 750 m) oxygen levels may fall into the range of 2.5–3.0 mg/L, which is tolerated by most species but which could be limiting to some. However, considering the variety of animals that live in or migrate through the oxygen minimum layer, such oxygen levels cannot be considered to be a major impediment to open-water marine life.

In most areas of the continental shelf, oxygen levels remain above 3.0 mg/L, but, as discussed in chapter 7, hypoxia (concentrations below 2.0 mg/L) is a regular seasonal occurrence in bottom waters of the nearshore shelf west of the Mississippi River Delta off Louisiana, and it may extend off east Texas as far as the lower end of Galveston Island (fig. 7.7). These hypoxic events, which occur primarily during the summer months, result from high levels of river-borne nutrients that stimulate massive phytoplankton blooms in a highly stratified water column. The phytoplankton cells and zooplankton fecal pellets (from heavy grazing) sink to the bottom. Here bacterial oxidation of the organic material removes oxygen from the lower layer of the water column, resulting in hypoxic conditions. The hypoxic bottom water, which may be 10–15 m thick, occurs in the depth range of 10–30 m, and at its maximum it may extend up to 50 km offshore and cover about 10,000 km² of ocean bottom surface (Giammona and Darnell 1990; Harper, Potts, et al. 1981; NOAA 1992; N. Rabalais and Harper 1992). As seen in figure 15.3, the onset of hypoxia is marked by a dramatic reduction in the number of species and individuals present in the toxic bottom water. Although some of the mobile forms such as larger invertebrates and fishes apparently detect and avoid such areas, a few of the polychaetes and other invertebrates may be able to tolerate at least mild hypoxic conditions (Harper 1991; N. Rabalais and Harper 1992). Recovery of the bottom fauna from such hypoxic events is fairly rapid, in large measure because of the settlement

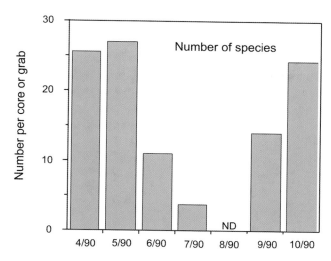

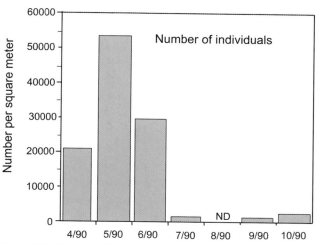

**Figure 15.3.** Seasonal distribution of species diversity and density (number of individuals per standard bottom sample) at a single station off the Louisiana coast characterized by summer hypoxia in 1990. During the months of April through September, bottom-water dissolved oxygen values remained below 2.0 mg/L. (From N. Rabalais and Harper 1992.)

occurs when oxygen returns to the bottom waters (Darnell 1992b; Darnell et al. 1976; Gunter 1947; Loesch 1960; May 1973; Schroeder and Wiseman 1988; J. Taylor and Saloman 1969).

In chapter 10 it was pointed out that marine bottom sediments are stratified. The topmost layer in contact with the water column contains free oxygen, the lowermost layer is anoxic, and these are separated by a thin transitional layer. The anoxic sediments charged with hydrogen sulfide, heavy metals in the chemically reduced state, and various organic acids, ketones, aldehydes, and so forth are toxic to most higher forms of plant and animal life. Such stratified sediments also underlie most coastal aquatic habitats. However, in marshes and swamplands the soils, which are rich in peat and other organic matter, may be totally anoxic, and marsh grasses and mangroves that grow in such soils have had to develop special mechanisms for providing their roots with oxygen. In some species this is accomplished by pumping oxygen down to the roots from the photosynthetic tissues above. In others, such as some of the mangroves, special woody breathing structures (*pneumatophores*) rise from the roots and transport oxygen directly from the atmosphere to the root systems. The roots then release some of this oxygen into the soil so that the environment adjacent to the roots is buffered from the anoxic surroundings. In somewhat parallel fashion, at all depths of the Gulf higher animals that live in marine sediments must pump oxygenated water into their burrows in order to avoid the toxic chemicals of the anoxic sediment environment. Exceptions are a few remarkable invertebrate species that harbor symbiotic bacteria and live in association with hydrocarbon and hydrogen sulfide seeps of the deep Gulf.

**Light**

The subject of light in relation to photosynthetic plants has been addressed in several of the previous chapters. Planktonic algae require light in sufficient quantity and at appropriate wavelengths to support photosynthesis. This takes place in the euphotic zone, the upper 200 m or less, depending on the amount of shading from suspended particulate matter. However, since very high light levels inhibit the photosynthetic process, the highest densities of phytoplankton cells and the maximum rates of photosynthesis do not occur at the sea surface but at some depth below, often around 40–50 m. Individual phytoplankton species themselves vary in respect to their needs and tol-

of larval forms from other areas. Hypoxic or anoxic brine pools, such as the Orca Basin, totally devoid of life, have been reported from the continental slope off Louisiana, but in this case it is the high salinity (260 ppt) rather than the absence of oxygen that precludes bacteria and other lower living forms. Hypoxic conditions have also been reported from various inshore coastal waters of the northern and eastern Gulf, including Galveston, Mobile, and Tampa Bays, as well as small bayous and canals. In each case this is a result of high organic loading, poor circulation, and elevated summer temperatures. The immediate effect on local populations is devastating, but recovery erances of light intensity, and some are considered light adapted whereas others are shade adapted. As in the case of the phytoplankton, the benthic algae are limited to the euphotic zone, as are the hermatypic corals and other invertebrates that depend on the photosynthetic products of the symbiotic intracellular algae (zooxanthellae) that live in their tissues. Light appears not to be limiting for the emergent and submerged rooted vegetation of the coastal marshes and swamps, but it does determine the maximum depth distribution of seagrass species.

Among marine animals sunlight provides for the vision of both predator and prey in the euphotic zone of the continental shelf and in the epipelagic and mesopelagic zones of the open Gulf. In less well-lighted areas and at night, bioluminescence provides important information for many species. In the deep sea, other animal species appear to be totally independent of light and rely on stimuli such as touch, vibrations, and chemical cues.

### Pressure

As noted earlier, the pressure of seawater increases about 1 atm (14.7 lb/in$^2$) for every increase of 10 m in depth. Thus, the pressure in the deep Gulf is vastly greater than that near the surface. An animal that spends its entire life within a narrow depth range can adjust its internal pressure to match that of its surroundings so that there is no differential, and the pressure factor essentially disappears. However, those species that undergo significant vertical migrations must take into account large changes in pressure, and for most marine migrators the morphological and physiological adjustments involved are largely unknown. What is known is that enzyme systems that work well at low pressures generally do not function well at high pressures. Hence, animals that regularly undergo large pressure changes apparently must either maintain alternate enzyme systems or make do with less efficient systems that function over a broad range of pressures.

Air-breathing vertebrates such as the leatherback turtle and some of the cetaceans, especially the toothed whales, regularly dive from the surface to depths of a thousand meters or more in search of food. The pressures at great depths could crush their bodies, and nitrogen and other gases could be forced into their circulatory systems if appropriate protective mechanisms were not available. Studies on sperm whales show that before a deep dive the animal respires at the surface for about 15 minutes and builds up a large oxygen reserve in the blood and body tissues. During the dive the rib cage and alveo-

lar sacs of the lungs collapse due to the outside pressure, preventing air under pressure in the lungs from entering the bloodstream. This prevents gases from being forced into the blood fluid and circumvents problems of nitrogen narcosis at depth and decompression sickness when the whale returns to the surface. Certainly, other mechanisms are involved, but this example does illustrate that in order to take advantage of a food supply, over the millennia marine animals have found ways of getting around the pressure problem.

### Suspended particulate material

Large quantities of suspended sediments are brought into the Gulf by rivers draining the surrounding land. The coarse fractions drop out near the river mouths, but finer silt and clay particles remain in suspension for some distance offshore and downcoast from the river mouths. Although they eventually settle out, the finer particles can be resuspended as bottom currents sweep along and roil the sediments. This is particularly true on the continental shelf of the northern Gulf around the mouths of the Mississippi River and neighboring streams from Alabama to central Texas. Layers of turbid water are referred to as *nepheloid* layers, and on the continental shelf off Louisiana it is not uncommon to encounter two nepheloid layers, one at the surface due to Mississippi River outflow and another below due to bottom roiling, with a layer of clear water between (fig. 15.4).

Suspended particulate material affects living organisms of the Gulf in several ways. By reducing light transmission, that is, through shading, turbid water restricts photosynthesis and greatly limits phytoplankton populations in the immediate vicinity of river mouths. It also reduces visibility, and some animals that rely on vision for feeding must avoid highly turbid areas. The suspended particles themselves may further interfere by adhering to planktonic species with sticky or gelatinous coverings, such as some planktonic algae, medusae, comb jellies, urochordates, larvae, and other forms. The particles may clog the gills or interfere with feeding by others. Sedentary benthic animals, such as bivalves, may not be able to survive repeated burial and reburial by the constant rain of particles, and the clogging of pores, gills, and feeding structures limits other benthic forms such as sponges, sea anemones, coral polyps, and various mollusks, echinoderms, bryozoans, and tunicates. When the suspended material settles out, it forms finely particulate sediments that further restrict the types of species that can occupy

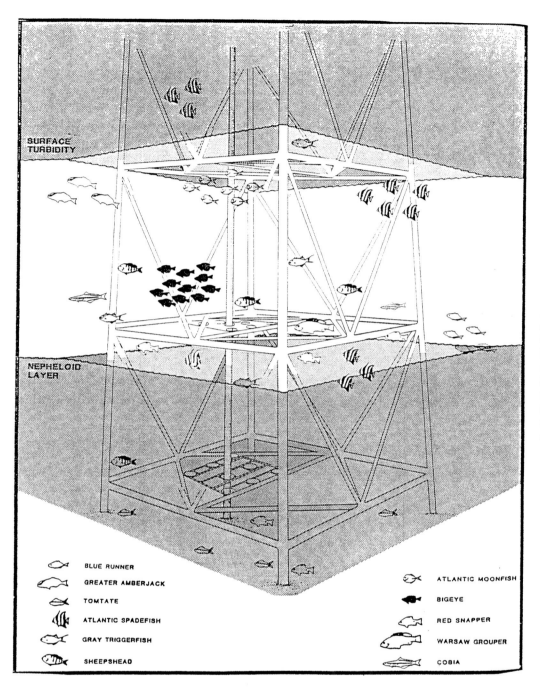

**Figure 15.4.**
Diagrammatic representation of fish distribution around legs of a petroleum platform in relation to the surface turbidity and bottom nepheloid layers. Fishes are identified as follows: blue runner (*Caranx crysos*), greater amberjack (*Seriola dumerili*), tomtate (*Haemulon aurolineatum*), Atlantic spadefish (*Chaetodipterus faber*), gray triggerfish (*Balistes capriscus*), sheepshead (*Archosargus probatocephalus*), Atlantic moonfish (*Selene setapinnis*), bigeye (*Priacanthus arenatus*), red snapper (*Lutjanus campechanus*), warsaw grouper (*Epinephelus nigritus*), and cobia (*Rachycentron canadum*).

the benthic habitats. On the other hand, certain species are well adapted to survive in turbid water and/or soft sediments, and the overall effect of suspended particulate matter is to create habitat for the tolerant species and to restrict others. In effect, the combination of turbid waters and soft sediment bottoms versus clear water and hard bottoms divides continental shelves of the Gulf into two rather distinct biological regions: the turbid-water, soft-sediment, terrigenous-bottom areas of the northern, western, and southwestern Gulf; and the clear-water, hard-bottom, carbonate shelves off peninsular Florida and northern Yucatán.

## Hard substrates

Many types of marine plants and animals are sessile species that must grow attached to substrates that can provide firm anchorage against the force of waves or water currents. Depending on the species involved, the substrate may be bare limestone, a shell, a piece of coral, a mangrove root, or a blade of seagrass, for example, and the availability of firm substrates limits the distribution of such forms. The list of species requiring firm substrates is quite long and includes various filamentous and macrophytic algae, protozoans, sponges, hydroids, sea pens, sea fans, corals, bivalves, barnacles, bryozoans, and tunicates, among others. Most forms often referred to as biofouling species are included here. Many species requiring firm substrates are typically found in clear waters associated with the carbonate platforms off Florida and northern Yucatán, but a sizable number are tolerant of fairly turbid conditions. For many years such sessile organisms were quite rare along the coast and on the shelf of the northern Gulf, but following the construction of coastal piers, seawalls, and jetties many attached species appeared, and with the establishment of arrays of petroleum drilling rigs and production platforms on the northern Gulf shelf, the variety and abundance of attached species greatly increased. Thus, it became clear that the rarity of such species in the past had been due, not to the turbid water, but to the absence of suitable hard substrates. Whereas many of the colonizers were derived from parents living in various places within the Gulf, others were introduced from outside the Gulf and came as biofouling "hitchhikers" on the hulls of commercial ships.

The attached flora and fauna play important roles in the biological economy of the Gulf of Mexico. Many are fast-growing colonizing species. Some are characterized by very high rates of primary production. As a group, they exhibit rapid rates of reproduction, with their larvae and disseminules adding greatly to the diversity, numbers, and biomass of the plankton. Associated with the attached species are various diatoms and many types of tiny invertebrates such as protozoans and wormlike species, as well as small polychaetes, snails, crustaceans, echinoderms, and even fishes (blennies and gobies), which find food and shelter here. The attached communities also attract more-mobile species such as larger crabs and fishes, which feed on the attached algae and associated invertebrate fauna.

## Water currents

The Gulf of Mexico is a very active place. Water currents of various velocities sweep around at all depths, but they are particularly active in the surface layers, which are most influenced by the Gulf Loop Current with its filaments, gyres, and eddies, and where wind-generated currents are most prominent. These currents are of great importance to marine life, which they influence in a variety of ways. By their continual movement and mixing they bring nutrients from deeper layers into the euphotic zone, making them available to the phytoplankton, and they ensure that virtually all portions of the shallow and deep Gulf remain well oxygenated. By sweeping soft sediments from the continental shelves, they aid in bringing living and dead organic materials to the deep Gulf, thereby providing food resources to the continental slope and abyss. Currents are necessary for the transport of plant disseminules and animal larvae throughout the Gulf. In their absence the flora and fauna of the Gulf would be vastly different. They regularly transport into the Gulf swarms of planktonic species and larval forms from the Caribbean, some of which remain and form local colonies. Various species of sedentary animals rely on water currents to sweep through bearing plankton and other suspended materials on which they feed. On the other hand, water currents create pressures that could dislodge organisms not firmly rooted or attached to solid substrates, and they limit certain species to environments where they can obtain firm anchorage. Currents also force many small species to live in sheltered habitats from which they cannot easily be dislodged. Exceptionally strong currents and surges, such as those generated by storms and hurricanes, can flush out bays and estuaries, reconfigure shoreline features, rip up seagrass meadows and attached benthic animals, and redistribute sediments of the continental shelf. Unlike air, water has a high density, and water in motion is a major factor to which marine species must be adapted and to which they must constantly adjust.

## Shelter

Many small polychaetes, mollusks, crustaceans, echinoderms, fishes, and so forth belong to a group known collectively as the cryptic fauna. These animals live in nooks and crannies around complex hard structures such as rocky outcrops and coral reefs or secreted among mangrove roots, lower portions of seagrass meadows, or biofouling mats. Subject to displacement by water currents and highly vulnerable to predation in the open, they carry out their life functions hidden away where they are hard to locate and capture. Larger forms, such as the common octopus (*Octopus vulgaris*), venture forth at night but must return to spend the day hidden away in the shelter of a rock pile or a large snail shell. Others, such as the hermit crabs, carry their shelters (discarded snail shells) with them. Although generally quite inconspicuous, these shelter-seeking cryptic species make up a significant percentage of the animal species inhabiting the Gulf.

## BIOLOGICAL FACTORS

In extreme environments physical factors often play a dominant role in determining which species can survive in an area, and as noted above, this is often the case in coastal waters. However, for most of the Gulf, biological factors are the primary determinants of species success. Such factors as food availability, interspecies competition, predation, parasitism, and disease, among others, really control the structure and functioning of most marine ecosystems. These factors are discussed below. Other biological concerns such as reproduction, larval life, and survival strategies will be taken up later.

### Food availability

All animals must have available food supplies of appropriate types and in sufficient abundance to provide the proteins, carbohydrates, lipids, vitamins, and other chemicals necessary for growth and reproduction and to supply the energy needed to carry out the required metabolic functions. From what has been said in previous chapters it is clear that the waters and sediments of the Gulf provide a great variety of potential food types. The various species of organisms making up the phytoplankton, zooplankton, nekton, marginal vegetation, and benthos, and the accompanying organic detritus together constitute a wide array of food resources on which the consumer species may draw. As seen in table 15.1, these resources are most abundant in coastal areas and in the euphotic zone of the open water, and for most of the Gulf they are more abundant in and near the sediments than they are in the open water column. In both the water column and the sediments the organic content tends to decrease with depth, and the ratio of biomass abundance in the shallow versus the very deep Gulf is on the order of 500 or 1,000 to 1.

Among the various consumer species, feeding methods vary widely from the sedentary forms that simply strain food from the passing water to the very active squids, fishes, and marine mammals that search for and chase down their prey. A generalized classification of feeding types and mechanisms is given in table 15.2. At the outset it must be pointed out that feeding types

**Table 15.1.** *Estimates of the primary production and standing crop of living biomass at various levels of the Gulf of Mexico.*

Water column of the open Gulf
- phytoplankton primary production
  - euphotic zone — up to 5.0 mg C/m³/day
  - below euphotic zone — 0.0 mg C/m³/day
- zooplankton
  - near surface — 100+ mg/m³
  - below 3,500 m — 0.2 mg/m³

Coastal rooted vegetation — > 1,000 g/m²

Benthos
- coastal (< 100 m) — 500+ g/m2
- deep Gulf (> 3,500 m) — ca. 1 g/m²

and food categories listed on paper are only approximations of what happens in the real world. Actually, feeding categories and food groups broadly overlap. Many of the smaller species consume any available organic materials (plant, animal, or detritus) that fall within the size range of particles they can apprehend and swallow. Within each category some forms are highly selective for particular target species or narrowly defined groups of species, whereas others are not so picky and will accept a wider range of items. Although an oversimplification of reality, the classification scheme is conceptually useful.

Passive suspension feeders are particularly numerous and widespread in waters relatively free of suspended inorganic particles and are found at all depths of the Gulf. They include such groups as the sponges, bryozoans, and tunicates, as well as many sedentary bivalves and some planktonic forms such as the pyrosomes and salps (tunicate relatives). Many of the sedentary passive feeders are colonial, with some members of the colony specialized for feeding, others for reproduction. Grazers, which feed largely on phytoplankton, include an array of small planktonic invertebrates such as many species of copepods and various larval forms. The "sweepers," like some polychaetes, copepods, barnacles, and others, use cilia, setae, or feather-like appendages to glean from the water all types of particles within a given size range, and these particles are then transferred to the mouth area, where some sorting takes place. Among the zooplankton are found species of polychaetes and chaetognaths, as well as small squids, crustaceans, and fishes that are actively predatory on other zooplanktonic species. Larger nektonic forms, of

**Table 15.2.** *A general classification and description of feeding types of marine animals of the Gulf of Mexico.*

| Feeding types | Food categories | Feeding mechanisms and behavior |
|---|---|---|
| **Suspension feeders** | | |
| – Passive | organic detritus, phytoplankton, and zooplankton | may be sedentary or planktonic; employ mucus, cilia, and/or nets to obtain food particles |
| – Active | | |
| • Grazers | phytoplankton | employ various appendages, etc. to obtain food |
| • "Sweepers" | organic detritus, phytoplankton, and zooplankton | employ cilia, setae, etc. to sweep particles from the water column |
| • Active predators | zooplankton | seek out and grasp individual prey organisms |
| **Deposit feeders** | | |
| – Selective | individual benthic animals | seek out and consume prey upon or within sediments |
| – Nonselective | organic detritus, carrion, benthic animals | swallow sediments; digest what they can and void the remainder |
| **Larger predators** | | |
| – Active hunters | larger animals | seek prey by active swimming or stalking |
| – Ambushers | large or small animals | often hidden or camouflaged; may use lures to attract prey |
| **Scavengers** | carrion | consume dead animals, mostly on the bottom |
| **Detritus feeders** | dead organic matter of all types | includes some planktonic, but mostly benthic sediment feeders |
| **Omnivores** | living and dead plants and animals | consume a wide range of organic materials depending on availability |

course, feed on all the above types, that is, on the passive feeders, grazers, sweepers, and smaller predators.

Animals that live upon or within the bottom sediments include many forms that feed selectively on individual bottom organisms. Others, particularly among the polychaetes and echinoderms, appear to simply swallow vast quantities of sediments, digesting what they can and voiding the indigestible material as large masses of fecal matter. Larger predators are encountered among the cephalopods, crustaceans, fishes, sea turtles, and marine mammals. Most of these hunt down and capture prey. Some such as toadfishes, scorpionfishes, and some flatfishes are camouflaged and ambush their food organisms, while others such as the anglerfishes and many deep-sea forms use baits or lures (often luminous) to attract prey within striking distance. Scavengers include some gastropods, polychaetes, amphipods, isopods, and fishes (notably, the hagfishes). These feed on carcasses

of dead animals that fall to the bottom. Other consumer groups include those that feed primarily on organic detritus (harvesting the associated microbes and small invertebrates as well as the matrix itself), and the omnivores, which consume a wide variety of living and dead plant and animal material.

From the above it can be concluded that whether in coastal areas or in the open Gulf, whether in the water column or in the benthic realm, consumer species are everywhere present that are capable of taking advantage of the available types of food resources. Within each realm some are feeding specialists, while others are generalists. Some species feed largely on phytoplankton or other vegetation, some on organic detritus, and some are primarily carnivores. Whatever is missed in the water column falls to the bottom to be taken by scavengers and other groups. Since most animals consume food items smaller than themselves, it is generally true that the meiofauna feed on the microbiota, the macrofauna on the meiofauna, and the megafauna on both the meiofauna and the macrofauna. High-quality food is generally plentiful in most coastal areas, in the benthos of the continental shelf, and in the euphotic zone of the open Gulf. In the deeper Gulf, in both the benthos and the water column, food may be quite scarce. This forces many deepwater species to make long daily vertical migrations to obtain food in the shallower but richer layers of the sea. Others remain behind to feed on the migrators and whatever falls from above. In order to locate a meal a deep-sea animal must search through large volumes of relatively empty ocean water or across large areas of bottom sediments, and even when food is located it is often of poor quality. Clearly, any animal must receive more energy from its food than it expends in obtaining it, and since deep-sea animals often must survive for long periods between meals, it seems probable that they have developed physiological mechanisms for lowering their metabolic rate, and thus energy expenditure, during their long periods of fasting. Many deep-sea animals must live at their nutritional edge, and starvation may not be a rare phenomenon in the deep Gulf. Animals living in nutritional poverty must be able to adjust their metabolism so that their first priority is their own individual survival, and only after their basic needs are satisfied can they allocate material and energy resources to reproduction. However, when food is abundant, their reproductive output increases accordingly.

## Interspecific competition

Competition between two species arises when both vie for the same scarce resource. Although most animal species of the open Gulf require dissolved oxygen, for this resource they are not really in competition since there is more than enough oxygen to satisfy all. Only when the combined demand exceeds the local supply of a resource does the problem really become serious, and then, although both species may suffer awhile, the better competitor must in the long run win out. The loser must either restrict its needs to resources not required by the other or become locally extinct. Sometimes accommodation is reached by both species reducing their needs in the area of overlap, and the species continue living side by side with restricted requirements. This process, often referred to as *competitive exclusion,* is considered to be one of the main driving forces of both the day-to-day ecology of individual organisms and, in the long run, of speciation and biological evolution. Competition is generally most severe be-

tween closely related species because their requirements are most nearly alike, and the outcome often depends on very subtle differences between the two species. Interspecific competition may take place with respect to food, space, shelter, and other biological resources.

Although simple to understand in principle, interspecific competition is often difficult to prove in nature. Overlap in requirements is often obvious, but demonstration of the critical scarcity of a resource is a greater challenge. A number of investigators have studied food habits of the fishes of estuaries and continental shelves of the northern and eastern Gulf, most notably W. Carr and Adams (1973), Darnell (1958, 1961, 1991a), Darnell, Defenbaugh, and Moore (1983), Lewis and Yerger (1976), R. Livingston (1982, 1984), Odum and Heald (1972), Overstreet and Heard (1982), Rogers (1977), Ross (1974, 1977), P. F. Sheridan (1979), P. F. Sheridan, Trimm, and Baker (1984), and Simons (1997). Although there is considerable overlap in the food items consumed by the various species, and they are undoubtedly competing for some types of food organisms, what the studies really show is partitioning of the food resources so that the basic requirements of the various consumer species are slightly different. Having been in contact over thousands of years, the species have mutually adjusted their needs, feeding methods, and seasonal timing so that the species we observe today have reduced their dependence on food resources required by others. Most species consume a wide spectrum of items so that if one resource is scarce they may shift to others.

Interspecific competition among marine organisms of the Gulf is perhaps most clearly seen among scleractinian corals that compete for space. Lang (1971, 1973) pointed out that in the crowded conditions of tropical reefs, when the polyps of one species touch those of another, along the margin of contact the more aggressive species extrudes digestive tissues over its less aggressive neighbor and digests away the polyps it can reach. Eventually the aggressive species overgrows and crowds out its neighbor. Other coral species grow high from their basal supports and branch out, shading lower species and reducing the ability of their zooxanthellae to carry on photosynthesis. This is really competition for light, but the result is the same. Whether directly through contact or indirectly through shading, the aggressive species are able to eliminate their neighbors, thereby securing the growing space for themselves. Within a given reef and from one reef to another there is a consistent hierarchy of species with greater or lesser aggressive tendencies. Although most of the research on this phenomenon has been conducted outside the Gulf, Bright, McGrail, et al. (1985) have shown such competitive interactions between coral species inhabiting the Flower Garden Banks of the northwestern Gulf. Crowding and competition for space by attached species may be observed on solid substrates throughout the Gulf, and it is particularly keen among biofouling species such as those attached to rock jetties, legs of oil and gas platforms, seagrass blades, mangrove roots, and so forth. Here the successful species may be the fastest growing, most aggressive, or most difficult to dislodge (barnacles), but success may also be determined by which species arrived first and were able to interfere with settlement and growth by later arrivals (see Sutherland 1974; Sutherland and Karlson 1977). Additional examples could be provided, but from the above it is clear that competition between species of one sort or another is a fact of daily life for many of the inhabitants of the Gulf of Mexico and a major force in shaping the structure of its marine communities.

## Predation

Throughout the previous chapters reference has been made to food habits of the various species, and from this information it can be concluded that predation is a major factor affecting populations of most animal species inhabiting the Gulf. In coastal waters and in the open Gulf, in the water column and around the bottom surface, predators regularly seek out and consume a great variety of prey species. Predators exist in eat-and-be-eaten hierarchies called food chains and webs. Copepods feed on bacteria, phytoplankton, protozoans, and so forth and are eaten by anchovies and other small fishes that themselves fall prey to cephalopods and larger fishes that, in turn, may be consumed by sharks, tunas, and marine mammals. The risk of being eaten is a constant fact of life for inhabitants of all the world oceans.

Predation benefits the predator population by providing the chemical and energy resources needed for growth and reproduction. In some situations it may benefit the prey population by eliminating the old, sick, or otherwise disabled members of the prey species, but these would die off anyway, and in many cases predation also eliminates large numbers of young and healthy individuals. What predation really does is to remove those members, healthy or otherwise, that are most vulnerable and thereby to select for survival those individuals less vulnerable to capture. Those most capable of avoiding predation

become parents of the next generation. Predation also selects for those predator types most capable of capturing prey. The result is an unceasing genetic battle to extend survival capabilities by the prey and capture capabilities by the predator species. However, it is not to the advantage of the predator species to run the prey to extinction, and long-standing predator-prey relationships generally wind up with a balance whereby predators harvest the excess prey, but enough survive to parent the next generation. When a particular prey species becomes too scarce, it behooves the predator to seek alternate food resources. Hence, most predator species tend to feed by food category, that is, they will take any of a number of different prey species falling within their capture capabilities, thereby avoiding systematic dependence on a single prey species. In parallel fashion, a number of different predator types may consume a given prey species so that when the prey is particularly (often seasonally) abundant it may temporarily become the primary food of several predator species. The reality is that each predator species consumes a number of prey species, and each prey species is consumed by a number of different predator species so that predation pressure falls broadly on a series of species and all survive. The dynamic balance between predator and prey species occurs throughout the food chains and webs of the sea, and this interaction is one of the chief processes regulating populations of animals in the Gulf of Mexico. Examples of adaptations to increase predator efficiency among fishes and marine mammals are provided in table 15.3.

Predator-prey interactions lend themselves to interesting exercises in mathematical modeling. The models often begin with the assumption of predation success by random encounters, that is, every time a predator meets a prey it consumes the prey. The probability of being consumed equals the probability of an encounter. This may be modified by increasing or decreasing the density of the predator or the prey species or both. In extension, the model may make assumptions about the vulnerability of the prey so that not every encounter results in a capture. It may also increase the efficiency of the predator by assuming that it is able to sense the prey at a greater distance, increasing the width of the search path and therefore the frequency of encounters. The model may be further expanded by the inclusion of additional predator and/ or prey species. Such models often track reality as prey populations develop better survival mechanisms and predators become more efficient. Mathematical models

**Table 15.3.** *Some examples of morphological, sensory, and behavioral adaptations among fishes and marine mammals of the Gulf of Mexico that enhance predator effectiveness.*

**Morphological adaptions**
- Whole body
  - Streamlining, fin placement, and muscle mass development to increase swimming speed and maneuverability
  - Mimetic body shapes, features, and color patterns as camouflage to escape visual detection by prey organisms
  - Special appendages that serve as rods and lures (illicia and escas), sometimes luminous, to attract prey
- Jaws
  - Protrusible jaws to aid in picking or sucking up small prey
  - Disarticulating jaws to permit swallowing of very large prey
- Dentition
  - Large incisor-like front teeth for scraping food organisms from solid substrates
  - Jaw teeth coalesced into plates in parrot-like mouth used for scraping and crushing corals and other hard substrates
  - Fang-like teeth for stabbing and holding large prey
  - Pads of platelike pharyngeal teeth for crushing mollusk shells

**Sensory adaptations**
- Vision
  - Extremely acute vision for seeing prey in dim light
  - Binocular vision to precisely locate prey
  - Telescopic eyes (directed forward or upward) to aid in locating prey in dim light
- Chemical senses
  - Acute sense of smell and/or taste to detect chemical cues (blood, pheromones, or other chemicals) released by prey species
- Hearing and detection of vibrations
  - Acute senses to detect sounds and water movements produced by prey
  - Ability to locate prey through echolocation
- Electrical sense
  - Ability to detect electrical fields produced by prey animals

**Behavioral adaptations**
- Bioluminescence to attract, startle, blind, or illuminate prey
- Vertical (diurnal) and horizontal (seasonal) migrations to fertile feeding areas
- Ability to hunt cooperatively in schools or packs to attack very large prey or schools of smaller prey organisms

---

of this type often provide deeper insights into the workings of species interactions and may provide numerical information about population pressures and survival probabilities under a variety of interaction conditions. Experimentation with natural populations in the open ocean is next to impossible, but simulation models help us understand what is really going on. Some references to predation in the Gulf of Mexico and surrounding areas include Dalby (1989), Frazer, Lindberg, and Stanton (1991), Heck and Coen (1995), Iversen, Jory, and Bannerot (1986),

Kurz (1995), Lizama and Blanquet (1975), Purcell (1985), and K. Wilson (1989).

**Parasitism and disease**

Parasitism and disease belong to a special class of interspecific relationships in which the infecting organisms benefit to the detriment of the host, and in this respect the interaction is similar to that of predation. Disease-causing organisms are often thought of as microbes, that is, single-celled or simple colonial forms such as viruses,

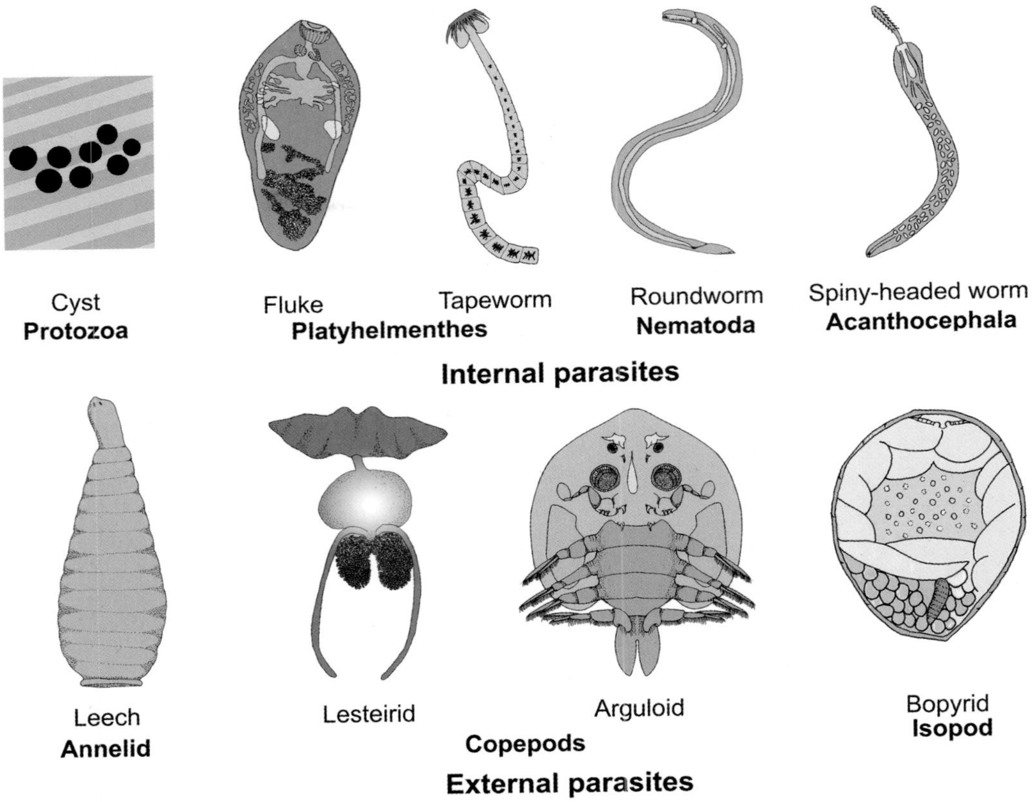

Cyst
**Protozoa**

Fluke    Tapeworm
**Platyhelminthes**

Roundworm
**Nematoda**

Spiny-headed worm
**Acanthocephala**

**Internal parasites**

Leech
**Annelid**

Lesteirid

Arguloid

Bopyrid
**Isopod**

**Copepods**

**External parasites**

**Figure 15.5.** Examples of some marine parasites. The protozoan *Perkinsus* (= *Dermocystidium*) *marinum*, is shown as spores embedded in tissues of the American oyster (*Crassostrea virginica*). The fluke, tapeworm, roundworm, and spiny-headed worm as adults all live in the internal organs (gut, liver, kidneys, etc.) of invertebrates and fishes. The lesteirid copepod is anchored in the host's muscles with only the posterior portion of the body, including the egg sacs, protruding externally. The arguloid copepod attaches itself to the outside surface of fishes by means of the two large round suction disks. The highly modified adult bopyrid isopod lives under the gill covers of caridean and penaeid shrimps.

rickettsias, bacteria, infectious protozoans, yeasts, and so forth, whereas parasites are considered to be multicellular organisms, especially various wormlike forms such as flukes (trematodes), tapeworms (cestodes), and spiny-headed worms (acanthocephalans) (fig. 15.5). Since the effects of parasites and disease agents on the hosts are much the same, in the present discussion they will all be referred to as parasites. These organisms may live within the host's body (endoparasites) or on the outside surface of the body (ectoparasites). Internally they may be found infecting almost any organ system, but a great many are specialized to live in the digestive tract. Although some plant and most animal phyla are represented among the parasites, they are most common among the viruses, bacteria, protozoans, flatworms, roundworms, spiny-headed worms, and various crustaceans (copepods, isopods, barnacles, etc.). Their hosts include practically all types of marine organisms from bacteria, which are parasitized by

viruses, up through marine mammals. Although less well known than their free-living hosts, parasites are probably the most numerous forms of organisms in the sea, and the number of species of parasites may far exceed that of all other types of marine organisms combined.

For simplicity, we may think of parasites as belonging to two general groups: the true parasites, which are specifically adapted for a life of parasitism; and the accidental parasites, which are adapted to live harmoniously within or upon the host (and often elsewhere) but which can become pathogenic when the host is injured or becomes weakened from stress. In nature the differences between these two groups are not sharply defined. Among the true parasites, some are host specific, but others are capable of infecting a variety of host species. Many of the wormlike parasites have complex life histories involving an egg, a free-swimming early larva, one or more parasitic larval stages that are passed in intermediate hosts, and a sexually reproducing adult that occurs in the final or definitive host species. The internal wormlike parasites and some others are transferred up the food chain and pass from infected prey to the predator hosts, but others, including most that attach to the outside of the body, are free living, generally lack parasitic larval stages, and locate their prey by other means. The types of parasites to which a host is subject depend a great deal on its mode of life. Planktonic and nektonic animals often host parasites with complicated life histories involving larval stages and several intermediate hosts, whereas bottom feeders more often obtain parasites with direct life histories and no intermediate hosts.

There is a fairly extensive world literature on parasites and diseases of marine organisms, and much of it relates to shallow-water species of commercial interest, but some work has been carried out on noncommercial species and even on parasites of deep-sea animals. Some general references to parasites and disease agents of marine organisms include Galtsoff (1964), Ho and Perkins (1985), Kinne (1983, 1984, 1985), Sindermann (1990), Snieszko (1970), and Vernberg (1974). Information on parasites of deep-sea animals has been summarized by Campbell (1983). Some references more specific to the Gulf of Mexico and the host groups (in parentheses) to which they relate include Ambroski, Hood, and Miller (1974), Chandler and Manter (1954), Overstreet (1978), Sprague (1954), and C. Wilson (1935) (general); Durako and Kuss (1994) (*Thalassia*); Gladfelter (1982) and Kuta and Richardson (1994) (corals); P. Phillips and Levin (1973) (medusae); Couch (1978) (shrimps); Armstrong (1974), Bullock (1957), D. Causey (1953), Ho (1969), Hutton and Sogandares-Bernal (1959), Korath (1955), Manter (1934, 1947), Nahas and Short (1965), Overstreet (1973), and Sanders (1964) (fishes). Overstreet (1978) discussed parasites of commercially important mollusks (oysters), crustaceans (shrimps and crabs), and fishes of the northern Gulf and provided many additional references.

Most marine animals are protected from harmful infections by an external skin, cuticle, exoskeleton, or scales as well as some form of internal immune system. When these systems are intact and fully functional the host is often little affected by other species that may live within or upon its body. Indeed, Skinner (1975) has noted that the parasites of adult striped mullets (*Mugil cephalus*) that feed on bottom detritus are essentially different species from those that infect the plankton-feeding young, and that older individuals tend to have a lower incidence of in-

fection than do the young. These and similar observations by other investigators suggest that in addition to a degree of natural immunity, healthy host individuals may develop some acquired resistance to parasites. On the other hand, regardless of how well healthy individuals respond to internal parasites, they still pay a metabolic price for having to provide food and other resources to their free-loading guests. This generally shows up as elevated host respiration, which in itself can cause stress in the host. Many of the so-called benign parasites and potential disease organisms that infect marine species can become pathogenic when the host is placed under stress. When faced with suboptimal physical conditions (high or low temperatures, low salinity, hypoxia, high levels of sedimentation, etc.), when crowded together (as in commercial rearing operations), or when weakened by heavy parasite loads or extreme exertion, the body loses some of its resistance and becomes more vulnerable and less able to cope with infectious agents. Weakened and stressed hosts become subject to invasion by additional species of parasites. Under such conditions, even if the hosts are not killed outright by the parasites and diseases, they become more vulnerable to predation. It is not generally to the parasite's advantage to kill the host and thereby to end its own life, but if the parasite is a larval stage residing in an intermediate host, then it is definitely to the parasite's advantage to weaken the host and make it more vulnerable to predation, thus increasing the parasite's chances of being transferred to a higher-level or definitive host. Marine animals that have suffered injurious lacerations, as from an unsuccessful attack by a predator, are of course subject to serious infections that can lead to debilitation and death.

The role of parasitism and disease in regulating populations of marine organisms of the Gulf is not clear. From what has been said above, parasitism plays a role in the enhancement of predator-prey relationships, but it may also play a more substantial and direct role in population regulation. When a virus invades a bacterium it causes the host to produce many new virus particles, which, upon death of the host, are released into the environment. In the absence of other nearby uninfected hosts, the particles simply disperse, but if the density of the host population is above a certain level the viruses can quickly spread from host to host, greatly reducing the density of the host population. Such density-dependent epidemics that spread by contagion could certainly act to regulate the host populations. Although this type of population control has not been demonstrated for Gulf species, it should be anticipated, especially in those species that sometimes occur in dense aggregations such as schools or breeding swarms.

## CATASTROPHIC EPISODIC EVENTS

So far as we can tell, life in the sea proceeds in a somewhat orderly fashion, with its rhythms set largely by changing physical factors. The day and night, tidal, and seasonal cycles of the environment oscillate in a rather regular and predictable manner, and to these physical patterns the populations have become genetically attuned. In the normal course of events variations in the major environmental parameters do occur, but for much of the sea these swings probably represent only minor deviations from the norm. However, in coastal areas the water is much shallower, the land is closer, and rivers enter the sea. Estuarine and related shallow shelf habitats, being more directly and strongly influenced by local weather phenomena and other physical factors, are especially variable with respect to the key forcing functions (Copeland 1970; Copeland and Bechtel 1974). Here extreme physical events sometimes occur, induced largely by meteorological factors, and these episodic occurrences can have catastrophic biological and ecological consequences. Such events appear suddenly and at sporadic intervals and are especially common in the Gulf of Mexico. Although no coastal section of the Gulf is immune, particularly affected are the bays, estuaries, and shallow shelf waters of the northern and eastern Gulf, which are most influenced by the Mississippi River and weather conditions of the large land mass to the north. Among the chief episodic events affecting the Gulf are major cold snaps, floods, large storms and hurricanes, hypoxic events, and noxious algal blooms, and, on the continental slope and abyssal plain of the deep Gulf, sediment slumps and turbidity flows. The Gulf is likely affected by other catastrophe-inducing events (associated with the Gulf Loop Current, etc.), but these are the ones we know about, and they will be addressed below. A summary of our knowledge about the distribution and effects of these events in the Gulf is presented in table 15.4, and some general references include Brongersma-Sanders (1957), Darnell (1992b), and Gunter (1947).

### Cold snaps

From time to time beginning in the late 1800s, newspaper articles and technical reports have appeared describing cold-induced mass mortalities of fishes and other animals in bays, estuaries, and shallow continental shelf waters extending from south Texas, along the northern and eastern Gulf coast, down to Key West, Florida (and along the southeast coast of the United States up to Cape Hatteras, North Carolina). In south Texas dead fishes include such shallow-water forms as anchovies, catfishes, pinfishes, sheepsheads, spotted seatrouts, mullets, and gobies, among others, and local estimates of dead fishes from a single event have ranged up to 90 million pounds. Some of the more important Gulf references include Beddingfield and McClintock (1994), Galloway (1941), Gunter (1941), Gunter and Hildebrand (1951), R. Moore (1976), Overstreet (1974), Rinckey and Saloman (1964), and Storey and Gudger (1936).

During normal winters, low temperature is not usually so harmful to shallow-water marine life. As the water temperature decreases in the fall and early winter months, the less tolerant mobile species leave the shallow inshore areas for deeper and warmer waters of the continental shelf. Those that remain are generally adapted to survive the cold water conditions of the normal winter. However, occasionally the coast is swept by a very strong cold front descending rapidly from the north, and this frigid air mass remains in place for several days. For example, in south Texas during such events wind speeds often exceed 42 km/hr (25 mi/hr), and air temperatures may hover below 0°C. In a matter of a few hours the temperature of the shallow bay waters may drop 5° to 15°C and then remain near the freezing mark, which is too fast and too cold for most of the remaining shallow-water fauna. They cannot leave, adjust to, or tolerate such conditions, and the result is mass mortality. These "blue northers," as they are called, may push down to the southern Gulf and beyond, but mortalities along the Mexican coast have not been well documented. Off Sanibel Island below Tampa, Florida, Storey (1937) noted that the degree of mortality among the fishes was related to their geographic affinities. She found that during a cold event all of the temperate species survived, but 45 percent of the subtropical and 82 percent of the tropical species were killed. This differential susceptibility strongly suggests that the occasional occurrence of sudden, extreme, and persistent cold snaps must be the primary factor limiting establishment of populations of tropical and subtropical species in coastal and shallow shelf waters of the northern Gulf.

**Table 15.4.** *Extreme episodic events that result in catastrophic mortality in biological populations of the Gulf of Mexico and their short-term consequences.*

**Cold snaps**
- Recorded from bays and estuaries along northern and eastern Gulf from south Texas to Key West, Florida.
- Can cause mass mortality of invertebrates and fishes of inshore waters. Probably limit distribution of tropical species from shallow shelf of the northern Gulf.

**Floods**
- Especially common in bays and estuaries and on continental shelf around mouth of the Mississippi River, but known from coastal waters around mouths of most major rivers. Floods bring much freshwater, sediment, and debris to estuaries and other coastal waters.
- In estuaries floods chase out mobile species and bury and induce low-salinity stress on benthic and less mobile species. By modifying current flow patterns floods can prevent larvae and young from entering estuaries. On continental shelf young invertebrates and fishes crowd nearshore waters, and older individuals are forced into deeper water.

**Major storms and hurricanes**
- Affect the entire coastline of Gulf. Damage results from high winds, elevated water level, pounding waves, strong currents, flooding, and heavy sedimentation.
- Storms physically tear up mangrove forests, salt marshes, seagrass meadows, coral reefs, barrier islands, and bottom habitats. They displace mobile species and fragment or bury nonmobile species.

**Hypoxic events**
- Reported from several coastal areas of the northern and eastern Gulf (Galveston Bay, Lake Pontchartrain, Mobile Bay, Tampa Bay, etc.) and from the shallow continental shelf off Louisiana and east Texas. They result from a coincidence of high organic matter, high temperature, and poor water circulation.
- Can cause mass mortality of invertebrates and fishes. Some invertebrates with high levels of tolerance may survive. Some mobile species may escape hypoxic areas.

**Noxious algal blooms**
- Known especially from the northern and eastern sectors of the Gulf, where they may affect ecological systems of bays, estuaries, and nearshore continental shelves. Caused by dinoflagellates, diatoms, and microalgae.
- Can result in massive invertebrate and fish kills, decreased seagrass production, and changes in species composition.

**Sediment slumps and turbidity flows**
- River-borne sediments deposited on the outer shelf and slope of the north-central Gulf periodically become unstable, fail, and send mudslides and turbidity flows downslope and onto the continental rise and abyssal plain.
- Such flows result in displacement and/or burial of slope and abyssal plain fauna. They also alter the composition of abyssal sediments from pelagic ooze to terrestrial clastic materials.

---

## Floods

Coastal flooding can be caused by locally heavy downpours, such as those associated with tropical storms and hurricanes, and it can also result from widespread precipitation or rapid snowmelt in upstream drainage basins, leading to high-volume river discharge at the coast. Flooding of bays, estuaries, and portions of the shallow shelf is particularly common around the Mississippi River Delta, where it has been most thoroughly studied, but it can also be encountered in other coastal shallows and around the mouths of most major rivers entering the Gulf. Although flooding may take place any time during the year, in the northern Gulf it is most common in the late spring, and farther south it is more frequent during the summer monsoonal season. Pertinent references include Butler (1952), Butler and Engle (1950), Dawson (1965), Goodbody (1961), Gunter (1953), Hawes and Perry (1978), R. Livingston and Duncan (1979), Poirrier and Mulino (1975), and Russell (1977). The effects may be local and of short duration, or they may be widespread and persist for many months. General effects of flooding include rapid lowering of salinity, a great increase in the load of suspended sediments accompanied by reduction in light penetration, and heavy sedimentation of at least a portion of the receiving basin. These are often accompanied by additional effects such as a rapid drop in water temperature, lower pH, lower levels of dissolved oxygen, a change in water color, and increased levels of nutrients such as nitrogen and phosphorus.

The biological effects of flooding are not as spectacular as in the case of cold-induced mortality because the beaches and nearshore waters are not littered with hordes of dead and dying fishes, but the results can be quite significant, nonetheless. These results depend on the extent, amount, and duration of the flooding, but they also relate to the tolerance characteristics of the various species and the seasonal timing of the event in relation to the life history details of the individual species. Conceptually, one may think of the flooding as having a two-phase effect, the initial shock associated with sudden changes in a number of physical and chemical parameters, and the longer-term, persistent phase. Faced with flooding conditions, many of the mobile species (squids, shrimps, crabs, and fishes) move out of the affected area, either to another part of the estuary or to continental shelf waters. Those that remain, including some of the mobile species and all of the less mobile benthic and attached flora and fauna, adjust as best they can. The impact of severe and prolonged flooding is most clearly illustrated by the effects of opening the Bonnet Carré spillway and permitting Mississippi River floodwaters to pass through Lake Pontchartrain into Lake Borgne and Mississippi Sound and from there to the continental shelf east of the Mississippi River Delta (Gunter 1953). Although this event results from human activities, it parallels natural flooding events that took place prior to the construction of artificial levees along the Mississippi River (Darnell 1962a). Severe and persistent flooding of the lake from Mississippi River discharge flushes the lake, removing the brackish-water plankton, which is ultimately replaced by species tolerant of fresher water. The mobile fauna either dies or moves out into the Gulf, the sedentary and attached species are largely wiped out because of their intolerance of the persistent fresher water, and essentially freshwater species invade from surrounding streams and marshes. The penaeid shrimps are replaced by river shrimps (*Macrobrachium ohione*) and

some crayfishes (*Cambarus* spp.). The brackish-water and saltwater fishes are replaced by freshwater catfishes, primarily the blue catfish (*Ictalurus furcatus*), largemouth bass (*Micropterus salmoides*), and other species of freshwater affinity (personal experience of the author). In nearby Lake Borgne, oyster mortality may approach 100 percent. The young of coastal fish species, which normally live in Lake Pontchartrain, accumulate on the inner continental shelf, and adults of such species as the Atlantic croaker (*Micropogonias undulatus*) move into deeper shelf waters (Russell 1977). Although it has not been specifically documented, it is clear that larval and juvenile invertebrates and fishes that normally make their way from spawning grounds into the bays and estuaries are prevented from doing so during the immediate period of flooding and for some time thereafter. The larvae and very young have limited powers of locomotion and are largely dependent on favorable water currents to enable them to traverse the passes into the estuaries.

Thus, the immediate biological effects of severe flooding include the displacement of many species, the death of others, and a great reduction of species diversity, and this is accompanied or followed by the invasion of species more tolerant of the fresher water. In the case of more localized or shorter-term flooding there is less displacement and mortality. Less severe flooding has been viewed as beneficial to oyster reefs in Lake Borgne because the oysters can often tolerate the conditions, but several pest species (boring sponges, boring clams, oyster drills, etc.) are wiped out. Regardless of the intensity and duration of the flooding event, normal salinity conditions are eventually reestablished, and the normal biota returns to the bays and estuaries. The following year or so, apparently as a result of the influx of nutrients, the biological production of the area may even exceed preflood levels.

## Storms and hurricanes

Each year the Gulf of Mexico is likely to be struck by several large storms, including one or more of hurricane force. Although occurring most frequently in the late summer and early fall in connection with tropical disturbances, severe storms can strike during any month and may be associated with intense winter cold fronts. All parts of the Gulf are subject to storm strikes, but the effects have been most thoroughly documented in the northern and eastern Gulf. Impacts are most intense in coastal lakes, estuaries, lagoons, and inner portions of the continental shelf. Important references include Ball, Shinn, and Stockman

(1967), Blair, McIntosh, and Mostkoff (1994), Bortone (1976), Chabreck and Palmisano (1973), Craighead and Gilbert (1962), McCoy et al. (1996), Meeder et al. (1994), R. Perkins and Enos (1968), Posey et al. (1996), Robins (1957), V. Springer and McErlean (1962), Tabb and Jones (1962), and Wright, Swaye, and Coleman (1970).

Physical effects of storms vary depending on such factors as season of occurrence, intensity, duration, where they strike the coast, and direction of approach. Typically they pack high winds, in extreme cases exceeding 300 km/hr (180 mi/hr). Since the winds of tropical storms rotate around a low-pressure center, or "eye," during the storm's approach they blow from one direction. After passage of the eye the winds may come from the opposite direction, and this shift can be quite abrupt. As the storm approaches, the sea level rises, and close to the center the storm surge may reach 5 m (16.4 ft) or more, sending large volumes of seawater over barrier islands and into the more protected inside waters of coastal lakes, estuaries, and lagoons. Waves are generally intense, and together with strong currents they may reshape shoreline features, cut new passes through the barrier islands, and erode beaches, causing them to recede, in extreme cases 30 m or more. Many storms are accompanied by heavy rainfall, and 25 cm (10 in) or more may fall in a few hours, leading to extensive flooding of coastal lowlands.

Within the water column the rotary motion of wave energy digs into bottom sediments, raising sand, shells, and other debris that may blast or smash into other objects in the water. Very strong currents transport this material and deposit it elsewhere. Bottom contours are reshaped, and sandbars and other features are rearranged. Salinity of the inshore waters may be greatly reduced (from rainfall and flooding) or increased (from seawater inflow with the storm surge). Turbidity levels increase, light penetration decreases, and the water temperature generally drops. As a result of the digging and resuspension of bottom sediments nutrient levels may rise, but hydrogen sulfide is also released into the water, sometimes in toxic concentrations.

Biological effects of strong storms are highly varied and often species specific. Rooted vegetation and benthic fauna may be dislodged from the bottom and crushed or smashed against other objects or thrown upon the beach. Attached flora and fauna may be torn off and dispersed. Mobile species seek shelter wherever they can among crevices in jetties and reefs or in deeper water, but many remain exposed. Here they are subject to physical dam-

age from buffeting and pounding. Some species suffer from clogging and choking by the suspended particles. Many fishes are killed by gill damage due to abrasion by suspended sand grains (Robins 1957). Others succumb to elevated levels of hydrogen sulfide.

The effects of a hurricane on marsh vegetation around the Mississippi River Delta have been documented by Chabreck and Palmisano (1973). Here there was a drastic reduction in marsh plants largely because of the sweeping action of wind and water, but some species fared worse than others. Vegetative cover was reduced by about a third, and marsh ponds and lakes were enlarged and deepened. More than four-fifths of the plant species suffered some decline. Floating vegetation (*Alternanthera, Eichhornia, Lemna,* and *Spirodela*) was simply swept away, and much of the rooted vegetation was uprooted or ripped apart and transported away by the current. Less affected were tough-stemmed rooted species such as *Bacopa, Phragmites,* and *Spartina.* After one year, recovery was well underway for the marsh species, but recovery in the lakes and ponds was slower, and the floating vegetation had not returned. This was in part because the lakes and ponds had become quite saline as a result of entrapment of salt water from the storm surge.

The effects of major storms and hurricanes on mangrove forests of south Florida have been discussed by Craighead and Gilbert (1962), McCoy et al. (1996), Odum, McIvor, and Smith (1982), and Tabb and Jones (1962), among others. Although these forests are adapted to survive minor storms, they can be devastated by severe storms and hurricanes. As a result of one hurricane heavy damage stretched over 40,000 ha (100,000 A) where the loss of trees ranged from 25 to 100 percent. Damage resulted from wind shearing of trunks 2–3 m above the ground, defoliation, and overwash mangrove islands being swept clean. Many tree deaths occurred months after the storm, apparently from damage sustained by the prop root systems. Shortly after passage of a hurricane it was determined that 60 percent of the trees were uprooted or broken, and 25 percent of the upright trees were dead. Only 14 percent of the upright trees were still well vegetated. Most severely affected were the white mangroves (*Laguncularia racemosa*), next were the black mangroves (*Avicennia germinans*), and least affected were the red mangroves (*Rhizophora mangle*). Since the latter tend to be lower and to grow largely in the subtidal zone, they are less affected by high winds and heavy siltation. Their root systems are flushed clean of silt and other debris by waves and tidal action. Fishes and invertebrates of mangrove swamps can be affected adversely by oxygen depletion resulting from the heavy accumulation of decomposing organic matter. Many species of wading birds and shorebirds nest in coastal mangroves, and their nests and fledglings can be destroyed by major storms. Fortunately, most of the larger storms arrive after the main nesting season is over, but some bird mortality has been reported. Since recovery of the mangrove forest involves reinvasion and growth of new trees over large areas, the full recovery process can take many years.

The effects of strong storms and hurricanes on submerged communities have also been studied. Meeder et al. (1994) documented some of the effects of a hurricane on the extensive seagrass beds of Florida Bay south of the Everglades. Here the *Thalassia* beds had been ripped up by heavy waves and strong water currents. Long D-shaped divots, some up to 8 m wide and even longer, had been dug out and overturned or transported away. In addition, much organic detritus and sedimentary material had been winnowed from the bottom surface within the beds. As a result of this breakup and erosion, the water depth over the beds had increased up to 45 percent. Zieman (1975) pointed out that the natural recolonization and development of such disturbed seagrass meadows is a very slow process requiring many years for complete recovery.

The effects of major storms and hurricanes on coral reefs of south Florida have been discussed by a number of authors including Ball, Shinn, and Stockman (1967), Blair, McIntosh, and Mostkoff (1994), Jaap (1984), Perkins and Enos (1968), and V. Springer and McErlean (1962). Jaap (1984) noted that between 1873 and 1966 the reefs were struck by 24 major hurricanes, an average of one every 3.6 years. All but one hit during the period of August–October. Since the primary physical damage results from the forces of wave energy and strong water currents, the shallower zones of the reef generally sustain the most damage. Smaller corals and other attached organisms are often dislodged, fragmented, and abraded by the severe physical pounding. Blair, McIntosh, and Mostkoff (1994) noted that off Biscayne Bay the most heavily impacted area was the foreshore slope of the offshore reef in the 17–29 m depth range. Here the algal community lost 40 to over 90 percent of the benthic cover. The sponge community lost 40–75 percent, soft corals 0–50 percent, and hard corals were least affected, with loss of up to 38 percent of the benthic cover. Fishes of the south Florida reefs were often dislocated some distance from their resident reef,

and smaller cryptic species were frequently left homeless as their sheltering sponges and corals were torn to pieces.

After passage of the storm the turbidity may remain high for a week or more. In succeeding months corals on some of the damaged fragments die, but others begin to regrow and regenerate new coral heads. Recovery of the south Florida coral reefs appears to take about five years or more, but for reefs of the Caribbean, which seem to suffer greater damage from hurricanes, full recovery has been estimated to take decades. In some areas where coral reefs are close to shore and near the mouths of large rivers (such as around Veracruz, Mexico) the reefs may be flooded by low-salinity water often loaded with suspended sediments washed down from the mountains by torrential rains. Such events can essentially wipe out the affected reefs.

Storm-induced forces of erosion, transport, and deposition also affect benthic communities of the open continental shelf. Posey et al. (1996) analyzed such effects at a study site 20 km off the coast of Cedar Key, Florida, at a depth of 13 m. After passage of the storm about a third of the species of common benthic animals declined in abundance. This was largely restricted to species that dwelled on or near the bottom surface (tube dwellers, juvenile bivalves, and epifauna), but no significant changes were observed in abundance of the deep-burrowing forms. The effects of wave energy on bottom sediments decline with depth of the water column, and it is probable that impacts on benthic animals diminish toward the middle and outer shelf. Nothing is known about the effects of major storms on animals of the water column, but following a large storm some medusas (cabbageheads, etc.), Portuguese man-of-wars, and sargassum are often found along the beaches, suggesting that the neuston and sargassum communities suffer considerable damage. One also wonders how sea turtles and marine mammals are able to surface and breathe air under violent storm conditions when very high winds remove wave tops and create a semiaquatic atmosphere for a few meters above the sea surface. For the most part, the effects of major storms and hurricanes on coastal communities around the Gulf appear to heal within a matter of months or a few years. However, some of the mangrove forests, seagrass meadows, and shallow coral reefs of south Florida may be kept in developmental stages of recovery as they are impacted by one storm after another. Since the devastation of the reefs off Veracruz is exacerbated by human activities in the watershed, they may never recover so long as the adjacent land is modified by agricultural and related land use practices.

## Hypoxic events

The distribution, causes, and effects of hypoxic events in bays and estuaries and on the continental shelf have been discussed in chapter 7 and earlier in the present chapter and need not be repeated here. Such events occur during the warm summer months in special localities characterized by poor water circulation and heavy loads of decomposing organic material. Most strongly impacted are certain bays and estuaries of the northern and eastern Gulf and nearshore waters of the continental shelf of Louisiana (west of the Mississippi River Delta) and east Texas. Although such events may result in considerable local mortality, including very visible fish kills, recovery takes place within a month or so following the reestablishment of good circulation and oxygenated bottom waters. No long-term damage to the local ecosystems results.

## Noxious algal blooms

Certain species of marine algae produce extremely potent chemical toxins, and when these algae are very abundant the toxins can cause massive mortalities in local plankton, invertebrate, and fish populations. Not all marine animals are equally susceptible, and some clams and oysters, for example, may concentrate the toxins in their cells and pass them up the food chains. Humans bathing in such waters can develop skin, eye, and respiratory irritations, and consumption of seafood contaminated by the algae can lead to serious neurological problems and death. Blooms of various species of toxic marine algae occur worldwide in coastal areas, where they often tinge the water with colors ranging from white through yellow, "tea-colored," red, green, and brown to black. Most of the really toxic forms are species of dinoflagellates, but diatoms have been implicated, and microalgae, although apparently not toxic, can become so dense that they disrupt the ecosystem. Within the Gulf of Mexico the two dinoflagellate species most responsible are *Karenia* (= *Gymnodinium* = *Ptychodiscus*) *brevis* and *Alexandrium* (= *Gonyaulax*) *monilata,* but other species including *Gymnodinium sanguineum* (= *G. splendens*), *Pyrodinium bahamense,* and others are likely involved in some cases. *Karenia brevis* and *A. monilata* have been found to be responsible for outbreaks of the notorious red tide. Toxic diatom blooms are known from the Atlantic coast of Canada and the west coast of the United States, and related and possibly toxic species including *Pseudo-nitzschia multiseries* (= *Nitzschia pungens*) and others are present in nearshore waters of the northern Gulf. They do not produce colored water. A species of microalga (*Aureoumbra lagunensis*), prevalent in the Laguna Madre of south Texas, imparts a brownish color to the water that is referred to as the "brown tide."

Over the past four decades much research has been carried out on toxic dinoflagellate species and their blooms, and more recently studies have also focused on toxic and potentially toxic diatoms and microalgae. Some of the pertinent references include Buskey et al. (1996), Connell and Cross (1950), Finucane and Dragovich (1959), Gunter (1952b), Gunter et al. (1948), Ingersoll (1882), Joyce (1975), D. Martin (1983), Perry, Stuck, and Howse (1979), Riley et al. (1989), Steidinger (1975), Steidinger and Joyce (1973), Stockwell, Buskey, and Whitledge (1993), Tester and Steidinger (1997), Wardle, Ray, and Aldrich (1975), J. Williams and Ingle (1972), and W. Wilson and Ray (1956).

In order to comprehend the factors that initiate, sustain, and finally dissipate blooms of noxious marine algae one needs to know intimate details of the physical and chemical environment in relation to life histories of the individual algal species, and at present much is still unknown. However, a great deal is understood about the red tide and, to a lesser extent, the brown tide blooms, and descriptions of these two blooms here will illustrate two contrasting types of noxious algal outbreaks.

The principal red tide organism, *Karenia brevis,* is a subtropical species present in low concentrations in waters of the Gulf Loop Current. It thrives in the temperature range of 16°–27°C (61°–81°F) but seldom survives below 10°C (50°F), and it prefers salinities above 24 ppt. Within the Gulf small outbreaks have occurred in many areas, but major blooms lasting two months or more have been recorded from the Bay of Campeche, Texas coast, and the southwest Florida shelf. The outer edge of the continental shelf off all these areas is characterized by intermittent or persistent upwelling, and all are bathed by the Gulf Loop Current or by meanders, filaments, or rings associated with or derived from the Loop Current.

Off southwest Florida blooms may develop any time of the year, but they most commonly occur in the late summer or fall. According to our present understanding (Tester and Steidinger 1997), here blooms are initiated when waters at the outer edge of the shelf are subject to major upwelling or downwelling processes that serve as pumps to bring nutrient-rich waters into the photic zone.

This stimulates growth and reproduction of the available phytoplankton species (many derived from Loop Current waters), among them *K. brevis*. However, as this dinoflagellate increases in numbers its powerful toxins inhibit the growth of other algal species, thereby reducing competition for the limited nutrient supply. As offshore waters are moved toward the middle shelf (i.e., to an area about 18–74 km from shore), cell densities increase from the initial value of a few hundred cells per liter to much higher concentrations, and within two to eight weeks they reach fish-killing concentrations of 100,000 to more than 500,000 cells per liter of seawater.

*Karenia brevis* is positively phototropic, and during the day it swarms near the surface, where it carries out photosynthesis. At night it disperses through the water column, where it has more access to nutrients. Unlike many algal species, it can obtain nitrogen from both inorganic and organic sources, further enhancing its competitive ability. After achieving very high densities at midshelf, populations may be transported to nearshore waters and then swept up and down the coast by alongshore currents, killing invertebrates and fishes and irritating surf-bathing humans.

Major coastal blooms may cover a surface area of up to 30,000 km² and last two months or more. Thus, the seasonal occurrence, initiation, buildup, and subsequent transport phases of the bloom all depend on the development of favorable physical conditions, and these factors also determine the persistence and ultimate dispersion of the bloom. Breakup occurs when the midshelf waters are swept out of the area or when the integrity of the water mass is weakened by mixing or dilution. Declining water temperature and wind stress can contribute to these processes.

The red tide scenario illustrates important facts about the ecology of planktonic organisms that inhabit the Gulf. First, the life history and success of the species is intimately tied to the prevailing physical and chemical conditions of the water, and these in turn are strongly influenced by prevailing meteorological conditions. Second, the species has developed a unique set of characteristics that enable it not only to survive but also, under special circumstances, to outcompete other species. Life histories of other toxic dinoflagellate species are somewhat different. *Gymnodinium sanguineum*, for instance, appears to prefer lower salinities, and blooms of this species may be initiated by fresher water and high nutrients associated with floods or derived from Mississippi River outflow and then spread downcoast by alongshore currents.

Powerful toxins are also produced by some benthic dinoflagellates, notably species of the genera *Coolia, Dinophysis, Gambierdiscus, Ostreopsis,* and *Prorocentrum.* These are basically tropical forms brought to the Gulf by the Yucatán current. Here they become resident in the reef tract of south Florida (Besada, Loeblich, and Loeblich 1982; de Sylva 1994), although some species have been reported elsewhere in the Gulf. When abundant these dinoflagellates may cause some disruption in the local ecosystems, but their main interest here stems from the fact that their toxins tend to become concentrated in species occupying higher levels of the food chains. Since the toxins are not destroyed by temperatures associated with normal cooking, humans consuming the flesh of certain predatory and other fish species (moray eels, snappers, groupers, dolphinfishes, jacks, wrasses, barracudas, mullets, mackerels, etc.) can receive high levels of dinoflagellate toxins, resulting in severe neurological symptoms (*ciguatera*) or death. Most cases of the disease have been reported from south Florida, but a single authenticated case is known from south Texas, and there is an unauthenticated report from Louisiana.

In January 1990 an outbreak of a then-unknown species of microalga was observed in Baffin Bay, an arm of the Laguna Madre of south Texas. Five months later it had spread into the upper reaches of the Laguna Madre proper, where it persisted in high concentrations for about nine years. Studies on water quality and other characteristics of the area had been underway prior to the outbreak, so the circumstances associated with the bloom initiation are fairly well known (Buskey et a1. 1996). Both the Laguna Madre and Baffin Bay are bounded inland by the arid lands of south Texas and on the Gulf side by a long barrier island (Padre Island) (fig. 15.6). During most years little freshwater flows into the bay and lagoon, and flushing is very poor. This, coupled with a high rate of evaporation, keeps the waters hypersaline most of the time. Prior to the bloom outbreak, in part because of a persistent drought, salinities in Baffin Bay had been high, mostly above 60 ppt. There had been several mass mortalities of invertebrates and fishes due to the high salinities and two sharp cold snaps, and as a result of the decomposition of dead organisms the water was charged with high levels of inorganic nitrogen, mostly in the form of ammonium ion. Under these conditions of low water

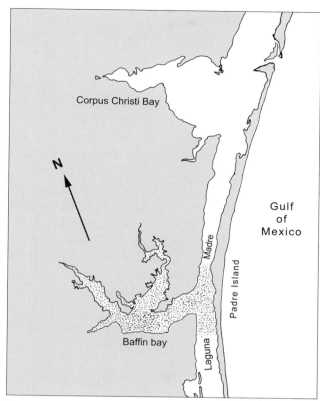

**Figure 15.6.** Map of Baffin Bay and the upper Laguna Madre of south Texas showing the known distribution of the brown tide–producing microalga *Aureoumbra lagunensis* (denoted by stippling), five months after its first appearance in the upper reaches of Baffin Bay. (From Buskey et al. 1996.)

a prebloom rate of 20–60 to fewer than 5 eggs/female/day. The microalgae were a poor-quality food for the zooplankton, which began feeding on other algal species, further reducing competition with the microalgae. The abundance, biomass, and diversity of the benthic animals also suffered a drastic reduction. Although the brown tide did not appear to affect adult shellfish and finfish directly, it did cause mortality in larval fishes, and their populations were reduced in heavy bloom areas. Because of the very high density of the tiny cells, water clarity was greatly reduced. In heavy blooms only about 50 percent of the photosynthetically available radiation penetrates to the seagrasses, and this resulted in local die-offs. By 1994 about 9 km² of seagrass meadows had been lost. These beds are important nursery areas for young invertebrates and fishes, and they provide essential winter food resources for migratory waterfowl. So, even though the species appears to be nontoxic, its impact on the ecosystem has been quite great. The brown tide outbreak persisted for nearly a decade. However, around 1998 heavy local rainfall reduced the salinity and helped flush out the bay and lagoon. This greatly reduced the density of the bloom, but in subsequent years there have been several short-term episodic outbreaks.

Circumstances surrounding the brown tide stand in sharp contrast against those associated with the red tide, discussed earlier. Both types of events involve complex relationships between the physical, chemical, and to some extent, biological factors of the environment, but the details vary considerably between the two species. One is a local endemic that inhabits coastal lagoons and breaks out under extreme environmental conditions. The other is associated with the Loop Current and erupts when nutrient and circulation patterns of the outer and middle shelf waters are appropriate. One species beats out the competition by being a superior competitor for nutrients, whereas the other eliminates potential competitors by producing powerful toxins. Although a great deal of painstaking scientific investigation has been required to work out the known connections between colored water and mass mortality of marine life, in both cases much yet remains to be learned.

## Sediment slumps and turbidity flows

As discussed in chapter 3, rivers of the northern, western, and southern sectors have deposited enormous volumes of land-derived sediments in the Gulf of Mexico. This pro-

temperature, very high salinity, and elevated inorganic nitrogen, the microalga (now known as *Aureoumbra lagunensis*) appeared in the arms of Baffin Bay and gradually increased in cell density from a normal concentration of around 100 to more than 2,000,000 cells/mL (and subsequently to more than 6,000,000 cells/mL).

Although the microalgae apparently do not produce toxic chemicals, their effect on the local ecosystem has been dramatic. They outcompete most other algal species, and under bloom conditions they occur in the bay and lagoon in almost pure culture. Because they are very small (4–5 µm), the individual cells have a large surface-area-to-volume ratio, which gives them an advantage over larger algae in obtaining nutrients. So, the reduction in other algal species results, in part, from its superior competitive ability in gaining nutrients. Populations of the dominant zooplankton species declined at the beginning of the bloom and remained low thereafter. The size of the adult copepods decreased, and egg production rates fell from cess has been particularly active in the northern Gulf, where the shelf has prograded seaward many miles and where post-Cretaceous deposits alone in some areas are now as much as 9 km (5.4 mi) thick (fig. 3.6). Much of this material has been contributed by the present-day and ancestral Mississippi River, whose deposits have extended the Mississippi Fan seaward all the way to the base of the Yucatán Scarp and for a considerable distance both east and west of the present delta (Bouma, Stelting, and Coleman 1985). To the west ancestral streams have also been quite active at one time or another, alternately filling the Houston and Rio Grande embayment areas (table 3.2).

During periods of high sea level stand, much riverborne material is deposited in the river valleys on land or, if brought to the coast, sequestered in coastal bays and lagoons, but some is transported to the Gulf proper, albeit at a relatively slow rate. That which lands on the continental shelf may subsequently be resuspended and deposited in deeper water on the continental slope. By contrast, during periods of low sea level stand, when rivers cut deep channels and erode deposits from the river valleys, they transport vast quantities of particulate material to the Gulf (Coleman, Prine, and Garrison 1980) (fig. 3.4). Here the streams empty at or near the outer edge of the continental shelf and release their sediment loads directly upon the outer shelf or continental slope (Feeley, Buffler, and Bryant 1985; Stow, Howell, and Nelson 1985). This rapid buildup of loose sediments leads to frequent episodes of sediment slumping and downslope turbidity flows. Whether during high or low sea level stands, these masses of loose sediments become mixed with water and race downslope to be deposited on the continental rise or abyssal plain. As they move they may be confined within preexisting channels or they may create new ones.

Benthic and demersal animals caught up in such submarine storms are likely to be tumbled and pummeled and rapidly transported downslope to unfavorable environments in much deeper water, where they and the local deepwater residents may become buried and smothered under layers of depositing sediments. After passage of a major gravity flow the track must be virtually devoid of higher marine life. Although the direct biological effects of such events have not been documented, there is ample evidence of past flows and some suggestion that they still occur occasionally if only on a reduced scale. In the deep Gulf, layers of fine clastic sediments are laminated between layers of pelagic ooze. Such processes must have been particularly active during the two million years or so of the Pleistocene epoch as the sea level repeatedly fell and rose in response to alternate periods of glacial advance and retreat on the North American continent.

## Lesser catastrophic events

In this review we have examined six major types of catastrophic events affecting biological populations of the Gulf (cold snaps, floods, major storms and hurricanes, hypoxic events, noxious algal blooms, and sediment slumps and turbidity flows). However, in such a large, diverse, and active environment as the Gulf of Mexico there are many other types of catastrophic events that could be cited, even if they sometimes occur on smaller scales or lack detailed documentation. A few of these are noted briefly here. Severe and prolonged droughts, which result in abnormally high salinities in estuaries and bays, modify the structure of the local ecosystems. By permitting entry of a wide assortment of marine predators into the otherwise protected coastal shallows, such conditions foster heavy predation of larval and postlarval shrimps, crabs, and fishes in the nursery areas.

Around the Florida Keys and Tortugas a combination of high summer water temperature and poor water circulation sometimes leads to thermal stress and mortality in corals (Jaap 1984). Certain infectious agents can cause significant mortality in susceptible species. Examples include slime mold disease (due to *Labyrinthula*) in seagrasses (*Thalassia*) of Florida Bay (Durako and Kuss 1994), black band disease (due to the cyanobacterium *Phormidium*) (Kuta and Richardson 1994) and white band disease (possibly due to a bacterium) (Gladfelter 1982) in corals of south Florida, and reef-fish disease (agents unknown) in several reef fish species of the Florida Keys (Landsberg 1995). Many coastal species depend on normal water current patterns to disperse their larvae to favorable sites for settlement or passage to nursery areas, but abnormal currents can sweep the larvae into the open Gulf, where they perish.

The Atlantic croaker (*Micropogonias undulatus*) is normally the dominant fish species in bays and estuaries adjacent to Mississippi Sound in the northeastern Gulf, but around 1990–1992 the species virtually disappeared from the coastal waters, and this event was preceded by abnormal meteorological and water circulation patterns (Darnell 1991b). Periodically the Gulf Loop Current spins off rotating rings of water that proceed in a westerly or southwesterly direction, where they ultimately spin down, either in the open water or along the edge of the continental slope. Each ring consists of an outer band of rotating Loop Current water and a core of Caribbean water trapped inside. As the rings and cores decay their physical and chemical characteristics change, and plankton and other species trapped in the rings and cores must undergo considerable mortality. Dead blue whales occasionally wash ashore on beaches of the western and northwestern Gulf, and one may visualize these animals beginning an epic journey toward Arctic feeding grounds only to become trapped in a Loop Current ring, where they swim around repeatedly in circles until finally, lost and exhausted, they expire and float ashore in the northwestern Gulf. As lower extensions of the rings brush along the continental slopes they must sweep countless numbers of small benthic animals away from the sediment surface into the open Gulf waters.

From the examples presented above, it is clear that catastrophic episodic events of various magnitudes represent a widespread phenomenon, particularly in shallow coastal environments, but they also occur in deeper areas of the Gulf, and they affect marginal vegetation as well as organisms of the water column and those associated with the benthos. Depending on the nature and severity of the events and where they occur, recovery time may be months to years, but in the absence of further disturbance the systems do eventually return to some sort of predisturbance equilibrium. However, such events are not without longer-term ecological significance. Disturbance areas are first invaded by algae as well as larval and adult invertebrates recruited from outside the disturbance zones, and many of these pioneering species are especially adapted to the task of quickly repopulating disturbed environments. The frequent occurrence of major natural perturbations ensures that these rapidly reproducing and mobile species remain available to begin the process of habitat wound healing whenever and wherever needed. Besides adding to the biological diversity, such species are necessary to the normal healthy functioning of marine ecosystems. In previous chapters we learned of the spatial heterogeneity of the various habitats around the Gulf, and we now gain insight into their temporal variability. It is worth noting that variable and heterogeneous environments, in turn, foster genetic variability within the individual species, which provides for adaptability and flexibility in dealing with the complex and ever-changing environmental milieu of the Gulf.

With a surface area of over 1.5 million square kilometers, the Gulf of Mexico is one of the world's major marine systems. Yet this vast area with its diverse habitats (including estuaries and coastal lagoons, marshes and swamps, seagrass meadows and coral reefs, broad terrigenous and carbonate shelves, continental slopes and abyssal plain, and large volume of deepwater habitat) is a single functional ecosystem bound together by physical and chemical processes, the exchange of materials and living plants and animals, and various activities of the organisms themselves. Earlier chapters described the individual subsystems of the Gulf, providing information on habitat distribution, species composition, biomass production, and so forth, and in the last chapter we examined how organisms have become adapted to the normally prevailing physical and biological factors of the environment and how they cope with erratic and occasionally extreme conditions. Here we will take an intimate look at life history patterns manifested by Gulf of Mexico species and consider how the species fit together into smaller multispecies aggregations and how these, in turn, are integrated into the several major subsystems that make up the total Gulf of Mexico ecosystem. Although there is much that we do not yet understand about the life histories and complex interrelationships of the organisms and subsystems, we do know enough to sketch, at least in outline, what is going on, how the systems function, and the factors governing these processes.

## INDIVIDUAL LIFE HISTORIES

Organic evolution may be thought of as a game in which the single goal is long-term survival of the species. Any group that played the game with a different goal is no longer with us. The species of plants and animals we observe on the earth today are those that have most successfully dealt with physical and biological problems, thereby enabling them to remain in the game. Through the processes of natural selection, only the most able competitors have survived. The life history of any species consists of a series of stages, each characterized by its own specialized morphology, physiology, and behavior, and one must assume that each of these stages and characteristics has

some special survival value for the species; otherwise it would have been eliminated long ago. For example, acute vision depends on the availability of a series of appropriate genes, and the maintenance of this vision entails constant positive selection for these genes. In the absence of such selection pressure, one or more of the genes may become modified or lost as selection works for other characteristics, and visual acuity is diminished or lost in the shuffle.

In the marine world one observes a great variety of animal species whose life histories include eggs, larvae, juveniles, and adults. We are struck by the array of bizarre morphologies and strange ways of doing things, but in the absence of information to the contrary we must assume that each shape, spine, and behavior has a specific function that enhances the survival capabilities of the species, even if such values are not always obvious to the observer. Nature often works in subtle ways, and we are still quite ignorant of the life histories of most marine animals. As seen in table 16.1, the life histories of marine animals consist of a number of components associated with reproduction, larval life, and postlarval life. Among the various species these components are put together in a variety of different ways to produce unique but successful combinations. It is customary to refer to the life history patterns of marine animals as their life history strategies, but this does not imply advance planning on the part of the species. Rather, the life history strategy is simply the sum total of adaptations that together constitute a formula for survival. The sections that follow provide information about the life history strategies of marine animals with special reference to those that inhabit the Gulf of Mexico.

## Reproductive strategies

### SEXUAL VERSUS ASEXUAL REPRODUCTION

The critical problem for each generation is to produce offspring for the following generation so as to keep the genetic line going into the future. This function involves two aspects: (1) production of sufficient numbers so that enough survive to reproductive age and (2) provision for genetic recombination and genetic diversity among the offspring. The latter aspect may be accomplished only by

**Table 16.1.** *Major components of marine animal life histories, which when properly adjusted provide for the survival of the various species.*

## Reproductive strategies

Sexual vs. asexual reproduction

One-shot vs. repeated spawning

Seasonality and migrations

Fecundity

  – Total number of eggs produced

  – Egg types

  – Larvae vs. no larvae

Genetic aspects

  – Importance of sexual reproduction

  – Importance of genetic variability

## Strategies of larval life

Problems of larval life

  – Obtaining food

  – Dispersal and migrations

    - Widespread dispersal

    - Targeted migrations

Types of larvae

  – Phylogenetic aspects

  – Planktonic vs. benthonic larvae

  – Short vs. long larval life

  – Number of larval stages

## Strategies of postlarval life

Intraspecific factors

  – Feeding

  – Growth and maturation

  – Intraspecific competition

  – Other population phenomena

Interspecific factors

  – Intimate species associations

  – Nonintimate associations

    - Feeding and predation

    - Shelter

    - Mobility and behavior

    - Defense and aggression

    - Interspecific competition

## Overall life history patterns

Life history types (*r* vs. *K* strategy)

General life history patterns

  – Benthos vs. water column

  – Estuarine and shelf species

  – Surface vs. deep-sea species

sexual reproduction, about which more will be said below. The former, the production of large numbers of offspring, may be accomplished through asexual or sexual reproduction. Species of the marine world, particularly those with planktonic eggs and larvae, face many uncertainties and potential sources of mortality (associated with transport to favorable environments, predation, location of suitable substrates, etc.), and the survival of such species often depends on the production of very large numbers of offspring.

Asexual reproduction by budding and other means is employed by many species of lower invertebrates to increase their numbers. For example, a single larva of a coral species that lands on a suitable hard substrate will grow into a mature polyp, which by budding provides new individuals and eventually all the thousands of polyps that make up the surface of a large coral head. Many species of hydroids and hydrozoans bud off new individuals constantly. This ability to reproduce asexually at least temporarily removes the necessity of finding a breeding partner, helps overcome mortality problems, and provides for wide dissemination of offspring. The disadvantage of asexual reproduction is that the offspring are all genetically identical with the parent and with each other. However, even those species that reproduce asexually must eventually carry out sexual reproduction in order to retain genetic diversity within the species.

Anyone who takes a ship out to the Flower Garden Banks off east Texas 7–10 days following the full moon in August (the dates may vary somewhat if the full moon occurs in early August) and makes a night dive down to the reef will encounter mass spawning by all the corals of several species (at least six species of the genera *Colpophyllia, Diploria, Montastrea,* and *Stephanocoenia*). At this time the warm water bathing the reef is thick with rising eggs and clouds of sperm (fig. 16.1), and as the fertilized eggs reach the surface they float off into the night toward an uncertain future (Gittings, Bolland, et al. 1992). This annual event, which may require several nights for completion, represents the only sexual reproductive activity of the species during the year. In a similar manner, one may anchor in the lee of the Chandeleur Islands east of the Mississippi River Delta in late August and by means of a night light suspended over the water observe the breeding swarm of a nereid polychaete worm. The water is alive with large fuzzy-looking females surrounded by numbers of small rapidly gyrating males excited by chemical stimuli (*pheromones*) released into the

**Figure 16.1.** Egg bundle release from brain coral (*Diploria strigosa*) at the East Flower Garden Bank. Photo by Gregory S. Boland.

water by the females. Many species of Gulf animals breed at a single time during the year in response to environmental cues such as water temperature, length of day, and lunar tides, and coordinated by species-specific signals including chemical cues, sound production, photophore patterns, and so forth. Other species exhibit more than one spawning peak, and yet others have long, extended breeding seasons or breed more or less continuously through the year. Whether one-shot, repeated, or continuous, each spawning pattern has its advantages and disadvantages, and each type is successful for certain species and in particular environments. When all the eggs are released at once, they are so abundant that the locally available predators can consume only a small fraction of them. However, unfavorable currents could sweep the entire year's reproductive products out to sea or to other unfavorable locations. For long-lived species such as corals, the loss of one year's reproductive effort may make little difference, but for short-lived species it may be safer to engage in repeated or continuous spawning. Species living in coastal areas where the environmental parameters vary widely often engage in repeated spawning. Many small planktonic and benthic invertebrates of the Gulf appear to reproduce throughout the year, or at least through long periods of the year.

FECUNDITY

Since each generation must produce enough offspring to survive to reproductive age and since great overproduction of young would be a waste of energy, the numbers of offspring produced should reflect, in a rough way at least,

the average mortality rates expected by a species during preadult life. In turn, this mortality must reflect the life history pattern of the species. To illustrate this relationship, table 16.2 provides information on the maximum recorded fecundity of adult females of a number of coastal species inhabiting the northern Gulf. The American oyster is a nonmobile bottom-dwelling mollusk that spends its entire postlarval life in the estuary, where it is subject to mass mortality due to flooding, siltation, and other extreme events as well as heavy predation by a number of invertebrate and fish species (boring sponges, oyster drills, black drums, etc.) and many types of parasites and diseases (Overstreet 1978). It spawns several times a year, and its larvae must eventually settle onto suitable hard substrates. Although most must settle within the same estuary where their parents live, it is important that some be swept into continental shelf waters, along the coast, and into distant estuaries to recolonize beds destroyed by episodic events and to maintain genetic diversity within the species. The chances of all this occurring successfully must be quite low, as suggested by the production of several hundred million eggs per year by the adult females. By contrast, species that spawn on the continental shelf and whose larvae must make the journey from the spawning grounds into the estuarine nursery areas generally produce several hundred thousand to a million eggs. They have to survive the hazards of transport into the estuaries as well as heavy predation along the way. Species that spawn on the shelf and remain there do not suffer the transport problems and have to contend only with predation. Such species produce only around forty thousand eggs per female. Finally, the hardhead catfish has eliminated both the transport and predation problems. In this species the adult male picks up the fertilized eggs and carries them in his mouth until they become free-living juveniles. In addition, the female produces a maximum of only around a hundred eggs. These few examples illustrate the strong correlation between female fecundity and life history pattern for coastal species. Similar relationships must apply to offshore species of the deep Gulf, but here data are not available for analysis.

Species in which there are free-living larval stages generally produce large numbers of tiny eggs containing little yolk. Upon hatching, the young must begin feeding immediately and nourish themselves as they develop and carry out their migrations. An alternative strategy is to produce fewer, larger eggs that contain much yolk. Such young do not begin feeding until they have reached a more advanced stage of development. Species with planktonic larvae are capable of wide dispersal and long migrations, allowing them to spread their genetic material and colonize distant areas. Those with large yolks and no larval stages have very limited powers of dispersal, genetic mixing, and colonization. Both life history patterns are successful in certain situations.

GENETIC ASPECTS

Among the wide array of life history patterns observed in marine plants and animals, one constant stands out. Whatever else may vary from species to species, all engage in sexual reproduction at some stage of their life histories. The universal occurrence of this phenomenon underscores its importance in the life of all species. As noted earlier, in the absence of sexual reproduction all individuals of the species are locked into a pattern of genetic identity in which the only change is due to occasional gene mutation (which is most likely to be detrimental, anyway). Long-term survival in the marine realm, as elsewhere, depends on the ability to change and adapt, and this in turn depends on the recombination of genetic material derived from two parents. Within the population as a whole there are a number of alternate gene types (*alleles*) that could occupy a given gene position (*locus*) on a chromosome. This reservoir of alleles provides for genetic variation and is the foundation for ecological adaptability, allowing for the testing of new combinations in each generation. In highly variable environments, such as coastal areas, the importance of adaptability is obvious, but what about the deep sea, where the physical environment appears to be much more constant? From what is known about deep-sea species, genetic variability is high here also. The reason appears to lie in the game of evolution. Adaptability is important in relation to not just physical factors, but biological factors, as well. As the competitors, predators, and parasites change, so must the species affected by these other forms, and this genetic race to stay in the game is not limited to variable coastal environments but rather pervades all situations wherever two or more species interact. Among marine animals there are many variations around the theme of sexual reproduction. In most species the sexes are separate, but some species are *hermaphroditic* (a given individual possesses both sexes at the same time, although self-fertilization is rare). In most species sex is constant throughout life, but in some species the sex changes with age. In *protandry* the male sex comes first and is later

**Table 16.2.** *Approximate fecundity of some coastal invertebrate and fish species of the northern Gulf of Mexico in relation to life history patterns. Fecundity is expressed as the maximum number of eggs produced per female per spawning. Note: The hardhead catfish is a mouth brooder. (Data from several sources.)*

| Species | Maximum number of eggs per female |
|---|---|
| **Spend entire life in estuaries** | |
| American oyster (*Crassostrea virginica*) | 115,000,000 |
| **Spawn in Gulf; use estuaries as nurseries** | |
| Pink shrimp (*Penaeus duorarum*) | 534,000 |
| White shrimp (*Penaeus setiferus*) | 860,000 |
| Gulf menhaden (*Brevoortia patronus*) | 128,000 |
| Sand seatrout (*Cynoscion arenarius*) | 325,000 |
| Silver seatrout (*Cynoscion nothus*) | 390,000 |
| Spot (*Leiostomus xanthurus*) | 514,000 |
| Atlantic croaker (*Micropogonias undulatus*) | 1,076,000 |
| Hardhead catfish (*Arius felis*) | 104 |
| **Spawn on shelf and remain there** | |
| Longspine porgy (*Stenotomus caprinus*) | 43,000 |
| Atlantic cutlassfish (*Trichiurus lepturus*) | 42,000 |

replaced by the female sex. In *protogyny* the reverse is true. Some species breed as single isolated pairs, while in others there are large breeding aggregations. In the deep sea where sexual partners are hard to locate, males of some species become parasitic on the females and remain attached throughout life, where they are available to fertilize her eggs. These various adaptations all underscore the importance of maintaining the process of genetic exchange and gene recombination generation after generation.

## Strategies of larval life

The larva is the earliest stage in the life history of many animal species following the developmental phases of the fertilized egg. In effect, it is an advanced embryo that is free living, has special adaptations for the free-living existence, and bears little resemblance to the adult or mature form of the species (as distinguished from the juvenile stage, which looks like a miniature adult). When scientists first encountered many larval forms, they described them as separate species and only later recognized them as being related to known adults. Even today many types of marine larvae are still not associated with the appropriate adult forms. Most species of marine invertebrates and some fishes, even many deep-sea benthic species, begin life as larvae (fig. 9.6). The great majority live up in the water column and are adapted for life in the plankton. Some of these develop into planktonic or nektonic adults, but many, especially among those found in continental shelf waters, ultimately descend to the bottom, where they undergo *metamorphosis* (change in body form) and grow into benthic adults.

PROBLEMS OF LARVAL LIFE

In order to survive and develop into juveniles, the larvae of marine animals must face and overcome a series of problems including the location and apprehension of sufficient food of the right types, protection from predators, transport to suitable areas, and for many species, the location of appropriate substrates for settlement and metamorphosis. To accomplish all this, the various types of larvae have developed many specialized features, the most obvious of which are morphological and behavioral, but they clearly have specialized sense organs and physiological and biochemical adaptations, as well.

The primary food of most early larvae is phytoplankton whose cells are small enough to be consumed, and this may be supplemented with bacteria, small invertebrates, and detritus particles in the appropriate size range. As the larvae develop and increase in size, zooplankton and other larger-sized particles often make up a greater percentage of the diet. So, for most planktonic larval forms the primary habitat is the euphotic zone where phytoplankton and zooplankton are most concentrated, although some types of larvae are known to undergo regular vertical migrations, being found in deeper layers during the day and becoming concentrated in surface waters at night. In order to obtain their food the larvae have developed a variety of feeding mechanisms. One of the most common involves bands of cilia to sweep food particles toward the mouth, and these may be supplemented by lateral appendages to aid in guiding the particles. Some produce strands of mucus-like fishing lines to which particles adhere. Others filter out food particles with fine-mesh nets. More carnivorous forms have claws for grasping and jaws for macerating prey. Planktonic marine larvae generally do not fare well off the mouths of large rivers, where there may be high concentrations of inorganic particles of no nutritive value that could clog or otherwise interfere with the feeding apparatus. In most coastal waters and upwelling areas where plankton is abundant, the larvae may have little trouble securing sufficient food, but many of those that get swept offshore into the relatively less productive areas of the open Gulf must face malnourishment or starvation.

Eggs, embryos, and the very young suffer the heaviest predation because their defense mechanisms are not yet fully developed. To avoid such losses the adult females of many species (notably lobsters, crabs, and caridean shrimps) carry the developing eggs on abdominal appendages, or *swimmerets,* and females of other groups carry the eggs and young in special brood pouches or chambers. The larvae are then released into the water in an advanced stage of development. In other forms the eggs are laid in protected areas and guarded by a parent until the larvae are released at an advanced stage. In yet other species, including some with large-yolked eggs, the planktonic larval stages are passed through very quickly, thus limiting the time of exposure to planktonic predators. Many larval forms avoid predation by being transparent and difficult to see. Many others bear an array of spines that can interfere with a predator's ability to swallow them or cause the predator to reject them if they are ingested, and such spines may be fixed in place or erected only when the larvae are molested. When attacked by a predator some polychaete larvae roll up into a ball with spines projecting in all directions. Many larvae can survive repeated attacks. In a few cases, such as in the tunicates, the larvae are toxic. To escape predation the larvae of some continental shelf species hide in the benthos during the day and become planktonic only at night. Defense mechanisms, such as those discussed above and others, serve to increase survival chances for the larvae of particular species, but overall it has been estimated that larval losses from predation may be as high as 10–20 percent per day. In coastal waters of the northern Gulf of Mexico major predators on larval forms include medusae, comb jellies, chaetognaths, a variety of crustaceans, and small fishes.

Considering the high mortality rates of marine planktonic larvae, one may wonder why it is advantageous to have such larval stages at all. As noted earlier, planktonic larvae, particularly those that can sustain themselves by feeding, confer on marine species two important benefits: the ability to achieve wide dispersal and the ability to carry out migrations to target habitats. Wide dispersal, especially among long-lived larvae, provides a high potential for colonizing newly available and often distant habitats, which allows the species to continue in spite of occasional local extinctions. It also provides for genetic exchange between widely separated populations of the same species, thereby maintaining high genetic variability within the populations as a hedge against inbreeding. This appears to be a powerful incentive to produce larvae with long planktonic lives. Although most larval forms remain planktonic for a few weeks to a few months, some are longer lived. Even those that last only a few months could cross the Atlantic, for example, and maintain genetic contact across this wide ocean barrier, but larvae of the spiny lobster, which can remain planktonic for at least two years, could, in theory, circumnavigate the North Atlantic Gyre at least twice before settling down. Spiny lobsters in the Gulf of Mexico may be offspring of parents residing in Bermuda, the Bahamas, the Caribbean, or even North Africa.

Many species of shrimps, crabs, and fishes spawn on continental shelves, but the larvae must carry out migrations to target nursery areas within bays and estuaries (fig. 16.2). Such migrations involve two phases. The first is transport to the inlets and passes, and the second entails movement through the passes to the protected inside waters. Eggs and early larvae tend to float passively, more or less like inert particles, and their dispersal from the spawning grounds must depend on the vagaries of wind and water currents. Losses at this stage as a result of unfavorable current regimes must account for much of the high year-to-year fluctuation in population levels of such coastal species. Older larvae, in which the sense organs and locomotory apparatuses are better developed, appear to be able to exert some control over their own destinies, and many can certainly change buoyancy, permitting them to remain planktonic, seek out appropriate current regimes, or sink to the bottom and wait. Their dispersal patterns must depend on the interaction of physical and biological factors.

For some species, transport to and through the passes follows what is often referred to as the "selective tidal transport" model (Lochmann, Darnell, and McEachran 1995). In this case the larvae can sense and be guided by chemical cues emanating from the estuary. They are also sensitive to tidal flow (or its timing) and possibly to light. During flood tide, when seawater flows into the estuary, they rise into the water column and are conveyed toward or through the estuary mouth, and on ebb tide they sink back to the bottom and wait out the falling tide when water leaves the estuary. Thus, by hitching rides on suc-

cessive incoming currents they "hop" their way into the estuaries. The larvae of many coastal species likely follow some variation of this mechanism to move from the continental shelf to the inside waters. Whether they are following the selective tidal transport or some other model, field observations reveal that larvae tend to be most concentrated in the passes during periods of incoming high tides at night (Epifanio 1988).

When it is time for them to settle, some types of marine larvae must locate an appropriate hard substrate. By means of various sensory mechanisms they inspect each surface to determine whether it is suitable, and if not they may become planktonic again and move on in search of the right material. Once satisfied, they affix themselves by means of "glue" produced by basal cement glands and undergo metamorphosis, a process that may take only a few hours. Thereafter, they grow into adults, which are generally much larger than the tiny larvae from which they were derived. There is a large technical literature dealing with various aspects of larval development, behavior, transport, dispersal, and settlement. A few of the more pertinent references include the following: Cameron (1986), Ditty and Shaw (1995, 1996), Ditty, Zieske, and Shaw (1988), Govoni (1993), Govoni, Hoss, and Colby (1989), Grimes and Finucane (1991), Grimes et al. (1990), Houde et al. (1979), Jackson (1986), Leming and Johnson (1985), Lochmann, Darnell, and McEachran (1995), Lyczkowski-Schultz et al. (1990), McGowan (1985), Perry, Eleuterius, et al. (1995), N. Rabalais et al. (1995), W. Richards et al. (1993), Scheltema (1971), R. Shaw, Cowan, and Tillman (1985), Weinstein (1988), and Young and Eckelbarger (1994). In shallow tropical and subtropical seas of the world, most invertebrate groups produce planktonic larvae, and many of those found in the Gulf are illustrated in C. Davis (1955). Drawings of larvae and early stages in the life histories of many fish species of the Gulf and western North Atlantic are presented in Fahay (1983).

## Strategies of postlarval life

After completing the larval stages and undergoing metamorphosis, whether in the water column or on soft or hard bottoms, the juveniles feed, grow into adults, and begin breeding to produce the next generation. During this period the growing individuals must contend with competition from members of their own species as well as pressure from other species. Since members of the same species have more or less identical requirements, intra-

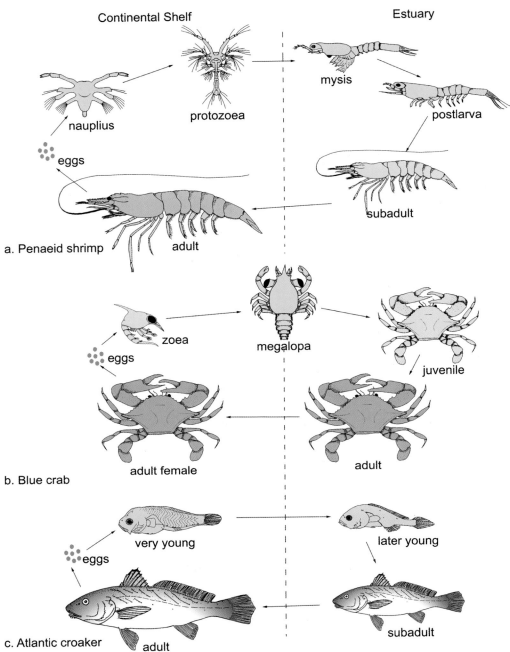

**Figure 16.2.** Life history types of some estuary-dependent species of the Gulf of Mexico. (a) Penaeid shrimps (*Penaeus aztecus, P. duorarum, and P. setiferus*) all spawn on the continental shelf. Eggs are released, and the young pass through a series of planktonic larval stages. The mysid stage, normally found in near-bottom waters, passes into the estuary, where it feeds and grows into the subadult. The subadults move back to the shelf, where they mature and spawn. (b) Adult female blue crabs (*Callinectes sapidus*) on the continental shelf of the northern Gulf release the young only after they have reached the megalops stage. These advanced larvae move into the estuaries, where they grow into mature adults. After mating, the male remains in the estuary, but the inseminated female migrates back to the continental shelf and remains near the mouth of the estuary, where she carries the developing eggs on her abdominal appendages until they are released as advanced larvae. (Note: As seen in Figure 16.6, the life history of *C. sapidus* in the eastern Gulf is somewhat different.) (c) Atlantic croakers (*Micropogonias undulatus*) and many other estuary-dependent fish species of the Gulf spawn on the continental shelf. There is no larval stage in most, exceptions being the Atlantic tarpon (*Megalops atlanticus*), ladyfish (*Elops saurus*), several menhadens (*Brevoortia* spp.), and a few others. The eggs develop into young that gradually find their way into the estuaries, where they feed and grow. As subadults they move back to the continental shelf, where spawning takes place.

specific competition for food, shelter, living space, and other resources is often intense. Territoriality, or defense of an area or resource against members of the same or different species, is common among tropical reef fishes, and it is observed in benthic environments elsewhere. Some cephalopods, crustaceans, and fishes defend burrows, home crevices, and other areas, especially if they contain eggs or developing young. We see competition in action when two hermit crabs struggle for possession of a favored snail shell to be used as a shelter or when a damselfish vigorously defends its patch of algae against greedy interlopers intent on harvesting their neighbor's food supply (Mahoney 1981; Myrberg and Thresher 1974), but often competition between members of the same species is more passive and subtle. In attacks on the common food supply it is generally "every individual for himself," and the best competitors survive to parent the next generation and pass on their genetically based survival traits.

Relations with other species take a variety of forms. In most instances these relationships involve free-living organisms obtaining food, competing for resources, and preying on one another, topics that have been addressed earlier. However, for many marine species the associations are more intimate, and sometimes species specific.

INTIMATE SPECIES ASSOCIATIONS
When two species regularly live together in close association, the relationship is referred to as *symbiosis.* For convenience, here we recognize three basic types of symbiosis: *commensalism, mutualism,* and *parasitism.* In nature these types of relationships are not always distinct from one another. They intergrade, sometimes one type can turn into another, and evolutionarily one type (commensalism) must have developed from or into the others. Since symbiotic relationships are widely distributed and important in the lives of many marine species, each of the types will be discussed and illustrated.

*Commensalism* refers to a relationship between two species in which one receives benefit, and the other receives neither benefit nor harm. The marine world is replete with examples of one species taking advantage of opportunities provided by another apparently without harming the host species, although the harm may be a matter of degree. We are not always sure. If a few colonies of filamentous algae grow upon a blade of seagrass, the algae benefit from having the substrate for attachment without apparent harm to the host, but if filamen-

tous algae coat the blade they can reduce light penetration and interfere with gas exchange processes to the detriment of the seagrass. In marine ecosystems the organisms that can be attached to other species include filamentous and other types of algae, as well as protozoans, sponges, hydroids, sea anemones, corals, polychaetes, snails, bivalves, barnacles, sea lilies, bryozoans, and tunicates, among others, and often these do not materially interfere with the host's ability to function normally. However, one wonders whether the host species sometimes pays a metabolic price for harboring the other species, or whether the host sometimes receives a benefit such as nutrition or protection. Sharksuckers, or remoras, attach themselves to sharks and other large fishes as well as sea turtles and whales (and even the hulls of ships). They are transported wherever the host travels, they are protected from predation by the host's presence, and they consume scraps of food from the host's meals. Nereid polychaetes and tiny pea crabs (*Pinnixa, Polyonyx,* etc.) live within the mantle cavities of bivalve mollusks and in the tubes or burrows of certain wormlike forms and crustaceans, receiving protection and food particles from the passing water currents apparently without bothering the host (Felder 1973; A. Williams 1965). Tiny gall crabs (*Opecarcinus* and *Troglocarcinus*), as larvae or very young, settle onto pits or other protected areas of certain types of corals (*Agaricia, Diploria, Favia, Montastrea, Oculina, Siderastrea,* etc.) and remain in place as the corals gradually grow around them and form a protective chamber with a tiny opening through which the crab is able to obtain food particles and oxygenated water (Kropp 1989; Kropp and Manning 1987), (fig. 16.3). The list of such associations is very long and involves various types of algae and most groups of the animal kingdom. In most cases the association is very loose and involves species that could well live apart, but in other cases, such as the pea crabs and gall crabs, the "guest" species is especially adapted for living with the host, and some may even be limited to a particular host species.

*Mutualism* is a type of intimate species association in which both partners receive benefit. In some instances one or both of the species can live alone, and the relationship is one of convenience ( *facultative mutualism*), but in other cases the relationship is obligatory for both partners. The classical example of mutualistic symbiosis in the sea involves tiny algae (zooxanthellae) that live within the tissues of corals and other marine animals (certain medusae, flatworms, bivalves, etc.). The algae

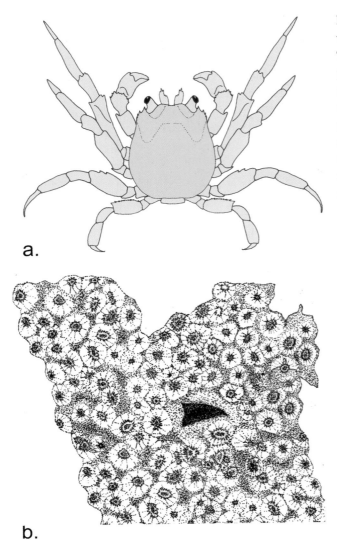

**Figure 16.3.** (a) The gall crab (*Troglocarcinus corallicola*) and (b) the opening of a gall crab home in coral (*Oculina varicosa*). (From Kropp and Manning 1987.)

heads or other prominent structures of the reefs and advertise their presence by conspicuous behavior patterns. The larger fishes line up to await their turn or jostle for position in the line. In this relationship the cleaners thoroughly inspect the external surface of the body as well as the inside lining of the mouth and gill arches, removing parasites and scraps of dead skin and cleaning tissues around wounds. In the process the cleaner species obtains food scraps, and the host fishes benefit from the removal of dead tissues and parasites. Other examples of facultative mutualism involve sea anemones attached to shells housing hermit crabs (placed there by the hermit crabs themselves). The hermit crabs receive protection from the stinging cells of the anemones, and the anemones receive transportation and scraps of food from the hermit crabs' meals. Certain types of small crabs, especially of the family Majidae, place live algae, hydroids, and other sessile organisms on the upper surfaces of their carapaces and hold them in position until they become firmly attached. These "decorator crabs," as they are called, receive the benefit of the camouflage, and the attached organisms receive transportation and food scraps. One summer afternoon on shipboard in the northern Gulf of Mexico shortly after the contents of a trawl had been dumped onto the deck, the author was surprised to observe a small clump of algae and hydroids moving slowly and deliberately away from the main mass of fishes, crabs, shrimps, mollusks, and other benthic marine life. The disguise was very effective, and the decorator crab would have gone completely unnoticed except for the fact that in its discomfort on the hot deck the crab was induced to crawl away, negating its efforts to appear inconspicuous. Facultative mutualism is probably a widespread phenomenon in the Gulf, and as the various commensalistic relationships are more thoroughly studied many will probably turn out to have mutualistic aspects.

*Parasitism* is a type of intimate species association in which one species benefits and the other is harmed. The parasite, normally the smaller of the two species, generally lives upon or within the host and often obtains nourishment from the tissues or body fluids of the host species. Parasitic relationships were discussed in some detail in the previous chapter. Here it is sufficient to note that some so-called parasites, currently considered to be harmful to their hosts, may eventually be found to bestow benefits. On the other hand, commensalistic and mutualistic relations can turn sour and act to the detriment of one of the partner species. Rules governing the

receive from their hosts living quarters, protection, and certain chemical nutrients (especially those containing nitrogen, phosphorus, and carbon). In turn, the algae carry out photosynthesis and supply their hosts with carbohydrates and other energy-rich compounds. In most cases the algae are capable of living independently and may abandon their hosts under stress conditions such as excessively high temperatures. Another example of mutualism is the cleaning symbiosis involving caridean shrimps (*Periclimenes* spp.) and small fishes (primarily wrasses and gobies), which act as the cleaners, and larger fishes (snappers, groupers, etc.), which avail themselves of the cleaning services. The cleaners, which are generally brightly colored, take up positions atop large coral relationships of species in nature may work well so long as the organisms are in good health and the environment is moderate, but they are flexible and subject to change as circumstances and environmental factors become stressful.

## AGGRESSION AND DEFENSE

Aggression involves the attempt of one individual to achieve dominance over another and to deprive the other of certain rights (such as possession of a territory) or to take its life (as in predation). Aggressive manifestations range from intimidation to attack and physical struggle. Intraspecific battles for dominance in a social hierarchy or for possession of a territory are generally nonviolent affairs involving ritualistic displays, feigned attacks, or brief physical skirmishes without injury to either party, but in some cases injury or death may result. During the breeding season the males of some marine mammal species may engage in physical conflicts over the right to breed with a female. An extreme example of aggression is provided by the tulip snail (*Fasciolaria*), where an encounter between two individuals inevitably results in a fight that could lead to simple capitulation or even death and cannibalism. Sometimes reef fishes defend their territories against all comers (within a given size range), including members of their own as well as other species. However, most interspecific aggression is related to predation, where one individual seeks to consume another.

Interspecific aggressors employ many means to locate, apprehend, and subdue their prey. Vision, touch, and sound and other vibrations are widely used to locate prey, as is sensitivity to chemical cues released into the water or chemical trails left along the bottom by prey animals. Sharks may home in on electrical fields set up by prey animals, and as noted in the previous chapter, in the deep sea certain types of fishes attract prey by means of lures and bioluminescent displays. To subdue their prey aggressive animals employ many types of offensive weapons involving an array of beaks, jaws, teeth, claws, and pincers, as well as toxins and venoms (table 16.3). Cnidaria stun their prey with stinging cells, cone snails (*Conus*) and some fishes immobilize prey by injecting venoms, and some deep-sea fishes may stun prey by brilliant light flashes. Most importantly, these various offensive mechanisms are orchestrated by appropriate behavior patterns that range from wait-and-ambush to hunt-and-chase-down the prey animals.

To defend themselves against attacks by predators,

prey animals have developed a range of anatomical features and behavior patterns to increase their chances of survival, and these involve means of avoidance and defense if they are encountered and attacked. To escape detection many species hide in holes in the bottom or among mangrove roots, seagrass blades, crannies in coral reefs, and similar out-of-the-way places. Many in open water undergo vertical migrations, hiding in deeper water or in bottom sediments during the day and ascending to surface waters only at night when they cannot be seen by visual predators. Many planktonic species are essentially transparent and nearly invisible in their seawater environments. Camouflage and protective coloration are employed by some forms living on reefs or among rafts of floating sargassum, climbing about sea fans and other soft corals, or resting on bottom sediments. To reduce their visibility in the water column, many types of marine fishes are countershaded, and the effect may be reinforced by the soft glow from rows of bioluminescent photophores. Many deep-sea demersal fishes have tapered tails to reduce water vibrations as they swim, and many deep-sea pelagic shrimps have extremely long antennae that can be used as feelers to warn of danger in advance of an attack.

Although the primary means of defense of most prey species seems to rest in avoidance of detection, many do have means of coping if they are recognized and beset by predators. In the complex environment of the reef many small invertebrates and fishes dart into safe havens. Within the open water column some species form large schools where it may be difficult for a predator to target a single individual. Others are capable of escape by very fast swimming and quick maneuvering, and flying fishes can take to the air and glide for long distances. Squids and a few species of fishes (tubeshoulders) startle the predator with a cloud of ink that may be bioluminescent, and some species of deep-sea cephalopods, crustaceans, and fishes may startle predators with brilliant flashes of light as they dart away into the darkness. Blowfishes take in large quantities of water, greatly increasing their volume and making them difficult to swallow. Many marine organisms possess an array of spines to discourage predators, and some species are able to inject powerful venoms into the would-be predators. Some exude toxic slimes, and in others the flesh or body fluids may be distasteful or toxic. One of the most potent toxins found in the animal kingdom (*tetrodotoxin*) is produced by pufferfishes and their relatives. Through the processes of continuing change

**Table 16.3.** *Types of toxic and venomous marine animals with examples of each. Toxic animals create problems when the flesh is ingested, and venomous animals create problems when the chemicals are injected. Known toxicities are based largely on mammalian studies, and the effects on other marine species are often unknown.*

**Toxic animals**

– Toxicity probably or definitely arises from consumption of toxic algae such as dinoflagellates. It is passed up through food chains and becomes most concentrated in the top predators. Included are ciguatera and other toxins. The chemicals are found in tissues of many types of fishes occurring around tropical reefs. Effects on marine animals are not entirely clear but may provide defense for some species.

Examples: fishes known to contain such toxins include moray eels (Muraenidae), snappers and groupers (Lutjanidae), jacks (Carangidae), butterflyfishes (Chaetodontidae), parrotfishes (Scaridae), wrasses (Labridae), surgeonfishes (Acanthuridae), barracudas (Sphyraenidae), and leatherjackets (Balistidae), among others.

– Toxic substances are produced by an animal and exuded from the skin when the animal is molested. Defensive weapon.

Examples: hagfishes (Myxinidae), soapfishes (Serranidae), boxfishes (Ostraciidae). (Note: The ink of some deep-sea cephalopods, such as *Heteroteuthis dispar*, may have toxic properties.)

– Toxins produced by fishes are present in muscles, body organs, or blood. Defensive weapon.

Examples: various eels (Congridae, Muraenidae, etc.), puffers (Tetraodontidae), ocean sunfishes (Molidae).

**Venomous animals**

– Venoms injected by stinging cells when animal is touched. Primarily offensive but may also serve as defensive weapon.

Examples: medusas, hydroids, sea anemones, corals, Portuguese man-of-wars (Cnidaria), hydrocorals (*Millepora*).

– Venoms injected by a bite. Offensive weapon.

Examples: certain octopuses (Cephalopoda) and moray eels (Muraenidae).

– Venoms injected by setae or spines. Mostly defensive weapon.

Examples: certain polychaete worms (Amphinomidae), snails (Conidae), many fishes including ratfishes (Chimaeridae), some sharks (Squalidae), stingrays (Dasyatidae, etc.), scorpionfishes (Scorpaenidae), toadfishes (Batrachoididae), stargazers (Uranoscopidae), surgeonfishes (Acanthuridae), and others.

---

and adaptation, we see the development of offensive mechanisms and strategies that are matched by counter-mechanisms and strategies, and these result in some successes and some failures by both predator and prey animals. There are some wins and some losses in the delicate balance that leads to the long-term survival of each player in this deadly game of evolution.

**Life history strategies**

The life history of an animal species is the summation of its stages from reproduction through larval and postlarval life. Having addressed each of these stages separately, we now examine how animals of the Gulf of Mexico have combined the stages into overall life history patterns, and this can most readily be achieved within the context of environmental stability and the selection process.

LIFE HISTORY TYPES

We begin with an analogy. If a species is introduced into a new environment containing a reasonable amount of resources and but few competitors, whether in the field or the laboratory, the population will increase in numbers according to a predictable pattern. Beginning with a small breeding population of a few individuals, the numbers increase rapidly at first, then (as intraspecific competition and other biological factors come into play) slow down, and finally level off (fig. 16.4). This is the classical S-shaped or sigmoid curve. The point at which the population achieves its maximum number is referred to as the equilibrium level or carrying capacity of the environment (designated by $K$). At any moment during the growth period or thereafter, the rate of increase of the population (designated by $r$) represents the difference between the birth and death rates. At the equilibrium level births

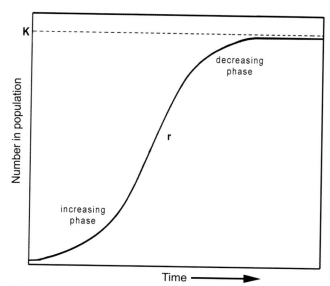

**Figure 16.4.** A typical growth curve showing the number of individuals in a population as a function of time since its inception. The quantity *r* represents the rate of population increase, and the upper asymptote *K* is the equilibrium level of the carrying capacity of the environment.

equal deaths, the value of *r* becomes zero, and the population remains stable. The sigmoid curve is also applicable to the development of ecological communities that begin with a few pioneering species invading a previously bare area and developing through a series of stages until the final climax community is reached. In this case the value of *r* would represent some measure of community metabolism, such as the rate of photosynthesis divided by the rate of respiration (P/R). Photosynthesis predominates during the developmental stages, and organic matter is being accumulated, but at the equilibrium or climax stage, photosynthesis and respiration balance one another, and the organic matter in the system remains more or less constant. We traditionally refer to the early developmental stages in the growth of a population or a community as the *r* phase and the equilibrium stage as the *K* phase. Within ecosystems species characteristic of the early successional stages, or *r* type, are adapted for colonization and rapid growth. Typically, they have high reproductive and dispersal capabilities, fast growth rates, and short life histories, but they are poor competitors and generally do not fare well under conditions of crowding. By contrast, *K*-adapted species tend to have lower reproductive potentials, lower dispersal capabilities, slower growth rates, and longer lives, but they are better adapted for handling competition, crowding, and survival in the presence of other species.

An overview of the life histories of species inhabiting the Gulf of Mexico reveals that most of those living in coastal environments are typical *r*-adapted species with repeated spawnings, very high reproductive rates, great dispersal capabilities, and short life histories. They are specifically adapted for colonization and survival in variable and physically unstable environments. On the other hand, around coral reefs certain species, including the corals themselves, are primarily *K*-adapted types with lower reproductive potentials, longer lives, and the ability to thrive under crowded conditions. Many other reef inhabitants, including invertebrates and reef fishes, have shorter lives but are otherwise primarily *K*-adapted species. The reef fishes and invertebrates have developed life history patterns designed for survival under conditions of high competitive interaction. Examples of commensalism and mutualism are common here. In the physically stable environment of the reef, biological interactions predominate.

Within the water column of the open Gulf, life histories tend to be fairly short, but there are exceptions in some of the larger fishes (sharks, tunas, billfishes, etc.), sea turtles, and marine mammals. Dispersal capabilities are high, with most producing planktonic larvae, but reproductive rates are generally not extraordinary. Exceptions include some of the larger solitary fish species, such as the louvar and giant mola, in which a single adult female may produce from a few million to more than 300 million eggs. Species of the deep Gulf also represent a mixed bag. Although we cannot be sure of their length of life, as a rule deep-sea organisms probably are not long lived. Fecundities are not extremely high, but some do produce planktonic larvae, providing for wide dispersal. Certainly, none of the open-water species are adapted to live under crowded conditions. Overall, it appears that in the basically subtropical environment of the Gulf, the necessity for planktonic larvae is almost a fact of life for invertebrates and fishes, and this vulnerable stage must be supported by a reasonably high rate of reproduction. However, beyond this, selection in coastal environments is definitely for *r*-type life history strategies, and around coral reefs it is largely for *K*-type strategies. In the open Gulf, both near the surface and in very deep water, although the case is not entirely clear, life histories tend toward the *r*-type strategy, which is what happens when the environment is variable, unpredictable, and not under the direct control of the organisms themselves.

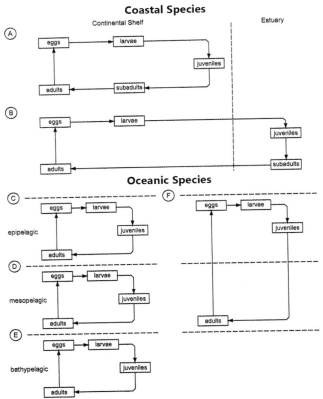

**Figure 16.5.** Six generalized life history patterns of Gulf of Mexico invertebrates and fishes. Patterns A and B are continental shelf types. Patterns C, D, E, and F are oceanic types. A = non–estuary dependent, B = estuary dependent, C, D, and E = patterns in which the larvae remain in the same basic depth zone as the adults, F = a pattern in which the eggs and larvae live in the epipelagic, but the adults live in a deeper zone of the sea.

GENERAL LIFE HISTORY PATTERNS

The bewildering array of life histories of the animal species inhabiting the Gulf of Mexico can be reduced to six basic types (fig. 16.5), each with a number of variations. These types apply whether development is direct (without a free-living larval stage) or indirect (with one or more free-living larval stages). In most cases the larvae and very young occupy habitats somewhat different from those of the adults, placing the young with favorable food supplies and reducing the possibility of accidental cannibalism, but in marine mammals the young must be present with the adults for nourishment, group protection, and development of learned behavior patterns.

Group A represents continental shelf species that do not depend on estuaries to complete their life histories. Some individuals may wind up there, but most remain in continental shelf waters. Many have planktonic eggs and/or larvae, whereas others are strictly demersal or benthic.

In some species the adults spawn over the outer shelf and the juveniles grow up in marine shallows and gradually move to deep water as they mature. Off Florida and Yucatán many of the Group A species utilize seagrass meadows or the more saline portions of mangrove swamps as nursery grounds. Such species include the pink shrimp, lane snapper, silver jenny, pigfish, grass porgy, pinfish, fringed and planehead filefishes, and striped burrfish, among others. In other forms such as the sea basses and most snappers and groupers, the young do not consistently appear in shallow waters but are found around the middle or outer shelf or broadly distributed across the continental shelf.

Group B includes those continental shelf species whose young regularly utilize estuaries as nursery areas. The larvae or young juveniles are especially adapted for traversing passes and inlets and for tolerating the low-salinity conditions of the inside waters. Included here are the brown and white shrimps, blue crabs, gulf and finescale menhadens, silver perch, sand and spotted seatrouts, spots, Atlantic croakers, and several species of flatfishes, among others. After growing up in the estuaries the subadults move back to the shelf to spawn, generally on the inner shelf, but in the case of the brown shrimp and a few others, on the middle or outer shelf.

Blue crabs mate in the estuaries, and the mated female normally moves out of the estuary but remains near the mouth of the pass, where she ultimately releases the larvae in a late stage of development. This species is notable in that its life history varies geographically within the Gulf. Off the west coast of Florida the pattern is different. There the great majority of mated females do not remain near the mouths of the passes but migrate northward to the Apalachicola Bay area at the base of the Florida panhandle, where they remain until the larvae are released. The larvae then move out of the bay into the open Gulf, where they are apparently transported southward by the descending arm of the Gulf Loop Current and ultimately shoreward toward the various estuaries by eddy currents (Oesterling and Evink 1977) (fig. 16.6). Remarkably, this deviation from the normal pattern must be a relatively new development. During the last Ice Age the west Florida continental shelf was largely exposed, and Apalachicola Bay did not exist. These features probably reached their present configurations only in the past 5,000 years or so as the sea level approached its present stand. Being genetically more or less isolated from other populations to the west, the Florida blue crabs appear to have developed

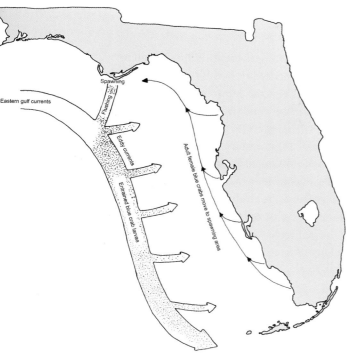

**Figure 16.6.** Graphic representation of the life history pattern of the blue crab in the eastern Gulf of Mexico. Adult females, inseminated in the estuaries, move out to nearshore waters of the shelf and then migrate northward to Apalachicola Bay, where they brood the eggs. Upon hatching, the planktonic larvae leave the bay and apparently become entrained in waters associated with the south-flowing arm of the Gulf Loop Current. From there they are somehow distributed to the various estuaries, where they transform into young crabs that feed and grow into a new generation of adults. (From Oesterling and Evink 1977.)

their own unique life history strategy in a remarkably short time.

The remaining life history groups all relate to oceanic species. Groups C, D, and E are alike in that the entire life history is passed within a single depth zone. Group C is confined to the epipelagic, D to the mesopelagic, and E to the bathypelagic zone. The various species may be completely restricted to the water column or the bottom, or they may have mixed patterns in which the adults live on the bottom while the eggs and/or larvae are planktonic. However, even for those species in which the adults reside in the water column, the younger stages are generally found at shallower depths than the adults. In some species the juveniles suddenly descend to the adult levels following metamorphosis, but in others they gradually descend as they mature.

Group F consists of open-ocean species in which the adults inhabit a depth zone different from, and deeper than, that of the young. In most cases the adults dwell in the mesopelagic, but sometimes they are bathypelagic. Their eggs and/or larvae are all epipelagic. In most in-

stances the adults reproduce at depth, and the eggs rise to the surface and develop into epipelagic young, but in a few cases the female herself releases the eggs in the epipelagic zone and returns to her resident depth. After a period of planktonic life and following metamorphosis, the young sink gradually or abruptly to the depth of the adults. Although somewhat oversimplified, these six patterns do provide considerable insight into the types of life history patterns developed by the various coastal and oceanic animal groups inhabiting the Gulf.

## MULTISPECIES AGGREGATIONS

In marine environments organisms are seldom distributed in random fashion. Rather they tend to cluster in favorable situations where they find shelter from the elements, substrates for attachment, food resources, protection from predators, or orientation features as for a home range or territory. Small multispecies aggregations often focus around rocks, large sponges, coral heads, seagrass blades, or other prominent objects on or near the bottom, gelatinous zooplankton and other organisms in the water column, and driftwood or floating rafts of sargassum at the surface. These aggregations vary in size, composition, and functional complexity and may involve from a few to hundreds of different species. Their biological connections with the system and its components vary from short-term and tenuous through longer-term and more intimate relationships involving commensalism, mutualism, parasitism, and predation. Since the major ecosystems of the Gulf are made up of or, at least include, small species aggregations, a brief treatment of these systems is in order.

### Small species aggregations

One often thinks of jellyfishes as being solitary forms floating around the sea, but a number of works have dealt with species associated with medusas, hydrozoans, ctenophores, and other gelatinous forms inhabiting waters of the Gulf and nearby Atlantic coast (Mansuetti 1963; P. Phillips 1972; P. Phillips and Levin 1973; P. Phillips, Burke, and Keener 1969). Among the associated species are hydroids, tapeworm larvae, floating marine snails (*Janthina*), parasitic isopods, hyperiid amphipods, crabs (*Callinectes, Libinia*), and a number of fish species including a pipefish (*Syngnathus*), several jacks (*Caranx, Chloroscombrus, Hemicaranx*), spadefish (*Chaetodipterus*), several species of butterfishes (*Nomeus, Peprilus*), flounder (*Paralichthys*), and filefish (*Monacanthus*). Most of the fishes are juveniles that are attracted to various floating objects, but some are adults, and one, the man-of-war fish (*Nomeus*), regularly associates with the Portuguese man-of-war. The invertebrate species are physically attached to or crawl around the surface of the host organisms or live within their bodies. Some are parasitic, and some of the snails, crabs, and fishes feed directly on tissues of the host species (as do louvars, molas, sea turtles, and other forms). As they move through the water these gelatinous hosts attract other species, and such groupings make up small but variable communities of marine organisms interacting in various ways. Such associations are not unique. Even marine copepods may represent small species aggregations. Ho and Perkins (1985) noted that copepods have both ectosymbionts and endosymbionts. Among the former are bacteria, fungi, diatoms, protozoans, and larval isopods. The endosymbionts include bacteria, dinoflagellates, protozoans, and flatworm larvae (flukes and tapeworms), as well as nematodes. Some are parasites, some are commensals, and a few may have mutualistic relations with their hosts. Small aggregations involving various types of interspecies relationships are widespread in the water column.

Rafts of sargassum that float on the surface of the Gulf, Caribbean, and North Atlantic also provide favorable environments for a variety of marine organisms that grow attached to the sargassum, clamber among its leaf- and branch-like structures, or swim beneath and dart for cover when danger threatens. From sargassum samples obtained off Cape Hatteras, North Carolina, Fine (1970) identified 38 species of invertebrates belonging to 6 different phyla. Represented were sea anemones, scyphozoans, gastropods, bivalves, polychaetes, pycnogonids, barnacles, stomatopods, tanaids, isopods, amphipods, decapods, and chaetognaths. Included were 3 species of shrimps and 8 species of crabs. From sargassum samples taken off Miami, Dooley (1972) identified 48 species of fishes, some of which were considered to be closely associated with the sargassum and the remainder loosely associated or incidental. Among the closely associated forms were 21 species belonging to 9 families (frogfishes, pipefishes, jacks, dolphinfishes, tripletails, sea chubs, damselfishes, butterfishes, and leatherjackets). The loosely associated and incidental groups included 27 species representing 11 families (requiem sharks, herrings, flying fishes, needlefishes, snipefishes, pipefishes, jacks, goatfishes, swordfishes, leatherjackets, and puffers). Most of the fishes were juveniles of species normally found in open water, but several such as the damselfishes, goatfishes, some leatherjackets, and puffers are more typical of coastal benthic environments. Here among the sargassum rafts they find food and some protection from predators. The sargassum itself provides surfaces for attachment of filamentous algae, protozoans, hydroids, and other small forms, and among the animals that shelter here are various grazers and their predators. The floating sargassum community apparently has been around for a long period of time, and a number of species specifically adapted for life here and seldom found elsewhere include a shrimp (*Latreutes fucorum*), two crabs (*Portunus anceps* and *P. sayi*), and two fishes (sargassumfish, *Histrio histrio*, and sargassum pipefish, *Syngnathus pelagicus*).

In marine benthic environments examples of small species aggregations abound. Those reported from waters of the Gulf and nearby areas of the Caribbean and south Atlantic states relate to a variety of hosts and habitat types (table 16.4). T. S. Hopkins (1981) gave lists of species found in association with sponges, corals, mollusks, and various habitats in the Florida Middle Ground area. Sponges, in particular, are known to provide living quarters and sheltering habitats for a great variety of animal types. Especially well represented here are mollusks, annelid worms, crustaceans, echinoderms, and fishes. Common host sponges of the Gulf and Caribbean area belong to the genera *Agelas, Callyspongia, Geodia, Haliclona, Ircinia, Microciona, Pseudoceratina, Spheciospongia, Verongia,* and *Xenospongia.* Within the Gulf and Caribbean area over two dozen fish species have been reported living in association with sponges (table 16.5). Most are facultative sponge dwellers that seek food and shelter here but are frequently found elsewhere. However, six species are considered to be obligate sponge dwellers that depend on the sponges for survival, and two species, the sponge goby and tusked goby, are morphologically adapted to live within the canal systems of the sponges. Habitats include the canals, central lumens, and crevices around the bases of the sponges. The commensalistic relationship in which one species lives within the body of another without being a parasite or causing the host serious harm has been referred to as *inquilism,* and the "guest" animals are known as *inquiline* species. Most of the sponge-dwelling fishes are small forms with slender bodies that can easily slide into and out of small cavities, and some of the facultative species are juveniles whose adults live among corals and elsewhere around the reef. In the West Indies more than 30 species of small fishes have been reported to dwell

**Table 16.4.** *Selected references to species associations in benthic environments of the Gulf of Mexico, Caribbean Sea, and Atlantic coast of the United States.*

| Host group or habitat | Reference |
|---|---|
| Seagrass epiphytes | Humm 1964; M. Sullivan 1979 |
| Sponges | Böhlke and Robins 1969; Gudger 1950; E. Livingston 1979; Pearse 1932, 1950; Tyler and Böhlke 1972 |
| Soft corals | Patton 1972 |
| Hard corals | Luckhurst and Luckhurst 1976; Scott 1987 |
| Sea anemones | Colin and Heiser 1973; Hanlon and Hixon 1986; Herrnkind, Stanton, and Conklin 1976 |
| Hermit crabs | Bruce 1989; Cutress, Ross, and Sutton 1970 |
| Shrimps and true crabs | Overstreet 1978 |
| Bryozoans | Lindberg and Stanton 1988 |
| Sea urchins | Stock and Gooding 1986 |
| Brittle stars | Hendler and Meyer 1982 |
| Coral rubble | Choi 1984 |
| Oyster beds | Wells 1961 |
| Invertebrates, fishes, and habitats of the Florida Middle Ground | T. S. Hopkins, Schroeder, et al. 1981 |
| Invertebrates and fishes of the Flower Garden Banks | Bright and Pequegnat 1974; Dennis and Bright 1988 |

**Table 16.5.** *Fishes recorded from sponges in the Gulf of Mexico and Caribbean Sea. (Data from Böhlke and Robins 1969; Gudger 1950; T. S. Hopkins 1981; Pearse 1932, 1950; Tyler and Böhlke 1972.)*

**Obligate sponge dwellers**

Apogonidae
  Sawcheek cardinalfish – *Apogon quadrisquamatus*
  Sponge cardinalfish – *Phaeoptyx xenus*
Gobiidae
  Sponge goby – *Evermannichthys spongicola*
  Yellowline goby – *Gobiosoma horsti*
  Yellowprow goby – *Gobiosoma xanthoprora*
  Tusked goby – *Risor ruber*

**Facultative sponge dwellers**

Chlopsidae
  Collared eel – *Kaupichthys nuchalis*
Gobiesocidae
  Skilletfish – *Gobiesox strumosus*
Apogonidae
  Bronze cardinalfish – *Astrapogon alutus*
Pomacentridae
  Beaugregory – *Pomacentrus leucostictus*
  Bicolor damselfish – *Pomacentrus partitus*
Labridae
  Slippery dick – *Halichoeres bivittatus*
  Bluehead – *Thalassoma bifasciatum*
Clinidae
  Banner blenny – *Emblemaria atlantica*
  Horned blenny – *Paraclinus grandicornis*
  Marbled blenny – *Paraclinus marmoratus*
  Checkered blenny – *Starksia ocellata*
Blenniidae
  Seaweed blenny – *Parablennius marmoreus*
Gobiidae
  Frilled goby – *Bathygobius soporator*
  Bridled goby – *Coryphopterus glaucofraenum*
  Tiger goby – *Gobiosoma macrodon*
  Code goby – *Gobiosoma robustum*
  Dwarf goby – *Lythrypnus elasson*
  Island goby – *Lythrypnus nesiotes*
  Green goby – *Microgobius thalassinus*
  Rusty goby – *Priolepis hipoliti*

among or near the tentacles of sea anemones, which provide protection from predators (Hanlon and Hixon 1986). Most of the fishes avoid the stinging tentacles, but some have adaptations that allow them to come into full contact without being harmed. Scott (1987) reported more than 30 species of invertebrates associated with corals of the Caribbean and nearby Atlantic. These included sponges, polychaetes, sipunculids, bivalves, barnacles, and decapods. Lindberg and Stanton (1988) listed 23 species of decapod crustaceans (shrimps, hermit crabs, and true crabs) from bryozoans of the south Florida seagrass meadows.

On a somewhat larger scale Wells (1961) found 303 species of animals associated with oyster beds in North Carolina estuaries, and in surveys of the Flower Garden coral/algal reefs off Texas, Bright and Pequegnat (1974) and Dennis and Bright (1988) listed 388 species of invertebrates and fishes (table 16.6). In both instances many groups of the animal kingdom were represented, and these organisms were associated not only with particular host species, but also with other habitat features, as well. Various types of intimate and nonintimate relationships were evident. Clearly, the oyster bed and coral/algal reef systems represent an order of magnitude increase in species diversity, and one may ask what factors operating in combination can account for the enhanced biological diversity. Part of the answer lies in the increased shelter and habitat stability. Whether in oyster beds, coral/algal reefs, salt marshes, mangrove swamps, or seagrass meadows, certain dominant species locally reduce the effects of waves and water currents and provide a more sheltered and stable physical environment where organic matter can accumulate, larvae can settle and grow, and the organisms can interact freely. Another important factor is the increase in habitat complexity in both a physical and biological sense. There are more surfaces for attachment, more nooks and crannies for the smaller cryptic species, and organic-rich bottom sediments for the development of benthic fauna. As shelter, habitat stability, and complexity increase, the colonizing organisms bring in their own suites of symbionts, and they create opportunities for additional species. Another factor leading to increased species diversity is food specialization. Long-term competition between coexisting species has resulted in prey specialization. The various habitats of the Gulf provide numerous examples of species with similar but slightly different food habits, and these differences permit the coexistence of larger numbers of species than would otherwise be possible. Finally, small species aggregations do not exist in a vacuum but are influenced by resources and species from neighboring ecosystems. These different systems may receive organic matter as well as larvae and young from one another, and many predatory species forage in different habitats of an area. In summary, we have seen how small species aggregations are based on various types of symbiotic and nonsymbiotic relationships and how aggregations, in turn, build up and become more diverse as the habitat becomes more stable and sheltered and increases in complexity, as species become more specialized in the use of food resources, and as the individual species aggregations interact with their neighbors.

**Table 16.6.** *Comparison of species diversity reported for a North Carolina oyster bed and the coral/algal reefs of the Flower Garden Banks of the northwestern Gulf of Mexico. Neither database represents all the species present. For the Flower Garden Banks the arthropod data are listed for the various subgroups. (Data from Bright and Pequegnat 1974; Dennis and Bright 1988; Wells 1961.)*

| Taxonomic group | | No. of species |
|---|---|---|
| **Oyster bed** | | |
| Protozoans | | 2 |
| Sponges | | 12 |
| Cnidarians | | 14 |
| Flatworms | | 8 |
| Nemerteans | | 4 |
| Mollusks | | 99 |
| Annelids | | 42 |
| Sipunculids | | 2 |
| Arthropods | | 76 |
| Bryozoans | | 20 |
| Echinoderms | | 5 |
| Chordates | | 19 |
| Total species | | 303 |
| **Coral/algal reefs** | | |
| Protozoans (Foraminiferans) | | 34 |
| Hydroids | | 17 |
| Corals | | 18 |
| Mollusks | | 65 |
| Polychaetes | | 17 |
| Sipunculids | | 1 |
| Arthropods | | 98 |
| Pycnogonids | 1 | |
| Ostracods | 33 | |
| Copepods | 1 | |
| Mysids | 1 | |
| Barnacles | 6 | |
| Stomatopods | 2 | |
| Amphipods | 5 | |
| Caridean shrimps | 28 | |
| Anomurans | 12 | |
| Macrurans | 2 | |
| Brachyurans | 7 | |
| Echinoderms | | 24 |
| Fishes | | 94 |
| Total species | | 368 |

## Major ecosystems

What constitutes a major as opposed to a minor ecosystem is a subjective matter since the two types intergrade with one another. However, for present purposes we will consider a major ecosystem to be one with the following general characteristics:

1. *Size.* It extends over a fairly large area, tens or hundreds of square miles.
2. *Composition.* It is made up of several subsystems.
3. *Physical gradients.* Because it is large, it generally includes gradients in physical factors.
4. *Internal patterns.* It has very complex structural and functional patterns for the capture, transfer, and transport of nutrients and energy.
5. *Exchange.* Through the exchange of nutrients, energy, and species it is functionally connected with other major ecosystems.
6. *Persistence.* It has a long-term persistence, hundreds or thousands of years.

For purposes of the present discussion we will divide the Gulf and its associated waters into five major types of ecosystems, each made up of several subsystems (table 16.7). In some cases the subsystems themselves could be considered to be major systems in their own right, but the present classification simplifies the discussion.

### ESTUARIES AND RELATED WATERS

Technically, an estuary is the expanded mouth of a river. It is a flow-through system that receives water as well as dissolved and suspended materials from the river, and after sedimentation, mixing, and other processes, it ultimately discharges the modified water to the Gulf (fig. 16.7). It is characterized by brackish water (with salinities intermediate between those of freshwater and seawater), tidal action (often reduced), and some protection from waves and water currents of the open sea. Coastal lagoons are saline water areas, frequently connected with estuaries that lie impounded behind barrier islands. Basically they are not flow-through systems but areas of poorer circulation and sometimes water stagnation. Although normally brackish, when they receive virtually no freshwater input or mixing with estuaries they may become hypersaline, with salinities well above those of seawater, as in the case of the Laguna Madre of south Texas and that of northeastern Mexico. Bays are coastal indentations of various sizes, and sounds are larger, partially protected, coastal waters. Within the Gulf all these water types are present,

**Table 16.7.** *Major ecological systems and subsystems within or associated with the Gulf of Mexico.*

**Estuaries (including lagoons, bays, and sounds)**
- Brackish and salt marshes
- Mangrove swamps
- Oyster beds
- Grass flats
- Intertidal mud and sand flats
- Subtidal bottoms
- Water column

**Continental shelf – terrigenous bottoms**
- Rocky outcrops
- Reefs
- Soft bottoms
- Water column

**Continental shelf – carbonate bottoms**
- Mangrove swamps
- Seagrass meadows
- Coral/algal reefs
- Live bottoms and patch reefs
- Soft bottoms
- Water column

**Open Gulf water column (beyond the continental shelves)**
- Epipelagic zone
- Mesopelagic zone
- Bathypelagic zone

**Deepwater benthos and near-bottom waters**
- Continental slope and rise
- Abyssal plain

and all exhibit some estuarine characteristics. These habitats dominate the coasts of the northern and northwestern sectors of the Gulf and south Florida, and they punctuate the coastlines elsewhere.

As seen in figure 16.8, the typical estuarine system embraces a series of functionally related subsystems. Shorelines and intertidal areas consist of marshes, swamps, and mud or sand flats. Estuarine bottoms are made up largely of mud, sand, shells, and shell fragments mixed with decomposing organic material derived from the various subsystems. Oyster reefs are often present, as are beds of submerged vegetation consisting of eelgrass (*Zostera*), widgeongrass (*Ruppia*), and other species. The open waters, although often extensive, are seldom more than three or four meters deep except in natural channels through the passes and in artificially dredged holes and navigation channels. The downstream salinity gradient is obvious, with lower salinities near the inland head of the estuary and higher salinities near the Gulf. Estuarine waters are generally stratified, with the denser, more saline water on the bottom and less dense, fresher water in the surface layer, and these waters are affected by the tidal-pump mechanism. On each rising tide saline Gulf

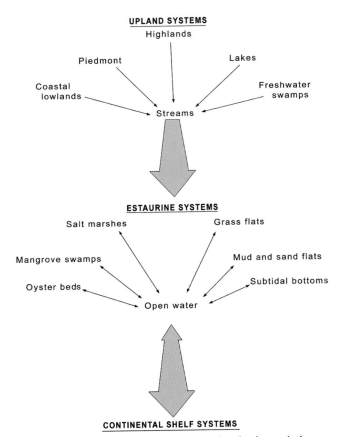

**Figure 16.7.** Downstream hydrologic series showing interrelations of a stream, its estuary, and the nearby continental shelf and of subsystems within the estuary. (From Darnell and Soniat 1979.)

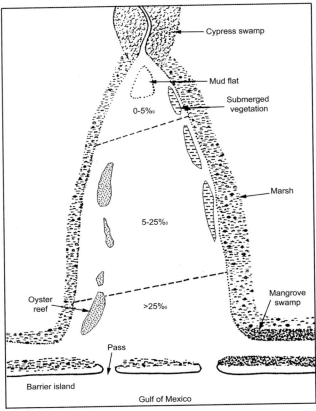

**Figure 16.8.** Diagram of an idealized estuary showing the distribution of various habitat types. Salinity is given in parts per thousand. (From Darnell 1992a.)

water moves into the estuary as a saltwater wedge, increasing salinities first along the bottom and some minutes later at the surface. Within the estuary this wedge spreads out but often flows inward as a river along the right-hand side of the estuary (as viewed from the Gulf). Inside the estuary a certain amount of mixing takes place between the surface and bottom water masses, and this mixing may be intensified by wind stress at the surface. On ebb tide the saltwater wedge reverses course and backs out of the estuary, often less saline than before, and the fresher water layer flows out along the surface. With the combined effects of wind, waves, tides, and water currents, estuarine waters are in motion much of the time. As a result, living, freshly dead, and decomposing organic material is regularly flushed from surrounding marshes, swamps, and intertidal flats and deposited on the estuarine bottom, a process intensified by stormy weather.

Functional relationships within the estuary are complex but fairly well understood. The system is considered to be a nutrient trap. Nutrients received from the river may be precipitated or taken up by the phytoplankton populations, and only a limited amount leaves the estuary through the outflowing surface water. Because of the general inflow of bottom water, nutrients associated with benthic particulate material are largely retained within the estuary. Photosynthesis is carried out within all the subsystems except in the deeper-lying bottoms and oyster beds. Marsh plants and mangrove trees, together with their populations of filamentous and other algae, produce large quantities of organic matter in the marginal areas. Diatoms and other microalgae are photosynthetically active on surfaces of the intertidal flats. Within the water column rooted vegetation and attached algae as well as the suspended phytoplankton populations are also highly productive. Because of the persistently high levels of available nutrients and the subtropical climate with its intense solar radiation, estuarine systems of the Gulf of Mexico are among the most photosynthetically productive biological systems on earth.

Either directly as living vegetation or indirectly through decomposition and organic detritus pathways, this great

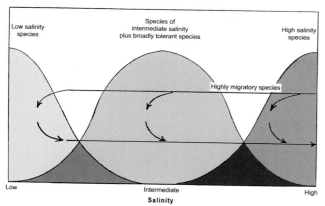

Figure 16.9. Different faunal groups normally encountered in estuaries of the Gulf of Mexico. Permanent residents include some widely distributed species of broad salinity tolerance as well as some restricted species of narrow salinity tolerance (which live in low, intermediate, or high salinity conditions). Also temporarily present in the estuary are a host of seasonally transient species, mostly young individuals of marine affinity, which may be found in low, intermediate, or high salinity conditions. (From Darnell 1992a.)

concentration of organic matter supports large populations of estuarine animals. As seen in figure 16.9, these consist of both local residents and seasonal transients. Among the true estuarine residents some have narrow tolerances and may be limited to particular salinity regimes, but most are tolerant of a wide range of salinities. The seasonal transients arrive primarily as larvae or early juveniles of marine species (penaeid shrimps, blue crabs, and a variety of fish species), and after several months of feeding and growth, as subadults or adults, they migrate back to the continental shelf, where spawning takes place. However, some of the transients are adults of marine species that are tolerant of the lower salinity conditions and that enter the estuaries and prey on the hordes of smaller animals residing there. Within the estuary two basic food chains are recognized, the plankton-based food chain of the water column and the benthos-based food chain of the bottom waters. These food chains are interconnected, and both depend heavily on the consumption of organic detritus (fig. 16.10).

The question of the export of nutrients and organic material to the continental shelves, sometimes referred to as "outwelling," has been examined in some detail by Darnell and Soniat (1979). Although quantitative data are sketchy, it is likely that most of the Gulf of Mexico estuaries are exporters, and for some the amount of such material transported to the shelf is probably quite high. The exported material consists of the nutrients, plankton, and

organic detritus particles present in the regularly outflowing surface water, the amount that leaves when the seasonal transients migrate back to the continental shelves, and the living and detrital material episodically flushed from the estuaries by floods, storms, and hurricanes. No one has been out in stormy weather monitoring this latter outflow, but some $^{13}C$ measurements of detrital material in sediments of the northern Gulf shelf suggest that the export from this source alone may be considerable. Large rivers such as the Mississippi have no estuaries and empty their waters directly onto the continental shelf. Some of the suspended material passes to the open Gulf, but much remains on the continental shelf. The freshwater and high levels of nutrients create estuary-like conditions on both sides of the Mississippi River Delta, particularly along the west Louisiana and upper Texas coasts.

CONTINENTAL SHELF SYSTEMS

Continental shelves, which underlie about a third of the surface area of the Gulf of Mexico, consist of two major types, referred to as the terrigenous and carbonate bottom provinces. From about the level of Mobile Bay in the northeast, the terrigenous province extends along the northern, western, and southern Gulf to the base of the Yucatán Peninsula. Bottoms here are largely soft sediments made up of sand, silt, and clay deposited by streams and subsequently sorted and redistributed by alongshore currents. In areas of the northern Gulf shelf influenced by fine sediments from the Mississippi River, and probably elsewhere, the bottom waters are often clouded by a benthic nepheloid layer that restricts the distribution of many types of benthic organisms, particularly photosynthetic plants and attached filter-feeding animals. Throughout the province isolated rocky outcrops protrude from the sediments, and these afford attachment and aggregation sites for a number of silt-tolerant species. On higher topographic features that extend above the nepheloid layer, various types of reef organisms thrive. Some of these higher banks, including the Flower Gardens off Texas and Louisiana and those near Tuxpan and Veracruz, Mexico, support hermatypic coral reefs. Also present on these and other banks may be communities of leafy and/or calcareous algae, sponges, hydroids, soft corals, bryozoans, sea lilies, and other forms. Estuaries and lagoons bordering the terrigenous-bottom shelves are often characterized by extensive areas of salt and brackish marshes, although mangrove swamps are more important toward the south.

The second major division of the continental shelves

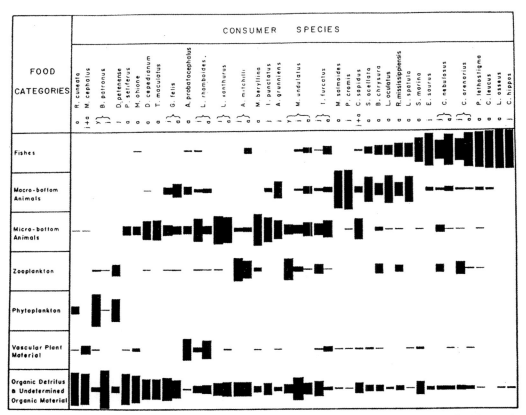

**Figure 16.10.** Trophic spectrum diagram for an estuarine community (Lake Pontchartrain, Louisiana) of the northern Gulf of Mexico showing the importance of various food categories in the diets of the most abundant large consumer species. Individual species are listed along the top (some with young, juvenile, and adult size groups separated out), and the food categories are listed along the left. The individual spectral bands represent the percentage of the food of a species falling within a given food category. For a given species (or life history stage) the total width of all bands would equal the vertical width of a given food category. For present purposes the names of the individual species are irrelevant. The important thing is to grasp how materials and energy are flowing through the ecosystem as evidenced by the food emphasis of the consumer species. In this case, note the heavy reliance on the lowermost category, organic detritus and undetermined organic material, and the relative unimportance of phytoplankton and recognizable vascular plant material. (From Darnell 1961.)

of the Gulf is the carbonate province. This includes the limestone platform off the southern and western flanks of the Florida Peninsula as well as the carbonate shelf off northern and western Yucatán. Each of these consists of a flat limestone floor carpeted by a veneer of soft sediments composed largely of carbonate rubble, mostly of biological origin. On the west Florida shelf a band of siliceous sand lies to a depth of 10–20 m, and a tongue of finer sediments extends southwestward from the Everglades. Unlike the terrigenous province, which receives considerable outflow of low-salinity water and fine particulate material from streams and estuaries, the carbonate province receives little surface stream flow and few fine sediments except locally off west and south Florida. However, both the Florida and Yucatán platforms do receive freshwater flow from underground streams that locally seep into the water column from subsurface sources. Within the car-

bonate province the nepheloid layer is absent, and the water column remains clear except immediately following storms. The numerous carbonate outcrops and bottom limestone exposures support patch reef and live-bottom communities of attached leafy algae, sponges, and soft and stony corals, as well as other sedentary species and their associated mobile fauna. Extensive seagrass meadows occupy the shallow-water bottoms in protected areas off west and south Florida and Yucatán. Especially well developed are the coral/algal reef tract off south Florida and the reefs around banks and islands of the Yucatán shelf. Estuaries punctuating the coasts of the carbonate province are bordered largely by mangrove swamps.

Terrigenous shelves of the Gulf receive nutrients and nonliving particulate as well as living organic materials from estuaries, the Mississippi River, and waters of the deep Gulf, much of the latter being associated with up-

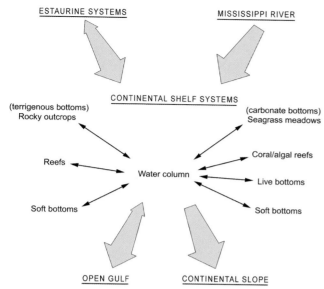

**Figure 16.11.** Conceptual diagram of the continental shelf systems and subsystems and their relationships and also interrelations between the shelves and estuaries and Mississippi River, on the one hand, and between the open Gulf and continental slope, on the other.

welling systems (fig. 16.11). These materials aid in supporting phytoplankton populations and food chains of pelagic consumer species. Organic material from the same sources as well as phytoplankton and fecal material sinking from above enrich the sediments and support dense populations of bottom animals living within and upon the soft sediments and in the suprabenthic waters. However, as in estuaries, the pelagic and benthic food chains are not separate but closely linked (fig. 16.12). Many of the pelagic consumer species also forage in bottom waters, and at night hordes of small benthic animals rise to feed in the water column. Because of the rapid decomposition of organic material and the regeneration of nutrients within the bottom sediments, and because of the vertical mixing of the water column, the nutrients are soon available for reuse by the phytoplankton populations. Since the vertical distance is not great, rapid recycling of nutrients and close linkage of pelagic and benthic food chains are characteristic features of continental shelf systems in general.

The same basic processes that occur in terrigenous shelf systems also take place in carbonate bottom areas but with somewhat different emphasis. Where estuaries are present, as along portions of west and south Florida,

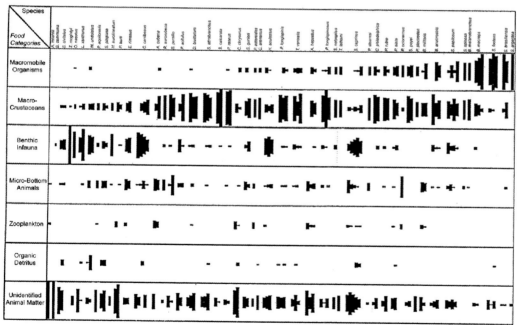

**Figure 16.12.** Trophic spectrum diagram for the Mississippi-Alabama continental shelf ecological community showing the importance of various food categories in the diets of major large consumer species. In contrast with the situation in the estuary (figure 16.10), in the food of continental shelf species organic detritus was seldom identifiable as such, but some was definitely included in the category "unidentifiable animal matter." Most of this material appeared to be the partially digested flesh of polychaetes and polychaete mucus containing a lot of very fine particulate organic detritus. (From Darnell 1991b.)

they export some nutrients and dead and living organic materials as in terrigenous areas, but since most of the carbonate shelves lack major contributions from streams and estuaries, they are more dependent on nutrients from the deep Gulf. Thanks to the Gulf Loop Current, which induces upwelling and sends small meanders and filaments across the Florida and Yucatán shelves, some carbonate areas at least seasonally are well supplied with oceanic nutrients. Since waters of the carbonate shelves are largely devoid of fine suspended particulate material, solar radiation penetrates to the bottom throughout most of the carbonate shelf province. Seagrasses together with their epiphytes, leafy and calcareous algae, and zooxanthellae within the tissues of corals and other animals can carry out photosynthesis at nearly all depths of the shelf. In addition, microalgae are probably active photosynthesizers in the bottom sediments. Thus, in the carbonate shelf areas photosynthesis is not restricted to the phytoplankton of the water column but is also carried out in various subsystems of the benthic realm including seagrass meadows, coral/algal reefs, patch reefs, live bottoms, and soft sediments. Most of the nutrients regenerated within the sediments are probably recycled locally through the benthic microalgae, with only limited amounts being released to the water column for reuse by the phytoplankton.

Within the terrigenous bottom province, aside from the pelagic-benthic coupling mentioned earlier, relations between subsystems are not significant, and they are mediated largely through the water column. Many animals produce planktonic eggs and/or larvae that are transported by water currents, and many attached species of the banks feed on phytoplankton and zooplankton. Among the benthic forms some, such as juvenile red snappers, appear to move around from one rocky outcrop to another. Within the carbonate province the situation is much different. Here relationships between subsystems are often complex and extensive. Both fringing mangrove swamps and seagrass meadows trap and retain sediments brought in by water currents, and both seasonally export large quantities of organic detritus to the water column and to neighboring benthic systems. Attached species of patch reefs, live bottoms, and reef tracts project into the water column where they glean plankton from the passing water, and many inhabitants of the various subsystems produce planktonic eggs and/ or larvae that are dispersed by the water currents. Mangrove swamps and seagrass meadows serve as major nursery areas for juvenile invertebrates and fishes whose adults live on soft bottoms, around coral/algal reefs, or in the water column. Some animals, such as sea urchins, which shelter in seagrass meadows during the day, graze on surrounding soft bottoms at night, and predators of the water column, soft-bottom areas, and coral/algal reefs often forage in seagrass meadows, some during the daytime and others at night. Where any combination of mangrove swamps, seagrass meadows, coral/algal reefs, and bare soft-bottom areas are present in close proximity to one another, as off south Florida, the combined subsystems can support a richer fauna than they would if the individual subsystems existed in isolation. This coupling of benthic subsystems is an important feature of carbonate bottom ecosystems. Both terrigenous and carbonate systems export nutrients, organic detritus, and living organisms to the open Gulf water column and to the continental slope benthic system.

### DEEP GULF WATER COLUMN AND OCEAN FLOOR

The deep Gulf water column receives input of nutrients and organic material primarily from the Caribbean Sea (via the Yucatán Current), outflow from the Mississippi River, and off-shelf flow from the continental shelves. Although most of the water passing through the Straits of Florida exits the Gulf, along the bottom a small backflow brings Atlantic water into the Gulf. The deep Gulf bottoms and near-bottom waters also contribute nutrients and organic material to the overlying water column, as will be discussed subsequently. In turn, the open Gulf contributes to the Atlantic Ocean, the Caribbean Sea (via a deep countercurrent), the continental shelves, and benthic regions of the deep Gulf. Subsystems of the water column include the euphotic, twilight, and aphotic (epipelagic, mesopelagic, and bathypelagic) zones as discussed in the previous chapters.

Nutrients become available in the euphotic zone of the open Gulf by upwelling from deeper layers and by outflow from continental shelves and the Mississippi River. Much of the upwelling is induced by the Gulf Loop Current and takes place along edges of the continental shelves off Yucatán and west Florida and along margins of the Loop Current in the intervening water, but upwelling also occurs in the western Gulf in association with counterclockwise rings and eddies and in response to wind-driven displacement of surface waters around continental shelf margins. Elevated levels of nutrients in the euphotic zone stimulate the growth of phytoplankton populations, which in turn support populations of zooplankton and higher levels of consumer species. These consumers include permanent residents of the euphotic zone, vertical migrators that live at greater depths during the day and ascend to feed in surface waters at night, and the larval and juvenile stages of numerous species whose adults live in deeper layers of the Gulf. Many of the larger consumer species of the open Gulf range widely across the continental shelves, especially during the warmer months, and some also forage in deeper waters and on the bottom at depths of several hundred meters or more.

Some of the phytoplankton that escapes consumption in the euphotic zone sinks to lower levels of the water column, and a portion of it apparently makes it all the way to the bottom. The sinking of individual cells by gravity is a very slow process, but sinking of larger phytoplankton cell aggregates is much more rapid. Organic material is also transferred to deeper layers of the water column and eventually to the bottom sediments by fecal material produced by various consumer species of the euphotic and other zones and by the rain of carcasses of animals that die in the upper layers. Whereas the various consumer species of the euphotic zone may feed on living phytoplankton, other animals, and/or decomposing organic material, below the euphotic zone carnivory and detritivory are the only options. As discussed earlier, many types of invertebrates and fishes living in the twilight zone and some in the aphotic zone migrate upward and feed in shallower zones and then return to their resident depths, where they in turn may be consumed by animals from yet greater depths. So, in addition to passive sinking, some organic material is transferred to deeper layers and even to the bottom of the Gulf by means of a living vertical "food ladder" of consumer species.

For the most part, organic material in the sea is transferred from the top down, that is, from shallower to deeper layers, but there is also some movement from deeper to shallower layers. Fats, oils, waxes, and any decomposing material with trapped gas bubbles tend to rise passively. Some of the larger euphotic-zone fishes forage in deeper water. The leatherback turtle is known to feed on gelatinous animals of the twilight zone, and many of the marine mammals, including species of dolphins, porpoises, and whales, regularly feed on squids, larger crustaceans, and fishes of the twilight zone. A few, such as the sperm whale, make it into the aphotic zone to feed. Finally, it should be mentioned that methane and other gaseous hydrocarbons and crude liquid petroleum emanating from numerous natural bottom seeps rise through the water column to the surface, where they provide energy resources for bacteria, which in turn are consumed by a variety of marine animals.

The deep Gulf bottom system includes the continental slope and rise, abyssal plain, and the near-bottom water extending a meter or so above the sediment surface. Because there is no usable light, photosynthetic plants are absent, and the ecosystem is composed entirely of consumer species. Locally, bottom seeps of hydrogen sulfide, gaseous hydrocarbons, and liquid petroleum support specialized communities, but mostly the deep-Gulf bottom ecosystem depends on organic material raining from above. This includes the phytoplankton, fecal material, and dead bodies that sink as well as any consumers involved in the vertical food ladder, mentioned earlier. To this are added patches of sargassum and seagrass blades, as revealed in bottom photographs. Much of the deep-Gulf benthic life consists of poorly known species living within the sediments. In addition to the microfauna and meiofauna, larger burrowing animals include primarily gastropods, bivalves, scaphopods, polychaetes, crustaceans, and echinoderms. Representatives of most of these groups as well as cephalopods and fishes forage along the surface or feed in the near-bottom water column. Attached to the bottom and extending into the near-bottom water are a number of types of sponges, hydroids, sea whips, and sea pens, as well as sea lilies and tunicates that feed by straining zooplankton and nonliving organic particles from the near-bottom waters. Within the Gulf the concentration of living organisms tends to decrease with depth, and in very deep water, whether in the water column or on the bottom, the concentration is only about a thousandth of what it is in the euphotic zone or on the continental shelf. The bottom sediments provide the ultimate sink for all organic material introduced into the Gulf by streams, estuaries, and water currents or produced by photosynthesis in the euphotic zone.

TROPHIC ORGANIZATION OF MAJOR ECOSYSTEMS
The basic trophic organization of any major marine ecosystem is illustrated in figure 16.13. Typically, each system imports and exports nutrients, living organisms, and particulate organic detritus, although the types and quantities vary from one system to another, and in a given system they may change seasonally and with local circumstances. Within each system there are normally three large pools of organic material, the photosynthesizers, several levels of consumer species, and the pool of organic detritus.

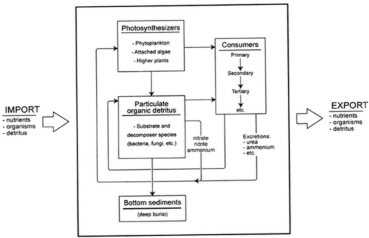

**Figure 16.13.** Simplified flow diagram of the basic trophic organization of any major marine ecosystem within the euphotic zone. The system imports dissolved nutrients, organisms, and organic detritus, and it exports materials in these same categories. Within the system there are three basic "pools" where materials are stored: the photosynthesizers (phytoplankton, attached algae, and/or higher plants), larger consumer species (herbivores and several levels of carnivores), and particulate organic detritus (the substrate itself plus a host of microscopic bacteria, fungi, and very small invertebrates). Operationally, organic and inorganic nutrients flow to the photosynthesizers. The consumer category receives from both the photosynthesizer and organic detritus pools, and organic detritus receives from the other two. Nutrients are released by both the consumer and detritus pools to be taken up by the photosynthesizers. Some of the organic detritus may become deeply buried in the sediments, where it remains out of circulation.

The latter includes substrates of dead organic material derived from the producers and consumers together with the attached or otherwise associated decomposer organisms. Most of the particulate organic detritus is located upon or within the bottom sediments, but at any one time a portion is suspended in the water column.

The photosynthesizers include the phytoplankton, benthic microalgae, larger attached algae, and vascular plants. Using nutrients, water, carbon dioxide, and radiant energy from the sun, they produce energy-rich organic compounds to support their own metabolism and growth. In turn, they are eaten by primary consumer species, or herbivores, and much of the unconsumed material dies and passes into the organic detritus pool. Phytoplankton is important in the water column of estuaries and continental shelves and in the euphotic zone of the open Gulf. Attached algae include many species of filamentous and nonfilamentous microalgae as well as various macroalgae. These flourish on the surfaces of vascular plants and other hard materials and on benthic substrates in

the intertidal and subtidal zones of estuaries and continental shelves. Because of their very rapid colonization and growth, they are important producers of organic carbon in estuaries, on carbonate shelves, and locally on terrigenous shelves where banks and other hard substrate exposures are not shaded or inundated by suspended inorganic particles. Vascular plants such as marsh grasses and mangroves grow along margins of estuaries and, in some cases, carbonate shelves, and beds of submerged vegetation are located in estuaries and in protected areas of carbonate shelves. Few marine animals consume vascular plants directly, and aside from providing food for some fishes (notably, the sparids), turtles, and manatees, their main contribution is to the pool of organic detritus.

Organic detritus has been defined as "all types of organic material in various stages of microbial decomposition which represent energy sources for consumer species" (Darnell 1967a, 1967b). Included here are whole bodies and body parts of plants and animals as well as fecal material produced by the latter. Upon death or release into marine waters, particulate organic material is quickly invaded by decomposer species, primarily bacteria and fungi (J. Turner 1979), whose enzymes begin the process of chemical degradation. Through a combination of chemical and mechanical action the larger materials are reduced to smaller and smaller particles, which may be deposited on the bottom locally or carried away by water currents. Ultimately, the substrate is reduced to carbon dioxide, water, and nutrients, which may be recycled by the photosynthesizing species to produce new organic material. During the process of decomposition the particles are invaded by protozoans, nematodes, and other small invertebrates, and together with the bacteria and fungi they enrich the particles and increase their nutritive value for consumer species. The organic detritus pool is important in all marine ecosystems, and it is the sole support for animal communities of the water column below the euphotic zone and in benthic habitats of the deep Gulf.

Decomposition rates vary widely depending on the nature of the original substrate, temperature, and other factors. Phytoplankton and other algae may degrade within a few days, and the bodies of zooplankton animals also break down rapidly. Larger animals such as fishes can take weeks or months, whereas the carcass of a whale requires more than a year. Fecal material degrades fairly rapidly depending on the temperature. For example, McCarthy (1973) found that after 20 days shrimp fecal material had lost 51.5 percent of its original organic carbon at 25°C but only 28.2 percent at 15°C. Vascular plant material takes much longer, and its breakdown occurs in two stages. The first is fairly rapid and entails the release of the liquid and semiliquid cell contents, leaving the fibrous structural portions of the plant, or *humus*. This material is composed largely of cellulose and lignins, which may require many months or years to break down completely. Rates of decomposition for coastal vascular plant species of the Gulf area have been summarized by a number of workers, including Stout (1984) for marsh grasses, Odum, McIvor, and Smith (1982) for mangrove leaves, and Zieman (1982) for seagrasses. Although decomposition rates for vascular plants vary considerably from one species to another, within about six months much of the material has been reduced to very small particles that can readily be ingested by small animals or transported away by water currents.

Other forms of organic detritus encountered in the sea include fecal material and organic aggregates. Some crustaceans and chaetognaths encapsulate the fecal material in a membrane that prevents the contents from disintegrating, but invading bacteria eventually destroy the membrane, releasing the contents. In copepod fecal pellets the membrane remains intact for several days at 25°C, and it may survive for 20–30 days at 5°C (J. Turner and Ferrante 1979). Numerous workers have observed fine filmy material ("marine snow") floating or slowly sinking in marine waters. Most of this appears to be a mucus-like substance produced by salps and other gelatinous animals and by biologically enhanced physical aggregation of smaller particles such as diatoms, fecal pellets, and so forth. This material is "sticky," and as it moves through the water it accumulates phytoplankton cells, fecal pellets, and other forms of organic as well as inorganic particles, forming *macroscopic aggregates* (Alldredge 1979; Alldredge and Jackson 1995). Such aggregates are often invaded by bacteria and a variety of small invertebrates (coelenterates, mollusks, annelids, a variety of crustaceans, echinoderm larvae, and urochordates) (E. Green and Dagg 1997). As the carcasses of dead animals, fecal material, and organic aggregates sink through the water column they may be ingested by various marine animals, but those materials that escape consumption eventually reach the bottom sediments.

In the open Gulf the rain of dead organic matter is particularly important in transferring organic material from the euphotic zone to deeper layers of the water column

**Table 16.8.** *Sinking rates of whole bodies and fecal material of selected marine organisms estimated from short-distance laboratory experiments. (Data from Robison and Bailey 1981.)*

| Detrital material | Whole bodies (m/day) | Fecal material (m/day) |
|---|---|---|
| Phytoplankton cells (single) | 1 – 151 | |
| Macroscopic aggregates | 91 | |
| Pteropods | | 400 – 800 |
| Copepods | 36 – 720 | 15 – 210 |
| Euphausiids | 1,555 | 43 – 86 |
| Urochordates | | |
|   Salps | | 450 – 2,700 |
|   Doliolids | | 41 – 208 |
| Midwater fishes | | 1,028 |

and to the benthic realm, and the amount that actually reaches the bottom depends in large measure on how fast the material descends. As seen in table 16.8, individual phytoplankton cells sink very slowly, but aggregates descend more rapidly. Likewise, larger carcasses such as euphausiids and the fecal material of midwater fishes sink much faster than the carcasses of smaller-bodied copepods or the fecal material of most small invertebrates. In warm surface waters decomposition rates are fairly rapid, but in colder waters beneath the euphotic zone the rate of decomposition is much slower. Fragmentation of fecal material and other particles reduces the rate of descent, and most decomposition and nutrient regeneration in the sea apparently takes place in the upper few hundred meters of the water column. These processes require oxygen, and the oxygen demand of decomposing organic material sinking from above is largely responsible for the oxygen minimum layer immediately below the euphotic zone. Rowe, Sibuet, et al. (1990) estimated the rate of organic matter reaching the bottom of the western Atlantic Ocean to range between 19 and 154 mg C/m²/day (= 6.9–56.2 g C/m²/yr), with higher values on the continental shelf and upper slope and lower values on the abyssal plain.

A considerable body of knowledge exists supporting the contention that organic detritus is consumed by marine animals and that it can be nutritious. The bodies of newly dead algae and the carcasses of animals are certainly consumed and should have much the same nutritional value as when they were alive. Digestion of food is seldom complete, and fecal material is often rich in undigested organic matter. In addition, the invading microbes and small invertebrates enrich the substrate with their own bodies so that consumers may obtain nutrients from the substrate and/or the invading organisms. Robison and Bailey (1981) noted that the fecal material of midwater fishes remains relatively rich in organic nutrients for some time after it is produced and that, because of luminous bacteria derived from the gut of the fish, the material may glow as it descends, attracting potential consumers. Some snail, clam, copepod, and shrimp species have been successfully raised in the laboratory on diets of their own fecal material, which was allowed to age for a few days to permit microbial invasion and growth (Johannes and Satomi 1966; Newell 1965; J. Turner and Ferrante 1979). Within estuaries vascular plant detritus is consumed by many species of fishes (Darnell 1958, 1961). Although they lack the enzymes necessary to break down cellulose and lignins, they probably harvest the associated microbes and small invertebrates, because stable carbon isotope studies do suggest that vascular plant material is an ultimate, if not a proximate, source of some of the organic carbon in estuarine animals (Conkright and Sackett 1986; Fry and Sherr 1984). However, Thayer, Govoni, and Conally (1983) found that zooplankton off the mouth of the Mississippi River had stable carbon isotope ratios indicative of a phytoplankton diet, and it is likely that on the continental shelf and in the open Gulf most vascular plant detritus is processed in the benthic sediments by microbes and small benthic consumers rather than in the water column.

We conclude that all forms of living and nonliving organic material present in estuarine and marine environments represent food resources for some consumer species, but the pathways by which nutrients, carbon, and energy flow through marine ecosystems are exceedingly complex and often understood only in broad outline. The classical "food chain" paradigm, inherited from studies of terrestrial ecology, holds that plants are consumed by herbivores (primary consumers), these by first-level carnivores, and so on up through discrete steps to top carnivores. This scenario applies fairly well to some marine systems, particularly those in the euphotic zone of the water column. However, difficulties arise when dealing with scavengers, detritus feeders, omnivores, and those species whose food habits change with life history stage

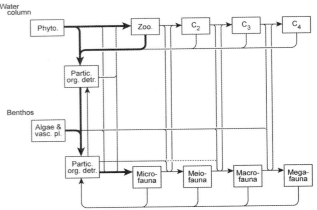

**Figure 16.14.** Simplified flow diagram showing the transfer of organic material through food chains in estuarine and continental shelf ecosystems of the Gulf of Mexico. Heavy lines denote the greatest presumed flows and dashed lines the least flows. Abbreviations are as follows: Phyto. = phytoplankton, Zoo. = zooplankton, $C_2$–$C_4$ = higher consumer levels, Partic. org. detr. = particulate organic detritus, and vasc. pl. = vascular plants. Note that the food chains of the water column and benthos are broadly connected.

(Darnell 1968; J. Turner and Roff 1993). An alternative view is that all marine animal species are omnivores and that they simply differ from one another in the relative percentages of the different food resources consumed. This view is also tenable, and it is especially useful when dealing with benthic food chains. One complication is that outside the estuary, organic detritus is difficult or impossible to identify in the guts of marine animals, and in any event, very little is known about the food consumed by the microfauna and meiofauna that live in bottom habitats and largely form the basis of benthic food chains.

In order to display functional relationships within complex ecosystems, workers put forth conceptual models. Although inevitably gross oversimplifications of real-world situations, such diagrams can be quite useful in providing an overview of the total system and depicting key functional relationships. Figure 16.14 represents the flow of nutrients, carbon, and energy through estuarine and continental shelf systems of the Gulf, as conceived by the author. Each trophic system consists of two recognizable food chains composed of five levels. That of the water column is based on phytoplankton but has some input from organic detritus. The benthic food chain is based on microalgae and macroalgae and vascular plants as well as a large pool of organic detritus and some input from living phytoplankton. All components of both food chains contribute to the organic detritus pool through carcasses

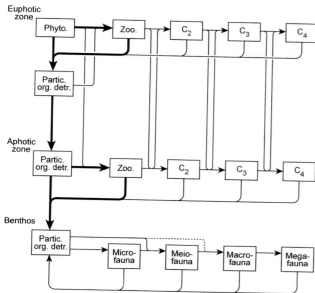

**Figure 16.15.** Simplified flow diagram showing the transfer of organic material through food chains of the euphotic and aphotic zones of the water column, and the benthic realm of the deep Gulf of Mexico. Heavy lines denote the greatest presumed flows and dashed lines the least flows. Abbreviations are as in figure 16.14. Featured are the broadly interconnected food chains of the euphotic and aphotic zones and the food chain of the benthic realm, which is based almost entirely on the "rain" of particulate organic detritus from above. Not shown is the presumed "food ladder" whereby organic material may be transferred from shallower layers to the benthos by a succession of living organisms.

and, in the case of the animals, fecal material. Most of the organic detritus produced in the water column eventually sinks to the bottom, but some of the benthic detritus becomes resuspended and enters the water column from below. An important feature of estuarine and shelf systems is that the water column and benthic food chains are closely bound, exchanging materials at every link of the chain. In nature these exchanges are most manifest in estuaries and shallow portions of the continental shelf, and they become weaker toward the outer shelf, where the vertical distance is greater.

Figure 16.15 is a comparable conceptual model representing the transfer of organic material in the water column and benthos of the open Gulf of Mexico. In this case three food chains are depicted, one each for the euphotic and aphotic zones and another for the benthic realm. The food chain of the euphotic zone is based on phytoplankton and to a lesser extent on recycled particulate detritus. Those of the aphotic zone and benthos are based solely on particulate organic detritus. Food chains of the euphotic and aphotic zones are bound together at every consumer

level by sinking organic detritus and by vertical migrators that live in one zone and feed in another or in which the adults live in one zone and the larvae and young inhabit the other. The benthic food chain is represented as being independent of that in the aphotic zone of the water column, but some, presumably minor, interaction must take place, including that of the vertical "food ladder." This model highlights the three food chains and the interactions between those of the two uppermost zones, and it stresses the importance of organic detritus in the trophic economy of the deep Gulf below the euphotic zone.

The third conceptual model, shown in figure 16.16, represents the ecosystem at the East Flower Garden Bank and depicts the trophic structure of the coral/algal reef and its associated species. This model is presented in more formal symbolic notation. All photosynthesizing compartments (producers) are bullet shaped, all consumer compartments are hexagons, and all storage compartments have rounded bottoms and pointed roofs. The respiratory loss of each consumer compartment is represented by a series of short horizontal lines of decreasing length connected to the bottom of the compartment. Also, the producer groups of the reef (corals, soft algae, and coralline algae) are enclosed in a box representing the domain of the main attached species of the reef. The composition of each compartment is explained in table 16.9. This model is presented to show an attached benthic system and its relations with components of the water column and with the various resident and transient organisms found in association with the attached species. In general form it represents all such benthic-attached systems of the Gulf. An interpretation of the model follows.

Water sweeping across the reef brings in nutrients, phytoplankton and zooplankton, and higher consumer species (advective imports) and ultimately carries away from the reef materials and organisms in the same general categories (advective exports). Attached producers of the reef take in light (L) and nutrients (N), and these are utilized at the ambient temperature (T) of the water bathing each depth level of the reef. Photosynthesizers of the reef proper include zooxanthellae within tissues of the corals, soft algae, and coralline algae. In their symbiotic relationship the zooxanthellae supply products useful to the corals, which in turn supply materials to the zooxanthellae through a feedback loop. Periodically, some of the zooxanthellae are released into the water column, where they take up life as free-living phytoplankton organisms. From the water column the coral polyps extract calcium

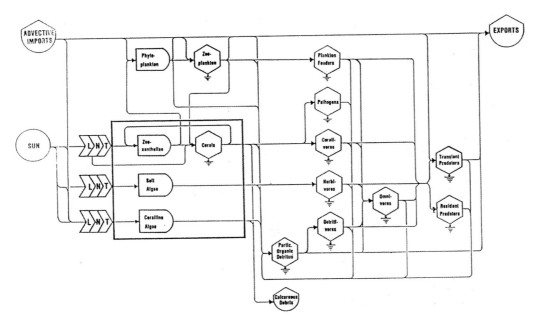

**Figure 16.16.**
Formalized flow diagram for the transfer of organic materials through food chains at the East Flower Garden Bank and showing relationships with components of the water masses that bathe the reef complex. Compositions of the various categories are given in table 16.9. For explanation of the model see the text. (L = Light, N = Nutrients, T = Temperature.)

(here treated as a nutrient) needed in the manufacture of their carbonate skeletons. They also glean from the passing water zooplankton used as food. Preying on the coral polyps are microscopic pathogens as well as certain polychaetes, echinoderms, and fishes (corallivores). The soft algae are grazed by some mollusks, crustaceans, and fishes (herbivores). Particulate organic detritus is produced by all plants and animals of the reef and water column, and this material becomes a food supply for the detritivores. Omnivores feed on all the previous compartments, and larger predators, both resident and transient, prey on the various other animal species inhabiting the reef. Much of the reef area is carpeted by particles of carbonate gravel (carbonate debris) produced by the corals and coralline algae of the reef, and this represents a major sink for the carbon metabolized by the reef inhabitants. Despite the seeming complexity of this flow diagram, in a simplified way it embodies the basic pathways for the transfer of nutrients and energy in the extremely complicated array of attached flora and fauna interacting with components of the water column. All the various species listed in table 16.9 as well as others not listed have a place in the system represented here by 4 producer groups, 11 consumer groups, and 1 storage compartment.

As noted earlier, the flow of materials and energy through marine ecosystems is vastly more complex than indicated by the conceptual models presented above. Complicated models have been put forth showing more compartments and many more pathways (see, for ex-

ample, Manickchand-Heileman, Soto, and Escobar 1998). These are quite useful for some purposes, but the gain in detail is accompanied by a loss in visual clarity, and unless some provision is made for distinguishing major versus minor pathways, such models can be extremely difficult to comprehend. Some of the models are supported by quantitative data for the evaluation of biomass stored within compartments and/or amounts flowing along the various pathways. Darnell and Bright (1983) provided a partial evaluation of the East Flower Garden Bank model given above, Dagg et al. (1991) gave information on the production and flow of materials in planktonic food webs of the northern Gulf continental shelf, and Manickchand-Heileman, Soto, and Escobar (1998) provided data for evaluation of their model of continental shelf food chains of the southwestern Gulf. Additional information of potential value for the evaluation of such models has been presented by a number of workers, as follows. The chemical and energy content of various marine animals has been published by Crabtree (1995), Stickney and Torres (1989), and Wissing et al. (1973). A nitrogen budget for a marine fish species has been presented by Darnell and Wissing (1975), and an energy budget for the brown shrimp has been developed by Ibrahim (1973).

Insight into the trophic structure of large marine ecosystems depends on the food analysis of large numbers of component groups. For continental shelves of the northern Gulf such information has been provided by Darnell (1991b), Rogers (1977), and Simons (1997). In the meso-

**Table 16.9.** *Definition and composition of categories in the conceptual model of the East Flower Garden coral/algal reef system. The compositional lists are suggestive rather than exhaustive, and generally only the most important elements are included.*

| Category | Composition |
|---|---|
| **Import/Export** | |
| Advective import | Nutrients (N, P, Ca), phytoplankton, zooplankton, transient predators, particulate organic detritus |
| Advective export | Nutrients, phytoplankton, zooplankton, transient predators, particulate organic detritus |
| **Components of Model** | |
| Producers | |
| Phytoplankton | Marine phytoplankton in water bathing the reef |
| Zooxanthellae | Symbionts within the hermatypic corals. Some are released periodically to become phytoplankton |
| Leafy algae | *Caulerpa, Chaetomorpha, Codium, Dictyota, Lobophora, Rhodymenia, Stypopodium, Valonia,* etc. |
| Coralline algae | *Archaeolithothamnion, Hydrolithon, Lithophyllum, Lithoporella, Lithothamnium, Mesophyllum, Prorolithon, Tenarea,* etc. |
| Consumers | |
| Zooplankton | Marine zooplankton in waters bathing the reef |
| Coral polyps | *Agaricia, Colpophyllia, Diploria, Helioseris, Madracis, Millepora, Montastrea, Porites, Stephanocoenia,* etc. |
| Plankton feeders | Sponges, sea anemones, bivalves, some polychaetes, crinoids, tunicates |
| Pathogens | Probably blue-green algae |
| Corallivores | *Hermodice, Diadema, Coralliophaga,* some crabs, some fishes (butterflyfishes, parrotfishes, and wrasses) |
| Herbivores | Some mollusks, small crustaceans, butterflyfishes, and parrotfishes |
| Detritivores | Meiofauna, small macrofauna, some annelids, mollusks, echinoderms, sea urchins, sea cucumbers, brittle stars, some starfishes, some fishes |
| Omnivores | Some mollusks, annelids, crustaceans, and fishes (butterflyfishes, damselfishes, filefishes, squirrelfishes, surgeonfishes, grunts, puffers, triggerfishes, and trunkfishes) |
| Transient predators | Sharks, some rays, dolphinfishes, jacks, barracudas, mackerels, tunas, etc. |
| Resident predators | Some mollusks (*Conus*), some polychaetes (nereids), some crustaceans (crabs), some fishes (eels, snappers, groupers, goatfishes, gobies) |
| Storage Compartments | |
| Particulate organic detritus | Decomposing organic material (including dead plants and animals plus secretions and fecal material). Bacteria, fungi, and other decomposing microbes are included in this group. |
| Calcareous debris | Reef rock and pieces of calcareous material derived from mechanical breakdown of stony corals, calcareous algae, mollusks, echinoderms, serpulid worms, and barnacles |

pelagic realm of the open Gulf such data have been put forth for cephalopods (Passarella and Hopkins 1991), shrimps (M. Flock and Hopkins 1992; Hefferman and Hopkins 1981; and Kinsey and Hopkins 1994), and fishes (T. L. Hopkins and Baird 1973, 1985a, 1985b; T. L. Hopkins and Gartner 1992; T. L. Hopkins and Sutton 1998; T. L. Hopkins, Flock, et al. 1994, T. L. Hopkins, Sutton, and Lancraft 1996; Lancraft, Hopkins, and Torres 1988; and T. Sutton and Hopkins 1996). The food of bottom fishes of the deep Gulf has been presented by Bright (1968, 1970), D. M. Martin (1978), and Rayburn (1975). The aim, of course, is to understand how carbon, nitrogen, energy, and so on flow through the various ecosystems and how such flows are influenced or controlled by physical factors (river outflow, Gulf Loop Current incursions, rings and eddies, upwelling, major meteorological events, etc.) and human intrusions (fisheries harvest, chemical pollution, degradation of estuarine nursery areas, etc.). Further along it will be of interest to ascertain how the systems can be manipulated and the fisheries harvest enhanced without jeopardizing the long-term sustainability of the Gulf ecosystems.

Every species inhabiting the Gulf of Mexico and its associated waters must contend with a series of environmental factors in order to complete its life history goals. Focusing on general patterns, we have examined some of the physical and biological factors and the ways in which species cope with the various environmental constraints. We have reviewed life history adaptations within a genetic, evolutionary, and ecological context. Finally, we have seen how each species is part of a larger multispecies aggregation and how such aggregations are integrated into larger functioning ecosystems that together constitute the Gulf of Mexico ecosystem. Simplified conceptual models provide some insight into how the complex systems function, but we must continue striving to grasp the intimate details if we are ever to understand how the systems really work.

# Part V

## HUMAN RELATIONS

In the preceding pages we have depicted the Gulf of Mexico as a natural ecosystem, an ecosystem as it once was, free of the hand of humanity. This, of course, only partially reflects the reality of the Gulf today. Everywhere we look we see evidence of human-induced changes. The greatest impacts are evident in the marginal bays, estuaries, marshes, swamps, and other coastal wetlands. Here environmental degradation and habitat loss have been rampant, especially in the east and north, and along the Mexican coast, as well. Habitat destruction and reduction in water quality are manifest from Florida Bay and the Everglades through Charlotte Harbor, Tampa Bay, Mobile Bay, Lake Pontchartrain, Louisiana coastal marshlands, Sabine Lake, and Galveston, Matagorda, San Antonio, and Corpus Christi Bays of Texas, and down the Mexican coast through the Laguna de Términos. Continental shelf areas have also been severely affected, most notably in the lower Keys and related areas of south Florida, the nearshore Louisiana and upper Texas shelf, and the reefs and banks in the vicinity of Veracruz, Mexico. These and other shelf areas have been impacted directly by heavy fishing pressure, mineral extraction, ocean dumping, and related activities, and indirectly by changes in upstream and coastal habitats.

Farther offshore in the open Gulf, petroleum and other chemical pollutants are still present, and the large predatory fish species are under heavy pressure from commercial fishing. Although human influences are felt throughout the Gulf, offshore at least, they have not yet become the dominant force. This is in large measure because of the vast size and dilution capacity of the Gulf waters and the power of the Gulf Loop Current and related gyres, eddies, and filaments that stir up the entire Gulf, sweep the shelves clean, distribute oxygen throughout the water column, and reduce the residence time of Gulf waters. The potential for future impacts from such human-induced changes as global warming is very great, and in this case the entire Gulf and its marginal areas would be severely and adversely affected.

The various aspects of the human impact problem are quite complex and the ramifications extensive, and together these merit the attention of a full volume. Here we can only touch on the highlights.

## HISTORICAL BACKGROUND

Five centuries ago when the first Europeans arrived in the New World, the Gulf was not entirely pristine. Already the indigenous peoples were harvesting biological resources from the bays and estuaries. In the southern Gulf the Mayas and Aztecs had developed intensive fisheries for mollusks, crustaceans, finfishes, sea turtles, and even manatees, and these creatures were taken both in protected coastal waters and on the continental shelves. The conquering Spaniards continued to make use of these resources and are known to have slaughtered large numbers of the native monk seals for the valuable oil that could be rendered from their fat. Throughout subsequent years, as the human population around the Gulf increased, so did the harvest from the bays and shallow continental shelves. Fishes, crabs, and shrimps were taken from the inside as well as nearshore shelf waters with long beach seines and by hook and line. Sea turtles, once common in coastal waters around much of the Gulf, suffered from habitat destruction as well as intensive market hunting for their meat and eggs. Ship traffic around the Gulf increased gradually, and the old wooden-hulled sailing vessels introduced into the Gulf a number of Old World biofouling organisms including two species of isopods, the "wharf roach" (*Ligia*) and the wood-boring "gribble" (*Limnoria*).

Following the development of steam-powered vessels in the early nineteenth century, ship traffic increased considerably. The first steamboat to visit New Orleans arrived at the docks in 1812. Until that time traffic on the Mississippi River had been limited to flatboats and keelboats propelled by the river's current and sometimes assisted by oars and sails, and all major ship traffic in the Gulf was wind driven. The steamboat provided a means for the farmers and merchants of Middle America to get their products to market and thus ushered in an era of heavy Mississippi River traffic. Export and import traffic increased all around the Gulf, but particularly at New

Orleans, which by the mid-1850s had become the major port of Middle America and, in terms of tonnage crossing its docks, the second largest port in the nation. During the late eighteenth and early nineteenth centuries, whalers sometimes entered the Gulf and harvested several species, including the great sperm whale. By the late nineteenth and early twentieth centuries, from the clear shallow waters of the Florida shelf, divers and tongers harvested several species of commercially valuable sponges, often depleting the local populations.

Despite the activities noted above, even through the late 1800s human impacts on the Gulf of Mexico and its ecosystems were still generally small, limited largely to coastal areas, and primarily local. But then things began to change. Starting around 1900 and during subsequent years, great oil discoveries in the coastal lowlands of east Texas and southwest Louisiana led to the development of coastal refineries, petrochemical plants, oil terminals, and major harbor facilities. Port cities grew in size and complexity, with manufacturing activities clustered on waterways and around harbors. Ship traffic increased greatly. During these and subsequent years much coastal marshland was damaged or destroyed by petroleum exploration and development operations, chemical pollution of coastal waters was increasing, and many coastal areas were awash in oil pollution. Dredge spoil, old construction materials, and other debris of civilization were being dumped indiscriminately in coastal wetlands or on the nearshore continental shelf. At this time environmental protection legislation was poorly developed and weakly enforced. During and following the Second World War these and related problems intensified because of a variety of factors, chief of which were expanded technology, great development and diversification of the chemical industry, widespread use of agricultural fertilizers and pesticides, increase and especially coastalization of the human population, offshore petroleum exploration and recovery operations, and greatly expanded commercial fisheries. These and related human activities and their impacts on the Gulf, many developed or intensified during the past half century, are taken up in the next section.

## HUMAN ACTIVITIES THAT AFFECT THE GULF
Major human activities that impact the coastal and marine areas of the Gulf of Mexico are listed in table 17.1.

## Modification of streams and watersheds
In pre-Columbian times the Mississippi River was a very sinuous stream characterized by numerous meanders as it coursed back and forth across the broad, flat floodplain valley on its way to the Gulf. Annually, with the spring thaw of snow and ice in its northern reaches, the river overflowed its banks, inundated the floodplain forests, and recharged the freshwater swamps and oxbow lakes with water, restocking them with fishes and other aquatic life. Much river-borne sediment was deposited here, elevating and enriching the floodplain soil. The wetlands and lowlands of the alluvial valley constituted a haven for wildlife, and they served as a vast temporary storage reservoir for the floodwaters. Following the spring flood the excess water gradually drained back into the main channel. At the river's mouth the peak flow from the spring flood was greatly dampened, and this peak was followed by a slow reduction in flow volume as the upstream floodplains gradually released their stored water (Darnell et al. 1976).

LEVEES AND CUTOFFS
Beginning with the settlement of New Orleans in the early 1700s, levees were constructed along the Mississippi River to contain its waters within the primary channel and to prevent inundation of the floodplain that now became available for farm fields and human settlements. Through the years these earthen walls have grown to impressive dimensions, and in all but the most severe floods they contain the waters within the prescribed channels. In order to hasten water transport to the Gulf and to facilitate river traffic, the US Army Corps of Engineers has straightened and shortened the river by cutting through the bases of many stream meanders. As a result of the levees and cutoffs, the spate associated with the spring thaw now shoots

**Table 17.1.** *Major human activities that impact the coastal and marine areas of the Gulf of Mexico.*

– Modification of streams and watersheds
– Dredging and channelization of coastal waters and wetlands
– Coastalization of the human population
– Commercial shipping
– Offshore mineral extraction
– Ocean dumping
– Commercial and recreational fishing

rapidly to the Gulf, giving short, high peak flows and relatively rapid return to low flow rates. A series of dams constructed in the headwater streams has further altered the normal seasonal pattern of water discharge and sediment loads transported to the Gulf.

Under prelevee circumstances, as it approached the Gulf of Mexico, the Mississippi River deposited great volumes of sediments, which built up a delta. When the level of one delta reached a certain height, the river abandoned this old delta and formed a new one (see fig. 4.12), and this process of alternate delta formation and abandonment went on for thousands of years. Delta sediments have a high water content, and over time the weight of overlying sediments squeezes out much of the water, a process referred to as compaction. As soils of a new delta become compacted they shrink, and the level of the soil drops, but the annual river overflow normally deposits new sediments that maintain the height of the land. Now that levees have been constructed, the Mississippi River no longer floods the lowland soils of southeast Louisiana, and because this annual sediment is no longer replenished, and also because of compaction and other factors to be discussed later, much of the land of coastal Louisiana is being lost at an alarming rate (R. Turner 1990).

Following the Revolutionary War, as the settlers moved westward they encountered the great deciduous forests of eastern North America and later the extensive tall- and short-grass prairies of the American Midwest. By the mid-1800s much of the forest had been cut over and converted to farms, and within a few more decades vast acreages of prairie land had fallen to the plow. Natural forests and prairies build up the soil and efficiently retain topsoil and the major chemical nutrients contained therein. For example, experiments and field observations have shown that the eastern deciduous forest receives more phosphorus in rainfall than it loses to the streams that drain it (Ryden, Syers, and Harris 1973). However, unless great care is taken, most farm fields lose large quantities of particulate sediments as well as water-soluble nutrients such as nitrates and phosphates. So, the development of Middle America was accompanied and followed by great loss of topsoil and nutrients to the drainage systems, and these were transported to the Gulf by the Mississippi River and other streams draining the area.

## NUTRIENT AND SEDIMENT LOADING OF STREAMS

To replace these losses, by 1900 some farmers were already applying limited amounts of fertilizer to their fields,

but widespread and heavy fertilizer use really began in the 1930s, when power from the TVA dams made cheap fertilizer widely available. Its use increased during subsequent years and peaked around 1980, remaining high thereafter (R. Turner and Rabalais 1991). Since most of the fertilizers are highly soluble in water, they are soon leached out of the fields, and during the past several decades nitrogen, in particular, has appeared regularly in waters of the Mississippi River in proportion to its annual application to farmland in the upstream drainage. Since most of these nutrients are carried to the Gulf by the spring flood, at a time when the Gulf is under the influence of strong and persistent winds from the southeast, much of the nutrient-rich Mississippi River water is transported westward along the inner continental shelf of Louisiana and east Texas. Here overfertilization (*eutrophication*) has regularly led to late spring phytoplankton blooms and summer hypoxia, as described earlier.

## CHEMICAL POLLUTION OF WATERWAYS

The westward expansion of the American population coincided with the early stages of the industrial revolution. Expanding agriculture was paralleled by increasingly sophisticated technology and industrial development and, in later stages, by the upsurge of the American chemical industry. It has been estimated that we now manufacture more than 100,000 different kinds of chemical compounds. Many are harmful to a wide array of living forms and are considered to be "biocides." Large numbers of the chemicals (such as plastics) have no natural counterparts and are nonbiodegradable. One way or another, many of these chemicals find their way into streams. Thus, over the past century and a half the Mississippi River, in particular, has brought ever-increasing loads of chemical pollutants to the Gulf. These include residues from upstream agricultural, municipal, and industrial sources. To a lesser extent the same is true for other streams entering the Gulf in both the United States and Mexico.

## FRESHWATER DIVERSION

The focus has been on the Mississippi River because human activities and impacts have been greatest and most varied there, but environmental effects elsewhere have also been important, most notably in south Florida. With respect to sea level, the land in this area is very low, that is, slightly above or below the level of the Gulf. In presettlement times water from Lake Okeechobee flowed southward as a sheet of freshwater, which ultimately emptied into Florida Bay. During its course this broad, shallow "river" passed through a variety of aquatic and semiaquatic habitat types including creeks, lakes, ponds, hammocks, and shallow grass flats, as well as fresh, brackish, and saline mangrove swamps and forests of the Everglades. During the past century dams and canals were constructed to divert much of the freshwater for use by coastal communities, chiefly Miami, and the low, fertile drained soil now became prime agricultural land for the raising of sugarcane, market vegetables, cattle, and other commodities. Thus, deprived of much of its natural freshwater inflow, the Everglades became subject to desiccation, wildfires, and saltwater intrusion, all to the detriment of the natural vegetation and wildlife. Lacking its normal freshwater input, Florida Bay and its flora and fauna also began to change. One complication is that mercury has now been found in soils and waters of the Everglades. Efforts have been underway to restore to the Everglades some of its natural freshwater flow and to determine the source and extent of the mercury pollution.

Thus, modifications of streams and watersheds concurrent with the agricultural and industrial development of Middle America have led to changes in annual discharge patterns and in the sediment loads, nutrient concentrations, and chemical pollutants brought to the Gulf by the Mississippi River and other streams draining the nation. The levees have also resulted in a great decrease in soil replenishment, especially in coastal lowlands of Louisiana, and they have reduced beach nourishment along the northern Gulf coast and caused overenrichment leading to hypoxia in nearshore shelf waters of the north-central Gulf. Chemical pollutants are now detectable in shelf sediments of the northern Gulf.

## Dredging and channelization of coastal waters and wetlands

### BACKGROUND

Dredging involves the digging up of bottom sediments either from open water or from land or semiaquatic habitats such as marshlands. If the dredging continues in an unbroken line it creates a channel or canal through which water may flow and through which boats and ships may travel. In the process of dredging, sediments are removed, creating holes or grooves in the bottom, and the removed sediment must be deposited somewhere. In coastal areas most of the dredging falls into one of three categories: creation and maintenance of navigation channels in open water such as bays or the continental shelf, creation and maintenance of navigation channels and canals through coastal lowlands and wetlands, and dredging for the removal of fossil shell deposits in coastal lagoons, bays, and wetlands. Some dredging also takes place to provide spoil material for beach nourishment and to serve as land fill to build up low-lying coastal areas for real estate or industrial development.

Although dredges of several different designs are in operation, they all produce the same basic types of environmental disturbances. These include effects associated with direct physical modification of the bottom, effects on more distant habitats resulting from the release of large volumes of suspended sediments and sometimes noxious chemicals into the water column, and effects associated with the material that has been deposited as spoil banks.

Since sediments of the Gulf coastal region are made up largely of fine particulate materials, the dredged channels are not permanent structures. They fill in rapidly and must be kept at the desired depth by a program of continuous dredging (called "maintenance dredging"). Thus, for navigation channels the environmental disruptions are not simply one-shot intrusions but long-term, chronic environmental disturbances. Some of the more specific problems are examined below.

### CHANNELS

Channels of appropriate depth in open water are of critical importance for the navigation of commercial vessels. They provide for the movement of local as well as oceanic ship traffic across the continental shelf to ports and harbors within the bays. Some may be beneficial to marine life by serving as conduits guiding larval and juvenile fishes and invertebrates from spawning grounds on the continental shelf to nursery areas within the bays and estuaries. However, these channels definitely trap sediments normally carried downcoast by alongshore currents, and they probably also interfere with alongshore movement of marine animals, as during spawning migrations. Within the bays and estuaries water circulation in deep channels is often restricted, leading to conditions of bottom-water hypoxia and death of animals trapped in the depths.

### CANALS

In order to facilitate the coastwise transport of barged materials, through the years the US Army Corps of Engineers has created and maintained a canal system extending along the northern Gulf coast essentially from the Florida panhandle to the Rio Grande. This canal, the *intracoastal waterway,* lies from one to a few miles inland from the open Gulf, and together with many feeder canals it traverses every available type of coastal habitat. The salt, brackish, and freshwater marshes as well as bays, estuaries, and coastal lagoons of the northern Gulf are all transited and are now interconnected by this extensive canal network. In addition, particularly in coastal Louisiana, the marshes are crisscrossed by large numbers of straight-line canals, many originally developed by private companies in connection with oil and gas exploration and production facilities such as drilling and production rigs, pipelines, boat maintenance facilities, and similar operations. Many of the canals are now abandoned, but once created they persist for many years, and they tend to enlarge and develop into very shallow stretches of open water, which are poor habitat for coastal aquatic life.

Despite their usefulness for society or for private interests, these coastal canals have come at an alarming environmental cost. To create the canals the dredgers have dug up and destroyed many square miles of prime coastal wetland habitat, and for the navigation canals the dredge spoil was generally placed in a continuous pile adjacent to the canal, thereby destroying a great deal of additional wetland habitat. Under natural conditions low coastal wetlands are drained by highly complex dendritic patterns of small creeks maintained by the gradual seaward flow of freshwater and the gentle ebb and flow of tidal currents. Straight-line canals and spoil banks that cut across the marshlands disrupt the normal flow patterns and result in more rapid drainage of the freshwater, intrusion of salt water deep into the marshlands, lowering of the water table, and often drying out of nearby wetlands. Because the canals have poor water circulation and are surrounded by soils of high organic content, they often become anoxic. The sulfates in the seawater are converted to sulfides, including the highly toxic hydrogen sulfide. Anoxic conditions and hydrogen sulfide are lethal to many inhabitants of the wetlands, and they interfere with the bayward and seaward migratory movement of marine life. Furthermore, in some cases the canals and spoil banks block the passage of aquatic and terrestrial animals alike and reduce the usefulness of the remaining habitat for wildlife. Frequent maintenance dredging prolongs these problems (Darnell et al. 1976).

## SHELL DREDGING

Many of the coastal waters and wetlands are underlain by deposits of oyster or clam shells, which can be recovered by special dredges. The shell material itself is a valuable source of high-quality lime, which is used in many industrial processes, in cement production, and in road construction as a substitute for gravel. Ground up, it is an important supplement to poultry feed, where it provides a useful form of calcium carbonate needed for egg shell production. Historically, the various coastal states issued permits that allowed private companies to harvest the shells, and the states received royalties for each ton of shells removed. However, as the extent of associated environmental damage has become evident, most states have now ceased issuing the permits. Some oyster reefs have persisted for thousands of years, and as the land has subsided or as the sea level has risen the reefs have continued to grow by the accumulation of new layers of shells on top of the old. When shell dredgers mine these deep deposits they create holes in the bottom that may be as much as several meters in depth, and such holes may persist for decades. They have poor water circulation and tend to become hypoxic. Over time they gradually fill in with very fine silts and clays, which provide poor habitat for small benthic animals (Rogers 1976).

In the process of dredging for shells, large volumes of sediments are taken up, and the spoil is simply released nearby. Larger particles such as sand settle out locally, but the fine clay particles remain in suspension for a considerable time and may be transported widely through the estuary. For example, coating of submerged vegetation beds has been recorded as far as a mile away from the site of an active dredge. Eventually, much of the bottom becomes covered with a layer of fine-particulate clay-based gel. Bottom fauna that is consumed by the dredge, deeply buried by spoil, or more lightly coated at a distance fares badly, and food chains dependent on the bottom fauna are disrupted.

Estuaries are known as sediment traps since they accumulate particulate material brought in by freshwater streams, on the one hand, and by bottom saltwater currents, on the other. As they build up particulate material, they also accumulate various types of chemical compounds in the bottom sediments. Dredging tends to release these chemicals back into the water column. Included are various nutrients as well as hydrogen sulfide, heavy metals, pesticides, various industrial chemicals, and so on. Many are toxic to marine life, and some are accumulated and passed up the food chains.

Lake Pontchartrain in southeastern Louisiana provides a dramatic example of the impacts of shell dredging compounded by the effects of other human activities in the area. This is a particularly important case because we know quite well what the lake was like prior to heavy human impact. Here the dredgers sought to recover shells of the small roundish clam *Rangia cuneata.* Unlike oysters, the rangias do not form solid banks of shells but normally live as solitary individuals scattered throughout the lake on top of the firm bottoms. Originally their populations were quite dense, numbering dozens of adults (and many young) per square meter of bottom. In order to mine the shells, the dredgers were obliged to dig up nearly the entire bottom of the lake (nearshore areas being off-limits to the dredgers). Since the lake occupies an area of about 1,645 km² (635 mi²), there was a great deal of bottom to be dredged, and the shell recovery operation was allowed to continue for several decades. The impacts are clear. Formerly the lake bottom was quite firm, and the sediments had a high content of organic matter derived from the heavy plankton blooms (primarily of blue-green algae including several species of *Anabaena, Lyngbya,* and *Microcystis*) and from organic detritus contributed largely by the extensive marginal marshlands that once bounded much of the southern, eastern, and parts of the northeastern margins of the lake. A large and varied benthic invertebrate fauna of all size groups was present throughout the lake, and this supported the most important food chains of the ecosystem (Darnell 1958, 1961). Now, after decades of shell dredging, the bottom consists of a gel-like, semifluid mud that is largely devoid of organic material and small benthic animals and is too thin to support the weight of adult rangia clams. Although larvae from other areas settle, they do not long survive on such bottoms. The fluid mud continues to release toxic chemicals, and the biomass of benthic organisms has been greatly reduced (W. Sikora, J. Sikora, and Prior 1981). Meanwhile, the former marshlands along the southern margin of the lake have largely disappeared as a result of drainage, levee construction, and land fill, thus depriving the lake of much of its former organic detritus input. Beds of shallow-water submerged vegetation have been significantly reduced (R. Turner, Darnell, and Bond 1980). The average salinity of the lake has risen, and areas of hypoxia have been reported. Chemical pollutants now enter the lake from the industrial canal on the south and from streams entering the north shore. Although various human activities have contributed to the habitat destruction and biological demise of this formerly productive coastal ecosystem, clearly the primary factor has been destruction of the bottom sediments by the long-term activity of the shell dredgers.

## Coastalization of the human population

### POPULATION GROWTH AND IMPACT

During most of the years following European settlement, population densities around coastlines of the Gulf were fairly low. Slowly at first, and then rapidly during the past few decades, population levels have increased, with most of the growth being associated with major coastal cities. According to US Census Bureau figures, in the 30-year interval between 1970 and 2000 the human population of the coastal zone counties bordering the Gulf of Mexico grew from 6.12 to 11.77 million, an increase of 92.3 percent. During the same period the population of the entire United States increased by only 38.7 percent, that of the Pacific coast by 59.8 percent, and that of the Atlantic coast by 35.1 percent. During this period the population of the northern Gulf increased about 2.5 times as fast as that of the nation, 1.5 times as fast as that of the Pacific coast, and 2.6 times as fast as that of the Atlantic coast. These figures demonstrate that the northern Gulf coast is now one of the fastest growing regions of the country, and this phenomenal growth has been accompanied by a series of environmental problems. The pressure is manifested as increasing demand for habitation, employment, and recreational sites; greater local demand for food and potable freshwater; production of ever greater volumes of liquid and solid waste materials; and release of increasing amounts of atmospheric pollutants including greenhouse gases. In turn, these have led to various alterations in coastal environments including loss and fragmentation of natural habitats, diversion of freshwater supplies for human use, pollution of the coastal waters and atmosphere, general reduction in habitat quality, and changes in native species composition and ecosystem functions.

### HABITAT LOSS

In order to supply real estate for houses, industrial development, harbors, marinas, and other human facilities, natural coastal systems around the Gulf have suffered widespread destruction. Coastal marshes and other wetlands have been drained and filled, mangrove forests have been felled, and portions of estuaries such as tidal and subtidal flats have been walled off by dikes and sea-walls and backfilled to elevate the land for development. Beaches and barrier islands have been built up and developed with buildings, roadways, seawalls, bulkheads, jetties, and other types of hard structures. Some of these developments simply eliminate natural habitat. Others disrupt the normal alongshore transport of sediments and contribute to beach erosion, which could threaten the existence of barrier islands and the shoreward habitats they protect.

FRESHWATER DIVERSION

The increasing human presence in coastal areas has created heavy demand for high-quality freshwater. To meet this demand engineers have dammed a number of coastal streams and diverted much of the impounded surface water for municipal and industrial use, wastewater processing, power plant cooling, and irrigation of agricultural cropland. As a result, much less freshwater now enters some bays, estuaries, and coastal lagoons, and salinity levels in these inside waters have been rising and inducing changes in the marshlands, submerged vegetation beds, and associated ecosystems. The pressure on freshwater resources is particularly acute during periods of extended drought. Many coastal communities are also actively pumping freshwater from underground aquifers, and in some areas this has already led to saltwater intrusion into the aquifers. The diversion of freshwater in south Florida was discussed earlier, but the Texas Water Plan calls for the diversion of large volumes of freshwater from streams of east and central Texas for use by large cities such as Houston and San Antonio and for irrigation of farm fields in arid south Texas.

WATER QUALITY REDUCTION

Rampant human population growth has inevitably led to a decline in water quality in coastal waters and wetlands. Runoff from agricultural fields has added fertilizer nutrients and various pesticide residues. Runoff from city streets and effluents from industrial plants add a variety of chemical wastes including heavy metals, oils, greases, and other organic and inorganic pollutants. Some, such as heavy metals, are especially toxic to marine life and to humans who consume seafood from contaminated areas. Municipal sewage treatment plants may release wastewater rich in nitrogen and phosphorus compounds, leading to overenrichment and hypoxia in estuarine waters (Neilson and Cronin 1981). Some of the municipal efflu-

ents may also contain harmful bacteria, viruses, and other human pathogens, leading to contaminated seafood and beach closures. Nonbiodegradable solid wastes of metal, glass, plastic, and so forth often appear in waters and along shorelines of bays and estuaries heavily used by people. The general picture emerges that where human populations are dense the quality of coastal waters tends to decline through the addition of nutrients, chemical pollutants, pathogens, and solid wastes, but the exact pattern of decline depends on local factors and is somewhat unique for each receiving body of water.

From the above it is clear that the great increase in human presence in coastal areas around the Gulf, particularly during the past few decades, has led to dramatic changes in the native habitats and coastal ecosystems. Among the most important are the loss and fragmentation of coastal habitats and their ecosystems and the reduction of freshwater inflow with consequent desiccation and saltwater intrusion. Other changes include the reduction in water quality through the addition of excess nutrients, chemical pollutants, human pathogens, and solid wastes. The coastal systems we see today are diminished in extent and much modified from those encountered by the early settlers.

## Commercial shipping

BACKGROUND

Commercial shipping is a major industry in the Gulf of Mexico and a critical link in the regional economy. Ports of the Gulf handle almost half the total US import and export shipping tonnage, and of the 19 major ports around the Gulf, 14 are in American waters (table 17.2). Two-thirds of the ships leaving US ports and about a quarter of those leaving Mexican ports are bound for destinations outside the Gulf. Roughly two-thirds of the large tanker and dry cargo ships leave the Gulf through the Straits of Florida, and the remainder exit through the Yucatán Channel. However, a large percentage of the commercial shipping traffic travels from one port to another within the Gulf. As seen in table 17.2, crude oil and refined petroleum products make up a large portion of the cargo. Other materials include agricultural products and foodstuffs, agricultural and industrial chemicals, and iron and steel products and equipment, as well as bulk materials such as ores, coal, and gravel.

In addition to cargo passing through the land-based ports referred to above, much of the imported crude oil is off-loaded at the Louisiana Offshore Oil Port (LOOP),

**Table 17.2.** *Major commercial shipping activities in the Gulf of Mexico.*

**Major ports of the Gulf**

United States

    Florida – Tampa, Pensacola

    Alabama – Mobile

    Mississippi – Pascagoula

    Louisiana – New Orleans, Lake Charles

    Texas – Port Arthur, Beaumont, Galveston, Texas City, Houston, Freeport, Corpus Christi, Brownsville

Mexico

    Tamaulipas – Tampico

    Veracruz – Tuxpan, Veracruz, Coatzacoalcos

    Tabasco – Dos Bocas

**Major imports and exports**

United States

    Imports – crude oil, iron and steel

    Exports – agricultural products (wheat, corn, soybeans), iron and steel products, industrial and agricultural chemicals, crude oil, petroleum products, coal

Mexico

    Imports – processed food products, equipment

    Exports – crude oil

**Materials shipped from port to port around the Gulf**

Crude oil, refined petroleum products, iron and steel products, industrial and agricultural chemicals, containerized cargo such as foodstuffs and equipment, bulk cargo (iron ore, coal, shell, sand, gravel)

---

located on the continental shelf off central Louisiana. The above figures do not include vessels in the commercial fishing fleets, the thousands of recreational boats that ply shallow coastal waters, or the various ships, boats, and barges associated with oil and gas recovery operations in the Gulf. From the standpoint of ship and boat operations the Gulf of Mexico is a very busy place, primarily in the coastal areas but in the offshore waters, as well.

The vast amount of ship and boat traffic in the Gulf requires port facilities for mooring, loading and unloading, storage, warehousing, processing, refining, and so forth. The ports also attract various types of industry, and all of these facilities must be housed on raised land adjacent to or near the water and protected by concrete walls. In addition, there must be navigation channels, turn-ing basins, and marinas. The navigation channels cut through estuaries and coastal lagoons and extend across shallow waters of the continental shelves. These must be kept at the desired depth by constant maintenance dredging, and the spoil must be placed somewhere. Too often the convenient solution is to dispose of the spoil in a line adjacent to the dredged channel. On the continental shelf the channel must be protected from filling by alongshore movement of sediments, and this is often accomplished by the construction of long stone and concrete jetties that project from the shoreline, sometimes a kilometer or more into the open Gulf.

COASTAL IMPACTS

The effects of commercial shipping and the smaller ship and boat traffic on the environment and biota are quite varied, and they relate to the shore facilities, navigation channels, and operation of vessels at sea. As mentioned earlier, the construction and development of docks, warehouses, and other shore facilities involves the loss of large acreages of prime lowland, wetland, and estuarine habitat. The channels and turning basins entail further habitat loss as well as chronic disturbance from maintenance dredging and spoil placement. Leakage, spillage, and effluent discharge into the channels add a variety of chemical pollutants, and because of the poor water circulation, hypoxic conditions are likely to develop. Within estuaries and on the continental shelf the channels may block cross-channel movement of organisms, and on the continental shelf the larvae and very young of estuary-dependent fishes and invertebrates must somehow find and negotiate the long channels into the estuaries.

OCEAN IMPACTS

On the busy continental shelf, collisions between boats, ships, barges, and rigs or platforms are inevitable, resulting in chronic low-level and occasional major spills of fuel, crude petroleum, and other substances. Ship anchors damage and sometimes rupture bottom pipelines, releasing more crude oil. A certain amount of garbage and trash is lost overboard or deliberately dumped, and among these materials plastic bags and plastic sheeting are particularly deadly for the sea turtles and marine mammals that ingest them. There is a danger of boat and ship collisions with sea turtles and cetaceans, from which the animals come out second best. All the ship and boat traffic generates a great deal of noise of many sound frequencies and volumes. This marine noise pollution may interfere with the normal behavior patterns of marine animals, and at high levels it may become pathological or lethal.

Tanker ships that carry liquid cargoes frequently wash out their bilges before taking on a new load, and this practice adds to the chemical pollution levels in Gulf waters. Of particular concern is the possibility of the introduction of noxious alien species from foreign waters into the Gulf. These may arrive as biofouling species attached to the outside hulls, or they may travel inside ballast tanks of ships entering the Gulf. Before undertaking a major ocean voyage, large ships regularly fill their ballast tanks with seawater, which acts to stabilize the ship during its trip at sea. As it approaches its destination, the ship traditionally discharges the ballast water and its living contents into nearshore waters of the destination port. Thus, these large oceangoing vessels are conveyers of invasive marine pest species from one ocean to another around the world.

## Offshore mineral extraction

Although some sand, shell, salt, and sulfur are mined from bottom and subbottom deposits of the Gulf of Mexico, the main products are petroleum and natural gas, and since hydrocarbon recovery operations have by far the greatest actual and potential impacts on the environment and biota of the Gulf, they will be the focus of the present section.

### BACKGROUND

In the Gulf of Mexico the first open-water oil field (Creole field) was established off central Louisiana in 1938, but significant development of offshore hydrocarbon deposits did not begin until 1947, when the Ship Shoal field was discovered about 20 miles off the western Louisiana coast. Subsequent development has been rapid and extensive. Four decades later, by the end of 1986, more than 26,750 oil and gas wells had been drilled on the continental shelf and slope, and more than 17,350 miles of bottom pipelines had been laid in US waters alone. At that time more than 86 million barrels of oil and 87 trillion cubic feet of natural gas had been produced from offshore Louisiana and Texas, and an additional 15 million tons of sulfur and 4.8 million tons of salt had been removed. Payments to the federal government for leases, rentals, royalties, and bonuses had exceeded $70 billion (Defenbaugh 1990). Recent years have seen the development of more sophisticated technology and the movement of drilling

activities from the continental shelf first to the upper and then to the middle and lower continental slope. As late as the early 1990s most drilling leases were for bottom tracts in water depths of 305 m (1,000 ft) or less, but by 2001 almost a third were in water depths of 1,524 m (5,000 ft) or more, and 11 percent were in water depths greater than 2,287 m (7,500 ft). In fact, new discoveries have already been made at a water depth of 2,964 m (9,727 ft), and technology is available to extract oil and gas from this and even greater depths. It is known that petroleum deposits are associated with the Sigsbee Knolls in about 3,750 m of water, so drilling is possible even in the deepest portions of the Gulf. Thus, the impacts of oil and gas development will be felt at all depths.

### ACTIVITIES ASSOCIATED WITH OIL AND GAS DEVELOPMENT

To understand the environmental effects of oil and gas development we must first examine how the subsurface deposits are located and developed. As seen in table 17.3, the operations fall into four sequential phases:

1. *Initial seismic survey.* Ascertains the likely location of subsurface oil and gas deposits.
2. *Site evaluation phase.* Exploratory drilling takes place to determine whether the deposits are actually there and worth developing.
3. *Main development and production phase.* May last for 20 years or more until the wells run dry.
4. *Postproduction phase.* Platforms are removed and the site is cleaned up.

These phases and their associated activities and events as well as their environmental effects are discussed below.

Once the responsible federal agency announces which tracts of submerged lands will be offered for lease, interested companies carry out detailed geophysical and geological surveys of each lease block to determine hydrocarbon potential, identify important drill sites, and map shallow drilling hazards such as unstable bottom sediments. In the process they use such equipment as sidescan sonar, air guns and sparkers, and subbottom profilers, and the sediments may be sampled with coring devices. Here the main points of environmental concern are increased ship traffic, increased background noise, and the very loud explosive sounds produced by air guns and sparkers. These matters are of growing concern as the activities move into waters most intensively utilized by marine mammals.

**Table 17.3.** *Offshore oil and gas operational phases, associated activities and events, and primary results of environmental concern. Derived effects are taken up later. Note that all phases result in increased ship and boat traffic and background noise, and most also involve some chemical pollution and seafloor disturbance.*

| Operational phase and associated activities/events | Primary results |
| --- | --- |
| **Seismic survey** | |
| Geophysical and geological surveying | – increased ship traffic |
| | – explosive sounds |
| **Site evaluation** | |
| Drill rig installation and removal | – increased ship traffic |
| | – seafloor disturbance |
| | – presence of structures |
| Routine drilling operations | – discharge of drilling mud and cuttings and other solid and liquid wastes |
| | – minor oil spills |
| Blowouts | – major or minor oil spills |
| **Development and production** | |
| Platform installation | – presence of structures |
| Routine drilling operations | – discharge of drilling mud and cuttings and other solid and liquid wastes |
| | – minor oil spills |
| Blowouts | – major or minor oil spills |
| Routine production operations | – discharge of produced water and other liquid wastes |
| | – discharge of solid debris |
| | – minor oil spills |
| Pipeline installation and routine operation | – dredging and channelization |
| | – presence of structures |
| Pipeline rupture | – major or minor oil spills |
| Tanker/barge accidents | – major or minor oil spills |
| **Postproduction cleanup** | |
| Platform removal | – explosions |
| Platform disposal | – presence of structures |

Once a company has obtained a lease to a particular tract, it carries out a site evaluation during which it drills one or more wells to determine the hydrocarbon potential of the area. This phase consists of two main groups of activities: drilling rig installation and removal, and routine drilling operations. These activities are accompanied by increased boat and ship traffic and background noise, seafloor disturbance (mainly from drilling rig installation and removal), presence of solid structures (drilling rigs, anchors, chains, cables, etc.), and discharge of various liquid, semisolid, and solid materials into the water or onto the bottom. Such discharges include drilling mud and its chemical additives, drill cuttings (rock fragments), various liquid wastes (cooling water, desalinization brine, sanitary and domestic wastes, etc.), and solid debris (pipes, cables, wrenches, and other objects dropped from the rig). Possible accidental occurrences during this phase include minor fuel and oil spills and (very rarely) blowouts resulting in major or minor oil spills.

During the development and production phase, platforms are installed to extract the oil and gas, which in turn may be transported directly to shore or to offshore terminals by pipelines, tankers, or barges. Drilling and long-term operation are also part of this phase. Environmental effects may result from increased boat and ship traffic and noise, seafloor disturbance (mainly from platform and pipeline installation), presence of structures, discharge of drilling muds and cuttings, discharge of produced water and other liquid wastes, and release of solid debris. Accidental occurrences of concern include blowouts, pipeline ruptures, and tanker and barge accidents, all of which can result in major or minor oil spills. The production life of an offshore platform is generally about 20–25 years.

The postproduction phase is short and consists of the removal and disposal of spent production platforms and the cleanup of any residual debris. In the late 1980s there were around 3,400 oil and gas platforms and related structures in offshore waters of the northern Gulf, and about 55 to 65 of these were being removed annually. Platform legs may be severed below the bottom surface by cutting devices or by explosives. Once removed, the retired rigs and platforms must be disposed of. Formerly they were brought to shore and cut up for scrap, an expensive procedure, but under a "rigs to reefs" program they are now being placed in special undersea "graveyards" for spent rigs and platforms. By the end of 2001 a total of 167 such structures had been converted to artificial reefs in the designated disposal areas. Effects of environmental concern include the pressure waves resulting from explosive devices used to sever the platform legs and the persisting hard substrates resulting from the rigs to reefs program.

ENVIRONMENTAL IMPACTS

From the primary effects listed in table 17.3 and discussed above, we can derive the various types of potential environmental impacts, and these generally apply whatever the water depth. However, petroleum and natural gas recovery operations in very deep water entail certain additional difficulties, risks, and potential environmental effects, and these will be addressed later. It should be noted that in marine waters under US jurisdiction all aspects of the oil and gas industry are highly regulated and closely monitored, and these rules take into account potential environmental damage. Although virtually all of the potential types of environmental impact discussed below have occurred in the Gulf of Mexico at one time or another, only a few are considered major in terms of severity, widespread occurrence, or long-term persistence. The following discussion relates primarily to oil and gas operations in US waters, and these are concentrated in the northern Gulf, primarily off Louisiana and

Texas, but most of the considerations must also apply to the oil and gas recovery activities in Mexican waters in the Bay of Campeche in the southern Gulf.

As mentioned earlier, throughout the years oil and gas exploration and development operations have caused much wetland habitat loss and water pollution in coastal environments of the northern Gulf of Mexico. Impacts of marine operations fall into seven general classes of special concern. These include: (1) increased background noise, (2) explosive sounds, (3) major and minor oil spills, (4) other chemical pollution, (5) discarded solid debris, (6) seafloor disturbance, and (7) presence of hard-substrate structures. Each of these impact categories is examined below.

EFFECTS OF BACKGROUND NOISE
AND EXPLOSIVE SOUNDS

Under provisions of the Marine Mammal Protection Act of 1972 it is illegal to harm or harass marine mammals or to otherwise induce changes in their natural behavior. Species considered to be endangered, such as the sperm whale and West Indian manatee, receive further protection under the US Endangered Species Act. In earlier years the possibility of marine mammal disturbance by surface ship and boat traffic was not considered to be a problem, but recent evidence suggests that normal behavioral patterns may be disrupted by low-frequency sounds associated with boat and ship traffic and especially by repetitive pinging noises from seismic surveys. Since whales and dolphins communicate by sound and use it for echolocation, the noise associated with surface boat and ship traffic is of some concern, especially as the oil and gas activities intensify over the continental slope, where marine mammal populations reach their greatest density and diversity.

Explosive sounds are mechanically produced by air guns and sparkers during seismic surveys and by actual explosives during the removal of production platforms. Resulting shock waves propagating through the surrounding water can kill or injure nearby fishes, sea turtles, and marine mammals, and at a greater distance they might interfere with communication and other behavior. Before detonating explosive charges on platform legs, efforts are made to ensure that sea turtles and marine mammals are not in the vicinity, but here again, as the oil and gas operations move into deeper water where many marine mammals feed, this is a matter of special concern.

## EFFECTS OF OIL SPILLS

Most phases of oil and gas operations entail spillage or leakage of minor amounts of fuel or crude oil, but surface ship accidents, pipeline ruptures, and oil well blowouts have the potential for releasing large volumes of crude petroleum into Gulf waters. A blowout may occur when a drill pipe penetrates a subsurface gas or oil deposit under extreme pressure. If this pressure exceeds the ability of the pipes, valves, and drilling fluids to contain the upward push, then the pressurized gas, petroleum, and formation waters simply blow upward through the drilling cavity, pushing ahead of them pipes, mud, and anything else in the way. At the surface the material gushes into the ocean from below or into the atmosphere, where the liquid falls back into the sea. Sometimes the spewing gas and petroleum ignite, and the top of the well becomes a spectacular flaming torch. If this raging fountain is not brought under control promptly, large volumes of natural gas and petroleum can be burned up, lost to the atmosphere, or spilled into the ocean. Although most such spills are of small magnitude, they have the potential for huge loss of gas and petroleum, as occurred in the southern Gulf in 1979 when the Ixtoc 1 well blew out (see chapter 7). It raged out of control for months and added to the Gulf an estimated 0.4–1.4 million tons of crude oil, which eventually spread throughout most of the Gulf waters.

Once released into the water, crude oil rises and floats on the surface. Water-soluble fractions, particularly many of the lower-molecular-weight compounds, tend to dissolve in seawater and disperse over time, and most of these ultimately evaporate into the overlying atmosphere. Insoluble compounds, primarily the higher-molecular-weight chemicals (oils, greases, paraffins, waxes, tars, asphaltenes, etc.), remain floating at or near the surface awhile and may become widely distributed around the Gulf, but the gradual incorporation of sand grains, silt particles, and other materials increases their density, and these masses ultimately sink to the bottom. Having lost the lighter and more toxic compounds, the weathered asphalt-like material is now prime substrate for attachment by biofouling species or soft material for penetration by a variety of marine borers and burrowers (W. Pequegnat and Jeffrey 1979).

Unlike the situation in the Arctic where the Exxon Valdez oil spill occurred, the temperate/subtropical waters of the Gulf are warm, and the dissipation and evaporation of lighter compounds proceed fairly rapidly. Furthermore, crude oil has been leaking from natural seeps in the bottom sediments of the Gulf for thousands (probably millions) of years, and Gulf waters contain large populations of bacteria capable of rapidly degrading most chemical components of crude petroleum. As indicated by the Ixtoc 1 event, a major oil spill in the Gulf of Mexico does not have anything like the long-term impact of one in a cooler climate.

Nevertheless, these matters are of concern. Some of the crude oil fractions (including heterocyclic and some aromatic compounds such as naphthalenes, phenanthrenes, fluorenes, dibenzothiophenes, etc.) are known to be highly toxic or carcinogenic, and chronic exposure even to low levels could prove hazardous to marine life. Furthermore, massive oil spills can devastate marine organisms caught up in the smelly, gunky mess before natural physical processes have a chance to reduce the damage. Heavy oiling of coastal marshes, mangrove swamps, seagrass meadows, and coral reefs can result in mass mortality in these systems, and their full recovery can take years or decades. The oiling of birds' feathers interferes with their ability to swim and fly, and heavy coating can lead to death as seabirds ingest the material in their attempts to clean themselves. Caught up in surface spills, sea turtles are known to ingest the oil with fatal consequences. Crude oil can also foul the baleen plates of baleen whales and thus interfere with their feeding activities. Würsig, Jefferson, and Schmidly (2000) reported that in a large oil spill off Galveston, Texas, in 1990 (from the *Mega Borg* oil tanker), bottlenose dolphins apparently did not know how to avoid extensive oil-covered areas. They remained and surfaced repeatedly to breathe in the very volatile fresh areas of the spill—areas in which humans became nauseated and sick to their stomachs in minutes. Under such conditions the lung tissue of dolphins rapidly becomes coated with petroleum-derived volatile compounds. These are known to enter the bloodstream and combine with hemoglobin, rendering the breathing apparatus useless and suffocating the individual. How widespread this problem is among the toothed whales is unknown, but the matter is of great concern as oil and gas activities move into deeper water.

## EFFECTS OF OTHER CHEMICAL POLLUTANTS

During routine oil and gas drilling and production operations, various types of liquid, semiliquid, and solid materials are released from the rigs and platforms, and these include drilling fluids, produced waters, and other liquid wastes as well as drill cuttings and other solid debris. Each of these is discussed briefly below. Drilling fluids are very dense, mud-like materials that are pumped down the central drill pipe during active drilling operations. They exit from holes in or near the drill bit and proceed back to the surface in the *annulus,* that is, in the space between the outside of the drill pipe and the inside of the well casing. Pumped down under high pressure, these fluids serve several functions. First, they cool and lubricate the drill bit itself. Second, they serve to remove the drill cuttings. As the bit grinds through layers of rock the resulting rock fragments, called "drill cuttings," become embedded in the dense mud and are transported up to the drilling platform, where they are removed so that the mud can be reused. Finally, by maintaining a strong positive pressure in the well bore they aid in preventing "formation" fluids (liquid trapped in subsurface deposits) from entering the annulus and diluting the drilling fluids.

Worldwide several types of drilling fluids are in use, and all consist of complex mixtures of chemical compounds. Water-based fluids are most commonly used in the Gulf of Mexico. These are the simplest, cheapest, and most environmentally friendly. Basically they consist of clay plus a variety of chemical additives. Commonly these include barium, caustic soda, lignite, lignosulfonates, water, and soluble polymers. Other additives may include diesel fuel, chromium salts, surfactants, a biocide (paraformaldehyde), and sometimes heavy metals such as copper, cadmium, mercury, and lead. Although considerable efforts are made to recover and reuse the drilling fluids, inevitably some are lost and discarded with the drill cuttings.

Oil and gas operations also generate large volumes of produced water, or "formation" water, as it is sometimes called. This liquid is essentially fossil seawater that is released from subbottom formations. It is a concentrated brine whose salinity may be several times that of seawater, and it has a slightly elevated temperature and a low oxygen content. It often contains low-level concentrations of heavy metals such as copper, chromium, manganese, and strontium. Some hydrocarbons (oil and gas) are always present. Of particular concern are the radionuclides (especially [226]radium) that are also generally present, as their concentrations may occasionally be fairly high. Following treatment, most of this water is simply released into the Gulf. The amount generated during production operations varies from one location to another, but Neff (1987) reported that in total the offshore platforms of the northern Gulf may discharge up to 1.5 million liters of produced water per day. This discharge affects mainly the local environment because the liquid rapidly becomes diluted in Gulf waters. Local effects, however, have been measured, and of greatest concern is the radioactive material released in the zone of pollution around the rigs and platforms where considerable commercial and recreational fishing takes place.

The rigs and platforms generate additional waste fluids. These include cooling water, desalinization brines, domestic and sanitary wastewater, and deck washings. Seawater used to cool deck engines is returned to the Gulf at a slightly elevated temperature. On the platform freshwater is extracted from seawater for domestic and other uses, and the resulting brine is returned to the Gulf. Domestic and sanitary wastes are generated by sinks, showers, toilets, galleys, and so forth, and after some treatment this material is discarded. Deck drainage results from deck washings, tank cleaning operations, drip pans, and runoff from work areas, curbs, and gutters. This miscellaneous waste fluid contains nutrient chemicals, some hydrocarbons and other organic material, and drilling muds. These are all present in low concentrations, and none are particularly toxic. Since such discharged wastes are carefully monitored and must meet strict standards, they are generally benign and are rapidly diluted in the Gulf water.

In summary, in addition to the crude petroleum lost to the Gulf, various other chemical pollutants are of concern, primarily as they affect areas near platforms. Highly soluble fluids released into the water are rapidly diluted, and by the time they have gone out from one to a few hundred meters concentrations are near background levels. Muds and other semisolid materials that fall to the bottom create a zone of pollution, which in some cases may extend a thousand meters or more from the platform. Here the abundance and diversity of marine life is generally depressed. In severe cases the effects may persist for years, but once the pollution ceases, recovery is generally rapid. Although these perturbations are local, the number of local sites affected indicates that in total these impacts must be widespread.

Polluted areas directly around platforms are sometimes heavily fished by commercial and recreational fishermen, raising the possibility of contaminated seafood. Many of these chemicals persist and add to the growing levels of chemical pollutants in the world oceans. Although their effects on distant marine organisms and ecosystems cannot yet be documented and assessed, this growing marine burden should be of concern to everyone.

### DISCARDED SOLID DEBRIS

Around the bases of drilling rigs and platforms a certain amount of solid debris falls and accumulates on the bottom surface. This includes the drill cuttings deliberately deposited as well as a certain amount of other materials that accidentally fall from the platforms or from boats in the vicinity. Drill cuttings are simply rock fragments from the limestones, sandstones, and siltstones encountered by the drill bit. The problem here is that when a well is drilled through 12,000 to 18,000 feet of bottom layers, quite a mass of cuttings accumulates. It has been determined that the drilling of a single exploratory well may generate 1,000–2,000 metric tons of cuttings. A production well generates somewhat less, but since a single platform may have up to 100 producing wells, the volume of cuttings produced is huge. These cuttings are always coated with a certain amount of oil and drilling fluids, and they are generally discarded through pipes that shunt the materials away from the immediate base of the rig or platform.

In addition to the drill cuttings, other solid debris accumulates around the bases of the platforms. Inevitably, some tools, cables, hard hats, work gloves, and other items accidentally fall into the water and sink. Barges and other work boats as well as commercial and recreational fishing vessels frequent the offshore structures, and occasional collisions may send solid materials to the bottom. Other waste debris (mostly lost or discarded by people in boats) includes fishing lines, netting, metal cans, and plastic materials, among other items. Although the petroleum industry and offshore boat operators' groups have gone to great lengths to reduce the volume of lost and discarded solid trash, a certain amount still finds its way to the bottom, where currents distribute the lighter material widely around the continental shelf. Besides adding to the general bottom clutter, some of the solid materials do have very specific adverse impacts on marine life. As mentioned earlier, plastic bags and sheeting can prove deadly for sea turtles and marine mammals that ingest them. The cuttings may leach out toxic materials and provide solid substrate for the attachment of biofouling species and niches for cryptic species, thus changing the nature of the bottom fauna. Fortunately, much of the solid material is removed from the bottom at the end of the petroleum production phase when the site is cleaned up, but the widely distributed material may persist for years.

### SEAFLOOR DISTURBANCE

Seafloor disturbance results from installing rigs and platforms, constructing pipelines, discarding solid and semisolid (toxic and nontoxic) waste materials, and various other activities such as the dragging of ships' anchors. All these activities disturb the bottom sediments and their inhabitants one way or another, perhaps by digging them up, dumping materials on top of them, or allowing toxic substances to accumulate. In any instance, the results are generally local, but in total the effects are widespread. Benthic organisms and communities are adversely affected and sometimes devastated. Most recover rapidly, but as mentioned earlier, some of the effects may persist for years.

### PRESENCE OF HARD-SUBSTRATE STRUCTURES

The final environmental impact category relates to the presence of hard-substrate structures. In prehistoric times few structures and very little hard substrate were present on the continental shelf of the north-central Gulf of Mexico, but by 1985 more than 4,000 oil and gas platforms had been installed, and it has been estimated that a single platform in 30 m of water provides about 8,000 m² of exposed hard substrate. By the end of 1987 more than 17,000 miles of bottom pipelines had been laid. Although on most of the continental shelf, pipelines are initially buried in the bottom sediments, a significant fraction subsequently becomes exposed on the seabed, where the pipes provide much additional hard substrate. Collectively, the drilling rigs, active and discarded production platforms, and bottom pipelines, together with various cables, anchoring devices, and other structures, represent a vast amount of new firm substrate where little was present before.

Much of this substrate is suitable for the attachment of biofouling species. Upon such surfaces grow thick mats of algae, sponges, mollusks, barnacles, sea squirts (urochordates), and other invertebrate species, and upon or through the mat many mobile and cryptic forms crawl or hide. Periodically portions of the mat break off and fall to the bottom, enriching the sediments and stimulating the development of bottom communities. Many fish species accumulate around the structures, some to feed on the

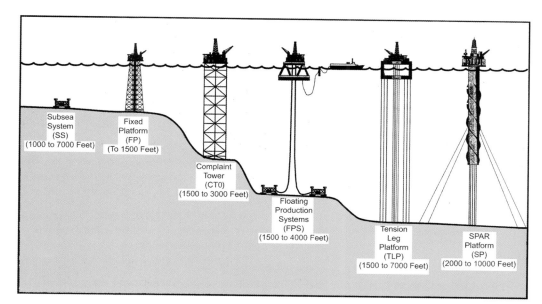

**Figure 17.1.**
Basic types of fixed and floating facilities for handling oil and gas production in offshore waters of the Gulf of Mexico. Fixed facilities include the subsea system (SS), fixed platform (FP), and compliant tower (CT). Floating facilities include the floating production system (FPS), tension leg platform (TLP), and SPAR platform (SP).

biofouling mat and its inhabitants or on the bottom communities, and some to prey on the resident invertebrate and fish populations. Other pelagic fish species, some of which forage elsewhere, appear to be attracted by the structures themselves. Sea turtles often settle on the bottom and rest near platform legs. Thus, for better or worse, the presence of the structures changes the nature of the northern Gulf shelf ecosystems in fundamental ways.

At the same time, these structures concentrate the flora and fauna within what we have called the zone of chemical pollution, and the fishes and crustaceans that humans harvest from such areas may contain chemical pollutants. Exposed bottom pipelines, some 20 inches or more in diameter, have the potential for interfering with normal migration patterns of benthic animals such as penaeid shrimps. Finally, it has been noted that the solid substrates provide settlement and colonization sites for invasive biofouling species from other parts of the world oceans. With the increase in marine ship traffic the introduction of such foreign invaders is inevitable, and without question some of these species will be unwanted pests. The existence of platform graveyards in the northern Gulf ensures that solid substrates will be available for such colonization long after the petroleum and natural gas resources have been depleted.

### DEEPWATER OPERATIONS

Through the years engineers have developed several types of fixed and floating facilities for handling oil and gas production in marine waters, and these fall into six basic cate-

gories, each with a number of variations. The decision to use one or another type is based on such considerations as cost, operating conditions, and water depth. The approximate maximum depth in which each type of facility may be used is given in figure 17.1.

In general, bottom-supported towers are used in shallow water, tension-leg platforms in somewhat deeper water, and in very deep water floating production and subsea systems are the facilities of choice. Most floating platforms are held in place by cables attached to bottom mooring anchors, but the FPSO system can be kept in position by a series of small propellers or thrusters. The FPSO itself is a ship-shaped structure containing tanks for the temporary storage of crude oil, which must be off-loaded periodically to tankers or oceangoing barges for transport to shore. Subsea systems can be used even in the greatest depths currently under operation. Such systems include both seafloor and surface components. The seafloor facilities include the producing wells, flowlines for the petroleum and natural gas, manifolds (with appropriate pipe connections, valves, etc.), and "umbilicals" containing wires to the control systems and production equipment on a platform that could be located many miles from the actual seafloor wells and manifolds.

### DEEPWATER HAZARDS

As mentioned earlier, drilling and production in deep water are attended by a number of special hazards. These have to do with current regimes in the water column, conditions at the seafloor and within the bottom sediments, gas hydrates, and other risks associated with operating at great depths. Variations in the Gulf Loop Current have been found to induce undesirable oscillations in some deepwater structures, and there is also concern about the unknown currents in deep water that could affect operations there and determine the direction and dispersal spread of any deepwater spills or blowouts. Seafloor stability is a constant concern for those working on the continental slope. Here sediment slumping is always a possibility, especially if the sediments contain pockets of gas or gas hydrates. The latter are encountered at depths of 500 m or more where the temperature is low and the pressure is high, and they may be present as mounds on the seafloor surface or as deposits within the sediments. Gas hydrates consist of ice-like water crystals that trap methane and other gases in a somewhat unstable physical state. Both natural and anthropogenic factors can cause a hydrate to dissociate, suddenly releasing the gas, which rapidly expands. This could trigger seafloor slumps or subaqueous landslides. The drilling itself could lead to dissociation, especially if the gas hydrate comes into contact with warm drilling fluids or hot hydrocarbons. At present the distribution, thickness, and extent of gas hydrate deposits on and within the continental slope and abyssal plain of the Gulf are unknown. Of further concern is that gas hydrates can form on drilling equipment and within pipelines, where they could stop fluid flow, increase pressure, and lead to pipeline rupture with consequent sea-bottom oil spills. Engineers are searching for ways to get cold petroleum to flow easily through deepwater pipelines or to heat the pipelines without creating gas hydrate problems. There are also concerns about how to deal with oil spills and blowouts if they occur in very deep water, where everything must be handled by submersibles or underwater television and remote control.

## Ocean dumping

Throughout recorded history humans have produced quantities of solid and liquid wastes, and the oceans have long been considered convenient places to dump these unwanted materials that might otherwise clutter, foul, or pollute the nearby land or fresh waters. Of course, the receiving capacity of the seas was considered to be virtually limitless. In the years following the Second World War, as the economies of the industrialized nations expanded, ever-greater volumes of waste were produced, and much of it was disposed of at sea. Unexploded military ordnance was placed in vast designated undersea "graveyards." Coastal cities piped their untreated or barely treated sewage directly to submarine outfalls. Large volumes of dredge spoils were placed onto continental shelves, often near navigation channels. Concrete and other solid construction debris were indiscriminately discarded offshore outside of the major shipping lanes. Around the world thousands of drums of highly toxic liquid industrial wastes were annually barged out of sight of land and simply dumped onto the sea bottom. Even drums of highly radioactive chemical wastes were disposed of at sea not far from land. It is not known how much of such material entered the world oceans, because no one was keeping track of the growing problem.

A significant backlash began in the late 1960s. In seafood-dependent Japanese cities (Minimata and Niigata), first the domestic cats and then the human population began suffering severe brain damage and dying as a result of heavy mercury pollution of the coastal waters. In Europe the semienclosed Baltic and North Seas became so polluted that fisheries declined, much of the contaminated seafood was no longer fit for human consumption, and seagoing stocks of salmon could no longer complete their life cycles. Americans awoke to the large volumes of radioactive wastes that were brewing in seafloor drums off California's Golden Gate, and to the marine birds (brown pelicans and others) that were experiencing reproductive failure due to the accumulation in their bodies of chlorinated hydrocarbons (DDT, DDE, etc.) and other toxic chemicals. Additional evidence of widespread pollution problems was becoming apparent in bays and estuaries and in offshore marine waters around the nation's coastlines. Added to this was the recurring news of frequent tanker collisions, massive oil spills, fouled beaches, and dead and dying wildlife. As the general public became aware and then aroused, there developed a widely recognized need for government institutions and policies to control and restrain the hitherto largely unregulated marine dumping practices. In 1972, the United States enacted the Marine Protection, Research, and Sanctuaries Act, which prohibited the dumping of all materials into coastal and oceanic waters under US jurisdiction except by permit from the US Army Corps of Engineers (for dredge spoils) or from the Environmental Protection Agency (for all other materials except the few already regulated by other statutes). Subsequently, a series of technical meetings held in Tokyo, London, and elsewhere under the auspices of the United Nations led to a series of international agreements to control and reduce ocean dumping and marine pollution worldwide.

Unfortunately, that is not the end of the story. Good laws and agreements are one thing; enforcement is another. Within US waters, enforcement is vested in the US Coast Guard, which is ill equipped to assess chemical pollution and which has many more-pressing duties. Although things have gotten much better, we know that some dumping still takes place because of the drums and other litter that continues to accumulate on the beaches of Padre Island (Texas) and other coastal areas. Of particular concern is the number of permits the "regulating" agencies issue for oil platform "graveyards" for places like the highly polluted Houston Ship Channel and the continental shelf. Of further concern are permits given for the establishment of "FADs" (fish-attracting devices) on the continental shelves. These are essentially artificial reefs created by sinking old cargo ships, streetcars, automobile bodies, chunks of concrete, and other dense materials to form elevated masses of solid substrate to which fouling organisms can attach and around which various fish species accumulate. Here also congregate large numbers of sport fishermen, and this has been used as justification for drastically changing the nature of the ecosystems of the northern Gulf and providing ample substrate for the attachment of unwanted invasive species from around the world. Within the Gulf of Mexico, where sport fishing is an economically important industry, the policy of establishing artificial reefs is widespread. Outside the more developed nations, antipollution laws are few and enforcement is lax, but on the other hand, these developing nations are not the world's major marine polluters.

## Commercial and recreational fishing

### COMMERCIAL FISHING

Throughout the nineteenth and early part of the twentieth centuries, commercial fishing using hook-and-line and long beach seines occurred in estuaries and nearshore Gulf waters. Bays and sandy beaches were scoured for adult sea turtles and their eggs, and within the bays there was a thriving oyster industry. In clear shallow waters off the west coast of Florida, especially around Tarpon Springs just north of Tampa, sponges were harvested from nearshore bottoms of the continental shelf. As a result of these activities, sea turtle populations in the bays of south Texas were greatly reduced, and nesting

along the northwestern Gulf became rare. Otherwise the effects on estuarine and marine populations were generally minor and localized.

### FISHING METHODS

However, following the Second World War, spurred by growing markets and advances in technology, the commercial fishing industry underwent a great expansion. Particularly important were the development of more powerful engines for the fishing boats and cable-pulling winches and the appearance of strong rot-resistant plastic cords and ropes for nets and fishing lines. In this postwar expansion fishermen were now using bottom trawls, purse seines, longlines, gill nets, hook-and-line, and other gear to harvest a wide spectrum of sea life including oysters and clams, squids, crabs, shrimps, tropical lobsters, bottom-dwelling (demersal) fishes, schooling pelagic fishes, large individual pelagics, and reef-related species. These fishing methods and their catch are discussed briefly below.

Bottom trawls were introduced into the Gulf prior to 1940. These consist of tapering, sack-shaped nets that are held open by kite-like wooden and metal "doors." These nets, spread wide, are dragged along the bottom behind powered boats. Trawls used in the Gulf are designed to catch primarily the very valuable penaeid shrimps, but they also take large quantities of nontarget species (the bycatch) including crabs and other invertebrates as well as slower-swimming demersal fishes (such as catfishes, members of the porgy and drum families, etc.) and occasional sea turtles.

Like its Atlantic counterpart, the gulf menhaden (*Brevoortia patronus*) is a sardine-like fish that passes its juvenile and subadult life in bays and estuaries. After moving back to the continental shelf the adults feed largely on phytoplankton and zooplankton and form enormous schools in the water column. Valuable oil can be rendered from their bodies, and the remaining flesh is used for cat food, organic fertilizer, and other products. Once the schools are located by onboard sonar, they can be surrounded by large, deep nets called "purse seines." A rope or cable along the bottom serves as a drawstring, and when this is pulled the net forms a pocket. The trapped fishes are then ladled into the hold of the mother ship. As in the case of the shrimp trawls, purse seines capture large numbers of nontarget fishes.

Longline fishing was developed by the Japanese to

harvest larger, free-swimming pelagic fishes of the open ocean. The longline consists of a long rope, often miles in length, suspended near the sea surface by spaced buoys, and from which hang many strong fishing lines with large baited hooks. For several decades longline fishing has been carried out in epipelagic waters of the deep Gulf of Mexico, where the catch includes primarily large sharks, mackerels, bluefin and yellowfin tunas, and billfishes (sailfishes, marlins, and swordfishes). The bycatch is made up of sharks and a few other fish species and occasional leatherback sea turtles.

Gill nets, which are constructed of strong, thin, nearly invisible nylon cord or monofilament line, have fairly large mesh openings. This allows the head of a streamlined fish to pass through, but not the body and fins. Nor can a fish back out because cords of the net get caught inside the gill cover. Once trapped, the fish struggles until it becomes exhausted and eventually dies or is extracted by the fisherman. Gill nets are both cruel and extremely efficient. Some oceanic gill nets stretch a mile or more in length, and if lost or unattended, they are free to float around surface waters of the sea for months or years, trapping various fishes and sometimes entangling dolphins and whales. Fortunately, gill nets are now banned from US Gulf of Mexico waters.

For at least two centuries commercial fishermen have visited natural offshore reefs, hard banks, and other structures to harvest snappers, groupers, and other reef-related species by hook-and-line. This practice has increased greatly during the past few decades, especially in response to the emplacement of artificial reefs and the proliferation of oil and gas rigs and platforms, which also attract reef-related as well as several types of pelagic fishes.

### COMMERCIAL FISHERY LANDINGS

Today the combined commercial fishery of the US portion of the Gulf brings in around 1.4 billion pounds of fishes and shellfishes annually and accounts for about 20 percent of the total US commercial fishery landings. In terms of dollar value, the shrimp fishery is the most valuable fishery of the US Gulf, and in 1999 this harvest accounted for 78 percent of the entire nation's shrimp landings. However, the largest catch by weight is brought in by the purse seine–based menhaden fishery. In addition, Gulf estuaries produce about 66 percent of the nation's annual oyster harvest. The total value of the commercial fishery landings of the Gulf has been increasing steadily and by

2003 accounted for $652 million (C. M. Adams, Hernandez, and Cato 2005). In terms of both poundage and dollar value of the landings, around 90 percent of the US Gulf fisheries are based on species that are estuary dependent during some stage of their lives. In the past few decades this percentage has been decreasing slowly as fishermen have shifted more attention to reef-related species. In 2003 the value distribution of the commercial fishery catch around the northern Gulf, given by state where the landings occurred and expressed as a percentage of the total value, was as follows: Louisiana (42.7%), Texas (24.4%), Florida (20.5%), Mississippi (6.7%), and Alabama (5.7%). Mexico has a large and valuable shrimp fishery as well as fisheries for mollusks, reef-related, and estuarine species, but figures on the poundage and dollar value of these landings are not readily available.

### IMPACTS OF THE COMMERCIAL FISHERY

The annual harvest of 2.5 billion pounds of fishes and shellfishes from Gulf waters by the several fisheries discussed above affects the Gulf in several ways, and since the fishing effort has been increasing steadily, the impacts are also on the rise. These impacts fall into three basic categories: (1) effects on the environment, (2) effects on target species, and (3) effects on nontarget species and the ecosystems in general.

#### Environmental effects

Clearly, the greatest environmental damage results from shrimp trawls. As these are dragged across the same bottoms over and over again, they remove all irregularities (small hills, rubble piles, small rocky outcrops, etc.), thus simplifying and homogenizing the sea-bottom habitat and removing refuges for juvenile fishes and smaller invertebrates. From the soft bottoms the trawls raise vast clouds of silt, which contribute greatly to the nepheloid layer. Much of this suspended material ultimately settles and coats rocky and sandy bottom habitats with layers of silt, further reducing bottom habitat heterogeneity. Although the trawls raise nutrients (nitrogen and phosphorus) into the water column, they also release various chemical pollutants otherwise safely locked up in the reduced layer of the bottom sediments. Hook-and-line fishing, especially around natural reefs and outcrops, inevitably results in some anchor damage to the bottom and reefs, and the loss of fishing lines and other gear further degrades the reef habitats. Purse seining and longlining do not appear to materially affect the environment, but all the commercial boat activity increases the level of background noise, and lost and discarded solid debris adds to the bottom clutter.

*Effects on target species*

Impacts on target species vary from one fishery to another, and they can best be assessed by periodic monitoring of the individual wild fishery stocks. In its 2001 report to the US Congress on the status of the nation's marine fishery stocks, the National Marine Fisheries Service (NMFS) provided information on about 75 target species of finfishes and invertebrates of special interest found in waters of the Gulf of Mexico under US jurisdiction, and it classified the species into three general groups: (1) not overfished, (2) overfished or approaching this condition, and (3) status unknown. Only 9 species were listed as still being in good condition (not overfished). These included the brown, pink, and white shrimps; spiny and smoothtail lobsters; and stone crab, dolphinfish, greater amberjack, and Spanish mackerel. The second group (overfished or approaching this condition) included 8 species: red and vermilion snappers; goliath, red, and Nassau groupers; and gag, red drum, and king mackerels. The remaining 50 or so species (i.e., about 75 percent of the reported species) were mostly reef related, and even though these are supposed to be managed fisheries, the status of these species was listed as "unknown."

Interestingly, the three penaeid shrimp species that are so heavily harvested appear not to be overfished. The secret of their sustainability is that these estuary-dependent species have very large population sizes, high rates of reproduction, rapid growth, and short life histories (ca. 18–24 months), and the populations are at least partially protected until they reach commercial size. This provides an opportunity for most to reproduce before being subjected to the full pressure of the commercial shrimp fleet. Although there is no evidence that the harvest of the shrimps and other species in this category has damaged the populations, this does not mean there has been no effect. Absence of evidence does not constitute proof. Rather, the matter simply has not been investigated in detail. The long-range genetic consequences of such heavy "predation pressure" are totally unknown, as are the potential and possibly subtle effects on other species and the ecosystem.

The menhaden populations appear not to be overfished at present, but whether or not they are approaching this condition is unknown. In order for many schooling species to complete the spawning process, there must be a critical population density. This lowest level could be reached without prior warning, leading to the complete collapse of the population and the fishery.

*Effects on nontarget species*

Effects on nontarget species occur primarily through the bycatch of the commercial shrimp fishery. Formerly this bycatch was quite large. It included sea turtles and 10 or more pounds of nontarget species for every pound of commercially valuable shrimps. In the past decade the bycatch has been reduced to about 4:1, and this has been brought about by the federally mandated use of turtle excluder devices (TEDs) and fish excluder devices (FEDs). These devices attached inside the trawl nets have bars that shunt sea turtles and larger fishes out through a hole in the top of the net while still retaining most of the valuable shrimp catch. This is a great improvement, but since the annual shrimp landings exceed 50 million pounds, at least 200 million pounds of other invertebrates and fishes are trashed each year. The bycatch is still far too high.

Darnell, Defenbaugh, and Moore (1983) analyzed the composition and seasonal distribution patterns of the trawl catch (including target species and bycatch) in the northwestern Gulf, and Darnell and Kleypas (1987) carried out a similar analysis for the northeastern and eastern Gulf. For the northwestern Gulf shelf, between the Rio Grande and the Mississippi River Delta, the catch consisted of 12 species of penaeid shrimps and 164 species of fishes. For the northeastern and eastern Gulf shelf there were 17 species of penaeid shrimps and 347 species of fishes. Most of the abundant species of the northern Gulf are estuary dependent and short lived (Atlantic croaker, spot, seatrout, etc.), or not estuary dependent but still short lived (hardhead catfish, longspine porgy, etc.). These appear able to withstand the heavy fishing pressure, but effects on populations of the other species are largely unknown.

The bycatch of the purse seine fishery represents only about 3 percent by weight of the total menhaden catch, but the latter amounts to 1.5 billion pounds annually. So the nontarget catch of this fishery runs around 45 million pounds per year, which is also a considerable waste. Longlines capture a variety of oceanic sharks and other species, all of which are essentially target species. But the fishery also takes numbers of the highly endangered leatherback sea turtle, which is a matter of great concern. Hook-and-line fishermen take mostly reef-related and some larger pelagic species (blackfin tunas, mackerels, jacks, etc.), but effects on nontarget species are largely unknown.

Long-term impacts of the various commercial fishing activities on the environments and ecosystems of the Gulf have not been investigated in detail, but some observations are possible. As we have already mentioned, the shrimp trawls reduce habitat diversity over vast stretches of the terrigenous-bottom continental shelves, and the hook-and-line fishery tends to degrade and destroy portions of the reef and hard-bank habitats.

At the population level, the heavy and persistent harvest will definitely induce major changes in populations of the target species and those making up the bycatch. The primary result will be a great reduction in genetic variability of the wild populations and consequent diminution in their ability to adapt to environmental changes. Other likely consequences will be increased growth rates (as the young no longer have to compete with adults) and attainment of sexual maturity at an earlier age (as the catch removes adults that spawn at later ages).

At the ecosystem level, we see truncation of marine food pyramids. As the hook-and-line fishery removes the top-level predators (sharks, mackerels, tunas, billfishes, etc.), it then moves down and targets the next lower level of carnivores (cobias, dolphinfishes, jacks, snappers, groupers, etc.). Around the world, removal of the predatory groups leads to proliferation of lower-trophic-level, often "weedy," species, a process referred to as "fishing down the food web." By any measure, the very heavy and persistent commercial fishing pressure leads to a great reduction in biodiversity in terms of genetic makeup, species composition, and the complexity of species interactions. There must also be specific adverse effects on sea turtle and marine mammal populations in terms of environmental disturbance and competition for food resources. The great harvest comes at a great environmental and ecological cost.

RECREATIONAL FISHING

Recreational fishing is a very important activity, which directly or indirectly adds an estimated $17 billion annually to the economy of the Gulf coastal states. According to the records, in 1998 recreational fishermen made about 16 million trips to the Gulf and caught 12.5 million fishes totaling 204 metric tons (nearly half a million pounds). Most of these fishermen use hook-and-line. Many fish from the shoreline or from piers or jetties, whereas others angle from party boats or private craft. The heaviest fishing occurs in bays and estuaries, but much recreational fishing also takes place offshore, particularly around oil rigs and platforms. Still other fishermen engage in offshore trolling for pelagic species. All told, the recreational catch involves a wide variety of species, a large percentage of which are estuary dependent. The total recreational harvest is only a small fraction of that taken by the commercial fishery, and impacts on the environment and fish populations are, for the most part, minimal. However, the catch of two species, the red drum and red snapper, is of some concern. These two species are under pressure from both the recreational and commercial fisheries as well as other factors.

The red drum, often referred to as "redfish," is a central ingredient in Cajun cooking, as it is used in "blackened redfish," court bouillon, and other dishes. It is much sought after as a target species by both commercial and recreational fishermen. Although an estuary-dependent member of the drum family, it is a relatively long-lived species that does not reproduce until at least its second or third year of life, and therefore it can be taken prior to reproduction. The young sometimes appear as bycatch in shrimp trawls. Since the species is now under pressure from overfishing, some states have been operating captive breeding and stocking programs in which a few parent individuals become responsible for producing a large percentage of the offspring of the next generation. As mentioned earlier, this is a form of inbreeding, which in the long run is bound to lead to a reduction in genetic diversity and ecological adaptability in wild populations. However, the red drum populations of the northwestern Gulf have been placed at even greater risk by the liquid natural gas (LNG) industry. Arriving in northern Gulf coastal waters from overseas, the LNG-laden tankers have to warm their cargoes so that the liquid can be converted to a gaseous state for transport through the pipelines. Here the engineering strategy is simply to pump hundreds of millions of gallons of warm nearshore Gulf water through pipes in the LNG to warm it up. Unfortunately, this simple solution certainly entrains and kills hundreds of millions of larvae and young individuals of various marine animals, most particularly those of the red drum, whose populations are already under great pressure from the fishery activities. Probably the simplest, although not the cheapest, solution would be for the warming to take place well offshore, beyond the area of greatest red drum larval density, and the natural gas could then be piped ashore from terminals located on the outer continental shelf.

The red snapper is another special case. It is a long-lived reef-related species whose young remain widely distributed around the open continental shelf before locating and finally settling down on a particular outcrop or reef. Like the red drum, this fish is highly sought after by both commercial and recreational fishermen, and the pressure on the species from this source is considerable. However, the real problem is that young prereproductive snappers are heavily represented in the shrimp trawl bycatch, and it has been estimated that the species will never recover to really sustainable levels until the bycatch mortality has been substantially reduced.

### POLLUTION-RELATED ISSUES

In the previous section we examined several types of human activities that affect the Gulf and discussed the major impacts of each. Here we focus specifically on environmental pollution, which, as we have seen, may be derived from a variety of human activities. The concept of pollution is interpreted broadly as any substance, factor, or agent added intentionally or unintentionally by human activity that degrades or reduces the quality of the environment. Included are (1) a host of chemical substances, (2) agents or factors that adversely modify the physical environment, and (3) biological agents such as parasites, disease organisms, and nuisance species. It should be noted that courts have upheld the contention that under terms of the Clean Water Act undesirable nonnative species, when introduced into US waters, constitute "biological pollution."

## Chemical pollution

A great deal of information is now available concerning the presence and concentrations of various chemical pollutants in waters, sediments, and living organisms of the US Gulf bays and estuaries and continental shelves. Less knowledge is on hand for coastal and shelf environments of the southern Gulf and the deep Gulf, in general. These data have been accumulated through numerous local studies as well as several broad-scale and long-term investigations carried out under the auspices of the National Oceanic and Atmospheric Administration (NOAA) and the US Environmental Protection Agency (EPA). Chemical pollutants may be thought of as falling into one of three categories: (1) those whose primary effect is to reduce the level of oxygen in the water, (2) those nutri-

ent chemicals that fertilize the waters and stimulate the growth of phytoplankton but that lead to eutrophication when present in excess quantity, and (3) those chemicals that are essentially toxic or carcinogenic (cancer causing), or that lead to metabolic disorders such as lowered fertility, partial or complete collapse of the immune system, or chronic stress. The complexity of the chemical pollution issue is illustrated by reference to a listing of the specific chemical pollutants (and related parameters) regularly monitored in bays and estuaries of the US Gulf coast by NOAA (table 17.4).

### OXYGEN-CONSUMING COMPOUNDS

Organic (carbon-containing) compounds such as sugars, amino acids, starches, fats, proteins, and a host of other substances exist in a chemically reduced state. Microorganisms including bacteria, yeasts, and fungi derive energy by breaking down these compounds to carbon dioxide, water, and other simple substances, in the process consuming a great deal of oxygen. Thus, it is convenient to think of these readily oxidizable compounds, when introduced into natural aquatic systems, as having a high biological oxygen demand (BOD). Garbage, untreated or poorly treated human sewage, and animal wastes all have a high BOD, and these are the primary factors leading to reduced or depleted oxygen resources in the bottom sediments and water columns of bays and estuaries of the Gulf coast. Many petroleum-derived and other industrial organic chemicals also have a high BOD. Certain inorganic chemicals such as some metals and sulfides exhibit a high chemical oxygen demand (COD). In most cases the oxygen-demanding material enters the water as particulate matter, and this tends to accumulate in muddy bottoms of dredge holes and channels and other areas where the water circulation is poor. Depleted oxygen levels resulting from the unregulated or poorly regulated dumping of such materials into coastal waters severely reduces the quality of the environment for aquatic life and decreases the ability of the waters to provide oysters, shrimps, crabs, fishes, and other products useful to humans. Most of the bays and estuaries of the US Gulf coast have, at one time or another, developed hypoxic or anoxic areas. Hypoxia resulting from the dumping of high BOD or COD chemicals is not known to have occurred anywhere on the continental shelves or in other open Gulf environments. Although oxygen-consuming chemicals are released in quantity during petroleum recovery operations, hypoxia is averted because of high ini-

**Table 17.4.** *Chemical pollutants regularly monitored by NOAA's National Status and Trends Program.*

Major elements – aluminum, iron, manganese, silicon

Trace elements – antimony, arsenic, cadmium, chromium, copper, lead, mercury, nickel, selenium, silver, tin, zinc

Polychlorinated biphenyls (PCBs) – 21 types

Mono-, di-, tri-, and tetra-butyltins

Chlorinated pesticides – DDT (and its metabolites), aldrin, dieldrin, chlordanes, heptachlor, heptachlor epoxide, lindanes, hexachlorobenzenes, tetrachlorobenzenes, mirex, endrin, endosulfans, chlorpyrifos, pentachloroanisole, chlorinated dioxins, chlorinated dibenzofurans

Polycyclic aromatic hydrocarbons (PAHs)

  2-ring – diphenyl, naphthalene, 1-methylnaphthalene, 2-methylnaphthalene, 2, 6-dimethylnaphthalene, 1, 6, 7-trimethylnaphthalene, $C_1 - C_4$ alkyl naphthalenes

  3-ring – fluorene, phenanthrene, 1-methylphenanthrene, anthracene, acenaphthylene, acenaphthene, dibenzothiophene, $C_1 - C_3$ alkylfluorenes, $C_1 - C_4$ alkylphenanthrenes, $C_1 - C_3$ alkyldibenzothiophenes

  4-ring – fluoranthene, pyrene, bena(a)anthracene, chrysene, $C_1 - C_4$ alkylchrysenes

  5-ring – benzo(a)pyrene, benzo(e)pyrene, perylene, dibenz(ah)anthracene, benzo(b)fluoranthene, benzo(k)fluoranthene

  6-ring – benzo(ghi)perylene, indeno (1, 2, 3 cd) pyrene

Related parameters – sediment grain size, sediment toxicity, total organic carbon, lipid, salinity, temperature, conductivity, dissolved oxygen, *Clostridium perfringens* spores

---

tial oxygen levels in the water column and good water circulation leading to dilution, mixing, and wide dispersion of the chemicals.

### NUTRIENT CHEMICALS

Nutrient chemicals are primarily those substances, rich in the elements nitrogen and phosphorus, that fertilize natural waters and stimulate the growth of phytoplankton, attached algae, and rooted vegetation. Additionally, the element silicon, when present as silicate, aids in the production of diatoms. Nutrient chemicals enter coastal waters as garbage, human sewage, animal wastes, and some industrial effluents, but a great deal also represent fertilizer added to upstream farm fields, which is picked up by runoff waters and transported to the coast by streams. When present in low to moderate concentrations the nutrients stimulate development of a wide variety of phytoplankton species, including diatoms, but often the concentrations of nitrogen and phosphorus greatly exceed that of silicate. In this event, diatom production lags, and other groups, particularly blue-green algae and dinoflagellates, become the dominant forms. If the rate of nutrient input is quite high, the system undergoes a process of overproduction known as eutrophication, and hypoxia or anoxia may result.

These effects are illustrated in figure 17.2. The upper diagram shows the flow of organic carbon through a natural aquatic system receiving low levels of nutrients. Processes of the system are essentially in balance, and each succeeding trophic level is able to handle the organic material passing to it without major bottlenecks or accumulations. Since the whole process proceeds at a relatively slow rate, mineralization of the organic material creates only a modest oxygen demand. The bottom diagram shows the same system receiving a high rate of nutrient input. Here the phytoplankton and benthic plants are stimulated to very high production levels, accumulating much more organic material than the zooplankton and benthic animals can process. Thus, a great deal of organic detritus, derived mostly from the massive phytoplankton bloom, accumulates in the bottom sediments for mineralization. As the microbes process this large amount of organic material, there arises a very high oxygen demand. This may lead to hypoxia or anoxia and to the release into the water column of unoxidized organic and other chemicals such as methane, ammonia, nitrites, hydrogen sulfide, organic acids, and so forth. Such an environment is hostile to most higher forms of life, and from the human standpoint it is a highly undesirable condition.

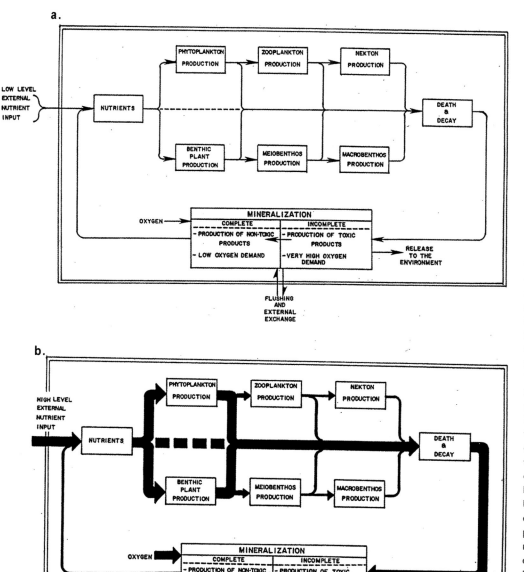

**Figure 17.2.**
The effects of low and high levels of nutrient loading on the flow of organic carbon through the various trophic levels of aquatic systems. a. A system receiving low levels of nutrient input. As the material flows through the system the organisms at each trophic level are able to handle the load without bottlenecks or accumulations. Mineralization proceeds with only modest oxygen demands. b. A system receiving high levels of nutrient input. The widths of the arrows indicate increases in the magnitude of flow through each pathway. The excess nutrients stimulate production overloads at each level, which results in great accumulation of organic material in the sediments. This creates an extremely high oxygen demand, leading ultimately to oxygen depletion. Toxic products of incomplete mineralization of the organic material rise into the water column, and the natural aerobic system collapses. (From Darnell and Soniat 1981.)

TOXIC, CARCINOGENIC, AND STRESSFUL CHEMICALS
Toxic, carcinogenic, and other metabolically stressful chemicals are derived from a variety of agricultural, municipal, and industrial sources, and they include a large series of organic compounds, a number of heavy metals, and a wide spectrum of industrial chemicals. Runoff from agricultural farm fields often contains high concentrations of insecticides and herbicides, mostly organic compounds, some with attached chlorine atoms, and others combined with heavy metals. Brought by streams to coastal waters, these biocides are lethal to a wide variety of marine organisms. Runoff from city streets may contain oils, greases, lead, and other substances derived from motor vehicles. Municipal trash, garbage, and sewage that enter coastal waters also contain many toxic and other undesirable chemicals.

ORGANIC COMPOUNDS

As a group, industrial and manufacturing wastes release into natural waters a bewildering array of chemical types including, among others, various organic chemicals and heavy metals. Organics include the less toxic aliphatic (carbon chain) compounds and the more toxic and often carcinogenic aromatic (carbon ring) compounds. Some of the carcinogenic forms, such as some naphthalenes, also occur in crude petroleum, and when these chemicals are released into the environment they induce tumor growth in aquatic animals. There is also a significant health risk for humans who consume seafood such as raw or cooked oysters laced with such chemical compounds. *Pyrogenic* organic compounds result from the incomplete combustion of organic materials by motor vehicles, power plants, and so forth, and some of these likewise may be toxic, carcinogenic, or otherwise stressful to marine life. Some of the organic compounds may be similar to natural animal hormones. Upon entering the body they have the potential for interfering with normal metabolic pathways, inhibiting reproduction, suppressing the immune system, and disrupting the chemical pathways of the body. The net result is less reproduction, stress, and often death.

In warm waters of the Gulf of Mexico low-molecular-weight organic compounds are quite volatile and are generally subject to fairly rapid evaporative loss. Included in this group are some fairly toxic components of petroleum. Background levels are generally low (less than 10 ng/L), with concentrations in sediments several orders of magnitude higher than in the water. Near chemical plant outfalls and oil platforms, concentrations are sometimes much higher (over 1,000 ng/L). Environmental concentrations decrease with distance from the source and with distance from shore. Less is known about concentrations in the biota.

Polyaromatic hydrocarbons include some very dangerous (carcinogenic) aromatic compounds. Concentrations vary greatly from one area to another, and they are highest near known sites of industrial chemical contamination. Concentrations in sediments range from less than 5 to 36,700 ppb, and in oysters they occur in concentrations from less than 20 to 18,620 ppb. The "hot spots" are Galveston Bay of Texas and Barataria Bay of Louisiana.

High-molecular-weight hydrocarbons are environmentally persistent, but most are not considered to be dangerous. However, as mentioned earlier, the naphthalenes and their derivatives are a major cause of concern. High-molecular-weight hydrocarbons tend to show high concentrations in bays and coastal lagoons as well as in sediments and organisms along the shelf of upper Texas, Louisiana, Mississippi, and Alabama, where tissue concentrations may range up to 190 µg/g. Concentrations tend to be low in sediments and organisms off south Texas, on the inner shelf of Florida, and in deeper water of the outer shelf generally. Naphthalene and its derivatives have been found in organisms around oil and gas platforms near the Mississippi River Delta.

Organochlorine compounds (chlorinated hydrocarbons), so dangerous to marine life, also become concentrated in seafood products used as human food. The most common forms found in organisms of the northern Gulf are DDT metabolites, PCBs (polychlorinated biphenyls), and dieldrin, although a variety of pesticides (toxaphene, methyl and ethyl parathion, carbophenothion, ethion, dacthal, and mirex) are also known. Concentrations of PCBs are highest in estuarine sediments and organisms found near industrial sites. Pesticides are most concentrated in estuarine sediments and organisms downstream from agricultural areas, and they are particularly high in Matagorda and Galveston Bays. Chlorinated hydrocarbons occur in high concentrations in Mississippi River sediments, and they are present in low concentrations throughout the estuarine and marine food chains.

TRACE METALS

Trace metals tend to be highest in estuaries near industrial sites, but they are sometimes abundant on the shelf, as well. The upper extension of Matagorda Bay (Lavaca Bay) of central Texas is noted for the high concentrations of mercury in its sediments as well as in blue crabs and other organisms. Some metals pass up through the food chains, where they become more concentrated, and in general, amounts in animal tissues exceed those in sediments. Oysters taken from the vicinity of industrial sites have high concentrations of many metals. Even on the continental shelf, concentrations of chromium, cadmium, and iron are considerably higher in tissues of red snappers than they are in the sediments. For example, chromium has a sediment concentration of 0.03 µg/g dry weight, whereas it is 0.72 µg/g dry weight in snapper tissue, a concentration factor of 24.

From this brief review we can see that some aspects of chemical pollution in the northern Gulf are causes for concern. The major problems tend to be located in bays and estuaries near industrial sites or downstream from agricultural areas. Other major sources of contamination include the Mississippi River and oil and gas platforms. Some pollutants become concentrated as they move up through food chains and are of concern as contaminants of seafood.

## Physical pollution

Pollution of the physical environment is brought about by the accumulation of semisolid and solid debris, the modification of physical properties of the water column (such as temperature), and the production of annoying or destructive underwater sounds or explosions. Humans are a messy lot. The debris that regularly accumulates along Gulf beaches attests to the littering that goes on at sea, and what is actually observed is by and large only the portion that floats. This flotsam and jetsam is generated by human beings involved in international ship traffic, by petroleum rigs and platforms and the associated boat and ship traffic, and by both commercial and recreational fishermen. Specific issues of the physical pollution problem have already been addressed in sections dealing with various human activities. Of particular concern is the accumulation of great quantities of nonbiodegradable plastic materials in the marine environment. Smaller particles interfere with the feeding of many marine animals, particularly the filter feeders and some marine birds. Plastic bags and other floating bubble-like materials resemble gelatinous prey, and when sea turtles or marine mammals consume these, a choking death results. Plastic rings (such as those from six-packs) and discarded fishing lines and nets may entangle and trap a variety of marine animals from fishes to birds. The list goes on and on. To these are added structures deliberately placed on the seafloor to act as fish-attracting devices (FADs) or artificial reefs and, of course, the oil rigs and platforms placed in special "graveyards" under the auspices of the rigs to reefs program. For better or worse, these large structures provide solid substrate habitat, often where little or none was present before. They attract and concentrate a foreign biota and therefore modify the original ecosystem. Despite whatever good they may do by creating fish habitat, they also provide substrates for attachment of newly introduced pest species.

Located near bays and estuaries of the Gulf coast, some power plants and industrial operations take in large volumes of water for cooling, desalinization, chemical recovery, and other purposes, and they return the water in modified form. In some cases the water is heated. In other instances it may be modified chemically, as to a high-salinity brine. This modified effluent generally reduces the quality of the habitat for numerous aquatic species. Warming of surface waters of the entire Gulf through the accumulation of atmospheric greenhouse gases will be discussed as a separate topic below.

Sound pollution in coastal and oceanic waters of the Gulf has also been discussed earlier. We actually have no idea of what effects the sounds made by routine boat and ship traffic may have on most marine animals, but without any supporting data it is widely assumed that the animals get used to it and eventually ignore it. Explosive sounds derived from seismic guns and actual detonations (used to sever the legs of free-standing petroleum platforms) create sudden, intense pressure waves that can kill nearby animals and startle or injure marine creatures at a distance. Since the endangered sea turtles and marine mammals are susceptible to such effects, special efforts are now being made to ensure that these animals are absent from the area before the explosions are set off. The problem of explosive sounds from seismic work has not yet been resolved. However, it is noted that by law marine mammals are not to be harassed, and their natural behavior is not to be modified in any way. The effects of certain other sounds on the behavior of marine mammals in the Gulf are currently under study.

## Biological pollution

Through the years human activities have led to the rise of dozens of noxious species in coastal and marine habitats around the Gulf of Mexico. In a few instances these represent native species essentially out of control. In other cases nonnative species introduced from around the world have settled into their new habitats to the detriment of the native species and humans. Four basic mechanisms are involved: (1) human modification of natural environmental factors, (2) discharge of pathogens in human sewage or animal wastes, (3) accidental or deliberate release of foreign organisms from cultivation or captivity, and (4) incidental species transport by foreign ship traffic.

ENVIRONMENTAL MODIFICATION
In the first case, humans modify the environment in such a way that desirable species are placed at a competitive disadvantage, allowing undesirable species to proliferate. A classic case of this involved duck farms on Long Island in which the overcrowded ducks loaded the estuary with high concentrations of nitrogen-rich organic excrement. This in turn provided toxic dinoflagellates with

a preferred nutrient and led to explosive outbreaks. Consumption of shellfish from contaminated areas resulted in many human illnesses and some deaths. Related dinoflagellates reside quietly in estuaries bordering the Gulf of Mexico, and there is concern that such outbreaks could occur there.

## RELEASE OF PATHOGENS

The release of untreated or inadequately treated human sewage from municipal systems or poorly maintained septic systems and the drainage of animal wastes from farm fields and stockyards inevitably run the risk of releasing human and animal pathogens into coastal waters. These include various fecal coliform bacteria as well as the bacterial agents for typhoid fever, cholera, salmonellosis, and other types of human gastroenteritis diseases. Also involved may be the agents for hepatitis A and other viral diseases. Public health officials carefully monitor the coastal waters, and when evidence of such pollution is encountered, they order the closure of all affected shellfish-harvesting areas. Unfortunately, this in itself does not guarantee protection of human health because bay shell fishermen are notorious for disregarding the regulations, and they collect and market shellfish from contaminated areas anyway. There is also concern regarding the effects of these and other pathogens on nonhuman species inhabiting the bays. Manatees die unexpectedly. Bottlenose dolphins appear to have succumbed to a distemper-like virus. Sea turtles develop skin papillomas. Little is known about possible effects on the fish and invertebrate populations. Although the human connections are still tenuous, such conditions in the Gulf and elsewhere are on the rise, and it is suspected that somehow we must be at least partially responsible.

## RELEASE OF ALIEN SPECIES

Americans import plants and animals from around the world as ornamentals for gardens, fish ponds, home aquaria, household pets, and as stock for commercial aquaculture ventures. After the plants and animals have served their decorative or otherwise useful purposes, they are sometimes released into the local environments by caring and well-meaning citizens. More often the alien organisms simply escape from captivity and become established in the wild. Additional species are brought in as biofouling organisms on the hulls of foreign ships or as "stowaways" in the ballast tanks. Oceangoing vessels regularly take in ballast water when they leave one port

and then discharge this water—and its contents—prior to entering their destination port. This load of water serves to stabilize the ship during its sea voyage. Whether they are released or have escaped from captivity or whether they are vagrant biofoulers or stowaways, many foreign species have been introduced into coastal and open marine waters of the Gulf of Mexico, and a list of those known to have become established here is provided in table 17.5.

## INTRODUCTION OF ALIEN SPECIES
## BY SEAGOING SHIPS

Beginning with the arrival of the wharf roach and gribble (isopods) on old wooden-hulled sailing vessels, through the years 2 species of algae, 7 species of vascular plants, and 18 species of animals are known to have been introduced to the Gulf from various parts of the world. The most recent introductions are the brown and green mussels (*Perna perna* and *P. viridis*) and the Australian spotted jellyfish (*Phyllorhiza punctata*). Introduced vascular plants have taken over many coastal waters, replacing native species, clogging navigable waterways, and requiring millions of dollars in cleanup and control activities. The introduced mollusks clog water intake pipes of power plants and other facilities, also necessitating costly cleanup and control operations. The spotted jellyfish, which may grow to a diameter of over two feet, is a voracious predator on zooplankton and micronekton, and it is capable of devastating coastal populations of larval fishes. The nutria, an aggressive and very fertile rodent from South America, has taken over much of the habitat formerly occupied by the native muskrat, and it has spread into and devastated many freshwater and brackish-water marshlands of the Gulf coast. However, the invasive pests with which we now contend are just the "tip of the potential iceberg." Within the Gulf of Mexico we simply do not need the deadly "sea-wasp" jellyfish of Australian waters. Nor do we need any of the highly venomous sea snakes of the Indo-Pacific region or the crown-of-thorns starfish that has decimated Pacific coral reefs. The list of undesirables is long. Potential invaders from around the world could seriously damage local ecosystems, endanger human lives, and/or occasion millions of dollars worth of damage and control operations.

Of immediate concern are four types of shrimp viruses currently at large in shrimp mariculture operations within the United States. Two of the viruses (infectious hypodermal and hematopoietic necrosis virus, or IHHNV, and

**Table 17.5.** *Invasive and introduced species now found in the Gulf of Mexico or adjacent bays and estuaries.*

| Common name | Scientific name | Group | Native range |
|---|---|---|---|
| **Plants** | | | |
| — | *Pseudonitzschia australis* | Diatom | Australasia |
| Caulerpa | *Caulerpa taxifolia* | Alga | South America, Africa, Asia |
| Alligatorweed | *Alternanthera philoxeroides* | Higher plant | South America |
| Asian marshweed | *Limophila sessiliflora* | Higher plant | Southeast Asia |
| Melaleuca | *Melaleuca quinquenervia* | Higher plant | Australia, Indian Ocean |
| Parrot feather | *Myriophyllum aquaticum* | Higher plant | South America |
| Water lettuce | *Pistia stratiotes* | Higher plant | ? |
| Curly-leaf pondweed | *Potamogeton crispus* | Higher plant | Eurasia, Africa, Australia |
| Japanese eelgrass | *Zostera japonica* | Higher plant | Asia |
| **Animals** | | | |
| Red beard sponge | *Microciona prolifera* | Sponge | Northwest Atlantic |
| Australian spotted jellyfish | *Phyllorhiza punctata* | Cnidarian | Australia |
| Soft-shelled clam | *Mya arenaria* | Mollusk | Northwest Atlantic |
| European melampus | *Mysosotella mysolis* | Mollusk | Europe, Mediterranean |
| Brown mussel | *Perna perna* | Mollusk | South America, Africa |
| Green mussel | *Perna viridis* | Mollusk | Indo-Pacific |
| False limpet | *Siphonaria pectinata* | Mollusk | Mediterranean (?) |
| Quilted melania | *Tarebia granifera* | Mollusk | Southeast Asia, Hawaii |
| Wharf roach | *Ligia exotica* | Isopod | North Atlantic, Mediterranean |
| Gribble | *Limnoria tripunctata* | Isopod | North Atlantic |
| — | *Ampithoe valida* | Amphipod | North Atlantic |
| — | *Corophium acherusicum* | Amphipod | Europe, West Africa |
| — | *Corophium insidiosum* | Amphipod | North Atlantic |
| — | *Melita nitida* | Amphipod | Northwest Atlantic |
| Striped barnacle | *Balanus amphitrite* | Barnacle | Indo-Pacific |
| Kuruma prawn | *Penaeus japonicus* | Shrimp | Indo-Pacific |
| Chinese mitten crab | *Eriocheir sinensis* | Crab | Southeast Asia |
| Nutria | *Myocastor coypu* | Rodent | South America |

Taura syndrome virus, or TSV) are from Latin America, and the other two (white spot syndrome virus, or WSV, and yellow head virus, or YHV) originated in Asia. All of these viruses cause 90–100 percent mortality rates in US mariculture shrimp populations, and they also affect crabs and other crustaceans. So far, these viruses have not been found in wild shrimp populations of the Gulf, but they could be spread by waters from seafood processing plants, escapes from mariculture ponds, infected shrimp used for fish bait, wastes released from shrimping boats, or possibly on the feet of birds. The threat increases as we import ever greater quantities of foreign-raised shrimps.

Clearly, the potential exists for devastating viral outbreaks in our native wild shrimp populations. In the face of these and other biological pollution threats, efforts are now underway to prevent the release of noxious invasive pathogens and other pests from mariculture populations and to control the release of ballast water in American coastal waters by vessels arriving from foreign ports.

## Atmospheric pollution and global warming

Little published information is available concerning air quality over the Gulf of Mexico in terms of particulates, aerosols, and noxious gases such as nitrous oxide. However, it does seem clear that at least periodically air quality over the northern Gulf must be quite poor. Atmospheric pollutants from industrial areas near Monterrey, Mexico, have been shown to reduce air quality in the Big Bend area of Texas. In addition, the sprawling Houston metropolitan area and the Houston-Galveston corridor with its many petrochemical plants and other industrial facilities are known to generate large volumes of atmospheric pollutants. Winds from the north and northwest must, from time to time, sweep this contaminated air over the Gulf, allowing the particulates and aerosols to enter and pollute Gulf waters.

During the past few decades much attention has been given to the matter of greenhouse gases and global warming, and specific information is available concerning the effects of global warming on the Gulf of Mexico region. Since the beginning of the industrial revolution about two centuries ago, the concentration of carbon dioxide in the earth's atmosphere has risen by more than 30 percent, reaching its highest level in almost half a million years. By trapping atmospheric heat, carbon dioxide and related gases have contributed to the increase in global temperature (a rise of 0.7°–1.4°F since 1900), with the highest temperatures occurring in the past two decades. The trend in global warming is accelerating, and it is predicted to continue during the next century. From historic climatic records and computer models, projections can be made concerning such phenomena as the rate of global warming, changes in the oceanic thermohaline circulation system, sea level rise, precipitation patterns, and intensity of tropical cyclones. Here we focus on predictions for the Gulf of Mexico region and their potential physical and biological consequences.

Potential changes in major environmental parameters of the Gulf of Mexico during the next century are summarized in table 17.6. According to the models, both atmospheric and surface water temperatures will definitely increase. The exact amounts will vary depending on the location above or within the Gulf and local circumstances. Precipitation patterns around the earth will change, but many of the details are still unclear. It appears that less moisture will fall in Texas and northeastern Mexico and that there will be an increase in precipitation in the Mississippi River watershed. Tropical storms entering the Gulf may or may not increase in frequency, but because of warming surface waters, those that do enter the Gulf will be more severe on average. It is not yet clear whether the

**Table 17.6.** *Predicted environmental changes in the Gulf of Mexico region during the next century as a result of projected global warming scenarios.*

**Atmospheric temperatures**
- Summer high to increase 1.7°–3.9°C (3°–7°F)
- Winter low to increase 2.8°C (5°F) in the east and 5.5°C (10°F) in the west

**Water temperatures**
- Warmer, the amount depending on location

**Sea level rise**
- Absolute rise (globally) = 0.2–0.5 m (8–20 in)
- Relative rise (locally)
  • Along most of the Gulf coast = 0.38 m (15 in)
  • In subsidence areas of the Louisiana coast = 1.1 m (44 in)

**Precipitation**
- In general, more rain in the tropics, less in the subtropics
- Texas and northeastern Mexico will probably receive less precipitation
- The Mississippi River watershed will possibly receive more precipitation

**Tropical storms affecting the Gulf of Mexico**
- Frequency: predictions are ambiguous
- Intensity: a greater percentage of the storms that do affect the Gulf should be intense (i.e., categories 3–5) due to the warmer surface water

**Gulf Loop Current**
- Projections are unclear; could be stronger or weaker

Gulf Loop Current will be stronger or weaker. The effects of these potential changes on the Gulf and its biological systems are discussed below.

Increased surface water temperatures throughout the Gulf and reduction in the severity of winter temperatures along the northern Gulf will modify the ecology in several ways. Warmer surface waters will strengthen stratification, and there will be less vertical mixing of surface and subsurface layers. As a result, upwelling will diminish, with a consequent reduction in primary production in waters of the open Gulf. Many of the corals of the Florida reef tract are already subject to summer heat stress, which often leads to loss of zooxanthellae, bleaching, and death. Even slight increases in shallow water temperature of the eastern Gulf will greatly expand this problem, causing major damage to the remaining corals and eventually to other reef species that depend on them. Around the Gulf many additional species will be placed under thermal stress, leading to weakened resistance to parasites and disease. Particularly affected will be the temperate species and Pleistocene relicts of the northern Gulf. As the surface waters continue to warm, subtropical species such as mangroves, most of which are now limited to the southern Gulf, will spread into the northern Gulf, and various tropical species from the Caribbean should move northward and become established in Gulf waters. Thus, there will be major disruption of the native ecosystems, which should be especially serious in the coral reefs of the eastern Gulf and the coastal systems of the northern Gulf. These systems will undergo species loss and changes in species composition, with a general reduction in stability and biodiversity.

Sea level rise coupled with storm surges and pounding from severe tropical cyclones will lead to extensive flooding of low-lying coastal areas around the Gulf and especially along subsiding coasts such as those of eastern and central Louisiana. Beaches, barrier islands, and other exposed coastal features will be subject to frequent and serious erosion and land loss. Coastal marshlands, mangrove swamps, seagrass meadows, and coral reefs will be devastated by pounding, flooding, and strong currents associated with hurricanes, resulting in loss of coastal wetlands and their ecosystems. The habitats will be further degraded by saltwater intrusion into bays, estuaries, and other low-salinity coastal areas. There is also danger of saltwater intrusion into coastal underground aquifers. Elevated salinities will create conditions favorable for blooms of noxious blue-green algae and dinoflagellates. The overall picture is one of coastal land loss, habitat degradation, ecosystem disruption, and species loss.

Increasingly arid conditions in Texas and northeastern Mexico will convert the coastal scrublands into veritable deserts. Stream flow will be diminished or halted, with reduced freshwater inflow into the bays and lagoons and long-term reduction in aquifer recharge. These factors, along with greater evaporation, will result in even higher salinity levels and wide expansion of hypersaline conditions along this arid coastline. Native aquatic species of the area will be placed under even greater heat and salinity stress, and competition will increase as tropical and subtropical species move up from the south.

The potential northerly shift of the frost line signals a winter warming of Middle America, and a warming in the headwaters of the Mississippi River should result in less snowmelt and a reduction in the spring peak flow. However, precipitation in the drainage basin is predicted to rise, resulting in a 20 percent increase in annual discharge of the Mississippi River. If this is truly the case, then there should be an increase in the amount of nutrients and sediment brought to the Gulf by the river. Whether or not these factors will have a major effect on hypoxic areas off the Louisiana and upper Texas coastline is an open question. Much depends on the seasonal distribution of the discharge.

It is quite clear that global warming will have major impacts on the ocean thermohaline circulation system that transfers surface waters to the ocean bottom and around the globe with a gradual rising back to the surface. Major circulation changes will occur, particularly in the North Atlantic, but precise effects on the western boundary current (Gulf Stream) and the Gulf Loop Current are not yet certain. If the Loop Current is strengthened, there should be an increase in the flushing of surface as well as deeper waters of the Gulf, and that could be beneficial. However, if the current becomes weaker, a number of adverse consequences could arise. In this event, the surface currents would be reduced, particularly in the western Gulf. Rings, eddies, and filaments shed by the Loop Current would be reduced in frequency and intensity (rotational speed, etc.). There would be less flushing of the continental shelves and less oxygenation of surface waters of the western Gulf. Upwelling and primary production would be reduced. There would be reduced flushing and oxygenation of the deep Gulf, and the residence time of deep Gulf basin waters would be lengthened. As a consequence of reduced flushing, chemical pollutants would tend to accumulate on the continental shelf and in the deep Gulf. The life histories of many marine plants and animals would be disrupted as the weaker surface currents failed to provide for effective larval transport. In the long range there would be less genetic transfer among isolated populations and slower recolonization of damaged or devastated areas. The degree to which these physical and biological consequences actually took place would depend largely on the extent of the weakening of the Gulf Loop Current.

Among the world's large marine ecosystems the Gulf of Mexico is unique in possessing long stretches of estuaries and coastal lagoons nourished by inland streams and fringed by extensive coastal marshlands, mangrove swamps and forests, and in some cases, seagrass meadows. It is also unique in that one-third of its surface is underlain by broad, shallow continental shelves. These coastal and shelf waters are highly productive of marine life, and although the open Gulf is a virtual biological desert, the coastal waters are so extensive and so productive that the Gulf provides nearly half the US annual commercial fishery landings of oysters, shrimps, crabs, and finfishes. The bottom surface and subsurface deposits yield vast quantities of mineral resources, primarily petroleum and natural gas, but also substantial amounts of salt, sulfur, sand, shell, and gravel. The Gulf is a haven for tourists and those interested in year-round outdoor recreational opportunities. It is also a highway for heavy marine commercial shipping traffic and a receiving system for large volumes of the nation's waste chemical pollutants. With high rates of flushing and transport, it is a self-cleansing system that removes pollutants through dilution, decontamination, burial, and export, and it carries oxygen throughout its surface and deeper layers. The Gulf strongly influences the climate of surrounding lands including eastern Mexico and southern and eastern portions of the United States, moderating winter temperatures and providing moisture for regional precipitation. In these and other respects the Gulf of Mexico is a major asset for both the United States and its neighbor, Mexico.

In previous chapters we have described the various environments of the Gulf of Mexico and the very diverse flora and fauna that inhabit the Gulf shorelines, waters, and sea bottoms. We have provided some insights into the ecological processes through which individual species carry out their life history patterns. We have discussed how the species interrelate with one another, from viruses and bacteria through small species aggregations around copepods, gelatinous zooplankton and nekton, sargassum rafts, oyster beds, salt marshes, mangrove forests, seagrass meadows, and coral reefs, and how these in turn are bound together in the larger ecosystems of the estuaries and lagoons, continental shelves, and deep Gulf. Although we can now glimpse, at least in outline form, the various dimensions of Gulf ecology, there is still much to be learned. We have only begun to understand the internal workings of this marvelous ecosystem. Yet with increasing human use various portions of the Gulf of Mexico, particularly the coastal areas, are coming under intense pressure that could unravel this complex and sensitive web of life. Ways must be found to manage the Gulf for the benefit of human society while providing for the long-term survival of the natural system.

## THE NEED FOR NEW MANAGEMENT PERSPECTIVES

Historically, human societies have approached the management of coastal and marine areas from a problem-oriented perspective. Through their legislatures and governmental agencies they have enacted laws and regulated human uses of the environment on a problem-by-problem or piecemeal basis. Each agency has had its own specific and limited mission to carry out while ignoring other problems, which presumably fell under the aegis of other agencies. Along with this narrow problem-oriented perspective there has been a general tendency to "externalize the costs of doing business." This laissez-faire policy allows that it is acceptable to destroy natural habitat areas for human use because there is plenty more or that it is all right to dump human trash, sewage, and other wastes into the environment and let natural processes take care of the cleanup problem. Coastal and marine waters turned out to be great places for industries, municipalities, and the US Army Corps of Engineers to dispose of solid and semisolid wastes, chemical residues, and dredge spoil material. There was no financial cost, most of the material was out of sight of the public, and the environmental processes often diluted the chemicals, transported them to other areas, and eventually decontaminated or buried them. The negative effects on coastal and marine environments and living systems were sometimes a result of calculated management decisions, but more often they resulted from benign neglect or absence of clear knowledge of the consequences of a given course of action. The management problem was further complicated by the "layering" of federal, state, and local agencies with overlapping and sometimes conflicting missions and authorities.

When human populations around the Gulf were small and vast areas of natural environment were available, these practices and policies worked fairly well, but during the past half century things have changed dramatically, particularly along the coastlines and on the shallow shelf areas of the Gulf. Increased human utilization of the coastal and marine resources has now reached the stage where conflicts between opposing uses are commonplace, and their impacts on biological species, natural processes, and environments are, in many cases, disastrous. As we move into the future these conflicts and pressures will continue to grow in intensity and complexity. Thus, it has become widely recognized that we simply cannot afford to continue in the old ways. The natural systems cannot handle the load. If the negative environmental effects are to be reduced or reversed and if human use and demands are to be met in an equitable way, then new perspectives, new policies, and new management systems and strategies will be required to cope with the changing situations.

Sensing the growing environmental degradation and seeing the need for cleaner land, air, and water, beginning in the mid-1960s the US Congress began passing major legislation designed to restructure governmental agencies and to focus on providing for improved environmental management, a trend that has continued to the present day. Much of the legislation has affected the way we manage human uses of coastal and marine resources, but very little of it has been directed specifically toward a thorough examination and revamping of our policies toward the oceans themselves. Finally, in the year 2000, the US Congress commissioned a study to examine the state of the oceans, to review the nation's policies concerning the oceans, and to recommend changes in the ways in which we manage the oceans and marine resources. The final report of the commission has now appeared (US Commission on Ocean Policy 2004), and this was preceded by another major report, dealing with management of the nation's marine fisheries (Pew Foundation 2003). At this writing, the various ocean management issues are under intense scrutiny at the highest levels of government. New management structures and environmental use and protection strategies are emerging, and many of the matters brought up in the present chapter will undoubtedly be addressed in the near future. Among

the new (and some old) paradigms of ecosystem health, sustainability, and management are such concepts as

- optimum human use,
- stewardship of the environment,
- sustainability of renewable resources,
- preservation of biodiversity,
- interrelatedness of ecosystems and their components,
- interrelatedness of humans and ecosystems,
- adaptive management,
- multiple use management,
- participatory management,
- comprehensive or holistic ecosystem-based management,
- a management regime for federal waters, and
- international responsibility.

Whether or not the agency personnel and environmental managers really comprehend the full implications, these phrases are now appearing frequently in the management-related literature. These and related topics will be examined in some detail in the following pages, but first we must address the question of who has jurisdiction over the Gulf of Mexico and therefore the authority to manage it. References particularly pertinent to the present chapter include the following: Belsky (1986), E. Chávez (1981), E. Chávez and Hidalgo (1988), H. Chávez and Tunnell (1993), J. Clark (1986, 1987, 1998), Darnell and Shimkin (1972), Darnell and Soniat (1981), Day, Culley, et al. (1979), Dyer and Orth (1994), GESAMP (1994), Gray (1997), Haskell, Lindelof, and Causey (1994), Holdgate and White (1977), Ketchum (1972), Kinne and Bulnheim (1980), Kumpf, Steidinger, and Sherman (1999), Library of Congress (1977), E. Livingston (1979), National Academy of Science (1995), NOAA Estuarine Programs Office (1989), Norse (1993), Office of National Marine Sanctuaries (2008), Pew Foundation (2003), G. Ray (1991), G. Ray and Grassle (1991), G. Ray and Gregg (1991), G. Ray and McCormick-Ray (1992), J. Richardson (1985), Sherman and Alexander (1986, 1989), Sherman, Alexander, and Gold (1990, 1991, 1993), R. Stickney (1984), Texas A&M University (2002), Tunnell and Dokken (2006), Tunnell and Judd (2002), Twilley et al. (2001), US Commission on Ocean Policy (2004), US Department of State (1982), Van Dyke, Zaelke, and Hewison (1993), Wenk (1972), and J. Wilson and Halcrow (1985). Considerable use has also been made of various articles in *Louisiana Coastal Law*, a publication of the Louisiana Sea Grant Program.

## INTERNATIONAL LAW OF THE SEA

Unlike laws of the land, which are enacted by legislatures, historically the law of the sea has been built up through time by the way maritime nations have customarily used the sea. When a certain usage became widely practiced, it was accepted as the normal way of doing things and was tacitly recognized as a *customary law of the sea.* Such customary laws could, of course, be formalized by written agreements between two or more maritime nations, in which case they became *treaty law.* However, until recent years the body of maritime law contained much unwritten customary law and local treaties complicated by the diverse and often conflicting jurisdictional claims of the various maritime nations.

During the past several decades international groups working under the auspices of the United Nations have devoted considerable effort to the standardization of seaward jurisdictional boundaries and to codification of customary laws of the sea. Since the UN is not a legislative body, its major reports and treaties, or "conventions," do not have the force of law until they are ratified by the member states. A brief sketch of the development of the presently recognized law of the sea is given below, and this is followed by an examination of US law as it relates to the oceans in general, and the Gulf of Mexico in particular.

### Historical background

Maritime nations of the ancient classical world generally exercised control over the sea adjacent to their own coastlines, but in waters beyond this coastal zone they permitted free trade by ships of other nations. This situation prevailed up through the Middle Ages, when rulers of some coastal nations felt the need to exercise jurisdiction over a zone of sea adjacent to their coastlines with a width variously defined as "a day's sailing," "as far as one can see on a clear day," "the distance of a cannon shot," and so on. This nearshore band of protected water eventually gave rise to the concept of a nation's *territorial sea,* over which a coastal nation has outright authority as it does on land. The *high seas* were considered to be all the vast ocean space seaward of the territorial sea.

Following discovery of the New World by Columbus in 1492, the nations of Europe, particularly Spain and Portugal, sent out exploratory voyages and claimed ownership of all lands discovered. In an effort to resolve some of the conflicting claims, in 1493 Pope Alexander VI divided the "undiscovered world" between the two Catholic nations, Spain and Portugal, by a longitudinal line located 100 leagues (ca. 300 miles) west of the Azores islands. These two nations then attempted to divide the New World and the known ocean (the Atlantic) between themselves and claimed control (i.e., a policy of "closed seas"). This did not sit well with the other nations, and ultimately, in 1580, Queen Elizabeth I of England openly repudiated this claim and asserted an "open seas" policy whereby all nations had free access and free passage to all of the high seas. When the Spanish ambassador protested Sir Francis Drake's intrusion into "Spanish" waters in the *Golden Hind,* the queen responded with a classic statement of what was to become a basic principle of the law of the sea: "The use of the sea and air is common to all; neither can any title to the ocean belong to any people or private man, for as much as neither nature nor regard of the public use permitted any possession thereof." This also provided an impetus for the buildup of British naval power. The matter was finally settled in 1588 by the destruction of the Spanish Armada.

Subsequently, in an article published in 1609 and in later writings, the Dutch scholar Hugo Grotius, defending the maritime rights of the Dutch East India Company, agreed with the British position and expanded on the concept of "freedom of the seas." While recognizing the right of each coastal nation to exercise national sovereignty over its adjacent nearshore waters, he argued that this should not impede free passage of other nations' vessels even within their territorial waters. However, beyond this zone of national sovereignty the ocean could be used by anyone. According to Grotius it would defy the very nature of the ocean to make it the sole property of any one nation. Seeking to limit the power of Spain and Portugal, most other maritime nations were quick to agree with this idea. The question then became how far to extend a coastal nation's jurisdiction from the shoreline, that is, the width of the territorial sea.

Shortly after the United States was founded, France and England were at war. As a neutral country the United States wished to trade with both nations, but warships of each of the warring nations would board US ships just out of harbor and seize trade goods as "contraband" bound for the other country. In response, in 1793 then Secretary of State Thomas Jefferson drafted a proclamation stating that foreign nations were not to engage in hostile actions within territorial waters of the United States (declared to extend one marine league, or about 3 miles, from the coastline). In 1794 Congress passed the Neutrality Act, formalizing the US territorial sea claim and extending the jurisdiction of coastal states for ship actions within this zone. The act also allowed the United States to capture foreign ships within this zone with just cause. The US Supreme Court upheld the 3-mile limit claim for the territorial sea, and Chief Justice John Marshall proclaimed that special jurisdiction could extend to 12 miles offshore to enhance enforcement of the nation's coastal laws. Other nations made similar claims, and a precedent was set for claiming jurisdiction out to the 12-mile limit. Eventually, some nations found it expedient to claim jurisdiction some distance seaward of the territorial sea for special purposes (to control fishing rights off the nation's shores and to control pollution, smuggling, petroleum reserves, etc.). In this so-called *contiguous zone* a nation exercised only limited control as necessary to achieve its special purpose.

## Modern law of the sea

In 1930 the League of Nations attempted to develop a set of principles on which an international law of the sea could be based. Although little progress was made at the time, some of the problems were defined, and the stage was set for real progress by working groups of the United Nations some decades later. In 1958 the first United Nations Conference on the Law of the Sea (UNCLOS I), or Geneva Convention, succeeded in drafting four treaties, or "conventions." These included the

- Convention on the Territorial Sea and Contiguous Zone,
- Convention on the High Seas,
- Convention on Fishing and Conservation of the Living Resources of the High Seas, and
- Convention on the Continental Shelf.

Although they failed to resolve the main jurisdictional questions, these conventions did define the continental shelf, territorial sea, contiguous zone, and so forth, and they recognized many different maritime rights (the right of innocent passage, the right to fishery resources, the right to defend, etc.). They also recognized the right of coastal nations to explore and exploit the mineral resources of their adjacent continental shelves (as some nations, including the United States, were already doing).

In 1960 UNCLOS II essentially ended in failure. In 1970 UNCLOS III passed a "Declaration of Principles," which, among other things, asserted that the resources of the seabed of the deep ocean were the "common heritage of all mankind," and not the exclusive property of those na-

tions with the technological capability to exploit them. At issue were the vast fields of mineral-rich manganese nodules littering certain areas of the floor of the deep sea. The declaration also reiterated the need for a comprehensive body of oceanic law that would finally codify centuries of confusing maritime law, which had been developed largely by custom. This would take another 12 years.

Finally, in 1982 the United Nations adopted (in draft form) the major Law of the Sea Treaty, which, with certain modifications, is the presently recognized law of the sea. Some of the salient points are given below.

- The *Territorial Sea* is recognized as extending up to 12 nautical miles from the coastline, and within this zone the coastal nation can exercise exclusive sovereignty over the air, sea, and seabed.
- A *Contiguous Zone* is recognized that extends seaward an additional 12 nautical miles (to a total of 24 nautical miles) and can be claimed by a coastal nation in order to accommodate the nation's immigration, fiscal, pollution control, and related concerns.
- In addition, the treaty recognizes that coastal nations may claim an *Exclusive Economic Zone* (EEZ) reaching out up to 200 nautical miles from the coastline. Here a nation can exercise jurisdiction over the exploitation of living and nonliving resources of the water column and upon or within the seabed.
- Marine scientific research on the continental shelf or in the exclusive economic zone of another nation should be carried out with the consent of that nation. This means that researchers must now apply for permits to work in such foreign waters, a process that generally requires six months or longer and could hamper the investigations.
- The Law of the Sea Treaty was found unacceptable by many of the developed nations, including the United States, primarily on the basis of some of the deep-sea mining provisions. Although most of the objectionable provisions have subsequently been removed, the treaty has still not been ratified by the United States.

## UNITED STATES' MARITIME JURISDICTION
### Post–World War II development

By the mid-1940s it had become clear that vast oil and gas reserves lay buried beneath the seabed of the continental shelves of the Gulf of Mexico and elsewhere. To lay claim to these resources, in 1945 President Harry Truman issued a major proclamation that said, in part, "The government of the United States regards the natural resources of the subsoil and the seabed of the continental shelf beneath the high seas but contiguous to the coasts of the United States as appertaining to the United States, and subject to its jurisdiction and control." This sudden extension of the territorial jurisdiction over 50 million square miles of submerged land was said not to affect the "character of the high seas" that cover the area, but it was a special-purpose jurisdiction related to the seabed and subsoil. This territorial extension by the United States set a precedent for other nations to do likewise, and eventually the United Nations recognized the rights of coastal nations to the mineral resources of their contiguous continental shelves.

This territorial extension by the federal government left open the question of the seaward extent of state jurisdiction, and a number of lawsuits erupted between the states and the federal government as the states sought to extend their jurisdiction well out onto the continental shelves. When the state of California attempted to regulate the leasing of oil and gas resources within its perceived territorial waters, the right was challenged by the US attorney general, and harking back to the Neutrality Act of 1794, the US Supreme Court eventually declared that the federal government, rather than the states, has paramount rights and power over the sea and seabed lying beyond the 3-mile limit granted in the Neutrality Act.

Subsequently, in 1953, Congress passed two major pieces of legislation relating to the continental shelves and their resources, the Submerged Lands Act and the Outer Continental Shelf Lands Act. These fortified the contents of the Truman Proclamation. They formally extended US jurisdiction over the continental shelves and authorized the secretary of the interior to lease the submerged lands for economic development of the mineral resources. They also reaffirmed state ownership of the seabed and its natural resources within the limit of 3 nautical miles from the coastline.

Following adoption in 1982 of the Law of the Sea Treaty by the United Nations, in 1983 President Ronald Reagan issued a presidential proclamation extending the US special jurisdiction to the 200-mile limit of the now recognized Exclusive Economic Zone, exercising US control over the living and nonliving resources of this vast area. In 1988 President Reagan issued a second proclamation extending the US territorial sea out to 12 nautical miles (from the original 3-mile limit) in accordance with the treaty (for control of the air, sea, and seabed).

## Jurisdiction within the Gulf of Mexico

### STATE BOUNDARIES

Lawsuits brought against the United States by Louisiana, Mississippi, and Alabama to extend their jurisdiction across the continental shelf were dismissed because congressional acts admitting them as states made no mention of any coastal boundaries beyond the traditional 3-mile territorial sea limit. However, the state of Texas, which was an independent republic before its admission, had a constitution that stated its boundaries extended to 3 leagues, or 9 miles, beyond its shores, and Congress had agreed to accept the Texas constitution upon its admission to the Union. Florida's claim was somewhat similar, since its constitution also referred to its outer boundary as being 3 leagues from the mainland, and Congress had approved it during Florida's readmission to the Union following the Civil War, but this extension applied only to the Florida west coast, that is, in the Gulf of Mexico. The result is that within the Gulf both Florida and Texas wound up with 3-league or 9-mile territorial seas, but the remaining states were restricted to the traditional 3-mile limit. Parenthetically, due to compaction, erosion, and other factors, in some areas the shoreline has been receding, especially along the Louisiana coast. Thus, it has been necessary to establish an immovable baseline from which the state (and federal) jurisdiction is measured.

### FEDERAL BOUNDARIES

Political boundaries defining US jurisdictions in the Gulf of Mexico are recognized as follows.

### TERRITORIAL SEA

Both the United States and Mexico claim a 12-nautical-mile territorial sea within which each nation exercises sovereignty but cannot deny the right of innocent passage to foreign nations.

### CONTIGUOUS ZONE

This is a band of high seas extending 12 miles beyond the territorial sea, or 24 miles from the coastal baseline. Within this zone a nation can exercise the control necessary to prevent infringement of its customs, fiscal, immigration, or environmental regulations.

## PROHIBITED OIL POLLUTION ZONE

Pursuant to the 1973 International Convention for the Prevention of Pollution from Ships (MARPOL), this is a band of coastal waters and high seas extending from the coastal baseline to 50 nautical miles from the nearest land areas. Oil tankers are prohibited from discharging oil in this zone (except under certain conditions spelled out in the MARPOL regulations). Ships other than tankers and greater than 400 gross tons are generally restricted from discharging oil within 12 nautical miles of the nearest land areas. Both the United States and Mexico are signatory to this Convention.

## EXCLUSIVE ECONOMIC ZONE

The Exclusive Economic Zone extends 200 nautical miles from the baseline of the territorial sea. Where the extent of the EEZs of two nations overlap, boundaries are determined on the basis of equitable principles. Consistent with international law, in 1983 the United States extended its jurisdiction over the EEZ, and within this area it claims (1) sovereign rights for the purposes of exploring, exploiting, conserving, and managing natural resources, both living and nonliving, of the seabed and subsoil and the superadjacent waters and with regard to other activities for the economic exploitation and exploration of the zone such as the production of energy from water, currents, and wind, and (2) jurisdiction with regard to the establishment and use of artificial islands and installations and structures having economic purposes, and the protection and preservation of the marine environment. This zone coincides with the US Fishery Conservation Zone, where the United States claims exclusive rights to manage fishery resources. Mexico claims similar rights and jurisdiction over its EEZ.

## US-MEXICO MARITIME BOUNDARY

Following negotiations, in 1978 the United States and Mexico reached agreement on a shared maritime boundary in the Gulf of Mexico. It provides that neither country shall claim exclusive sovereign rights or jurisdiction over the water or seabed on the other country's side of the maritime boundary. In 1997 the United States finally ratified the treaty, thereby establishing a permanent boundary between the two countries extending out up to 200 nautical miles from the coastlines. Although most of the Gulf was thereby divided, two gaps remained, the "western gap" and the "eastern gap." Both represented high seas areas beyond the 200-mile jurisdiction of either nation. However, the western gap was totally bounded by US and Mexican waters, and here a division could be made between the two nations for the presumed petroleum reserves lying beneath this area of 5,092 nautical miles. In April 2000, the United States and Mexico reached agreement on a boundary treaty that divided the western gap between them. This treaty established a buffer zone of 1.4 nautical miles on either side of the actual boundary in which no oil or gas exploration or drilling could take place for a period of 10 years following ratification of the treaty. The eastern gap boundary could be resolved only by participation of the Republic of Cuba, as well, and no such negotiations are currently planned. However, the eastern gap does not appear to have major petroleum-producing potential and is of no immediate concern. For much of its distance the US-Mexican boundary roughly follows the 26°N parallel. Its position and the eastern gap are shown in figure 18.1. The upshot of all this is that between them, the United States and Mexico share the mineral and fishery resources of the Gulf of Mexico. Perhaps more importantly, these two nations have tacitly agreed to jointly manage the resources, environment, and even the Gulf of Mexico ecosystem in a responsible and sustainable manner. If this works out well, it could become a model for the management of all the rest of the world's oceans.

## FEDERAL INVOLVEMENT IN MANAGING THE GULF
### Major federal legislation concerning the environment and marine resources

Following the Second World War there was a great expansion in the American economy. Domestic and foreign markets needed goods of all types. Industries and agriculture were booming. The population was increasing, cities were growing, and there was a significant demographic movement of the human population into coastal areas of the nation. However, during this same period chemical and other pollutants were building up on the nation's land and in the air and surface waters. In less than two decades after the war, this buildup had reached crisis proportions in some areas, and pollution-related problems began to hit front pages of local newspapers. Citizens became alarmed by the news and by the writings of such authors as Rachel Carson, who persuasively argued in her book *Silent Spring* that agricultural and other chemicals were killing off the native wildlife. Added to this were the widely publicized court cases that proved her point, especially the case against DDT and related chemicals, which

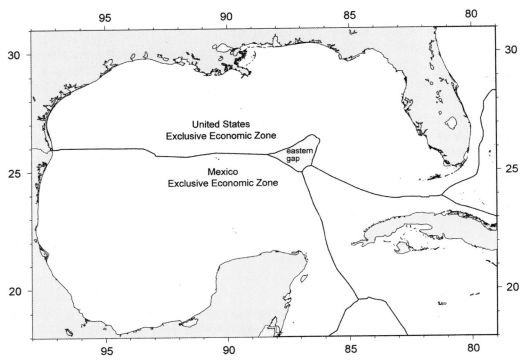

**Figure 18.1.**
The negotiated boundary line between the Exclusive Economic Zones of the United States and Mexico showing approximate boundary extensions in neighboring marine waters. The location of the "eastern gap" is also shown. For explanation see text.

were shown to be systematically destroying segments of the native American bird life. Responding to the public outcry and in light of increasing scientific evidence about the extent of the environmental degradation, during the period 1966–1976 the US Congress passed a series of important legislative acts that serve as the foundation for much of our present national policy and law concerning the environment in general, and the coastal zone and marine areas in particular.

### ENVIRONMENTAL REGULATION

Dealing primarily with the environment are five pieces of major federal legislation: the National Environmental Policy Act (NEPA, 1969), Environmental Quality Improvement Act (1970), Clean Water Act (1972), Endangered Species Act (1973), and United Nations Environment Program Participation Act (1973). Environmental management at the federal level was also greatly affected by portions of the Government Reorganization Plan (1970). In addition to treating the environment in general, all these laws and actions also directly affected how we deal with coastal areas and the oceans.

The first of these laws (NEPA) declared a national environmental policy that would reduce environmental damage, increase our knowledge and understanding of ecosystems, and encourage harmony between humans and nature. The act established the Council on Environmental Quality (CEQ) within the Executive Office of the President, and for each federal action likely to have a significant effect on the environment, it requires the preparation of a detailed *environmental impact statement*. CEQ was ordered to keep track of the status and conditions of the natural and human-altered environments as well as trends in environmental management and resource use. CEQ was also to report at least annually on these matters and to set forth ways of limiting resource deficiencies and to recommend any additional federal legislation needed to accomplish these goals. The second law provided staff and financial support for CEQ so that it could assist other federal agencies in environmental matters and coordinate environmental management among the various federal agencies. It also provided support for the collection and analysis of environmental data and for the study of the effects of technology on the environment.

The Government Reorganization Plan recommended establishment of the US Environmental Protection Agency (EPA) as a separate independent government agency to consolidate federal efforts to deal with environmental problems (especially pollution), to establish and enforce national environmental pollution standards, to conduct research on the effects of pollution and methods of pollution abatement, and to assist other agencies and groups in dealing with pollution issues. The agency would also assist CEQ in developing and recommending national policies for dealing with environmental pollution.

The Clean Water Act (CWA) specifically mandated that EPA establish and enforce national standards for the quality of the nation's navigable waters. Through a permitting program it was to control all pollutants entering the waters except dredge spoils (to be permitted by the US Army Corps of Engineers) and a few other substances already regulated by other agencies. Pursuant to this goal EPA was to aid in developing new technologies for pollution control and abatement and to provide financial assistance for pollution management planning and for construction of certain waste treatment facilities.

The Endangered Species Act announced that the federal government recognized the importance of preserving our native plant and animal species. It provided means of protecting those species already endangered or threatened by preserving the ecosystems of which they were a part and by developing other programs for the protection and restoration of such species. The program was to be administered by the US Department of the Interior. The United Nations Environment Program Participation Act announced a US policy of participation in coordinated international efforts to solve environmental problems of global and international concern and to provide funds in support of such efforts.

Thus, during the late 1960s and early 1970s the US government announced policies and established agencies and programs to deal with the complex and growing environmental problems. EPA was to clean up and control the mess, and CEQ was to report regularly to the president and the nation on how things were going. There was much optimism, and it was announced that the nation should be pollution free by 1985. Steps were also taken to preserve the native plants and wildlife, and the United States decided to participate with the United Nations in international efforts to attack global environmental problems.

## THE COASTS

Two of the laws enacted during the 1966–1976 period affected mainly coastal areas: the National Sea Grant College and Program Act (1966) and the Coastal Zone Management Act (1972). The first of these established Sea Grant colleges and Sea Grant programs in coastal states around the nation. These programs were intended to stimulate research and development of new technolo-

gies for utilization of the nation's coastal and marine resources (with a focus on mariculture, or the raising of marine plants and animals for human use). They were also to provide educational opportunities and widely disseminate new coastal and marine-related knowledge. The Coastal Zone Management Act established partnerships between the federal government and coastal states to manage the nation's coastal zone. It provided for the establishment of a series of *estuarine sanctuaries* where natural and human processes could be studied, and it provided funds for coastal land acquisition for public access to beaches and for protection of barrier islands. This act involved a major overhaul in the way the federal government dealt with the states and the way in which it handled environmental management. The act will be discussed in some detail below.

## THE OCEANS

Regarding the oceans, US policy and management were affected by the Government Reorganization Plan (1970) and by three laws enacted during 1966–1976: the Marine Mammal Protection Act (1972); Marine Protection, Research, and Sanctuaries Act (1972); and the Fishery Conservation and Management Act (1976). The Reorganization Plan pointed out the need to establish, within the US Department of Commerce, the National Oceanic and Atmospheric Administration (NOAA) in order to bring together in one organization oceanic and atmospheric initiatives previously scattered among many different agencies. NOAA was to become the major federal agency to focus on and provide understanding of the oceans and atmosphere. Among the agencies to be transferred into NOAA would be the Sea Grant Program (from the National Science Foundation) and most of the marine aspects of the US Bureau of Commercial Fisheries (from the US Department of the Interior). Some specific functions of NOAA include: observation and interpretation of the global environment from orbiting satellites, weather forecasting and prediction of severe and potentially damaging weather events, storage and retrieval of environmental data, preparation of appropriate maps and charts, and research on the oceans and atmosphere including the maintenance and use of data buoys.

The Marine Mammal Protection Act, to be administered by NOAA, established a moratorium on the taking of marine mammals from marine waters within 200 nautical miles of the US coastline and the importation of marine mammals or their parts (with exceptions for certain Native Americans and for scientific purposes). This law protects whales, porpoises, and certain other marine mammals not already protected by other laws. The act also authorizes establishment of the *Marine Mammal Commission* to study marine mammal stocks and to make recommendations concerning any needed management action.

The Marine Protection, Research, and Sanctuaries Act prohibits the dumping of all materials into coastal and oceanic waters except by permit from the US Army Corps of Engineers (for dredged materials) or from the EPA (for all other materials not regulated by other laws). It also directs the US Department of Commerce to carry out studies on the effects of dumping into coastal and marine waters within US jurisdiction, and it authorizes the US Department of Commerce to designate, establish, and manage a series of *marine sanctuaries* in waters off the coasts of the United States and its territories for the purpose of preserving or restoring such areas for their conservation, recreational, and ecological values.

The Fishery Conservation and Management Act declares that coastal and highly migratory oceanic fishes as well as those that dwell on US-controlled continental shelves and anadromous species that spawn in US rivers and estuaries are valuable and renewable natural resources. Since some of the stocks were depleted and foreign fleets were interfering with domestic fishing efforts, there was a need for a national fishery management program. The act itself

- established a *Fishery Conservation Zone* (FCZ) extending seaward 200 nautical miles from the coastline;
- announced that within the FCZ the United States would exercise management authority over all fish species and other fishery resources except highly migratory species (which are addressed in certain international treaties and conventions). These resources include commercially important species of sponges, soft and hard corals, mollusks, crustaceans, and fishes (but not plants, birds, sea turtles, and marine mammals, these being covered by other laws);
- obliged US agencies to work with the international community to manage the highly migratory species;
- provided for the preparation and implementation of *Fishery Management Plans* designed to achieve

and maintain, on a continuing basis, the optimum yield for each fishery; and

- established within the US Department of Commerce a national marine fishery management program to carry out the mandates of the act. The program was to be administered by an agency of the US Department of Commerce (the National Marine Fisheries Service) whose duties would be to
  - prepare and implement a Fishery Management Plan for each major species of commercial importance designed to achieve and maintain, on a continuing basis, the optimum yield for the fishery;
  - set, monitor, and enforce fishery regulations designed to achieve the goals of the act;
  - review foreign applications for entry into the fishery of the US FCZ; and
  - conduct research in support of the act.

The act also calls for the establishment of a series of regional Fishery Management Councils to advise the NMFS on these matters. Enforcement would be carried out by the US Coast Guard and, as necessary, by the US Department of Defense and the coastal states. The management of coastal and marine fisheries will be examined in greater detail in a later section.

During the critical period under discussion, one of the most significant federal actions was the establishment within the US Department of Commerce of the National Oceanic and Atmospheric Administration, which was to become the lead federal agency in dealing with the coastal zone, the oceans, and the atmosphere. Eventually placed within NOAA were the following programs: Sea Grant, Coastal Zone Management, Estuarine Sanctuaries, Marine Sanctuaries, Marine Mammal Commission, and programs dealing with coastal and marine commercial fisheries (initially called the National Marine Fisheries Service but recently renamed NOAA Fisheries). It was thought by some that NOAA would become a "wet NASA," but NOAA has never achieved funding levels comparable with those accorded to the National Aeronautics and Space Administration. The movement of commercial fishery administration from the Department of the Interior to the Department of Commerce (with its historic emphasis on exploitation rather than preservation) was met with some skepticism, and in some quarters it was felt that "the fox has been set to work guarding the chickens." As it has turned out, this concern was not entirely unfounded.

Through the Coastal Zone Management Act, the federal government established partnerships with the various coastal states to jointly manage the nation's coastal zone. Through the Sea Grant Office, funding programs would be developed for utilizing coastal and marine resources. A ban was placed on the dumping of materials into coastal and marine waters except by permit from the Army Corps of Engineers or from the EPA. Special protection was extended to marine mammals, and marine fishery resources were to be managed for optimum yield and on a sustainable basis. Both estuarine and marine sanctuaries were to be established to protect species and ecosystems and to study the effects of human perturbations. During subsequent years some of these acts would be amended and other related legislation would be passed, but the tone had been set for federal protection of the environment and for special attention to management of coastal and marine resources.

It should be noted that the Bureau of Land Management (BLM) (and later the Minerals Management Service, [MMS], which subsequently became the Bureau of Ocean Energy Management [BOEM] and the Bureau of Safety and Environmental Enforcement [BSEE]) was already busy leasing areas for the exploitation of mineral resources, primarily petroleum and natural gas, but also salt and sulfur, on continental shelves off the US coastlines, particularly in the Gulf of Mexico. Prior to establishment of the MMS, mineral resource activities on these leased lands were regulated by the US Geological Survey (USGS). What changed was that after 1969, as mandated by the National Environmental Policy Act, detailed environmental impact statements had to be prepared prior to each oil and gas lease sale, and thereafter the BLM was obliged to sponsor and fund considerable marine research in order to prepare the impact statements. Through subsequent years the BLM- and MMS-sponsored research was to provide most of the baseline information for our present understanding of the environments, biological composition, and ecosystems in marine waters off our nation's coastlines, and this is especially true for the Gulf of Mexico.

## Federal-state relations in the coastal zone

COASTAL ZONE MANAGEMENT ACT

In 1972 Congress passed the Coastal Zone Management Act (CZMA) to address the increasing degradation of the nation's estuaries and other coastal systems. It also hoped to provide a decision-making mechanism for resolving the increasingly bitter disputes between those who wished to expand commercial development of the coastal zone and those who wanted to preserve for future generations the fishery, wildlife, ecological, and aesthetic values of the natural environment. Basically, the act encouraged coastal states to develop plans, with federal oversight, for managing the resources of their respective coastal zones. Matching funds would be made available to aid the states in developing the plans and implementing the programs once the plans had been approved at the federal level. To accomplish all this, the act created a uniquely cooperative management scheme between federal and state governments to address both local and national interests in preservation and development of the coastal zone.

Nominally, the act encouraged states to exercise their constitutional power over land and water uses in the coastal zone and promised that once the plans had been accepted at the federal level, the actions of federal agencies would be consistent with the state programs. It stated: "Each federal agency activity within or outside the coastal zone that affects any land or water use or natural resource of the coastal zone shall be carried out in a manner that is consistent, to the maximum extent practicable, with the enforceable policies of approved state management programs."

Within the United States resource management normally takes place at three levels of government (federal, state, and local). The overlapping jurisdictions and the complex and sometimes conflicting regulations often make for a cumbersome way of doing business. This entire process has been simplified by the "consistency" provisions of the CZMA.

The focus and power for management was placed at the state level because: (1) the mix of competing problems was considered to be too complex for local governments to handle, (2) commercial interests often dominate local decision-making processes, and (3) adequate protection of natural resources often involves areas not fully under the control of local governments. For these reasons the state was considered to be the best level for resolving coastal management issues.

In formulating their initial plans the states had to yield to a number of federal requirements, that is, they essentially had to make their own plans comply with certain federal standards, processes, and regulations before the plans would be accepted at the federal level. Among these provisions was the requirement that the state develop "unified policies, criteria, standards, methods, and processes for dealing with land and water use decisions of other than local significance." Another provision was the requirement that the state establish procedures to designate areas of significant conservation, recreational, ecological, or aesthetic value for purposes of preservation or restoration.

In the act the coastal zone was defined to include the shore lands and related waters and wetland environments. The zone extends inland only to the extent necessary to control human activities that have a direct and significant effect on the coastal waters. Seaward it extends to the outer limit of the US territorial sea (12 nautical miles). In this respect the act extends some state powers seaward beyond the normal limits of state jurisdiction. The act further authorizes the establishment of a series of estuarine sanctuaries about which more will be said later.

## FISHERY CONSERVATION AND MANAGEMENT ACT (1976)

As a fundamental management principle, this act announces that to the extent practicable each individual stock of fish will be managed as a unit throughout its range, and interrelated stocks of fish should be managed as a unit or in close coordination. Of course, individual species of fish often range across jurisdictions, but regulations by the different authorities must have the same goals of stock conservation. Therefore, the principle obliges federal and state agencies to cooperate in managing stocks of coastal fish species. In general, states still manage marine fisheries within the limits of state jurisdiction, and the National Marine Fisheries Service manages fisheries seaward of state authority. However, the act also specifically allows for federal preemption of state regulations in certain instances within the state's own territorial waters. This would occur only if an action or inaction by the state in its territorial waters "substantially and adversely affects" the application of an approved Fishery Management Plan that concerns species that exist primarily in the Exclusive Economic Zone. In both of these acts the federal government invites state participation in the management of certain resources in waters seaward of the normal state territorial sea.

## Federal interests in the oceans

Table 18.1 lists the federal departments and agencies that have some jurisdiction over coastal or marine affairs, and most of these have direct interests in the Gulf of Mexico. The 38 agencies are scattered through 11 cabinet-level departments and 4 independent agency groups. This is a prime example of federal piecemeal management. No one agency has been in charge, and no one has been coordinating or even keeping track of the activities of the different groups. During the past half century the United States has progressively extended federal jurisdiction beyond its coastlines, but as a nation we have yet to develop a clear policy or unified plan for the management of our ocean resources. Thus, there has developed a strong need for the creation of a high-level *National Ocean Council* whose duties would be to

1. keep the executive branch of government informed of the progress and interactions of our various marine management operations,
2. formulate a unified national marine policy for ocean governance,
3. develop a comprehensive ocean management plan that would embrace both the coastal zone and open ocean, and
4. inform Congress of additional legislation needed to implement the management plan.

This is particularly important at a time when commercial interests are planning to establish on the continental shelves of the Gulf of Mexico major structures for carrying out mariculture operations and for deriving energy from the winds and water currents. Our intrusion into the marine environment is still in its infancy.

In 2004, by executive order, a cabinet-level Commission on Ocean Policy was established to address the above and related issues. This committee, to be chaired by the head of the Council on Environmental Quality, includes representatives from the various federal departments and agencies listed in table 18.1. Hopefully, from this group a coherent and sustainable national ocean policy will emerge.

There has been a tendency to concentrate certain aspects of ocean investigation and management into a single agency (NOAA). The National Marine Fisheries Service has shown considerable initiative in approaching the problems of broadscale marine fisheries management from a holistic perspective, and the estuarine and marine sanctuaries programs are major steps in the direction of preservation of coastal and marine biodiversity. Some of these developments are discussed below.

**Table 18.1.** *Federal departments and agencies that have some jurisdiction over coastal or marine affairs.*

**Cabinet-level departments**

Agriculture
- Agricultural Research Service
- Animal and Plant Health Inspection Service
- National Resources Conservation Service

Commerce
- National Environmental Satellite, Data, and Information Service
- National Marine Fisheries Service (NMFS)
- National Oceanic and Atmospheric Administration (NOAA)
- National Ocean Service
- National Weather Service
- Office of Oceanic and Atmospheric Research
- Office of Program Planning and Information

Defense
- Defense Advanced Research Projects Agency (DARPA)
- Department of the Navy
- National Geospatial-Intelligence Agency
- US Army Corps of Engineers

Energy
- Office of Fossil Energy
- Office of Science

Health and Human Services
- National Institutes of Health
- US Food and Drug Administration

Homeland Security
- Federal Emergency Management Agency (FEMA)
- Transportation Security Administration
- US Coast Guard

Interior
- Minerals Management Service (MMS, now BOEM and BSEE)
- National Park Service (NPS)
- US Fish and Wildlife Service (USFWS)
- US Geological Survey (USGS)

Justice
- Environment and Natural Resources Division

Labor
- Occupational Safety and Health Administration (OSHA)

State
- Bureau of Oceans and International Environmental and Scientific Affairs

Transportation
- Saint Lawrence Seaway Development Corporation
- US Maritime Administration

**Independent agencies**

Environmental Protection Agency (EPA)
- Office of Air and Radiation
- Office of Research and Development
- Office of Waters

National Aeronautics and Space Administration (NASA)
- Office of Earth Science

National Science Foundation (NSF)
- Directorate for Biological Sciences
- Directorate for Geosciences
- Office of Polar Programs

US Agency for International Development

---

MINERALS MANAGEMENT SERVICE (MMS)
OPERATIONS

The Minerals Management Service (now BOEM and BSEE, as mentioned above) was responsible for the development of the mineral resources of the continental shelves and deep oceans outside of state territorial waters and to the outer limits of the 200-mile Exclusive Economic Zone. Geographically, MMS divided the US Gulf into three major areas (Eastern, Central, and Western Planning Areas), and within each area the bottom was sectioned into squares of approximately 100 km² each (referred to as "lease blocks"). Periodically, MMS offered certain promising lease blocks for sale, and oil and gas com-

panies engaged in competitive bidding for the right to explore and exploit the subsurface mineral resources of the blocks they purchased. The financial stakes were quite high, and a single lease sale event could provide the federal government with revenues of well over a billion dollars. So, there were only around 200 companies or fewer involved in the bidding, and often several of these would bind together in joint ventures for bidding and then developing the resources.

Since the fieldwork was so hazardous and many human lives were at stake, all aspects of the exploration and development processes were heavily regulated. On-site inspections occurred regularly. These regulations not only dealt with operations on the rigs and platforms but also extended to the laying of pipelines and the operation of servicing vessels as well as site cleanup after production ended. Industry officials and workers understood the need for such regulations and inspections, and compliance was a top priority.

When the Bureau of Land Management began its operations in the Gulf in the 1950s, little was known about the marine environment and its biological inhabitants. However, through the years BLM sponsored many large-scale but detailed environmental studies, so that now the Gulf of Mexico is one of the best known marine areas in the world. The agency is no longer managing an unknown resource, and the environmental impacts, which are kept to a minimum, are also quite well understood.

## MARINE FISHERIES MANAGEMENT
## IN THE GULF OF MEXICO

Following the Second World War, with the advent of larger boats, more powerful engines, and better nets, commercial fishing in the Gulf expanded greatly. It spread from the bays and estuaries well onto the continental shelves, where fishermen began harvesting the lucrative shrimp and menhaden populations and intensified hook-and-line fishery operations around offshore reefs and banks. Perceiving the need for interstate communication in the virtual absence of integrating federal programs, in 1949 the fishery managers of the five Gulf states created the Gulf States Marine Fisheries Commission, an informal group that convened periodically to discuss common problems and work out differences between the state programs.

This situation prevailed until 1976, when the Fishery Conservation and Management Act extended US fishery jurisdiction out 200 nautical miles beyond the coastline. It mandated the establishment of the Gulf of Mexico Fishery Management Council to advise the National Marine Fisheries Service on matters related to the management of fishery resources of this vast area (essentially, the northern half of the Gulf of Mexico). The council was to be composed of 17 voting members selected as follows: 5 members would be the principal state fishery officials (as affirmed by the governors of each of the five Gulf states), 11 members would be appointed by the secretary of commerce from lists of candidates submitted by the governors, and one member would be the regional administrator of the National Marine Fisheries Service for the Gulf area. The list of potential candidates provided by the governors could include representatives from indus-

try, commercial or recreational fishermen, or even members of the general public known to be knowledgeable about fishery issues. In drafting the act, Congress hoped that the councils would include administrators, lawyers, fishermen, economists, scientists, and others and would provide broad-based input for management decisions.

The act further mandated that for every species of commercial importance the council should prepare and submit to NMFS for approval a Fishery Management Plan designed to achieve and maintain, on a continuing basis, the optimum yield for the fishery. Upon approval of the plan, the secretary of commerce would set, monitor, and enforce regulations necessary to achieve the goals of the act.

As seen in table 18.2, for the Gulf of Mexico seven Fishery Management Plans have been accepted. These include 54 named species plus various unnamed types of hard and soft corals. In addition, the NMFS is monitoring fisheries for 7 additional species. Notably absent from the Fishery Management Plans are species such as the Atlantic croaker; spot; sand, silver, and spotted seatrouts; and several other species of considerable commercial and/or recreational interest that inhabit bays and estuaries and the nearshore continental shelf and that are generally less important in waters beyond state jurisdiction. By this omission the NMFS apparently decided not to manage certain fisheries of the coastal zone (where major fishery problems really exist). On the other hand, it does have management plans for the brown, pink, and white shrimps, two stone crabs, and the red drum, all of which are essentially coastal species.

Despite the intentions of Congress, council members often represent vested interests, and contentious advocacy (commercial vs. recreational interests, shrimper vs. red snapper interests, etc.) often dominates discussions and voting of the council members. At council meetings turf wars, not conservation issues, rule the day. Furthermore, by law the National Marine Fisheries Service is to manage for optimum (not maximum possible) yields. In environments with very high natural variability that are also subject to episodic catastrophic events, wise management would be precautionary management (i.e., in determining the allowable catch, as a hedge against negative uncertainties, management should take a conservative approach and establish catch limits well short of the calculated maxima). As a result of present practices, many of the Gulf fishery stocks are overfished, some severely so, and the condition of others is largely unknown. Clearly,

**Table 18.2.** *Species of fishes and invertebrates in the seven Fishery Management Plans (FMPs) through which the National Marine Fisheries Service manages the fisheries of the US Gulf of Mexico Exclusive Economic Zone. The species may be listed as "in the management unit" (subject to federal fishing regulations) or "in the management unit for data collection only" (which means the landings are monitored, but the species are not subject to federal fishing regulations at this time). The list is subject to change but was accurate as of July 2003.*

**Species in the management unit**

1. Coastal migratory pelagics FMP
    cobia – *Rachycentron canadum*
    king mackerel – *Scomberomorus cavalla*
    Spanish mackerel – *Scomberomorus maculatus*
2. Red drum FMP
    red drum – *Sciaenops ocellatus*
3. Reef fish FMP
    – Snappers (family Lutjanidae)
    queen snapper – *Etelis oculatus*
    mutton snapper – *Lutjanus analis*
    schoolmaster – *Lutjanus apodus*
    red snapper – *Lutjanus campechanus*
    cubera snapper – *Lutjanus cyanopterus*
    gray (mangrove) snapper – *Lutjanus griseus*
    dog snapper – *Lutjanus jocu*
    mahogany snapper – *Lutjanus mahogoni*
    lane snapper – *Lutjanus synagris*
    silk snapper – *Lutjanus vivanus*
    yellowtail snapper – *Ocyurus chrysurus*
    wenchman – *Pristipomoides aquilonaris*
    vermilion snapper – *Rhomboplites aurorubens*
    – Groupers (family Serranidae)
    rock hind – *Epinephelus adscensionis*
    speckled hind – *Epinephelus drummondhayi*
    yellowedge grouper – *Epinephelus flavolimbatus*
    red hind – *Epinephelus guttatus*
    goliath grouper (jewfish) – *Epinephelus itajara*
    red grouper – *Epinephelus morio*
    misty grouper – *Epinephelus mystacinus*
    warsaw grouper – *Epinephelus nigritus*
    snowy grouper – *Epinephelus niveatus*
    Nassau grouper – *Epinephelus striatus*
    black grouper – *Mycteroperca bonaci*
    yellowmouth grouper – *Mycteroperca interstitialis*
    gag – *Mycteroperca microlepis*
    scamp – *Mycteroperca phenax*
    yellowfin grouper – *Mycteroperca venenosa*
    – Tilefishes (family Malacanthidae)
    goldface tilefish – *Caulolatilus chrysops*
    blackline tilefish – *Caulolatilus cyanops*

anchor tilefish – *Caulolatilus intermedius*
blueline tilefish – *Caulolatilus microps*
tilefish – *Lopholatilus chamaeleonticeps*
– Jacks (family Carangidae)
    greater amberjack – *Seriola dumerili*
    lesser amberjack – *Seriola rivoliana*
    banded rudderfish – *Seriola zonata*
– Triggerfishes (family Balistidae)
    gray triggerfish – *Balistes capiscus*
– Wrasses (family Labridae)
    hogfish – *Lachnolaimus maximus*
– Sand perches (family Serranidae)
    dwarf sand perch – *Diplectrum bivittatum*
    sand perch – *Diplectrum formosum*
4. Shrimp FMP
    brown shrimp – *Farfantepenaeus aztecus*
    pink shrimp – *Farfantepenaeus duorarum*
    royal red shrimp – *Hymenopenaeus robustus*
    white shrimp – *Litopenaeus setiferus*
5. Spiny lobster FMP
    spiny lobster – *Panulirus argus*
    spiny lobster – *Scyllarides nodifer*
6. Stone crab FMP
    stone crab (Cedar Key - north) – *Menippe adina*
    stone crab (Florida peninsula) – *Menippe mercenaria*
7. Corals and coral reefs FMP
    corals of the class Hydrozoa (stinging and hydrocorals)
    corals of the class Anthozoa (sea fans, sea whips, sea pens, precious coral, stony corals)

**Species in the management unit for data collection only**

1. Coastal migratory pelagics FMP
    dolphinfish – *Coryphaena hippurus*
    little tunny – *Euthynnus alletteratus*
    bluefish – *Pomatomus saltatrix*
    cero – *Scomberomorus regalis*
2. Spiny lobster FMP
    spotted spiny lobster – *Panulirus guttatus*
    smooth tail lobster – *Panulirus laevicauda*
    Spanish slipper lobster – *Scyllarides aequinoctialis*

major changes are needed in the way catch allocations are made. We must get the fox out of the henhouse.

There are also basic problems with the fishing industry itself. In the first place, there are far too many fishermen after too few fishes, or as is often said, the industry is overcapitalized. In US Gulf coastal waters the commercial fishing fleet involves several thousand boats, and the recreational fleet represents at least 10 times this number. It is difficult to monitor the actions of so many boats. Profits are marginal, and temptations to cheat are high. In recent years the fishermen have brought many lawsuits against the governing agency, mainly over estimates of stock size and the magnitude of the allowable catch, and even if these suits are eventually dismissed, enforcement of recommended catch limits can be delayed for months or years. A long-range solution to the overfishing problem in the Gulf is to reduce the number of licensed fishermen by the establishment of a "limited entry" fishery. If only a limited number of permits are issued, the size of the exploiting fleet can be controlled. Other measures include restrictions on gear type as well as the seasons and areas in which fishing can take place.

In 1996, Congress, concerned about problems of overfishing, habitat destruction, excessively large bycatch, and the relative lack of scientific knowledge, passed the federal Sustainable Fisheries Act (often referred to as the Magnuson-Stevens Act), which amends and updates the earlier Fishery Conservation and Management Act in a number of ways. Among other things, the act highlights the necessity for the identification and long-term protection of *essential fish habitat,* defined as "the waters and substrate necessary to fish for spawning, breeding, feeding, or growth to maturity." To implement this provision, the council and the National Marine Fisheries Service must (1) describe and identify essential habitat for each species, (2) identify adverse impacts caused by fishing and other activities, and (3) consider measures to conserve and enhance essential fish habitat. All of this turns out to be a very formidable task, which is to some extent beyond the capabilities of the council. Clearly, the greatest problems lie in the coastal zone, where commercial development, pollution, and habitat degradation and loss threaten nursery area habitat for many species. In tackling this set of problems the Gulf council would have to deal with matters in state waters and consult with the EPA, Army Corps of Engineers, local governments, and so on. The NMFS is helping out in these matters.

Other problems are associated with essential fish habi-

tat seaward of the coastal zone, and because there are so many species covered in the management plans, by some reckoning, essential fish habitat could include the entire Gulf of Mexico. However, some portions of the continental shelves are clearly more important than others for the commercially important fish and invertebrate species, and these are being scrutinized carefully. An interesting development is the establishment on the continental shelves of *fishery reserves,* where fishing activity is restricted or prohibited. The essential point here is that broad authority has been given to really plan and manage living resources of the entire US portion of the Gulf of Mexico. This level of management authority transcends present capabilities of the Gulf council, and it must rely heavily on NOAA, which has considerable information-gathering and environmental-monitoring capabilities and a much better perspective on the larger picture.

## SANCTUARIES, RESERVES, AND OTHER AREAS OF MANAGEMENT INTEREST

Any agency concerned with protection of essential fish habitat must first ascertain what is already protected. Then it can proceed to determine what yet remains to be afforded protection status. Table 18.3 lists those coastal and marine areas of the US Gulf that are already legally protected, have special national programs, or are of particular concern to the National Marine Fisheries Service. Two of the areas are National Marine Sanctuaries, the Florida Keys and the Flower Garden Banks. The Office of National Marine Sanctuaries (2008) has given a condition report on the Flower Garden Banks. Four of the areas are National Estuarine Research Reserves (formerly called Estuarine Sanctuaries). Both groups are administered by NOAA. Seven coastal areas have National Estuary Programs under the aegis of EPA as estuaries of national significance. These are now eligible to receive special federal funding for study, cleanup, and protection. An additional 15 coastal and marine sites are listed as Areas of Particular Concern to the NMFS for possible designation and protection as essential fish habitat. Some of these sites are already protected, but most are not.

If all the localities on the list eventually receive protection status, then 27 areas around the US Gulf will have some form of legal protection. Nearly half the sites are in bays and estuaries, where they would afford protection for nursery habitats of estuary-related species. Most of the remainder are nearshore or offshore banks and reefs, which would provide protection for the unique fish and

**Table 18.3.** *Coastal and marine areas of the US Gulf of Mexico that are legally protected or that are of particular management interest. The list does not include national parks and wildlife refuges or coastal areas under state management.*

**National Marine Sanctuaries (NOAA)**

| | |
|---|---|
| Florida Keys, FL | Flower Garden Banks, TX/LA |

**Estuarine Research Reserves (NOAA)**

| | |
|---|---|
| Rookery Bay, FL | Weeks Bay, AL |
| Apalachicola Bay, FL | Mission-Aransas Bay, TX |

**National Estuary Programs (EPA)**

| | |
|---|---|
| Charlotte Harbor, FL | Barataria-Terrebonne Bays, LA |
| Sarasota Bay, FL | Galveston Bay, TX |
| Tampa Bay, FL | Coastal Bend Bay and Estuaries |
| Mobile Bay, AL | Program, TX |

**NMFS – Habitat Areas of Particular Concern (NOAA)**

Tortugas North and South Ecological Reserves, FL
Pulley Ridge, FL
Madison-Swenson Marine Reserve, FL
Florida Middle Grounds, FL
Flower Garden Banks, TX/LA
Reefs and banks of the northwestern Gulf, TX/LA

| | | |
|---|---|---|
| Stetson | Geyer | Alderdice |
| McNeil | McGrail | Jakkula |
| Bright | Bouma | |
| Rezak | Sonnier | |

invertebrate populations that inhabit such areas. Missing are some of the mangrove habitats of the Everglades, the seagrass meadows of Florida Bay and the Big Bend area of Apalachee Bay, the unique banks and "pinnacles" off Alabama and Mississippi, the vast stretches of soft (sand, silt, and/or clay) bottoms of the open continental shelves, and special habitats associated with the shelf break and upper continental slope. However, this is a good beginning for what must become a continuing effort to locate and preserve habitats critical for protection of the living resources of the Gulf.

After passage of the federal Sustainable Fisheries Act (1996) mandating protection of essential fish habitats, and in response to a presidential executive order, the US Department of Commerce established within NOAA the National Marine Protected Areas Center, whose ultimate goal is to create and sustain a national system with re- gional networks of marine protected areas (MPAs). To aid in this endeavor, in the year 2000 the agency set up a 30-member Federal Advisory Committee made up of indi- viduals representing diverse backgrounds including sci- ence, conservation, natural resources, and governmental agencies. The committee has submitted its first recom- mendations, and it continues to provide advice and guid- ance on how to bring the system to reality and make it succeed in the long run.

What we observe here are early steps in another stage of the planning and zoning of the nation's coastal and marine waters and seabed. First, there was the establish- ment of designated shipping lanes for marine transporta- tion traffic and safe anchorage areas for idle ships. Next, there were specified areas related to national defense (tar- get areas for gunnery practice, submarine fairways, and designated sea-bottom areas for disposal of munitions and other materiel left over from the Second World War). Then, areas were set aside for various energy-related functions, including offshore facilities for off-loading petroleum and liquid natural gas (LNG), "graveyards" for spent oil rigs and platforms, and so forth. Now, finally, we are beginning to set aside areas as sanctuaries, reserves, and essential habitat for the living resources of the coast and the open Gulf.

However, this effort is definitely different from those of previous stages. Although the present focus is on site- specific areas, the approach is much broader. Through the National Marine Fisheries Service NOAA is now focusing on "total ecosystem management" of the US portion of the Gulf of Mexico. As pointed out by Belsky (1986, 1989) and others, the United States clearly has the international authority for broadscale marine ecosystem management within its Exclusive Economic Zone, and the two previ- ously mentioned fishery management acts provide NOAA and the fishery management councils with the federal authority for such management, at least in a preliminary way. Conceived broadly, total marine ecosystem man- agement would include concern for the various environ- mental factors as well as all the biological components and ecosystem processes—a very tall order. Where all this might eventually lead we can only guess, but in any event, it does go well beyond the setting aside of a small number of critical habitat areas, as important as this is. To manage the entire Gulf ecosystem we would need the cooperation of the Mexican and probably the Cuban gov- ernments, but a coordinated management effort would be to the benefit of all.

## FUTURE PROSPECTS

The Gulf of Mexico, which for millions of years has persisted as a relatively undisturbed natural system, is now coming under progressively more intense human utilization and disturbance. This comes at a time when our knowledge of the environmental processes and biological components is beginning to come into focus and when legal authority and management systems are being set in place. Looking ahead, we must ponder the problems to be faced and the management structure and processes needed to cope with the human–marine ecosystem complex. Our real concern is to develop an overview without becoming bogged down in the details.

### Major issues

Here we recognize six key issues to be addressed. These deal with coastal habitat loss, water pollution, invasive species, freshwater inflow to estuaries, biodiversity, and renewable resources.

#### REVERSAL OF COASTAL HABITAT LOSS

Historically, the Gulf of Mexico was a very heterogeneous place, and the coastal zone, in particular, displayed great habitat diversity. In some areas marshlands and other coastal habitats were quite extensive, and each played a particular and significant role in the cycling of nutrients, production of biological resources, and maintenance of sustainable yields over long periods. Beaches, swamps, marshlands, seagrass meadows, oyster reefs, and mudflats, together with the open-water areas, channels, and passes were all integrated into sustainable functional ecosystems. As documented in the previous chapter, the increased presence and the industrial and other activities of human populations in the coastal zone have been accompanied by massive and progressive loss of wetlands and other coastal habitats. Through dredged channels and canals, spoil banks and landfills, concrete walls and buildings humans have eliminated native habitats to make way for industrial, municipal, residential, and harbor facilities and structures.

However, environmental ignorance is no longer an issue. We have now begun to realize the biological price of the march of civilization. Future management decisions must be made with full knowledge of the environmental costs. We must ask whether a proposed facility has to be placed on a sensitive coastal habitat or whether it can be located elsewhere, thus saving scarce habitat. We must advance the sciences of habitat enhancement, habitat restoration, and mitigation measures that can be brought into play when further habitat loss is unavoidable. Detailed environmental knowledge, sensitivity, and a sense of stewardship must guide future decisions dealing with natural habitats and environmental heterogeneity, in general.

#### WATER POLLUTION CONTROL

The maintenance of healthy coastal and marine ecosystems is first of all an exercise in pollution control, and here we are concerned about the management of pathogens and parasites as well as excess nutrients and toxic chemicals. The US Environmental Protection Agency (EPA) is the primary federal regulatory authority, and together with state and local agencies it is largely responsible for control of the biological and chemical pollutants that still pour into our coastal and marine waters. By issuing permits to point-source polluters, the agency actually sanctions present levels of environmental degradation.

To alleviate this situation EPA should require tertiary treatment of all human and animal wastes emanating from sewage treatment plants, septic systems, animal feedlots, and other high-nutrient organic input sources. In addition, the agency should definitely set maximum limits on the levels of nutrients and toxic chemical pollutants allowed to enter natural waters, and over time the maximum levels should be progressively lowered. Permits for discharges should be issued only up to the point where the maximum level is achieved. This would undoubtedly create hardship for some municipalities and polluting industries and would have consequent political repercussions, but in the public interest hard decisions must be made. EPA has now been in business for nearly four decades, and by its own admission, the estuaries and other coastal waters around the Gulf are still in deplorable condition. The agency must be made to stand firm as guardian of the public and ecosystem health. Controlling pollution from upstream and non–point sources is a much more difficult problem, but working with other federal as well as state and local agencies, over time EPA could also gradually bring down pollution levels from these sources.

#### CONTROL OF INVASIVE SPECIES

Exotic invasive species can be introduced into coastal and marine waters of the Gulf by release from home aquaria, escape from mariculture operations, and transport on outside hulls or in ballast water of seagoing ships. Once established in Gulf waters, invasive species are virtually impossible to eradicate, and they may outcompete and eventually replace some of the native species. They also sometimes interfere in major ways with human activities, and their control can be quite costly.

In the past two decades Congress has passed several laws to aid in the control of exotic species and to reduce the chances of further introductions. Some of this legislation deals with safer aquacultural practices, while other portions prohibit the release of ballast water in coastal areas. Here there are no simple solutions. Private citizens must be mindful of the dangers associated with releasing aquarium flora and fauna originating elsewhere, and scientists and governmental authorities must remain alert to the arrival of exotic species so that eradication measures can be attempted before the invaders gain a firm foothold.

## MAINTENANCE OF ADEQUATE FRESHWATER INFLOW TO ESTUARIES AND OTHER COASTAL HABITATS

It has been well documented that the health and productivity of bays, estuaries, and coastal lagoons along the margins of the Gulf depend, in large measure, on the continuance of reasonable levels of freshwater inflow. Therefore, any major upstream diversion of freshwater for other purposes must be considered deleterious to the biological health of the coastal systems. As discussed in the previous chapter, the Florida Everglades have been devastated by the diversion of freshwater for municipal and agricultural use. The Texas Water Plan calls for the diversion of water from the streams of east and central Texas to provide freshwater for Houston and other cities and agricultural lands of arid south Texas.

Local management of freshwater resources generally falls under the jurisdiction of state governments, although federal agencies are also involved, and all of the agencies are well aware of the environmental costs of the various freshwater diversion trade-offs. Strong pressure from coalitions of concerned citizens' groups may be required to force the agencies to make and stick with environment-friendly freshwater allocation decisions. Meanwhile, a great deal can be done to conserve freshwater resources. In some of the areas of concern, citizens and municipalities still waste great quantities watering lawns, using old-fashioned, high-volume flush toilets, and so forth. Important water conservation lessons could be learned from places like Phoenix and Tucson in Arizona, where freshwater is really scarce.

## CONSERVATION OF BIODIVERSITY

The great tragedy of our time is the worldwide loss of biodiversity as local populations and whole species become extinct due to the degradation and destruction of natural habitats, overfishing, and other factors. The great challenge, indeed the ecological imperative, for managers of coastal or marine areas or resources is to preserve the biodiversity of the ecosystems within their jurisdiction in the face of other demands of advancing civilization. Sensitive management for sustainable natural resources requires a thorough understanding of the basic concepts, associated problems, and potential solutions to the biodiversity dilemma.

Biodiversity has been defined as the variability among living systems from all sources (i.e., local and interpopulation genetics, species composition, functional relationships, and community/ecosystem-level diversity). Within local sexually reproducing populations each individual is genetically unique and different from all others. Populations of the same species, geographically isolated from one another (as, for example, blue crabs from Tampa Bay, Mobile Bay, Lake Pontchartrain, and Galveston Bay), are even more genetically distinct from one another due to local mutations, different environmental selective pressures, genetic drift, and other factors. Likewise, as noted in earlier chapters, the species compositions and abundances vary from one locality to another.

Functional diversity refers to the range of functions and species interactions occurring within natural ecosystems (producers vs. consumers, predators vs. prey, suspension vs. deposit feeders, etc.). Community or ecosystem diversity expresses the number of species (or some other quantitative index of diversity) present in a given community or habitat area, and it is sometimes referred to as "habitat diversity." On a broader scale the variety of ecosystems or habitat types present in a given region together forms a mosaic with even greater biodiversity.

Field studies have revealed that species diversity is greater in the sea than on land or in freshwater. In marine areas it is generally greater in the benthos than in the water column, and in the pelagic realm it is greater in coastal than in oceanic areas. Genetic diversity tends to be higher in stressed environments, that is, areas such as the coastal zone where over evolutionary time the species have had to adapt to variable environmental conditions. In turn, high species diversity (or species redundancy) leads to greater efficiency in the transfer of chemical elements through the ecosystem (biogeochemical cycling).

Without question, the best way to conserve genetic and species diversity is to preserve habitats. We do not just preserve alligators, we preserve swamps. Then we have the alligators and all the other plant and animal inhabitants of the area in a self-sustaining system at no cost to society. It is even better, wherever possible, to preserve habitat mosaics (such as mangrove swamps, seagrass meadows, and coral reefs together), which would include more biodiversity than conserving each habitat type in isolation. In any case, protection of specific areas should be part of a larger overall strategy that involves the following considerations:

- Protect individual local areas as part of a planned regional system of coastal and marine reserves covering all the basic habitat types.
- At the regional level strive to protect habitat mosaic areas, light-use areas, and corridors between protected areas (which could allow for some interbreeding and exchange of genetic material between otherwise fragmented and isolated populations).
- Lower the level of pollution and other environmental stresses both within and outside the primary protected areas.
- Periodically inventory the protected areas to ensure that the protective measures are effective.
- Develop and, where necessary, implement strategies and methods for the enhancement and/ or restoration of degraded habitats and their living populations.

The protection of natural habitats, particularly in the coastal zone, carries many rewards for human society, some of them clearly of great economic value. For example, protection of wetlands and related coastal areas stabilizes the coastline and aids in the protection of inland areas from hurricanes and storm surges. It facilitates the recharge of groundwater resources and inhibits saltwater intrusion into underground aquifers. It provides habitat for birds, mammals, and other types of wildlife and critical nursery areas for many species of fishes and invertebrates, some of great commercial importance. Protected natural habitats also aid in cleansing the waters and reducing the loads of excess nutrients and toxic pollutants. Natural areas can create pleasant scenic vistas and may increase the level of ecotourism. They can be used effectively as natural laboratories and outdoor classrooms for student and public education about the native biota

and conservation issues. Properly planned, operated, and periodically assessed, regional systems of habitat reserves should aid in sustaining high levels of native genetic and ecological resources for the use and enjoyment of future human generations. They must become integral elements of all coastal and marine management planning.

## IMPLEMENTATION OF AN ECOSYSTEM-BASED APPROACH TO THE MANAGEMENT OF COASTAL AND MARINE FISHERIES

Traditionally, fishery scientists have focused on the management of single target species. Wild fish stocks would be assessed and then allowable catch limits established with little concern about the effects of the fishery on other target and nontarget species or on the environment. For such simple systems, mathematical models have proven to be useful tools for predicting outcomes of alternative management strategies. Recently, some groups have attempted to manage marine fisheries on a multispecies basis, that is, to manage several species together as a coherent unit. In such cases the species of interest might be closely related in an ecological sense as, for example, several predator species competing for more or less the same prey or both predator and prey species managed together.

Over the past decade and a half much attention has been focused on the possibility of managing large numbers of species together, giving full consideration to the relationships of those species to each other and to the broader ecosystems of which they are a part. Such ecosystem-based management also considers the fishery within the context of the human socioeconomic system. Fully implemented, this approach would require the input of prodigious amounts of information, much of it unavailable at the present time, and it would entail a quantum leap in the complexity with which management must contend.

These three approaches (single species, multispecies, and ecosystem based) are not mutually exclusive. Each might be appropriate for a particular species or group of species under a given set of circumstances, and in the world of practical fisheries management it is possible that all three could be implemented (for different species groups) simultaneously. Ecosystem-based management practices are currently being brought into play around the world, and since it is clearly the intent of Congress that the management of our own nation's coastal and ocean fisheries should involve the progressive introduction of ecosystem-based management practices, this matter requires some explanation.

**Table 18.4.** *A summary of recommendations made by the Ecosystem Principles Advisory Panel in its report to Congress. (EPAP 1999.)*

**Develop a Fishery Ecosystem Plan** (for each fisheries management area) that would require the Council to take the following steps:

– Delineate the geographic extent of the ecosystem(s) that occur(s) within the Council's authority.

– Develop a conceptual model of the food web.

– Describe the habitat needs of different life history stages for all plants and animals that represent the "significant food web" and how they are considered in conservation and management measures.

– Calculate total removals, including incidental mortality, and show how they relate to standing biomass, production, optimum yields, natural mortality, and trophic structure.

– Assess how uncertainty is characterized and what kinds of buffers against uncertainty are included in conservation and management actions.

– Develop indices of ecosystem health as targets for management.

– Describe long-term monitoring data and how they are used.

– Assess the ecological, human, and institutional elements of the ecosystem that most significantly affect fisheries and are outside Council/Department of Commerce authority. Included should be a strategy to address these influences in order to achieve both Fishery Management Plan and Fishery Ecosystem Plan objectives.

**Take measures to implement the Fishery Ecosystem Plan.**

– Encourage the Council to apply stated ecosystem principles, goals, and policies.

– Provide basic ecological training for Council members and staff.

– Prepare guidelines for preparation of the Fishery Ecosystem Plans.

– Develop demonstration Fishery Ecosystem Plans.

– Provide oversight to ensure development of and compliance with the Fishery Ecosystem Plans.

– Enact legislation requiring Fishery Ecosystem Plans.

**Carry out research required to support management needs.**

– Determine the ecosystem effects of fishing.

– Monitor trends and dynamics in heavily exploited marine ecosystems.

– Explore ecosystem-based approaches to fisheries governance.

---

A few years ago, at the request of Congress, the National Marine Fisheries Service convened a panel of experts to, among other things, recommend how best to integrate ecological principles into future fisheries management and research (EPAP 1999). This group pointed out that a comprehensive ecosystem-based fisheries management approach "would require managers to consider all interactions that a target fish stock has with predators, competitors, and prey species; the effects of weather and climate on fisheries biology and ecology; the complex interactions between fishes and their habitat; and the effects of the fishing on fish stocks and their habitat." Fortunately, not all of this has to be done at once. This approach could be implemented gradually. An important first step would be for managers to consider how

the harvesting of one species might impact other species in the ecosystem. The panel recommended that for each large marine ecosystem (such as the Gulf of Mexico) there should be developed an overall *Fishery Ecosystem Plan* (FEP) to provide Fishery Management Council members with a clear description of the biological and institutional context within which the fisheries are managed, direct how the information should be used, and set policies through which management options could be developed.

A brief summary of the panel's recommendations is provided in table 18.4. The basic recommendations are to develop a meaningful Fishery Ecosystem Plan, begin implementing it, and carry out the scientific research needed to support the plan and its implementation. Here we see some of the nation's top environmental scientists steering management so as to permit sustained harvests without compromising the supportive ecosystems. This marriage of science and management is the linchpin of ecosystem-based management, and its continuance is the real hope for sustainable yields, retention of biodiversity, maintenance of ecosystem health, and long-term ecosystem viability and productivity. However, none of the approaches to the management of coastal and ocean fisheries is likely to achieve great success unless strong steps are taken to prevent overfishing, protect critical natural habitats and biodiversity, and support expanded research and environmental monitoring programs.

## Management problems and process

During the past several decades it has become apparent that the life-support system of this planet cannot survive without active human assistance. Broad appreciation of this fact has stimulated interest in the science of ecosystem management. However, ecosystems cannot be managed without a sophisticated input of technical information about the system being managed, characteristics of the norm, symptoms of perturbation, levels of tolerance, and carrying capacities. The necessity of supplying this information in a usable form has placed a great burden on the scientist. Proper understanding and use of this information transfers some of this burden to the shoulders of the manager, but in truth, there must be a scientist-manager partnership.

MANAGEMENT OF THE ECOSYSTEM

Any ecosystem involves a group of organisms of various sizes living together in a defined area and interacting with factors of the environment and with each other to maintain, over long periods, the flow of nutrients and energy through a characteristic trophic structure in which a portion of the nutrients is continually recycled. Aspects of coastal and marine ecosystems important to management include the following:

- Biodiversity is extremely important to natural ecosystem functioning.
- Components and processes are linked in complex and often subtle ways.
- Ecosystems exist in various sizes ranging from small (an estuary) to large (the Gulf of Mexico).
- Large ecosystems consist of various subsystems linked together through the transfer of nutrients, energy, and biological species.

- Ecosystem boundaries are open, and ecosystems may receive from or release to other ecosystems nutrients and biological species.
- Because of the vicissitudes of weather, water currents, activities of the organisms themselves, and the laws of probability, ecosystems are subject to high degrees of natural variability.
- Ecosystems have real thresholds and limits which when exceeded can result in major system restructuring or collapse.
- Once thresholds and limits have been exceeded, the changes can be irreversible.

In dealing with ecosystems, management must handle vast amounts of information, but often there are not enough data to support specific decisions. Nevertheless, management cannot await scientific certainty and often must act on the basis of expert judgment plus whatever information is available.

Imposed on the natural variability is that induced by human activities. This additional level of variation must somehow be sorted out from the natural variability. This is possible, to some extent, if monitoring of environmental phenomena is carried out at appropriate time and space scales. However, even at best the ability to predict ecosystem behavior is limited, and therefore ecosystem management must be carried out on a precautionary basis. As a hedge against unforeseen variability, a significant margin of safety should be built into environmental decisions.

In dealing with marine fisheries, the entire ecosystem is the management unit. There are so many species of interest that managing the whole system may be the best way to manage most of the stocks. Optimizing allowable catch levels for all species at the same time is a logical impossibility, and in some cases the managers must attempt to work out smaller segments of the larger ecosystem and thereby focus on a few species of interest at a time.

MANAGEMENT AND ECOSYSTEM HEALTH

In our world today we must distinguish between relatively natural unmodified ecosystems, on the one hand, and those heavily modified and managed for optimum human use, on the other. These two environmental states are generally incompatible. To manage for naturalness will preclude most human uses; to manage for heavy human use will preclude naturalness. Herein lies the basic dilemma. We cannot afford to eliminate the native species and ecosystems of the planet, yet we must use them to the benefit of society. Therefore, we need a practical definition of ecosystem health that includes the needs of nature while recognizing legitimate, nondestructive uses by society. Accordingly, ecosystem health has been defined as "that state in which the components and processes remain well within specified limits of system integrity selected to assure that there is no diminution in the capacity of the system to render its basic services to society throughout the indefinite future" (Darnell and Soniat 1981).

Critical to this definition are three factors: (1) dedication to the proposition, espoused in the National Environmental Policy Act of 1969, that unborn generations shall not inherit a ravaged biosphere; (2) existence of a knowledge base sufficient for the determination of critical limits of tolerance; and (3) willingness of society to establish limits on human intrusion well short of the absolute limits of system tolerance. Thus, ecosystem health must be defined in terms of two limits: the one that cannot be transgressed under any circumstances without incurring long-range system damage (the *categorical limit*), and the other that might be temporarily violated without major long-term damage to the system (the *conditional limit*).

Categorical limits are set by those species that inhabit the ecosystems. Major loss of habitat, species, and locally adapted genetic strains is irreversible. As pointed out earlier, the preservation of natural habitats in sufficient variety, quantity, and quality is the best way to ensure the long-range perpetuation of most of the local species and genetic stocks. However, not all species can be retained in special reserves, and this is particularly true in the case of estuarine and other coastal species, which may utilize several different habitats during the course of their complicated life histories. Hence, all coastal and marine ecosystems, even those heavily utilized by society, must serve a secondary function of maintaining the native genetic stocks.

Conditional limits mark the beginning of ecosystem stress. Loss of genetic material is relatively minor and tolerable, and the environmental damage can be repaired once the pressure is removed. Critical factors here are what levels of environmental quality we are willing to pay for, what levels of resource harvest we can be satisfied with, and how much habitat we can agree to set aside in our natural area reserves.

## A SYSTEMS APPROACH TO COASTAL AND OCEAN MANAGEMENT

The proper management of our coastal and ocean resources will require the input of large volumes of data of various types, and we must develop the capacity for reaching rational decisions once the necessary information is available. Systems approaches provide powerful means of organizing and summarizing vast bodies of information and, by means of algorithms, mathematical models, and other techniques, of translating the information into different levels of perception, analysis, and action (Darnell and Shimkin 1972). With this information and these models, the general problem for the manager is to determine how to proceed from the present state (of the natural and social environment) toward some desired future state with a high degree of probability of attaining the goal. In moving from the present to the desired future state the manager must remain cognizant of the two levels of constraint, the categorical limits (which are absolute) and the conditional limits (which are related to some cost-effectiveness calculation). The distinction between these two limits is quite important because the line separating them defines the break between areas controlled by natural ecological and genetic constraints (or by legal regulations) and those open for bargaining. Thus, the environmental manager must be knowledgeable about the present state and must also be able to define a desirable future state as well as an environmentally and socially tolerable pathway between the two.

As shown in figure 18.2, the coastal and ocean management problem can be reduced to an operation in quality control in which human utilization is balanced against environmental effects. It is a problem of operation under constraints where future time must be phased into foreseeable horizons for short-term planning. So, as we proceed into the future by a successive series of appropriately informed and rational decisions, we can periodically adjust coefficients of the human use–environmental quality equations for achievement of successively perceived optimal states. Thus, the progressive introduction of systems approaches should bring orderly criteria evaluation into standard management practice, develop better models and other methods to support complex management decisions, and train professionals for higher levels of competence in the management of coastal and ocean resources. This is especially true for the coastal zone, where human use is greatest and the problems are most complex.

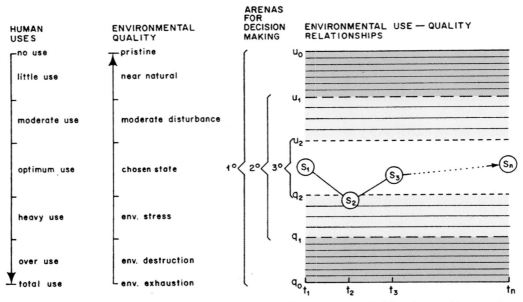

**Figure 18.2.** The pathway into the indefinite future for successive management decisions balanced between human use, on the one hand, and environmental quality, on the other. In this figure, u ($u_0$, $u_1$, $u_2$) represents different levels of human use, q ($q_0$, $q_1$, $q_2$) represents different levels of environmental quality, and t ($t_1$, $t_2$, $t_3$, . . . $t_n$) represents different time horizons from the present ($t_1$) into the indefinite future ($t_n$). The term s ($s_1$, $s_2$, $s_3$, . . . $s_n$) represents successive environmental states from the present state ($s_1$) to the indefinite future state ($s_n$). Management must strive to maintain the environmental state within the clear area, that is, the optimum use pathway between $u_2$ and $q_2$ where human use is optimized and the environment is not placed under major stress. The state may be allowed to temporarily exceed the limits of $u_2$ or $q_2$ (into the light-shaded area) without incurring permanent damage to the system. However, management must at all costs avoid allowing the state to enter the dark-shaded areas beyond $u_1$ (where the environment is little used and resources are wasted) and $q_1$ (where there is major environmental damage and irretrievable loss of natural resources). (From Darnell and Shimkin 1972.)

LINKING COASTAL ZONE AND OCEAN MANAGEMENT
As our attention turns more and more toward natural resource and ecosystem-based management, we must shift our focus from local areas and existing political boundaries to larger systems and boundaries established by nature. Although there are ample practical reasons to have separate management regimes for the coastal zone and the open Gulf, there are also compelling reasons why these two management thrusts should be integrated or at least closely coordinated. From what has been said in previous chapters, it is clear that the bays, estuaries, and coastal lagoons with their open waters, marshes, seagrass meadows, and swamps are simply subsystems of the larger Gulf of Mexico ecosystem. These inside coastal and open Gulf waters are bound together through the exchange of water, salts, nutrients, organic matter, and biological species, and this exchange is critical to the life histories of many of the coastal species. The individual species themselves know no political boundaries, and during their life histories they may move back and forth between state and federal jurisdictions. Long-term sustainment of high levels of commercial fishery production in the Gulf depends largely on the maintenance of ample natural habitats and reasonable levels of water quality in the estuaries and coastal lagoons.

Both the coastal zone and ocean management entities must deal with natural resource planning and management under societal and ecological constraints. However, intensive coastal zone management has been carried out for several decades, and its experience in dealing with various contentious issues, strongly held opinions, and consensus building could prove useful to managers of ocean systems. In turn, the latter might provide insight into the handling of large systems and the science-based resource management approach to the ocean areas where environmental factors play a dominant role. However it is done, both management regimes must enlarge their scope to include geographically larger areas and the various subsystems that together make up the larger Gulf of Mexico ecosystem. The commonality of their goals and areas of interest is great, and proper management of the larger system requires the expertise and attention of both.

As a step toward development of the coastal zone–

ocean management linkage, the federal government must develop an overall ocean policy backed up by regionally specific management plans in which the regions coincide with the fisheries management areas. In other words, there should be a specific federal management plan for the Gulf of Mexico. The regional plans should include the open-ocean areas and should extend landward through the coastal zone. In turn, the five Gulf states should develop their own ocean management policies designed to interface with the federal plan, and these should become part of their amended coastal zone management plans. In order to facilitate management of the entire Gulf of Mexico ecosystem, the federal plan should also include provisions for the cooperative and coordinated management of the Gulf with Mexico and Cuba.

SIMPLIFICATION OF THE MANAGEMENT PROCESS
In dealing with coastal and ocean problems, a guiding principle should be that the best governance is the simplest and least required to do the job effectively. In coastal areas under state jurisdiction three levels of government (federal, state, and local) all have some management authority. In former years there were three sets of regulations and three arenas for decision making about the same basic issues. Laws and regulations overlapped and were often in conflict. To simplify the situation the Coastal Zone Management Act (1972) established the "consistency" principle. Before a state coastal zone management plan was accepted by the federal government, it had to meet federal standards and conform to federal regulations. However, once the plan had been accepted, when issues came up, if they met requirements of the state plan, then they were automatically in compliance with federal regulations. This harmonization of federal and state requirements has greatly simplified the coastal zone management process.

When two or more federal agencies share responsibility for action, through conference and consultation they can work out in advance what the individual agency responsibilities are and which should function as the *lead agency*. For example, both the Minerals Management Service and the US Coast Guard were involved in dealing with major marine oil spills. By a memorandum of understanding between the two agencies, the Coast Guard was designated the lead agency for immediate response, control, and cleanup.

Likewise, when federal and state agencies share authority for managing natural resources, through consul-

tation the individual responsibilities can be allotted to one or the other agency. For example, in dealing with the management of stocks of coastal fish species that utilize estuarine and continental shelf waters beyond normal state jurisdiction, both the state and federal fishery management agencies (National Marine Fisheries Service) or their appointees (Fishery Management Councils) must be involved. However, by conference they have divided the species between them and determined which species will be managed by the coastal state and which by the NMFS and the councils. These agreements have been formalized through interagency memoranda of understanding.

By means of these three mechanisms (consistency requirements, memoranda of understanding between federal agencies, and memoranda of understanding between federal and state agencies), the management of coastal and ocean resources of the Gulf of Mexico area has been greatly simplified, but there is still much room for further simplification. These and related mechanisms should be explored for additional ways of streamlining the process of coastal and ocean management.

CORNERSTONES OF COASTAL AND
OCEAN MANAGEMENT
Informed, intelligent, and sensitive management of our coastal and ocean resources must be founded on a set of guiding principles that reflect knowledge, practicality, and an altruistic view of management's responsibility. Such a set of principles has been expressed in the report of the US Commission on Ocean Policy (table 18.5). Here the 13 principles have been grouped into three categories: decision-making considerations, management considerations, and focus on the resources. In large measure, this compilation summarizes points made earlier in the present chapter, and it brings them together here for management consideration and guidance. Of all the principles, the most important is the one that relates to stewardship of the environment and natural resources. Unless the manager feels a strong ethical commitment toward the protection of nature and natural resources, the other principles do not matter much. Without such a commitment, management is reduced to the role of a simple mechanical steering device.

**GENERAL RECOMMENDATIONS**
Recognizing the importance of the nation's coastal and ocean resources, the declining quality of these environments, and the inadequacy of current institutions of gov-

**Table 18.5.** *Cornerstones of intelligent, sensitive, and informed management of our coastal and ocean resources. These are based on guiding principles stated in the report of the US Commission on Ocean Policy (2004).*

**Decision-making considerations**

Participatory governance. Governance of ocean and coastal uses should ensure widespread participation by all citizens on issues that affect them.

Best available science and information. Ocean and coastal policy decisions should be based on the best available information on natural and social processes that affect these environments.

Ocean, land, and atmosphere connections. Ocean and coastal policies and management decisions should take into account the complex interactions between the oceans, land, and atmosphere.

Understandable laws and clear decisions. To facilitate public understanding and compliance, the laws and management decisions relating to ocean and coastal issues and the reasons behind them should be made clear and available to all parties.

Accountability. Decision makers should be identifiable and accountable for their actions that affect the ocean and coastal resources.

Timeliness. Ocean and coastal governance systems should operate with as much efficiency and predictability as possible.

International responsibility. The United States should act cooperatively with other nations in developing and implementing international ocean policy.

**Management considerations**

Multiple use management. Ocean and coastal resources should be managed in a way that balances competing uses while preserving and protecting the overall integrity of the systems.

Ecosystem-based management. Ocean and coastal management should reflect the relationships among all ecosystem components, including humans and nonhuman species, and the environments in which they live. Management areas must be defined on the basis of ecosystems, not political boundaries.

Adaptive management. Ocean and coastal management programs should be designed to provide new information that can be used to continually improve the scientific basis for future management. Periodic reevaluation of past measures and incorporation of new information should continually improve future management decisions.

**Focus on the resources**

Preservation of marine and coastal biodiversity. Efforts should be made to maintain or recover the natural levels of biodiversity and ecosystem services provided by our marine and coastal systems. To this end downward trends in biodiversity must be reversed.

Sustainability of renewable resources. The harvest of renewable natural resources should be carried out in such a fashion as to meet the needs of the present generation without compromising the options of future generations.

Stewardship of the environment and renewable resources. Since the federal government holds the ocean and coastal resources in public trust for present and future generations, it must balance different uses for the benefit of all Americans. The public itself must recognize the value of the ocean and coasts and must support appropriate policies and actions that permit use while minimizing environmental impacts.

---

ernance, in the year 2000 Congress passed the National Ocean Policy Act, which mandated the establishment of a blue-ribbon Commission on Ocean Policy. This group was to assess the overall situation (and particularly the ocean governance and related agency structure) and recommend to the president, Congress, and the nation measures designed to remedy existing problems. The commission, which began its work in September 2001, held many public hearings, visited numerous sites, reviewed the latest scientific and technical literature, and received input from hundreds of experts. On the basis of all this information the commission issued a preliminary report in April 2004 and its final report in September of the same year. The administration's response was published in December 2004.

The commission's thoughtful and inclusive report put forth 282 recommendations detailing ways of improving our governance of coastal and ocean resources. The various recommendations are directed toward Congress, the executive branch of leadership, nine federal departments (Agriculture, Commerce, Defense, Health and Human Services, Homeland Security, Interior, Labor, State, and Transportation), three independent agencies (EPA, NASA, and NSF), and intergovernmental groups, as well as regional bodies. The recommendations range from new federal legislation needed and the establishment of a National Ocean Council to advise the president to such specific matters as studies of the bycatch of the commercial and recreational fisheries. Ten of the recommendations apply to the National Marine Fisheries Service, 20 to the US Environmental Protection Agency.

Although it is not possible to discuss all of these recommendations here, interested readers are encouraged to obtain a copy of the report and examine them on their own. Table 18.6 provides those general recommendations appearing in the report's executive summary. Among these are major themes that have been stressed in earlier pages of the present chapter, including the following:

- establishment of a National Ocean Council in the Executive Office of the President,
- restructuring and strengthening of government for more effective management of coastal and ocean resources,
- reduction of chemical pollution and enhancement of environmental quality of the coastal and ocean waters,
- restructuring and improvement of marine fisheries management,
- exploration of the use of dedicated access (limited entry) privileges in marine commercial fisheries,
- great improvement in our scientific knowledge of the coastal and ocean environments and ecosystems,
- increased use of scientific knowledge in environmental and natural resource decision making,
- broadening management's focus to embrace ecosystem-based decision making, and
- cultivation of a strong stewardship ethic.

The recommendations also place heavy emphasis on ocean education and the preparation and training of a new generation of leaders in the ocean sciences.

In 2005 members of the US Commission on Ocean Policy (the Joint Ocean Commission Initiative) issued a "Report Card" grading the government's response to various sections of the report, and these grades are given below.

A–  Initial response to the commission report
D+  National ocean governance reform
B–  Regional and state ocean governance reform
F   International leadership
D   Research, science, and education
C+  Fisheries management reform
F   New funding for ocean policy and programs

This brings up the general question of what can be done when through hostility, inertia, or misplaced priorities the government at any level fails to act or acts ineffectively on important environmental or natural resource matters. Fortunately, there is an effective remedy. A few decades ago Galveston Bay in Texas was widely recognized as one of the most polluted and environmentally degraded bodies of water in the nation. After passage of the Water Quality Act in 1987, EPA established the National Estuary Program with the goal of identifying and upgrading environmental quality in "estuaries of national significance." The governor of the state of Texas then nominated Galveston Bay to be recognized as an estuary of national significance. Shortly thereafter, several environmentally concerned citizens organized the Galveston Bay Foundation to ensure a strong public advocate for environmental concerns relating to the bay. They educated themselves on environmental matters, obtained much publicity, and recruited a large membership (now numbering hundreds of individuals from all walks of life—truly a grassroots organization). Their stated mission is to preserve and enhance the bay for its multiple uses through programs in education, conservation, research, and environmental advocacy.

In 1987 the US Army Corps of Engineers issued a draft environmental impact statement and held public hearings on its plans to widen and deepen the Houston Ship Channel (to a width of 600 ft and a depth of 50 ft) and to dump the dredge spoil, some of it contaminated, onto the bay bottom, thus destroying an estimated 11,000 acres of productive habitat including some oyster reefs. The newly formed Galveston Bay Foundation mounted strong oppo-

**Table 18.6.** *Executive Summary recommendations of the US Commission on Ocean Policy.*

**A. Use ecosystem-based management.**
– US ocean and coastal resources should be managed to reflect the relationships among all ecosystem components, including human and nonhuman species, and the environments in which they live. Applying this principle will require refining relevant geographic management areas based on ecosystem, rather than political, boundaries.

**B. Create a new national ocean policy framework.**
– Improve federal leadership and coordination.
– Strengthen federal agency structure to enable effective implementation of national ocean policy, and enhance the ability of agencies to address links among ocean, land, and air.

**C. Strengthen science and meet information needs.**
– Improve scientific understanding of the ocean and coastal environment and ensure effective science-based measures to use, safeguard, and restore ocean and coastal resources.
– Enhance the nation's ability to observe, monitor, and forecast ocean and coastal conditions to better understand and respond to the interactions among oceanic, atmospheric, and terrestrial processes.

**D. Enhance ocean education.**
– Improve decision-makers' understanding of the ocean.
– Cultivate a broad stewardship ethic.
– Prepare a new generation of leaders on ocean issues.

**E. Take these critical actions recommended by the US Commission on Ocean Policy.**
– Establish a national Ocean Council, chaired by an assistant to the president, and create a Presidential Council of Advisors on Ocean Policy in the Executive Office of the President.
– Strengthen NOAA and improve the federal agency structure.
– Develop a flexible and voluntary process for creating regional ocean councils, facilitated and supported by the National Ocean Council.
– Double the nation's investment in ocean research.
– Implement the National Integrated Ocean Observing System.
– Increase attention to ocean education through coordinated and effective formal and informal programs.
– Strengthen the link between coastal and watershed management.
– Create a coordinated management regime for federal waters.
– Create measurable water pollution reduction goals, particularly for non–point sources, and strengthen technical assistance and other management tools to reach these goals.
– Reform fisheries management by separating assessment and allocation, improving the Regional Fishery Management Council system, and exploring the use of dedicated access privileges.
– Accede to the United Nations Convention on the Law of the Sea.
– Establish an Ocean Policy Trust Fund based on revenue from offshore oil and gas development and other new and emerging offshore uses to pay for implementing the recommendations.

---

sition to this outrageous proposal. During the next several months, through a series of discussion meetings between representatives of the Corps, several other federal and state agencies, port authorities, and environmental groups, a final plan was worked out to reduce the width and depth of the Ship Channel (to 530 ft wide and 45 ft deep), greatly reducing the amount of material to be dredged. Contaminated sediments were to be disposed of elsewhere in safe areas on land, and the uncontaminated material was to be treated as a resource and used to enhance the quality of habitats around the bay. It would serve to create 4,500 acres of new intertidal marshland

and a 12-acre bird island, and restore shoreline protection structures around the bay. Further mitigation measures included the restoration of oyster reefs and the planting of trees around the shoreline as habitat for local and migratory birds. To everyone's benefit and enhancement of the environment, confrontation had turned into cooperation.

Through subsequent years the Galveston Bay Foundation has continued to play an important role in elevating the quality of the bay environment and educating the local citizens about environmental matters. Similar groups of concerned and dedicated citizens have sprung up in several areas around the Gulf coast, and their success stories are many. So, the final recommendation of this chapter is for concerned citizens to form or join local environmental action groups, get involved, become educated about environmental issues, and participate in the governance of their own local coastal and ocean areas. The concerned citizen is the last and best hope for protection of the environment and sustainment of its natural resources. If the public does not become involved, the job may not get done. The future of the Gulf of Mexico ecosystem ultimately rests in the hands of its coastal residents.

This discussion cannot be brought to an end without brief mention of an important development on the coast of south Texas, the recent establishment of the Harte Research Institute for Gulf of Mexico Studies on the campus of Texas A&M University–Corpus Christi. Endowed by a $46 million gift, this institute is moving to provide international leadership in generating and disseminating knowledge about the Gulf of Mexico ecosystem and its critical role in the economics of the North American region. From its inception it has been bringing together representatives from the United States, Mexico, and Cuba. Its focus is on the Gulf of Mexico as a coherent large marine ecosystem. It plans to periodically monitor and issue reports on the state of the Gulf and to interface with agencies at all levels of federal, state, and local government as well as industry, private organizations, and the general public. Its personnel will include an interdisciplinary group of specialists in the fields of coastal and marine policy and law, marine biodiversity and conservation science, ecosystem studies and modeling, geographic information science, socioeconomics, and ocean and human health. One significant conference has already been held (Tunnell and Dokken 2006), and a number of major projects are under way. With its vision, missions, and methods, this institute bodes well for the future of the Gulf of Mexico, and it may eventually serve as a model for the development, management, and protection of marine resources in other areas of our small planet.

# LITERATURE CITED

Abbott, R.E. 1975. The faunal composition of the algal-sponge zone of the Flower Garden Banks, northwest Gulf of Mexico. M.S. thesis. Texas A&M University, College Station. 205 pp.

———. 1979. Ecological processes affecting the reef coral population at the East Flower Garden Bank, northwest Gulf of Mexico. Ph.D. dissertation. Texas A&M University, College Station. 154 pp.

——— and T.J. Bright. 1975. *Benthic Communities Associated with Natural Gas Seeps on Carbonate Banks in the Northwestern Gulf of Mexico.* Report, Study of Naturally Occurring Hydrocarbons in the Gulf of Mexico. Oceanography Dept., Texas A&M University. 191 pp.

Abele, L.G. 1970. The marine decapod crustaceans of the northeastern Gulf of Mexico. M.S. thesis. Florida State University, Tallahassee. 138 pp.

——— and J.W. Martin. 1989. American species of the deep-sea shrimp genus *Bythocaris* (Crustacea, Decapoda, Hippolytidae). *Bull. Mar. Sci.* 45(1):26–51.

Adams, C.C. 1996. *Species Composition, Abundance, and Depth Zonation of Sponges (Phylum Porifera) on an Outer Continental Shelf Gas Production Platform, Northwestern Gulf of Mexico.* Center for Coastal Studies, Texas A&M–Corpus Christi. TAMU-CC-9601-CCS. 130 pp.

Adams, C.M., E. Hernandez, and J. Cato. 2005. The economic significance of the Gulf of Mexico related to population, income, employment, minerals, fisheries and shipping. *Ocean Coast. Manage.* 47:565–80.

Adams, J.A. 1960. A contribution to the biology and postlarval development of the sargassumfish, *Histrio histrio* (Linnaeus), with a discussion of the *Sargassum* complex. *Bull. Mar. Sci.* 10(1):55–82.

Adelman, H.C. 1967. The taxonomy and summer and fall vertical distribution of the chaetognaths off Galveston, Texas. Ph.D. dissertation. Texas A&M University, College Station. 108 pp.

Aladro-Lubel, A. 1984. Algunos ciliados intersticiales de Isla de Enmedio Veracruz, Mexico. *Anales del Instituto de Biología, Serie Zoología, Universidad Nacional Autónoma de México* 1:1–59.

Albers, C.C. 1966. Foraminiferal ecological zones of the Gulf coast. *Gulf Coast Assoc. Geol. Socs., Trans.* 5(16): 345–54.

Alexander, S.K., S.J. Schropp, and J.R. Schwartz. 1982. Spatial and seasonal distribution of hydrocarbon utilizing bacteria of sediment from the northwestern Gulf of Mexico. *Contr. Mar. Sci.* 25:13–19.

Alldredge, A.L. 1979. The chemical composition of macroscopic aggregates in two neritic seas. *Limnol. Oceanogr.* 24(5):855–66.

——— and G.A. Jackson. 1995. Aggregation in marine systems. *Deep-Sea Res.* II 42:1–273.

Alvarez, W., J. Smit, W. Lowrie, F. Asaro, S.V. Margolis, P. Claeys, M. Kastner, and A.R. Hildebrand. 1992. Proximal impact deposits at the Cretaceous-Tertiary boundary in the Gulf of Mexico: A restudy of DSDP leg 77 sites 536 and 540. *Geology* 20(8):697–700.

Alvarez-Cadena, J. N. and L. Segura-Puertas. 1997. Zooplankton variability and copepod species assemblages from a tropical coastal lagoon. *Gulf Res. Rep.* 9(4):345–55.

Ambroski, R.L., M.A. Hood, and R.R. Miller (eds.). 1974. *1974 Proceedings of Gulf Coast Regional Symposium on Diseases of Aquatic Animals.* Louisiana State University, Center for Wetland Resources, Baton Rouge, LA. LSU-SG-74-05.

Anderson, M.E., R.E. Crabtree, H.J. Carter, K.J. Sulak, and M.D. Richardson. 1985. Distribution of demersal fishes of the Caribbean Sea found below 2,000 meters. *Bull. Mar. Sci.* 37(3):794–807.

Andrews, A.P. 1998. El comercio marítimo de los mayas del Posclásico. *Arqueología Mexicana* 6(3):16–23.

Armstrong, H.W. 1974. A study of the helminth parasites of the family Macrouridae from the Gulf of Mexico and Caribbean Sea: Their systematics, ecology, and zoogeographical implications. Ph.D. dissertation. Texas A&M University, College Station. 350 pp.

Arnold, E.L., Jr. 1958. *Gulf of Mexico Plankton Investigations: 1951–53.* US Fish and Wildlife Service, Special Scientific Report, Fisheries no. 269. 53 pp.

Austin, H.M. 1971a. The characteristics and relationships between the calculated geostrophic current component and selected indicator organisms in the Gulf of Mexico Loop Current System. Ph.D. dissertation. Florida State University, Tallahassee. 369 pp.

———. 1971b. Ecology of fishes on Florida's Middle

Ground. M.S. thesis. Florida State University, Tallahassee. 56 pp.

Baca, B.J., L. O. Sorensen, and E.R. Cox. 1979. Systematic list of the seaweeds of south Texas. *Contr. Mar. Sci.* 22:179–92.

Bahr, L.M. and W.P. Lanier. 1981. The ecology of intertidal oyster reefs of the South Atlantic coast: A community profile. US Fish and Wildlife Service. FWS/OBS/-81/15. 105 pp.

Balech, E. 1967a. Dinoflagellates and tintinnids in the northeastern Gulf of Mexico. *Bull. Mar. Sci.* 17(2):280–98.

———. 1967b. *Microplankton of the Gulf of Mexico and Caribbean Sea.* Texas A&M University Research Foundation. 67-10-T. 144 pp.

Ball, M.M., E.A. Shinn, and K.W. Stockman. 1967. The geologic effects of Hurricane Donna on south Florida. *J. Geol.* 75(5):583–97.

Banner, A.H. 1954. The Mysidacea and Euphausiacea. pp. 447–48. In: P.S. Galtsoff (ed.). *Gulf of Mexico: Its Origin, Waters, and Marine Life.* US Fish and Wildlife Service Fishery Bulletin 89. 604 pp.

Bard, E., B. Hamelin, and R.G. Fairbanks. 1990. U-Th ages obtained by mass spectrometry in corals from Barbados: Sea level during the past 130,000 years. *Nature* 346 (6283):456–58.

Barnes, R.D. 1968. *Invertebrate Zoology.* W.B. Saunders, Philadelphia, PA. 743 pp.

Barriero-Gilemes, M.T. and J. Balderas-Cortes. 1991. Evaluación de algunas communidades de productores primarios de la Laguna de la Mancha, Veracruz. *Anales del Instituto de Ciencias del Mar y Limnología* 18(2):229–45.

Baughman, J.L. 1952. The marine fisheries of the Mayas as given in Diego de Landa's "Relacion de las Cosas de Yucatan" with notes of the probable identification of fishes. *Texas J. Sci.* 4(4):432–59.

Bayer, F.M. 1954. Anthozoa: Alcyonaria. pp. 279–84. In: P.S. Galtsoff (ed.). *Gulf of Mexico: Its Origin, Waters, and Marine Life.* US Fish and Wildlife Service Fishery Bulletin 89. 604 pp.

Beddingfield, S.D. and J.B. McClintock. 1994. Environmentally-induced catastrophic mortality of the sea urchin *Lytechinus variegatus* in shallow seagrass habitats of Saint Joseph's Bay, Florida. *Bull. Mar. Sci.* 55(1):235–40.

Bedinger, C.A., Jr. (ed.). 1981. *Ecological Investigations of Petroleum Production Platforms in the Central Gulf of Mexico.* Part 6. *Benthic Biology.* Final report by the Southwest Research Institute to the US Dept. of the Interior, Bureau of Land Management, New Orleans, LA, contract no. AA551-CT8-17. 3 vols.

Behre, E.H. 1954. Decapoda of the Gulf of Mexico. pp. 451–55. In: P.S. Galtsoff (ed.) *Gulf of Mexico: Its Origin, Waters, and Marine Life.* US Fish and Wildlife Service Fishery Bulletin 89. 604 pp.

Bekker, V.E., Y.N. Shcherbachev, and V.M. Tchuvasov. 1975. Deep sea pelagic fishes of the Caribbean Sea, Gulf of Mexico, and Puerto-Rican Trench. *Tr. Inst. Okeanol. Akad. Nauk SSSR* 100:289–336.

Belsky, M.H. 1986. Legal constraints and options for total ecosystem management of large marine ecosystems. Chapt. 12, pp. 241–60. In: K. Sherman and L.M. Alexander (eds.). *Variability and Management of Large Marine Ecosystems.* AAAS Selected Symposium Series, 99. Westview Press, Boulder, CO. 319 pp.

———. 1989. The ecosystem model mandate for a comprehensive United States ocean policy and Law of the Sea. *San Diego Law Rev.* 26(3):417–95.

Bender, E.S. 1971. Studies on the life history of the stone crab, *Menippe mercenaria* (Say), in the Cedar Key area. M.S. thesis. University of Florida, Gainesville. 110 pp.

Bennett, J.L. and T.L. Hopkins. 1989. Aspects of the ecology of the calanoid copepod genus *Pleuromamma* in the eastern Gulf of Mexico. *Contr. Mar. Sci.* 31:119–36.

Bennington, J.P. 1979. Luminescent bacteria and deep-sea macrourids (Pisces: Gadiformes) from the Gulf of Mexico. Ph.D. dissertation. Texas A&M University, College Station. 152 pp.

Berkowitz, S.P. 1976. A comparison of the neuston and near-surface zooplankton in the northwest Gulf of Mexico. M.S. thesis. Texas A&M University, College Station. 148 pp.

Berry, S.S. 1934. List of marine mollusca of the Atlantic coast from Labrador to Texas, by C.W. Johnson. *Proc. Boston Soc. Nat. Hist.* 40(1):160–65.

Bert, T.M. and H.J. Humm. 1979. Checklist of the marine algae on the offshore oil platforms of Louisiana. *Rice University Studies* 65(4):437–46.

Besada, E.G., L.A. Loeblich, and A.R. Loeblich III. 1982. Observations on tropical, benthic dinoflagellates from Ciguatera areas: *Coolia, Gambierdiscus,* and *Ostreopsis. Bull. Mar. Sci.* 32(3):723–35.

Bianchi, T.S., J.R. Pennock, and R.R. Twilley (eds.). 1999. *Biogeochemistry of Gulf of Mexico Estuaries.* John Wiley & Sons, New York. 428 pp.

Bierbaum, T.J. and J.A. Zischke. 1979. Changes in barnacle population structure along an intertidal community gradient in the Florida Keys. *Mar. Biol.* 53(4):345–51.

Biggs, D.C., R.R. Bidigare, and D.E. Smith. 1981. Population density of gelatinous macro-zooplankton: In situ estimation in oceanic surface waters. *Biol. Oceanogr.* 1:157–73.

———, G.S. Fargion, P. Hamilton, and R. Leben. 1996. Cleavage of a Gulf of Mexico Loop Current eddy by a deep water cyclone. *J. Geophys. Res.* 101(C9):20629–41.

——— and P.H. Ressler. 2000. Water column biology. Chapt. 6, pp. 141–87. In: *Continental Shelf Associates, Inc. Deepwater Program: Gulf of Mexico Deepwater Information Resources Data Search and Literature Synthesis.* Vol. 1. *Narrative Report.* US Dept. of the Interior, Minerals Management Service, Gulf of Mexico OCS Region, New Orleans, LA. OCS Study MMS 2000-049. 340 pp.

——— and P.H. Ressler. 2001. Distribution and abundance of phytoplankton, zooplankton, ichthyoplankton, and micronekton in the deepwater Gulf of Mexico. *Gulf of Mexico Sci.* 2001(1):7–29.

———, D.E. Smith, R.R. Bidigare, and M.A. Johnson. 1984. In situ estimation of the population density of gelatinous planktivores in Gulf of Mexico surface waters. *Mem. Univ. Newfoundland, Occas. Pap. Biol.* 9:17–34.

Bird, D.E., K. Burke, S.A. Hall, and J.F. Casey. 2005. Gulf of Mexico tectonic history: Hot spot tracks, crustal boundaries, and early salt distribution. *Amer. Assoc. Petrol. Geol. Bull.* 89(3): 311–28.

Bishof, D.E. 1980. The ecology of molluscan infauna on the southwestern continental shelf of Florida. M.S. thesis. University of South Florida, Tampa. 150 pp.

Björnberg, T.K.S. 1971. Distribution of plankton relative to the general circulation system in the area of the Caribbean Sea and adjacent regions. pp. 343–56. In: *Symposium on Investigations and Resources of the Caribbean Sea and Adjacent Regions.* Willemstad, Curacao, Netherlands Antilles, 18–26 Nov. 1968. UNESCO, Paris.

Bjorndal, K.A. (ed.). 1982. *Biology and Conservation of Sea Turtles.* Smithsonian Institution Press, Washington, DC.

Blackburn, J.B., Jr. 2004. *The Book of Texas Bays.* Texas A&M University Press, College Station. 290 pp.

Blair, S.M., T.L. McIntosh, and B.J. Mostkoff. 1994. Impacts of Hurricane Andrew on the offshore reef systems of central and northern Dade County, Florida. *Bull. Mar. Sci.* 54(3):961–73.

Blake, N.J. 1979. Infaunal macromolluscs of the eastern Gulf of Mexico. Vol. 2, Chapt. 14. In: Dames and Moore (eds.). *The Mississippi, Alabama, Florida Outer Continental Shelf Baseline Environmental Survey 1977/1978.* Report to the US Dept. of the Interior, Bureau of Land Management, Washington, DC, contract no. AA550-CT7-34.

Blalock, R.A., J.W. Hain, L.J. Hansen, D.L. Palka, and G.T. Waring. 1995. *U.S. Atlantic and Gulf of Mexico Marine Mammal Stock Assessments.* NOAA, National Marine Fisheries Service, Technical Memorandum 363. 211 pp.

Bock, W.D. 1979. Foraminifera of the MAFLA area. Vol. 2, Chapt. 12, pp. 626–39. In: Dames and Moore (eds.). *The Mississippi, Alabama, Florida Outer Continental Shelf Baseline Environmental Survey, 1977/1978.* Report to the US Dept. of the Interior, Bureau of Land Management, Washington, DC, contract no. AA550-CT7-34.

Boesch, D.F. and N.N. Rabalais. 1991. Effects of hypoxia on continental shelf benthos: Comparisons between the New York Bight and the northern Gulf of Mexico. pp. 27–34. In: R.V. Tyson and T.H. Pearson (eds.). *Modern and Ancient Continental Shelf Anoxia.* Geol. Soc. Spec. Publ. no. 58.

Bogdanov, D.V., V.A. Sokalov, and N.S. Khromov. 1968. Regions of high biological and commercial productivity in the Gulf of Mexico and Caribbean Sea. *Oceanology* 8(3):371–81.

Böhlke, J.E. and C.C.G. Chaplin. 1993. *Fishes of the Bahamas and Adjacent Waters, with Nomenclatural Changes and Additions by E.B. Böhlke and W.F. Smith-Vaniz.* University of Texas Press, Austin. 771 pp.

——— and C.R. Robins. 1969. West Atlantic sponge-dwelling gobies of the genus *Evermannichthys:* Their taxonomy, habits, and relationships. *Proc. Acad. Nat. Sci. Philadelphia* 121(1):1–24.

Booker, R.B. 1971. Some aspects of the biology and ecology of the deep-sea echinoid *Phormosoma placenta* Wyv. Thomson. M.S. thesis. Texas A&M University, College Station. 86 pp.

Borror, A.C. 1962. Ciliate protozoa of the Gulf of Mexico. *Bull. Mar. Sci.* 12(3):333–49.

Bortone, S.A. 1976. Effects of a hurricane on the fish fauna at Destin, Florida. *Florida Scientist* 39:245–48.

Boschung, H.T., Jr. 1992. *Catalog of Freshwater and Marine Fishes of Alabama.* Bull. Ala. Mus. Nat. Hist. 14. 266 pp.

——— and G. Gunter. 1966. A new species of lancet, *Branchiostoma bennetti* (Order Amphioxi), from Louisiana. *Copeia* 1966(3):485–89.

——— and R.F. Shaw. 1988. Occurrence of planktonic lancets from Louisiana's continental shelf, with a review of pelagic *Branchiostoma* (Order Amphioxi). *Bull. Mar. Sci.* 43(2):229–40.

———, J.D. Williams, D.W. Gorshall, D.K. Caldwell, and M.C. Caldwell. 1983. *The Audubon Society Field Guide to North American Fishes, Whales, and Dolphins.* Alfred Knopf, New York. 848 pp.

Bouma, A.H., G.T. Moore, and J.M. Coleman (eds.). 1978. *Framework, Facies, and Oil-Trapping Characteristics of the Upper Continental Margin.* Amer. Assoc. Petrol. Geol., Stud. in Geol. 7. 326 pp.

———, C.E. Stelting, and J.M. Coleman. 1985. Mississippi Fan, Gulf of Mexico. pp. 143–50. In: A.H. Bouma, W.R. Normark, and N.E. Barnes (eds.). *Submarine Fans and Related Turbidite Systems.* Springer-Verlag, New York. 351 pp.

Bralower, T.J., C.K. Paull, and R.M. Leckie. 1998. The Cretaceous-Tertiary boundary cocktail: Chicxulub impact triggers margin collapse and extensive sediment gravity flows. *Geology* 26(4):331–34.

Bright, T.J. 1968. A survey of the deep-sea bottom fishes of the Gulf of Mexico below 350 meters. Ph.D. dissertation. Texas A&M University, College Station. 217 pp.

———. 1970. Food of deep-sea bottom fishes. pp. 245–52. In: W.E. Pequegnat and F.A. Chace Jr. (eds.). *Contributions on the Biology of the Gulf of Mexico.* Texas A&M University Oceanographic Studies. Vol. 1. Gulf Publishing, Houston. 270 pp.

——— and C. Cashman. 1974. Fishes. pp. 339–409. In: T.J. Bright and L.H. Pequegnat (eds.). *Biota of the West Flower Garden Bank.* Gulf Publishing, Houston. 435 pp.

———, F. Ferrari, D. Martin, and G.A. Franceschini. 1972. Effects of a total solar eclipse on the vertical distribution of certain oceanic zooplankton. *Limnol. and Oceanogr.* 17(2):296–301.

———, S.R. Gittings, and R. Zingula. 1991. Occurrence of Atlantic reef corals on offshore platforms in the northwestern Gulf of Mexico. *Northeast Gulf Sci.* 12(1):55–60.

———, G.P. Kraemer, G.A. Minnery, and S.T. Viada. 1984. Hermatypes of the Flower Garden Banks, northwestern Gulf of Mexico: A comparison to other western Atlantic reefs. *Bull. Mar. Sci.* 34:461–76.

———, D.W. McGrail, R. Rezak, G.S. Boland, and A.R. Trippett. 1985. *The Flower Gardens: A Compendium of Information.* US Dept. of the Interior, Minerals Management Service. OCS Study MMS 85–0024. 103 pp.

——— and L.H. Pequegnat (eds.). 1974. *Biota of the West Flower Garden Bank.* Gulf Publishing, Houston. 435 pp.

Britton, J.C. and B. Morton. 1989. *Shore Ecology of the Gulf of Mexico.* University of Texas Press, Austin. 387 pp.

Brongersma-Sanders, M. 1957. Mass mortality in the sea. pp. 941–1010. In: J. Hedgpeth (ed.). *Treatise on Marine Ecology and Paleoecology.* Geol. Soc. Amer. Mem. 67. Vol. 1. 1296 pp.

Bruce, L. 1989. Invertebrates associated with the thinstripe hermit crab *Clibonarius vittatus* (Bosc) (Crustacea: Decapoda: Diogenidae) from the barrier islands of Mississippi. *Gulf Res. Rep.* 8(2):213–17.

Bryant, W.R., J. Antoine, M. Ewing, and B. Jones. 1968. Structure of Mexican continental shelf and slope, Gulf of Mexico. *Amer. Assoc. Petrol. Geol. Bull.* 52(7):1204–28.

———, J. Lugo, C. Córdova, and A. Salvador. 1991. Physiography and bathymetry. pp. 13–30. In: A. Salvador (ed.). *The Gulf of Mexico Basin.* Geological Society of America, Boulder, CO. Vol. J. 568 pp.

———, A.A. Meyerhoff, N.K. Brown Jr., M.A. Furrer, T.E. Pyle, and J.W. Antoine. 1969. Escarpments, reef trends, and diapiric structures, eastern Gulf of Mexico. *Amer. Assoc. Petrol. Geol. Bull.* 53:2506–42.

Buffler, R.T. 1991. Seismic stratigraphy of the deep Gulf of Mexico basin and adjacent margins. pp. 353–88. In: A. Salvador (ed.). *The Gulf of Mexico Basin.* Geological Society of America, Boulder, CO. Vol. J. 568 pp.

Bullis, H.R., Jr. 1956. Preliminary results of deep-water exploration for shrimp in the Gulf of Mexico by the R/V OREGON (1950–1956). *Comm. Fish. Rev.* 18(12):1–12.

——— and J.R. Thompson. 1965. Collections made by the exploratory fishing vessels *Oregon, Silver Bay, Combat,* and *Pelican,* made during 1956–60 in the southwestern North Atlantic. US Dept. of the Interior, Special Scientific Report, Fisheries no. 510. 130 pp.

Bullock, W.L. 1957. The acanthocephalan parasites of fishes of the Texas coast. *Publ. Inst. Mar. Sci., Univ. Texas* 4(2):278–83.

Burke, T.E. 1974a. Echinoderms. pp. 311–31. In: T.J. Bright and L.H. Pequegnat (eds.). *Biota of the West Flower Garden Bank.* Gulf Publishing, Houston. 435 pp.

———. 1974b. Echinoderms of the West Flower Garden Reef Bank. M.S. thesis. Texas A&M University, College Station. 165 pp.

Burkenroad, M.D. 1934. The Penaeidae of Louisiana. *Bull. Amer. Mus. Nat. Hist.* 68:61–143.

Busby, R.F. 1965. Sediments and reef corals of Cayo Arenas, Campeche Bank, Yucatan, Mexico. M.S. thesis. Texas A&M University, College Station. 106 pp.

———. 1966. *Sediments and Reef Corals of Cayo Arenas, Campeche Bank, Yucatan, Mexico.* US Naval Oceanographic Office, Washington, DC. 58 pp.

Busch, W.N., B.L. Brown, and G.F. Mayer (eds.). 2003. *Strategic Guidance for Implementing an Ecosystem-Based Approach to Fisheries Management.* Prepared by the Ecosystem Approach Task Force for the Marine Fisheries Advisory Committee. Silver Spring, MD. 34 pp.

Bushnell, V.C. (ed.). 1972. *Chemistry, Primary Productivity and Benthic Algae of the Gulf of Mexico.* American Geographical Society, New York.

Buskey, E.J., S. Stewart, J. Peterson, and C. Collumb. 1996. *Current Status and Historic Trends of Brown Tide and*

*Red Tide Phytoplankton Blooms in the Corpus Christi Bay National Estuary Program Study Area.* Texas Natural Resource Conservation Commission Report CCBNEP-07. 174 pp.

Butler, P.A. 1952. *Effects of Floodwaters on Oysters in Mississippi Sound in 1950.* US Fish and Wildlife Service Research Report 31. 20 pp.

————. 1954. Summary of our knowledge of the oyster in the Gulf of Mexico. pp. 479–89. In: P.S. Galtsoff (ed.). *Gulf of Mexico: Its Origin, Waters, and Marine Life.* US Fish and Wildlife Service Fishery Bulletin 89. 604 pp.

———— and J.B. Engle. 1950. *The 1950 Opening of the Bonnet Carré Spillway: Its Effects on Oysters.* US Fish and Wildlife Service, Special Scientific Report, Fisheries no. 14. 10 pp.

Cairns, S.D. 1976. Cephalopods collected in the Straits of Florida by the R/V GERDA. *Bull. Mar. Sci.* 26:233–72.

————. 1977. Stony corals. I. *Caryophyllina* and *Dendrophyllina* (Anthozoa:Scleractinia). Florida Marine Research Laboratory, *Mem. Hourglass Cruises* 3(4):1–27.

————. 1978. A checklist of the ahermatypic Scleractinia of the Gulf of Mexico, with the description of a new species. *Gulf Res. Rep.* 6:9–15.

Caldwell, D.K. and M.C. Caldwell. 1973. Marine mammals of the eastern Gulf of Mexico. pp. III-I-1 through III-I-23. In: J.I. Jones, R.E. Ring, M.O. Rinkel, and R.E. Smith (eds.). *A Summary of Knowledge of the Eastern Gulf of Mexico, 1973.* State University System of Florida, Gainesville.

Cameron, R.A. (ed.). 1986. Proceedings of invertebrate larval biology workshop, Friday Harbor Laboratories, University of Washington, 26–30 March 1985. *Bull. Mar. Sci.* 39(2):145–622.

Camp, D.K. 1973. Stomatopod crustacea. Florida Marine Research Laboratory, *Mem. Hourglass Cruises* 3(2):1–100.

Campbell, R.A. 1983. Parasitism in the deep sea. pp. 473–552. In: G.T. Rowe (ed.). *Deep-Sea Biology.* Vol. 8. *The Sea.* John Wiley & Sons, New York. 560 pp.

Capurro, L.R.A. and J.L. Reid (eds.). 1972. *Contributions to the Physical Oceanography of the Gulf of Mexico.* Texas A&M University Oceanographic Studies. Vol. 2. Gulf Publishing, Houston. 288 pp.

Carlgren, O. and J.W. Hedgpeth. 1952. Actinaria, Zoantharia, and Ceriantharia from shallow water in the northwestern Gulf of Mexico. *Publ. Inst. Mar. Sci., Univ. Texas* 2(2):141–71.

Carlton, J.M. 1975. A guide to common Florida salt marsh and mangrove vegation. *Florida Mar. Res. Publ.* 6:1–30.

Carney, R.S. 1971. Some aspects of the ecology of *Mesothuria lactea* Theel, a common bathyal holothurian

in the Gulf of Mexico. M.S. thesis. Texas A&M University, College Station. 96 pp.

————, R.L. Haedrich, and G.T. Rowe. 1983. Zonation of the fauna in the deep sea. pp. 371–98. In: G.T. Rowe (ed.). *Deep-Sea Biology.* Vol. 8. *The Sea.* John Wiley & Sons, New York. 560 pp.

Carr, A. 1952. *Handbook of Turtles.* Comstock Publishing Associates, Ithaca, NY. 542 pp.

————. 1967. *So Excellent a Fishe.* Natural History Press, Garden City, NY. 248 pp.

———— and D.K. Caldwell. 1956. The ecology and migrations of sea turtles. 1. Results of field work in Florida, 1955. *Amer. Mus. Novitiates* 1793:1–23.

Carr, W.E.S. and C.A. Adams. 1973. Food habits of juvenile marine fishes occupying seagrass beds in the estuarine zone near Crystal River, Florida. *Trans. Amer. Fish. Soc.* 102:511–40.

Casey, R.E. 1979a. Ciliated protozoa. L-1–65. In: W. Flint and N.N. Rabalais (eds.). *Environmental Studies, South Texas Outer Continental Shelf.* Final report, 1977. Vol. 5, appendices J–M. Report by the University of Texas Marine Science Institute to the Bureau of Land Management, Washington, DC, contract no. AA550-CT7-11.

————. 1979b. Shelled microzooplankton and general microplankton. K-1–38. In: W. Flint and N.N. Rabalais (eds.). *Environmental Studies, South Texas Outer Continental Shelf.* Final report, 1977. Vol. 5, appendices J–M. Report by the University of Texas Marine Science Institute to the Bureau of Land Management, Washington, DC, contract no. AA550-CT7-11.

————, J.M. Spaw, F. Kunze, R. Reynolds, T. Duis, K. McMillen, D. Pratt, and V. Anderson. 1979. Radiolarian ecology and the development of the radiolarian component in Holocene sediments, Gulf of Mexico and adjacent seas with potential paleontological implications. *Gulf Coast Assoc. Geol. Socs., Trans.* 29:228–37.

Cashman, C.W. 1973. Contributions to the ichthyofaunas of the West Flower Garden Reef and other reef sites in the Gulf of Mexico and western Caribbean. Ph.D. dissertation. Texas A&M University, College Station. 248 pp.

Castro-Aguirre, J.L. 1978. *Catálogo sistemático de los peces marinos que penetran a las aguas continentales de México, con aspectos zoogeográphicos y ecológicos.* Departamento de Pesca, Dirección General del Instituto Nacional de Pesca, Serie Cien. 19:1–298.

———— and A. Márquez-Espinoza. 1981. *Contribución al*

*conocimiento de la ictiofauna de la Isla de Lobos y zonas adyacentes, Veracruz, Mexico.* Departamento de Pesca, Dirección General del Instituto Nacional de Pesca, Serie Cien. 22:1–85.

Causey, B.D. 1969. The fish of Seven and One-Half Fathom Reef. M.S. thesis. Texas A&I University, Kingsville. 110 pp.

Causey, D. 1953. Parasitic copepods of Texas coastal fishes. *Publ. Inst. Mar. Sci., Univ. Texas* 3(1):5–16.

Chabreck, R.H. 1970. Marsh zones and vegetative types in the Louisiana coastal marshes. Ph.D. dissertation. Louisiana State University, Baton Rouge. 113 pp.

———. 1972. *Vegetation, water, and soil characteristics of the Louisiana coastal region.* Louisiana State University, Agricultural Experiment Station Bulletin No. 644. 72 pp.

——— and G. Linscombe. 1978. *Vegetative type map of the Louisiana coastal marshes.* Louisiana Dept. of Wildlife and Fisheries, New Orleans.

——— and A.W. Palmisano. 1973. The effects of Hurricane Camille on marshes of the Mississippi River Delta. *Ecology* 54(5):1118–23.

Chace, F.A., Jr. 1940. Report on the scientific results of the *Atlantis* expeditions to the West Indies under the joint auspices of the University of Havana and Harvard University: The brachyuran crabs. *Torreia* 3:3–67.

———. 1954. Stomatopoda. pp. 449–50. In: P.S. Galtsoff (ed.). *Gulf of Mexico: Its Origin, Waters, and Marine Life.* US Fish and Wildlife Service Fishery Bulletin 89. 604 pp.

Chamberlain, C.K. 1966. Some Octocorallia of Isla de Lobos, Veracruz, Mexico. *Brigham Young Univ. Geol. Stud.* 13:47–54.

Chandler, A.C. and H.W. Manter. 1954. Parasitic helminths. pp. 331–58. In: P.S. Galtsoff (ed.). *Gulf of Mexico: Its Origin, Waters, and Marine Life.* US Fish and Wildlife Service Fishery Bulletin 89. 604 pp.

Chapman, V.J. 1960. *Salt Marshes and Salt Deserts of the World.* Leonard Hill Books, London. 392 pp.

———. 1976a. *Coastal Vegetation.* Pergamon Press, New York. 292 pp.

———. 1976b. *Mangrove Vegetation.* J. Cramer, Germany. 447 pp.

Chávez, E.A. 1973. Observaciones generales sobre las comunidades del Arrecife de Lobos, Veracruz. *Anales de la Escuela Nacional de Ciencias Biológicas, Mexico* 20:13–21.

———. 1981. Toward a rational management of western Gulf of Mexico shore fisheries. pp. 2018–23. In: G.E. Lasker (ed.). *Applied Systems and Cybernetics.* Vol. 4. Pergamon Press, Oxford, UK.

——— and E. Hidalgo.1988. Los arrecifes corallinos de Caribe Noroccidental y Golfo de México en el contexto socioeconómico. *Anales del Instituto de Ciencias del Mar y Limnología* 15(1):167–75.

———, E. Hidalgo, and M.A. Izaguirre. 1985. A comparative analysis of Yucatan coral reefs. *Proceedings of the Fifth International Coral Reef Congress, Tahiti* 6:355–61.

———, E.Y. Sevilla, and M.L. Hidalgo. 1970. Datos acerca de las comunidades bentonicas del Arrecife de Lobos, Veracruz. *Revista de la Sociedad Mexicana de Historia Natural* 31:211–80.

Chávez, H. 1966. Peces colectados en el arrecife Triángulos Oeste y en Cayo Arenas, Sonde de Campeche, Mexico. *Acta Zool. Mex.* 8(1):1–12.

———. 1969. Tagging and recapture of the lora turtle (*Lepidochelys kempii*). *Intern. Turtle and Tortoise Soc. Jour.* 3:14–19, 32–36.

———, M. Contreras G., and T.P.E. Hernandez D. 1968. On the coast of Tamaulipas. *Intern. Turtle and Tortoise Soc. Jour.* 2(5):16–19, 27–34.

——— and R. Kaufmann. 1974. Información sobre la tortuga marina *Leipidochelys kempii* (Garman), con referencia a un ejemplar marcado en México y observade en Colombia. *Bull. Mar. Sci.* 24(2):372–77.

——— and J.W. Tunnell, Jr. 1993. Needs for management and conservation of the southern Gulf of Mexico. pp. 2040–53. In: *Coastal Zone '93.* Proceedings of the Eighth Symposium on Coastal and Ocean Management, New Orleans, LA.

Cheney, D.P. and J.P. Dyer III. 1974. Deep-water benthic algae of the Florida Middle Ground. *Mar. Biol.* 27:185–90.

Cheng, L. and J.H. Wormuth. 1992. Are there separate populations of *Halobates* in the Gulf of Mexico? *Bull. Mar. Sci.* 50(2):307–19.

Chitwood, B.G. 1951. North American marine nematodes. *Texas J. Sci.* 3(4):617–72.

———. 1954. Tardigrades of the Gulf of Mexico. p. 325. In: P.S. Galtsoff (ed.), *Gulf of Mexico: Its Origin, Waters, and Marine Life.* US Fish and Wildlife Service Fishery Bulletin 89. 604 pp.

——— and R.W. Timm. 1954. Free-living nematodes of the Gulf of Mexico. pp. 313–23. In: P.S. Galtsoff (ed.). *Gulf of Mexico: Its Origin, Waters, and Marine Life.* US Fish and Wildlife Service Fishery Bulletin 89. 604 pp.

Choi, D.R. 1984. Ecological succession of reef-cavity-dwellers (coelobites) in coral rubble. *Bull. Mar. Sci.* 35(1):72–79.

Chrétiennot-Dinet, M.-J., A. Sournia, M. Ricard, and C. Billard. 1993. A classification of the marine phytoplankton of the world from class to genus. *Phycologia* 32(3):159–79.

Christmas, J.Y. and W.W. Langley. 1973. Estuarine

invertebrates, Mississippi. pp. 255–319. In: J.Y. Christmas (ed.). *Cooperative Gulf of Mexico Estuarine Inventory and Study.* State of Mississippi, Gulf Coast Research Laboratory. 434 pp.

Clapp, R.B. 1983. *Marine Birds of the Southeastern United States and Gulf of Mexico.* Pt. III. *Charadriiformes.* US Fish and Wildlife Service FWS/OBS-83/30. 853 pp.

———, R.C. Banks, D. Morgan-Jacobs, and W.A. Hoffman. 1982. *Marine Birds of the Southeastern United States and Gulf of Mexico.* Pt. I. *Gaviiformes through Pelecaniformes.* US Fish and Wildlife Service FWS/OBS-82/01. 637 pp.

———, D. Morgan-Jacobs, and R.C. Banks. 1982. *Marine Birds of the Southeastern United States and Gulf of Mexico.* Pt. II. *Anseriformes.* US Fish and Wildlife Service FWS/OBS-82/20. 492 pp.

Clark, A.H. 1954. Echinoderms (other than holothurians) of the Gulf of Mexico. pp. 373–79. In: P.S. Galtsoff (ed.) *Gulf of Mexico: Its Origin, Waters, and Marine Life.* US Fish and Wildlife Service Fishery Bulletin 89. 604 pp.

Clark, J.R. (ed.) 1985. *Coastal Resources Management: Development Case Studies.* Renewable Resources Information Series Coastal Management Publ. No. 3. Prepared by Research Planning Inst., Inc., Columbia, SC, for National Park Service, US Dept. of the Interior, and US Agency for International Development. 749 pp.

———. 1986. *Coastal Zone Management Handbook.* CRC Press/Lewis Publishers, New York. 704 pp.

———. 1987. *Coastal Ecosystem Management.* John Wiley & Sons, New York.

———. 1998. *Coastal Seas: The Conservation Challenge.* Blackwell Publishing, Oxford. 134 pp.

Clark, S.T. and P.B. Robertson. 1982. Shallow water marine isopods of Texas. *Contr. Mar. Sci.* 25:45–59.

Cobb, S.P., C.R. Futch, and D.K. Camp. 1973. The rock shrimp, *Sicyonia brevirostris* Stimpson, 1871 (Decapoda, Penaeidae). Florida Marine Research Laboratory, *Mem. Hourglass Cruises* 3(1):1–38.

Cochrane, J.D. and F.J. Kelly. 1986. Low frequency circulation on the Texas-Louisiana continental shelf. *J. Geophys. Res.* 91:10645–59.

Coe, W.R. 1954. The nemertean fauna of the Gulf of Mexico. pp. 303–9. In: P.S. Galtsoff (ed.). *Gulf of Mexico: Its Origin, Waters, and Marine Life.* US Fish and Wildlife Service Fishery Bulletin 89. 604 pp.

Coleman, J.M., D.B. Prine, and L.E. Garrison. 1980. *Subaqueous Sediment Instabilities in the Offshore Mississippi River Delta.* US Dept. of the Interior, Minerals Management Service, Gulf of Mexico OCS Region, Metairie, LA. 49 pp.

Colin, P.L. and J.B. Heiser. 1973. Association of two species of cardinalfishes (Apogonidae: Pisces) with sea anemones in the West Indies. *Bull. Mar. Sci.* 23(3):521–24.

Collard, S.B. and C.N. D'Asaro. 1973. Benthic invertebrates of the eastern Gulf of Mexico. pp. III G1–27. In: J.I. Jones, R.E. Ring, M.O. Rinkel, and R.E. Smith (eds.). *A Summary of Knowledge of the Eastern Gulf of Mexico.* State University System of Florida, Institute of Oceanography.

———, H.H. Roberts, and W.R. Bryant. 1991. Late Quaternary sedimentation. In: A. Salvador (ed.). *The Gulf of Mexico Basin.* Geological Society of America, Boulder, CO. Vol. J. 568 pp.

Conger, P.S., G.A. Fryxell, and S.Z. El-Sayed. 1972. Diatom species reported from the Gulf of Mexico. pp. 18–23. In: S.Z. El-Sayed, W.M. Sackett, L.M. Jeffrey, A.D. Fredericks, R.P. Saunders, P.S. Conger, G.A. Fryxell, K.A. Steidinger, and S.A. Earle. *Chemistry, Productivity, and Benthic Algae of the Gulf of Mexico.* American Geographical Society Serial Atlas of the Marine Environment, Folio 22. 29 pp.

Conkright, M.D. and W.M. Sackett. 1986. A stable carbon isotope evaluation of the contribution of terrigenous carbon to the marine food web in Bayboro Harbor, Tampa Bay, Florida. *Contr. Mar. Sci.* 29:131–39.

Connell, C.H. and J.B. Cross. 1950. Mass mortality of fish associated with the protozoan *Gonyaulax* in the Gulf of Mexico. *Science* 112(2909):359–63.

Continental Shelf Associates, Inc. 2000. *Deepwater Program: Gulf of Mexico Deepwater Information Resources Data Search and Literature Synthesis.* 3 vols. US Dept. of the Interior, Minerals Management Service, Gulf of Mexico OCS Regional Office, New Orleans, LA. OCS Study MMS 2000-2004.

Continental Shelf Associates, Inc., and Geochemical and Environmental Research Group, Texas A&M University. 2002. *Mississippi/Alabama Pinnacle Trend Ecosystem Monitoring.* Final synthesis report. US Dept. of the Interior, Geological Survey, Biological Resources Division, USGS BSR 2001-0007 and Minerals Management Service, Gulf of Mexico OCS Region, New Orleans, LA. OCS Study MMS 2001-080. 415 pp.

Contreras, F. 1930. Contribución al conocimiento de las jaibas de México. *Anales del Instituto de Biología, Universidad Nacional Autónoma de México* 1(3):227–41.

Cooper, G.A. 1954. Brachiopoda occurring in the Gulf of Mexico. pp. 363–65. In: P.S. Galtsoff (ed.). *Gulf of Mexico: Its Origin, Waters, and Marine Life.* US Fish and Wildlife Service Fishery Bulletin 89. 604 pp.

———. 1973. Brachiopods (recent). Florida Marine Research Laboratory, *Mem. Hourglass Cruises* 3(3):1–17.

Copeland, B.J. 1965. Fauna of the Aransas Pass Inlet, Texas. I. Emigration as shown by tide trap collections. *Publ. Inst. Mar. Sci., Univ. Texas* 10:9–21.

———. 1967. Environmental characteristics of hypersaline lagoons. *Contr. Mar. Sci.* 12:207–81.

———. 1970. Estuarine classification and response to disturbances. *Trans. Amer. Fish. Soc.* 99(4):826–35.

——— and T.J. Bechtel. 1974. Some environmental limits of six Gulf coast estuarine organisms. *Contr. Mar. Sci.* 18:169–204.

Costello, T.J. and D.M. Allen. 1966. Migrations and geographic distribution of the pink shrimp, *Penaeus duorarum,* of the Tortugas and Sanibel grounds, Florida. *US Fish and Wildlife Service Fishery Bulletin* 66(3):491–502.

Couch, J.A. 1978. Diseases, parasites, and toxic responses of commercial penaeid shrimps of the Gulf of Mexico and south Atlantic coasts of North America. *Fish. Bull.* 76(1):1–44.

Crabtree, R.E. 1995. Chemical composition and energy content of deep-sea demersal fishes from tropical and temperate regions of the western North Atlantic. *Bull. Mar. Sci.* 56(2):434–49.

Craighead, F.C. and V.C. Gilbert. 1962. The effects of Hurricane Donna on the vegetation of southern Florida. *Quart. J. Fla. Acad. Sci.* 25:1–28.

Croley, F.C. and C.J. Dawes. 1970. Ecology of the algae of a Florida key. Part I. A preliminary checklist, zonation, and seasonality. *Bull. Mar. Sci.* 20(1):165–85.

Cropper, D.A. 1973. Living cheilostome Bryozoa of West Flower Garden Bank, Northwest Gulf of Mexico. M.S. thesis. Texas A&M University, College Station. 89 pp.

Cross, R.D. and D.L. Williams. 1981. *Proceedings of the National Symposium on Freshwater Inflow to Estuaries.* US Fish and Wildlife Service, Office of Biological Services FWS/OBS-81/04. 2 vols.

Cruise, J. 1971. The planktonic shrimp genus *Lucifer:* Its distribution and use as an indicator organism in the eastern Gulf of Mexico. M.S. thesis. Florida State University, Tallahassee. 185 pp.

Cruz-Ábrego, F.M., F. Flores-Andolais, and Y. Solis-Weiss. 1991. Distribución de moluscos y caracterización ambiental en zonas de descarga de aquas continentales del Golfo de México. *Anales del Instituto de Ciencias del Mar y Limnología* 18(2):247–59.

Cruz-Kaegi, M.E. 1992. Microbial abundance and biomass in sediments of the Texas: Louisiana shelf. M.S. thesis. Texas A&M University, College Station. 59 pp.

——— and G.T. Rowe. 1992. Benthic biomass gradients on the Texas-Louisiana shelf. pp. 145–49. In: *Nutrient Enhanced Coastal Ocean Productivity.* NECOP Workshop Proceedings, Oct. 1991. NOAA Coastal Ocean Program. Texas A&M Sea Grant Publication TAMU-SG-92-109. 153 pp.

Culver, S.J. and M.A. Buzas. 1981. *Distribution of Recent Benthic Foraminifera in the Gulf of Mexico.* Smithsonian Contrib. Mar. Sci. 2(8). 898 pp.

Cummings, J.A. 1982. Vertical distribution patterns of calanoid copepods in the western Gulf of Mexico. Ph.D. dissertation. Texas A&M University, College Station. 130 pp.

Curtis, D.M. 1960. Relation of environmental energy levels and ostracod biofacies in east Mississippi Delta area. *Amer. Assoc. Petrol. Geol. Bull.* 44(4):471–94.

Cutress, C., D.M. Ross, and L. Sutton. 1970. The association of *Calliactis tricolor* with its pagurid, calappid, and majid partners in the Caribbean. *Canadian J. Zool.* 48:371–76.

Cuzon du Rest, R.P. 1963. Distribution of the zooplankton in the salt marshes of southeastern Louisiana. *Publ. Inst. Mar. Sci., Univ. Texas* 9:132–55.

Dagg, M., C. Grimes, S. Lorenz, B. McKee, R. Twilley, and W. Wiseman Jr. 1991. Continental shelf food chains of the northern Gulf of Mexico. pp. 329–45. In: K. Sherman, L.M. Alexander, and B.D. Gold (eds.). *Food Chains, Yields, Models, and Management of Large Marine Ecosystems.* Westview Press, Boulder, CO.

Dalby, J.E., Jr. 1989. Predation of ascidians by *Melongena corona* (Neogastropoda: Melongenidae) in the northern Gulf of Mexico. *Bull. Mar. Sci.* 45(3):708–12.

Darcy, G.H. and E.J. Gutherz. 1984. Abundance and density of demersal fishes on the West Florida shelf, January 1978. *Bull. Mar. Sci.* 34(1):81–105.

Dardeau, M.R. 1984. Synalpheus shrimps (Crustacea: Decapoda: Alpheidae). I. The Gambarelloides group, with a description of a new species. Florida Marine Research Laboratory, *Mem. Hourglass Cruises* 7(2):1–125.

——— and R.W. Heard Jr. 1983. Crangonid shrimps (Crustacea: Caridea) with a description of a new species of *Pontocaris.* Florida Marine Research Laboratory, *Mem. Hourglass Cruises* 6(2):1–39.

Darling, J.D., C. Nicklin, K.S. Norris, H. Whitehead, and B. Würsig. 1995. *Whales, Dolphins and Porpoises.* National Geographic Society, Washington, DC. 232 pp.

Darnell, R.M. 1958. Food habits of fishes and larger invertebrates of Lake Pontchartrain, Louisiana, an estuarine community. *Publ. Inst. Mar. Sci., Univ. Texas* 5:353–416.

———. 1959. Studies on the life history of the blue crab (*Callinectes sapidus* Rathbun) in Louisiana waters. *Trans. Amer. Fish. Soc.* 88:294–304.

———. 1961. Trophic spectrum of an estuarine community, based on studies of Lake Pontchartrain, Louisiana. *Ecology* 42(3):553–68.

———. 1962a. Ecological history of Lake Pontchartrain, an estuarine community. *Amer. Midl. Nat.* 68(2):434–44.

———. 1962b. Fishes of the Río Tamesí and related coastal lagoons in east-central Mexico. *Publ. Inst. Mar. Sci., Univ. Texas* 8:299–365.

———. 1967a. Organic detritus in relation to the estuarine ecosystem. pp. 376–82. In: G.H. Lauff (ed.). *Estuaries.* AAAS Publ. 83. 757 pp.

———. 1967b. The organic detritus problem. pp. 374–75. In: G.H. Lauff (ed.). *Estuaries.* AAAS Publ. 83. 757 pp.

———. 1968. Animal nutrition in relation to secondary production. *Amer. Zool.* 8:83–93.

———. 1991a. Demersal fish food analysis. pp. 9–1 through 9–76. In: J.M. Brooks (ed.). *Mississippi-Alabama Continental Shelf Ecosystem Study: Data Summary and Synthesis.* Vol. 2. *Technical Narrative.* US Dept. of the Interior, Minerals Management Service, Gulf of Mexico OCS Region, New Orleans, LA. OCS Study MMS 91–0063. 862 pp.

———. 1991b. Summary and synthesis. pp. 15–1 through 15–147. In: J.M. Brooks (ed.). *Mississippi-Alabama Continental Shelf Ecosystem Study: Data Summary and Synthesis.* Vol. 2. *Technical Narrative.* US Dept. of the Interior, Minerals Management Service, Gulf of Mexico OCS Region, New Orleans, LA. OCS Study MMS 91–0063. 862 pp.

———. 1992a. Biology of the estuaries and inner continental shelf of the northern Gulf of Mexico. pp. 161–72. In: G. Flock (ed.). *The Environment and Economic Status of the Gulf of Mexico.* Report of a symposium sponsored by EPA, NOAA, and other agencies. 186 pp.

———. 1992b. Ecological history, catastrophism, and human impact on the Mississippi/Alabama continental shelf and associated waters. *Gulf Res. Rep.* 8(4):375–86.

——— and T.J. Bright. 1983. Ecosystem dynamics. pp. 401–67. In: R. Rezak, T.J. Bright, and D.W. McGrail (eds.). *Reefs and Banks of the Northwestern Gulf of Mexico: Their Geological, Biological, and Physical Dynamics.* Northern Gulf of Mexico Topographic Features Synthesis. Final report to US Dept. of the Interior, Minerals Management Service, Gulf of Mexico OCS Region, New Orleans, LA, contract no. AA851-CT1-55.

——— and R.E. Defenbaugh (eds.). 1990. Ecology of the Gulf of Mexico. *Amer. Zool.* 30(1):1–105.

———, R.E. Defenbaugh, and D. Moore. 1983. *Northwestern Gulf Shelf Bio-Atlas: A Study of the Distribution of Demersal Fishes and Penaeid Shrimp of Soft Bottoms of the Continental Shelf from the Rio Grande to the Mississippi River Delta.* US Dept. of the Interior, Minerals Management Service. Open File Rept. 82-04. 438 pp.

——— and J.A. Kleypas. 1987. *Eastern Gulf Shelf Bio-Atlas: A Study of the Distribution of Demersal Fishes and Penaeid Shrimp of Soft Bottoms of the Continental Shelf from the Mississippi River Delta to the Florida Keys.* US Dept. of the Interior, Minerals Management Service, New Orleans, LA. OCS Study MMS 86–0041. 548 pp.

———, W.E. Pequegnat, B.M. James, F.J. Benson, and R.E. Defenbaugh. 1976. *Impacts of Construction Activities in Wetlands of the United States.* US Environmental Protection Agency, Ecological Research Series EPA-600/3-76-045. 393 pp.

——— and D.J. Schmidly. 1988. Marine Biology. Chapt. 6, pp. 203–325. In: N.W. Phillips and B.M. James (eds). *Offshore Texas and Louisiana Marine Ecosystems Data Synthesis.* Vol. 2. *Synthesis Report.* US Dept. of the Interior, Minerals Management Service, Gulf of Mexico OCS Region, New Orleans, LA. OCS Study/MMS 88–0067. 477 pp.

——— and D.B. Shimkin. 1972. A systems view of coastal zone management. Chapt. 12, pp. 346–64. In: B.H. Ketchum (ed.). *The Water's Edge: Critical Problems of the Coastal Zone.* MIT Press, Cambridge, MA. 393 pp.

——— and T.M. Soniat. 1979. The estuary/continental shelf as an interactive system. pp. 487–525. In: R.J. Livingston (ed.). *Ecological Processes in Coastal and Marine Systems.* Plenum Press, New York. 548 pp.

——— and ———. 1981. Nutrient enrichment and estuarine health. pp. 225–45. In: B.J. Neilson and L.E. Cronin (eds.). *Estuaries and Nutrients.* Humana Press, Clifton, NJ. 643 pp.

——— and T.E. Wissing. 1975. Nitrogen turnover and food relationships of the pinfish *Lagodon rhomboides* in a North Carolina estuary. pp. 81–110. In: F.J. Verberg (ed.). *Physiological Ecology of Estuarine Organisms.* University of South Carolina Press, Columbia. 397 pp.

Davis, C.C. 1955. *The Marine and Freshwater Plankton.* Michigan State University Press, East Lansing. 562 pp.

Davis, G.E. 1982. A century of natural change in coral distribution at the Dry Tortugas: A comparison of reef maps from 1881 and 1976. *Bull. Mar. Sci.* 32(2):608–23.

Davis, J.H. 1940. The ecology and geological role of mangroves in Florida. *Carneg. Inst. Wash. Publ.* 517:303–412.

Davis, R.A. 1964. Foraminiferal assemblages of Alacran Reef, Campeche Bank, Mexico. *J. Paleont.* 38(2):417–20.

Davis, R.W., G.S. Fargion, N. May, T.D. Leming, M. Baumgartner, W.E. Evans, L.J. Hansen, and

K. Mullin. 1998. Physical habitats of cetaceans along the continental slope in the north central and western Gulf of Mexico. *Mar. Mamm. Sci.* 14:490–507.

———, G. Scott, B. Würsig, W. Evans, G. Fargion, L. Hansen, R. Benson, et al. 1994. *Distribution and Abundance of Marine Mammals in the North-Central and Western Gulf of Mexico.* US Dept. of the Interior, Minerals Management Service, OCS Study Rept. 94-0003. 129 pp.

Davita, R., M. Creel, and P.F. Sheridan. 1983. Food of coastal fishes during brown shrimp, *Penaeus aztecus,* migration from Texas estuaries (June–July 1981). *Fish. Bull.* 81:396–404.

Dawes, C.J., S.A. Earle, and F.C. Croley. 1967. The offshore benthic flora of the southwest coast of Florida. *Bull. Mar. Sci.* 17(1):211–31.

——— and J.F. Van Breedveld. 1969. Benthic marine algae. Florida Marine Research Laboratory, *Mem. Hourglass Cruises* 1(2):1–47.

Dawson, C.E. 1965. Rainstorm induced mortality of lancets, *Branchiostoma,* in Mississippi Sound. *Copeia* 1965(4):505–6.

Day, J.W., Jr., W.H. Conner, F. Ley-Lou, R.H. Day, and A. Machado Navarro. 1987. The productivity and composition of mangrove forest, Laguna de Términos, Mexico. *Aquat. Bot.* 27:267–84.

———, D.D. Culley Jr., R.E. Turner, and A.J. Mumphrey Jr. (eds.). 1979. *Coastal Marsh and Estuary Management.* Louisiana State University, Division of Continuing Education, Baton Rouge. 511 pp.

———, W.G. Smith, P.R. Wagner, and C.S. Wilmer. 1973. *Structure and Carbon Budget of a Salt Marsh and Shallow Bay Estuarine System in Louisiana.* Louisiana State University, Baton Rouge. Sea Grant Publ. No. LSU-SG-72-04. 80 pp.

Deevey, E.S. 1950. Hydroids from Louisiana and Texas, with remaks on the Pleistocene biogeography of the western Gulf of Mexico. *Ecology* 31:334–67.

———. 1954. Hydroids of the Gulf of Mexico. pp. 267–72. In: P.S. Galtsoff (ed.). *Gulf of Mexico: Its Origin, Waters, and Marine Life.* US Fish and Wildlife Service Fishery Bulletin 89. 604 pp.

Defenbaugh, R.E. 1974. Hydroids. pp. 93–112. In: T.J. Bright and L.H. Pequegnat (eds.). *Biota of the West Flower Garden Bank.* Gulf Publishing, Houston. 435 pp.

———. 1976. A study of the benthic macro-invertebrates of the continental shelf of the northern Gulf of Mexico. Ph.D. dissertation. Texas A&M University, College Station. 476 pp.

———. 1990. The Gulf of Mexico: A management perspective. *Amer. Zool.* 30(1):7–13.

——— and S.H. Hopkins. 1973. *The Occurrence and Distribution of the Hydroids of the Galveston Bay Area.* Texas A&M University, Sea Grant Publ. No. TAMU-SG-73-210. 202 pp.

Deichmann, E. 1954. The holothurians of the Gulf of Mexico. pp. 381–410. In: P.S. Galtsoff (ed.). *Gulf of Mexico: Its Origin, Waters, and Marine Life,* US Fish and Wildlife Service Fishery Bulletin 89. 604 pp.

de la Cruz, A.A. 1973. The role of tidal marshes in the productivity of coastal waters. *ASB Bull.* 20(4):147–56.

de Laubenfels, M.W. 1953. Sponges from the Gulf of Mexico. *Bull. Mar. Sci.* 2(3):511–57.

den Hartog, C. 1970. *The Seagrasses of the World.* North-Holland Publishing, Amsterdam. 275 pp.

Dennis, G.D. 1985. Reef fish assemblages on hard banks in the northwestern Gulf of Mexico. M.S. thesis. Texas A&M University, College Station. 184 pp.

——— and T.J. Bright. 1988. Reef fish assemblages on hard banks in the northwestern Gulf of Mexico. *Bull. Mar. Sci.* 43(2):280–307.

de Sylva, D.P. 1975. Nektonic food webs in estuaries. pp. 420–47. In: L.E. Cronin (ed.). *Estuarine Research.* Academic Press, New York. Vol. 1. 738 pp.

———. 1994. Distribution and ecology of ciguatera fish poisoning in Florida, with emphasis on the Florida Keys. *Bull. Mar. Sci.* 54(3):944–54.

DiMarco, S.F., W.D. Nowlin, and R.O. Reid. 2005. A statistical description of the velocity fields from upper ocean drifters in the Gulf of Mexico. pp. 101–10. In: W. Sturges and A. Lugo-Fernandez (eds.). *Circulation in the Gulf of Mexico: Observations and Models.* American Geophysical Union, Geophysical Monograph Series 161. 360 pp.

Ditty, J.G. and R.F. Shaw. 1995. Seasonal occurrence, distribution, and abundance of larval bluefish, *Pomatomus saltatrix* (Family Ponatomidae), in the northern Gulf of Mexico. *Bull. Mar. Sci.* 56(2):592–601.

——— and ———. 1996. Spatial and temporal distribution of larval striped mullet (*Mugil cephalus*) and white mullet (*M. curema,* family: Mugilidae) in the northern Gulf of Mexico, with notes on mountain mullet, *Agonostomus monticola. Bull. Mar. Sci.* 59(2):271–88.

———, G.G. Zieske, and R.F. Shaw. 1988. Seasonality and depth distribution of larval fishes in the northern Gulf of Mexico above latitude 26°00′ N. *Fish. Bull.* 86:811–23.

Dooley, J.K. 1972. Fishes associated with the pelagic sargassum complex, with a discussion of the sargassum community. *Contr. Mar. Sci.* 16:1–32.

Dörjes, J.D. and J.D. Howard. 1975. Fluvial-marine transition indicators in an estuarine environment,

Ogeechee River-Ossabaw Sound. *Senckenbergiana Marit.* 7:137–80.

Dorman, C.E. and R.H. Bourke. 1981. Precipitation over the Atlantic Ocean, 30°S to 70°N. *Mon. Weath. Rev.* 109:554–63.

Dortch, Q. 1994. Changes in phytoplankton numbers and species composition. pp. 46–49. In: M.J. Dowgiallo (ed.). *Coastal Oceanographic Effects of Summer 1993 Mississippi River Flooding.* US Dept. of Commerce, Special NOAA Report.

———. 1996. Phytoplankton survey. pp. 74–90. In: S.P. Murray and J. Donley (eds.). *Mississippi River Plume Hydrography: Second Annual Report.* US Dept. of the Interior, Minerals Management Service, Gulf of Mexico OCS Region, New Orleans, LA. OCS Study MMS 96-0022. 175 pp.

Downey, M.E. 1973. *Starfishes from the Caribbean and the Gulf of Mexico.* Smithson. Contrib. Zool., no. 126. 158 pp.

Drennan, K.L. 1968. *Hydrographic Studies in the Northeast Gulf of Mexico.* Gulf South Research Institute, New Iberia, LA. Ref. 68-0-1.

Dubois, R. 1975. A comparison of the distribution of the Echinodermata of a coral community with that of a nearby rock outcrop on the Texas continental shelf. M.S.thesis. Texas A&M University, College Station. 153 pp.

Durako, M.J. and K.M. Kuss. 1994. Effects of *Labyrinthula* infection on the photosynthetic capacity of *Thalassia testudinum. Bull. Mar. Sci.* 54(3):727–32.

Dyer, K.R. and R.J. Orth (eds.). 1994. *Changes in Fluxes in Estuaries: Implication from Science to Management.* Olsen and Olsen, Fredensborg, Denmark. 485 pp.

Earle, S.A. 1969. Phaeophyta of the eastern Gulf of Mexico. *Phycologia* 7(2):71–254.

———. 1972. Benthic algae and sea grasses. pp. 15–18. In: S.Z. El-Sayed, W.M. Sackett, L.M. Jeffrey, A.D. Fredericks, R.P. Saunders, P.S. Conger, G.A. Fryxell, K.A. Steidinger, and S.A. Earle. *Chemistry, Productivity, and Benthic Algae of the Gulf of Mexico.* American Geographical Society Serial Atlas of the Marine Environment, Folio 22. 29 pp.

Ebbs, N.K., Jr. 1966. The coral-inhabiting polychaetes of the northern Florida reef tract. Part 1. Aphroditidae, Polynoidae, Amphinomidae, Eunicidae, and Lysaretidae. *Bull. Mar. Sci.* 16:485–555.

Edwards, G.S. 1969. Distibution of shelf sediments, offshore from Anton Lizardo and the Port of Veracruz, Veracruz, Mexico. M.S. thesis. Texas A&M University, College Station. 89 pp.

———. 1971. *Geology of the West Flower Garden Bank.* Texas

A&M University, Sea Grant Publ. No. TAMU-SG-71-215. 199 pp.

Edwards, P. 1969. Field and cultural studies on the seasonal periodicity of growth and reproduction of selected Texas benthic marine algae. *Contr. Mar. Sci.* 14:59–114.

———. 1976. *Illustrated Guide to the Seaweeds and Seagrasses in the Vicinity of Port Aransas, Texas.* University of Texas Press, Austin. 126 pp.

——— and D.F. Kapraun. 1973. Benthic marine algal ecology in the Port Aransas, Texas area. *Contr. Mar. Sci.* 17:15–52.

Eiseman, N.J. and S.M. Blair. 1982. New records and range extensions of deepwater algae from East Flower Garden Bank, northwestern Gulf of Mexico. *Contr. Mar. Sci.* 25:21–26.

Ekdale, A.A. 1974. Marine mollusks from shallow water environments (0 to 60 meters) off the northeast coast, Mexico. *Bull. Mar. Sci.* 24:638–88.

Eleuterius, L.N. 1973. The marshes of Mississippi. pp. 147–90. In: J.Y. Christmas (ed.). *Cooperative Gulf of Mexico Estuarine Inventory and Study, Mississippi.* Mississippi Gulf Coast Research Laboratory, Ocean Springs. 434 pp.

———. 1976. The distribution of *Juncus roemerianus* in the salt marshes of North America. *Chesapeake Sci.* 17(4):289–92.

——— and C.K. Eleuterius. 1979. Tide levels and salt marsh zonation. *Bull. Mar. Sci.* 29(3):394–400.

Elliot, B.A. 1979. Anticyclonic rings and the energetics of the circulation of the Gulf of Mexico. Ph.D. dissertation. Texas A&M University, College Station. 205 pp.

El-Sayed, S.Z. 1972. Primary production and standing crop of phytoplankton. pp. 8–13. In: S.Z. El-Sayed, W.M. Sackett, L.M. Jeffrey, A.D. Fredericks, R.P. Saunders, P.S. Conger, G.A. Fryxell, K.A. Steidinger, and S.A. Earle. *Chemistry, Productivity and Benthic Algae of the Gulf of Mexico.* American Geographical Society Serial Atlas of the Marine Environment, Folio 22. 29 pp.

———, W.M. Sackett, L.M. Jeffrey, A.D. Fredericks, R.P. Saunders, P.S. Conger, G.A. Fryxell, K.A. Steidinger, and S.A. Earle. 1972. *Chemistry, Productivity, and Benthic Algae of the Gulf of Mexico.* American Geographical Society Serial Atlas of the Marine Environment, Folio 22. 29 pp.

Emery, K.O. 1963. Coral reefs off Veracruz, Mexico. *Geofisica Internacional* 3:11–17.

Environmental Science and Engineering, Inc.; LGL Ecological Research Associates, Inc.; and Continental Shelf Associates, Inc. 1987. *South Florida Shelf Ecosystems Study Data Synthesis.* Report to the US Dept. of the

Interior, Minerals Management Service, Gulf of Mexico OCS Region, New Orleans, LA, contract no. 14-12-0001-30276. 2 vols.

EPAP (Ecosystem Principles Advisory Panel). 1999. *Ecosystem-Based Fishery Management: A Report to Congress.* US Dept. of Commerce, NOAA, NMFS, Silver Spring, MD. 55 pp.

Epifanio, C.E. 1988. Transport of invertebrate larvae between estuaries and the continental shelf. pp. 104-14. In: M.P. Weinstein (ed.). *Larval Fish and Shellfish Transport through Inlets.* American Fisheries Society Symposium 3. 165 pp.

Ernst, C.H. and R.W. Barbour. 1972. *Turtles of the United States.* University Press of Kentucky, Lexington. 347 pp.

Ernst, W.G. and J.G. Morin. 1982. *Ecosystem Processes in the Deep Oceans.* Prentice-Hall, Englewood Cliffs, NJ.

Escobar-Briones, E. and L.A. Soto. 1997. Continental shelf benthic biomass in the western Gulf of Mexico. *Cont. Shelf Res.* 17:585-604.

Espinoza-Avalos, J. 1996. Distribution of sea grasses in the Yucatan Peninsula, Mexico. *Bull. Mar. Sci.* 59(2):449-54.

Etter, P.C. 1975. A climatic heat budget study of the Gulf of Mexico. M.S. thesis. Texas A&M University, College Station. 87 pp.

———. 1983. Heat and freshwater budgets of the Gulf of Mexico. *J. Phys. Oceanogr.* 13(11):2058-69.

Every, M.G. 1968. The taxonomy and areal distribution of the Chaetognatha in the oceanic Gulf of Mexico. M.S. thesis. Texas A&M University, College Station. 67 pp.

Fahay, M.P. 1983. *Guide to the Early Stages of Marine Fishes Occurring in the Western North Atlantic Ocean, Cape Hatteras to the Southern Scotian Shelf. J. Northwest. Atl. Fish. Sci.* 4. 423 pp.

Fairbanks, R.G. 1989. A 17,000-year glacio-eustatic sea level record: Influence of glacial melting rates on the Younger Dryas event and deep-ocean circulation. *Nature* 342(6250):637-42.

Fairchild, R.R. and L.O. Sorensen. 1985. Sea urchins from the Brazos-Santiago Pass jetty, South Padre Island, Texas. *Texas J. Sci.* 37:383-85.

Farrell, D.H. 1979. Benthic molluscan and crustacean communities in Louisiana. pp. 401-36. In: C.H. Ward, M.E. Bender, and D.J. Reish (eds.). *The Offshore Ecology Investigation: Effects of Oil Drilling and Production in a Coastal Environment.* Rice University Studies 65 (4 and 5), Houston. 589 pp.

Farrell, T.M., C.F. D'Elia, L. Lubbers III, and L.J. Pastor Jr. 1983. Hermatypic coral diversity and reef zonation at Cayo Arcas, Campeche, Gulf of Mexico. *Atoll Res. Bull.* 270:1-13.

Feeley, M.H., R.T. Buffler, and W.R. Bryant. 1985. Depositional units and growth pattern of the Mississippi Fan. pp. 253-57. In: A.H. Bouma, W.R. Normark, and N.E. Barnes (eds.). *Submarine Fans and Related Turbidite Systems.* Springer-Verlag, New York. 350 pp.

Felder, D.L. 1971. The decapod crustaceans of Seven and One-Half Fathom Reef. M.S. thesis. Texas A&M University, Kingsville. 101 pp.

———. 1973. *An Annotated Key to Crabs and Lobsters (Decapoda: Reptantia) from Coastal Waters of the Northwestern Gulf of Mexico.* Louisiana State University, Sea Grant Publ. No. LSU-SG-73-02. 103 pp.

——— and A.C. Chaney. 1979. Decapod crustacean fauna of Seven and One-Half Fathom Reef, Texas: Species composition, abundance, and species diversity. *Contr. Mar. Sci.* 22:1-29.

Fenchel, T. and R. Riedl. 1970. The sulfide system: A new biotic community underneath the oxidized layer of marine sand bottoms. *Mar. Biol. (Berl.).* 7:255-68.

Ferrari, F.D. 1973. Some Corycaeidae and Oncaeidae (Copepoda: Cyclopoida) from the epipelagic waters of the Gulf of Mexico. Ph.D. dissertation. Texas A&M University, College Station. 215 pp.

Fine, M.L. 1970. Faunal variation on pelagic *Sargassum. Mar. Biol.* 7(2):112-22.

Finucane, J.H. 1976. Ichthyoplankton. pp. 20-31. In: *Environmental Studies, South Texas Outer Continental Shelf, 1975.* Vol. 1. Report by NOAA, NMFS to the Bureau of Land Management, New Orleans, LA.

———. 1977. Ichthyoplankton. In: *Environmental Studies, South Texas Outer Continental Shelf,* 1976. Report by NOAA, NMFS to the Bureau of Land Management, New Orleans, LA. 528 pp.

———, L.A. Collins, L.E. Barger, and J.D. McEachran. 1977. Ichthyoplankton/mackerel eggs and larvae. pp. 1-504. In: *Environmental Studies, South Texas Outer Continental Shelf, 1977.* Report by NOAA, NMFS to the Bureau of Land Management, New Orleans, LA.

——— and A. Dragovich. 1959. *Counts of the Red Tide Organisms,* Gymnodinium breve, *and Associated Oceanographic Data from Florida West Coast.* US Fish and Wildlife Service, Special Scientific Report, Fisheries no. 289. 220 pp.

Firth, R.W., Jr. 1971a. Polychelidae. In: W. Pequegnat, L. Pequegnat, R.W. Firth Jr., B.M. James, and T.W. Roberts. *Gulf of Mexico Deep-Sea Fauna.* American Geographical Society Serial Atlas of the Marine Environment, Folio 20. 12 pp.

———. 1971b. A study of the deep-sea lobsters of the families Polychelidae and Nephropidae (Crustacea,

Decapoda). Ph.D. dissertation. Texas A&M University, College Station. 194 pp.

Fisk, H.N. 1944. *Geological Investigations of the Alluvial Valley of the Lower Mississippi River.* US Army Corps of Engineers Mississippi River Commission, Vicksburg, MS. 78 pp.

Fitzhugh, K. 1984. Temporal and spatial patterns of the polychaete fauna on the central Northern Gulf of Mexico continental shelf. pp. 211–26. In: P.A. Hutchings (ed.) *Proceedings of the First International Polychaete Conference, Linnean Society of New South Wales.*

Fleminger, A. 1956. Taxonomic and distributional studies on the epiplanktonic calanoid copepods (crustacea) of the Gulf of Mexico. Ph.D. dissertation. Harvard University, Cambridge, MA. 317 pp.

Flint, R.W. 1981. Gulf of Mexico outer continental shelf benthos: Macroinfaunal-environmental relationships. *Biol. Oceanogr.* 1:135–55.

——— and N.N. Rabalais. 1980. Polychaete ecology and niche patterns: Texas continental shelf. *Mar. Ecol. Prog. Ser.* 3:193–202.

——— and ——— (ed.). 1981. *Environmental Studies of a Marine Ecosystem: South Texas Outer Continental Shelf.* University of Texas Press, Austin. 240 pp.

Flock, G. (ed.). 1990. *The Environmental and Economic Status of the Gulf of Mexico.* Gulf of Mexico Program Office, John C. Stennis Space Ctr., Stennis, MS. 186 pp.

Flock, M.E. 1989. Vertical distribution and feeding ecology of sergestid shrimp (Decapoda: Natantia) of the eastern Gulf of Mexico. M.S. thesis. University of South Florida, Tampa. 47 pp.

——— and T.L. Hopkins. 1992. Species composition, vertical distribution, and food habits of the sergestid shrimp assemblage in the eastern Gulf of Mexico. *J. Crust. Biol.* 12:210–23.

Flores-Coto, C. 1974. Contribución al conocimiento de las apendicularias del arrecife "La Blanquilla," Veracruz, Mexico, con descripción de una nueva especie. *Anales del Centro de Ciencias del Mar y Limnología* 1(1):41–59.

Folk, R.L. and R. Robles. 1964. Carbonate sands of Isla Pérez, Alacran reef complex, Yucatan. *Jour. Geol.* 72(3):255–92.

Fosberg, F.R. 1962. A brief study of the cays of Arrecife Alacran, a Mexican atoll. *Atoll Res. Bull.* 93:1–25.

Fotheringham, N. 1981. Observations on the effects of oil field structures on their biotic environment: Platform fouling community. pp. 179–208. In: B.S. Middleditch (ed.). *Environmental Effects of Offshore Oil Production: The Buccaneer Gas and Oil Field Study.* Plenum Press, New York. 446 pp.

Franceschini, G.A. 1961. Hydrological balance of the Gulf of Mexico. Ph.D. dissertation. Texas A&M University, College Station. 71 pp.

Frankel, C. 1999. *The End of the Dinosaurs, Chicxulub Crater and Mass Extinctions.* Cambridge University Press, Cambridge. 223 pp.

Franks, J.S., J.Y. Christmas, W.L. Siler, R. Combs, R. Waller, and C. Burns. 1972. A study of nektonic and benthic faunas of the shallow Gulf of Mexico off the State of Mississippi as related to some physical, chemical, and geological factors. *Gulf Res. Rep.* 4(1):1–148.

Frazer, T.K., W.J. Lindberg, and G.R. Stanton. 1991. Predation on sand dollars by gray triggerfish, *Balistes capriscus,* in the northeastern Gulf of Mexico. *Bull. Mar. Sci.* 48(1):159–64.

Fredericks, A.D. 1972. Distribution of organic carbon. pp. 6–7. In: S.Z. El-Sayed, W.M. Sackett, L.M. Jeffrey, A.D. Fredericks, R.P. Saunders, P.S. Conger, G.A. Fryxell, K.A. Steidinger, and S.A. Earle. *Chemistry, Productivity, and Benthic Algae of the Gulf of Mexico.* American Geographical Society Serial Atlas of the Marine Environment, Folio 22. 29 pp.

Friend, J.H., M. Lyon, N. Garrett, J.L. Borom, J. Ferguson, and G.C. Lloyd. 1981. *Alabama Coastal Region Ecological Characterization.* Vol. 3. *A Socioeconomic Study.* US Fish and Wildlife Service, Office of Biological Services FWS/OBS-81/41. 367 pp.

Fritts, T.H., W. Hoffman, and M.A. McGehee. 1983. The distribution and abundance of marine turtles in the Gulf of Mexico and nearby Atlantic waters. *J. Herpetology* 17:327–44.

———, A.B. Irvine, R.D. Jennings, L.A. Collum, W. Hoffman, and M.A. McGehee. 1983. *Turtles, Birds, and Mammals in the Northern Gulf of Mexico and Nearby Atlantic Waters.* US Fish and Wildlife Service, Division of Biological Services, Washington, DC. FWS/oB5-82/65. 455 pp.

Fry, B. and E.B. Sherr. 1984. $^{13}$C measurements as indicators of carbon flow in marine and freshwater ecosystems. *Contr. Mar. Sci.* 27:13–47.

Futch, C.R. and S.E. Dwinell. 1977. Nearshore marine ecology at Hutchinson Island, Florida. IV. Lancets and fishes. *Fla. Mar. Res. Publ.* 24. 23 pp.

Gaardner, K.A. and G.R. Hasle. 1971. Coccolithophorids of the Gulf of Mexico. *Bull. Mar. Sci.* 21(2):519–44.

Gabriel, B.C. and A.A. de la Cruz. 1974. Species composition, standing stock and net primary production of a salt marsh community in Mississippi. *Chesapeake Sci.* 15(3):72–77.

Gage, J.D. and P.A. Tyler. 1991. *Deep-Sea Biology: A Natural*

*History of Organisms at the Deep-Sea Floor.* Cambridge University Press, Cambridge. 504 pp.

Gagliano, S.M. and J.L. van Beek. 1970. *Hydrologic and Geologic Studies of Coastal Louisiana.* Report no. 1 to the US Army Corps of Engineers, contract no. DACW 29–69-C-0092. 140 pp.

Gallaway, B.J. 1980. *Pelagic Reef Demersal Fishes and Macrocrustaceans/Biofouling Communities.* Vol. 2, W.B. Jackson and E.P. Wilkens, (eds.). *Environmental Assessment of the Buccaneer Gas and Oil Field in the Northwestern Gulf of Mexico, 1975–1980.* NOAA Tech. Memo. NMFS-SEFC-48.

———— and G.S. Lewbel. 1982. *The Ecology of Petroleum Platforms in the Northwestern Gulf of Mexico: A Community Profile.* US Fish and Wildlife Service, Office of Biological Services, Washington, DC. FWS/OBS-82/27. 92 pp.

————, L.R. Martin, and R.L. Howard. 1988. *Northern Gulf of Mexico Continental Slope Study. Annual Report, Year 3.* Report to Minerals Management Service, New Orleans, LA, contract no. 14–12–0001–30212. OCS Study MMS 87–0060.

————, ————, ————, G.S. Boland, and G.D. Dennis. 1981a. Effects on artificial reef and demersal fish and macrocrustacean communities. pp. 237–93. In: B.S. Middleditch (ed.). *Environmental Effects of Offshore Oil Production: The Buccaneer Gas and Oil Field Study.* Plenum Press, New York. 446 pp.

————, ————, ————, ————, and ————. 1981b. *The Artificial Reef Studies.* Vol. 2, C.A. Bedinger Jr. and L.Z. Kilby (eds.). *Ecological Investigations of Petroleum Production Platforms in the Central Gulf of Mexico.* Report by the Southwest Research Institute to the Bureau of Land Management, New Orleans, LA, project 01–5245. 199 pp.

Galloway, J.C. 1941. Lethal effect of the cold winter of 1939–40 on marine fishes at Key West, Florida. *Copeia* 1941(2):118–19.

Galtsoff, P.S. (ed.). 1954a. *Gulf of Mexico: Its Origin, Waters, and Marine Life.* US Fish and Wildlife Service Fishery Bulletin 89. 604 pp.

————. 1954b. Historical sketch of the explorations in the Gulf of Mexico. pp. 3–36. In: P.S. Galtsoff (ed.). *Gulf of Mexico: Its Origin, Waters, and Marine Life.* US Fish and Wildlife Service Fishery Bulletin 89. 604 pp.

————. 1964. *The American Oyster* Crassostrea virginica *Gmelin.* US Fish and Wildlife Service Fishery Bulletin 64. 480 pp.

García-Montes, J.P., L.A. Soto, and A. Garcia. 1988. Cangrejos portúnidos del suroeste del Golfo de México.

Aspectos pesqueros y ecológicos. *Anales del Instituto de Ciencias del Mar y Limnología* 15(1):135–50.

George, R.Y. and P.J. Thomas. 1979. Biofouling community dynamics in Louisiana shelf oil platforms in the Gulf of Mexico. pp. 533–74. In: C.H. Ward, M.E. Bender, and D.J. Reish (eds.). *The Offshore Ecology Investigation: Effects of Oil Drilling and Production in a Coastal Environment.* Rice University Studies 65 (4 and 5), Houston. 589 pp.

Geraci, J.R. and V.L. Lounsbury. 1993. *Marine Mammals Ashore: A Field Guide for Strandings.* Texas A&M University Sea Grant Publ. 305 pp.

GESAMP (Group of Experts on the Scientific Aspects of Marine Environmental Protection). 1994. *Guidelines for Marine Environmental Assessments.* GESAMP-IMO/FAO/UNESCO-IOC/WMO/WHO/IAEA/UN/UNEP, Joint Rept. No. 54. 40 pp.

Gettleson, D.A. 1976. An ecological study of the benthic meiofauna of a soft bottom area on the Texas outer continental shelf. Ph.D. dissertation. Texas A&M University, College Station. 257 pp.

Giammona, C.P., Jr. 1978. Octocorals in the Gulf of Mexico: Their taxonomy and distribution, with remarks on their paleontology. Ph.D. dissertation. Texas A&M University, College Station. 260 pp.

———— and R.M. Darnell. 1990. Environmental effects of the Strategic Petroleum Reserve Program on Louisiana continental shelf communities. *Amer. Zool.* 30(1):37–43.

Gitschlag, G.R., B.A. Herczeg, and T.R. Barcak. 1997. Observations of sea turtles and other marine life at the explosive removal of offshore oil and gas structures in the Gulf of Mexico. *Gulf Res. Rep.* 9(4):247–62.

Gittings, S.R. 1983. Hard bottom macrofauna of the East Flower Garden brine seep: Impact of a long term, point-source brine discharge. M.S. thesis. Texas A&M University, College Station. 72 pp.

————. 1985. Notes on barnacles (Cirripedia: Thoracica) from the Gulf of Mexico. *Gulf Res. Rep.* 8(1):35–41.

————, G.S. Boland, K.J.P. Deslarzes, C.L. Combs, B.S. Holland, and T.J. Bright. 1992. Mass spawning and reproductive viability of reef corals at the East Flower Garden Bank, northwest Gulf of Mexico. *Bull. Mar. Sci.* 51(3):420–28.

———— and T.J. Bright. 1987. Mass mortality of *Diadema antillarum* at the Flower Garden Banks, northwest Gulf of Mexico: Effects on algae and coral cover. Benthic Ecology Meeting, Raleigh, NC (abstract).

————, ————, and E.N. Powell. 1984. Hard bottom macrofauna of the East Flower Garden brine seep: Impact of a long-term sulfurous brine discharge. *Contr. Mar. Sci.* 27:105–25.

———, ———, and W.W. Schroeder. 1991. Topographic features characterization–biological. Chapt. 13, pp. 1–117. In: J.M. Brooks (ed.). *Mississippi-Alabama Marine Ecosystem Study.* Final report to US Dept. of the Interior, Minerals Management Service, New Orleans, LA. MMS 89–0095.

———, ———, ———, W.W. Sager, J.S. Laswell, and R.Rezak. 1992. Invertebrate assemblages and ecological controls on topographic features in the northeast Gulf of Mexico. *Bull. Mar. Sci.* 50(3):435–55.

———, G.D. Dennis, and H.W. Harry. 1986. *Annotated Guide to the Barnacles of the Northern Gulf of Mexico.* Texas A&M University, Sea Grant Publ. TAMU-SG-86-402. 36 pp.

———, K.J.P. Deslarzes, D.K. Hagman, and G.S. Boland. 1992. Reef coral populations and growth on the Flower Garden Banks, northwest Gulf of Mexico. *Proceedings of the Seventh International Coral Reef Symposium, Guam* 1:90–96.

Gladfelter, W.B. 1982. White-band disease in *Acropora palmata:* Implications for the structure and growth of shallow reefs. *Bull. Mar. Sci.* 32(2):639–43.

González, J.G. 1957. The copepods of the Mississippi Delta region. M.S. thesis. Texas A&M University, College Station. 132 pp.

Goodbody, I. 1961. Mass mortality of a marine fauna following tropical rains. *Ecology* 42(1):150–55.

Goode, G.B. and T.H. Bean. 1896. *Oceanic Ichthyology, a Treatise on the Deep-sea and Pelagic Fishes of the World Based Chiefly upon the Collections Made by the Steamers, Blake, Albatross, and Fish Hawk in the Northwestern Atlantic, with an Atlas containing 417 figures.* Spec. Bull. U.S. Nat. Mus. 2. 553 pp.; atlas, 123 pls.

Gore, R.H. and L.E. Scotto. 1979. Crabs of the family Parthenopidae (Crustacea: Brachyura: Oxyrhyncha) with notes on specimens from the Indian River region of Florida. Florida Marine Research Laboratory, *Mem. Hourglass Cruises* 3(6):1–98.

Gosselink, J.G. 1984. *The Ecology of Delta Marshes of Coastal Louisiana: A Community Profile.* US Fish and Wildlife Service, Office of Biological Services FWS/OBS-84-09. 134 pp.

———, E.P. Odum, and R.M. Pope. 1973. *The Value of the Tidal Marsh.* Louisiana State University, Sea Grant Publ. No. LSU-SG-74-03. 30 pp.

Govoni, J.J. 1993. Flux of larval fishes across frontal boundaries: Examples from the Mississippi River plume front and the western Gulf Stream front in winter. *Bull. Mar. Sci.* 53(2):538–66.

———, D.E. Hoss, and D.R. Colby. 1989. The spatial distribution of larval fishes above the Mississippi River plume. *Limnol. Oceanogr.* 34:178–87.

Gray, J.S. 1997. *Marine Biodiversity: Patterns, Threats, and Conservation Needs.* International Marine Organization, GESAMP Rep. Stud. 62. 24 pp.

Green, E.P. and M.J. Dagg. 1997. Mesozooplankton associations with medium to large marine snow aggregates in the northern Gulf of Mexico. *J. Plankton Res.* 19(4):435–47.

Green, G. 1977. Sinopsis taxonómica de trece especies de esponjas del arrecife La Blanquilla, Veracruz, México. *Anales del Centro de Ciencias del Mar y Limnología* 4(1):79–98.

Greiner, G.O.G. 1970. Distribution of major benthonic foraminiferal groups on the Gulf of Mexico continental shelf. *Micropaleontol.* 16:83–101.

Grice, G.D. 1960. Calanoid and cyclopoid copepods collected from the Florida Gulf coast and Florida Keys in 1954 and 1955. *Contr. Mar. Sci.* 10(2):217–26.

———. 1969. Calanoid copepods from the Caribbean Sea and Gulf of Mexico. 1. New species and new records from midwater trawl samples. *Bull. Mar. Sci.* 19(2):446–55.

Grimes, C.B. and J.H. Finucane. 1991. Spatial distribution and abundance of larval and juvenile fish, chlorophyll, and macrozooplankton around the Mississippi River discharge plume, and the role of the plume in fish recruitment. *Mar. Ecol. Prog. Ser.* 75:109–19.

———, ———, L.A. Collins, and D.A. DeVries. 1990. Young king mackerel, *Scomberomorus cavalla,* in the Gulf of Mexico, a summary of the distribution and occurrence of larvae and juveniles, and spawning dates for Mexican juveniles. *Bull. Mar. Sci.* 46(3):640–54.

Grimm, D.E. 1978. The occurrence of the Octocorallia (Coelenterata: Anthozoa) on the Florida Middle Ground. M.S. thesis. University of Alabama, Tuscaloosa. 85 pp.

Gudger, E.W. 1950. Fishes that live as inquilines (lodgers) in sponges. *Zoologica* 35(2):121–26.

Gunter, G. 1941. Death of fishes due to cold on the Texas coast: January, 1940. *Ecology* 22(2):203–8.

———. 1942. Further miscellaneous notes on American manatees. *J. Mammal.* 23(1):89–90.

———. 1947. Catastrophism in the sea and its paleontological significance, with special reference to the Gulf of Mexico. *Amer. J. Sci.* 245(11):669–76.

———. 1952a. Historic changes in the Mississippi River and the adjacent marine environment. *Publ. Inst. Mar. Sci., Univ. Texas* 2(2):119–39.

———. 1952b. The import of catastrophic mass mortality for marine fisheries along the Texas coast. *J. Wildl. Manag.* 16:63–69.

———. 1953. The relationship of the Bonnet Carré spillway to oyster beds in Mississippi Sound and the "Louisiana Marsh" with a report on the 1950 opening. *Publ. Inst. Mar. Sci., Univ. Texas* 3(1):15–71.

———. 1954. Mammals of the Gulf of Mexico. pp. 543–51. In: P.S. Galtsoff (ed.). *Gulf of Mexico: Its Origin, Waters, and Marine Life.* US Fish and Wildlife Service Fishery Bulletin 89. 604 pp.

——— and R.A. Geyer. 1955. Studies on the fouling organisms of the northwest Gulf of Mexico. *Publ. Inst. Mar. Sci., Univ. Texas* 4(1):37–67.

——— and H.H. Hildebrand. 1951. Destruction of fishes and other organisms on the south Texas coast by the cold wave of January 28–Februry 3, 1951. *Ecology* 32(4):731–36.

———, R.H. Williams, C.C. Davis, and F.G.W. Smith. 1948. Catastrophic mass mortality of marine animals and coincident phytoplankton bloom on the west coast of Florida, November 1946 to August 1947. *Ecol. Monogr.* 18:309–24.

Haagensen, D.A. 1976. Thecosomata. Part II. pp. 551–712. In: *Caribbean Zooplankton.* Office of Naval Research Report.

Hackney, C.T. and A.A. de la Cruz. 1981. Some notes on the macrofauna of an oligohaline tidal creek in Mississippi. *Bull. Mar. Sci.* 31(3):658–61.

Haedrich, R.L., G.T. Rowe, and P. Polloni. 1975. Zonation and faunal composition of epibenthic populations on the continental slope south of New England. *J. Mar. Res.* 33:191–212.

———, ———, and ———. 1980. The megabenthic fauna in the deep sea south of New England. *J. Mar. Res.* 57:165–79.

Hagman, D.K. and S.R. Gittings. 1992. Coral bleaching on high latitude reefs at the Flower Garden Banks, NW Gulf of Mexico. *Proceedings of the Seventh International Coral Reef Symposium,Guam* 1:38–43.

Halpern, J.A. 1970. Goniasteridae (Echinodermata, Asteroidea) of the straits of Florida. *Bull. Mar. Sci.* 20(1):193–286.

Hamilton, W.J., Jr. and J.O. Whittaker Jr. 1979. *Mammals of the Eastern United States.* 2nd ed. Cornell University Press, Ithaca, NY. 345 pp.

Hanlon, R.T. 1985. Cephalopods of the northwestern Gulf of Mexico. *Texas Conchologist* 21(3):90–93.

——— and R.F. Hixon. 1986. Behavioral associations of coral reef fishes with the sea anemone, *Condylactis gigantea,* in the Dry Tortugas, Florida. *Bull. Mar. Sci.* 39(1):130–34.

Hann, R.W., Jr., C.P. Giammona, and R.E. Randall (eds.). 1985. *Offshore Oceanographic and Environmental Monitoring Services for the Strategic Petroleum Reserve. Annual Report for the West Hackberry Site from November 1983 through November 1984.* Report to US Dept. of Energy. NTIS No. DOE-PO-10850-5.

Happ, G., J.G. Gosselink, and J.W. Day. 1977. The seasonal distribution of organic carbon in a Louisiana estuary. *J. Estuarine Coastal Mar. Sci.* 5:695–705.

Harding, J.L. 1964. Petrology and petrography of the Campeche Lithic Suite, Yucatan shelf, Mexico. Ph.D. dissertation. Texas A&M University, College Station. 140 pp.

Harp, J.C. 1980. The meiofaunal communities associated with two salt marshes on Dauphin Island, Alabama, with emphasis on the benthic harpacticoid copepods. M.S. thesis. University of South Alabama, Mobile. 82 pp.

Harper, D.E., Jr. 1968. Distribution of *Lucifer faxoni* (Crustacea: Decapoda: Sergestidae) in neritic waters off the Texas coast, with a note on the occurrence of *Lucifer typus. Contr. Mar. Sci., Univ. Texas* 13:1–16.

———. 1970. Ecological studies of selected level-bottom macroinvertebrates off Galveston, Texas. Ph.D. dissertation. Texas A&M University, College Station. 300 pp.

———. 1991. Macroinfauna and macroepifauna. pp. 7–1 to 7–43. In: J.M. Brooks (ed.). *Mississippi-Alabama Continental Shelf Ecosystem Study.* Vol. 2. *Technical Narrative.* Report by Texas A&M University to US Dept. of the Interior, Minerals Management Service, New Orleans, LA.

———, L.D. McKinney, and J.M. Nance. 1985. Benthos. In: R.W. Hann, C.P. Giammona, and R.E. Randall (eds.). *Offshore Oceanographic and Environmental Monitoring Services for the Strategic Petroleum Reserve. Annual Report for the Bryan Mound Site from September 1983 through August 1984.* Report to US Dept. of Energy, contract no. DOE-P010850-4.

———, ———, ———, and R.R. Salzer. 1991. Recovery responses of two benthic assemblages following an acute hypoxic event on the Texas continental shelf, northwestern Gulf of Mexico. pp. 49–64. In: *Modern and Ancient Continental Shelf Anoxia.* R. Tyson and T.H. Pearson (eds.). Geological Society of London Special Publ. No. 58.

———, D.L. Potts, R.R. Salzer, R.J. Case, R.L. Jaschek, and C.M. Walker. 1981. Distribution and abundance of macrobenthic and meiobenthic organisms. pp. 133–

77. In: B.S. Middleditch (ed.). *Environmental Effects of Offshore Oil Production: The Buccaneer Gas and Oil Field Study.* Plenum Press, New York. 446 pp.

Hartman, D.S. 1979. Ecology and behavior of the manatee (*Trichechus manatus*) in Florida. American Society of Mammalogists Special Publ. No. 5.

Hartman, O. 1954. Polychaetous annelids of the Gulf of Mexico. pp. 413–17. In: P.S. Galtsoff (ed.). *Gulf of Mexico: Its Origin, Waters, and Marine Life.* US Fish and Wildlife Service Fishery Bulletin 89. 604 pp.

———. 1957. The littoral marine annelids of the Gulf of Mexico. *Publ. Inst. Mar. Sci., Univ. Texas* 2(1):7–124.

Hartman, W.D. 1955. A collection of sponges from the west coast of the Yucatan Peninsula with descriptions of two new species. *Bull. Mar. Sci.* 5(3):161–89.

Harvey, E.N. 1952. *Bioluminescence.* Academic Press, New York. 649 pp.

Haskell, B.D., E. Lindelof, and B.D. Causey. 1994. Monitoring the health of the Florida Keys National Marine Sanctuary: Research needs. *Bull. Mar. Sci.* 54(3):1077.

Hastings, J.W. 1983. Chemical control of bioluminescence in marine organisms. *Bull. Mar. Sci.* 33(4):818–28.

Hastings, R.W., L.H. Ogren, and M.T. Mabry. 1976. Observations on the fish fauna associated with offshore platforms in the northeastern Gulf of Mexico. *Fish. Bull. US Nat. Mar. Fish. Serv.* 74:387–401.

Hawes, S.R. and H.M. Perry. 1978. Effects of 1973 floodwaters on plankton populations in Louisiana and Mississippi. *Gulf Res. Rep.* 6(2):109–24.

Healy, P.F., H.I. McKillop, and B. Walsh. 1984. Analysis of obsidian from Moho Cay, Belize: New evidence on classic Maya trade routes. *Science* 225 (4660):414–16.

Heard, R.W. 1979a. *Guide to Common Tidal Marsh Invertebrates of the Northeastern Gulf of Mexico.* Mississippi-Alabama Sea Grant Consortium Publ. No. MASGP-79-004. 82 pp.

———. 1979b. Macroinfaunal crustaceans. Chapt. 16. In: Dames and Moore (ed.). *The Mississippi, Alabama, Florida Outer Continental Shelf Baseline Environmental Survey 1977/1978.* Report to the US Dept. of the Interior, Bureau of Land Management, Washington, DC, contract no. AA550-CT7-34.

Heck, K.L., Jr. and L.D. Coen. 1995. Predation and the abundance of blue crabs: A comparison of selected east and Gulf coast (USA) studies. *Bull. Mar. Sci.* 57(3):877–83.

Hecker, B. 1985. Fauna from a cold sulfur-seep in the Gulf of Mexico: Comparisons with hydrothermal vent communities and evolutionary implications. *Bull. Biol. Soc. Wash.* 6:465–74.

Hedgpeth, J.W. 1953. An introduction to the zoogeography of the northwestern Gulf of Mexico with reference to the invertebrate fauna. *Publ. Inst. Mar. Sci., Univ. Texas* 3(1):108–224.

———. 1954a. Anthozoa: The anemones. pp. 285–90. In: P.S. Galtsoff (ed.). *Gulf of Mexico: Its Origin, Waters, and Marine Life.* US Fish and Wildlife Service Fishery Bulletin 89. 604 pp.

———. 1954b. Bottom communities of the Gulf of Mexico. pp. 203–14. In: P.S. Galtsoff (ed.). *Gulf of Mexico: Its Origin, Waters, and Marine Life.* US Fish and Wildlife Service Fishery Bulletin 89. 604 pp.

———. 1954c. Enteropneusta. p. 369. In: P.S. Galtsoff (ed.). *Gulf of Mexico: Its Origin, Waters, and Marine Life.* US Fish and Wildlife Service Fishery Bulletin 89. 604 pp.

———. 1954d. The lancets. p. 499. In: P.S. Galtsoff (ed.). *Gulf of Mexico: Its Origin, Waters, and Marine Life.* US Fish and Wildlife Service Fishery Bulletin 89. 604 pp.

———. 1954e. Miscellaneous vermes. pp. 419–20. In: P.S. Galtsoff (ed.). *Gulf of Mexico: Its Origin, Waters, and Marine Life.* US Fish and Wildlife Service Fishery Bulletin 89. 604 pp.

———. 1954 f. Pycnogonida. pp. 425–27. In: P.S. Galtsoff (ed.). *Gulf of Mexico: Its Origin, Waters, and Marine Life.* US Fish and Wildlife Service Fishery Bulletin 89. 604 pp.

———. 1954g. Scyphozoa. pp. 277–78. In: P.S. Galtsoff (ed.). *Gulf of Mexico: Its Origin, Waters, and Marine Life.* US Fish and Wildlife Service Fishery Bulletin 89. 604 pp.

———. 1954h. Xiphosura. p. 423. In: P.S. Galtsoff (ed.). *Gulf of Mexico: Its Origin, Waters, and Marine Life.* US Fish and Wildlife Service Fishery Bulletin 89. 604 pp.

Heezen, B.C. and C.D. Hollister. 1971. *The Face of the Deep.* Oxford University Press, New York. 672 pp.

Hefferman, J.J. and T.L. Hopkins. 1981. Vertical distribution and feeding of the shrimp genera (*Gennadas* and *Bentheogennema* (Decapoda: Penaeidae) in the eastern Gulf of Mexico. *J. Crust. Biol.* 1:461–73.

Heilprin, A. 1890. The corals and coral reefs of the western waters of the Gulf of Mexico. *Proc. Acad. Nat. Sci. Philadelphia* 42:303–16.

Hendler, G. and D.L. Meyer. 1982. An association of a polychaete, *Branchiosyllis exilis,* with an ophiuroid, *Ophiocoma echinata,* in Panama. *Bull. Mar. Sci.* 32(3):736–44.

———, J.E. Miller, D.L. Pawson, and P.M. Kier. 1995. *Sea Stars, Sea Urchins, and Allies: Echinoderms of Florida and the Caribbean.* Smithsonian Institution Press, Washington, DC. 390 pp.

Henkel, D.H. 1982. Echinoderms of Enmedio Reef,

southwestern Gulf of Mexico. M.S. thesis. Corpus Christi State University. 78 pp.

Henry, D.P. 1954. Cirripedia: The barnacles of the Gulf of Mexico. pp. 443–46. In: P.S. Galtsoff (ed.). *Gulf of Mexico: Its Origin, Waters, and Marine Life.* US Fish and Wildlife Service Fishery Bulletin 89. 604 pp.

Henry, W.K. and A.H. Thompson. 1976. An example of polar air modification over the Gulf of Mexico. *Mon. Weath. Rev.* 104(10):1324–27.

Heremans, K. 1982. High pressure effects on proteins and other molecules. *Ann. Rev. Biophys. Bioeng.* 11:1–21.

Herring, P.J. (ed.). 1978. *Bioluminescence in Action.* Academic Press, New York. 570 pp.

Herrnkind, W., G. Stanton, and E. Conklin. 1976. Initial characterization of the commensal complex associated with the anemone, *Lebrunia danae,* at Grand Bahama. *Bull. Mar. Sci.* 26(1):65–71.

Hildebrand, A.R., M. Pilkington, M. Connors, C. Ortiz-Aleman, and R.E. Chávez. 1995. Size and structure of the Chicxulub crater revealed by horizontal gravity gradients and cenotes. *Nature* 376(6539):415–17.

Hildebrand, H.H. 1954. A study of the fauna of the brown shrimp (*Penaeus aztecus* Ives) grounds in the western Gulf of Mexico. *Publ. Inst. Mar. Sci., Univ. Texas* 3(2):233–366.

———. 1955. A study of the fauna of the pink shrimp (*Penaeus duorarum* Burkenroad) grounds in the Gulf of Campeche. *Publ. Inst. Mar. Sci., Univ. Texas* 4(1):169–232.

———. 1963. Hallazgo del área de anidación de la tortuga marina "lora," *Lepidochelys kempi* (Garman) en la costa occidental del Golfo de México. *Ciencia* 22(4):105–12.

———, H. Chávez, and H. Compton. 1964. Aporte al concocimiento de los peces del Arrecife Alacranes, Yucatán (México). *Ciencia* 23(3):107–34.

Ho, J. 1969. Copepods of the family Taeniacanthidae (Cyclopoida) parasitic on fishes in the Gulf of Mexico. *Bull. Mar. Sci.* 19(1):111–30.

——— and P.S. Perkins. 1985. Symbionts of marine copepoda: An overview. *Bull. Mar. Sci.* 37(2):586–98.

Hochachka, P.W. and G.N. Somero. 1973. *Strategies of Biochemical Adaptation.* Saunders, Philadelphia.

Hoese, H.D. 1972. Invertebrates. pp. 33–42. In: H.D. Hoese and J.M. Valentine Jr. (eds.). *University of Southwestern Louisiana Studies on the Chandeleur Islands.* University of Southwestern Louisiana, Lafayette. Res. Ser. No. 10. Biology. 60 pp.

——— and R. H. Moore. 1998. *Fishes of the Gulf of Mexico, Texas, Louisiana, and Adjacent Waters.* 2nd ed. Texas A&M University Press, College Station. 422 pp.

Holdgate, M.W. and G.F. White (eds.). 1977. *Environmental*

*Issues.* Rept. 10, Scientific Committee on Problems of the Environment (SCOPE), International Council of Scientific Unions (ICSU). John Wiley & Sons, New York. 224 pp.

Holland, J.S., J. Holt, S. Holt, R. Kalke, and N. Rabalais. 1980. Benthic invertebrates, macroinfauna and epifauna. Chapt. 15, pp. 515–89. In: R.W. Flint and N. Rabalais (eds.). *Environmental Studies, South Texas Outer Continental Shelf, 1975–77.* Vol. 1. *Ecosystem Description.* Report to the US Dept. of the Interior, Bureau of Land Management, Washington, DC, contract no. PB80–181506.

Holthuis, L.B. 1974. Biological results of the University of Miami deep-sea expeditions. 106. The lobsters of the superfamily Nephropidea of the Atlantic Ocean (Crustacea: Decapoda). *Bull. Mar. Sci.* 24(4):723–884.

——— and W.R. Mikulka. 1972. Biological results of the University of Miami deep-sea expeditions. 91. Notes on the deep-sea isopods of the genus *Bathynomus* A. Milne Edwards, 1879. *Bull. Mar. Sci.* 22(3):575–91.

Hooks, A.T., K.L. Heck Jr., and J.R. Livingston. 1976. An inshore marine invertebrate community: Structure and habitat associations in the northwestern Gulf of Mexico. *Bull. Mar. Sci.* 26(1):99–109.

Hopkins, T.L. 1966. Plankton of the St. Andrew Bay system of Florida. *Publ. Inst. Mar. Sci., Univ. Texas* 11:12–64.

———. 1973. Zooplankton. pp. III F 1–10. In: J.I. Jones, R.E. Ring, M.O. Rinkel, and R.E. Smith (eds.). *A Summary of Knowledge of the Eastern Gulf of Mexico.* State University System of Florida, Institute of Oceanography, St. Petersburg.

———. 1982. The vertical distribution of zooplankton in the eastern Gulf of Mexico. *Deep Sea Res.* 29:1069–83.

——— and R.C. Baird. 1973. Diet of the hatchetfish *Sternoptyx diaphana. Mar. Biol.* 21:34–46.

——— and ———. 1985a. Aspects of the trophic ecology of the mesopelagic fish *Lampanyctus alatus* (family Myctophidae) in the eastern Gulf of Mexico. *Bull. Oceanogr.* 3:285–313.

——— and ———. 1985b. Feeding ecology of four hatchetfishes (Sternoptychidae) in the eastern Gulf of Mexico. *Bull. Mar. Sci.* 36:260–77.

———, M.E. Flock, J.V. Gartner Jr., and J.J. Torres. 1994. Structure and trophic ecology of a low latitude midwater decapod and mysid assemblage. *Mar. Ecol. Prog. Ser.* 109:143–56.

——— and J.V. Gartner Jr. 1992. Resource-partitioning and predation impact of a low-latitude myctophid community. *Mar. Biol.* 114:185–97.

———, ———, and M.E. Flock. 1989. The caridean shrimp

(Decapoda: Natantia) assemblage in the mesopelagic zone of the eastern Gulf of Mexico. *Bull. Mar. Sci.* 45(1):1–14.

———— and T.M. Lancraft. 1984. The composition and standing stock of mesopelagic micronekton at 27ºN 86ºW in the eastern Gulf of Mexico. *Contr. Mar. Sci.* 27:143–58.

———— and T.T. Sutton. 1998. Midwater fishes and shrimps as competitors and resource partitioning in low latitude oligotrophic systems. *Mar. Ecol. Prog. Ser.* 164:37–45.

————, ————, and T.M. Lancraft. 1996. The trophic structure and predation impact of a low latitude midwater fish assemblage. *Progr. Oceanogr.* 38:205–39.

Hopkins, T.S. 1974a. *In Situ and Dredging Studies on the Epifauna and Flora at 30 Stations in the MAFLA Area of Eastern Gulf of Mexico*. Final report to US Dept. of the Interior, Bureau of Land Management, contract no. 08550-CT4–11.

————. 1974b. Observations on the Florida Middle Ground through the use of open-circuit SCUBA. pp. 227–28. In: R.E. Smith (ed.). *Proceedings of Marine Environmental Implications of Offshore Drilling Eastern Gulf of Mexico*. State University System of Florida, Institute of Oceanography, St. Petersburg.

————. 1976. *Epifaunal and Epifloral Benthic Communities in the MAFLA Year 02 Lease Area*. Final report by State University System of Florida Institute of Oceanography to Bureau of Land Management, contract no. 08550-CT5–30.

————. 1979. Macroepifauna. pp. 789–835. In: Dames and Moore (ed.). *The Mississippi, Alabama, Florida Outer Continental Shelf Baseline Environmental Survey 1977–1978*. Report to the US Dept. of the Interior, Bureau of Land Management, Washington, DC, contract no. AA550-CT7–34.

————. 1981. Biology. Sect. 4, pp. 61–131. In: R. Norman (ed.). *Northern Gulf of Mexico Topographic Features Study*. Final report by Texas A&M University to US Dept. of the Interior, Bureau of Land Management, contract AA551-CT8–35.

————, D.R. Blizzard, S.A. Brawley, S.A. Earle, D.E. Grimm, D.K. Gilbert, P.G. Johnson, et al. 1977. A preliminary characterization of the biotic components of composite strip transects on the Florida Middle Ground, northeastern Gulf of Mexico. *Proceedings, Third International Coral Reef Symposium, University of Miami Rosenstiel School of Marine and Atmospheric Science* 1:31–37.

————, ————, and D.K. Gilbert. 1977. The molluscan fauna of the Florida Middle Ground with comments on its zoogeographical affinities. *Northeast Gulf Sci.* 1(1):39–47.

————, W. Schroeder, T. Hilde, L. Doyle, and J. Steinmetz. 1981. Florida Middle Ground. Chapt. 21. In: R. Norman (ed.). *Northern Gulf of Mexico Topographic Features Study*. Report by Texas A&M University to US Dept. of the Interior, Bureau of Land Management, contract AA551-CT8–35.

Hopper, B.E. 1961a. Marine nematodes from the coast line of the Gulf of Mexico, I. *Can. J. Zool.* 39:183–99.

————. 1961b. Marine nematodes from the coast line of the Gulf of Mexico, II. *Can. J. Zool.* 39:359–65.

————. 1963. Marine nematodes from the coast line of the Gulf of Mexico, III. Additional species from Gulf Shores, Alabama. *Can. J. Zool.* 41:841–63.

Hoskins, C.M. 1963. *Recent Carbonate Sedimentation on Alacran Reef, Yucatan, Mexico*. National Academy of Sciences National Research Council Publ. 1089:1–160.

Houde, E.D., J.C. Leak, C.E. Dowd, S.A. Berkeley, and W.J. Richards. 1979. *Ichthyoplankton Abundance and Diversity in the Eastern Gulf of Mexico*. Report to the Bureau of Land Management, contract no. 11550-CT7–28. 546 pp.

Howard, R.L., G.S. Boland, B.J. Gallaway, and G.D. Dennis. 1980. *Effects of Oil and Gas Field Structures and Effluents on Fouling Community Production and Function*. Vol. 6, W.B. Jackson and P.Wilkins (eds.). *Environmental Assessment of an Active Field in the Northwestern Gulf of Mexico*. Annual report by the National Marine Fisheries Service to the US Environmental Protectection Agency, Washington, DC, proj. no. EPA-IAG-D5-E693-EO.

Howell, A.H. 1932. *Florida Bird Life*. Florida Dept. of Game and Fish, Cowan-McCann, New York. 579 pp.

Hubbard, G.F. 1977. A quantitative analysis of benthic polychaetous annelids from the northwestern Gulf of Mexico. M.S. thesis. Texas A&M University, College Station. 84 pp.

————. 1995. Benthic polychaetes from the northern Gulf of Mexico continental slope. Ph.D. dissertation. Texas A&M University, College Station. 265 pp.

Huerta, L. and A.G. Barrientos. 1965. Algas marinas de la barra de Tuxpan y de los arrecifes Blanquilla y Lobos. *Anales de la Escuela Nacional de Ciencias Biológicas, México* 13(1–4):5–21.

————, M.L. Chávez, and M.E. Sánchez. 1977. Algas marinas de la Isla de Enmedio, Veracruz. pp. 313–24. In: F.A. Manique, (ed.). *Memorias de V Congreso Nacional de Oceanografía, Guaymas, Sonora, Mexico*. 724 pp.

Huff, J.A. and S.P. Cobb. 1979. Penaeoid and sergestoid shrimps (Crustacea: Decapoda). Florida Marine Research Laboratory, *Mem. Hourglass Cruises* 5(4):1–102.

Hughes, W.A. 1968. The thecosomatous pteropods of the Gulf of Mexico. M.S. thesis. Texas A&M University, College Station. 59 pp.

Hulings, N.C. 1961. The barnacle and decapod fauna from the near shore area of Panama City, Florida. *Quart. J. Fla. Acad. Sci.* 24(3):215–22.

———. 1967. A review of the recent marine podocopid and platycopid ostracods of the Gulf of Mexico. *Contr. Mar. Sci., Univ. Texas* 12:80–100.

Humann, P. 1994. *Reef Fish Identification: Florida, Caribbean, Bahamas.* 2nd ed. New World Publications, Jacksonville, FL. 396 pp.

Humm, H.J. 1952. Marine algae from Campeche Banks. *Fla. State Univ. Stud.* 7:27.

———. 1956. Sea grasses of the northern Gulf coast. *Bull. Mar. Sci.* 4:305–8.

———. 1964. Epiphytes of the seagrass, *Thalassia testudinum,* in Florida. *Bull. Mar. Sci.* 14(2):306–41.

———. 1973. The biological environment: Salt marshes, benthic algae of the east Gulf of Mexico, seagrasses, mangroves. In: J.I. Jones, R.E. Ring, M.O. Rinkel, and R.E. Smith (eds.). *A Summary of Knowledge of the Eastern Gulf of Mexico.* State University System of Florida Institute of Oceanography.

——— and R.L. Caylor. 1957. The summer marine flora of Mississippi Sound. *Publ. Inst. Mar. Sci., Univ. Texas* 4(2):228–64.

——— and R.M. Darnell. 1959. A collection of marine algae from the Chandeleur Islands. *Publ. Inst. Mar. Sci., Univ. Texas* 6:265–76.

——— and H.H. Hildebrand. 1962. Marine algae from the Gulf coast of Texas and Mexico. *Publ. Inst. Mar. Sci., Univ. Texas* 8:227–68.

——— and S.E. Taylor. 1961. Marine chlorophyta from the upper west coast of Florida. *Bull. Mar. Sci.* 11(3):321–80.

Humphrey, W.D. 1979. Diversity, distribution, and relative abundance of benthic fauna in a Mississippi tidal marsh. Ph.D. dissertation. Mississippi State University, Mississippi State. 93 pp.

Hutton, R.F. 1964. A second list of parasites from marine and coastal animals of Florida. *Trans. Amer. Micros. Soc.* 83:439–47.

——— and F. Sogandares-Bernal. 1959. *Studies on the Trematode Parasites Encysted in Florida Mullets.* Florida State Board of Conservation, Spec. Sci. Rept. 1:1–88.

Hyman, L.H. 1954. Free-living flatworms (Turbellaria) of the Gulf of Mexico. pp. 301–2. In: P.S. Galtsoff (ed.). *Gulf of Mexico: Its Origin, Waters, and Marine Life.* US Fish and Wildlife Fishery Bulletin 89. 604 pp.

Ibrahim, M.A. 1973. The energetics of growth, respiration, and egestion of the brown shrimp *Penaeus aztecus aztecus* Ives. Ph.D. dissertation. Texas A&M University, College Station. 82 pp.

Ichiye, T., H. Kuo, and M.R. Carnes. 1973. *Assessment of Currents and Hydrography of the Eastern Gulf of Mexico.* Texas A&M University Oceanography Dept., Contribution No. 601.

Imhoff, T.A. 1976. *Alabama Birds.* University of Alabama Press, Tuscaloosa. 445 pp.

Ingersoll, E. 1882. On the fish-mortality in the Gulf of Mexico. *Proc. U.S. Nat. Mus.* 4:74–80.

International Symposium. 1981. Second international symposium on biology and management of tropical shallow water communities (coral reefs, bays, and estuaries) 20 July–2 August 1980. Papua New Guinea. *Bull. Mar. Sci.* 31(3):477–815.

Iversen, E.S., D.L. Jory, and S.P. Bannerot. 1986. Predation on queen conchs, *Strombus gigas,* in the Bahamas. *Bull. Mar. Sci.* 39(1):61–75.

Iverson, R.L. and T.L. Hopkins. 1981. A summary of knowledge of plankton production in the Gulf of Mexico: Recent phytoplankton and zooplankton research. pp. 147–211. In: *Proceedings of a Symposium on Environmental Research Needs in the Gulf of Mexico (GOMEX), Key Biscayne, FL, 30 Sept.–5 Oct. 1979.* Vol. 2A.

Ivester, M.S. 1979. Analysis of benthic meiofauna from the MAFLA/eastern Gulf of Mexico. pp. 641–64. In: Dames and Moore (eds.). *The Mississippi, Alabama, Florida Outer Continental Shelf Baseline Environmental Survey, 1977/1978.* Vol. 2. *A Compendium of Work Element Reports.* Report to the US Dept. of the Interior, Bureau of Land Management, New Orleans, LA, contract no. AA550-CT7-34.

Jaap, W.C. 1984. *The Ecology of the South Florida Coral Reefs: A Community Profile.* US Fish and Wildlife Service, FWS/OBS-82/08. 138 pp.

Jackson, G.A. 1986. Interaction of physical and biological processes in the settlement of planktonic larvae. *Bull. Mar. Sci.* 39(2):202–12.

Jaenicke, R. 1981. Enzymes under extremes of physical conditions. *Ann. Rev. Biophys. Bioeng.* 10:1–67.

James, B.M. 1966. The Euphausiacea of the Gulf of Mexico and northwestern Caribbean. M.S. thesis. Texas A&M University, College Station. 75 pp.

———. 1970. Euphausiacean crustacea. pp. 205–30. In: W.E. Pequegnat and F.A. Chace Jr. (eds.). *Contributions on the Biology of the Gulf of Mexico.* Texas A&M University Oceanographic Studies. Vol. 1. Gulf Publishing, Houston. 270 pp.

———. 1972. Systematics and biology of the deep-water

Palaeotaxodonta (Mollusca: Bivalvia) from the Gulf of Mexico. Ph.D. dissertation. Texas A&M University, College Station. 135 pp.

Jefferson, T.A. 1995. Distribution, abundance, and some aspects of the biology of cetaceans in the offshore Gulf of Mexico. Ph.D. dissertation. Texas A&M University, College Station. 232 pp.

——— and A.J. Schiro. 1997. Distribution of cetaceans in the offshore Gulf of Mexico. *Mammal Review* 27:27–50.

Jeffrey, L.M. 1972. Organic chemistry. pp. 5–6. In: S.Z. El-Sayed, W.M. Sackett, L.M. Jeffrey, A.D. Fredericks, R.P. Saunders, P.S. Conger, G.A. Fryxell, K.A. Steidinger, and S.A. Earle. *Chemistry, Productivity, and Benthic Algae of the Gulf of Mexico.* American Geographical Society Serial Atlas of the Marine Environment, Folio 22. 29 pp.

Jochens, A.E., L.C. Bender, S.F. DiMarco, J.W. Morse, M.C. Kennicutt II, M.K. Howard, and W.D. Nowlin Jr. 2005. *Understanding the Processes that Maintain the Oxygen Levels in the Deep Gulf of Mexico: Synthesis Report.* US Dept. of the Interior, Minerals Management Service, Gulf of Mexico OCS Region, New Orleans, LA. OCS Study MMS 2005-032. 142 pp.

Johannes, R.E. and M. Satomi. 1966. Composition and nutritive value of fecal pellets of a marine crustacean. *Limnol. Oceanogr.* 11(2):191–97.

Jones, J.I. 1968. The relationship between planktonic foraminiferal populations to water masses in the western Caribbean and lower Gulf of Mexico. *Bull. Mar. Sci.* 18(4):946–82.

Jordan, C.L. 1973. The physical environment. pp. IIA1-IIA14. In: J.I. Jones, R.E. Ring, M.O. Rinkel, and R.E. Smith (eds.). *A Summary of Knowledge of the Eastern Gulf of Mexico, 1973.* State University System of Florida Institute of Oceanography, St. Petersburg.

Joyce, E.A., Jr. 1975. *Proceedings of the Florida Red Tide Conference, 10–12 October 1974, Sarasota, Florida.* Florida Dept. of Natural Resources Res. Publ. 8. 18 pp.

——— and J. Williams. 1969. Rationale and pertinent data. Florida Marine Research Laboratory, *Mem. Hourglass Cruises* 1 (1):1–50.

Kapraun, D.F. 1970. Field and cultural studies of *Ulva* and *Enteromorpha* in the vicinity of Port Aransas, Texas. *Contr. Mar. Sci.* 15:205–83.

———. 1974. Seasonal periodicity and spatial distribution of benthic marine algae in Louisiana. *Contr. Mar. Sci.* 18:139–68.

———. 1980. Summer aspects of algal zonation on a Texas jetty in relation to wave exposure. *Contr. Mar. Sci.* 23:101–9.

Keefe, C.W. 1972. Marsh production: A summary of the literature. *Contr. Mar. Sci.* 16:163–81.

Keith, D.E. and N.C. Hulings. 1965. A quantitative study of selected nearshore infauna between Sabine Pass and Bolivar Point, Texas. *Publ. Inst. Mar. Sci., Univ. Texas* 10:33–40.

Keller, C.E., J.A. Spendelow, and R.D. Greer. 1984. *Atlas of Wading Bird and Seabird Nesting Colonies in Coastal Louisiana, Mississippi, and Alabama: 1983.* US Fish and Wildlife Service, FWS/OBS-84/13. 127 pp.

Kelly, F.J. 1988. Physical oceanography and meteorology. pp. 45–73. In: N.W. Phillips and B.M. James (eds.). *Offshore Texas and Louisiana Marine Ecosystems Data Synthesis.* Vol. 2. *Synthesis Report.* Report to the US Dept. of the Interior, Minerals Management Service, New Orleans, LA. 477 pp.

———. 1991. Physical oceanography/water mass characterization. pp. 10–1 to 10–151. In: J.M. Brooks (ed.). *Mississippi-Alabama Continental Shelf Ecosystem Study: Data Summary and Synthesis.* Vol. 2. *Technical Narrative.* US Dept. of the Interior, Minerals Management Service, Gulf of Mexico OCS Region, New Orleans, LA. OCS Study MMS 91-0063. 862 pp.

Kennedy, E.A., Jr. 1959. A comparison of the molluscan fauna along a transect extending from the shoreline to a point near the edge of the continental shelf of the Texas coast. M.S. thesis. Texas Christian University, Fort Worth. 136 pp.

———. 1976. A distribution study of deep-sea macrobenthos collected from the western Gulf of Mexico. Ph.D. dissertation. Texas A&M University, College Station. 201 pp.

Kennicutt, M.C., II 1991. Sediment hydrocarbon and bulk organic matter distributions. Chapt. 4, pp. 4–1 to 4–44. In: J.M. Brooks (ed.). *Mississippi-Alabama Continental Shelf Ecosystem Study: Data Summary and Synthesis.* Vol. 2. *Technical Narrative.* US Dept. of the Interior, Minerals Management Service, Gulf of Mexico OCS Region, New Orleans, LA. OCS Study MMS 91-0063. 862 pp.

———, 2000. Chemical oceanography. Chapt. 5, pp. 123–39. In: Continental Shelf Associates, Inc. *Deepwater Program: Gulf of Mexico Deepwater Information Resources Data Search and Literature Synthesis.* Vol. 1. *Narrative Report.* US Dept. of the Interior, Minerals Management Service, Gulf of Mexico OCS Region, New Orleans, LA. OCS Study MMS 2000-049. 340 pp.

———, J.M. Brooks, E.L. Atlas, and C.S. Giam. 1988. Organic compounds of environmental concern in the Gulf of Mexico: A review. *Aquat. Toxicol.* 11:191–212.

———, ———, R.R. Bidigare, and G.J. Denoux. 1988. Gulf of Mexico hydrocarbon seep communities: I. Regional distribution of hydrocarbon seepage and associated fauna. *Deep Sea Res.* 35:1639–51.

———, ———, ———, R.R. Fay, T.L. Wade, and T.J. MacDonald. 1985. Vent-type taxa in a hydrocarbon seep region on the Louisiana slope. *Nature* 317:351–53.

———, J.L. Sericano, T.L. Wade, F. Alcazar, and J.M. Brooks. 1987. High molecular weight hydrocarbons in Gulf of Mexico continental slope sediments. *Deep Sea Res.* 34(3):403–24.

Ketchum, B.H. (ed.). 1972. *The Water's Edge: Critical Problems of the Coastal Zone.* MIT Press, Cambridge, MA. 393 pp.

Khromov, N.S. 1965. Distribution of plankton in the Gulf of Mexico and some aspects of its seasonal dynamics. pp. 36–56. In: A.S. Bogdanov (ed.). *Soviet-Cuban Fishery Research.* Israel Program for Scientific Translations, 1969.

Kilby, J.D. 1955. The fishes of two gulf coastal marsh areas of Florida. *Tulane Stud. Zool.* 2:175–247.

Kim, C.S. 1964. Marine algae of Alacran Reef, southern Gulf of Mexico. Ph.D. dissertation. Duke University, Durham, NC. 212 pp.

King, C.E. 1962. Some aspects of the ecology of psammolittoral nematodes in the northwestern Gulf of Mexico. *Ecology* 43:515–23.

Kinne, O. (ed.). 1983. *Diseases of Marine Animals.* Vol. 2, *Introduction, Bivalves to Scaphopoda.* Biologische Anstalt Helgoland, Hamburg, Germany. 571 pp.

——— (ed.). 1984. *Diseases of Marine Animals.* Vol. 4, Pt. 1. *Introduction, Pisces.* Biologische Anstalt Helgoland, Hamburg, Germany. 541 pp.

——— (ed.). 1985. *Diseases of Marine Animals.* Vol. 4, Pt. 2. *Introduction, Reptilia, Aves, Mammalia.* Biologische Anstalt Helgoland, Hamburg, Germany. 884 pp.

———, and H.P. Bulnheim (eds.).1980. *Protection of Life in the Sea.* 14th European Marine Biology Symposium. *Helgoländer Meeresunters* 33(1–4).

Kinsey, S.T. and T.L. Hopkins. 1994. Trophic strategies of euphausiids in a low latitude ecosystem. *Mar. Biol.* 118:651–61.

Kirby-Smith, W.W. 1976. The detritus problem and the feeding and digestion of an estuarine organism. pp. 469–79. In: M. Wiley (ed.). *Estuarine Processes.* Academic Press, New York. Vol. 1. 541 pp.

Klima, E.F. and D.A. Wickham. 1971. Attraction of coastal pelagic fishes with artificial structures. *Trans. Amer. Fish. Soc.* 100(1):86–99.

Kolb, C.R. and J.R. Van Lopek. 1958. *Geology of the Mississippi River Delta Plain, Southeastern Louisiana.* US Army Corps of Engineers, Waterways Exp. Sta. Tech. Rept. 483 pp. 2 vols.

Korath, K.J. 1955. Studies on the monogenetic trematodes of the Texas coast. I. Results of a survey of marine fishes at Port Aransas, with a review of Monogenea reported from the Gulf of Mexico and notes on euryhalinity, host-specificity, and relationship of the remora and the cobia. *Publ. Inst. Mar. Sci., Univ. Texas* 4(1):233–50.

Kornicker, L.S. 1983. *Rutidermatidae of the Continental Shelf of Southeastern North America and the Gulf of Mexico (Ostracoda: Myodocopina).* Smithson. Contrib. Zool., no. 371. 86 pp.

———. 1984. *Cypridinidae of the Continental Shelves of Southeastern North America, the Northern Gulf of Mexico, and the West Indies (Ostracoda: Myodocopina).* Smithson. Contrib. Zool., no. 401. 36 pp.

———. 1986. *Sarsiellidae of the Western Atlantic and Northern Gulf of Mexico, and Revision of the Sarsiellinae (Ostracoda: Myodocopina).* Smithson. Contrib. Zool., no. 415. 217 pp.

———, F. Bonet, R. Cann, and C.M. Hoskins. 1959. Alacran Reef, Campeche Bank, Mexico. *Publ. Inst. Mar. Sci., Univ. Texas* 6:1–22.

——— and D.W. Boyd. 1962. Shallow water geology and environments of Alacran Reef complex, Campeche Banks, Mexico. *Amer. Assoc. Petrol. Geol. Bull.* 46:640–73.

Kraemer, G.P. 1982. Population levels and growth rates of scleractinian corals within the *Diploria-Montastrea-Porites* zone of the East and West Flower Garden Banks. M.S. thesis. Texas A&M University, College Station. 139 pp.

Kropp, R.K. 1989. A revision of the Pacific species of gall crabs, genus *Opecarcinus* (Crustacea: Cryptochiridae). *Bull. Mar. Sci.* 45(1):98–129.

——— and R.B. Manning. 1987. *The Atlantic Gall Crabs, Family Cryptochiridae (Crustacea: Decapoda: Brachyura).* Smithson. Contrib. Zool., no. 462. 21 pp.

Krutak, P.R. and S.E. Rickles. 1979. Equilibrium in modern coral reefs, western Gulf of Mexico: Role of ecology and ostracod microfauna. *Gulf Coast Assoc. Geol. Socs., Trans.* 29:263–74.

———, ———, and R. Gio Argaez. 1980. Modern ostracod species diversity, dominance, and biofacies pattern in Veracruz-Anton Lizardo reefs, Mexico. *Anales del Centro de Ciencias del Mar y Limnología* 7(2):181–97.

Kuhlmann, D.H.H. 1975. Characterisierung der Korallenriffe von Veracruz, Mexico. *International Revue der gesamten Hydrobiologie* 60(4):495–521.

Kumpf, H., K. Steidinger, and K. Sherman (eds.). 1999.

*The Gulf of Mexico Large Marine Ecosystem: Assessment, Sustainability, and Management.* Blackwell Science, Malden, MA. 704 pp.

Kurz, R.C. 1995. Predator-prey interactions between gray triggerfish (*Balistes capriscus* Gmelin) and a guild of sand dollars around artificial reefs in the northeastern Gulf of Mexico. *Bull. Mar. Sci.* 56(1):150–60.

Kushlan, J.A. 1974. Effects of a natural fish fill on the water quality, plankton, and fish population of a pond in Big Cypress Swamp, Florida. *Trans. Amer. Fish. Soc.* 103:235–43.

———— and D.A. White. 1977. Nesting wading bird populations in southern Florida. *Fla. Sci.* 49:65–72.

Kuta, K.G. and L.L. Richardson. 1994. Distribution and frequency patterns of black band disease in the northern Florida Keys. *Bull. Mar. Sci.* 54(3):1078.

Ladd, H.H. 1951. Brackish-water and marine assemblages of the Texas coast, with special reference to mollusks. *Publ. Inst. Mar. Sci., Univ. Texas* 2(1):125–63.

Lagaaij, R. 1963. New additions to the bryozoan fauna of the Gulf of Mexico. *Publ. Inst. Mar. Sci., Univ. Texas* 9:162–236.

Lancraft, T.M., T.L. Hopkins, and J.J. Torres. 1988. Aspects of the ecology of the mesopelagic fish *Gonostoma elongatum* (Gonostomatidae, Stomiiformes) in the eastern Gulf of Mexico. *Mar. Ecol. Prog. Ser.* 49:27–40.

Landsberg, J.H. 1995. Tropical reef-fish disease outbreak and mass mortalities in Florida, USA: What is the role of dietary biological toxins? *Dis. Aquat. Org.* 22:83–100.

Lang, J. 1971. Interspecific aggression by scleractinian corals. 1. The rediscovery of *Scolymia cubensis* (Milne Edwards and Haime). *Bull. Mar. Sci.* 21(4):952–59.

————. 1973. Interspecific aggression by scleractinian corals. 2. Why the race is not only to the swift. *Bull. Mar. Sci.* 23(2):260–79.

Layne, J.N. 1974. The land mammals of south Florida. *Miami Geol. Soc. Mem.* 2:386–413.

Leary, S.P. 1967. *The Crabs of Texas.* Tex. Parks Wild. Dept. Bull. 43 (Ser. 7). 37 pp.

Leatherwood, S., D.K. Caldwell, and H.E. Winn. 1976. *Whales, Dolphins, and Porpoises of the Western North Atlantic: A Guide to their Identification.* NOAA, National Marine Fisheries Service, Circ. 396. 176 pp.

Leavesley, A., M. Bauer, K. McMillen, and R. Casey. 1978. Living shelled microzooplankton (radiolarians, foraminiferans, and pteropods) as indicators of oceanographic processes in water over the outer continental shelf of south Texas. *Gulf Coast Assoc. Geol. Socs., Trans.* 28:229–38.

Lecroy, S.E. 1995. Amphipod crustacea. III. Family

Colomastigidae. Florida Marine Research Laboatory, *Mem. Hourglass Cruises* 11(2):1–138.

Lehman, R.L. and J.W. Tunnell Jr. 1992. Species composition and ecology of the macroalgae of Enmedio Reef, Veracruz, Mexico. *Texas J. Sci.* 44(4):445–57.

Leipper, D.F. 1954a. Marine meteorology of the Gulf of Mexico, a brief review. pp. 89–98. In: P.S. Galtsoff (ed.). *Gulf of Mexico: Its origin, Waters and Marine Life.* US Fish and Wildlife Service Fishery Bulletin 89. 604 pp.

————. 1954b. Physical oceanography of the Gulf of Mexico. pp. 119–37. In: P.S. Galtsoff (ed.). *Gulf of Mexico: Its Origin, Waters, and Marine Life.* US Fish and Wildlife Service Fishery Bulletin 89. 604 pp.

Lemaitre, R. 1991. To the memory of Gilbert L. Voss: *Portunus vossi,* a rare new species of swimming crab (Decapoda: Brachyura: Portunidae) from the west coast of Florida. *Bull. Mar. Sci.* 49(1, 2):546–51.

Leming, T.D. and D.R. Johnson. 1985. Application of circulation models to larval dispersement and recruitment. *MTS Jour.* 19(2):34–41.

Leuterman, A. 1979. The taxonomy and systematics of the Gymnolaemate and Stenolaemate Bryozoa of the northwest Gulf of Mexico. Ph.D. dissertation. Texas A&M University, College Station. 309 pp.

Lewis, T.C. and R.W. Yerger. 1976. Biology of five species of searobins (Pisces, Triglidae) from northeastern Gulf of Mexico. *Fish. Bull.* 74(1):93–103.

Ley-Lou, F. 1985. Aquatic primary productivity, nutrient chemistry, and oyster community metabolism in mangrove bordered tidal channel, Laguna de Terminos, Mexico. M.S. thesis. Louisiana State University, Baton Rouge. 59 pp.

Library of Congress, Congressional Research Service. 1977. *A Compilation of Federal Laws Relating to Conservation and Development of Our Nation's Fish and Wildlife Resources, Environmental Quality, and Oceanography.* Ser. No. 95-B. US Govt. Printing Office. 933 pp.

Lidz, B. H., E. A. Shinn, J. H. Hudson, H. G. Multer, R. B. Halley, and D. M. Robson. 2008. Controls on Late Quaternary coral reefs of the Florida Keys. Chapt. 2, pp. 9–74. In: B. M. Riegl and R. E. Dodge (eds.). *Coral Reefs of the USA.* Springer Science & Business Media B. V., Netherlands.

Lidz, L. and B. Lidz. 1966. Foraminiferal biofacies of Veracruz reefs. *Amer. Assoc. Petrol. Geol. Bull.* 50(7):1514–17.

Limbaugh, C.L., H. Pederson, and F.A. Chace Jr. 1961. Shrimps that clean fishes. *Bull. Mar. Sci.* 11(2):237–57.

Lin, S. and J.W. Morse. 1991. Sulfate reduction and iron

sulfide mineral formation in Gulf of Mexico anoxic sediments. *Amer. J. Sci.* 291:55–89.

Lindberg, W.J. and G. Stanton. 1988. Bryozoan-associated decapod crustaceans: Community patterns and a case of cleaning symbiosis between a shrimp and a crab. *Bull. Mar. Sci.* 42(3):411–23.

Lindner, M.J. and W.W. Anderson. 1954. Biology of commercial shrimps. pp. 457–61. In: P.S. Galtsoff (ed.). *Gulf of Mexico: Its Origin, Waters, and Marine Life.* US Fish and Wildlife Service Fishery Bulletin 89. 604 pp.

Lipka, D.A. 1970. The systematics and distribution of Enoploteuthidae and Cranchiidae (Cephalopoda: Oegopsida) from the Gulf of Mexico. M.S. thesis. Texas A&M University, College Station. 134 pp.

———. 1974. Mollusks. pp. 141–96. In: T.J. Bright and L.H. Pequegnat (eds.). *Biota of the West Flower Garden Bank.* Gulf Publishing, Houston. 435 pp.

———. 1975. Systematics and zoogeography of cephalopods from the Gulf of Mexico. Ph.D. dissertation. Texas A&M University, College Station. 347 pp.

Livingston, E.H. 1979. Observations on sponge-dwelling fishes on the Florida Middle Grounds. M.S. thesis. University of Alabama, Tuscaloosa. 64 pp.

Livingston, R.J. 1982. Trophic organization of fishes in a coastal seagrass system. *Mar. Ecol. Prog. Ser.* 7:1–12.

——— (ed.). 1979. *Ecological Processes in Coastal and Marine Systems.* Plenum Press, New York. 548 pp.

———. 1984. Trophic response of fishes to habitat variability in coastal seagrass systems. *Ecology* 65(4):1258–75.

——— and J.L. Duncan. 1979. Climatological control of a north Florida coastal system and impact due to upland forestry management. pp. 339–81. In: R.J. Livingston (ed.). *Ecological Processes in Coastal and Marine Systems.* Plenum Press, New York. 548 pp.

Lizama, J. and R.S. Blanquet. 1975. Predation on sea anemones by the amphinomid polychaete *Hermodice carunculata. Bull. Mar. Sci.* 25:442–43.

Lochmann, S.E., R.M. Darnell, and J.D. McEachran. 1995. Temporal and vertical distribution of crab larvae in a tidal pass. *Estuaries and Coasts* 18(18):255–63.

Lockhart, F.D., W.J. Lindberg, N.J. Blake, R.B. Erdman, H.M. Perry, and R.S. Waller. 1990. Distributional differences and population similarities for two deep-sea crabs (family Geryonidae) in the northeastern Gulf of Mexico. *Can. J. Fish. Aquat. Sci.* 47(11):2112–22.

Loesch, H.C. 1960. Sporadic mass shoreward migrations of demersal fish and crustaceans in Mobile Bay, Alabama. *Ecology* 41:292–98.

Logan, B.W. 1962. Coral reef and bank communities of the Campeche Shelf, Yucatan, Mexico. *Geol. Soc. Amer., Spec. Paper* no. 68:218 (abstract).

———. 1969. Coral reefs and banks, Yucatan Shelf, Mexico (Yucatan Reef Unit). pp. 129–98. In: *Carbonate Sediments and Reefs, Yucatan Shelf, Mexico.* Amer. Petrol. Inst., Mem. 11. 355 pp.

———, J.L. Harding, W.M. Ahr, J.D. Williams, and R.G. Snead. 1969. Late Quaternary carbonate sediments of Yucatan Shelf, Mexico. pp. 5–128. In: *Carbonate Sediments and Reefs, Yucatan Shelf, Mexico.* Amer. Petrol. Inst., Mem. 11. 355 pp.

Longley, W.H. and S.F. Hildebrand. 1941. *Systematic Catalog of the Fishes of Tortugas, Florida, with Observations on Color, Habits and Local Distribution.* Carnegie Institution of Washington Publ. 535. 331 pp.

Lot-Helgueras, A. 1971. Estudios sobre fanerógamas marinas en las cercanías de Veracruz, Veracruz. *Anales del Instituto de Biología, Universidad Nacional Autónoma de México Ser. Bot.* 42(1):1–48.

———, C. Vásquez-Yañez, and F. Menéndez. 1975. Physiognomic and floristic changes near the northern limits of mangroves in the Gulf coast of Mexico. In: G.E. Walsh, S.C. Snedaker, and H.J. Teas (eds.). *Proceedings of the International Symposium on Biology and Management of Mangroves.* IFAS, University of Florida, Gainesville. 846 pp.

Lowe, G.C. and E.R. Cox. 1978. Species compositon and seasonal periodicity of marine benthic algae of Galveston Island, Texas. *Contr. Mar. Sci.* 21:9–24.

Lowery, G.H., Jr. 1974. *Louisiana Birds.* Louisiana State University Press, Baton Rouge. 651 pp.

——— and R.J. Newman. 1954. The birds of the Gulf of Mexico. pp. 519–42. In: P.S. Galtsoff (ed.). *Gulf of Mexico: Its Origin, Waters, and Marine Life.* US Fish and Wildlife Service Fishery Bulletin 89. 604 pp.

Luckhurst, B.E. and K. Luckhurst. 1976. Some infaunal fishes associated with the scleractinian coral *Madracis mirabilis. Can. J. Zool.* 54:1395–97.

Lugo, A.E., G. Evink, M.M. Brinson, A. Broce, and S.C. Snedaker. 1975. Diurnal rates of photosynthesis, respiration, and transpiration in mangrove forests in south Florida. pp. 335–50. In: F. Golley and G. Medina (eds.). *Tropical Ecological Systems.* Springer-Verlag, New York.

———, M. Sell, and S.C. Snedaker. 1976. Mangrove ecosystem analysis. pp. 113–45. In: B.C. Patten (ed.). *Systems Analysis and Simulation in Ecology.* Academic Press, New York.

——— and S.C. Snedaker. 1974. The ecology of mangroves. *Ann. Rev. Ecol. Syst.* 5:39–64.

Lyczkowski-Schultz, J., D.L. Ruple, S.L. Richardson, and J.H. Cowan Jr. 1990. Distribution of fish larvae relative to time and tide in a Gulf of Mexico barrier island pass. *Bull. Mar. Sci.* 46(3):563–77.

Lynch, S.A. 1954. Geology of the Gulf of Mexico. pp. 67–86. In: P.S. Galtsoff (ed.). *Gulf of Mexico: Its Origin, Waters, and Marine Life.* US Fish and Wildlife Service Fishery Bulletin 89. 604 pp.

Lyons, W.G. 1970. Scyllarid lobsters (Crustacea, Decapoda). Florida Marine Research Laboratory, *Mem. Hourglass Cruises* 1(4):1–74.

———. 1979. Molluscan communities of the west Florida shelf. *Bull. Amer. Malacol. Union, Inc.* 37–40.

———, S.P. Cobb, D.K. Camp, J.A. Mountain, T. Savage, L. Lyons, and E.A. Joyce Jr. 1971. *Preliminary Inventory of Marine Invertebrates Collected near the Electrical Generating Plant, Crystal River, Florida, in 1969.* Florida Dept. of Natural Resources Laboratory, Prof. Pap. 14. 45 pp.

——— and S.B. Collard. 1974. Benthic invertebrate communities of the eastern Gulf of Mexico. pp. 157–65. In: R.F. Smith (ed.). *Proceedings of Conference/Workshop on Implications of Offshore Drilling in the Eastern Gulf of Mexico.* State University System of Florida Institute of Oceanography, St. Petersburg.

MacDonald, A.G. 1975. *Physiological Aspects of Deep Sea Biology.* Monogr. of the Physiol. Soc., no. 31. Cambridge University Press, Cambridge.

MacDonald, I.R. 2000. Chapt. 8, pp. 209–30. In: Continental Shelf Associates, Inc. *Deepwater Program: Gulf of Mexico Deepwater Information Resources Data Search and Literature Synthesis.* Vol. 1. *Narrative Report.* US Dept. of the Interior, Minerals Management Service, Gulf of Mexico OCS Region, New Orleans, LA. OCS Study MMS 2000–049. 340 pp.

———, G.S. Boland, J.S. Baker, J.M. Brooks, M.C. Kennicutt II, and R.R. Bidigare. 1989. Gulf of Mexico chemosynthetic communities: II. Spatial distribution of seep organisms and hydrocarbons at Bush Hill. *Mar. Biol.* 10:235–47.

———, W.R. Callender, J. Burke, R.A. MacDonald, and S.J. MacDonald. 1990. Fine scale distribution of methanotrophic mussels at a Louisiana slope cold seep. *Prog. Oceanogr.* 25:15–24.

——— and C. Fisher. 1996. Life without Light. *Nat. Geograph.* 190(4):86–97.

———, W.W. Schroeder, and J.M. Brooks. 1995. *Chemosynthetic Ecosystems Study.* Final report by the Geochemical and Environmental Research Group, Texas A&M University, to the US Dept. of the Interior, Minerals Management Service, Gulf of Mexico OCS Region, New Orleans, LA. OCS Study MMS 95–0023. 338 pp.

Machain-Castillo, M.L. 1989. Ostracod assemblages in the southern Gulf of Mexico: An overview. *Anales del Instituto de Ciencias del Mar y Limnología* 16(1):119–34.

Macias, G.G. 1968. Contribución al conocimiento de la sistemática y ecología de las esponjas del Arrecife la Blanquilla, Veracruz, Ver. M.S. thesis. *Universidad Nacional Autónoma de México.* 102 pp.

Macnae, W. 1968. A general account of the fauna and flora of mangrove swamps and forests of the Indo-West-Pacific region. *Adv. Mar. Biol.* 6:73–270.

Maddocks, R.F. 1974. Ostracodes. pp. 199–229. In: T.J. Bright and L.H. Pequegnat (eds.). *Biota of the West Flower Garden Bank.* Gulf Publishing, Houston. 435 pp.

Mahoney, B.M. 1981. An examination of interspecific territoriality in the dusky damselfish, *Eupomacentrus dorsopunicans* Poey. *Bull. Mar. Sci.* 31(1):141–46.

Manickchand-Heileman, S., L.A. Soto, and E. Escobar. 1998. A preliminary trophic model of the continental shelf, southwestern Gulf of Mexico. *Est. Coast. Shelf Sci.* 46:885–99.

Manning, R.B. 1959. A checklist of stomatopod crustaceans of the Florida-Gulf of Mexico area. *Quart. J. Fla. Acad. Sci.* 22(1):14–24.

Mansuetti, R. 1963. Symbiotic behavior between small fishes and jellyfishes, with new data on that between the stromateid, *Peprilus alepidotus,* and the scyphomedusa, *Chrysaora quinquecirrha. Copeia* 1963(1):40–80.

Manter, H.W. 1934. Some digenetic trematodes from deep-water fish of Tortugas, Florida. *Pap. Tortugas Lab., Carnegie Inst. Wash. Publ.* 28:257–345.

———. 1947. The digenetic trematodes of marine fishes of Tortugas, Florida. *Amer. Midl. Nat.* 38:257–416.

Marcus, E. and E. Marcus. 1959. Some opisthobranchs from the northwestern Gulf of Mexico. *Publ. Inst. Mar. Sci., Univ. Texas* 6:250–64.

Marmer, H.A. 1954. Tides and sea level in the Gulf of Mexico. pp. 101–18. In: P.S. Galtsoff (ed.). *Gulf of Mexico: Its Origin, Waters, and Marine Life.* US Fish and Wildlife Service Fishery Bulletin 89. 604 pp.

Márquez, M. 1978. *Sea Turtles.* Vol. 6, *FAO Species Identification Sheets for Fishery Purposes. West Central Atlantic (Fishing Area 31).* FAO, Rome.

Marshall, N.B. 1954. *Aspects of Deep Sea Biology.* Hutchinson, London. 380 pp.

———. 1980. *Deep Sea Biology: Developments and Perspectives.* Garland STPM Press, New York. 566 pp.

Martin, D. 1983. Why don't we have more red tides in Florida? *J. Environ. Sci.* 18:685–700.

Martin, D.M. 1978. Distribution and ecology of the Synaphobranchidae of the Gulf of Mexico. M.S. thesis. Texas A&M University, College Station. 113 pp.

———. 1984. Distribution and ecology of the Synaphobranchidae of the Gulf of Mexico. *Gulf Res. Rep.* 7(4):311–24.

Martin, R.G. and A.H. Bouma. 1978. Physiography of Gulf of Mexico. pp. 3–19. In: A.H. Bouma, G.T. Moore, and J.M. Coleman (eds.). *Framework, Facies, and Oil-Trapping Characteristics of the Upper Continental Margin.* Amer. Assoc. Petrol. Geol., Stud. in Geol. 7. 326 pp.

Marum, J.P. 1979. Significance of distribution patterns of planktonic copepods in Louisiana coastal waters and relationships to oil drilling and production. pp. 355–77. In: C.H. Ward, M.E. Bender, and D.J. Reish (eds.). *Effects of Oil Drilling and Production in a Coastal Environment.* Rice University Studies 65 (4 and 5), Houston. 589 pp.

Mathieson, A.C. and C.J. Dawes. 1975. Seasonal studies of Florida sublittoral marine algae. *Bull. Mar. Sci.* 25(1):46–65.

Mauchline, J. and L.R. Fisher. 1969. The biology of euphausiids. In: S. Russell and M. Young (eds.). *Advances in Marine Biology.* Vol. 7. Academic Press, London. 454 pp.

May, E.B. 1973. Extensive oxygen depletion in Mobile Bay, Alabama. *Limnol. Oceanogr.* 18:353–66.

McCarthy, F.D. 1973. Microbial degradation of the feces of the pink shrimp *Penaeus duorarum,* and the effects of temperature and oxygen limitation. Ph.D. dissertation. Texas A&M University, College Station. 70 pp.

McCarty, D.M. 1974. Polychaetes of Seven and One-Half Fathom Reef. M.S. thesis. Texas A&I University, Kingsville. 213 pp.

McCoy, E.D., H.R. Mashinsky, D. Johnson, and W.E. Meshaka Jr. 1996. Mangrove damage caused by Hurricane Andrew on the southwestern coast of Florida. *Bull. Mar. Sci.* 59(1):1–8.

McEachran, J.D. and J.D. Fechhelm. 1998. *Fishes of the Gulf of Mexico.* Vol. 1. *Myxiniformes to Gasterosteiformes.* University of Texas Press, Austin. 1112 pp.

——— and ———. 2005. *Fishes of the Gulf of Mexico.* Vol. 2. *Scorpaeniformes to Tetraodontiformes.* University of Texas Press, Austin. 1004 pp.

McGowan, M.F. 1985. Ichthyoplankton of the Flower Garden Banks, Northwest Gulf of Mexico. Ph.D. dissertation. University of Miami, Florida. 376 pp.

McKinney, L.D. 1977. The origin and distribution of shallow water Gammaridean Amphipoda in the Gulf of Mexico and Caribbean Sea, with notes on their ecology. Ph.D.

dissertation. Texas A&M University, College Station. 401 pp.

———, J.M. Nance, and D.E. Harper. 1985. Benthos. pp. 6–1 to 6–98. In: R.W. Hann Jr., C.P. Giammona, and R.E. Randall (eds.). *Offshore Oceanographic and Environmental Monitoring Services for the Strategic Petroleum Reserve. Annual Report for the West Hackberry site.* Vol. 1. Report to the US Dept. of Energy, Washington, DC.

McLelland, J.A. 1984. Observations on chaetognath distributions in the northeastern Gulf of Mexico during the summer of 1974. *Northeast Gulf Sci.* 7(1):49–59.

———. 1989. An illustrated key to the Chaetognatha of the northern Gulf of Mexico with notes on their distribution. *Gulf Res. Rep.* 8(2):145–72.

——— and H.M. Perry. 1989. Records of deep-water chaetognaths from the northern Gulf of Mexico. *Gulf Res. Rep.* 8(2):181–87.

McMillen, K.J. and R.E. Casey. 1978. Distribution of living polycistine radiolarians in the Gulf of Mexico and Caribbean Sea, and comparison with the sedimentary record. *Marine Micropaleont.* 3(2):121–45.

McNulty, J.K., R.C. Work, and H.B. Moore. 1962. Some relationships between the infauna of the level bottom and the sediment in South Florida. *Bull. Mar. Sci.* 12:322–32.

McRae, E.D., Jr. 1950. An ecological study of the Xanthidae crabs of the Cedar Key area. M.S. thesis. University of Florida, Gainesville. 73 pp.

McRoy, C.P. and C. Helfferich (eds.) 1980. *Seagrass Ecosystems: A Scientific Perspective.* Marcel Dekker, New York.

Meeder, J.F., R. Jones, J.J. O'Brien, M.S. Ross, R.J. Sawicki, and A.M. Strong. 1994. Effects of Hurricane Andrew on *Thalassia* ecosystem dynamics and the stratigraphic record. *Bull. Mar. Sci.* 54(3):1080.

Menzel, D.W. and J.H. Ryther. 1960. The annual cycle of primary production in the Sargasso Sea off Bermuda. *Deep-Sea Res.* 6(4):351–67.

Menzies, R.J., R.Y. George, and G.T. Rowe. 1973. *Abyssal Environment and Ecology of the World Oceans.* John Wiley & Sons, New York. 488 pp.

——— and W.L. Kruczynski. 1983. Isopod crustacea (exclusive of Epicaridea). Florida Marine Research Laboratory, *Mem. Hourglass Cruises* 6(1):1–126.

Merrell, W.J. and J.M. Morrison. 1981. On the circulation of the western Gulf of Mexico with observations from April 1978. *J. Geophys. Res.* 86(CS):4181–85.

Michel, H.B. 1984. *Chaetognatha of the Caribbean Sea and Adjacent Areas.* NOAA Tech. Rept. NMFS 15, US Dept. of Commerce. 33 pp.

———— and M. Foyo. 1976. Siphonophora, Heteropoda, Copepoda, Euphausiacea, Chaetognatha, and Salpidae. Part I. In: *Caribbean Zooplankton.* Office of Naval Research Report. 549 pp.

Mille-Pagaza, S., R. Reyes-Martínez, and E. Sánchez-Salazar. 1997. Distribution and abundance of Chaetognatha on the Yucatan shelf during May, 1986. *Gulf Res. Rep.* 9(4):263–75.

Miller, J.E. and D.L. Pawson. 1984. Holothurians (Echinodermata: Holothuroidea). Florida Marine Research Laboratory, *Mem. Hourglass Cruises* 7(1):1–79.

Minnery, G.A. 1984. Distribution, growth rates, and diagenesis of coralline algal structures on the Flower Garden Banks, northwestern Gulf of Mexico. Ph.D. dissertation. Texas A&M University, College Station. 177 pp.

————. 1990. Crustose coralline algae from the Flower Garden Banks, northwestern Gulf of Mexico: Controls on distribution and growth morphology. *J. Sed. Pet.* 69(6):992–1007.

————, R. Rezak, and T.J. Bright. 1985. Depth zonation and growth form of crustose coralline algae, Flower Garden Banks, northwestern Gulf of Mexico. pp. 237–46. In: D.F. Toomey and M.H. Nitecki (eds.). *Paleoalgology: Contemporary Research and Applications.* Springer-Verlag, Berlin.

Monniot, C. and F. Monniot. 1987. Abundance and distribution of tunicates on the northern continental slope of the Gulf of Mexico. *Bull. Mar. Sci.* 41(1):36–44.

Montagna, P.A. and D.E. Harper Jr. 1996. Benthic infaunal long term response to offshore production platforms in the Gulf of Mexico. *Can. J. Fish. Aquat. Sci.* 53:2567–88.

Moody, C.L. 1967. Gulf of Mexico distributive province. *Amer. Assoc. Petrol. Geol. Bull.* 51(2):179–99.

Moore, D., H.A. Brusher, and L. Trent. 1970. Relative abundance, seasonal distribution, and species composition of demersal fishes off Louisiana and Texas, 1962–1964. *Contr. Mar. Sci.* 15:45–70.

Moore, D.R. 1958. Notes on Blanquilla Reef, the most northerly coral formation in the western Gulf of Mexico. *Publ. Inst. Mar. Sci., Univ. Texas* 5:151–55.

————. 1961. The marine and brackish water mollusca of the State of Mississippi. *Gulf Res. Rep.* 1(1):1–58.

————. 1963. Distribution of the sea grass, *Thalassia,* in the United States. *Bull. Mar. Sci.* 13(2):329–42.

Moore, G.T., G.W. Starke, L.C. Bonham, and H.O. Woodbury. 1978. Mississippi Fan, Gulf of Mexico: Physiography, stratigraphy, and sedimentational patterns. pp. 155–91. In: A.H. Bouma, G.T. Moore, and J.M. Coleman (eds.). *Framework, Facies, and Oil-Trapping Characteristics of the Upper Continental Margin. Amer. Assoc. Petrol. Geol.,* Stud. Geol. 7. 326 pp.

Moore, K.A. and R.L. Wetzel. 1988. The distribution and productivity of seagrass in the Terminos Lagoon. pp. 207–20. In: A. Yáñez-Arancibia and J.W. Day Jr. (eds.). *Ecology of Coastal Ecosystems in the Southern Gulf of Mexico: The Terminos Lagoon Region.* UNAM, Coastal Ecol. Inst., Louisiana State University, Editorial Universitaria, Mexico, D.F.

Moore, R.H. 1976. Observations on fishes killed by cold at Port Aransas, Texas, 11–12 January 1973. *Southwest Nat.* 20(4):461–66.

Morée, M.D. 1979. The vertical distribution of Phoronimid amphipods at two stations in the central Gulf of Mexico. M.S. thesis. Texas A&M University, College Station. 77 pp.

Morin, J.G. 1983. Coastal bioluminescence: Patterns and functions. *Bull. Mar. Sci.* 33(4):787–817.

Morris, E.H., J. Charlot, and A.A. Morris. 1931. *The Temple of the Warriors at Chichén Itzá, Yucatán.* Carnegie Institution of Washington vol. 1., publ. 406.

Morton, J.W. 1977. *Ecological Effects of Dredging and Dredge Spoil Disposal: A Literature Review.* US Fish and Wildlife Service Technical Paper 94. 33 pp.

Mukai, L.S. 1974. The species composition and distribution of the asteroids of the Gulf of Mexico and the Caribbean. M.S. thesis. Texas A&M University, College Station. 72 pp.

Müller-Karger, F.E., J.J. Walsh, R.H. Evans, and M.B. Meyers. 1991. On the seasonal phytoplankton concentration and sea surface temperature cycles of the Gulf of Mexico as determined by satellites. *J. Geophys. Res.* 96(C7):12645–65.

Mullin, K.D., W. Hoggard, C. Roden, R. Lohoefener, C. Rogers, and B. Taggert. 1994. Cetaceans on the upper continental slope in the north-central Gulf of Mexico. *Fish. Bull.* 92:773–86.

Murdy, E.O. 1983. *Saltwater Fishes of Texas: A Dichotomous Key.* Texas A&M University, Sea Grant Publ. No. TAMU-SG-83-607. 220 pp.

Murrell, M.C. and J.W. Fleeger. 1989. Meiofauna abundance on the Gulf of Mexico continental shelf affected by hypoxia. *Cont. Shelf Res.* 9(12):1049–62.

Myers, A.A. 1981. Amphipod crustacea. I. Family Aoridae. Florida Marine Research Laboratory, *Mem. Hourglass Cruises* 5(5):1–75.

Myrberg, A.A., Jr. and R.E. Thresher. 1974. Interspecific aggression and the relevance of the concept of territoriality in reef fishes. *Amer. Zool.* 14:81–96.

Nahas, F.M. and R.B. Short. 1965. Digenetic trematodes of marine fishes from Apalachee Bay, Gulf of Mexico. *Tulane Stud. Zool.* 12:39–50.

Nakamura, I. 1976. *Catálogo de Peces Marinos de México.* Secretaría de Industria y Comercio, Subsec. Pesca., Instituto Nacional de Pesca. 462 pp.

National Academy of Science. 1995. *Understanding Marine Biodiversity: A Research Agenda for the Nation.* National Academies Press, Washington, DC. 114 pp.

Neff, J.M. 1987. Biological effects of drilling fluids, drill cuttings, and produced waters. In: D.F. Boesch and N.N. Rabalais (eds.). *The Long-Term Effects of Offshore Oil and Gas Development.* Elsevier Applied Science, New York. 708 pp.

Neilson, B.J. and L.E. Cronin (eds.). 1981. *Estuaries and Nutrients.* Humana Press, Clifton, NJ. 643 pp.

Nelson, J.S. 1994. *Fishes of the World.* John Wiley & Sons, New York. 600 pp.

Nelson, T.J., T.L. Stinnett, and J.W. Tunnell. 1988. Comparative assessment of an unusually dense octocoral community in the southwestern Gulf of Mexico. *Proceedings of the Sixth International Coral Reef Symposium, Australia* 2:791–96.

Nelson, W.R. and J.S. Carpenter. 1968. Bottom longline explorations in the Gulf of Mexico: A report on "*Oregon II*'s" first cruise. *Comm. Fish. Rev.* 30(10):57–62.

Nessis, K.N. 1975. Cephalopods of the American Mediterranean Sea. *Trudy Inst. Okeanol., Acad. Sci. USSR* 100:259–89.

———. 1987. *Cephalopods of the World.* TFH Publications, Neptune City, NJ. 351 pp.

Neumann, C.J., B.R. Jarvinen, C.J. McAdie, and J.D. Elms. 1993. *Tropical Cyclones of the Atlantic Ocean, 1871–1992.* NOAA, National Weather Service, Histor. Climat. Ser. 6–2. 3rd revis. 193 pp.

Newell, R. 1965. The role of detritus in the nutrition of two marine deposit feeders, the prosobranch *Hydrobia ulvae* and the bivalve *Macoma baltica. Proc. Zool. Soc. London* 144:25–45.

NOAA. 1986. *Gulf of Mexico Coastal and Ocean Zones Strategic Assessment: Data Atlas.* US Dept. of Commerce, NOAA, National Ocean Service, and National Marine Fisheries Service.

NOAA. 1992. *Nutrient Enhanced Coastal Ocean Productivity.* NECOP Workshop Proceedings, Oct. 1991. NOAA Coastal Ocean Program. Texas A&M University, Sea Grant Publ. No. TAMU-SG-92–09. 153 pp.

NOAA Estuarine Programs Office. 1989. *Galveston Bay: Issues, Resources, Status and Management.* US Dept. of Commerce, NOAA, Estuary Program Office, NOAA Estuary of the Month Seminar Ser. No. 13. 114 pp.

Norman, R. (ed.). 1981. *Northern Gulf of Mexico Topographic Features Study.* Report by the Dept. of Oceanography,

Texas A&M University, to the US Dept. of the Interior, Bureau of Land Management, contract no. AA551-CT8–35. Tech. Rept. No. 81-2-T. 5 vols.

Norse, E.A. (ed.). 1993. *Global Marine Biodiversity: A Strategy for Building Conservation into Decision Making.* Island Press, Washington, DC. 383 pp.

Nowlin, W.D. 1972. Winter circulation patterns and property distributions. pp. 3–52. In: L.R.A. Capurro and J.L. Reid (eds.). *Contributions on the Physical Oceanography of the Gulf of Mexico.* Gulf Publishing, Houston. 288 pp.

———, A.E. Jochens, S.F. DiMarco, and R.O. Reid. 2000. Physical oceanography. Chapt. 4, pp. 61–121. In: Continental Shelf Associates, Inc. *Deepwater Program: Gulf of Mexico. Deepwater Information Resources Data Search and Literature Synthesis.* Vol. 1. *Narrative Report.* US Dept. of the Interior, Minerals Management Service, Gulf of Mexico OCS Region, New Orleans, LA. OCS Study MMS 2000–049. 340 pp.

———, A.E. Jochens, S.F. DiMarco, R.O. Reid, and M.K. Howard. 2005. Low-frequency circulation over the Texas-Louisiana continental shelf. pp. 219–40. In: W. Sturges and A. Lugo-Fernandez (eds.). *Circulation in the Gulf of Mexico: Observations and Models.* American Geophysical Union, Geophys. Monogr. Ser. 161. 360 pp.

——— and H. McLellan. 1967. A characterization of the Gulf of Mexico waters in winter. *J. Mar. Res.* 25:29–59.

Oberholser, H.C. and E.B. Kincaid Jr. 1974. *The Bird Life of Texas.* Vol. 1. University of Texas Press, Austin. 530 pp.

Ode, H. 1973. A survey of the molluscan fauna of the northwest Gulf of Mexico: Preliminary report. *Texas Conchol.* 9(4):73–83.

Odum, E.P. 1974. Halophytes, energetics, and ecosystems. pp. 599–602. In: R.J. Reimold and W.R. Queen (eds.). *Ecology of Halophytes.* Academic Press, New York.

Odum, W.E. 1970. Insidious alteration of the estuarine environment. *Trans. Amer. Fish. Soc.* 99(4):836–46.

———, J.S. Fisher, and J. Pickral. 1979. Factors controlling the flux of particulate organic carbon from estuarine wetlands. pp. 69–80. In: R.J. Livingston (ed.). *Ecological Processes in Coastal and Marine Systems.* Plenum Press, New York. 548 pp.

——— and E.J. Heald. 1972. Trophic analysis of an estuarine mangrove community. *Bull. Mar. Sci.* 22(3):671–738.

——— and ———. 1975. The detritus-based food web of an estuarine mangrove community. pp. 265–86. In: L.E. Cronin (ed.). *Estuarine Research.* Academic Press, New York.

———, C.C. McIvor, and T.J. Smith III. 1982. *The Ecology of the Mangroves of South Florida: A Community Profile.* US

Fish and Wildlife Service, Office of Biological Services, FWS/OBS 81/24. 144 pp.

Oesterling, M.L. and G.L. Evink. 1977. Relationship between Florida's blue crab population and Apalachicola Bay. *Florida Mar. Res. Publ.* 26:101–21.

Office of National Marine Sanctuaries. 2008. *Flower Garden Banks National Marine Sanctuary Condition Report 2008.* US Dept. of Commerce, National Oceanic and Atmospheric Administration, Silver Spring, MD. 49 pp.

O'Neil, P.E. and M.F. Metee. 1982. *Alabama Coastal Region Ecological Characterization.* Vol. 2. *A Synthesis of Environmental Data.* US Fish and Wildlife Service, Office of Biological Services, FWS/OBS 82/42. 346 pp.

Onuf, C.P. 1996. Biomass patterns in seagrass meadows of the Laguna Madre, Texas. *Bull. Mar. Sci.* 58(2):404–20.

Opresko, D.M. 1972. Redescriptions and reevaluations of the antipatharians described by L.F. de Pourtales. *Bull. Mar. Sci.* 22(4):950–1017.

Opresko, L., R. Thomas, and F.M. Bayer. 1976. *A Guide to the Larger Marine Gastropods of Florida, the Gulf of Mexico, and the Caribbean Region.* University of Miami Sea Grant Program, Sea Grant Field Guide Ser. 5:1–54.

Ortiz, M. 1991. Amphipod crustacea. II. Family Bateidae. Florida Marine Research Laboratory, *Mem. Hourglass Cruises* 8(1):1–31.

Orton, R.B. 1964. *The Climate of Texas and the Adjacent Gulf Waters.* US Dept. of Commerce, Weather Bureau, US Govt. Printing Office, Washington, DC. Report no. 0–734–017. 195 pp.

Osburn, R.C. 1954. The Bryozoa of the Gulf of Mexico. pp. 361–62. In: P.S. Galtsoff (ed.). *Gulf of Mexico: Its Origin, Waters, and Marine Life.* US Fish and Wildlife Service Fishery Bulletin 89. 604 pp.

O'Shea, T.J., B.B. Ackerman, and H.F. Percival (eds.). 1995. *Population Biology of the Florida Manatee.* US Fish and Wildlife Service, Tech. Rept. 1. 289 pp.

Overstreet, R.M. 1973. Some species of *Lecithaster* Lühe, 1901 (Digenea: Hemiuridae) and related genera from fishes in the northern Gulf of Mexico. *Trans. Amer. Micros. Soc.* 92:231–40.

———. 1974. An estuarine low-temperature fish-kill in Mississippi, with remarks on restricted necropsies. *Gulf Res. Rep.* 4(3):328–50.

———. 1978. *Marine Maladies: Worms, Germs, and Other Symbionts from the Northern Gulf of Mexico.* Mississippi-Alabama Sea Grant Consortium (MASGP-78–021), Gulf Coast Research Laboratory, Ocean Springs, MS. 140 pp.

——— and R.W. Heard. 1982. Food contents of six commercial fishes from Mississippi Sound. *Gulf Res. Rep.* 6(2):137–50.

Park, E.T. 1970. Calanoid copepods from the Caribbean Sea and Gulf of Mexico. 2. New species and new records from plankton samples. *Bull. Mar. Sci.* 20(2):472–546.

———. 1975a. Calanoid copepods of the genera *Aetideopsis, Pseudaetideus,* and *Chiridius* from the Gulf of Mexico. *Bull. Mar. Sci.* 25(2):272–90.

———. 1975b. Calanoid copepods of the genera *Gaetanus* and *Gaidius* from the Gulf of Mexico. *Bull. Mar. Sci.* 25(1):9–34.

———. 1975c. Zooplankton project. pp. 93–306. In: G.P. Pfeiffer (ed.). *Environmental Assessment of the South Texas Outer Continental Shelf: Chemical and Biological Components.* Report to the Bureau of Land Management, contract no. 08550-CTS-17. 730 pp.

Parker, J.C. 1965. An annotated checklist of the fishes of the Galveston Bay System, Texas. *Publ. Inst. Mar. Sci., Univ. Texas* 10:201–20.

Parker, R.H. 1956. Macro-invertebrate assemblages as indicators of sedimentary environments in east Mississippi Delta region. *Amer. Assoc. Petrol. Geol. Bull.* 40:295–376.

———. 1960. Ecology and distributional patterns of marine macro-invertebrates, northern Gulf of Mexico. pp. 302–37. In: F.P. Shepard, F.B. Phleger, and T.H. van Andel (eds.). *Recent Sediments, Northwest Gulf of Mexico.* American Association of Petroleum Geologists, Tulsa, OK. 394 pp.

Parr, A.E. 1939. Quantitative observations on the pelagic sargassum vegetation of the western North Atlantic. *Bull. Bingham Oceanogr. Coll.* 6(7):1–94.

Passarella, K.C. 1990. Oceanic cephalopod assemblage in the eastern Gulf of Mexico. M.S. thesis. University of South Florida, Tampa. 50 pp.

——— and T.L. Hopkins. 1991. Species composition and food habits of the micronektonic cephalopod assemblage in the eastern Gulf of Mexico. *Bull. Mar. Sci.* 49(1,2):638–59.

Pattengill-Semmens, C.V. and S.R. Gittings. 2003. A rapid assessment of the Flower Garden Banks National Marine Sanctuary (stony corals and fishes). pp. 500–11. In: J.C. Lang (ed.). *Status of Coral Reefs in Western Atlantic: Results of Initial Surveys, Atlantic and Gulf Rapid Relief Assessment (AGRRA) Program.* Atoll Research Bulletin, no. 496.

Patton, W.K. 1972. Studies on the animal symbionts of the gorgonian coral, *Leptogorgia virgulata* (Lamarck). *Bull. Mar. Sci.* 22:419–31.

Paull, C.K., B. Hecker, R. Commeau, R.P. Freeman-Lynde, C. Neumann, W.P. Corso, S. Golubic, J.E. Hook, E. Sikes, and J. Curray. 1984. Biological communities at the

Florida Escarpment resemble hydrothermal vent taxa. *Science* 226:965–67.

———, A.J.T. Jull, L.J. Toolin, and T. Linick. 1985. Stable isotope evidence for chemosynthesis in an abyssal seep community. *Nature* 317:709–11.

——— and A.C. Neumann. 1987. Continental margin brine seeps: Their geological consequences. *Geology*15:545–48.

Peak, D.E. 1999. Distribution and relative abundance of pelagic seabirds of the northern Gulf of Mexico. pp. 236–47. In H. Kumpf, K. Steidinger, and K. Sherman (eds.). *The Gulf of Mexico Large Marine Ecosystem: Assessment, Sustainability, and Management.* Blackwell Science, Malden, MA.

Pearse, A.S. 1932. Inhabitants of certain sponges at Dry Tortugas. *Pap. Tortugas Lab.* 7:119–24.

———. 1950. Notes on the inhabitants of certain sponges at Bimini. *Ecology* 31(1):149–51.

Penfound, W. and E. Hathaway. 1938. Plant communities in the marshlands of southeastern Louisiana. *Ecol. Monogr.* 8:1–56.

Pequegnat, L.H. 1970a. Deep-sea caridean shrimps with descriptions of six new species. pp. 21–57. In: W.E. Pequegnat and F.A. Chace Jr. (eds.). *Contributions on the Biology of the Gulf of Mexico.* Texas A&M University Oceanographic Studies. Vol. 1. Gulf Publishing, Houston. 270 pp.

———. 1970b. A study of deep-sea caridean shrimps (Crustacea: Decapoda: Natantia) of the Gulf of Mexico. Ph.D. dissertation. Texas A&M University, College Station. 225 pp.

———. 2000. Lists of mysid, penaeid, and caridean shrimps collected in the Gulf of Mexico (personal communication).

——— and W. E. Pequegnat. 1970. Deep-sea anomurans of superfamily Galatheidea with descriptions of three new species. pp. 125–70. In: W.E. Pequegnat and F.A. Chace Jr. (eds.). *Contributions on the Biology of the Gulf of Mexico.* Texas A&M University Oceanographic Studies. Vol. 1. Gulf Publishing, Houston. 270 pp.

——— and J.P. Ray. 1974. Crustaceans and other arthropods. pp. 231–88. In: T.J. Bright and L.H. Pequegnat (eds.). *Biota of the West Flower Garden Bank.* Gulf Publishing, Houston. 435 pp.

——— and M. K. Wicksten. 2006. Oplophorid shrimps (Decapoda: Caridea: Oplophoridae) in the Gulf of Mexico and Caribbean Sea from the collections of the research vessels *Alaminos, Oregon* and *Oregon II. Crustacean Res.* 35:92–107.

——— and J.H. Wormuth. 1977. Neuston project. pp. 7–1 to 7–83. Vol. 1, *Environmental Studies of the South Texas Outer Continental Shelf, 1976. Biology and Chemistry.* Report by the University of Texas Marine Science Institute to the Bureau of Land Management, New Orleans, LA.

Pequegnat, W.E. 1964. *Biofouling Studies off Panama City, Florida.* 1. Texas A&M Research Foundation, Ref. 66–17T. Texas A&M University, Dept. of Oceanography, College Station. 33 pp.

———. 1970. Deep-water brachyuran crabs. pp. 171–204. In: W.E. Pequegnat and F.A. Chace Jr. (eds.). *Contributions on the Biology of the Gulf of Mexico.* Texas A&M University Oceanographic Studies. Vol. 1. Gulf Publishing, Houston. 270 pp.

———. 1983. *The Ecological Communities of the Continental Slope and Adjacent Regimes of the Northern Gulf of Mexico.* Report by TerEco Corp. to the US Dept. of the Interior, Minerals Management Service, contract no. AA851-CT1–12. 398 pp.

——— and F.A. Chace Jr. (eds.). 1970. *Contributions on the Biology of the Gulf of Mexico.* Texas A&M University Oceanographic Studies. Vol. 1. Gulf Publishing, Houston. 270 pp.

———, R.M. Darnell, B.M. James, E.A. Kennedy, L.H. Pequegnat, and J.T. Turner. 1976. *Ecological Aspects of the Upper Continental Slope of the Gulf of Mexico.* Report by TerEco Corp. to the US Dept. of the Interior, Bureau of Land Management, contract no. 08550-CT4–12. 360 pp.

———, B.J. Gallaway, and L.H. Pequegnat. 1990. Aspects of the ecology of the deep-water fauna of the Gulf of Mexico. *Amer. Zool.* 30(1):45–64.

———, B.M. James, A.H. Bouma, W.R. Bryant, and A.D. Fredericks. 1972. Photographic study of deep-sea environments of the Gulf of Mexico. pp. 67–218. In: V.H. Henry and R. Rezak (eds.). *Contributions on the Geological Oceanography of the Gulf of Mexico.* Texas A&M University Oceanographic Studies. Vol. 3. Gulf Publishing, Houston. 303 pp.

——— and L.M. Jeffrey. 1979. Petroleum in deep benthic ecosystems of the Gulf of Mexico and Caribbean Sea. *Contr. Mar. Sci.* 22:63–75.

——— and L.H. Pequegnat. 1968. *Ecological Aspects of Marine Fouling in the Northeastern Gulf of Mexico.* Texas A&M Research Foundation, Ref. 68–22T. Texas A&M University, Dept. of Oceanography, College Station. 80 pp.

——— and ———. 1971. New species and new records of *Munidopsis* (Decapoda: Galatheidae) from the Gulf of Mexico and Caribbean Sea. Suppl., pp. 1–24. In: W.E. Pequegnat and F.A. Chace Jr. (eds.). *Contributions on the Biology of the Gulf of Mexico.* Texas A&M University

Oceanographic Studies. Vol. 1. Gulf Publishing, Houston. 270 pp.

———, ———, R.W. Firth Jr., B.M. James, and T.W. Roberts. 1971. *Gulf of Mexico Deep-Sea Fauna, Decapoda and Euphausiacea.* American Geographical Society Serial Atlas of the Marine Environment, Folio 20. 12 pp.

———, ———, J.A. Kleypas, B.M. James, E.A. Kennedy, and G.F. Hubbard. 1983. *The Ecological Communities of the Continental Slope and Adjacent Regimes of the Northern Gulf of Mexico.* US Dept. of the Interior, Minerals Management Service, Gulf of Mexico OCS Region, New Orleans, LA. OCS Study MMS 1983-22. 398 pp.

——— and W.B. Sikora. 1977. Meiofauna project. pp. 8-1 to 8-55. In: *Enviromental Studies, South Texas Outer Continental Shelf, Biology and Chemistry, 1976.* Vol. 1. Report by the University of Texas Marine Science Institute to the US Dept. of the Interior, Bureau of Land Management, New Orleans, LA, contract PB80-181506.

——— and C. Venn. 1980. Meiofauna project. pp. 469–514. In: R.W. Flint and N.N. Rabalais (eds.). *Environmental Studies, South Texas Outer Continental Shelf, 1975–1977.* Vol. 1. *Ecosystem Description.* Report to the US Dept. of the Interior, Bureau of Land Management, Washington, DC, contract no. PB80-181506.

Pérez-Farfante, I. 1969. Western Atlantic shrimps of the genus *Penaeus. US Fish and Wildlife Service Fishery Bulletin* 67(3):461–591.

———. 1971. *Western Atlantic Shrimps of the Genus* Metapenaeopsis *(Crustacea, Decapoda, Penaeidae), with Descriptions of Three New Species.* Smithson. Contrib. Zool., no. 79. 37 pp.

———. 1977. American solenocerid shrimps of the genera *Hymenopenaeus, Haliporoides, Pleoticus, Hadropenaeus* new genus, and *Mesopenaeus* new genus. *Fish. Bull.* 75(2):261–346.

——— and H.R. Bullis Jr. 1973. *Western Atlantic Shrimps of the Genus* Solenocera *with Description of a New Species (Crustacea, Decapoda, Penaeidae).* Smithson. Contrib. Zool., no. 153. 33 pp.

Perkins, R.D. and P. Enos. 1968. Hurricane Betsy in the Florida-Bahamas area: Geological effects and comparison with Hurricane Donna. *J. Geol.* 76:710–17.

Perkins, T.H. and T. Savage. 1975. A bibliography and checklist of polychaetous annelids from Florida, the Gulf of Mexico, and the Caribbean region. Florida Marine Research Laboratory, *Fla. Marine Res. Publ.* 14:1–62.

Perry, H.M. and J.Y. Christmas. 1973. Estuarine zooplankton, Mississippi. pp. 198–254. In: J.Y. Christmas (ed.). *Cooperative Gulf of Mexico Estuarine Inventory*

*and Study.* State of Mississippi, Gulf Coast Research Laboratory, Ocean Springs. 434 pp.

———, C.K. Eleuterius, C.B. Trigg, and J.R. Warren. 1995. Settlement patterns of *Callinectes sapidus* megalopae in Mississippi Sound: 1991, 1992. *Bull. Mar. Sci.* 57(3):821–33.

———, K.C. Stuck, and H.D. Howse. 1979. First record of a bloom of *Gonyaulax monilata* in coastal waters of Mississippi. *Gulf Res. Rep.* 6(3):313–16.

———, R. Waller, C. Trigg, J. McBee, R. Erdman, and N. Blake. 1995. A note on bycatch associated with deepwater trapping of *Chaceon* in the northcentral Gulf of Mexico. *Gulf Res. Rep.* 9(2):139–42.

Pew Foundation. 2003. *America's Living Oceans: Charting a Course for Sea Change.* Pew Oceans Commission, Arlington, VA. 144 pp.

Pflum, C.E. and W.E. Frerichs. 1976. *Gulf of Mexico Deep-Water Foraminifers.* Cushman Foundation for Foraminiferal Research, Spec. Publ. No. 14. 125 pp.

Phillips, N.W., D.A. Gettleson, and K.D. Spring. 1990. Benthic biological studies of the southwest Florida shelf. *Amer. Zool.* 30(1):65–75.

——— and B.M. James (eds.). 1988. *Offshore Texas and Louisiana Marine Ecosystems Data Synthesis.* US Dept. of the Interior, Minerals Management Service, New Orleans, LA. OCS Study MMS 88-0067. 3 vols.

Phillips, P.J. 1972. The pelagic cnidaria of the Gulf of Mexico. Ph.D. dissertation. Texas A&M University, College Station. 212 pp.

———. 1973. The occurrence of the remarkable scyphozoan, *Deepstaria enigmatica,* in the Gulf of Mexico and some observations on cnidarian symbionts. *Gulf Res. Rep.* 4(2):166–68.

———, W.D. Burke, and E.J. Keener. 1969. Observations on the trophic significance of jellyfishes in Mississippi Sound with quantitative data on the associative behavior of small fishes with medusae. *Trans. Amer. Fish. Soc.* 98:703–12.

——— and N.L. Levin. 1973. Cestode larvae from scyphomedusae in the Gulf of Mexico. *Bull. Mar. Sci.* 23(3):574–84.

Phillips, R.C. 1978. Seagrasses and the coastal marine environment. *Oceanus* 21(3):30–40.

——— and C.P. McRoy (eds.). 1979. *Handbook of Seagrass Biology.* Garland STPM Press, New York.

Phleger, F.B. 1951. Ecology of Foraminifera, northwest Gulf of Mexico. Pt. I. Foraminifera distribution. *Geol. Soc. Amer. Mem.* 46:1–88.

———. 1960. Sedimentary patterns of microfaunas in northern Gulf of Mexico. pp. 267–301. In: F.P. Shepard,

F.B. Phleger, and T.H. van Andel (eds.). *Recent Sediments, Northwest Gulf of Mexico.* American Association of Petroleum Geologists, Tulsa, OK. 394 pp.

———— and F.L. Parker. 1954a. Ecology of Foraminifera, northwest Gulf of Mexico. Pt. 2. Foraminifera species. *Geol. Soc. Amer. Mem.* 46:1–64.

———— and ————. 1954b. Gulf of Mexico Foraminifera. pp. 235–41. In: P.S. Galtsoff (ed.), *Gulf of Mexico: Its Origin, Waters, and Marine Life.* US Fish and Wildlife Service Fishery Bulletin 89. 604 pp.

Pierce, E.L. 1954. Notes on the Chaetognatha of the Gulf of Mexico. pp. 327–29. In: P.S. Galtsoff (ed.). *Gulf of Mexico: Its Origin, Waters, and Marine Life.* US Fish and Wildlife Service Fishery Bulletin 89. 604 pp.

————. 1965. The distribution of lancets (Amphioxi) along the coasts of Florida. *Bull. Mar. Sci.* 15(2):480–94.

Plotkin, P.T., M.K. Wicksten, and A.F. Amos. 1993. Feeding ecology of the loggerhead sea turtle (*Caretta caretta*) in the northwestern Gulf of Mexico. *Mar. Biol.* 115:1–15.

Plough, H.H. 1978. *Sea Squirts of the Atlantic Continental Shelf from Maine to Texas.* Johns Hopkins University Press, Baltimore, MD. 118 pp.

Poag, C.W. and W.E. Sweet. 1972. Claypile Bank, Texas continental shelf. pp. 223–61. In: V.J. Henry and R. Rezak (eds.). *Contributions to the Geological and Geophysical Oceanography of the Gulf of Mexico.* Texas A&M University Oceanographic Studies. Vol. 3. Gulf Publishing, Houston. 303 pp.

Poirrier, M.A. and M.M. Mulino. 1975. Effects of the 1973 opening of the Bonnet Carré Spillway upon epifaunal invertebrates in southern Lake Pontchartrain. *Proc. La. Acad. Sci.* 38:36–40.

Polloni, P., R. Haedrich, G. Rowe, and C.H. Clifford. 1979. The size-depth relationship in deep ocean animals. *Int. Rev. Ges. Hydrobiol.* 64(1):39–46.

Ponnamperuma, P.N. 1972. The chemistry of submerged soils. *Adv. Agron.* 24:29–96.

Pool, D.J., S.C. Snedaker, and A.E. Lugo. 1977. Structure of mangrove forests in Florida, Puerto Rico, Mexico, and Costa Rica. *Biotropica* 9:195–212.

Posey, M.H., W. Lindberg, T. Alphin, and F. Vosse. 1996. Influence of storm disturbance on an offshore benthic community. *Bull. Mar. Sci.* 59(3):523–29.

Post, H.A. and A.A. de la Cruz. 1977. Litterfall, litter decomposition, and flux of particulate organic material in a coastal plain stream. *Hydrobiologica* 55:201–7.

Potts, D.T. and J.S. Ramsey. 1987. *A Preliminary Guide to Demersal Fishes of the Gulf of Mexico Continental Slope (100 to 600 Fathoms).* Alabama Sea Grant Publ. MASGP-86-009. 95 pp.

Powell, E.H., Jr. and G. Gunter. 1968. Observations on the stone crab *Menippe mercenaria* (Say), in the vicinity of Port Aransas, Texas. *Gulf Res. Rep.* 2:285–99.

Powell, E.N., T.J. Bright, A. Woods, and S. Gittings. 1983. Meiofauna and the thiobios in the East Flower Garden brine seep. *Mar. Biol.* 73:269–83.

Powell, P.E. and P.J. Szaniszlo. 1980. Benthic mycology. pp. 395–437. In: R.W. Flint and N.N. Rabalais (eds.). *Environmental Studies, South Texas Outer Continental Shelf, 1975–1977.* Vol. 2. Report to the US Dept. of the Interior, Bureau of Land Management, Washington, DC, contract no. PB80-181522.

Powers, L.W. 1977. A catalogue and bibliography to the crabs (Brachyura) of the Gulf of Mexico. *Contr. Mar. Sci.* 20 (suppl.). 190 pp.

Prescott, W.H. 1964. *The Conquest of Mexico.* Bantam Books, New York. 740 pp. (Reprint of 1843 edition entitled *History of the Conquest of Mexico.*)

Presley, B.J. 1991. Trace metals and distributions. Chapt. 5, pp. 5-1 to 5-20. In: J.M. Brooks (ed.). *Mississippi-Alabama Continental Shelf Ecosystem Study: Data Summary and Synthesis.* Vol. 2. *Technical Narrative.* OCS Study MMS 91-0063. US Dept. of the Interior, Minerals Management Service, Gulf of Mexico OCS Region, New Orleans, LA. 862 pp.

———— and J.M. Brooks. 1988. Marine chemistry. Chapt. 5, pp. 133–201. In: N.W. Phillips and B.M. James (eds.). *Offshore Texas and Louisiana Marine Ecosystems Data Synthesis.* Vol. 2. *Synthesis Report.* OCS Study MMS 88-0067. US Dept. of the Interior, Minerals Management Service, Gulf of Mexico OCS Region, New Orleans, LA. 477 pp.

Price, W.A. 1954. Geology. pp. 39–67. In: P.S. Galtsoff (ed.). *Gulf of Mexico: Its Origin, Waters, and Marine Life.* US Fish and Wildlife Fishery Bulletin 89. 604 pp.

Price, W.W. 1976. The abundance and distribution of Mysidacea in the shallow waters off Galveston Island, Texas. Ph.D. dissertation. Texas A&M University, College Station. 207 pp.

————, A.P. McAlister, R.M. Tousley, and M. DelRe. 1986. Mysidacea from continental shelf waters of the northwestern Gulf of Mexico. *Contr. Mar. Sci.* 29:45–58.

Provenzano, A.J., Jr. 1959. The shallow-water hermit crabs of Florida. *Bull. Mar. Sci.* 9(4):349–420.

Pulley, T.E. 1952. A zoogeographic study based on the bivalves of the Gulf of Mexico. Ph.D. dissertation. Harvard University, Cambridge, MA. 215 pp.

Purcell, J.E. 1985. Predation on fish eggs and larvae by pelagic cnidarians and ctenophores. *Bull. Mar. Sci.* 37(2):739–55.

Putt, R.E., D.A. Gettleson, and N.W. Phillips. 1986. Fish assemblages and benthic biota associated with natural hard-bottom areas in the northwestern Gulf of Mexico. *Northeast Gulf Sci.* 8(1):51–63.

Quayle, R.G. and D.C. Fulbright. 1977. Wind and wave statistics for the North American Atlantic and Gulf coasts. *Mariner's Weather Log.* 21:13–14.

Rabalais, N.N. 1982. The ascidians of the Aransas Pass inner jetties, Port Aransas, Texas. pp. 65–74. In: B.R. Chapman and J.W. Tunnell Jr. (eds.). *South Texas Fauna: A Symposium Honoring Dr. Allan H. Chaney.* Texas A&M University Caesar Kleberg Wildlife Research Institute, Kingsville.

———. 1990. Biological communities of the South Texas continental shelf. *Amer. Zool.* 30:77–87.

———, F.R. Burditt Jr., L.D. Coen, B.E. Cole, C. Eluterius, K.L. Heck Jr., T.A. McTigue, et al. 1995. Settlement of *Callinectes sapidus* megalopae on artificial collectors in four Gulf of Mexico estuaries. *Bull. Mar. Sci.* 57(3):855–76.

———, R.S. Carney, and E.G. Escobar-Briones. 1999. Overview of continental shelf benthic communities of the Gulf of Mexico. Chapt. 10, pp. 171–95. In: H. Kumpf, K. Steidinger, and K. Sherman (eds.). *The Gulf of Mexico Large Marine Ecosystem: Assessment, Sustainability, and Management.* Blackwell Science, Malden, MA. 704 pp.

——— and D.E. Harper Jr. 1992. Studies of benthic biota in areas affected by moderate and severe hypoxia. pp. 150–53. In: *Nutrient Enhanced Coastal Ocean Productivity Workshop Proceedings,* Texas A&M University, Sea Grant Publ. No. TAMU-SG-92-109.

Rabalais, S.C. 1978. Fouling communities of selected artificial substrates in the northwestern Gulf of Mexico. M.S. thesis. Texas A&I University, Kingsville. 99 pp.

——— and N.N. Rabalais. 1980. The occurrence of sea turtles on the south Texas coast. *Contr. Mar. Sci.* 23:123–29.

Randall, J.E. 1968. *Caribbean Reef Fishes.* T.F.H. Publications, Jersey City, NJ. 318 pp.

Rannefeld, J.W. 1972. The stony corals of Enmedio Reef off Veracruz, Mexico. M.S. thesis. Texas A&M University, College Station. 104 pp.

Rass, T.S. 1971. Deep sea fishes in the Caribbean Sea and the Gulf of Mexico (the American Mediterranean region). pp. 509–26. In: *UNESCO-FAO Symposium on Investigations and Resources of the Caribbean Sea and Adjacent Regions.* UNESCO, Paris.

Ray, G.C. 1991. Coastal zone biodiversity patterns. *BioScience* 41(7):490–98.

——— and J.F. Grassle. 1991. Marine biological diversity. *BioScience* 41(7):453–61.

——— and W.P. Gregg Jr. 1991. Establishing biosphere reserves for coastal barrier ecosystems. *BioScience* 41(5):301–9.

——— and M.G. McCormick-Ray. 1992. Functional coastal-marine biodiversity. pp. 384–97. In: R.E. McCabe (ed.). *Transactions of the 57th North American Wildlife and Natural Resources Conference.* Wildlife Management Institute, Washington, DC.

Ray, J.P. 1974. A study of the coral reef crustaceans (Decapoda and Stomatopoda) of two Gulf of Mexico reef systems: West Flower Garden, Texas and Isla de Lobos, Veracruz, Mexico. Ph.D. dissertation. Texas A&M University, College Station. 323 pp.

Rayburn, R. 1975. Food of deep-sea demersal fishes of the northwestern Gulf of Mexico. M.S. thesis. Texas A&M University, College Station. 119 pp.

Rebel, T.P. 1974. *Sea Turtles and the Turtle Industry of the West Indies, Florida, and the Gulf of Mexico.* University of Miami Press, Coral Gables, FL. 250 pp.

Rehder, H.A. 1954. Mollusks. pp. 469–74. In: P.S. Galtsoff (ed.). *Gulf of Mexico: Its Origin, Waters and Marine Life.* US Fish and Wildlife Service Fishery Bulletin 89. 604 pp.

Reid, G.K., Jr. 1954. An ecological study of the Gulf of Mexico fishes, in the vicinity of Cedar Key, Florida. *Bull. Mar. Sci.* 4(1):1–94.

Reimold, R.J. and W.H. Queen (eds.). 1974. *Ecology of Halophytes.* Academic Press, New York.

Rezak, R., T.J. Bright, and D.W. McGrail. 1983. *Reefs and Banks of the Northwestern Gulf of Mexico: Their Chemical, Biological, and Physical Dynamics.* Final report to US Dept. of the Interior, Minerals Management Service, New Orleans, LA, contract no. AA851-CT1-55. 501 pp.

———, ———, and ———. 1985. *Reefs and Banks of the Northwestern Gulf of Mexico: Their Geological, Biological, and Physical Dynamics.* John Wiley & Sons. New York. 259 pp.

——— and G.S. Edwards. 1972. Carbonate sediments of the Gulf of Mexico. pp. 263–80. In: R. Rezak and V.J. Henry (eds.). *Contributions on the Geological and Geophysical Oceanography of the Gulf of Mexico.* Texas A&M University Oceanographic Studies. Vol. 3. Gulf Publishing, Houston. 303 pp.

———, S.R. Gittings, and T.J. Bright. 1990. Biotic assemblages and ecological controls on reefs and banks of the northwest Gulf of Mexico. *Amer. Zool.* 30(1):23–35.

——— and V.J. Henry (eds.). 1972. *Contributions on the Geological and Geophysical Oceanography of the Gulf of*

*Mexico.* Texas A&M University Oceanographic Studies. Vol. 3. Gulf Publishing, Houston. 303 pp.

———, W.W. Sager, J.S. Laswell, and S. Gittings. 1989. Seafloor features on the Mississippi-Alabama outer continental shelf. *Gulf Coast Assoc. Geol. Socs., Trans.* 39:511–14.

Rice, W.H. and L.S. Kornicker. 1962. Mollusks of Alacran Reef, Campeche Bank, Mexico. *Publ. Inst. Mar. Sci., Univ. Texas* 8:366–403.

——— and ———. 1965. Mollusks from the deeper waters of the northwestern Campeche Bank, Mexico. *Publ. Inst. Mar. Sci., Univ. Texas* 10:108–72.

Richards, F.A. 1957. Oxygen in the ocean. pp. 185–238. In: J.W. Hedgpeth (ed.). *Treatise on Marine Ecology and Paleoecology.* Vol. 1. *Ecology.* Geological Society of America Memoir 67.

Richards, W.J., M.G. McGowan, T. Leming, J.T. Lamkin, and S. Kelley. 1993. Larval fish assemblages at the Loop Current boundary in the Gulf of Mexico. *Bull. Mar. Sci.* 53(2):475–537.

Richardson, J. (ed.). 1985. *Managing the Ocean: Resources, Research, Law.* Lomond Publications, Mt. Airy, MD. 407 pp.

Richardson, W.J., C.R. Greene Jr., C.I. Malme, and D.H. Thomson. 1995. *Marine Mammals and Noise.* Academic Press, San Diego, CA. 576 pp.

Rico-Gray, V. and A. Lot-Helgueras. 1983. Producción de hojasca del manglar de la Laguna de la Mancha, Ver., México. *Botanica* 8:295–301.

Rigby, J.K. and W.G. McIntyre. 1966. Isla de Lobos and associated reefs, Veracruz, Mexico. *Brigham Young Univ. Geol. Stud.* 13:1–46.

Riley, C.M., S.A. Holt, G.J. Holt, E.J. Buskey, and C.R. Arnold. 1989. Mortality of larval red drum (*Sciaenops ocellatus*) associated with a *Ptychodiscus brevis* red tide. *Contr. Mar. Sci.* 31:137–46.

Rinckey, G.R. and C.H. Saloman. 1964. Effects of reduced water temperature on fishes of Tampa Bay, Florida. *J. Fla. Acad. Sci.* 27(1):9–16.

Roberts, H., R. Carney, M. Kupchik, C. Fisher, K. Nelson, E. Becker, L. Goehring, et al. 2007. *Alvin* explores the deep Gulf of Mexico slope. *EOS* 88(35):341–42.

Roberts, T.W. 1970. A preliminary study of the family Penaeidae and their distribution in the deep water of the Gulf of Mexico. M.S. thesis. Texas A&M University, College Station. 97 pp.

———. 1977. An analysis of deep-sea benthic communities in the northeast Gulf of Mexico. Ph.D. dissertation. Texas A&M University, College Station. 258 pp.

——— and W.E. Pequegnat. 1970. Deep-water decapod shrimps of the family Penaeidae. pp. 21–57. In: W.E. Pequegnat and F.A. Chace (eds.). *Contributions to the Biology of the Gulf of Mexico.* Texas A&M University Oceanographic Studies. Vol. 1. Gulf Publishing, Houston. 270 pp.

Robertson, W.B., Jr. 1955. An analysis of the breeding bird populations of tropical Florida in relation to vegetation. Ph.D. dissertation. University of Illinois, Urbana.

——— and J.A. Kushlan. 1974. The southern Florida avifauna. *Miami Geol. Soc. Mem.* 2:414–52.

Robins, C.R. 1957. Effects of storms on the shallow-water fish fauna of southern Florida with new records of fishes from Florida. *Bull. Mar. Sci.* 7(3):266–75.

———, R.M. Bailey, C.E. Bond, J.R. Brooker, E.A. Lachner, R.N. Lea, and W.B. Scott. 1991. *Common and Scientific Names of Fishes from the United States and Canada.* American Fisheries Society, Spec. Publ. 20. 183 pp.

———, G.C. Ray, J. Douglass, and R. Freund. 1986. *A Field Guide to Atlantic Coast Fishes of North America.* Peterson Field Guide Series, 32. Houghton Mifflin, Boston. 324 pp.

Robison, B.H. and T.G. Bailey. 1981. Sinking rates and dissolution of midwater fish fecal matter. *Mar. Bio.* 65(2):136–42.

Rogers, R.M., Jr. 1976. Distribution of meiobenthic organisms in San Antonio Bay in relation to season and habitat disturbance. pp. 337–44. In: A.H. Bouma (ed.). *Shell Dredging and Its Influence on Gulf Coast Environments.* Gulf Publishing, Houston. 454 pp.

———. 1977. Trophic relationships of selected fishes on the continental shelf of the northern Gulf of Mexico. Ph.D. dissertation. Texas A&M University, College Station. 229 pp.

Romero R., M.E. 1998. La navegación Maya. *Arqueología Mexicana* 6(33):6–15.

Rooker, J.R., Q.R. Dokken, C.V. Pattengill, and G.J. Holt. 1997. Fish assemblages on artificial reefs in the Flower Garden Banks National Marine Sanctuary, USA. *Coral Reefs* 16:83–92.

Ross, S.T. 1974. Resource partitioning in searobins (Pisces: Triglidae) on the west Florida shelf. Ph.D. dissertation. University of South Florida, St. Petersburg.

———. 1977. Patterns of resource partitioning in searobins (Pisces: Triglidae). *Copeia* 1977:561–71.

Rowe, G.T. 1966. A study of the deep water benthos of the northwestern Gulf of Mexico. M.S. thesis. Texas A&M University, College Station. 95 pp.

———. 1998. Organic carbon cycling in abyssal benthic food chains: Numerical simulations of bioenhancement by sewage sludge. *J. Mar. Syst.* 14:337–54.

——— and M.C. Kennicutt II. 2002. *Deepwater Program:*

*Northern Gulf of Mexico Continental Slope Habitat and Benthic Ecology.* Year 2: Interim Report. US Dept. of the Interior, Minerals Management Service, Gulf of Mexico OCS Region, New Orleans, LA. OCS Study MMS 2002-063. 158 pp.

———— and D.W. Menzel. 1971. Quantitative benthic samples from the deep Gulf of Mexico. *Bull. Mar. Sci.* 21:556–66.

————, P.T. Polloni, and G.W. Horner. 1974. Benthic biomass estimates from the northwestern Atlantic Ocean and the northern Gulf of Mexico. *Deep Sea Res.* 21:641–50.

————, M. Sibuet, J. Deming, J. Tietjen, and A. Khripounoff. 1990. Organic carbon turnover time in deep-sea benthos. *Progr. Oceanogr.* 24:141–60.

————, R. Theroux, W. Phoel, H. Quinby, R. Wilke, D. Koschoreck, T. Whitledge, P. Falkowski, and C. Fray. 1988. Benthic carbon budgets for the continental shelf south of New England. *Cont. Shelf Res.* 8:511–27.

Ruppert, E.E., R.S. Fox, and R. D. Barnes. 2004. *Invertebrate Zoology: A Functional Evolutionary Approach.* 7th ed. Thomson Brooks/Cole, Belmont, CA.

Russell, M. 1977. Apparent effects of flooding on distribution and landings of industrial bottomfish in the northern Gulf of Mexico. *Northeast Gulf Sci.* 1(2):77–82.

Ryden, J.C., J.K. Syers, and R.F. Harris. 1973. Phosphorus in runoff and streams. pp. 1–45. In: N.C. Brady, (ed.). *Advances in Agronomy* 25. Academic Press, New York. 400 pp.

Ryder, G., D. Fastovsky, and S. Gartner (eds.). 1996. *The Cretaceous-Tertiary Event and Other Catastrophies in Earth History.* Geological Society of America, Spec. Paper 307. 569 pp.

Sackett, W.M. 1972. Chemistry. pp. 1–5. In S.Z. El-Sayed, W.M. Sackett, L.M. Jeffrey, A.D. Fredericks, R.P. Saunders, P.S. Conger, G.A. Fryxell, K.A. Steidinger, and S.A. Earle. *Chemistry, Productivity, and Benthic Algae of the Gulf of Mexico.* American Geographical Society Serial Atlas of the Marine Environment, Folio 22. 29 pp.

Sage, W.W. and M.J. Sullivan. 1978. Distribution of blue green algae in a Mississippi gulf coast salt marsh. *J. Phycol.* 14:333–37.

Salcedo-Vargas, M.A. 1991. Checklist of the cephalopods from the Gulf of Mexico. *Bull. Mar. Sci.* 49 (1,2):216–20.

Salvador, A. (ed.). 1991a. *The Gulf of Mexico Basin.* Geological Society of America, Boulder, CO. Vol. J. 568 pp.

————. 1991b. Introduction. In: A. Salvador (ed.). *The Gulf of Mexico Basin.* Geological Society of America, Boulder, CO. Vol. J. 568 pp.

————. 1991c. Origin and development of the Gulf of Mexico basin. In: A. Salvador (ed.). *The Gulf of Mexico Basin.* Geological Society of America, Boulder, CO. Vol. J. 568 pp.

Sánchez, J.A. and L.A. Soto. 1987. Camarones de la superfamilia Penaeoidea (Rafinesque, 1815) distribuidos en la plataforma continental del suroeste del Golfo de México. *Anales del Instituto de Ciencias del Mar y Limnología* 14(2):157–80.

Sánchez, R. and M. Elana. 1963. Datos relativos a los manglares de México. *Anales de la Escuela Nacional de Ciencias Biológicas, México* 12:61–72.

Sánchez-Gil, P., A. Yáñez-Arancibia, and F. Amezcua Linares. 1981. Diversidad, distribución, y abundancia de las especies y poblaciones de peces demersales de la Sonda de Campeche (verano 1978). *Anales del Instituto de Ciencias del Mar y Limnología* 8(1):209–39.

Sanders, D.C. 1964. Blood parasites of marine fish of southwest Florida, including a new hemogregarine from the menhaden (*Brevoortia tyrannus* Latrobe). *Trans. Amer. Micros. Soc.* 83:218–25.

Santiago, V. 1977. Algunos estudios sobre las madréporas del Arrecife "La Blanquilla" Veracruz, México. Tesis profesional, Facultad de Ciencias, Universidad Nacional Autónoma de México. 103 pp.

Saunders, R.P. and G.A. Fryxell. 1972. Diatom distribution. pp. 13–14. In: S.Z. El-Sayed, W.M. Sackett, L.M. Jeffrey, A.D. Fredericks, R.P. Saunders, P.S. Conger, G.A. Fryxell, K.A. Steidinger, and S.A. Earle. *Chemistry, Productivity and Benthic Algae of the Gulf of Mexico.* American Geographical Society Serial Atlas of the Marine Environment, Folio 22. 29 pp.

———— and D.A. Glenn. 1969. Diatoms. Florida Marine Research Laboratory, *Mem. Hourglass Cruises* 1(3):1–119.

Savilov, A.I. 1967. Oceanic insects of the genus *Halobates* (Hemiptera, Gerridae) in the Pacific. *Oceanology* 7:252–60.

Scheltema, R.S. 1971. The dispersal of the larvae of shoal-water benthic invertebrate species over long distances by ocean currents. pp. 7–28. In: D. Crisp (ed.) *Fourth European Marine Biology Symposium.* Cambridge University Press, London.

Schink, D.R., P.H. Santschi, O. Corapcioglu, P. Sharma, and U. Fehn. 1995. $^{129}$I in Gulf of Mexico waters. *Earth and Planet. Sci. Letters* 135:131–38.

Schmidly, D.J. 1981. *Marine Mammals of the Southeastern United States Coast and the Gulf of Mexico.* US Fish and Wildlife Service, Office of Biological Services, FWS/OBS-80/41. 163 pp.

———— and S.H. Shane. 1978. *A Biological Assessment of the*

*Cetacean Fauna of the Texas Coast.* US Marine Mammal Commission Report MMC-74/05. 38 pp.

Schmitt, W.L. 1954. Copepoda. pp. 439–42. In: P.S. Galtsoff (ed.). *Gulf of Mexico: Its Origin, Waters, and Marine Life.* US Fish and Wildlife Service Fishery Bulletin 89. 604 pp.

Schmitz, W.J., Jr. 2005. Cyclones and westward propogation in the shedding of anticyclonic rings from the Loop Current. In: W. Sturges and A. Lugo-Fernandez (eds.). *Circulation in the Gulf of Mexico: Observations and Models.* Geophysical Monograph Series Vol. 161. 360 pp.

Schneider, C.W. and R.B. Searles. 1991. *Seaweeds of the Southeastern United States.* Duke University Press, Durham, NC. 553 pp.

Schomer, N.S. and R.D. Drew. 1982. *An Ecological Characterization of the Lower Everglades, Florida Bay and the Florida Keys.* US Fish and Wildlife Service, Office of Biological Services, FWS/OBS 82/58.1.

Schroeder, W.W. 1971. The distribution of euphausiids in the oceanic Gulf of Mexico, the Yucatan Strait, and the northwest Caribbean. Ph.D. dissertation. Texas A&M University, College Station. 174 pp.

——— and W.J. Wiseman. 1988. Mobile Bay estuary: Stratification, oxygen depletion, and jubilees. pp. 41–52. In: *Hydrodynamics of Estuaries.* Vol. 2. *Estuarine Case Studies.* CRC Press, Boca Raton, FL.

Schuchert, C. 1985. Historical geology of the Antillean-Caribbean region, or the lands bordering the Gulf of Mexico and the Caribbean Sea: John Wiley & Sons, New York. 811 pp.

Schwartz, J.R., S.K. Alexander, C.S. Giam, and H.S. Chan. 1980. Analysis of benthic bacteria. pp. 439–68. In: R.W. Flint and N.N. Rabalais (eds.). *Environmental Studies, South Texas Outer Continental Shelf, 1975–1977.* Vol. 1. *Ecosystem Description.* Report to the US Dept. of the Interior, Bureau of Land Management, Washington, DC, contract no. PB80–181506.

Scott, P.J.B. 1987. Associations between corals and macroinfaunal invertebrates in Jamaica, with a list of Caribbean and Atlantic coral associates. *Bull. Mar. Sci.* 40(2):271–86.

Sears, M. 1954a. Ctenophores in the Gulf of Mexico. p. 297. In: P.S. Galtsoff (ed.). *Gulf of Mexico: Its Origin, Waters, and Marine Life.* US Fish and Wildlife Service Fishery Bulletin 89. 604 pp.

———. 1954b. Hydromedusae of the Gulf of Mexico. pp. 273–74. In: P.S. Galtsoff (ed.). *Gulf of Mexico: Its Origin, Waters, and Marine Life.* US Fish and Wildlife Service Fishery Bulletin 89. 604 pp.

———. 1954c. Siphonophores in the Gulf of Mexico. pp. 275–76. In: P.S. Galtsoff (ed.). *Gulf of Mexico: Its*

*Origin, Waters, and Marine Life.* US Fish and Wildlife Service Fishery Bulletin 89. 604 pp.

Segura-Puertas, L. 1992. Medusae (Cnidaria) from the Yucatan Shelf and Mexican Caribbean. *Bull. Mar. Sci.* 51(3):353–59.

Serafy, D.K. 1979. Echinoids (Echinodermata: Echinoidea). Florida Marine Research Laboratory, *Mem. Hourglass Cruises* 5(3):1–120.

Shaw, J.K., P.G. Johnson, R.M. Ewing, C.E. Comiskey, C.C. Brandt, and T.A. Farmer. 1982. *Benthic Macroinfauna Community Characterization in Mississippi Sound and Adjacent Waters.* US Army Corps of Engineers, Mobile District, Mobile, AL. 442 pp.

Shaw, R.F., J.H. Cowan Jr., and T.L Tillman. 1985. Distribution and density of *Brevoortia patronus* (Gulf menhaden) eggs and larvae in the continental shelf waters of western Louisiana. *Bull. Mar. Sci.* 36(1):96–103.

Shelton, R.C. and P.B. Robertson. 1981. Community structure of intertidal macrofauna on two surf-exposed Texas sandy beaches. *Bull. Mar. Sci.* 31(4):833–42.

Shemitz, S.D. 1974. Populations of bear, panther, alligator, and deer in the Florida Everglades. *Fla. Sci.* 37:157–67.

Shepard, F.P. and D.G. Moore. 1955. Central Texas coast sedimentation: Characteristics of sedimentary environment, recent history, and diagenesis. *Amer. Assoc. Petrol. Geol. Bull.* 39(8):1463–593.

——— and ———. 1960. Bays of central Texas coast. pp. 117–52. In: F.P. Shepard, F.B. Phleger, and T.H. van Andel (eds.). *Recent Sediments, Northwest Gulf of Mexico: A Symposium Summarizing the Results of Work Carried On in Project 51 of the American Petroleum Institute 1951–1958.* American Association of Petroleum Geologists. 394 pp.

———, F.B. Phleger, and T.H. van Andel (eds.). 1960. *Recent Sediments, Northwest Gulf of Mexico: A Symposium Summarizing the Results of Work Carried On in Project 51 of the American Petroleum Institute 1951–1958.* American Association of Petroleum Geologists. 394 pp.

Sheridan, P.F. 1979. Trophic resource utilization by three species of sciaenid fishes in a northwest Florida estuary. *Northeast Gulf Sci.* 3(1):1–14.

———, D.L. Trimm, and B.M. Baker. 1984. Reproduction and food habits of seven species of northern Gulf of Mexico fishes. *Contr. Mar. Sci.* 27:175–203.

Sheridan, P.T. 1992. Comparative habitat utilization by estuarine macrofauna within the mangrove ecosystem of Rookery Bay, Florida. *Bull. Mar. Sci.* 50(1):21–39.

Sherman, K. and L.M. Alexander (eds.). 1986. *Variability and Management of Large Marine Ecosystems.* AAAS Selected

Symposium Series 99. Westview Press, Boulder, CO. 319 pp.

———— and ———— (eds.). 1989. *Biomass Yields and Geography of Large Marine Ecosystems.* AAAS Selected Symposium Series 3. Westview Press, Boulder, CO. 493 pp.

————, ————, and B.D. Gold (eds.). 1990. *Large Marine Ecosystems: Patterns, Processes, and Yields.* AAAS Publ. No. 90-30S. AAAS Press, Washington, DC. 242 pp.

————, ————, and ———— (eds.). 1991. *Food Chains, Yields, Models, and Management of Large Marine Ecosystems.* Westview Press, Boulder, CO. 320 pp.

————, ————, and ———— (eds.). 1993. *Large Marine Ecosystems: Stress, Mitigation, and Sustainability.* AAAS Publ. No. 92-395. AAAS Press, Washington, DC. 376 pp.

Sherrod, C.L. and C. McMillan. 1985. The distributional history and ecology of mangrove vegetation along the northern Gulf of Mexico coastal region. *Contrib. Mar. Sci.* 28:129-40.

Shier, C.F. 1965. A taxonomic and ecological study of shallow water hydroids of the northeastern Gulf of Mexico. M.S. thesis. Florida State University, Tallahassee. 120 pp.

Shier, D.E. 1964. Marine Bryozoa from northwest Florida. *Bull. Mar. Sci.* 14(4):603-62.

Shifflett, E. 1961. Living, dead, and total foraminiferal faunas, Heald Bank, Gulf of Mexico. *Micropaleontol.* 7(1):45-54.

Shinn, E.A. 1980. Geological history of Grecian Rocks, Key Largo Coral Reef Marine Sanctuary. *Bull. Mar. Sci.* 30(3):646-56.

————, J.H. Hudson, R.B. Halley, and B. Lidz. 1977. Topographic control and accumulation rate of some Holocene coral reefs: South Florida and Dry Tortugas. pp. 1-7. In: D. Taylor (ed.). *Proceedings of the Third International Coral Reef Symposium, University of Miami, FL.*

Shipp, L.P. 1977. The vertical and horizontal distribution of decapod larvae in relation to some environmental conditions within a salt marsh area of the north central Gulf of Mexico. M.S. thesis. University of South Alabama, Mobile. 129 pp.

Shipp, R.L. and T.S. Hopkins. 1978. Physical and biological observations of the northern rim of the DeSoto Canyon made from a research submersible. *Northeast Gulf Sci.* 2(2):113-21.

Shirley, J.C. 1974. The echinoderms of Seven and One-Half Fathom Reef. M.S. thesis. Texas A&I University, Kingsville. 82 pp.

Sikora, W.B., J.P. Sikora, and A.M. Prior. 1981. *Environmental Effects of Hydraulic Dredging for Clam Shells in Lake Pontchartrain, Louisiana.* LSU Center for Wetland Resources Publ. LSU-CEL-81-18. 140 pp.

Simberloff, D. 1976. Experimental zoogeography of islands: Effects of island size. *Ecology* 57:629-48.

———— and E.O. Wilson. 1969. Experimental zoogeography of islands: The colonization of empty islands. *Ecology* 50:278-96.

Simons, J.D. 1997. Food habits and trophic structure of the demersal fish assemblages on the Mississippi-Alabama continental shelf. Ph.D. dissertation. Texas A&M University, College Station. 309 pp.

Sindermann, C.J. 1990. *Principal Diseases of Marine Fish and Shellfish.* Academic Press, San Diego, CA. 2 vols.

Skinner, R. 1975. Parasites of the striped mullet, *Mugil cephalus,* from Biscayne Bay, Florida, with descriptions of a new genus and three new species of trematodes. *Bull. Mar. Sci.* 25(3):318-45.

Smith, D.E. 1982. Abundance and foraging ability of Physonect Siphonophores in subtropical oceanic surface waters. Ph.D. dissertation. Texas A&M University. College Station. 205 pp.

Smith, F.G.W. 1954a. Biology of the commercial sponges. pp. 263-66. In: P.S. Galtsoff (ed.). *Gulf of Mexico: Its Origin, Waters, and Marine Life.* US Fish and Wildlife Service Fishery Bulletin 89. 604 pp.

————. 1954b. Biology of the spiny lobster. pp. 463-65. In: P.S. Galtsoff (ed.). *Gulf of Mexico: Its Origin, Waters, and Marine Life.* US Fish and Wildlife Service Fishery Bulletin 89. 604 pp.

————. 1954c. Gulf of Mexico Madreporaria. pp. 291-95. In: P.S. Galtsoff (ed.). *Gulf of Mexico: Its Origin, Waters, and Marine Life.* US Fish and Wildlife Service Fishery Bulletin 89. 604 pp.

————. 1954d. Taxonomy and distribution of sea turtles. pp. 513-15. In: P.S. Galtsoff (ed.). *Gulf of Mexico: Its Origin, Waters, and Marine Life.* US Fish and Wildlife Service Fishery Bulletin 89. 604 pp.

————. 1971. *Atlantic Reef Corals.* University of Miami Press, Coral Gables, FL. 164 pp.

Smith, G.B. 1976. *Ecology and Distribution of Eastern Gulf of Mexico Reef Fishes.* Florida Marine Research Publ. 19. 78 pp.

————, H.M. Austin, S.A. Bartone, R.W. Hastings, and L.H. Ogren. 1975. *Fishes of the Florida Middle Ground with Comments on Ecology and Zoogeography.* Florida Marine Research Publ. 9. 14 pp.

Snider, J.E. 1975. Quantitative distribution of shelled pteropods in the Gulf of Mexico including related sampling studies. Ph.D. dissertation. Texas A&M University, College Station. 213 pp.

Snieszko, S.F. (ed.). 1970. *A Symposium on the Diseases of Fishes and Shellfishes*. American Fisheries Society, Spec. Publ. No. 5.

Snyder, S.W. 1978. Distribution of planktonic foraminifera in surface sediments of the Gulf of Mexico. *Tulane Studs. in Geol. and Paleontol.* 14(1):1–80.

Soley, J.C. 1914. *The Gulf Stream in the Gulf of Mexico*. Pilot Chart for the North Atlantic Ocean. US Navy, Washington, DC.

Somero, G.N. 1982. Physiological and biochemical adaptations of deep-sea fishes: Adaptive responses to the physical and biological characteristics of the abyss. pp. 257–78. In: W.G. Ernst and J. Morin (eds.). *Ecosystem Processes in the Deep Oceans*. Prentice-Hall, Englewood Cliffs, NJ.

———, J.F. Siebenaller, and P.W. Hochachka. 1983. Biological and physiological adaptations of deep-sea animals. pp. 261–330. In: G.T. Rowe (ed.). *Deep-Sea Biology*. John Wiley & Sons, New York. 560 pp.

Sonnier, F., J. Teerling, and H.D. Hoese. 1976. Observations on the offshore reef and platform fish fauna of Louisiana. *Copeia* 1976(1):105–11.

Soto, L.A. 1972. Decapod shelf-fauna of the northeastern Gulf of Mexico, distribution and zoogeography. M.S. thesis. Florida State University, Tallahassee. 129 pp.

———. 1979. Decapod crustacean shelf fauna of the Campeche Bank: Fishery aspects and ecology. *Gulf Carib. Fish. Inst.* 32:66–81.

———. 1980. A decapod crustacea shelf-fauna of the Northeastern Gulf of Mexico. *Anales del Instituto de Ciencias del Mar y Limnología* 7(2):79–109.

———. 1985. Distributional patterns of deep-water brachyuran crabs in the Straits of Florida. *J. Crust. Biol.* 5(3):480–99.

———. 1986. Deep-water brachyuran crabs of the Straits of Florida (Crustacea, Decapoda). *Anales del Instituto de Ciencias del Mar y Limnología* 13(1):1–68.

———. 1991. Faunal zonation in the deep-water brachyuran crabs in the Straits of Florida. *Bull. Mar. Sci.* 49(1,2):623–37.

——— and E. Escobar-Briones. 1995. Coupling mechanisms related to benthic production in the SW Gulf of Mexico. pp. 79–118. In: E. Eleftheriou, A.D. Ansell, and C.J. Smith (eds.). *Biology and Ecology of Shallow Coastal Waters*. 28th European Marine Biology Symposium. Olsen & Olsen, Fredensborg, Denmark.

——— and A. Garcia. 1987. Evaluación de los efectos de hidrocarburos fósiles sobre las poblaciones de camarones peneidos en el Banco de Campeche. *Anales del Instituto de Ciencias del Mar y Limnología* 14(2):133–46.

Spivey, H.R. 1981. Origins, distribution, and zoogeographic affinities of the Cirripedia (Crustacea) of the Gulf of Mexico. *J. Biogeogr.* 8:153–76.

Sprague, V. 1954. Protozoa. pp. 243–56. In: P.S. Galtsoff (ed.). *Gulf of Mexico: Its Origin, Waters, and Marine Life*. US Fish and Wildlife Service Fishery Bulletin 89. 604 pp.

Springer, S. and H.R. Bullis Jr. 1956. *Collections by the Oregon in the Gulf of Mexico*. US Fish and Wildlife Service, Special Scientific Report, Fisheries no. 196. 134 pp.

Springer, V.G. 1961. Notes and additions to the fish fauna of the Tampa Bay area. *Copeia* 1961(4):480–82.

——— and A. McErlean. 1962. A study of the behavior of some tagged south Florida coral reef fish. *Amer. Midl. Nat.* 67(2):386–97.

——— and K.D. Woodburn. 1960. *An Ecological Study of the Fishes of the Tampa Bay Area*. Florida State Board of Conservation, Marine Laboratory, Prof. Pap. Ser. 1. 104 pp.

Stanton, R.J. and I. Evans. 1971. Environmental controls of benthic macrofaunal patterns in the Gulf of Mexico adjacent to the Mississippi Delta. *Gulf Coast Assoc. Geol. Socs., Trans.* 21:371–78.

Steidinger, K.A. 1972a. Dinoflagellate distribution. pp. 14–15. In: S.Z. El-Sayed, W.M. Sackett, L.M. Jeffrey, A.D. Fredericks, R.P. Saunders, P.S. Conger, G.A. Fryxell, K.A. Steidinger, and S.A. Earle. *Chemistry, Productivity, and Benthic Algae of the Gulf of Mexico*. American Geographical Society Serial Atlas of the Marine Environment, Folio 22. 29 pp.

———. 1972b. Dinoflagellate species reported from the Gulf of Mexico and adjacent coastal areas. pp. 23–25. In: S.Z. El-Sayed, W.M. Sackett, L.M. Jeffrey, A.D. Fredericks, R.P. Saunders, P.S. Conger, G.A. Fryxell, K.A. Steidinger, and S. A. Earle. *Chemistry, Productivity and Benthic Algae of the Gulf of Mexico*. American Geographical Society Serial Atlas of the Marine Environment, Folio 22. 29 pp.

———. 1973. Phytoplankton. pp. III E 1–10. In: J.I. Jones, R.E. Ring, M.O. Rinkel, and R.E. Smith (eds.). *A Summary of Knowledge of the Eastern Gulf of Mexico*. State University System of Florida, Institute of Oceanography.

———. 1975. Basic factors influencing red tides. pp. 153–62. In: V.R. LoCicero (ed.). *Proceedings of the First International Conference on Toxic Dinoflagellate Blooms*. Marine Science and Technology Foundation, Wakefield, MA.

————— and E.A. Joyce. 1973. *Florida Red Tides.* Florida Dept. of Natural Resources Educational Series, no. 17. 26 pp.

————— and J.F. Van Breedveld. 1971. Benthic marine algae from waters adjacent to the Crystal River electric power plant (1969 and 1970). *Florida Marine Research Laboratory, Prof. Pap. Ser.* 16:1–46.

————— and J. Williams. 1970. Dinoflagellates. Florida Marine Research Laboratory, *Mem. Hourglass Cruises* 2:1–251.

Stephenson, T.A. and A. Stephenson. 1950. Life between tide-marks in North America: The Florida Keys. *J. Ecol.* 38(2):354–402.

Stickney, D.G. and J.J. Torres. 1989. Proximate composition and energy content of mesopelagic fishes from the eastern Gulf of Mexico. *Mar. Biol.* 103:13–24.

Stickney, R.R. 1984. *Estuarine Ecology of the Southeastern United States and Gulf of Mexico.* Texas A&M University Press, College Station. 310 pp.

Stock, J.H. 1986. Pycnogonida from the Caribbean and the Straits of Florida. *Bull. Mar. Sci.* 38(3):399–441.

————— and R.U. Gooding. 1986. A new siphonostomatoid copepod associated with the West Indian sea urchin, *Diadema antillarum. Bull. Mar. Sci.* 39(1):102–9.

Stockwell, D.A., E.J. Buskey, and T.E. Whitledge. 1993. Studies on conditions conducive to the development and maintenance of a persistent "brown tide" in Laguna Madre, Texas. pp. 693–98. In: T.J. Smayda and Y. Shimizu (eds.). *Proceedings of the Fifth International Conference on Toxic Marine Phytoplankton.* Elsevier Press, Amsterdam.

Storey, M. 1937. The relation between normal range and mortality of fish due to cold at Sanibel Island, Florida. *Ecology* 18(1):10–26.

————— and E.W. Gudger. 1936. Mortality of fishes due to cold at Sanibel Island, Florida, 1886–1936. *Ecology* 17(4):640–48.

Storr, J.F. 1964. *Ecology of the Gulf of Mexico Commercial Sponges and Its Relation to the Fishery.* US Fish and Wildlife Service, Special Scientific Report, Fisheries no. 466:1–73.

Stout, J.P. 1984. *The Ecology of Irregularly Flooded Salt Marshes of the Northeastern Gulf of Mexico: A Community Profile.* US Fish and Wildlife Service, Biological Report 85(7.1). 98 pp.

Stow, D.A.V., D.G. Howell, and C.H. Nelson. 1985. Sedimentary, tectonic, and sea-level controls. pp. 15–22. In: A.H. Bouma, W.R. Normark, and N.E. Barnes (eds.). *Submarine Fans and Related Turbidite Systems.* Springer-Verlag, New York. 351 pp.

Street, G.T. and P.A. Montagna. 1996. Loss of genetic diversity in Harpacticoida near offshore platforms. *Mar. Biol.* 126:271–82.

Stuck, K.C., H.M. Perry, and R.W. Heard. 1979a. An annotated key to the Mysidacea of the northcentral Gulf of Mexico. *Gulf. Res. Rep.* 6(3):225–38.

—————, —————, and —————. 1979b. Records and range extensions of Mysidacea from coastal and shelf waters of the eastern Gulf of Mexico. *Gulf Res. Rep.* 6(3):239–48.

Sturges, W. and A. Lugo-Fernandez (eds.). 2005. *Circulation in the Gulf of Mexico: Observations and Models.* Geophysical Monographs 161. American Geophysical Union, Washington, DC. 347 pp.

Suárez, M. and R. Gasca-S. 1992. Pterópodos (Gastropoda: Thecosomata: Pseudothecosomata) de aguas someras (0–50 m) del sur del Golfo de México. *Anales del Instituto de Ciencias del Mar y Limnología* 19(2):201–9.

Subrahmanyam, C.B. and C.L. Coultas. 1980. Studies on the animal communities in two north Florida salt marshes. Part III. Seasonal fluctuations of fish and macroinvertebrates. *Bull. Mar. Sci.* 30(4):790–818.

————— and S.H. Drake. 1975. Studies of the animal communities in two north Florida salt marshes. Part I. Fish communities. *Bull. Mar. Sci.* 25:445–65.

—————, W.L. Kruczynski, and S.H. Drake. 1976. Studies on the animal communities in two north Florida salt marshes. Part II. Macroinvertebrate communities. *Bull. Mar. Sci.* 26(2):172–95.

Sullivan, J.R. 1979. The stone crab, *Menippe mercenaria,* in the Southwest Florida fishery. *Florida Mar. Res. Publ.* 36:1–37.

Sullivan, M.J. 1979. Epiphytic diatoms on three seagrass species in Mississippi Sound. *Bull. Mar. Sci.* 29(4):459–64.

Sutherland, J.P. 1974. Multiple stable points in natural communities. *Amer. Nat.* 108:859–73.

————— and R.H. Karlson. 1977. Development and stability of the fouling community at Beaufort, North Carolina. *Ecol. Monogr.* 47:425–46.

Sutton, G.M. 1951. *Mexican Birds: First Impressions.* University of Oklahoma Press, Norman.

Sutton, T.T. and T.L. Hopkins. 1996. Trophic ecology of the stomiid (Pisces: Stomiidae) fish assemblage of the Eastern Gulf of Mexico: Strategies, selectivity, and impact of a top mesopelagic predator group. *Mar. Biol.* 127:179–92.

Sweeney, M.J. and C.F.E. Roper. 1998. Classification, type localities, and type repositories of recent cephalopoda. pp. 561–95. In: N.A. Voss, M. Veccione, R.B. Tole, and M.J. Sweeney (eds.). *Systematics and Biogeography of*

*Cephalopods.* Smithson. Contr. Zool., no. 586. Vol. 1. 599 pp.

Tabb, D.C. and A.C. Jones. 1962. Effects of Hurricane Donna on the aquatic faunas of North Florida Bay. *Trans. Amer. Fish. Soc.* 91(4):375–78.

Tattersall, W.M. 1951. A review of the Mysidacea of the United States National Museum. *U.S. Natl. Mus. Bull.* 201:1–292.

Taylor, D.D. 1969. The distribution of the heteropods of the Gulf of Mexico. M.S. thesis. Texas A&M University, College Station. 61 pp.

———— and L. Berner. 1970. The Heteropoda (Mollusca: Gastropoda). pp. 231–44. In: W.E. Pequegnat and F.A. Chace Jr. (eds.). *Contributions on the Biology of the Gulf of Mexico.* Texas A&M University Oceanographic Studies. Vol. 1. Gulf Publishing, Houston. 270 pp.

Taylor, J.L. and C.H. Saloman. 1969. Some effects of hydraulic dredging and coastal development in Boca Ciega Bay, Florida. *US Fish and Wildlife Service Fishery Bulletin* 67(2):213–41.

Taylor, W.R. 1954a. Distribution of marine algae in the Gulf of Mexico. *Pap. Mich. Acad. Sci. Arts Lett.* 39:85–109.

————. 1954b. Sketch of the character of the marine algal vegetation of the shores of the Gulf of Mexico. pp. 177–92. In: P.S. Galtsoff (ed.). *Gulf of Mexico: Its Origin, Waters, and Marine Life.* US Fish and Wildlife Service Fishery Bulletin 89. 604 pp.

————. 1960. *Marine Algae of the Eastern Tropical and Subtropical Coasts of the Americas.* University of Michigan Press, Ann Arbor. 870 pp.

Teerling, J. 1975. A survey of sponges from the northwestern Gulf of Mexico. Ph.D. dissertation. University of Southwestern Louisiana, Lafayette. 186 pp.

Temple, R.F., D.L. Harrington, and J.A. Martin. 1977. *Monthly Temperature and Salinity Measurements of Continental Shelf Waters of the Northwestern Gulf of Mexico.* NOAA Tech. Rept., NMFS, Special Scientific Report, Fishery Bulletin 707. 26 pp.

Tester, P.A. and K.A. Steidinger. 1997. *Gymnodinium breve* red tide blooms: Initiation, transport and consequences of surface circulation. *Limnol. Oceanogr.* 42(5, pt. 2): 1039–51.

Tett, P.B. and M.G. Kelly. 1973. Marine bioluminescence. *Oceanogr. Mar. Biol.* 11:89–173.

Texas A&M University. 2002. *A Sustainable Gulf of Mexico: Research, Technology, and Observations 1950 to 2050.* Sustainable Coastal Margins Program, Texas A&M University, Conference Proceedings. TAMU-SG-02-102. 105 pp.

Thayer, G.W., J.J. Govoni, and D.W. Conally. 1983. Stable carbon isotope ratios of the planktonic food web in the northern Gulf of Mexico. *Bull. Mar. Sci.* 33(2):247–56.

Thayer, P.A., A. La Roque, and J.W. Tunnell Jr. 1974. Relict lacustrine sediments on the inner continental shelf, southeast Texas. *Gulf Coast Assoc. Geol. Socs., Trans.* 24:337–47.

Thomas, L.P. 1962. The shallow water amphiurid brittle stars (Echinodermata, Ophiuroidea) of Florida. *Bull. Mar. Sci.* 12(4):623–94.

Thomas, P.J. 1975. The fouling community on selected oil platforms off Louisiana, with special emphasis on the Cirripedia fauna. M.S. thesis. Florida State University, Tallahassee. 129 pp.

Thorne, R.F. 1954. Flowering plants of the waters and shores of the Gulf of Mexico. pp. 193–202. In: P.S. Galtsoff (ed.). *Gulf of Mexico: Its Origin, Waters and Marine Life.* US Fish and Wildlife Service Fishery Bulletin 89. 604 pp.

Tierney, J.O. 1954. The Porifera of the Gulf of Mexico. pp. 259–61. In: P.S. Galtsoff (ed.). *Gulf of Mexico: Its Origin, Waters, and Marine Life.* US Fish and Wildlife Service Fishery Bulletin 89. 604 pp.

Toral-Almazán, S. and A. Reséndez-Medina. 1974. Los ciclidos (Pisces: Perciformes) de la Laguna de Términos y sus afluentes. *Rev. Biol. Trop.* 21:254–74.

Tozzer, A.M. 1941. *Landa's Relación de las Cosas de Yucatán.* Pap. Peabody Mus. 18. 394 pp.

Trabant, P.K. and B.J. Presley. 1978. Orca Basin, anoxic depression on the continental slope, northwest Gulf of Mexico. pp. 303–11. In: A.H. Bouma, G.T. Moore, and J.M. Coleman (eds.). *Framework, Facies, and Oil-Trapping Characteristics of the Upper Continental Margin.* Amer. Assoc. Petrol. Geol., Stud. in Geol. 7. 326 pp.

Treece, G.D. 1980. Bathymetric records of marine shelled Mollusca from the northeastern shelf and upper slope of Yucatan, Mexico. *Bull. Mar. Sci.* 30:552–70.

Trefry, J.H. and B.J. Presley. 1976. Heavy metals in sediments from San Antonio Bay and the northwest Gulf of Mexico. *Environ. Geol.* 1:282–94.

Tresslar, R.C. 1974a. Corals. pp. 115–39. In: T.J. Bright and L.H. Pequegnat (eds.). *Biota of the West Flower Garden Bank.* Gulf Publishing, Houston. 435 pp.

————. 1974b. Foraminifers. pp. 67–91. In: T.J. Bright and L.H. Pequegnat (eds.). *Biota of the West Flower Garden Bank.* Gulf Publishing, Houston. 435 pp.

Tressler, W.L. 1954. Marine Ostracoda. pp. 429–37. In: P.S. Galtsoff (ed.). *Gulf of Mexico: Its Origin, Waters, and Marine Life.* US Fish and Wildlife Service Fishery Bulletin 89. 604 pp.

Tunnell, J.W., Jr. 1973. Molluscan population of a

submerged reef off Padre Island, Texas. *Bull. Amer. Malacol. Union* 38:25–26.

———. 1974. Ecological and geographical distribution of Mollusca of Lobos and Enmedio coral reefs, southwestern Gulf of Mexico. Ph.D. dissertation. Texas A&M University, College Station. 172 pp.

———. 1985. Environmental stresses of the Veracruz coral reefs (Southwestern Gulf of Mexico). *Proceedings of the Fifth International Coral Reef Congress, Tahiti* 2:384.

———. 1988. Regional comparison of southwestern Gulf of Mexico to Caribbean Sea coral reefs. *Proceedings of the Sixth International Coral Reef Symposium, Australia* 3:303–8.

———. 1992. Natural versus human impacts to southern Gulf of Mexico coral reef resources. *Proceedings of the Seventh International Coral Reef Symposium, Guam* 1:300–306.

——— and B.D. Causey. 1969. Vertebrate Pleistocene fossils from the continental shelf, northwestern Gulf of Mexico. *Texas A&I Univ. Stud.* 2(1):75–76.

——— and A.H. Chaney. 1970. A checklist of the mollusks of Seven and One-Half Fathom Reef, northwestern Gulf of Mexico. *Contr. Mar. Sci.* 15:193–203.

——— and B.R. Chapman. 1988. First record of red-footed boobies nesting in the Gulf of Mexico. *Amer. Birds.* 42(3):380–81.

———, E.A. Chávez, and K. Withers (eds.). 2007. *Coral Reefs of the Southern Gulf of Mexico.* Texas A&M University Press, College Station. 194 pp.

——— and Q.R. Dokken (eds.). 2006. *Proceedings of the State of the Gulf of Mexico Summit.* Corpus Christi, TX. 28–30 March, 2006. Harte Research Institute for Gulf of Mexico Studies, Texas A&M University, Corpus Christi. 44 pp.

——— and F.W. Judd (eds.). 2002. *The Laguna Madre of Texas and Tamaulipas.* Texas A&M University Press, College Station. 346 pp.

——— and T.J. Nelson. 1989. A high density-low diversity octocoral community in the southwestern Gulf of Mexico. *Diving for Science* 1989:325–35.

Turner, J.T. 1979. Microbial attachment to copepod fecal pellets and its possible ecological significance. *Trans. Amer. Micros. Soc.* 98(1):131–35.

——— and J.G. Ferrante. 1979. Zooplankton fecal pellets in aquatic ecosystems. *BioScience* 29(11):670–77.

——— and J.C. Roff. 1993. Trophic levels and trophospecies in marine plankton: Lessons from the microbial food web. *Mar. Microbial Food Webs* 7(2):225–48.

Turner, R.E. 1976. Geographic variations in salt marsh macrophyte production: A review. *Contr. Mar. Sci.* 20:47–68.

———. 1977. Intertidal vegetation and commercial yields of penaeid shrimp. *Trans. Amer. Fish. Soc.* 106(5):411–16.

———. 1990. Landscape development and coastal wetland losses in the northern Gulf of Mexico. *Amer. Zool.* 30(1):89–105.

———, R.M. Darnell, and J. Bond. 1980. Changes in the submerged macrophytes of Lake Pontchartrain (Louisiana): 1954–1973. *Northeast Gulf Sci.* 4(1):44–49.

——— and J.G. Gosselink. 1975. A note on standing crop of *Spartina alterniflora* in Texas and Florida. *Contr. Mar. Sci.* 19:113–18.

——— and N.N. Rabalais. 1991. Changes in Mississippi River water quality this century: Implications for coastal food webs. *BioScience* 41:140–47.

Twilley, R.R., E.J. Barron, H.L. Gholz, M.A. Harwell, R.L. Miller, D.J. Reed, J.B. Rose, E.H. Siemann, R.G. Wetzel, and R.J. Zimmerman. 2001. *Confronting Climate Change in the Gulf Coast Region.* Report of the Union of Concerned Scientists and the Ecological Society of America. 82 pp.

Tyler, J.C. and J.E. Böhlke. 1972. Records of sponge-dwelling fishes, primarily of the Caribbean. *Bull. Mar. Sci.* 22(3):601–42.

Uchupi, E. and K.O. Emery. 1968. Structure of continental margin off Gulf coast of United States. *Amer. Assoc. Petrol. Geol. Bull.* 52(7):1162–93.

Uebelacker, J.M. and P.G. Johnson (eds.). 1984. *Taxonomic Guide to the Polychaetes of the Northern Gulf of Mexico.* Final report by Barry A. Vittor & Assoc. to the US Dept. of the Interior, Minerals Management Service, New Orleans, LA, contract no. 14–12–001–29091. 7 vols.

United Nations. 1984. *Convention for the Protection and Development of the Marine Environment in the Wider Caribbean.* Geneva.

US Commission on Ocean Policy. 2004. Preliminary report of the US Commission on Ocean Policy, governor's draft, Washington, DC. 413 pp.

US Department of Commerce. 1959. *Climatological and Oceanographic Atlas for Mariners.* Vol. 1. *North Atlantic Ocean.* US Govt. Printing Office, Washington, DC.

———. 1968. *Climatic Atlas of the United States.* US Govt. Printing Office, Washington, DC.

US Department of State. 1982. *Proceedings of the U.S. Strategy Conference on Biological Diversity.* 16–18 Nov. 1981. US Dept. of State Publ. 9262. US Govt. Printing Office, Washington, DC. 126 pp.

US Naval Oceanographic Office. 1963. *Oceanographic*

*Atlas of the North Atlantic Ocean, Section IV.* US Naval Oceanographic Office, Publ. 700. Washington, DC.

————. 1967. *Oceanographic Atlas of the North Atlantic Ocean, Section II.* US Naval Oceanographic Office, Publ. 700, Washington, DC.

US Naval Weather Service Command. 1970. *Summary of Synoptic Meteorological Observations: North American Coastal Marine Areas.* Vol. 5. NTIS, Springfield, VA.

US Navy Hydrographic Office 1959. *Climatological and Oceanographical Atlas for Mariners.* Vol. 1. *North Atlantic Ocean.*

Van Dyke, J.M., D. Zaelke, and G. Hewison (eds.). 1993. *Freedom for the Seas in the 21st Century: Ocean Governance and Environmental Harmony.* Island Press, Washington, DC. 504 pp.

Van Name, W. 1954. The Tunicata of the Gulf of Mexico. pp. 495–97. In: P.S. Galtsoff (ed.). *Gulf of Mexico: Its Origin, Waters, and Marine Life.* US Fish and Wildlife Service Fishery Bulletin 89. 604 pp.

Vargas Maldonado, I., A. Yáñez-Arancibia, and F. Amezcuna Linares. 1981. Ecología y estructura de las comunidades de peces en áreas de *Rhizophora mangle* y *Thalassia testudinum* de la Isla del Carmen, Laguna de Términos, sur del Golfo de México. *Anales del Instituto de Ciencias del Mar y Limnología* 8(1):241–65.

Vargo, G.A. and T.L. Hopkins. 1990. Plankton. Chapt. 6, pp. 195–230. In: *Synthesis of Available Biological, Geological, Chemical, Socioeconomic, and Cultural Resource Information for the South Florida Area.* US Dept. of the Interior, Minerals Management Service. OCS Study MMS 90–0019.

Vázquez de la Cerda, A.M., R.O. Reid, S.F. DiMarco, and A.E. Jochens. 2005. Bay of Campeche circulation: An update. pp. 279–93. In: W. Sturges and A. Lugo-Fernandez (eds.). *Circulation in the Gulf of Mexico: Observations and Models.* American Geophysical Union, Geophysical Monograph Ser. 161. 360 pp.

Venn, C. 1980. Studies on harpacticoid copepod populations of two transects across the south Texas outer continental shelf. M.S. thesis. Texas A&M University, College Station. 151 pp.

Vernberg, W.B. (ed.). 1974. *Symbiosis in the Sea.* University of South Carolina Press, Columbia.

Viada, S.T. 1980. Species composition and population levels of scleractinian corals within the *Diploria-Montastrea-Porites* zone of the East Flower Garden Bank, northwest Gulf of Mexico. M.S. thesis. Texas A&M University, College Station. 96 pp.

Vidal, V.M.V., F.V. Vidal, and J.M. Perez-Molero. 1989. *Atlas Oceanográfico del Golfo de México.* Vol. 1. Instituto de Investigaciones Electricas, Cuerna Vaca, México. 415 pp.

Villalobos, A. 1971. Estúdios ecológicos en un arrecife coralino en Veracruz, México. pp. 332–45. In: *Symposium on Investigations and Resources of the Caribbean Sea and Adjacent Regions.* Organized by UNESCO and FAO.

Vittor, B.A. 1979. Abundance, diversity, and distribution of benthic polychaetous annelids in the eastern Gulf of Mexico. Vol. 2A, Chapt. 15. In: Dames and Moore (ed.). *The Mississippi, Alabama, Florida Outer Continental Shelf Baseline Environmental Survey, 1977/1978.* Report to the US Dept. of the Interior, Bureau of Land Management, Washington, DC, contract no. AA550-CT7-34.

Vittor, Barry & Associates, Inc. 1985. *Tuscaloosa Trend Regional Data Search and Synthesis Study.* Final report to the US Dept. of the Interior, Minerals Management Service, New Orleans, LA, contract no. 14-12-0001-30048. 2 vols.

Vokes, H.E. and E.H. Vokes. 1983. *Distribution of Shallow Water Marine Mollusca, Yucatan Peninsula, Mexico.* Middle America Research Institute, Publ. 54. 183 pp.

Voss, G.L. 1954. Cephalopoda of the Gulf of Mexico. pp. 475–78. In: P.S. Galtsoff (ed.). *The Gulf of Mexico: Its Origin, Waters, and Marine Life.* US Fish and Wildlife Service Fishery Bulletin 89. 604 pp.

————. 1956. A review of the cephalopods of the Gulf of Mexico. *Bull. Mar. Sci.* 6(2):85–178.

———— and M. Solis Ramirez. 1966. *Octopus maya,* a new species from the Bay of Campeche, Mexico. *Bull. Mar. Sci.* 16(3):615–25.

Voss, N.A., M. Vecchione, R.B. Tole, and M.J. Sweeney (eds.). 1998. *Systematics and Biogeography of Cephalopods.* Smithson. Contr. Zool., no. 586. 599 pp. 2 vols.

Waisel, Y. 1972. *Biology of Halophytes.* Academic Press, New York. 395 pp.

Waller, R., H. Perry, C. Trigg, J. McBee, R. Erdman, and N. Blake. 1995. Estimates of harvest potential and distribution of the deep sea red crab, *Chaceon quinquedens,* in the northcentral Gulf of Mexico. *Gulf Res. Rep.* 9(2):75–84.

Walls, J.G. 1975. *Fishes of the Northern Gulf of Mexico.* T.F.H. Publications, Jersey City, NJ. 432 pp.

Walsh, G.E. 1974. Mangroves: A review. pp. 51–174. In: R. Reimhold and W. Queen (eds.). *Ecology of Halophytes.* Academic Press, New York.

————, S. Snedaker, and H. Teas (eds.). 1974. *Proceedings of the International Symposium on Biology and Management of Mangroves.* IFAS, University of Florida, Gainesville.

Walsh, J.J., D.A. Dieterle, M.B. Myers, and F.E. Muller-Karger. 1989. Nitrogen exchange at the continental margin: A numerical study of the Gulf of Mexico. *Prog. Oceanogr.* 23:245–301.

Wardle, W.J., S.M. Ray, and A.S. Aldrich. 1975. Mortality of marine organisms associated with offshore summer blooms of the toxic dinoflagellate *Gonyaulax monilata* Howe at Galveston, Texas. pp. 257–63. In: V.R. LoCicero (ed.). *Proceedings of the First International Conference on Toxic Dinoflagellate Blooms.* Wakefield, MA.

Wass, M. 1955. The decapod crustaceans of Alligator Harbor and adjacent inshore areas of northwestern Florida. *Quart. J. Fla. Acad. Sci.* 18(3):129–76.

Weinstein, M.P. (ed.). 1988. *Larval Fish and Shellfish Transport through Inlets.* American Fisheries Society Symposium 3. 165 pp.

Wells, H.W. 1961. The fauna of oyster beds, with special reference to the salinity factor. *Ecol. Monogr.* 31:239–66.

———. 1966. Barnacles of the northeastern Gulf of Mexico. *Quart. J. Fla. Acad. Sci.* 29(2):81–95.

Wenk, E., Jr. 1972. *The Politics of the Ocean.* University of Washington Press, Seattle.

Whitten, H.L., F. Rosene, and J.W. Hedgpeth. 1950. The invertebrate fauna of Texas coast jetties: A preliminary survey. *Publ. Inst. Mar. Sci., Univ. Texas* 1(2):53–87.

Wickham, D.A., J.W. Watson Jr., and L.H. Ogren. 1973. The efficacy of midwater artificial structures for attracting pelagic sport fish. *Trans. Amer. Fish. Soc.* 102:563–72.

Wiley, G.N., R.C. Circe, and J.W. Tunnell Jr. 1982. Mollusca of the rocky shores of east central Veracruz State, Mexico. *Nautilus* 96(2):55–61.

Williams, A.B. 1965. *Marine Decapod Crustaceans of the Carolinas. Fish. Bull.* 65(1). 298 pp.

———. 1966. The western Atlantic swimming crabs *Callinectes ornatus, C.danae* and a new related species (Decapoda, Portunidae). *Tulane Stud. Zool.* 13(3):83–93.

———. 1974. The swimming crabs of the genus *Callinectes. Fish. Bull.* 72 (3):685–798.

Williams, J. and R.M. Ingle. 1972. *Ecological Notes on Gonyaulax monilata Blooms on the West Coast of Florida.* Florida Dept. of Natural Resources, Marine Research Laboratory, Leaflet Ser. 1 (Pt. 1, No. 5). 12 pp.

Wills, J.B. and T.J. Bright. 1974. Worms. pp. 291–309. In: T.J. Bright and L.H. Pequegnat (eds.). *Biota of the West Flower Garden Bank.* Gulf Publishing, Houston. 435 pp.

Wilson, C.B. 1935. Parasitic copepods from the Dry Tortugas. *Pap. Tortugas Lab.* 29:327–47.

Wilson, J.G. and W. Halcrow. 1985. *Estuarine Management and Quality Assessment.* Plenum Press, New York. 225 pp.

Wilson, K.A. 1989. Ecology of mangrove crabs: Predation, physical factors, and refuges. *Bull. Mar. Sci.* 44(1):263–73.

Wilson, W.B. and S.M. Ray. 1956. The occurrence of

*Gymnodinium brevis* in the western Gulf of Mexico. *Ecology* 37:385.

Wissing, T.E., R.M. Darnell, M.A. Ibrahim, and L. Berner Jr. 1973. Caloric values of marine animals from the Gulf of Mexico. *Contr. Mar. Sci.* 17:1–7.

Wolff, G.A. and J.H. Wormuth. 1984. Zooplankton. pp. 8–1 through 8–106. In: R.W. Hann Jr., C.P. Giammona, and R.E. Randall (eds.). *Offshore Oceanographic and Environmental Monitoring Services for the Strategic Petroleum Reserve.* Annual report for the West Hackberry Site from November 1983 to November 1984. US Dept. of Energy, contract no. DE-AC96–83P010850.

Wood, C.E. 1974. A key to the Natantia (Crustacea, Decapoda) of the coastal waters on the Texas coast. *Contr. Mar. Sci.* 18:35–56.

Woodmansee, R.A. 1966. Daily vertical migration of *Lucifer:* Planktonic numbers in relation to solar and tidal cycles. *Ecology* 47(5):847–50.

Wormuth, J.H. 1979. Determination of average dry weights of important neuston species collected in 1976. In: R.W. Flint (ed.). *Environmental Studies, South Texas Outer Continental Shelf, Biology and Chemistry.* Report by the University of Texas Marine Science Institute to the Bureau of Land Management, New Orleans, LA. Supplement to 1976 report.

———, J.D. McEachran, and L.H. Pequegnat. 1980. Analysis of a two year study of neuston. pp. 285–346. In: R.W. Flint and N.N. Rabalais (eds.). *Environmental Studies, South Texas Outer Continental Shelf, 1975–1977.* Vol. 3. Report by the University of Texas Marine Science Institute to the Bureau of Land Management, New Orleans, LA.

Wright, L.D., F.J. Swaye, and J.M. Coleman. 1970. *Effects of Hurricane Camille on the Landscape of the Brenton-Chandeleur Island Chain and the Eastern Portion of the Lower Missisippi Delta.* Coastal Studies Institute, Louisiana State University, Technical Report 76. 34 pp.

Würsig, B., T.A. Jefferson, and D.J. Schmidly. 2000. *The Marine Mammals of the Gulf of Mexico.* Texas A&M University Press, College Station. 232 pp.

Yáñez-Arancibia, A. (ed.). 1985a. *Ecología de Comunidades de Pescas en Estuarios y Lagunas Costeras.* UNAM, Instituto de Ciencias del Mar y Limnología. 653 pp.

———. 1985b. *Recursos Pesqueros Potenciales de Mexico: La Pesca Acompañante del Camarón.* UNAM, Instituto de Ciencias del Mar y Limnología. 748 pp.

——— and J.W. Day Jr. 1982. Ecological characterization of Terminos Lagoon, a tropical lagoon-estuarine system in the southern Gulf of Mexico. pp. 431–40. In: *Oceanologica Acta. Proceedings of the International*

*Symposium on Coastal Lagoons.* SCOR/IABO/UNESCO, Bordeaux, France, 1981.

———, F.A. Linares, and J.W. Day Jr. 1980. Fish community structure and function in Terminos Lagoon, a tropical estuary in the southern Gulf of Mexico. pp. 465–82. In: V.S. Kennedy (ed.). *Estuarine Perspectives.* Academic Press, New York. 533 pp.

——— and P. Sánchez-Gil. 1986. *Los Peces Demersales de la Plataforma Continental del Sur del Golfo de México. 1. Caracterización Ambiental, Ecología, y Evaluación de las Especies, Poblaciones y Comunidades.* UNAM, Instituto de Ciencias del Mar y Limnología. Publ. Esp. 9. 230 pp.

Yerger, R.W. 1965. The leatherback turtle on the Gulf coast of Florida. *Copeia* 1965(3):365–66.

Yingst, J.Y. and D.C. Rhoads. 1985. The structure of soft-bottom benthic communities in the vicinity of the Texas Flower Garden Banks, Gulf of Mexico. *Est., Coast., and Shelf Sci.* 20:569–92.

Young, C.M. and K.J. Eckelbarger (eds.). 1994. *Reproduction, Larval Biology, and Recruitment of Deep-Sea Benthos.* Columbia University Press, New York. 336 pp.

Young, R.E. 1983. Oceanic bioluminescence: An overview of general functions. *Bull. Mar. Sci.* 33(4):829–45.

Zieman, J.C. 1975. Quantitative and dynamic aspects of the ecology of turtle grass, *Thalassia testudinum.* pp. 241–62. In: L.E. Cronin (ed.). *Estuarine Research.* Vol. 1. Academic Press, New York.

———. 1982. *The Ecology of the Seagrasses of South Florida: A Community Profile.* US Fish and Wildlife Service, Office of Biological Services, FWS/OBS-82/25. 158 pp.

——— and R.T. Zieman. 1989. The ecology of the seagrass meadows of the west coast of Florida: A community profile. US Fish and Wildlife Service Biological Report 85(7.25). 155 pp.

Zimmer, B., W.F. Precht, E.L. Hickerson, and J. Sinclair. 2006. Discovery of *Acropora palmata* at the Flower Garden Banks National Marine Sanctuary, northwestern Gulf of Mexico. *Coral Reefs* 25:192.

Zimmerman, M.S. and R.J. Livingston. 1976. Seasonality and physico-chemical ranges of benthic macrophytes from a north Florida estuary (Apalachee Bay). *Contr. Mar. Sci.* 20:33–45.

——— and R.J. Livingston. 1979. Dominance and distribution of benthic macrophyte assemblages in a north Florida estuary (Apalachee Bay, Florida). *Bull. Mar. Sci.* 29(1):27–40.

ZoBell, C.E. 1954. Marine bacteria and fungi in the Gulf of Mexico. pp. 217–22. In: P.S. Galtsoff (ed.). *Gulf of Mexico: Its Origin, Waters, and Marine Life.* US Fish and Wildlife Service Fishery Bulletin 89. 604 pp.

# INDEX

Page numbers in *italics* refer to figures and tables.